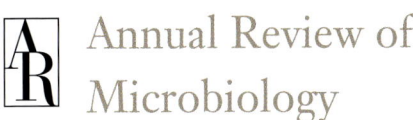
Annual Review of Microbiology

Editorial Committee (2012)

Katherine A. Borkovich, University of California, Riverside
Lynn W. Enquist, Princeton University
Susan Gottesman, National Cancer Institute
Stephen L. Hajduk, University of Georgia
Caroline S. Harwood, University of Washington
Jung-Hye Roe, Seoul National University
Thomas M. Schmidt, Michigan State University
Olaf Schneewind, University of Chicago

Responsible for the Organization of Volume 66 (Editorial Committee 2010)

Peggy A. Cotter
Daniel DiMaio
N. Louise Glass
Susan Gottesman
Stephen L. Hajduk
Caroline S. Harwood
Nancy A. Moran
Katherine A. Borkovich (Guest)
Thomas M. Schmidt (Guest)

Production Editor: Cleo X. Ray
Managing Editor: Linley E. Hall
Bibliographic Quality Control: Mary A. Glass
Electronic Content Coordinators: Suzanne K. Moses, Erin H. Lee
Illustration Editor: Glenda Lee Mahoney

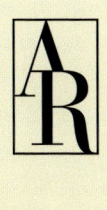

Annual Review of Microbiology

Volume 66, 2012

Susan Gottesman, *Editor*
National Cancer Institute, Bethesda

Caroline S. Harwood, *Associate Editor*
University of Washington

Olaf Schneewind, *Associate Editor*
University of Chicago

www.annualreviews.org • science@annualreviews.org • 650-493-4400

Annual Reviews
4139 El Camino Way • P.O. Box 10139 • Palo Alto, California 94303-0139

Annual Reviews
Palo Alto, California, USA

COPYRIGHT © 2012 BY ANNUAL REVIEWS, PALO ALTO, CALIFORNIA, USA. ALL RIGHTS RESERVED. The appearance of the code at the bottom of the first page of an article in this serial indicates the copyright owner's consent that copies of the article may be made for personal or internal use, or for the personal or internal use of specific clients. This consent is given on the condition that the copier pay the stated per-copy fee of $20.00 per article through the Copyright Clearance Center, Inc. (222 Rosewood Drive, Danvers, MA 01923) for copying beyond that permitted by Section 107 or 108 of the U.S. Copyright Law. The per-copy fee of $20.00 per article also applies to the copying, under the stated conditions, of articles published in any *Annual Review* serial before January 1, 1978. Individual readers, and nonprofit libraries acting for them, are permitted to make a single copy of an article without charge for use in research or teaching. This consent does not extend to other kinds of copying, such as copying for general distribution, for advertising or promotional purposes, for creating new collective works, or for resale. For such uses, written permission is required. Write to Permissions Dept., Annual Reviews, 4139 El Camino Way, P.O. Box 10139, Palo Alto, CA 94303-0139 USA.

International Standard Serial Number: 0066-4227
International Standard Book Number: 978-0-8243-1166-7
Library of Congress Catalog Card Number: 49-432

All Annual Reviews and publication titles are registered trademarks of Annual Reviews.

∞ The paper used in this publication meets the minimum requirements of American National Standards for Information Sciences—Permanence of Paper for Printed Library Materials, ANSI Z39.48-1992.

Annual Reviews and the Editors of its publications assume no responsibility for the statements expressed by the contributors to this *Annual Review*.

TYPESET BY APTARA
PRINTED AND BOUND BY FRIESENS CORPORATION, ALTONA, MANITOBA, CANADA

Preface

The year 2011 was the fiftieth anniversary of a number of the founding discoveries of molecular biology, and this volume of the *Annual Review of Microbiology* contains a number of chapters (particularly those of Ullmann, Lengyel, and Hazelbauer) to help us appreciate how we got to those landmark papers, what didn't work as well as what worked, and how we got from there to here. I hope you will find them instructive as well as entertaining.

Among the lessons I learned or was reminded of:

- As annoying as rules and regulations and travel complications are now, as stringent as grant support may be, for most of us, doing science doesn't require being smuggled out of our home countries. A reread of Agnes Ullmann's chapter may be a good antidote for a tendency to complain.

- The best science has always been fostered by intense and frequent communication, with a bit of competition helping it to move along. Peter Lengyel describes the role of a Cold Spring Harbor meeting in fomenting ideas about breaking the code, and Gerald L. Hazelbauer describes the role that interactions at meetings play in the field of chemotaxis.

As usual, my sincere thanks to the Editorial Committee for Volume 66, to Cleo X. Ray for keeping us all on track, and particular thanks to Stephen L. Hajduk as his term on our committee ends. Katherine A. Borkovich and Thomas M. Schmidt were welcome guests for the planning meeting for this volume and have now been added to our Editorial Committee.

Susan Gottesman
Editor

Annual Review of Microbiology

Volume 66, 2012

Contents

A Fortunate Journey on Uneven Grounds
Agnes Ullmann 1

Memories of a Senior Scientist: On Passing the Fiftieth Anniversary
of the Beginning of Deciphering the Genetic Code
Peter Lengyel 27

Yeast ATP-Binding Cassette Transporters Conferring
Multidrug Resistance
Rajendra Prasad and Andre Goffeau 39

'Gestalt,' Composition and Function of the
Trypanosoma brucei Editosome
H. Ulrich Göringer 65

Physiology and Diversity of Ammonia-Oxidizing Archaea
David A. Stahl and José R. de la Torre 83

Bacterial Persistence and Toxin-Antitoxin Loci
Kenn Gerdes and Etienne Maisonneuve 103

Activating Transcription in Bacteria
David J. Lee, Stephen D. Minchin, and Stephen J.W. Busby 125

Herpesvirus Transport to the Nervous System and Back Again
Gregory Smith 153

A Virological View of Innate Immune Recognition
Akiko Iwasaki 177

DNA Replication and Genomic Architecture in Very Large Bacteria
Esther R. Angert 197

Large T Antigens of Polyomaviruses: Amazing Molecular Machines
Ping An, Maria Teresa Sáenz Robles, and James M. Pipas 213

Peroxisome Assembly and Functional Diversity
in Eukaryotic Microorganisms
Laurent Pieuchot and Gregory Jedd 237

Microbial Population and Community Dynamics on Plant Roots and
Their Feedbacks on Plant Communities
James D. Bever, Thomas G. Platt, and Elise R. Morton 265

Bacterial Chemotaxis: The Early Years of Molecular Studies
Gerald L. Hazelbauer 285

RNA Interference Pathways in Fungi: Mechanisms and Functions
Shwu-Shin Chang, Zhenyu Zhang, and Yi Liu 305

Evolution of Two-Component Signal Transduction Systems
Emily J. Capra and Michael T. Laub 325

The Unique Paradigm of Spirochete Motility and Chemotaxis
*Nyles W. Charon, Andrew Cockburn, Chunhao Li, Jun Liu,
Kelly A. Miller, Michael R. Miller, Md. A. Motaleb,
and Charles W. Wolgemuth* 349

Vaginal Microbiome: Rethinking Health and Disease
Bing Ma, Larry J. Forney, and Jacques Ravel 371

Electromicrobiology
Derek R. Lovley 391

Origin and Diversification of Eukaryotes
Laura A. Katz 411

Genomic Insights into Syntrophy: The Paradigm
for Anaerobic Metabolic Cooperation
Jessica R. Sieber, Michael J. McInerney, and Robert P. Gunsalus 429

Structure and Regulation of the Type VI Secretion System
Julie M. Silverman, Yannick R. Brunet, Eric Cascales, and Joseph D. Mougous 453

Network News: The Replication of Kinetoplast DNA
Robert E. Jensen and Paul T. Englund 473

Pseudomonas aeruginosa Twitching Motility: Type IV Pili in Action
Lori L. Burrows 493

Postgenomic Approaches to Using Corynebacteria as Biocatalysts
Alain A. Vertès, Masayuki Inui, and Hideaki Yukawa 521

Index

Cumulative Index of Contributing Authors, Volumes 62–66 551

Errata

An online log of corrections to *Annual Review of Microbiology* articles may be found at
http://micro.annualreviews.org/

Related Articles

From the *Annual Review of Biochemistry*, Volume 81 (2012)

Toward the Single-Hour High-Quality Genome
Patrik L. Ståhl and Joakim Lundeberg

Mass Spectrometry–Based Proteomics and Network Biology
Ariel Bensimon, Albert J.R. Heck, and Ruedi Aebersold

Structural Perspective of Peptidoglycan Biosynthesis and Assembly
Andrew L. Lovering, Susan S. Safadi, and Natalie C. J. Strynadka

Discovery, Biosynthesis, and Engineering of Lantipeptides
Patrick J. Knerr and Wilfred A. van der Donk

Structure Unifies the Viral Universe
Nicola G.A. Abrescia, Dennis H. Bamford, Jonathan M. Grimes, and David I. Stuart

From the *Annual Review of Genetics*, Volume 45 (2011)

Uncovering the Mystery of Gliding Motility in the Myxobacteria
Beiyan Nan and David R. Zusman

Toxin-Antitoxin Systems in Bacteria and Archaea
Yoshihiro Yamaguchi, Jung-Ho Park, and Masayori Inouye

The Rules of Engagement in the Legume-Rhizobial Symbiosis
Giles E.D. Oldroyd, Jeremy D. Murray, Philip S. Poole, and J. Allan Downie

DNA Elimination in Ciliates: Transposon Domestication and Genome Surveillance
Douglas L. Chalker and Meng-Chao Yao

CRISPR-Cas Systems in Bacteria and Archaea: Versatile Small RNAs for Adaptive Defense and Regulation
Devaki Bhaya, Michelle Davison, and Rodolphe Barrangou

Sex in Fungi
Min Ni, Marianna Feretzaki, Sheng Sun, Xuying Wang, and Joseph Heitman

Genomic Analysis at the Single-Cell Level
Tomer Kalisky, Paul Blainey, and Stephen R. Quake

From the *Annual Review of Immunology*, Volume 30 (2012)

Adaptive Immunity to Fungi
Marcel Wüthrich, George S. Deepe, Jr., and Bruce Klein

Microbial Translocation Across the GI Tract
Jason M. Brenchley and Daniel C. Douek

Tolerance of Infections
Janelle S. Ayres and David S. Schneider

microRNA Regulation of Inflammatory Responses
Ryan M. O'Connell, Dinesh S. Rao, and David Baltimore

The Microbiome in Infectious Disease and Inflammation
Kenya Honda and Dan R. Littman

From the *Annual Review of Medicine*, Volume 63 (2012)

New Paradigms for HIV/AIDS Vaccine Development
Louis J. Picker, Scott G. Hansen, and Jeffrey D. Lifson

Emerging Concepts on the Role of Innate Immunity in the Prevention and Control of HIV Infection
Margaret E. Ackerman, Anne-Sophie Dugast, and Galit Alter

Vaccines for Malaria: How Close Are We?
Mahamadou A. Thera and Christopher V. Plowe

Novel Therapies for Hepatitis C: Insights from the Structure of the Virus
Dahlene N. Fusco and Raymond T. Chung

From the *Annual Review of Phytopathology*, Volume 50 (2012)

Stagonospora nodorum: From Pathology to Genomics and Host Resistance
Richard P. Oliver, Timothy L. Friesen, Justin D. Faris, and Peter S. Solomon

Pathogenomics of the *Ralstonia solanacearum* Species Complex
Stéphane Genin and Timothy P. Denny

Genome-Enabled Perspectives on the Composition, Evolution, and Expression of Virulence Determinants in Bacterial Plant Pathogens
Magdalen Lindeberg

Diversity and Natural Functions of Antibiotics Produced by Beneficial and Pathogenic Soil Bacteria
Jos M. Raaijmakers and Mark Mazzola

The Role of Secretion Systems and Small Molecules in Soft-Rot *Enterobacteriaceae* Pathogenicity
Amy Charkowski, Carlos Blanco, Guy Condemine, Dominique Expert, Thierry Franza, Christopher Hayes, Nicole Hugouvieux-Cotte-Pattat, Emilia Lopez Solanilla, David Low, Lucy

Moleleki, Minna Pirhonen, Andrew Pitman, Nicole Perna, Sylvie Reverchon, Pablo Rodriguez Palenzuela, Michael San Francisco, Ian Toth, Shinji Tsuyumu, Jacquie van der Walls, Jan van der Wolf, Frederique Van Gijsegem, Ching-Hong Yang, and Iris Yedidia

From the *Annual Review of Plant Biology*, Volume 63 (2012)

The Evolution of Flavin-Binding Photoreceptors: An Ancient Chromophore Serving Trendy Blue-Light Sensors
  Aba Losi and Wolfgang Gärtner

Plant Innate Immunity: Perception of Conserved Microbial Signatures
  Benjamin Schwessinger and Pamela C. Ronald

Quantitative Imaging with Fluorescent Biosensors
  Sakiko Okumoto, Alexander Jones, and Wolf B. Frommer

Annual Reviews is a nonprofit scientific publisher established to promote the advancement of the sciences. Beginning in 1932 with the *Annual Review of Biochemistry*, the Company has pursued as its principal function the publication of high-quality, reasonably priced *Annual Review* volumes. The volumes are organized by Editors and Editorial Committees who invite qualified authors to contribute critical articles reviewing significant developments within each major discipline. The Editor-in-Chief invites those interested in serving as future Editorial Committee members to communicate directly with him. Annual Reviews is administered by a Board of Directors, whose members serve without compensation.

2012 Board of Directors, Annual Reviews

Richard N. Zare, *Chairperson of Annual Reviews, Marguerite Blake Wilbur Professor of Natural Science, Department of Chemistry, Stanford University*
Karen S. Cook, *Vice-Chairperson of Annual Reviews, Director of the Institute for Research in the Social Sciences, Stanford University*
Sandra M. Faber, *Vice-Chairperson of Annual Reviews, University Professor of Astronomy and Astrophysics, Astronomer, University of California Observatories/Lick Observatory, University of California, Santa Cruz*
John I. Brauman, *J.G. Jackson-C.J. Wood Professor of Chemistry, Stanford University*
Peter F. Carpenter, *Founder, Mission and Values Institute, Atherton, California*
Susan T. Fiske, *Eugene Higgins Professor of Psychology, Princeton University*
Eugene Garfield, *Emeritus Publisher*, The Scientist
Samuel Gubins, *President and Editor-in-Chief, Annual Reviews*
Steven E. Hyman, *Director, Stanley Center for Psychiatric Research, Broad Institute of MIT and Harvard*
Roger D. Kornberg, *Professor of Structural Biology, Stanford University School of Medicine*
Sharon R. Long, *Wm. Steere-Pfizer Professor of Biological Sciences, Stanford University*
J. Boyce Nute, *Palo Alto, California*
Michael E. Peskin, *Professor of Particle Physics and Astrophysics, SLAC, Stanford University*
Claude M. Steele, *Dean of the School of Education, Stanford University*
Harriet A. Zuckerman, *Senior Fellow, The Andrew W. Mellon Foundation*

Management of Annual Reviews

Samuel Gubins, President and Editor-in-Chief
Paul J. Calvi Jr., Director of Technology
Steven J. Castro, Chief Financial Officer and Director of Marketing & Sales
Jennifer L. Jongsma, Director of Production
Laurie A. Mandel, Corporate Secretary
Lisa R. Wucher, Director of Human Resources

Annual Reviews of

Analytical Chemistry
Anthropology
Astronomy and Astrophysics
Biochemistry
Biomedical Engineering
Biophysics
Cell and Developmental Biology
Chemical and Biomolecular Engineering
Clinical Psychology
Condensed Matter Physics
Earth and Planetary Sciences
Ecology, Evolution, and Systematics
Economics
Entomology
Environment and Resources
Financial Economics
Fluid Mechanics
Food Science and Technology
Genetics
Genomics and Human Genetics
Immunology
Law and Social Science
Marine Science
Materials Research
Medicine
Microbiology
Neuroscience
Nuclear and Particle Science
Nutrition
Pathology: Mechanisms of Disease
Pharmacology and Toxicology
Physical Chemistry
Physiology
Phytopathology
Plant Biology
Political Science
Psychology
Public Health
Resource Economics
Sociology

SPECIAL PUBLICATIONS
Excitement and Fascination of Science, Vols. 1, 2, 3, and 4

Agnes Ullmann

A Fortunate Journey on Uneven Grounds

Agnes Ullmann

Institut Pasteur, 75015 Paris, France; email: ullmann@pasteur.fr

Keywords

β-galactosidase complementation, cAMP in *E. coli*, adenylate cyclase toxin

Abstract

I was surprised to be invited to write a prefatory chapter for the *Annual Review of Microbiology*. Indeed, I did not feel that I belonged to that class of eminent scientists who had written such chapters. Perhaps it is because I am a kind of mutant: In spite of having experienced war, both German and Soviet occupations, repeated bombardments, dictatorships, and a revolution, I managed nonetheless to engage in scientific research, thus realizing a childhood dream. After having obtained my Doctor Rerum Naturalium degree in Budapest, Hungary, I was fortunate to meet Jacques Monod at the Pasteur Institute, and this became a turning point in my scientific career. In his laboratory, I contributed to the definition of the lactose operon promoter, uncovered intracistronic complementation in β-galactosidase, and investigated the role of cAMP in *Escherichia coli*. In my own laboratory, together with many gifted students and collaborators, I studied the role of adenylate cyclase in bacterial virulence. This allowed the engineering of recombinant adenylate cyclase toxin from *Bordetella pertussis* for the development of protective and therapeutic vaccines.

Contents

BEGINNINGS 2
FIRST STEPS IN RESEARCH 4
   Actin 4
   Erythrocyte ATPase 4
   In Vitro Protein Synthesis 4
INTERMISSION 5
MEETING JACQUES MONOD 6
BACK IN BUDAPEST 7
THE ESCAPE 8
AT THE PASTEUR INSTITUTE FOR GOOD 9
   Lactose Repressor 9
   Lactose Promoter 10
β-GALACTOSIDASE COMPLEMENTATION 11
THE TALE OF GLYCOGEN PHOSPHORYLASE 13
   Act I 13
   Act II 13
   Act III 14
   Act IV 14
BACK TO THE LACTOSE OPERON 14
   β-Galactosidase Fusions as a Tool 15
CYCLIC AMP EFFECTS IN BACTERIA 15
   Transcription Regulation 16
   Glucose Effects 16
   cAMP and Catabolite Repression 16
   Escherichia coli Adenylate Cyclase 17
FROM *ESCHERICHIA COLI* TO PATHOGENS 17
THE SAGA OF *BORDETELLA PERTUSSIS* ADENYLATE CYCLASE 18
   Structure-Function Relationships 19
   Antigen Delivery 19
   Reporter for Protein Targeting 20
   Bacterial Two-Hybrid System 20

BEGINNINGS

I was born in 1927 in Transylvania, a region in the Carpathian Basin, which now belongs to Romania. To many people, this region is associated with the bloodthirsty vampire Count Dracula. Transylvania was always a multiethnic and multilingual region; it belonged intermittently to Hungarians, Romanians, and Germans and was even part of the Ottoman Empire. It was considered one of the cultural centers in Central Europe. Most cities in Transylvania carried three names in three different languages, the last one used depending on the last occupant.

    As a child, I was trilingual, speaking Hungarian, Romanian, and German. According to distinguished linguists, when a child begins talking by speaking three languages, she ends up speaking none of them correctly. Alas, in my case, they were right. I attended public schools and, from age eleven, I was more interested in sports than in what was taught in school. I was doing athletics, rowing and playing tennis and ping-pong (I even earned medals!). Sports helped me in many ways. I was twelve when World War II began. I spent much of my time on sport fields (except during bombardments), which kept me away from the daily war problems. I had an excellent athletic trainer who taught me how to practice correctly. Once, having lost a 100-m sprint during a competition, I had asked him whose example I should follow to be successful. He answered that I didn't have to compare myself to anybody; all I had to do was to carry a stopwatch while running and try to do better than the day before. Maybe that lesson helped me later in science.

After finishing college, I decided to study science at the University of Cluj in Romania. My father, who had always hoped that I would become a doctor like my grandfather and my aunts, was not entirely surprised by my choice. Apparently, as a child, I constantly pestered him with odd questions such as "How come cows who eat green grass give white milk?"

My first real interest in science began when I was 14, after my father had given me for my birthday a Hungarian translation of Paul de Kruif's *Microbe Hunters*, a beautifully bound book with many pictures of microbes and scientists such as Antoni van Leeuwenhoek, Louis Pasteur, Robert Koch, and Paul Ehrlich. I had dreamed of becoming a microbe hunter and had chosen Pasteur as my hero because of his incredible accomplishments in so many fields. Seventeen years later, my dream was fulfilled: I began working at the Pasteur Institute in Paris. But many things happened in the years in between.

I earned my high school diploma (baccalauréat) in 1945 from a Romanian public school in Arad, Transylvania (now Romania), just as the war ended. I then attended the Faculty of Science at the University of Cluj. During World War II, Cluj, a charming city in the hills of Transylvania, belonged to Hungary and still bore its Hungarian name of Kolozsvár (Klausenburg in German) when I arrived there. Transylvania returned to Romania after the Paris Peace Treaties in 1947. During my first two years at the University, Cluj was a kind of no-man's-land. The civil population had fled Transylvania, not knowing to whom it might belong next. The Opera House was running. Attendance was almost free for students; furthermore, it was heated during our ice-cold winters. Artists and poets came from many Western countries: Jacques Thibaud and Yehudi Menuhin played violin sonatas for the students, and Tristan Tzara lectured on the Dada movement and surrealism, which I had never even heard of. Furthermore, even though the postwar period was not easy and we were often hungry and cold, we were totally immersed in a true climate of culture. By contrast, science instruction at the university was so inadequate that, after a couple of years, I decided to transfer to the University of Budapest in Hungary. But then, of course, I had to switch from Romanian to Hungarian.

In 1949, I obtained my chemistry diploma from the University of Science in Budapest. Predictably, in addition to the scientific subjects, I had to pass an exam in Marxism. Then, I started to work in the Department of Medical Chemistry, directed by Bruno F. Straub, a former student of the Nobel Prize winner Albert Szent-Györgyi. Everyday life was difficult not only for political reasons but also because salaries were very low. The Soviet Army had occupied Hungary since the end of the war, and a real Communist dictatorship was established in 1947. Later, Hungary became the Hungarian People's Republic. Political purges and show trials began; around 100,000 persons were arrested or deported to concentration camps and several thousands were executed. For several years, Hungary experienced the harshest dictatorship in Europe. I was also arrested but miraculously released after 48 hours. My schoolmate, who was arrested at the same time as me, was sentenced for no reason and, as I later learned, executed.

FIRST STEPS IN RESEARCH

Actin

Since the discoveries of Szent-Györgyi on muscle contraction, research at the Department of Medical Chemistry in the late 1940s had been dedicated to muscle proteins. Just before I arrived in 1949, Straub had discovered that actin contained bound ATP and that, during its polymerization into filamentous F-actin, ATP was hydrolyzed to ADP, yielding ADP-F-actin. I was charged with studying the mechanism of this polymerization as a function of ATP dephosphorylation, the hypothesis being that the energy of polymerization was provided by the hydrolysis of ATP. By studying the kinetics of polymerization as a function of temperature, I found that at 0°C F-actin still contained bound ATP instead of ADP. To explain this result, I suggested that actin itself might exhibit some ATPase activity. This idea was rejected offhand as an unorthodox hypothesis and the project was brought to an end. Nevertheless, many years later, it was shown by others that actin, indeed, had intrinsic ATPase activity.

It may be interesting to recall how we measured ATP concentrations when no spectrophotometer was available. We used a method set up by Szent-Györgyi himself based on the observation that a highly polymerized actomyosin gel dissociates and depolymerizes in the presence of ATP. So, we measured the kinetics of depolymerization as a function of known concentrations of ATP in a viscosimeter kept in ice. Once the reaction was calibrated, it was easy to determine the experimental ATP values.

Erythrocyte ATPase

A small group in the laboratory was studying the potassium permeability of erythrocytes. They found that the intracellular concentration of K^+ was dependent on the level of ATP. This was in the 1950s, when assays for measuring the concentration of K^+ were extremely time-consuming. We thought that an ATPase might be involved in potassium permeability; therefore, I decided to purify this hypothetical ATPase because the inorganic phosphate generated could easily be determined colorimetrically. At that time, one could not purchase ATP. We had to prepare the material ourselves from rabbit skeletal muscle, which was time-consuming. When I accumulated enough ATP, I started to purify the enzyme from different blood sources, but to my surprise, I could never purify it more than approximately 50-fold, nor could I obtain it in soluble form. Fortunately, an electron microscope was available in the building, and to my amazement, all the purified preparations were revealed as beautiful erythrocyte ghosts. Clearly, the enzyme was membrane bound and I could show that treatments that damaged the membrane integrity led to a loss of enzymatic activity. Due to my inexperience, I was not able to demonstrate any correlation between potassium transport and ATPase activity; I also missed the Na^+ and K^+ activation of the enzyme. Nevertheless, to my knowledge, this was the first observation of a membrane-bound ATPase. The results were published in German in 1952 in a practically unknown journal, *Acta Physiologica Hungarica*, with the obligatory abstract in Russian.

In Vitro Protein Synthesis

In the early 1950s, I became interested in protein biosynthesis. One had to choose a model system compatible with our limited working possibilities. We found a publication from Fritz Lipmann's laboratory describing the in vitro synthesis of cytochrome *c*. Even though I strictly followed their experimental conditions, I could not for months duplicate their results and finally gave up. Straub thought that I was totally incompetent if I couldn't even repeat an experiment coming from Lipmann's lab, but because I carried heavy teaching duties, he needed me around and accepted that I choose another project.

The cytochrome c story came to an end when many years later I met Fritz Lipmann and asked him what I could have done wrong for not having been able to reproduce his experiments. He said, "Thank God you could not reproduce them." It turned out that one of his collaborators had falsified the data and the paper had to be retracted. Lipmann was genuinely sorry for me and, from then on, each time I was in New York, arranged for me to stay in the guesthouse of Rockefeller University. Several years later, I told that story to Albert Szent-Györgyi, and he remarked that cheating was unfair toward Fritz because "Fritz was a person who only perceived the music of nature without listening to the background noise."

After my failure with cytochrome c, a small group of us in Straub's laboratory decided to study the formation of α-amylase under a variety of in vitro conditions with the hope of obtaining some information about the mechanism of its biosynthesis. L.E. Hokin had described in 1951 an increase of amylase activity in pancreas slices in vitro. These data were perfectly reproducible and the amylase system kept me busy for about four years. I set up different cell-free systems and studied the requirements to obtain an increase in amylase activity. In retrospect, much work was invested but the cell-free systems we had set up were ill-defined and some of the conclusions reached were not straightforward. When we finally obtained radioactive glycine, thanks to a microscale one-step isolation procedure we had devised, we were able to demonstrate a specific incorporation of glycine into amylase. The microscale method consisted of adsorbing amylase-containing extracts at 0°C on insoluble starch we had prepared, washing it exhaustively in the cold, and finally dissociating the starch-amylase complex in buffer at 37°C. The method yielded practically pure amylase. When we incubated pancreas slices in the presence of radioactive glycine and measured the incorporation of radioactivity in purified amylase, we observed a considerable time lag in the labeling of the enzyme compared with that of total proteins. This finding suggested the presence of some intermediary steps (a precursor?) in the synthesis of amylase. In cell-free systems, we could observe an increase in amylase activity but no specific labeling of the enzyme. Because ribonuclease inhibited this reaction, we concluded that RNA was somehow implicated in the process. The precursor hypothesis was never substantiated, and at that time, we were not able to define the intermediary steps in enzyme formation. This work led to eight publications, seven of them in *Acta Physiologica Hungarica*, and formed the substance of my doctoral thesis.

INTERMISSION

As mentioned above, when I arrived in Budapest in 1947, the Communist dictatorship was already installed and the Soviet Army occupied Hungary. From 1949 onward, political trials followed by death sentences were daily events. Those who were not arrested or purged lived in constant fear.

University studies were totally corrupted and the teaching of genetics was forbidden. Only the theories of the Russian agronomist Trofim Lysenko were taught. He rejected the science of genetics, believed in the heritability of acquired characteristics, and claimed that their inheritance played a major role in evolution. He considered Mendel's principles incompatible with dialectic materialism, which pleased Stalin. He not only succeeded in ruining Soviet agriculture but also caused the elimination of the best Soviet geneticists. It was during the Lysenko affair, in 1948, that I first heard of Jacques Monod. Western newspapers were banned in Hungary but a friend of mine succeeded in smuggling a newspaper called *Combat*, directed by Albert Camus and dated September 19, 1948. In it was an article by a certain Dr. Monod entitled "The Victory of Lyssenko

Has No Scientific Basis." To me the article came as a revelation and I decided on the spot that I would meet this Dr. Monod one day. Indeed I did 10 years later.

In 1949, when I started research and teaching, the political climate was not at all conducive to quiet scientific pursuits (though this was certainly not my only excuse for having been unsuccessful in research). Teaching was not easy: One was compelled to tell the students that all the main discoveries in science were made by Russian scientists. Once, at the end of a general chemistry lecture, a student quietly asked me who was the Western scientist who had really discovered the Mendeleev Periodic Table. He couldn't believe me when I told him that it really was Mendeleev. Everything associated with the Russians was summarily dismissed.

After Stalin's death in 1953, Hungary's Communist regime started to crumble. The voice of the opposition was no longer silenced; many of the persons who had been deported or arrested and who were still alive were rehabilitated. Public debates occurred everywhere (I participated in many of them) and justice was demanded for those guilty of murders. But we did not anticipate the 1956 Revolution. It actually started on October 23 during the course of a huge demonstration. We were about 50,000 marching peacefully to pay homage to the heroes of the 1848 Hungarian Revolution. Toward its end, when the crowd began to disperse and a few of the participants went to the radio station to deliver an appeal for freedom, the men of the secret police began to shoot. This happened next to the university while we were holding a meeting to discuss the follow-up of our afternoon protest. That evening, one of the symbols of the dictatorship, a giant bronze statue of Stalin, was toppled and smashed to pieces. The next day, groups of young people with newly acquired weapons roamed Budapest, and in the following days, the uprising spread throughout the country. On October 27, the Revolutionary Committee of Intellectuals was founded and established itself in one of the university buildings. I joined them there and I helped them coordinate their daily activities. It was an unbelievably exciting period of time because, for once, we felt free. We stayed there day and night because nobody owned cars and public transportation had stopped. Young people requisitioned Soviet tanks (unbelievably, in those days the Russians let them do that). Once, they even picked me up and drove me home. That was the noisiest ride I ever experienced.

We were full of enthusiasm and totally naïve. We really believed the Russian promises that they would withdraw from Hungary. But at 4 AM on November 4, 15 Russian armored divisions equipped with 6,000 tanks launched their attack on the unsuspecting country, putting an end to the revolution. Fighting, however, went on for many days and chaos continued to reign for many months. Hundreds of people were arrested and more than 1% of the population fled the country. By the end of the year, I was back at the bench and resumed teaching, but times were hard because, once again, we were living in constant fear.

MEETING JACQUES MONOD

In 1957, I received an invitation to attend a Symposium of the Society for Experimental Biology in London. Obtaining a passport was like a lottery. I was 30 years old and I had never left the country. Unexpectedly, I received my passport but I still needed an English entry visa plus transit visas to travel through Austria, Switzerland, and France. For unknown reasons, the British refused to grant me a visa. Fortunately, I knew the French consul in Budapest and he provided me with a French visa. So, I left for Paris (with only the five dollars I was allowed to carry with me), with the hope that I would meet the famous Dr. Monod who had been fighting against Lysenko's theories.

After arriving in Paris in January 1958, I immediately asked for an appointment with Professor Monod, which I rapidly obtained. I was nervously waiting in his secretary's office at the Pasteur Institute, fully expecting to meet a gray-bearded, distinguished old gentleman. To my surprise, a young fellow came whistling down the corridor and introduced himself as Jacques Monod. He was friendly but looked very busy. In fact, that was when the famous Pardee–Jacob–Monod (the PaJaMa) experiment, which led to the operon model, was being carried out. The fastest way to get rid of me was to invite me to present a seminar the next day. Following that, Monod politely asked me what I was doing in Paris and how long I intended to stay. Gathering all my courage, I told him that if he allowed me to work in his lab, I could extend my stay to six weeks; otherwise, I would have to return to Budapest very shortly. The next morning, I started working with François Gros, who at that time was interested in the mechanism by which chloramphenicol inhibited protein synthesis. I learned, for example, how to prepare bacterial extracts and how to measure aminoacyl-tRNA complexes, but I did not obtain any conclusive results during my stay (the inhibitory effect of chloramphenicol on peptidyl-transferase was discovered many years later).

One Saturday morning, the Spinco centrifuge in the corridor refused to start. A group of people was standing around it, rather perplexed, trying to figure out what was going on. I approached them asking if I could be of help. They looked at me incredulously, wondering how somebody coming from an Eastern country could have the technical know-how to repair a Spinco. I solved the problem in two minutes: The needle that makes the electric contact with the rotor had fallen off. That was for me an enormous success. They didn't know that, in Hungary, many such instruments were embargoed and could only be delivered in separate parts; we had to put everything together ourselves. Before coming to Paris, we had just reconstructed a Spinco but at the end of putting it together we found a needle in a small box. After working our way through the instructions for another hour, we finally discovered that it was an essential item that had to be installed last.

My collaboration with François Gros developed into a close friendship to the point that I felt I could speak to him openheartedly. One day, I confided to him that I wanted to leave Hungary for good, come to France, and go on working forever in the lab. Somewhat puzzled, he advised me to discuss this problem with Monod, but since I did not dare do it, he did it for me. The next day Jacques Monod invited me for dinner at his home. After a few hours during which I recounted what had happened during the revolution, the unending menace of persecution under which we were living, constantly fearing for our security, Jacques told me that he was willing to do everything in his power to help me. When I asked him why, he simply answered: "It is a matter of human dignity."

BACK IN BUDAPEST

After almost two months in total bliss, I had to leave Paris, which I did with a heavy heart. I had to go back to Budapest; otherwise I would have endangered my husband, my colleagues, and my friends. But I knew Jacques Monod would try to help. Indeed, a few weeks later, I received a letter from him indicating that I had to return to Pasteur for a few months to finish some crucial experiments. Of course, it was only a pretext for me to renew my passport.

Meanwhile, I wanted to take advantage of the techniques I had learned with François Gros and decided to study the mechanism of action of streptomycin. My husband, Tom Erdös, a former student of Szent-Györgyi who had been jailed after the revolution for a few months for political reasons, was working in a large tuberculosis institute where nonpathogenic strains of mycobacteria, including some streptomycin-sensitive, resistant and dependent strains of *Mycobacterium friburgiensis*, were available. The strains were certainly not isogenic but behaved as expected in bacterial cultures. We prepared cell extracts and looked for the incorporation of radioactive tyrosine (the only labeled amino acid we had available at that time) in the presence or absence of streptomycin. We observed no effect of streptomycin on the formation of the aminoacyl-tRNA complex but, to our great satisfaction, found that streptomycin did inhibit the incorporation of labeled tyrosine into a subcellular fraction. It was the first time that streptomycin had been shown to inhibit protein synthesis. We were even more satisfied when we could show that in streptomycin-resistant strains, streptomycin did not inhibit amino acid incorporation whereas in streptomycin-dependent strains, the antibiotic was required for incorporation. Although we could not characterize our "subcellular fraction" (later identified as ribosomes), we sent two short letters to *Nature* that were rapidly published (8, 9).

Around February 1959 I obtained for the second time a passport that would allow me to return to Paris. In no time, I secured the needed French and transit visas and left Budapest with the hope of never coming back. No return meant that something would have to be organized for my husband, Tom, to leave Hungary illegally. That was Jacques Monod's original idea: He figured that it would be easier to smuggle one person out rather than two, and thought that we would have plenty of time to plan for his escape during the six months I would be in Paris. With Tom, we agreed on a simple telephone code by which I would let him know how to proceed if his escape could be organized. This was essential because, since his arrest, he was under constant police surveillance. When my train departed Budapest, I was hoping that everything would work out as we had planned.

THE ESCAPE

François Gros was waiting for me at the Paris railway station and drove me directly to Pasteur. At that time, François was trying to identify messenger RNA, whose existence had been suggested by the PaJaMa experiment (37). He proposed that I share this task with him. For many weeks, we labeled *Escherichia coli* cultures with several millicuries of ^{32}P. After centrifugation, we ground the bacteria with alumina powder in a cooled mortar, suspended the mixture in buffer, and loaded the centrifuged suspension on a sucrose gradient. After eight hours of centrifugation, we counted the drops. The ribosomes were well separated but the radioactivity was never where it was expected to be found. When I asked how long I had to grind the bacteria, someone told me until I heard "the bacteria crying." All these procedures would seem quite folkloric today. A year later, when François spent some time in Jim Watson's laboratory, he finally succeeded in identifying the messenger (16).

Meanwhile, with the help of a friend (a Hungarian refugee who was living in Vienna, close enough to Hungary), Jacques Monod was trying to arrange for Tom's escape. Several strategies were considered, such as constructing a special car with a hiding place or stowing away aboard one of the Danube tankers. That latter plan nearly went through but for the first time in 50 years, the Danube was so shallow that all river traffic was interrupted. Jacques had collected money from many friends and colleagues but all our successive plans collapsed. We finally ran out of time and I had to return home. But I never doubted Jacques's total commitment to get us both out of Hungary, one way or another.

Before leaving, I had an idea of how we should correspond with Jacques and his secretary, Madeleine Brunerie, to avoid the surveillance by the secret police. For the first time, my work with amylase paid off. Knowing that one can detect starch with iodine, I proposed to write my messages with a starch solution that Jacques or Madeleine would reveal with an iodine solution. I demonstrated how this would work and Jacques liked the idea because it reminded him of his time in the Resistance. François Jacob and his wife, Lise, accompanied me to the railway station. Lise bought me books by Albert Camus, Jean-Paul Sartre, and Simone de Beauvoir, among others, most of which were prohibited in Hungary. So I spent the 20 hours of travel time reading those books, throwing away all those that were banned before reaching the border. I was particularly sorry to get rid of the books by George Orwell. But in Budapest, a colleague working in the zoology department succeeded in ordering a single copy of Orwell's *Animal*

Farm, along with other scientific books, and it was then circulated among several hundreds of us. Another friend in the history department never dared to order Orwell's *1984*: It was simply too dangerous.

Between September 1959 and April 1960, nothing happened in Budapest, except some exchanges of messages written with starch on the cover sheets of records by Béla Bartók or, more often, Mstislav Rostropovich (Jacques Monod was a cello player). Most of the strategies for getting us out of Hungary had failed. All Monod's correspondence concerning these plans, their organization, and attempts to raise the needed money can now be found in several folders in the *Fonds Monod* at the Archives of the Pasteur Institute. Once in Paris, it was a breathtaking experience for me to go over them.

In May 1960, Jacques arrived in Budapest with another carefully prepared scheme for our escape. Officially, he had been invited by the Hungarian Academy of Sciences to present some lectures. His first lecture on regulation was a tremendous success: It was the first time since 1948 that the word "gene" had been pronounced. His entire visit was a memorable event for the Hungarian scientific community. The new escape plan he presented seemed feasible to us: We were to meet at a given place with an Austrian "tourist" owning a trailer modified to contain an appropriate hiding place. Once inside it, he would drive us to Austria. At that time, Hungarian borders were tightly sealed and crossing could be nightmarish. Jacques asked us what would happen if the police caught us? We answered "prison for twenty years." When we assured Jacques that we were definitely willing to take that risk, we drove along the Danube to find an adequately isolated place where the rigged trailer could pick us up.

Our escape was scheduled for a Saturday in June 1960. The trailer arrived as planned and we crammed ourselves under a bathtub. At the border, two customs officers searched every corner of the vehicle with a flashlight for more than one hour. It was an absolute miracle that they did not catch us. The officers finally left the trailer and, after a few more controls, we were allowed to proceed. Shortly afterward, I smelled gasoline. Because there were no gas stations on the Hungarian side of the border, I knew we were in Austria.

We became political refugees in Vienna, and our only possessions were the clothes on our backs. Monod and other friends made arrangements to send some money and thanks to Philippe Monod, Jacques's brother, a diplomat who would later be named ambassador to Australia, we rapidly obtained all the official documents that would permit us to travel to France.

AT THE PASTEUR INSTITUTE FOR GOOD

Having participated emotionally in this rescue operation for two years, Madeleine Brunerie was waiting for us at the Paris train station and offered us a cup of coffee and a "liberty croissant." Philippe Monod kindly lent us his apartment. Jacques secured for each of us a Rockefeller Fellowship. I stayed at the Pasteur Institute while Tom worked at a research institute outside Paris.

Lactose Repressor

I wanted to begin working immediately. Jacques Monod told me that a few months earlier François Gros and Georges Cohen had attempted to isolate the Lac repressor but had abandoned that project. He suggested that I take it over. At that time, the Lac repressor was thought to be RNA that would bind inducer. I started to induce wild-type, constitutive, and repressor-negative bacteria with labeled ^{14}C-methyl-thiogalactoside (TMG) and succeeded in isolating labeled RNA. However, we still obtained stable TMG-RNA complexes in strains that were deleted for the lactose region and furthermore inducers other than TMG failed to bind to RNA. Therefore, we had to relinquish the idea that the repressor could form a stable complex with inducer. Later, when I analyzed certain TMG-labeled RNA fractions, I found to my great surprise that it was the methyl group, not TMG, that was bound to RNA: TMG served as a methyl-donor to yield methyl-adenine. In short, I had discovered RNA methylation before Ernest Borek! But because our goal was to isolate the repressor, I never investigated this phenomenon further.

At the same time, a number of other pieces of evidence came to light suggesting that the repressor might be a protein and that its inactivation would not necessarily involve a covalent interaction. Because the repressor was supposed to exhibit a strong affinity for both DNA and the inducer, I thought of making use of this property to purify it. I set up DNA-affinity columns (the first: it was 1960) and tried to elute specific fractions with inducer. The results were consistently negative. We then turned to a genetic approach, i.e., measured β-galactosidase basal levels in diploid strains harboring varying numbers of operator and regulatory genes. The data indicated that the number of repressor molecules was no higher than 10 per cell. This was so discouraging that I asked François Jacob if it were possible to obtain a strain that would make ten or a hundred times more repressor? He said that it was hopeless. Well, a few years later, Wally Gilbert & Benno Müller-Hill (13) did obtain those mutants, which allowed them to purify the Lac repressor and show that it was indeed a protein.

Lactose Promoter

By the end of 1963, most of the major concepts regarding the structure of the Lac operon had been delineated by Jacob and Monod. One question that remained to be solved was the site at which the transcription of the structural genes would commence. Genetic evidence suggested that transcription would not start at the operator. If an additional site existed, then joining structural genes of one operon to the control system of another would result in the subordination of those structural genes to the regulatory mechanism of the other operon. To check this prediction, without running the risk of deleting essential genes, François Jacob isolated a series of deletions on *Flac* episomes. The deletions extended on one side over various lengths of the *lacZ* gene (they were selected for the presence of a functional permease); on the other side, we expected them to extend into the alkaline phosphatase region. I participated in this project and was hoping that François would isolate a *pho-lac* fusion in which the Lac permease would be repressed by inorganic phosphate. But after having analyzed several thousand clones without success, François obtained deletions extending into *purE*. We proceeded like workers on an assembly line: François analyzed the petri dishes, his technician cultured the clones, and I analyzed the cultures to determine whether the permease would be regulated by purine (not an easy assay, by the way). I usually arrived at the lab late but also left late. Every morning, François would be pacing the corridor, growing increasingly impatient, and greet me with such remarks as "It is quite late to start." In the evening, he left early to have dinner with his children, but before leaving, he passed by my lab to ask whether there was anything new. "Not yet," I would always reply. Each time he mumbled something like, "Had you started earlier... call me if there is anything new."

One evening around 9 PM I found a clone in which purine repressed both permease and transacetylase synthesis. The *purE-lacZY* fusion had just created a new operon. Instead of calling François right away, I went to a movie. I called him when I came back, around midnight. Obviously I woke him up. All he said, curtly, was "Merci." The next day I arrived even later than usual. François did not make any remarks; he only asked to see the results. He then told me that next time he would prefer that I call him before 10 PM. It was the last time I ever heard him comment on my working hours.

These experiments represented the first example of in vivo genetic engineering (21). Gene fusions were to become a major tool to study genetic analysis and regulation. I sent a telegram to Monod, who was at the Salk Institute in La Jolla, California, to tell him of the result and took off to go skiing for 10 days.

β-GALACTOSIDASE COMPLEMENTATION

Early studies by Melvin Cohn in Jacques Monod's laboratory in the 1950s demonstrated that the *E. coli* β-galactosidase was a tetramer composed of four identical polypeptides (36). Later, David Perrin's work (38) on complementation between various *lacZ* point mutants favored the idea that the *lacZ* gene consisted of a single cistron.

Looking for gene fusions while working on the *lac* promoter, François Jacob obtained a large collection of deletions. Some of those covering the *lac* regulatory region and extending to various sites of the Z gene did complement certain promoter-distal point mutants. This raised the possibility that the Z gene might consist of several cistrons. François refined the in vivo complementation tests while I set up an in vitro system with crude extracts of different deletion mutants. Fortunately, the two approaches converged with the conclusion that all deletions not extending beyond a certain point in the gene (which we called the ω-barrier) would complement all promoter-distal point mutants or deletion mutants of the Z gene. We called this phenomenon ω-complementation. I purified the ω-peptide from the largest deletion still displaying complementation and showed that the ω-region represented about one-third of the total genetic length of *lacZ* (53). In 1994, Matthews and colleagues (22) determined the three-dimensional structure of β-galactosidase and showed that the polypeptide chain folded into five sequential domains. They concluded that the C-terminal third of β-galactosidase could fold independently from the rest and probably corresponded to the ω-fragment. It was most rewarding to see a crystallographic confirmation of our results 30 years after they had been obtained by genetic and biochemical approaches.

In the main corridor of the Institute, called *avenue de l'opéron*, discussions whether the *lacZ* represented one or several cistrons were reawakened when Gary Craven in Chris Anfinsen's laboratory presented some evidence that the β-galactosidase monomer might consist of several peptides. Chris invited me to spend a few months in his laboratory at the National Institutes of Health to settle this point, but by the time I arrived in 1965, the *lacZ* had once again reverted to a single cistron. That was my first visit to the United States: I learned some protein chemistry, met many new colleagues, and gave a few seminars at different institutions. However, my trips were restricted and I could only travel with the permission of the US Attorney General's office (they granted most of my requests). Indeed, after an investigation that lasted a few months, the visa office of the US Embassy in Paris had concluded that I was an Eastern European spy, so I was issued only a "restrictive" visa that did not allow me to move freely throughout the United States.

In fact, I remained subject to these idiotic restrictions for the next 20 years. One of my later trips took a turn to the absurd. Around 1970, I had gone to the US Embassy for a visa to attend a scientific Battelle Conference in Seattle, Washington, to which I had been invited. The civil servant involved insisted in examining my travel reservations. When I told him that I was flying from London to Seattle, he asked how I would travel from Paris to London. Exasperated, I told him I would swim. He became furious and refused to issue the visa until I showed him my Paris-London ticket, but that was not the end of it. The London flight was delayed such that we arrived in Seattle a few minutes past midnight, and I was refused entry because my restrictive entry visa had expired at midnight. I tried to explain what had happened to no avail. Fortunately, I had met Manfred Eigen at Heathrow Airport; he was also flying to Seattle to attend the same meeting.

Manfred had been awarded the Nobel Prize in Chemistry three years earlier and he told them that if they did not allow me in, he wouldn't enter either. The greatly embarrassed customs officer finally relented and we were both allowed to proceed in the company of Eddy Fischer, who had come to pick us up.

Once back at the Pasteur Institute, I tried to understand the mechanism of ω-complementation. I purified both in vivo– and in vitro–complemented enzymes, studied their properties, and filled up several notebooks. The conclusion was straightforward: ω-Complementation involved noncovalent association between peptides corresponding to different fragments of the wild-type chain. This finding was confirmed by immunological studies carried out with Franco Celada. Having shown that antibodies directed against the ω-peptide reacted with the wild-type enzyme, we concluded that ω was able to fold itself into the correct wild-type structure, indicating that it had to contain an independent nucleation center (4). This appeared evident after examining Matthews's X-ray structure (22).

Once ω-complementation was settled, I wondered whether other combinations of deletion mutants would generate an active enzyme. When I mixed extracts containing partial deletions of the promoter proximal segment of *lacZ* with extracts of β-galactosidase-negative mutants whose promoter-proximal segments were intact, I recovered enzymatic activity: α-Complementation was born! Jacques liked it, but François was somehow reluctant to accept those results because he had not observed those effects previously on eosin-methylene blue (EMB)-lactose plates. After a while, by choosing the appropriate heterozygous strains, he eventually confirmed the in vitro results but considered that in vivo α-complementation was very inefficient (49). Today, when α-complementation has become the basis for the commonly used blue/white screen to identify recombinant DNA (insertion of foreign DNA into plasmid-borne *lacZα* results in the abolition of α-complementation), one would smile at those barely positive in vivo results. The reasons for the EMB data became clear later: *Flac* episomes used by François were not multicopy vectors and EMB is much less sensitive than X-gal.

I showed that the α-peptide could be obtained in a practically pure state by boiling a crude extract in 6M guanidine in a pressure cooker. While studying the mechanism of α-complementation, several pieces of convergent data suggested that the role of α-peptide was conformational rather than structural (50). This was confirmed later by the crystal structure of β-galactosidase. It showed that the polypeptide chain contains an extended segment at the N terminus (the α-peptide), which participates in subunit interaction and stabilizes it (22).

Encouraged by the discovery of α-complementation, I went on to look for a β- or γ-complementation by mixing three different extracts of deletion strains but never succeeded in recovering enzymatic activity. One day, at lunchtime, I was complaining to André Lwoff about my failure to detect complementation with three partners. "Nothing surprising about that," he said. "You are not yet French enough to be ready for a ménage à trois." So I gave up the project.

I enjoyed the fact that α-complementation had an unexpectedly long career, even though no reference was ever made to our work in subsequent papers reporting new cloning vectors based on α-complementation. An amusing story happened in the early 1980s at a Cold Spring Harbor meeting on bacteria and phage. A nice young man presented a new cloning vector based on a screening method involving the appearance of Lac$^+$ clones. In short, he explained what α-complementation was. Not wanting to intervene publicly, I approached him during the coffee break and told him that the phenomenon he just described had been discovered more than 15 years ago. "By whom?" he asked. "By me, it so happens." He turned pale and, panic-stricken, asked, "Did you patent it?" "No," I said. "I only published it." He didn't even ask if it was in a high-impact-factor journal.

THE TALE OF GLYCOGEN PHOSPHORYLASE

Late one evening in 1961, Jacques walked into my lab. His tie was loose and he looked tired and worried. He stood silently at my bench and after a few minutes he said, "I think I have discovered the second secret of life." I looked at him rather alarmed and I suggested that he sit down and have a drink. After downing his second or third glass of scotch, he started to explain his discovery of a phenomenon he had already named "allostery." He then pointed out that the regulatory role of allosteric proteins was absolutely fundamental, arguing that the "invention" of indirect allosteric interactions in the course of evolution opened the way to an infinite number of possible regulations. During the following month, one spoke only of allostery in the lab. In fact, it couldn't be helped because whenever we discussed whatever topic, we came back to allostery within minutes. During this period, Jacques desperately searched for a suitable experimental model with the idea of doing some bench work himself. He asked me one day whether it was difficult to prepare muscle phosphorylase and whether it was easy to assay. I answered that it was a matter of just a few days; I had done it myself in Budapest.

Act I

Around that time, P. Roy Vagelos arrived to work on bacterial genetics for a postdoctoral year. (He had isolated a deletion of the whole lactose region, called RV X74, which biotech companies sell as X74 even today.) He was, however, willing to help me perform sucrose gradient experiments to determine whether 5′-AMP, the activator of dimeric phosphorylase b, acted as an allosteric effector or whether it converted the inactive dimer to the active tetrameric form as proposed earlier by Carl and Gerty Cori. Our data clearly showed that the S values of phosphorylase b were identical whether the centrifugation was carried out in the presence or absence of 5′-AMP. This strongly suggested that the activation of phosphorylase b by 5′-AMP was not due to its association into the tetrameric a form (55). Jacques was pleased with the phosphorylase b system and set up a dye-binding procedure inspired by the observations of Eraldo Antonini that the effect of bromthymol blue (BTB) on hemoglobin was due to an allosteric heme-heme interaction (1). It therefore seemed of particular interest to test whether phosphorylase would exhibit a similar behavior, which it did. The dye associated with phosphorylase b was significantly increased in the presence of 5′-AMP. Interestingly, phosphorylase a bound the same amount of BTB as did AMP/phosphorylase b, but the nucleotide had no additional effect. We concluded that in the presence of its allosteric effector, 5′-AMP, phosphorylase b underwent a reversible conformational change that increased its capacity to bind certain ligands (55).

Act II

For various reasons, we had to determine the molecular weight of the subunits (protomers) of oligomeric proteins. I thought that it would be easy, in 1966, to do so in 6M guanidine using an analytical ultracentrifuge. By carrying out the centrifugation with proteins whose molecular structures were well known, we could devise a calibration curve. Michel Goldberg, whose knowledge in physical chemistry far exceeded mine, warned me that the experiment could be done, but a number of specific parameters would have to be taken into consideration for a correct evaluation of the results. Because the ultracentrifuge used schlieren optics, photos had to be taken every 6 hours. Michel took the first picture at 6 AM and I took the last at midnight. We ran six different proteins whose subunit molecular weight varied from 14,000 to 135,000 Da and found the expected values for all proteins except phosphorylase b, which repeatedly gave a value 92,500 Da instead of 125,000 Da, published by the Coris 12 years earlier. I was terribly disappointed because after several months of hard work I thought that the method had to be abandoned. When we showed those results to Jacques, he was not discouraged at all and simply said, "Believe in your experiment; for once, Cori might have been wrong." I called Eddy Fischer in Seattle for his advice. He told me that they had just submitted a manuscript in which they had revised the molecular weight of phosphorylase b and found a value of 185,000 Da for the dimeric enzyme (43). That corresponded almost exactly to what we had found, so we rapidly sent in a paper to the same journal (48).

Act III

Henri Buc, who was working on the allosteric properties of phosphorylase b, came in one day worried, saying that if our new molecular weight value was correct, the number of bound 5'-AMP would no longer be 1 per monomer, as had been calculated for a molecular weight of 250,000. With Marie-Hélène Buc and Michel Goldberg, we decided to reexamine several parameters to solve this discrepancy. We measured molecular weights in nondenaturing conditions by analytical ultracentrifugation and light scattering. From the different measurements we obtained values between 190,000 and 198,000 Da. We then measured the molar extinction coefficient at 280 nm by five different methods and, to our satisfaction, found a value of 13.2 rather than 11.8, as published earlier. With these new values in hand, we established that there are precisely 2.0 stereospecific 5'-AMP sites per phosphorylase b dimer (3).

Act IV

When Maxime Schwartz joined the lab in 1963, Jacques suggested that he start working with me to learn how to handle rabbit phosphorylase b, with the idea that he would later study its allosteric conformational changes. But Maxime rapidly switched to a bacterial phosphorylase with the hope—at least Jacques's hope—that the bacterial enzyme would also be allosteric and that it would be amenable to genetic analysis. Maxime did indeed find a polysaccharide phosphorylase in *E. coli*. He showed that the enzyme was inducible by maltose, but to Jacques's disappointment, displayed no allosteric properties. Nevertheless, Maxime later discovered the positive regulation of the maltose regulon.

BACK TO THE LACTOSE OPERON

Since the early 1960s, several reports had appeared showing that clinical isolates of enterobacteria resistant to antibiotics carried plasmids that endowed them with the ability to utilize lactose. Normally, those bacteria are unable to ferment this sugar. With Nicole Guiso, we were interested in the origin of the Lac$^+$ character of these antibiotic-resistant plasmids. After having transferred the plasmids to the *E. coli* RV X74 Lac deletion strain, we compared their characteristics to those of the *E. coli* Lac operon. The proteins essential for lactose utilization, i.e., β-galactosidase and lactose permease, were biochemically and functionally similar to those of the *E. coli* Lac operon. The first surprise was that none of the plasmids carried enzymatically detectable thiogalactoside transacetylase. (Guy Cornelis later showed by hybridization techniques that the gene was also missing.) If the plasmids had acquired the lactose genes from an *E. coli* chromosome, then selection must have gotten rid of the transacetylase, an enzyme with no known physiological function in bacteria. The next and most unexpected surprise was that all repressors carried by the plasmids exhibited affinities for the inducer that were 10 to 30 times higher than that for the

E. coli Lac repressor. This might have given a selective advantage to the bacteria carrying these genes when growing in low lactose media. I might have been more successful in isolating the Lac repressor if I had been aware of these plasmids earlier (18).

β-Galactosidase Fusions as a Tool

The observation that the N-terminal 23 residues of β-galactosidase could be replaced with other amino acids without affecting enzymatic activity suggested that hybrid genes coding for hybrid proteins could be constructed. At that time, the purification of the hybrid proteins required several steps and the yields were poor. To avoid such low yields and protein degradation, I devised in 1984 a one-step procedure for the isolation of hybrid proteins exhibiting β-galactosidase activity (46). The procedure was based on affinity chromatography using a β-galactosidase inhibitor of high affinity. Starting with crude bacterial extracts, several milligrams of near-homogeneous proteins could be obtained within a few hours with an overall yield of around 90%. Some of these hybrid proteins contained up to 500 residues of the foreign sequence; they were nevertheless able to fold into a conformation leading to a tetrameric protein endowed with β-galactosidase activity. The method was a great success; for several years, colleagues from different laboratories arrived in the morning with their bacterial extracts and left in the afternoon with the pure protein. Since the discovery of the His-tag purification method, the β-galactosidase affinity chromatography procedure has been abandoned. *Sic transit gloria mundi*! (Thus passes the glory of the world!)

CYCLIC AMP EFFECTS IN BACTERIA

When Jacques Monod was appointed professor at the Collège de France in 1967, he devoted his first lectures to allosteric interactions. At the time, he was so convinced that only noncovalent interactions were important that, at a Gordon Conference on Proteins organized by Eddy Fischer and Dan Koshland, he proclaimed dismissively that, after all, peptide bonds were irrelevant, to which Fred Richards immediately countered by saying, "But Jacques, if you were right, proteins should be gases." Someone else in the back remarked that "the only person I know of who wasn't held together by covalent bonds was Lot's wife, after she had been converted into pillar of salt."

When Jacques asked me to give a seminar as part of a course on allosteric interactions at the Collège de France, I decided to play devil's advocate and talk about the cascade of covalent regulations of glycogen phosphorylase discovered earlier by Fischer et al. (12). The process involved a hormonal regulation, amplified by the second-messenger cAMP, which activates a cAMP-dependent protein kinase. This kinase then phosphorylates a second kinase, which in turn phosphorylates glycogen phosphorylase b, transforming it into active phosphorylase a (26). Reviewing the literature on this second messenger, I discovered a recent paper by Makman & Sutherland (33) describing the presence of cAMP in bacteria. I therefore decided to look for covalent modifications in *E. coli* triggered by cAMP.

François Jacob discouraged me from undertaking this project because, as he said, "There are no hormones in *E. coli*." Jacques also thought that it wasn't a good idea.

In those days, cAMP was not commercially available. With the help of Mic Michelson at the Institut de Biologie Physico-Chimique (IBPC), rue Pierre Curie (it had not yet been renamed rue Pierre et Marie Curie), on a Friday evening we synthesized a few grams of cAMP. The next day I grew *E. coli* in the presence of different carbon sources and found that addition of cAMP relieved bacteria from catabolite repression (CR) (52). On Monday morning Mic asked me to come to the IBPC. He took me into the dark room and showed me a petri dish that looked like a sky full of stars: He had discovered that bacterial luciferase synthesis was dramatically enhanced by cAMP. When it turned out that cAMP also relieved diauxic growth, which Jacques had discovered

25 years earlier (35), he started to become really interested. One day, Gerard Buttin entered the lab and, after looking at the diauxic curves, spontaneously remarked, "How nice, Jacques. Now you can at last complete your thesis."

It became obvious that covalent modifications were involved neither in CR nor in the relief of diauxic growth. On the contrary, evidence from us and others indicated that cAMP was an allosteric effector. By binding to CAP (the catabolite gene activator protein, its receptor protein), cAMP exerts a positive regulation on the otherwise negatively regulated *lac* operon (25). For 18 years, I worked on the role of cAMP in regulatory mechanisms in bacteria (47), and with my collaborators, we published approximately 50 papers on the subject.

Transcription Regulation

With François Gros's group, we were the first to show a role of cAMP in transcription initiation. Later on, with Antoine Danchin, Evelyne Joseph, and Chantal Guidi-Rontani, we studied its role in transcription termination. Our first observation concerned natural polarity, that is, the decreased expression of promoter distal genes as opposed to proximal genes. We showed that cAMP relieved the degree of natural polarity and interpreted this result by assuming that it was interfering with the transcription termination protein Rho. Indeed, direct measurements of mRNA synthesis corresponding to the promoter proximal and distal regions of the *lac* and *gal* operons showed that the relative synthesis of the promoter distal mRNAs was significantly increased in the presence of cAMP (17).

One of our manuscripts on the functional relationship between the cAMP-CAP complex and Rho was communicated by Roger Stanier to the *Proceedings of the National Academy of Sciences* (51). One referee requested minor modifications while the second just wrote, "I don't believe it." Because I did not want to argue with this referee about the role belief played in science, I simply quoted André Lwoff, who said "Once, the author of a discovery was subjected to three successive attacks. First, he was told that his findings were wrong; then, that they were already known. Lastly, when the poor fellow finally succeeded in demonstrating that his findings were right and original, he was told: 'That was obvious.'" The referee wrote to Roger: "Nice quotation; accept the manuscript."

Glucose Effects

After my first experiments on the relief of CR by cAMP, Jacques Monod's comment was that it was bad to get involved with a phenomenon that was far too complicated. Within a few years, I realized that he was right, but I was already too much involved in this unending story.

The terms catabolite repression and inducer exclusion were coined by Boris Magasanik (32) to describe a phenomenon called the glucose effect discovered in 1900 by Dienert (7). This phenomenon accounts for the "diauxic growth" observed by Monod in 1942 (35). Studying glucose inhibition of different catabolic enzymes, I succeeded in dissociating the relative contributions of CR and inducer exclusion. Surprisingly, I found that for some systems (*lac*, *tna*), glucose brought about mainly CR, whereas in other systems (*mal*, *gal*) glucose brought about inducer exclusion.

cAMP and Catabolite Repression

In the 1970s, the accepted model postulated that the extent of CR depended exclusively on the intracellular concentration of cAMP. This model was challenged when subsequent results indicated that it could be modulated independently of cAMP. They were based mainly on the observations (of Alain Dessein and Chantal Guidi-Rontani) that mutants deficient in adenylate cyclase and/or CAP were still subjected to CR (6). Searching for inhibitors that specifically affected catabolite-sensitive operons without interfering with overall cellular mechanisms, we found with Brigitte Sanzey that low concentrations of urea that did not inhibit bacterial growth specifically repressed the expression of catabolite-sensitive operons, a repression that was not relieved by cAMP (40). These and other results led us to propose that CR could be accounted for by a negative regulation afforded by a mediator: a specific metabolite or a class of metabolites that would accumulate under growth conditions that led to CR, and was degraded when CR was relieved (47). Searching for such modulators with Françoise Tillier, we purified from boiled extracts of *E. coli* a low-molecular-weight fraction that exerted a strong repressive effect on catabolite-sensitive operons, without showing any effect on catabolite-insensitive systems. The chemical nature of the active compound has not been elucidated. The compound, designated catabolite modulator factor, was actively metabolized by the bacteria, and cAMP only partially reversed its repression effect. In all respects, the repression obtained by catabolite modulator factor was similar to physiological CR (54). The paper describing these results was the last one signed by Jacques Monod; when it appeared in the *Proceedings of the National Academy of Sciences* (communicated by Boris Magasanik) Jacques had already passed away.

Escherichia coli Adenylate Cyclase

In the early 1980s, very little was known about the expression, structure, and activity of the *E. coli* adenylate cyclase. Some reports had suggested that the enzyme was loosely bound to the membrane and extremely labile. Antoine Danchin succeeded in cloning the *cya* gene and demonstrated that the bacterial adenylate cyclase consisted of two functional domains: an amino-terminal domain exhibiting cyclase catalytic activity and a carboxy-terminal domain having regulatory functions. Antoine then constructed a *cya-lacZ* fusion that coded for both adenylate cyclase and β-galactosidase activities (39).

Using the one-step purification procedure mentioned above, we obtained homogeneous Cya-LacZ fusion protein with which we could confirm that the adenylate cyclase activity was indeed associated with the amino-terminal domain of the wild-type protein. From the specific activity of the purified protein we calculated that wild-type *E. coli* contains about 1,000 molecules of adenylate cyclase per cell (39).

FROM *ESCHERICHIA COLI* TO PATHOGENS

Having spent many years on the role of cAMP in *E. coli*, I wanted to switch to a different area, i.e., study the role of this nucleotide in pathogenesis. It was already known that two bacterial toxins (secreted by *Vibrio cholerae* and by *Bordetella pertussis*) acted by increasing cAMP levels of the infected organisms. This increase was due to an ADP ribosylation (from NAD) of one of the regulatory subunits of the eukaryotic adenylate cyclase. But at the time, this field seemed very crowded and competitive so I decided to turn to something else.

Upon reviewing the relevant literature, I discovered reports on the existence of adenylate cyclase activity in *B. pertussis*, the etiologic agent of whooping cough (20). It took a few more years for Wolff et al. (57) to realize that it was a secreted protein that exhibited the unusual property of being activated by calmodulin, a eukaryotic protein par excellence. Confer & Eaton (5) were the first to demonstrate in 1982 that *B. pertussis* adenylate cyclase is a toxin. After entering animal cells, it elicits an unregulated increase in cAMP, thereby disrupting normal cellular functions. More importantly, Alison Weiss (of Stanley Falkow's laboratory) had opened the way to *B. pertussis* genetic analysis by showing that mutants deficient in adenylate cyclase were avirulent (56). Around the same time, Leppla (31) had reported that *Bacillus anthracis* toxin (the edema factor) was also a calmodulin-activated adenylate cyclase.

At the beginning, I was tempted to choose the anthrax system, because when I was a student, for an end-of-the-year diploma, my organic chemistry professor, V. Bruckner, asked me to purify the capsule of *B. anthracis* (poly-D-γ-glutamic acid), the structure of which he was trying to establish. To start the purification, I had to centrifuge large amounts of bacteria. Because we had only glass centrifuge tubes without cover, I asked whether the bacteria were dangerous. I was told that they could be lethal and that I had better be careful. I thought it was the first step one had to take to become a microbe hunter. I purified large amounts of capsule and survived; I obtained my diploma.

For a while, I worked with Michèle Mock on the *B. anthracis* system but finally decided to work on the adenylate cyclase of *B. pertussis*. Michèle and I could not understand how two taxonomically distant organisms with such different genomic G+C content (30% for *B. anthracis* versus 66% for *B. pertussis*) could produce adenylate cyclase toxins that required the same eukaryotic activator for their enzymatic activity, namely calmodulin (34). We never did solve that evolutionary puzzle, but both our groups uncovered many aspects of the structure-function relationships of the two systems.

THE SAGA OF *BORDETELLA PERTUSSIS* ADENYLATE CYCLASE

When we decided in 1984 to work on the *B. pertussis* adenylate cyclase (with Octavian Bârzu, Nicole Guiso, Isabelle Crenon, and Daniel Ladant), very little was known about the enzyme itself. No homogeneous preparation had been obtained. Various groups had described both low- and high-molecular-weight forms that were activated to various extents by calmodulin. Therefore, our first priority was to purify the enzyme in order to study its structure-function relationship.

From supernatants of a virulent *B. pertussis* strain, Daniel Ladant succeeded in purifying a 45-kDa form of adenylate cyclase with high affinity for calmodulin (0.2 nM). Limited proteolysis with trypsin yielded two fragments of 25 and 18 kDa (T25 and T18) that could be separated by gel-filtration chromatography and easily detected by SDS-PAGE (sodium dodecyl sulfate polyacrylamide gel electrophoresis). At that point, Daniel performed an experiment that only an uninhibited and imaginative young scientist would do: He excised the two fragments from the gel, eluted them in the presence of calmodulin, and recovered enzymatic activity. This showed that T25 and T18 represented separate domains of the adenylate cyclase that could complement one another in the presence of calmodulin, leading to a structure resembling the native enzyme (27). Fifteen years later, Daniel had the idea of setting up a bacterial two-hybrid screening system based on the T25 and T18 complementation (24).

We obtained antibodies against the 45-kDa form of the adenylate cyclase, which brought us two surprises. The first surprise came when Daniel isolated the enzyme from urea-treated bacterial extracts: He obtained mainly the ~200-kDa form that was fully active. Moreover, antibodies directed against the low-molecular-weight form recognized perfectly well the high-molecular-weight enzyme. The obvious conclusion was that the 45-kDa form resulted from a proteolytic cleavage of the high-molecular-weight form. Clearly, the 45-kDa species could represent the catalytic moiety of the native protein. The nature of the rest of the molecule was identified later thanks to antibodies raised against *E. coli* hemolysin (a gift from Werner Goebel): It recognized the same high-molecular-weight protein that our antibodies did. This result validated and extended Weiss & Falkow's (56) findings that several *Tn5*-induced nonhemolytic mutants were also adenylate cyclase deficient. The cloning and sequencing of the *cya* gene coding for adenylate cyclase (by Philippe Glaser and Antoine Danchin) confirmed the biochemical results and showed that the 5'-terminal end of the gene represented the catalytic moiety and the 3'-terminal end exhibited strong homology to the *E. coli* hemolysin (14).

The second surprise arose from a collaboration with Michèle Mock's group. We found that the antibodies directed against *B. pertussis* adenylate cyclase cross-reacted with the *B. anthracis* enzyme. Later, sequence analysis indicated that only the calmodulin-activated catalytic domain shared those structural similarities (34). In 2005, Tang et al. (19) established the three-dimensional structures of both adenylate cyclases and concluded that their catalytic regions shared similar structures and mechanisms of activation. By contrast, the interactions with calmodulin were completely different, providing molecular details of how two structurally homologous bacterial toxins could have undergone divergent evolution to bind calmodulin (19).

In the late 1980s, I attended a Gordon Conference on Microbial Toxins and Pathogenicity also attended by several colleagues involved in pertussis research. I reported our results on adenylate cyclase and that afternoon, while we were discussing our projects with Rino Rappuoli, John Coote, and Roy Gross, I told them that I had learned from Jacques Monod that it was "better to collaborate than to compete." Shouldn't we, for instance, join forces and initiate a collaborative project and apply for a common international grant? They agreed, provided that I would be the coordinator. After several months of hard work, I sent in a grant request to the Human Frontier Science Program (HFSP) entitled "Molecular and Cellular Mechanisms Underlying Regulation of Virulence Factors of *Bordetella pertussis*." Even though I knew that our chance of success was less than 1%, we were funded, which was a bonanza for our research. This grant radically changed our everyday lives: For once, I could support postdocs, which was not possible with French grants. The HFSP provided us with many facilities and, more importantly, sealed a very close and fruitful collaboration among our groups.

Structure-Function Relationships

The genetic organization of the *cya* locus revealed, in addition to the structural gene (*cyaA*) encoding for adenylate cyclase, the existence of three additional genes (*cyaB*, *cyaD*, and *cyaE*) required for the secretion of CyaA. Analysis of the transcriptional organization of the *cya* locus showed that *cyaA*, *cyaB*, *cyaD*, and *cyaE* were organized in a single operon transcribed from the *cyaA* promoter. Adjacent to the *cyaA* gene, *cyaC* was transcribed in the opposite direction. Its product catalyzed the palmitoylation of CyaA, a prerequisite for the conversion of the protein to the active toxin.

Transcriptional regulation of *B. pertussis* virulence factors, achieved by a two-component system (BvgA/BvgS), was studied extensively by the Falkow and Rappuoli groups (2). My own group was interested mainly in some molecular mechanisms underlying specific promoter activations. We elucidated the transcriptional organization of the *cya* locus (with Brid Laoide) (30), characterized the specificity of BvgA binding to virulence promoter regions (with Gouzel Karimova) (23), carried out functional analysis of the *cya* promoter (with Sophie Goyard) (15), and studied the in vitro transcriptional activation of *bvg*-regulated promoters as well as the specific role of *B. pertussis* RNA polymerase subunits (with Pierre Steffen and Sophie Goyard) (45).

For several years our research was focused on the structure-function relationship of CyaA. The 1,706-residue CyaA is a bifunctional protein: It exhibits both adenylate cyclase and hemolytic activities and is constructed in a modular fashion. The calmodulin-activated catalytic domain is located within the 400 amino-proximal residues, whereas the 1,306 carboxy-terminal residues are responsible for the hemolytic activity, for the binding of the toxin to eukaryotic cells, and for its internalization. In addition, it contains the secretion signals (29).

Antigen Delivery

To study the mechanism of CyaA secretion and its entry into eukaryotic cells, we had to reconstruct in *E. coli* a high level of expression of the *cya* genes, which was achieved mainly by Peter Sebo (42). Using the reconstituted system, he and Daniel Ladant, Jacques Bellalou, and Hiroshi Sakamoto elucidated several aspects of the secretion mechanism. Daniel Ladant carried out the next important experiment for our future understanding of the multiple capabilities of the CyaA toxin. He set up an insertional mutagenesis procedure for the introduction of as many as 16 codons into the *cyaA* gene and identified within the primary structure of the catalytic domain several permissive sites where insertions altered neither the activity nor the stability of the protein (28). This finding suggested that if we could insert foreign antigenic determinants into those permissive sites, we might be able to use the resulting recombinant CyaA toxin as a Trojan horse to deliver specific epitopes into antigen-presenting cells. Because I was not an immunologist, I approached Claude Leclerc at the Pasteur Institute. To my satisfaction, she agreed to test the idea, and it worked! Daniel Ladant and Peter Sebo constructed genetically detoxified (i.e., lacking adenylate cyclase catalytic activity) recombinant CyaA toxins carrying different $CD8^+$ epitopes. Then Leclerc's group, using different mouse models, showed that a recombinant CyaA toxin carrying a viral $CD8^+$ epitope could elicit epitope-specific cytotoxic T-cell immune responses and, more importantly, protect mice against infections triggered by high doses of the virus (11, 41). Also, recombinant CyaA toxin carrying specific tumor epitopes could protect mice against grafts of tumor cells (10). Further work fully confirmed that recombinant CyaA toxin represented an attractive nonreplicative vector for antigen delivery and could be used for the development of protective and therapeutic vaccines. After my mandatory retirement in 1995 (French law), the Pasteur Institute generously offered me an office so I could closely follow the development of a project entitled "Recombinant CyaA Toxins for Vaccinal Purposes" funded by the European Union. The results obtained so far have fully validated the vaccine potential of CyaA: Two of these recombinant molecules are currently under Phase I clinical trials.

Reporter for Protein Targeting

The structural flexibility of the CyaA catalytic domain proved to be very useful not only for antigen delivery but also for protein targeting and the characterization of protein-protein interactions. Guy Cornelis, who studied the *Yersinia* type III secretion system, which allows direct delivery of bacterial virulence proteins into the host cell cytosol, once mentioned that it was difficult to detect the proteins delivered because of their very low concentration. I suggested he try to construct a fusion between the C-terminal section of his protein and the catalytic domain of CyaA. The hybrid protein would exhibit no adenylate cyclase activity in *Yersinia* because bacteria do not produce calmodulin. By contrast, once internalized in eukaryotic cells, the hybrid enzyme would become activated by calmodulin and generate cAMP. As the CyaA catalytic domain alone is unable to penetrate into eukaryotic cells, appearance of cAMP in the target cell would be a direct measure of virulence protein internalization. After constructing the appropriate fusion proteins from the *cya*-containing plasmid we provided, Sory & Cornelis (44) were able to define the mechanism of internalization of the *Yersinia* virulence factors. The successful utilization of cAMP production as a selective reporter of eukaryotic environment inspired groups working on *Shigella* or *Salmonella* type III secretion systems to adopt this approach.

Bacterial Two-Hybrid System

Taking advantage of the modular structure of the catalytic domain of CyaA, Daniel Ladant designed a genetic system that would allow easy screening and selection of functional interactions between two proteins in vivo. The system is based on complementation between the T25 and T18 fragments of the catalytic domain that can be easily tested in an *E. coli cya* strain. If cAMP is produced, catabolic operons are activated (as I had shown 30 years earlier), thus giving rise to a selectable phenotype. T25 and T18 expressed in *E. coli cya* as separate entities are unable to associate and reconstitute a functional enzyme. On the other hand, when T25 and T18 are fused to peptides or proteins that interact with one another, heterodimerization of those chimeric polypeptides results in a functional complementation of the two fragments and cAMP synthesis. The bacteria are thus able to ferment carbohydrates; therefore, it is easy to detect positive clones on petri dishes containing lactose or maltose (24). The system functions as a versatile reporter of protein associations and has been used successfully to reveal interactions between bacterial, viral, and eukaryotic proteins. It proved to be an attractive complementary approach to the yeast two-hybrid system.

Today, when scores of scientists work on different aspects of *B. pertussis* adenylate cyclase, it is gratifying to remember that it started from a shot-in-the-dark weekend experiment with homemade cAMP. At that time, I could never have dreamed that many years later it would lead to versatile reporters for protein targeting and protein association and, more importantly, to a useful and innovative vaccine.

DISCLOSURE STATEMENT

The author is not aware of any affiliations, memberships, funding, or financial holdings that might be perceived as affecting the objectivity of this review.

ACKNOWLEDGMENTS

Looking back on my research career of the past 60 years, I was fortunate to witness the early days of molecular biology and to meet outstanding scientists and wonderful human beings. My encounter with Jacques Monod not only shaped my scientific career; more importantly, I owe him my freedom. I feel immensely privileged to have worked with him, and also to have had the chance to cruise with him on his sailing boat and thus discover the Mediterranean. After he passed away, my longstanding collaboration with Antoine Danchin was particularly beneficial. I would like to express my deep appreciation to all those who have collaborated with me and have helped me in several teaching adventures, including heroic ones in Brazil and India. I want to thank Daniel Ladant and Gouzel Karimova for helping me to still keep in touch with everyday bench life. For 15 years I served as Scientific Director of Development at the Pasteur Institute and I greatly appreciated my friendly collaboration with the two General Directors of the Institute, Raymond Dedonder and Maxime Schwartz. In addition, Maxime provided me in many instances friendly support. I am particularly grateful to those who nominated me for the awards that pleased me the most: the French Légion d'Honneur, the Robert Koch Gold Medal, and the Doctorate Honoris Causa from the Sapienza University of Rome. I am deeply indebted to Eddy Fischer, who, as an old friend of Jacques Monod, followed his tradition. Once, when I had shown Jacques a manuscript I had just written, he said it wasn't too bad but that he would have to translate it "from Agnes to French." This chapter has greatly benefited from Eddy's translation "from Agnes to English."

LITERATURE CITED

1. Antonini E, Wyman J, Moretti R, Rossi-Fanelli A. 1963. The interaction of bromthymol blue with hemoglobin and its effect on the oxygen equilibrium. *Biochem. Biophys. Acta* 71:124–32
2. Aricò B, Scarlato V, Monack DM, Falkow S, Rappuoli R. 1991. Structural and genetic analysis of the *bvg* locus in *Bordetella* species. *Mol. Microbiol.* 5:2481–91

3. Buc MH, Ullmann A, Goldberg ME, Buc H. 1971. Masse moléculaire et coefficient d'extinction de la glycogène phosphorylase b du muscle de lapin. *Biochimie* 53:283–89
4. Celada F, Ullmann A, Monod J. 1974. An immunological study of complementary fragments of β-galactosidase. *Biochemistry* 13:5543–47
5. Confer DL, Eaton JW. 1982. Phagocyte impotence caused by the invasive bacterial adenylate cyclase. *Science* 217:948–95
6. Dessein A, Schwartz M, Ullmann A. 1978. Catabolite repression in *Escherichia coli* mutants lacking cyclic AMP. *Mol. Gen. Genet.* 162:83–87
7. Dienert F. 1900. Sur la fermentation du galactose et sur l'accoutumance des levures à ce sucre. *Ann. Inst. Pasteur Paris* 19:139–89
8. Erdos T, Ullmann A. 1959. Effect of streptomycin on the incorporation of amino acids labelled with carbon-14 into ribonucleic acid and protein in a cell-free system of a *Mycobacterium*. *Nature* 183:618–19
9. Erdos T, Ullmann A. 1960. Effect of streptomycin on the incorporation of tyrosine labelled with carbon^{14} into protein of *Mycobacterium* cell fractions in vivo. *Nature* 185:100–101
10. Fayolle C, Ladant D, Karimova G, Ullmann A, Leclec C. 1999. Therapy of murine tumors with recombinant *Bordetella pertussis* adenylate cyclase carrying a cytotoxic T cell epitope. *J. Immunol.* 162:4157–62
11. Fayolle C, Sebo P, Ladant D, Ullmann A, Leclerc C. 1996. In vivo induction of CTL responses by recombinant adenylate cyclase of *Bordetella pertussis* carrying viral CD8$^+$ T-cell epitopes. *J. Immunol.* 156:4697–706
12. Fischer EH, Appleman MM, Krebs EG. 1964. The structure of phosphorylases. In *Ciba Foundation Symposium – Control of Glycogen Metabolism*, ed. H Whelan, MP Cameron, pp. 94–106. Chichester, UK: Wiley
13. Gilbert W, Müller-Hill B. 1966. Isolation of the *lac* repressor. *Proc. Natl. Acad. Sci. USA* 56:1891–98
14. Glaser P, Ladant D, Sezer O, Pichot F, Ullmann A, Danchin A. 1988. The calmodulin-sensitive adenylate cyclase of *Bordetella pertussis*: cloning and expression in *Escherichia coli*. *Mol. Microbiol.* 2:19–30
15. Goyard S, Ullmann A. 1993. Functional analysis of the *cya* promoter of *Bordetella pertussis*. *Mol. Microbiol.* 7:693–704
16. Gros F, Gilbert W, Hiatt HH, Attardi G, Spahr PF, Watson JD. 1961. Molecular and biological characterization of messenger RNA. *Cold Spring Harb. Symp. Quant. Biol.* 26:111–13
17. Guidi-Rontani C, Danchin A, Ullmann A. 1984. Transcriptional control of polarity in *Escherichia coli* by cAMP. *Mol. Gen. Genet.* 195:96–100
18. Guiso N, Ullmann A. 1976. Expression and regulation of lactose genes carried by plasmids. *J. Bacteriol.* 127:691–97
19. Guo Q, Shen Y, Lee YS, Gibbs CS, Mrksich M, Tang W-J. 2005. Structural basis for the interaction of *Bordetella pertussis* adenylate cyclase toxin with calmodulin. *EMBO J.* 24:3190–201
20. Hewlett EL, Wolff J. 1976. Soluble abenylate cyclase from the culture medium *Bordetella pertussis*: purification and characterization. *J. Bacteriol.* 127:890–98
21. Jacob F, Ullmann A, Monod J. 1965. Délétions fusionnant l'opéron lactose et un opéron purine chez *Escherichia coli*. *J. Mol. Biol.* 31:704–19
22. Jacobson RH, Zhang XJ, DuBose RF, Matthews BW. 1994. Three-dimensional structure of β-galactosidase of *E. coli*. *Nature* 369:761–66
23. Karimova G, Bellalou J, Ullmann A. 1996. Phosphorylation-dependent binding of BvgA to the upstream region of the *cyaA* gene of *Bordetella pertussis*. *Mol. Microbiol.* 20:489–96
24. Karimova G, Pidoux J, Ullmann A, Ladant D. 1998. A bacterial two-hybrid system based on a reconstituted signal transduction pathway. *Proc. Natl. Acad. Sci. USA* 95:5752–56
25. Kolb A, Busby S, Buc H, Garges S, Adhya S. 1993. Transcriptional activation by cAMP and its receptor protein. *Annu. Rev. Biochem.* 62:749–95
26. Krebs EG, Fischer EH. 1962. Molecular properties and transformations of glycogen phosphorylase in animal tissues. *Adv. Enzymol.* 24:263–90
27. Ladant D. 1988. Interaction of *Bordetella pertussis* adenylate cyclase with calmodulin: identification of two separate calmodulin binding domains. *J. Biol. Chem.* 263:2612–18

28. Ladant D, Glaser P, Ullmann A. 1992. Insertional mutagenesis of *Bordetella pertussis* adenylate cyclase. *J. Biol. Chem.* 267:2244–50
29. Ladant D, Ullmann A. 1999. *Bordetella pertussis* adenylate cyclase: a toxin with multiple talents. *Trends Microbiol.* 7:172–76
30. Laoide BM, Ullmann A. 1990. Virulence dependent and independent regulation of the *Bordetella pertussis cya* operon. *EMBO J.* 9:999–1005
31. Leppla SH. 1982. Anthrax toxin edema factor: a bacterial adenylate cyclase that increases cyclic AMP concentrations in eukaryotic cells. *Proc. Natl. Acad. Sci. USA* 79:3162–66
32. Magasanik B. 1970. Glucose effects: inducer exclusion and repression. In *The Lactose Operon*, ed. JR Beckwith, D Zipser, pp. 189–219. Cold Spring Harbor, NY: Cold Spring Harbor Lab. Press
33. Makman RS, Sutherland EW. 1965. Adenosine 3′,5′-phosphate in *Escherichia coli*. *J. Biol. Chem.* 240:1309–14
34. Mock M, Ullmann A. 1993. Calmodulin-activated bacterial adenylate cyclases as virulence factors. *Trends Microbiol.* 1:187–92
35. Monod J. 1958. *Recherche sur la croissance des cultures bactériennes*. Paris: Hermann et Cie. 210 pp.
36. Monod J, Cohn M. 1952. La biosynthèse induite des enzymes (adaptation enzymatique). *Adv. Enzymol.* 13:67–119
37. Pardee AB, Jacob F, Monod J. 1959. The genetic control and cytoplasmic expression of "inducibility" in the synthesis of β-galactosidase in *Escherichia coli*. *J. Mol. Biol.* 1:165–78
38. Perrin D. 1963. Complementation between products of the β-galactosidase structural gene of *Escherichia coli*. *Cold Spring Harbor Symp. Quant. Biol.* 28:529–31
39. Roy A, Danchin A, Joseph E, Ullmann A. 1983. Two functional domains in adenylate cyclase of *Escherichia coli*. *J. Mol. Biol.* 165:197–202
40. Sanzey B, Ullmann A. 1976. Urea, a specific inhibitor of catabolite sensitive operons. *Biochem. Biophys. Res. Commun.* 71:1062–68
41. Saron MF, Fayolle C, Sebo P, Ladant D, Ullmann A, Leclerc C. 1997. Anti-viral protection conferred by recombinant adenylate cyclase toxins from *Bordetella pertussis* carrying a CD8[+] T cell epitope from lymphocytic choriomeningitis virus. *Proc. Natl. Acad. Sci. USA* 94:3314–19
42. Sebo P, Glaser P, Sakamoto H, Ullmann A. 1991. High-level synthesis of active adenylate cyclase toxin of *Bordetella pertussis* in a reconstructed *Escherichia coli* system. *Gene* 10:19–24
43. Seery VL, Fischer EH, Teller DC. 1967. A reinvestigation of the molecular weight of glycogen phosphorylase. *Biochemistry* 6:3315–27
44. Sory MP, Cornelis GR. 1994 Translocation of a hybrid YopE-adenylate cyclase from *Yersinia enterocolitica* into HeLa cells. *Mol. Microbiol.* 14:583–94
45. Steffen P, Goyard S, Ullmann A. 1996. Phosphorylated BvgA is sufficient for transcriptional activation of virulence-regulated genes in *Bordetella pertussis*. *EMBO J.* 15:102–109
46. Ullmann A. 1984. One-step purification of hybrid proteins which have β-galactosidase activity. *Gene* 29:27–31
47. Ullmann A, Danchin A. 1983. Role of cyclic AMP in bacteria. In *Advances in Cyclic Nucleotide Research*, ed. P Greengard, GA Robison, 15:1–53. New York: Raven Press
48. Ullmann A, Goldberg ME, Perrin D, Monod J. 1968. On the determination of molecular weight of proteins and protein subunits in 6M guanidine. *Biochemistry* 7:261–65
49. Ullmann A, Jacob F, Monod J. 1967. Characterization by in vitro complementation of a peptide corresponding to an operator proximal segment of the β-galactosidase structural gene of *Escherichia coli*. *J. Mol. Biol.* 24:339–43
50. Ullmann A, Jacob F, Monod J. 1968. On the subunit structure of wild-type versus complemented β-galactosidase of *Escherichia coli*. *J. Mol. Biol.* 32:1–13
51. Ullmann A, Joseph E, Danchin A. 1979. Cyclic AMP as a modulator of polarity in polycistronic transcriptional units. *Proc. Natl. Acad. Sci. USA* 76:3194–97
52. Ullmann A, Monod J. 1968. Cyclic AMP as an antagonist of catabolite repression in *Escherichia coli*. *FEBS Lett.* 2:57–60
53. Ullmann A, Perrin D, Jacob F, Monod J. 1965. Identification par complémentation in vitro et purification d'un segment peptidique de la β-galactosidase d'*Escherichia coli*. *J. Mol. Biol.* 12:918–23

54. Ullmann A, Tillier F, Monod J. 1976. Catabolite modulator factor: a possible mediator of catabolite repression in bacteria. *Proc. Natl. Acad. Sci. USA* 73:3476–79
55. Ullmann A, Vagelos PR, Monod J. 1964. The effect of 5′ adenylic acid upon the association between bromthymol blue and muscle phosphorylase b. *Biochem. Biophys. Res. Commun.* 17:86–92
56. Weiss AA, Falkow S. 1984. Genetic analysis of phase change in *Bordetella pertussis*. *Infect. Immun.* 43:263–69
57. Wolff J, Cook GH, Goldhammer AR, Berkowitz SA. 1980. Calmodulin activates prokaryotic adenylate cyclase. *Proc. Natl. Acad. Sci. USA* 77:3840–44

Peter Lengyel

Memories of a Senior Scientist: On Passing the Fiftieth Anniversary of the Beginning of Deciphering the Genetic Code

Peter Lengyel

Department of Molecular Biophysics & Biochemistry, Yale University School of Medicine, New Haven, Connecticut 06520; email: peter.lengyel@yale.edu

Keywords

codon, codon nucleotide composition, codon nucleotide sequence, homopolyribonucleotide, polynucleotide phophorylase, random copolyribonucleotide

Abstract

2011 marked the fiftieth anniversary of breaking the genetic code in 1961. Marshall Nirenberg, the National Institutes of Health (NIH) scientist who was awarded the Nobel Prize in Physiology or Medicine in 1968 for his role in deciphering the code, wrote in 2004 a personal account of his research. The race for the code was a competition between the NIH group and Severo Ochoa's laboratory at New York University (NYU) School of Medicine, where I was a graduate student and conducted many of the experiments. I am now 83 years old. These facts prompt me to recall how I, together with Joe Speyer, an instructor in the Department of Biochemistry at NYU, unexpectedly became involved in deciphering the code, which also became the basis of my PhD thesis. Ochoa won the Nobel Prize in Physiology or Medicine in 1959 for discovering polynucleotide phosphorylase (PNP), the first enzyme found to synthesize RNA in the test tube. The story of how PNP made the deciphering of the code feasible is recalled here.

Contents

FROM READING ABOUT POLYNUCLEOTIDE PHOSPHORYLASE IN BUDAPEST TO BECOMING SEVERO OCHOA'S GRADUATE STUDENT IN NEW YORK 28
EXPERIMENTS WITH POLYNUCLEOTIDE PHOSPHORYLASE AND FURTHER PROJECTS AT NYU 28
GETTING UNEXPECTEDLY INVOLVED IN DECIPHERING THE GENETIC CODE 29
SLOW START OF THE CODE WORK WITH JOE SPEYER 30
FIRST DECODING OF THE NUCLEOTIDE COMPOSITIONS OF CODONS USING RANDOM COPOLYNUCLEOTIDES AT NYU 31
MEETING MARSHALL NIRENBERG AND CONTINUING CODE DECIPHERING WITH RANDOM COPOLYNUCLEOTIDES AT BOTH NYU AND NIH 33
   From Cold War to Code War 34
SUBMISSION OF MY PHD THESIS: NUCLEOTIDE COMPOSITIONS OF 18 CODONS, CODING CHARACTERISTICS OF THE INOSINE AND XANTHINE NUCLEOTIDES, AND THE RIBOSOME AS A TARGET OF THE ANTIBIOTIC STREPTOMYCIN 34
AN OVERVIEW OF THE PHASES IN DECIPHERING THE GENETIC CODE 35
POSTSCRIPT 36

FROM READING ABOUT POLYNUCLEOTIDE PHOSPHORYLASE IN BUDAPEST TO BECOMING SEVERO OCHOA'S GRADUATE STUDENT IN NEW YORK

I recall reading with excitement in Budapest in 1955 the first report concerning the polynucleotide phosphorylase (PNP) enzyme synthesizing RNA by Grunberg-Manago & Ochoa (11). With a diploma in chemical engineering, I was a graduate student of Brunó Ferenc Straub, an outstanding biochemist at the Semmelweiss University Medical School in Budapest, Hungary. After the defeat of the Hungarian uprising by the Soviet troops in 1956, I fled the country and immigrated to the United States.

It was my interest in PNP that prompted me (just one day after arriving in New York as a refugee) in January 1957 to get in touch with its discoverer, Severo Ochoa. Dr. George Bosworth Brown (Sloan Kettering Division of Cornell Medical School), whom I had met the same morning, kindly introduced me to him. This happened after Ochoa had finished a lecture on PNP at the New York Academy of Medicine. It was probably his sympathy for the cause of Hungary, together with my enthusiasm to become his student, that prompted him to accept me as a graduate student in his laboratory.

EXPERIMENTS WITH POLYNUCLEOTIDE PHOSPHORYLASE AND FURTHER PROJECTS AT NYU

PNP degrades RNA by phosphorolysis (11). This requires phosphate and it results in the formation of nucleoside diphosphates, as well as polyribonucleotides shortened at their 3' end. The

Polynucleotide phosphorylase (PNP): first enzyme found to synthesize polyribonucleotides in the test tube

enzyme can also synthesize RNA, e.g., polyribonucleotides containing as many as four types of ribonucleotides in a random sequence. The synthesis requires ribonucleoside diphosphates and produces, in addition to polyribonucleotides, phosphate (12).

My first project at New York University (NYU) assigned by Dr. Ochoa was to isolate RNAs from various plant leaves and compare their rates of PNP-catalyzed phosphorolysis with that of RNAs from various natural sources and polyA (25). At that time it was not known whether PNP was involved in vivo in the synthesis of RNA or only its degradation. It had been reported, however, that the pyrimidine analog 2-thiouracil is present in the RNA of *Tobacco mosaic virus* and *Bacillus megaterium* (22 and references therein). It occurred to me that, if PNP were involved in RNA synthesis in vivo, then 2-thiouridine 5′-diphosphate should serve as a substrate for the enzyme in vitro. I proposed to explore this question. Bob Chambers, a bioorganic chemist and instructor in the Department of Biochemistry, guided me in the eight-step chemical synthesis of 2-thiouridine 5′-diphosphate. This compound was converted by PNP into polythiouridylic acid, and it could also be incorporated into a copolyribonucleotide with the four natural nucleotides (22). These results did not prove that PNP was involved in RNA synthesis in vivo, but made it conceivable that, at least in principle, it might be. The above-outlined two projects provided me with much experience working with PNP.

My next project, suggested by Dr. Ochoa, was a component of his extensive exploration of propionic acid metabolism in animal tissues (3). The goal was to characterize methylmalonyl isomerase from sheep kidney. This enzyme converts methylmalonyl coenzyme A (CoA) (a branched-chain compound) to succinyl CoA (its straight-chain isomer) (3). I noted the similarity of the above conversion to that of β-methylaspartate (another branched-chain compound) to glutamate (its straight-chain isomer) by glutamate isomerase (1). Recognizing this similarity, I expected that methylmalonyl isomerase may require for activity the same vitamin B_{12} cobamide coenzymes as glutamate isomerase (48). In collaboration with Rajarshi Mazumder, a postdoctoral fellow, I started to search for vitamin B_{12} coenzymes in our enzyme. But, prior to completing our search, three research papers reported finding vitamin B_{12} coenzyme in methylmalonyl isomerase (for citations, see Reference 24). Although disappointed, we completed the study, working for many days in the dark, cold room because of vitamin B_{12} coenzyme's high sensitivity to light. We found vitamin B_{12} coenzyme in the sheep kidney methylmalonyl isomerase and established that of the three types of vitamin B_{12} coenzymes, i.e., dimethylbenzimidazolyl cobamide, benzimidazolyl cobamide, and adenosyl cobamide, only the first and second appeared to fully activate the enzyme (24).

After finishing this study, I was supposed to join an ongoing project in the department devoted to the problem of the deoxyribonucleic acid–dependent enzymatic synthesis of RNA. However, my participation in the project was cut short because we began to study the genetic code.

polyA: polyadenylic acid

Genetic code: list of sequences of ribonucleotide triplets encoding the various amino acids as well as chain initiation and chain termination signals

Homopolypeptide: polypeptide consisting of a series of the same amino acid linked in a chain

Homopolyribonucleotide: polyribonucleotide consisting of a single type of nucleotide

GETTING UNEXPECTEDLY INVOLVED IN DECIPHERING THE GENETIC CODE

It helps me to remember this old story that, after it began (on June 6, 1961 in Cold Spring Harbor, Long Island, New York), I described it in a letter to Severo Ochoa, who was away in Europe. (This letter is available online as supplemental material. Follow the **Supplemental Material link** from the Annual Reviews home page at **http://www.annualreviews.org**.)

To cite one sentence from my letter dated August 19, 1961, "On the very day [June 6, 1961] when you left for the summer and when I attended the lectures on messenger RNA at the Cold Spring Harbor Symposium on Cellular Regulatory Mechanisms (7) the idea occurred to me that homopolyribonucleotides should be tested as messengers which, if they would act at all, should induce the ribosomes to synthesize homopolypeptides (e.g.: poly C could induce the formation

Messenger ribonucleic acid (mRNA): binds to ribosomes and encodes the proteins synthesized

Cell-free amino-acid-incorporating system: cell extract with the ability to incorporate added (often radioactive) amino acids into proteins

polyU: polyuridylic acid

Random copolyribonucleotide: polyribonucleotide consisting of a random sequence of two or more kinds of ribonucleotides

of polyleucine)." I mentioned this idea there and then to my colleagues from the Ochoa laboratory who attended the Cold Spring Harbor Symposium with me: Charlie, Joe, Hans, and Yoshito (Charles Gilvarg, later professor at Princeton University; Joseph Speyer, later professor at the University of Connecticut; Hans Kroeger, later professor and Director of the Robert Koch Institute in Berlin; and Yoshito Kaziro, later professor at the Institute of Medical Science of the University of Tokyo).

My idea was evoked by listening to the fascinating presentation given by Sydney Brenner at the symposium on messenger RNA (mRNA) (4). He and his colleagues demonstrated that in vivo ribosomes are unspecific structures that synthesize at a given time the protein dictated by the mRNA that is attached to the ribosomes. This demonstration confirmed the validity of the mRNA concept by Jacob & Monod (14).

Prior to the discovery of mRNA, ribosomal RNA, a stable component of ribosomes, was thought to be the template for the synthesis of proteins. This (erroneous) belief provided no approach for code deciphering. The discovery of mRNA as a new intermediate in information transfer between DNA and protein that transiently binds to ribosomes and encodes on them the proteins they synthesize, however, suggested a novel approach.

The triggers evoking in me the idea of this novel approach were (*a*) Brenner's lecture at the symposium on messenger RNA as the transient template of protein synthesis on unspecific ribosomes (4), and (*b*) my work in the Ochoa laboratory with PNP, the first enzyme capable of producing synthetic RNA (11). The approach that occurred to me was to use simple synthetic messengers (e.g., homopolyribonucleotides or copolyribonucleotides of two, three, or four types of ribonucleotides of random sequence) for the deciphering of the code. These synthetic messengers could be produced by PNP. This enzyme, the discovery of which earned Ochoa the Nobel Prize in 1959 (36), thus became the magic wand for deciphering the genetic code.

For code deciphering, the various types of synthetic messengers had to be tested for their ability to drive in a cell-free amino-acid-incorporating system the uptake of each of the 20 radioactively labeled amino acids into an insoluble product whose radioactivity was determined. Our system was based on the pioneering research of Zamecnik (49) and Zamecnik & Keller (50).

On June 6, 1961, in Cold Spring Harbor, I thought only of the use of homopolyribonucleotides (of which only four kinds exist, polyA, polyC, polyG, and polyU) for the deciphering. Utilizing these as synthetic messengers could reveal nucleotide sequences encoding, at most, only four amino acids. The use of random copolyribonucleotides (which could reveal sequences encoding all 20 amino acids) occurred to me only somewhat later. However, my letter to Ochoa briefly mentions an example of this: "Continued code breaking, e.g., inserting A into polyU which should lead to three new incorporated amino acids, if the amount of inserted A will be low enough to prevent it from occupying neighboring positions in significant amount."

SLOW START OF THE CODE WORK WITH JOE SPEYER

As noted above, I had mentioned the idea of attempting to use synthetic homopolyribonucleotides for code deciphering to my four colleagues who accompanied me to Cold Spring Harbor on June 6, 1961. Joe, Hans, and Yoshito liked the idea and found it worthwhile to test. Charlie was in agreement with them; however, with brilliant insight and foresight, he added that in case the translation initiation signal included more than one kind of nucleotide, none of the four homopolyribonucleotides would be active. (Charlie's foresight turned out to be prophetic. Nevertheless, the slightly unphysiological conditions we used enabled our experiments to work.)

Around the end of June 1961, I asked Joe, who has been developing a cell-free amino-acid-incorporating system, whether we could try the experiments on coding together. He agreed. I synthesized polyA and polyU for the experiments. In Dr. Ochoa's absence, I had to ask Bob Warner, a professor in the department, for permission to order the labeled amino acids for the study. Joe was registered to attend the Phage Course at Cold Spring Harbor, and we decided that, upon his return, we would perform the experiments right away.

On the evening of July 31, a friend from Rockefeller University called to let me know that there was a breakthrough in research. I remember well my first sentence to him, "Don't tell me that the genetic code was broken!" He continued, "Yes, somebody at MIT has found that polyU encodes polyphenylalanine." I will never forget my great disappointment upon hearing that the first part of our planned project had been accomplished by somebody else. Early the next morning, I called Joe about the bad news. We decided to postpone the beginning of the experiments to consider whether and how to proceed.

What prompted us to go ahead with the study was that homopolyribonucleotides could reveal at most only codons for four amino acids, and that we had already planned to use random copolynucleotides to identify the nucleotide composition of codons for all 20 amino acids. Joe and I decided to first test whether the rumor concerning a polyU experiment was true. We attempted to reproduce the experiment starting at 11 AM on Monday, August 14. By 2 AM on August 15, we had successfully repeated what later turned out to be Marshall Nirenberg's and Heinrich Matthaei's classical experiment (35).

In our first experiment, for which we had already obtained or prepared all the necessary ingredients in advance, polyU stimulated the uptake of labeled phenylalanine to an insoluble product 32-fold. On August 17, when we repeated the experiment with slightly improved conditions, the stimulation was 55-fold.

The next day we ran into Jerry Hurwitz, a colleague at NYU, who had just returned from the Moscow International Congress of Biochemistry. He informed us about Marshall Nirenberg's exciting talk in which he had described his discovery with Heinrich Matthaei that polyU promotes the incorporation of phenylalanine into polyphenylalanine in a cell-free system.

On August 19, I reported to Dr. Ochoa, still away at the time, in a letter (from which I have already quoted earlier in this manuscript), the results of our above-mentioned experiments between August 14 and 17, together with the origin of my idea for code breaking using PNP, and the plans to extend the research using random copolynucleotides to explore codons for all 20 amino acids. On August 20, my wife, Suzanna, and I left for a long-planned 12-day vacation. As I reconsider the situation 50 years later, it would have been more appropriate to skip this vacation.

Nucleotide composition of codons: specifies the types of nucleotides linked in codons

FIRST DECODING OF THE NUCLEOTIDE COMPOSITIONS OF CODONS USING RANDOM COPOLYNUCLEOTIDES AT NYU

After Dr. Ochoa and I returned to New York on September 2, we had a long meeting with Joe Speyer. From this time on, Dr. Ochoa strongly and enthusiastically supported the code-deciphering project Joe and I had started and he took over its guidance. Ochoa must have rightfully considered that the project was only made possible by his earlier discovery (with Marianne Grunberg-Manago) of PNP (11, 36).

Nirenberg & Matthaei's classical experiment (35) (which we had repeated) proved that the first part of our working hypothesis (i.e., that a homopolynucleotide could direct the synthesis of a homopolypeptide in a cell-free system) was correct. Thus, we had to proceed to explore the second part of my hypothesis, namely, that a random copolynucleotide synthesized by PNP can dictate the incorporation of various amino acids into copolypeptides. The essence of this hypothesis was stated in my letter to Dr. Ochoa dated August 19: "continued code breaking, e.g., inserting A into polyU which should lead to three new incorporated amino acids, if the amount of inserted A will be low enough to prevent it from occupying neighboring positions in a significant amount."

To aid Joe and me in this exploration, Dr. Ochoa assigned Pete Lozina, an experienced technician. We first performed many experiments to improve the cell-free system that Joe had developed and, specifically, to show that added transfer RNA increased the polyU-promoted incorporation of phenylalanine into an acid-insoluble product. PolyU did promote polyphenylanine formation, indicating (using a triplet code) that UUU encodes phenyalanine. We synthesized our first random copolynucleotides on September 14. They were polyUC-5:1 and polyUA-5:1.

The ratio of UUU to UUC (or UCU or CUU) triplets in a random polyUC copolymer is the same as the UC ratio of the copolymer, i.e., in our case 5:1. In turn, the ratio of UUU to UCC (or CUC or CCU) triplets in the same copolymer should be 25:1. The validity of this conclusion was based on the randomness of the nearest-neighbor distributions of several copolymers synthesized by PNP (13, 38). Later studies also verified that the nucleotide ratio of the copolymers closely reflects the nucleotide compositions of the incubation mixtures in which they were prepared (8).

Our experiments revealed that the polyUC 5:1 copolymer promoted the incorporation of phenylalanine as well as serine, and the ratio of the phenylalanine to serine incorporation was 4.4 (i.e., close to 5). This ratio suggested that the codon for serine is UUC, UCU, or CUU, i.e., consists of two U and one C. Furthermore, the polyUA (UA-5:1) copolymer promoted the incorporation of phenylalanine as well as tyrosine, and the ratio of the phenylalanine to tyrosine incorporation was 4.0 (i.e., close to 5). This suggested that the codon for tyrosine is UUA, UAU, or AUU, i.e., consists of two U and one A (27).

Watching the counter of radioactivity with Dr. Ochoa and Joe, and witnessing that polyUC 5:1 and polyUA 5:1, our two random copolymers, promote the uptake of other particular amino acids in addition to phenylalanine, were among the greatest joys of my scientific career (see also Reference 37). The manuscript containing these and other results was submitted to the *Proceedings of the National Academy of Sciences* (*PNAS*) on October 25, 1961 (27). In its introduction we specified the important findings that triggered the idea of using synthetic polyribonucleotides, i.e., the concept of mRNA as the actual template of protein synthesis, and the observation by Brenner et al. that, after infection of *Escherichia coli* with T2 bacteriophage, the newly synthesized phage mRNA is bound to ribosomes. We concluded the introduction by citing Nirenberg & Matthaei's classical experiment that had revealed the polyU-dependent synthesis of polyphenylalanine in a system of *E. coli* extract supernatant fraction and ribosomes. The article describing their experiment had just appeared in the 1961 October issue of *PNAS* (35). We ended our manuscript by stating that "...the results reported in this paper would appear to open up an experimental approach to the study of the coding problem in protein synthesis" (27).

The subsequent identifications of the nucleotide compositions of the codons specifying the various other amino acids were based on the same considerations and principles as above (27, 44). This required the determination of the ratios in which different amino acids were incorporated into cell-free systems in many experiments involving the use of various random copolymers containing two or three kinds of nucleotides in different ratios.

I gave my first seminar about the successful use of synthetic copolyribonucleotides of random sequence for deciphering the nucleotide composition of codons at the Department of Chemistry at the Roswell Park Cancer Institute and the University of Buffalo in Buffalo, New York, on October 23, 1961. At about that time, our research group was increased by Carlos Basilio, a postdoctoral fellow from the University of Chile, Santiago. For the continuation of the code deciphering, we prepared a series of random copolyribonucleotides, including, for the first time, some copolymers with three different nucleotides, e.g., UAC, UAG, and UCG.

MEETING MARSHALL NIRENBERG AND CONTINUING CODE DECIPHERING WITH RANDOM COPOLYNUCLEOTIDES AT BOTH NYU AND NIH

Around the middle of October 1961, I noticed on the departmental bulletin board an announcement from the Massachusetts Institute of Technology (MIT) that Marshall Nirenberg would be presenting a seminar. I attended his talk, presented to an enthusiastic audience in a large and crowded lecture hall at MIT. During the question-and-answer period following the talk, I asked for permission to make an announcement. The following lines are from a relevant paragraph taken from Nirenberg's "Historical Review: Deciphering the Genetic Code—A Personal Account" (32):

> Peter Lengyel came up to the podium and told the audience that he and others in Ochoa's laboratory had used randomly ordered synthetic polynucleotides that contained several different nucleotide residues to direct the incorporation of other amino acids [i.e., other than phenylalanine] into protein. I flew back to Washington feeling very depressed because, although I had taken only two weeks to show that aminoacyl-tRNA is an intermediate in protein synthesis, I should have spent the time focusing on the more important problem of deciphering the genetic code. Clearly, I had to either compete with the Ochoa laboratory or stop working on the problem (text in brackets author's own).

Nirenberg continues his historical review describing that the next morning at the National Institutes of Health (NIH) he read up on the synthesis and characterization of randomly ordered synthetic polyribonucleotides by PNP. In the library, he met NIH scientist Robert Martin. Upon hearing Nirenberg's story about what happened at MIT on the previous day, Martin volunteered to stop his own research and synthesize random polynucleotides the same week. Martin was only the first of numerous NIH scientists (e.g., Leon Heppel and Maxine Singer) who helped Nirenberg determine the nucleotide composition of codons (31, 32).

On November 9, sometime after Nirenberg's MIT talk and my announcement, there was a meeting at the New York Academy of Medicine. At this meeting, Nirenberg reported his pioneering study revealing that polyU encodes polyphenylalanine. Ochoa announced our successful use of synthetic polynucleotides with a random sequence for deciphering the nucleotide compositions of 11 codons. That same afternoon, Nirenberg visited our laboratory. He talked to both Ochoa and me for the first time. The meeting was friendly but did not result in a collaboration.

We submitted our second paper concerning the deciphering of the nucleotide compositions of further codons to *PNAS* on November 20 (42). The next day, we sent preprints of both our first and second papers to Nirenberg.

The first paper, which involved the use of random copolynucleotides for codon deciphering from the Nirenberg laboratory, was submitted to *Biochemical and Biophysical Research Communications* on December 4, 1961, and was published on December 16 (31). It reported the nucleotide compositions of codons corresponding to 15 amino acids.

From this time on, for the next two years, the nucleotide compositions of codons were determined in experiments involving random copolynucleotides with two or three kinds of nucleotides synthesized by PNP, and were published from both the Ochoa (44) and Nirenberg (34) laboratories. The results were, in general albeit with some (temporary) exceptions, in good agreement. The race between the two research groups probably accelerated the establishment of the nucleotide compositions of the codons.

From Cold War to Code War

TMV: *Tobacco mosaic virus*

Nitrous acid deamination: process that results in the conversion of cytidylate to uridylate, adenylate to inosinate, and guanylate to xanthylate

Some of Nirenberg's collaborators in code deciphering were very unhappy by the involvement of an NYU team in this effort. I cite the intriguing paper entitled "A Revisionist View of the Breaking of the Genetic Code" by Robert G. Martin, a leading member of Nirenberg's team (30): "We at NIH were terribly angry with Ochoa and his colleagues for jumping in on Marshall and Heinrich's discovery. But of course, they too were working on protein synthesis at the time. Peter Lengyel, Ochoa's director of operation, says they had planned those experiments before hearing of Marshall's work. I am sure he is right."

I wrote this prefatory chapter in part to describe, for the first time, how and when the idea that became the basis of the NYU project on code deciphering was born. We have, perhaps mistakenly, never reported the story until now. As described in the section "Getting Unexpectedly Involved in Genetic Code Deciphering," it was at the Cold Spring Harbor Symposium in 1961 that the idea of an approach for deciphering the genetic code using synthetic polyribonucleotides occurred to me. This happened on June 6, whereas I heard the rumor only on July 31, i.e., almost eight weeks later, that somebody at MIT (this was obviously wrong; it should have been NIH) had performed the classical polyU experiment.

The scientific community was well aware that the two laboratories were racing to decipher the genetic code. Rollin Hotchkiss, the eminent Rockefeller University scientist, even coined a witty saying about this race. According to Hotchkiss, "The U-2 incident started the cold war, the U3 incident started the code war." This was based in part on the well-known U-2 incident, which had occurred one year earlier, in 1960. U-2 was an American high-flying spy plane that was shot down by the Soviet military high above the Soviet Union. This resulted in a large increase in the tensions between the United States and the Soviet Union.

SUBMISSION OF MY PHD THESIS: NUCLEOTIDE COMPOSITIONS OF 18 CODONS, CODING CHARACTERISTICS OF THE INOSINE AND XANTHINE NUCLEOTIDES, AND THE RIBOSOME AS A TARGET OF THE ANTIBIOTIC STREPTOMYCIN

I remain grateful to Joe Speyer for suggesting in April 1962 that I interrupt my lab work and submit the research on code deciphering as my PhD thesis. At the time, I had been in the Ochoa laboratory for over five years and had worked on several topics. Dr. Ochoa consented. The deadline for submitting the thesis was less than a month away.

That all the code work performed by that time had already been published or was in press facilitated the task of writing the thesis. The thesis was entitled "Use of Synthetic Polynucleotides in the Deciphering of the Genetic Code" (17). It included, following a historical introduction, the first five papers of the series entitled "Synthetic Polynucleotides and the Amino Acid Code" (2, 26, 27, 42, 43). These described, for the first time, the idea of an approach using synthetic random copolynucleotides containing two or three kinds of nucleotides, radioactively labeled amino acids and a cell-free amino-acid-incorporating system from *E. coli* to decipher the nucleotide composition of codons. The thesis reported the correct nucleotide composition of codons for 18 amino acids.

The thesis also described the coding properties of nucleotides containing the modified purine base hypoxanthine or xanthine in place of guanine (2, 43). Our studies were prompted, in part, by articles that reported the generation of *Tobacco mosaic virus* (TMV) mutants by treatment of TMV-RNA with nitrous acid (10). We used nitrous acid deamination to modify the coding characteristics of synthetic polynucleotides. The change in coding characteristics of polyUA (5:1) due to conversion of A to I by deamination with nitrous acid was readily apparent. PolyUA (5:1) promoted the incorporation of Ile and Tyr, among others, but not of Cys or Val. Deamination eliminated the ability to stimulate incorporation of Ile and Tyr and generated stimulation of the incorporation of Cys and Val. Evidently, on deamination by nitrous acid treatment, the UA polynucleotide acquired the coding characteristics of a UG polynucleotide. Because the deamination of a UA polynucleotide leads to a UI polynucleotide, it follows that hypoxanthine can replace guanine in the genetic code. This conclusion was tested with polyUI (5:1) synthesized by PNP. Although the activity of this polynucleotide was somewhat lower than that of polyUG (5:1), the Phe/Cys, Phe/Val, Phe/Gly, and Phe/Tri incorporation ratios were in reasonable agreement with the corresponding ratios produced by polyUG (5:1) (2, 43). Treatment of polyUG (5:1) with nitrous acid (resulting in a guanine to xanthine conversion), however, virtually eliminated the coding activity of this polynucleotide. Thus, xanthine is unable to substitute for guanine in the genetic code (2).

TCA: trichloroacetic acid

Another publication included in the thesis was entitled "Ribosomal Localization of Streptomycin Sensitivity" (41). This study was based on (*a*) the finding that the antibiotic streptomycin inhibits the incorporation of amino acids into protein in a cell-free preparation from streptomycin-sensitive strains of bacteria, but not from resistant strains (6), and on (*b*) the hypothesis that ribosomes from sensitive bacteria have a high affinity for streptomycin, whereas ribosomes from resistant strains have no affinity (46).

The in vitro polyU-dependent polyphenylalanine synthesis system was a convenient tool to establish whether the target of streptomycin action is indeed the ribosome. For this purpose, we fractionated cell-free extracts from streptomycin-sensitive and streptomycin-resistant *E. coli* strains into a fast-sedimenting fraction including the ribosomes and a supernatant fraction. Streptomycin inhibited the polyU-dependent synthesis of polyphenylalanine in reaction mixtures with ribosomes from sensitive cells combined with supernatants from either sensitive or resistant cells, whereas it inhibited polyphenylalanine synthesis much less so, if at all, in mixtures of ribosomes from resistant cells combined with supernatants from either sensitive or resistant cells. These results established that the streptomycin-sensitive component of an *E. coli* protein synthesis system is in the ribosomal fraction (41). Later studies of others revealed the remarkable fact that a large variety of antibiotics inhibits the functioning of ribosomes (47).

AN OVERVIEW OF THE PHASES IN DECIPHERING THE GENETIC CODE

Prior to the use of synthetic polynucleotides in cell-free systems, it was hoped that mutagenesis (especially of TMV by nitrous acid) (10) would reveal the genetic code. However, Nirenberg & Matthaei's classical experiment with polyU revealed the successful use of a synthetic homopolynucleotide for code breaking, i.e., phase 1 in code deciphering.

In his historical review (32), Nirenberg mentions learning from Mirko Beljanski (a former postdoctoral fellow of Ochoa) that Ochoa had suggested to Beljanski to check whether a synthetic homopolyribonucleotide can promote amino acid incorporation in an *E. coli* extract. Beljanski tested polyA with no success. Dr. Ochoa never mentioned Beljanski's experiments to me. I learned about them only by reading Nirenberg's historical review (32). Nirenberg also notes that Alfred Tissières tested polyA in an *E. coli* extract in Jim Watson's laboratory and was equally unsuccessful.

That polyA can promote amino acid incorporation in an *E. coli* extract was discovered only in 1962 by Bob Gardner, a gifted medical student who joined our team at NYU. He noticed that polyA caused a slight stimulation of lysine incorporation into a trichloroacetic acid (TCA)-insoluble product. Polylysine had been reported to be TCA soluble, but insoluble in a TCA-tungstate solution. Using the latter for precipitation, Bob established that polyA strongly promoted polylysine formation in the *E. coli* extract (8). The use of this precipitant facilitated the deciphering of codons not containing U residues.

Phase 2 in code deciphering (the central topic of this write-up) was establishing the nucleotide composition of codons by using random copolyribonucleotides containing two or three kinds of nucleotides synthesized by PNP and labeled amino acid incorporation in a cell-free system. Phase 3 was the determination of the nucleotide sequence of codons using an approach developed by Marshall Nirenberg and Philip Leder (16, 33). This was based on their finding that each triplet codon specifically promotes the binding to ribosomes of the (labeled) aminoacyl-tRNA specified by the codon. A further approach in the Khorana laboratory was based on the chemical synthesis of oligodeoxyribonucleotides that were amplified (and/or linked) by DNA polymerase, and then transcribed by RNA polymerase into long ribopolynucleotides of known sequences. These were used as messengers in cell-free systems to reveal features of the genetic code (15). Ribotrinucleotides prepared by chemical approaches were also used in the Khorana laboratory (15, 39) in the assay developed by Nirenberg & Leder (33). Both Nirenberg and Khorana were awarded the Nobel Prize in Physiology or Medicine in 1968 for their outstanding research.

The determination of the nucleotide sequence of the three nonsense codons (i.e., UAA, UAG, and UGA) triggering translation termination, by Alan Garen and Sidney Brenner, involved mutagenesis in *E. coli*. This could alter a single nonsense triplet (triggering the premature termination of the translation of the protein encoded) to various sense triplets encoding a series of different amino acids. A comparison of the codons encoding these amino acids allowed the sequence of the nonsense triplet from which they arose to be identified (9; see also 5).

Nucleotide sequence of a codon: specifies the order in which two or three types of nucleotides are linked in a codon

POSTSCRIPT

After finishing his work on the genetic code at NYU, Joe Speyer moved to the Cold Spring Harbor Laboratory. While there he discovered that a temperature-sensitive mutant of bacteriophage T4 DNA polymerase was mutagenic (40). This revealed that DNA polymerase may affect the selection of the nucleotides in DNA replication. Subsequently, as a professor in the Department of Molecular and Cellular Biology at the University of Connecticut, Joe studied the mechanisms of action of proteins with a mutator or antimutator phenotype (29, 45).

After concluding my studies on the genetic code at NYU, I spent one year (1963–1964) in the laboratory of the eminent French scientist Jacques Monod at the Pasteur Institute in Paris. In 1965, I joined the Department of Molecular Biophysics and Biochemistry at Yale University and became Professor Emeritus in 2001. During the 45+ years I spent at Yale, I studied the mechanisms of protein synthesis (18, 19), the interferon system (20, 21), and, finally, the multifunctional p200 family genes and proteins (23, 28). The references cited are reviews of the above studies.

DISCLOSURE STATEMENT

The author is not aware of any affiliations, memberships, funding, or financial holdings that might be perceived as affecting the objectivity of this review.

ACKNOWLEDGMENTS

I remain most grateful to Severo Ochoa, who accepted me as his graduate student when I arrived in New York as a refugee in 1957. I spent about seven years with Dr. Ochoa at NYU, during the course of which I rose from graduate student to assistant professor. Ochoa created an atmosphere in his department that promoted a constant exchange of scientific information and stimulated the generation of ideas. I also wish to express my gratitude to the late Joseph Speyer. The two of us performed most of the experiments on code deciphering together. I much appreciate the valuable suggestions of Dr. Daniel DiMaio and Dr. Alan Garen concerning the manuscript.

LITERATURE CITED

1. Barker HA, Weissbach H, Smyth RD. 1958. A coenzyme containing pseudovitamin B_{12}. *Proc. Natl. Acad. Sci. USA* 44:1093–97
2. Basilio C, Wahba AJ, Lengyel P, Speyer JF, Ochoa S. 1962. Synthetic polynucleotides and the amino acid code. V. *Proc. Natl. Acad. Sci. USA* 48:613–16
3. Beck WS, Flavin M, Ochoa S. 1957. Metabolism of propionic acid in animal tissues. III. Formation of succinate. *J. Biol. Chem.* 229:997–1010
4. Brenner S. 1961. RNA, ribosomes, and protein synthesis. *Cold Spring Harb. Symp. Quant. Biol.* 26:101–10
5. Brenner S, Stretton AO, Kaplan S. 1965. Genetic code: the 'nonsense' triplets for chain termination and their suppression. *Nature* 206:994–98
6. Erdos T, Ullmann A. 1959. Effect of streptomycin on the incorporation of amino-acids labelled with carbon-14 into ribonucleic acid and protein in a cell-free system of a Mycobacterium. *Nature* 183:618–19
7. Frisch L, ed. 1961. *Cellular Regulatory Mechanisms*. Cold Spring Harb. Symp. Quant. Biol. Vol. 26. Cold Spring Harbor, NY: Cold Spring Harb. Press
8. Gardner RS, Wahba AJ, Basilio C, Miller RS, Lengyel P, Speyer JF. 1962. Synthetic polynucleotides and the amino acid code. VII. *Proc. Natl. Acad. Sci. USA* 48:2087–94
9. Garen A. 1968. Sense and nonsense in the genetic code. Three exceptional triplets can serve as both chain-terminating signals and amino acid codons. *Science* 160:149–59
10. Gierer A, Mundry KW. 1958. Production of mutants of tobacco mosaic virus by chemical alteration of its ribonucleic acid in vitro. *Nature* 182:1457–58
11. Grunberg-Manago M, Ochoa S. 1955. Enzymatic synthesis and breakdown of polynucleotides: polynucleotide phosphorylase. *J. Am. Chem. Soc.* 77:3165–66
12. Grunberg-Manago M, Ortiz PJ, Ochoa S. 1955. Enzymatic synthesis of nucleic acidlike polynucleotides. *Science* 122:907–10
13. Heppel LA, Ortiz PJ, Ochoa S. 1957. Studies on polynucleotides synthesized by polynucleotide phosphorylase. II. Structure of polymers containing a mixture of bases. *J. Biol. Chem.* 229:695–710
14. Jacob F, Monod J. 1961. Genetic regulatory mechanisms in the synthesis of proteins. *J. Mol. Biol.* 3:318–56
15. Khorana HG. 1965. Polynucleotide synthesis and the genetic code. *Fed. Proc.* 24:1473–87
16. Leder P, Nirenberg M. 1964. RNA codewords and protein synthesis. 3. Nucleotide sequence of a valine RNA codeword. *Proc. Natl. Acad. Sci. USA* 52:420–27
17. Lengyel P. 1962. *Use of synthetic polynucleotides in the deciphering of the genetic code*. PhD thesis. New York Univ. 65 pp.
18. Lengyel P. 1969. The process of translation as seen in 1969. Summary of the Symposium on the Mechanism of Protein Synthesis. *Cold Spring Harbor Symp. Quant. Biol.* 34:828–41
19. Lengyel P. 1976. Ten years in protein synthesis. In *Reflections in Biochemistry*, ed. A Kornberg, BL Horecker, L Cornudella, J Oro, pp. 309–16. Oxford: Pergamon
20. Lengyel P. 1982. Biochemistry of interferons and their actions. *Annu. Rev. Biochem.* 51:251–82
21. Lengyel P. 2008. From RNase L to the multitalented p200 family proteins: an exploration of the modes of interferon action. *J. Interf. Cytokine Res.* 28:273–81
22. Lengyel P, Chambers RW. 1960. Preparation of 2-thiouridine-5′-diphosphate and the enzymatic synthesis of polythiouridylic acid. *J. Am. Chem. Soc.* 82:752
23. Lengyel P, Liu CJ. 2010. The p200 family protein p204 as a modulator of cell proliferation and differentiation: a brief survey. *Cell Mol. Life Sci.* 67:335–40
24. Lengyel P, Mazumder R, Ochoa S. 1960. Mammalian methylmalonyl isomerase and vitamin B(12) coenzymes. *Proc. Natl. Acad. Sci. USA* 46:1312–18
25. Lengyel P, Ochoa S. 1958. Phosphorolysis of leaf ribonucleic acids. *Biochim. Biophys. Acta* 28:200–1

26. Lengyel P, Speyer JF, Basilio C, Ochoa S. 1962. Synthetic polynucleotides and the amino acid code. III. *Proc. Natl. Acad. Sci. USA* 48:282–84
27. Lengyel P, Speyer JF, Ochoa S. 1961. Synthetic polynucleotides and the amino acid code. *Proc. Natl. Acad. Sci. USA* 47:1936–42
28. Luan Y, Lengyel P, Liu CJ. 2008. p204, a p200 family protein, as a multifunctional regulator of cell proliferation and differentiation. *Cytokine Growth Factor Rev.* 19:357–69
29. Lyons SM, Speyer JF, Schendel PF. 1985. Interaction of an antimutator gene with DNA repair pathways in *Escherichia coli* K-12. *Mol. Gen. Genet.* 198:336–47
30. Martin RG. 1984. A revisionist view of the breaking of the genetic code. In *NIH: An Account of Research in Its Laboratories and Clinics*, ed. J DeWitt Stetten, pp. 282–95. Orlando, FL: Academic
31. Martin RG, Matthaei JH, Jones OW, Nirenberg MW. 1962. Ribonucleotide composition of the genetic code. *Biochem. Biophys. Res. Commun.* 6:410–14
32. Nirenberg M. 2004. Historical review: deciphering the genetic code—a personal account. *Trends Biochem. Sci.* 29:46–54
33. Nirenberg M, Leder P. 1964. RNA codewords and protein synthesis. The effect of trinucleotides upon the binding of sRNA to ribosomes. *Science* 145:1399–407
34. Nirenberg MW, Jones OW, Leder P, Clark BFC, Sly WS, Pestka S. 1963. On the coding of genetic information. *Cold Spring Harbor Symp. Quant. Biol.* 28:549–57
35. Nirenberg MW, Matthaei JH. 1961. The dependence of cell-free protein synthesis in *E. coli* upon naturally occurring or synthetic polyribonucleotides. *Proc. Natl. Acad. Sci. USA* 47:1588–602
36. Ochoa S. 1960. Enzymatic synthesis of ribonucleic acid. Nobel Lecture 1959. pp. 146–64. Stockholm
37. Ochoa S. 1980. The pursuit of a hobby. *Annu. Rev. Biochem.* 49:1–30
38. Ortiz PJ, Ochoa S. 1959. Studies on polynucleotides synthesized by polynucleotide phosphorylase. IV. P32-labeled ribonucleic acid. *J. Biol. Chem.* 234:1208–12
39. Soll D, Ohtsuka E, Jones DS, Lohrmann R, Hayatsu H, et al. 1965. Studies on polynucleotides, XLIX. Stimulation of the binding of aminoacyl-sRNA's to ribosomes by ribotrinucleotides and a survey of codon assignments for 20 amino acids. *Proc. Natl. Acad. Sci. USA* 54:1378–85
40. Speyer JF. 1965. Mutagenic DNA polymerase. *Biochem. Biophys. Res. Commun.* 21:6–8
41. Speyer JF, Lengyel P, Basilio C. 1962. Ribosomal localization of streptomycin sensitivity. *Proc. Natl. Acad. Sci. USA* 48:684–86
42. Speyer JF, Lengyel P, Basilio C, Ochoa S. 1962. Synthetic polynucleotides and the amino acid code. II. *Proc. Natl. Acad. Sci. USA* 48:63–68
43. Speyer JF, Lengyel P, Basilio C, Ochoa S. 1962. Synthetic polynucleotides and the amino acid code. IV. *Proc. Natl. Acad. Sci. USA* 48:441–48
44. Speyer JF, Lengyel P, Basilio C, Wahba AJ, Gardner RS, Ochoa S. 1963. Synthetic polynucleotides and the amino acid code. *Cold Spring Harb. Symp. Quant. Biol.* 28:559–67
45. Speyer JF, Rosenberg D. 1968. The function of T4 DNA polymerase. *Cold Spring Harb. Symp. Quant. Biol.* 33:345–50
46. Spotts CR, Stanier RY. 1961. Mechanism of streptomycin action on bacteria: a unitary hypothesis. *Nature* 192:633–37
47. Vazquez D. 1979. *Inhibitors of Protein Biosynthesis*. Berlin/New York: Springer-Verlag. 312 pp.
48. Weissbach H, Toohey J, Barker HA. 1959. Isolation and properties of B(12) coenzymes containing benzimidazole or dimethylbenzimidazole. *Proc. Natl. Acad. Sci. USA* 45:521–25
49. Zamecnik PC. 1962. Unsettled questions in the field of protein synthesis. *Biochem. J.* 85:257–64
50. Zamecnik PC, Keller EB. 1954. Relation between phosphate energy donors and incorporation of labeled amino acids into proteins. *J. Biol. Chem.* 209:337–54

Yeast ATP-Binding Cassette Transporters Conferring Multidrug Resistance

Rajendra Prasad[1] and Andre Goffeau[2]

[1]School of Life Sciences, Jawaharlal Nehru University, New Delhi, 110067 India; email: rp47@mail.jnu.ac.in

[2]Institut des Sciences de la Vie, Université Catholique de Louvain, Louvain-la-Neuve, 1349 Belgium; email: agoffeau@hotmail.com

Keywords

transmembrane domain, nucleotide binding domain, pleiotropic drug resistance, Pdr5p, Cdr1p

Abstract

Overexpression of the ATP binding cassette (ABC) drug transporter P-glycoprotein (P-gp) is often responsible for the failure of chemotherapy as a treatment for human tumors. The presence of proteins homologous to P-gp in organisms ranging from prokaryotes to eukaryotes indicates that drug export is a general mechanism of multidrug resistance. Yeasts are no exception. They have developed a large subfamily of ABC exporters involved in pleiotropic drug resistance (PDR) and in the cellular efflux of a wide variety of drugs. The PDR transporters Pdr5p of *Saccharomyces cerevisiae* and Cdr1p of *Candida albicans* are important members of this PDR subfamily, which comprises up to 10 phylogenetic clusters in fungi. Here, we review current achievements concerning the structure, molecular mechanism, and physiological functions of yeast Pdr transporters.

Contents

INTRODUCTION	40
INVENTORY OF THE ATP-BINDING CASSETTE SUPERFAMILY	41
Saccharomyces cerevisiae Proteins	41
Candida albicans Proteins	42
STRUCTURE AND FUNCTION OF PLEIOTROPIC DRUG RESISTANCE PROTEINS	42
Heterologous Expression	42
Distinct Domains of PDR Proteins	43
Nucleotide-Binding Domains	44
Uncoupled ATPase Activity	47
Transmembrane Domains	50
SUBSTRATE PROMISCUITY OF YEAST PLEIOTROPIC DRUG RESISTANCE PROTEINS	52
PHYSIOLOGICAL ROLES OF YEAST PLEIOTROPIC DRUG RESISTANCE PROTEINS	53
Protection Against Cellular Toxins/Metabolites	53
Transport of Sterols	54
Translocation of Membrane Phospholipids	54
Ion Transport	55
Other Functions	55
CONCLUSIONS	56

INTRODUCTION

The large ATP-binding cassette (ABC) superfamily comprises diverse classes of transport proteins that are classified into nine families (A to I) according to the nomenclature adopted by the Human Genome Organization (HUGO) (12, 108). These proteins, which were first discovered in bacteria as high-affinity nutrient importers, rose to prominence when their ability to confer multidrug resistance (MDR) to cancer cells was identified (33). Among several mechanisms that contribute to MDR, the overexpression of drug efflux pumps belonging to the ABC superfamily is the most frequent cause of resistance to antifungals, herbicides, anticancer drugs, and other cytotoxic drugs (8, 74, 89). This review is focused on the characterization of the subfamily of full-sized multidrug exporters from yeast designated Pdr (pleiotropic drug resistance) proteins.

In 1973, Rank & Bech-Hansen (81) described the first point mutant of *Saccharomyces cerevisiae* that was resistant to a variety of inhibitors with different structures and different modes of action. This mutant, *pdr1*, was mapped to a single locus near the centromere of chromosome VII (9). The *PDR1* gene was sequenced by Balzi et al. (3) and was shown to be a novel member of the large Zn2/Cys6 transcription factor family. Disruption of *PDR1* conferred multidrug sensitivity. Several allelic drug-resistant mutants, subsequently designated *pdr1-1* to *pdr1-12*, were shown to be point mutations located in the inhibitory or terminal regions of Pdr1p (10). Together with *PDR1*, a second transcription factor located on chromosome II, *PDR3* (17, 101), was also found to control MDR. It was therefore proposed that *PDR1* and *PDR3* both control the transcription of several genes encoding drug exporters in the plasma membrane (4, 43, 44, 46).

ABC: ATP-binding cassette

Superfamily: a protein class of related proteins composed of one or more protein families

MDR: multidrug resistance

PDR: pleiotropic drug resistance

A first target of *PDR1* in *S. cerevisiae* was identified on a yeast DNA fragment that conferred MDR when amplified on a high-copy plasmid (57). This fragment of chromosome XV included a gene named *PDR5*, the deletion of which conferred drug hypersensitivity (61). This finding indicated that the *PDR5* gene is a target of the transcription factor encoded by *PDR1*. In 1994, three independent studies (5, 6, 34) reported the complete sequence of *PDR5* and showed that the encoded 160-kDa protein is a new member of the large ABC superfamily. Deletion of this ABC protein could sensitize cells to various inhibitors (48). Moreover, Balzi & Goffeau (4) showed that activation of *PDR1* by the *pdr1-3* and *pdr1-6* mutations led to considerable overexpression of Pdr5p in the plasma membrane. This demonstration of the interaction between *PDR1* and *PDR5* triggered an exhaustive search for genes involved in the PDR phenotype. To date, over 20 genes belonging to the PDR network of *S. cerevisiae* have been identified, including nine ABC proteins located in the plasma membrane (13, 68), six transcription factors (*PDR1*, *PDR3*, *YAP1*, *YRR1*, *RDR1*, and *STB5*) (56, 59), metabolic regulators (such as the casein kinases *YCK1/2* and the phosphatase *SIT4*), and several other interacting proteins (63).

A *PDR5*-like gene in a species other than *S. cerevisiae* was first identified by Prasad et al. (75), who transformed a *PDR5*-disrupted strain of *S. cerevisiae* with DNA fragments from the pathogenic yeast *Candida albicans*. The sequence of complementing DNA revealed the presence in *C. albicans* of a gene homologous to *PDR5* that was named *CDR1* (*Candida* drug resistance). This was an important discovery, as Cdr1p (*Candida* drug resistance protein 1) was later found to be a major determinant of antifungal resistance in *C. albicans* (89). Like Pdr5p, Cdr1p and its close homolog Cdr2p export not only azoles and their derivatives but also a variety of structurally unrelated compounds. Two other highly similar genes, *CDR3* and *CDR4*, have also been identified (2, 27). This review focuses only on the structure and function of full-sized Pdr protein transporters and avoids discussion of their regulatory circuitry, which has been recently reviewed (64, 95).

Pdr5p: pleiotropic drug resistance 5 protein

Cdr1p: *Candida* drug resistance 1 protein

Pdr12p: pleiotropic drug resistance 12 protein

Pdr11p: pleiotropic drug resistance 11 protein

INVENTORY OF THE ATP-BINDING CASSETTE SUPERFAMILY

Saccharomyces cerevisiae Proteins

In *S. cerevisiae*, 31 ABC proteins have been identified and classified into five phylogenetic subfamilies (13). The Pdr protein subfamily comprises nine full-sized members: Pdr5p, Pdr10p, Pdr15p, Snq2p, Pdr12p, Aus1p, Pdr11p, Pdr18p, and YOL075c. A recent analysis of 349 Pdr proteins from 55 fungal species belonging to the Ascomycota and Basidiomycota phyla identified a total of 10 phylogenetic clusters: A, B, C, D, E, F, G, H1a, H1b, and H (54). All Pdr proteins have a characteristic reverse topology (see below), but one can distinguish the sensu lato Pdr proteins belonging to clusters E and F from the sensu stricto Pdr proteins belonging to other clusters. Cluster E includes Aus1p and Pdr11p, which are influx pumps for sterols. Cluster F, which contains the *S. cerevisiae* protein of unknown function YOL075c, is the only cluster conserved in all tested fungal species. The members of clusters F and E may be close relatives of a common ancestor of all fungal Pdr proteins. Clusters A and D contain proteins only from the subphylum Saccharomycotina and are the only well-characterized Pdr proteins. Cluster A comprises Pdr5/10/15p, which are efflux pumps for cationic amphiphilic drugs such as rhodamine G. Cluster D comprises Snq2p and Pdr18p, which are efflux pumps for anionic amphiphilic drugs such as rhodamine B (47). Pdr12p-like pumps that export medium-chain fatty acids are loosely associated with cluster D (71, 94). It is striking that cluster A, which is present only in Saccharomycotina, shares a common ancestor with clusters B and C, which contain members only from the Pezizomycotina and Basidiomycotina subphyla. Unfortunately, the members of clusters B, C, G, H1a, H1b, and H2 have not yet been thoroughly characterized but are generally considered to be multidrug export pumps.

NBD: nucleotide-binding domain

Candida albicans Proteins

Using TBlastn searches, together with domain analysis of the complete genome sequence of *C. albicans* (**http://www.candidagenome.org/**), Prasad and colleagues (30) identified 81 nucleotide-binding domains (NBDs) belonging to 51 different putative open-reading frames encoding full-sized ABC proteins. This number is much greater than the 31 known ABC proteins in the *S. cerevisiae* genome (13). One reason for the greater number is that the *C. albicans* diploid genome sequence contains two copies of each gene, some of which are not identical to each other (30). Taking into consideration that each allelic pair represents a single ABC protein encoded by the *C. albicans* genome, the total number of putative ABC superfamily members is 28, including 12 half transporters that remain uncharacterized and are outside the scope of this review. By employing neighbor-joining tree and self-organizing-map-based clustering methods, these 28 putative ABC proteins can be grouped into five known subfamilies, including *C. albicans* Pdr protein (CaPdrp), and a sixth "others" category that includes soluble ABC nontransporter proteins unrelated to the existing fungal subfamilies. The Pdr protein subfamily of *C. albicans* comprises seven full-sized members: Cdr1p (75), Cdr2p (87), Cdr3p (2), Cdr4p (27), Cdr11p (Ca918, assembly #20 **http://www.candidagenome.org/download/Assembly20notes/**), CaSnq2p, and Ca4531 (30). The Cdr1/2/3/4p proteins are included in the phylogenetic cluster A together with the *S. cerevisiae* Pdr5/10/15p proteins. The *C. albicans* Cdr1p and Cdr2p proteins are active multidrug transporters; Cdr3p and Cdr4p play no apparent role in the development of antifungal resistance (2, 27, 74). CaSnq2p belongs to cluster D, containing the *S. cerevisiae* protein Snq2p, and Ca4531 belongs to cluster F, containing the *S. cerevisiae* protein YOL075c.

STRUCTURE AND FUNCTION OF PLEIOTROPIC DRUG RESISTANCE PROTEINS

Heterologous Expression

From phylogenetic and alignment analyses it is clear that sensu stricto PDR genes drive multidrug efflux functions in all plants and fungi, with the possible exceptions of the Microsporidia species *Encephalitozoon cuniculi* and the Chytridiomycotina species *Batrachochytrium dendrobatidis*, both of which are members of phyla that arose early after fungal emergence (49). For comparative studies of Pdr proteins from different species, it is necessary to express the PDR genes in *S. cerevisiae* under the same promoter. However, the use of strong promoters must be avoided, as overexpressed membrane proteins often accumulate in proliferating endoplasmic reticulum structures and can cause toxicity or are degraded (22, 70, 72, 102).

An early protocol for heterologous expression of yeast PDR genes was developed by Decottignies et al. (14) and improved by Nakamura et al. (66). It is based on the use of *S. cerevisiae* strain AD1-8, which is up to 200 times more sensitive to a variety of antifungals than the control strain. AD1-8 harbors multiple deletions in eight genes encoding major drug pumps: *YOR1*, *SNQ2*, *PDR5*, *YCF1*, *PDR10*, *PDR11*, *PDR15*, and the regulator *PDR3*. The gene to be tested is put on a low-copy vector under the constitutive *PDR5* promoter, which is activated by the *pdr1-3* mutation in the genome of the host *PDR1* gene. This system allows the correct localization of overexpressed Pdr proteins and has been successful for quantifying the drug sensitivity of the major Pdr protein pumps from *C. albicans* and *C. glabrata* (55, 97).

Constitutive *PDR5* promoter-driven expression in *S. cerevisiae* has been widely used to express a variety of ABC transporter proteins; the system tolerates N- and C-terminal tagging of overexpressed proteins with green fluorescent protein (GFP)/polyhistidine (His)/octapeptide marker sequence (FLAG) for purification and functional studies (55, 97). However, when constitutive *PDR5* promoter-based expression fails because the cloned membrane protein is toxic or degraded, expressing under the inducible Met25 promoter of *S. cerevisiae* (7), or expression in insect cells, can allow cloning of several functional ABC proteins (50). Recently, the major ABC multidrug transporter of *C. glabrata*, CgCdr1p, has been tagged with GFP and cloned into azole-sensitive (AS) and azole-resistant (AR) clinical isolates. After integration of the gene at its native chromosomal locus into AS and AR backgrounds, CgCdr1p-GFP was specifically overexpressed in AR isolates owing to the hyperactive native promoter in AR strains. This protocol prevents artifacts related to the heterologous expression of membrane proteins (77).

TMD: transmembrane domain

Sequence motif: a nucleotide or amino acid sequence pattern that is widespread and has a biological significance

TMS: transmembrane segment

ECL: extracellular loop

ICL: intracellular loop

Distinct Domains of PDR Proteins

Full ABC proteins are made up of two (or three) transmembrane domains (TMDs) and two cytoplasmic NBDs. Notably, the yeast genome reveals the existence of several ABC proteins that are putative half transporters. On the basis of biochemical and crystallographic evidence, these half proteins, which have only one NBD and one TMD, appear to function as homo- or heterodimers. In the forward topology, the TMDs precede the NBDs (TMD-NBD), whereas the NBDs come first in the reverse topology (NBD-TMD) (8, 73, 75). Unlike all other ABC subfamilies, the members of the yeast and plant Pdr protein subfamily possess reverse topology. In each NBD of Pdr proteins, at least five amino acid sequence motifs can be distinguished, and each TMD can be divided into 11 structural domains (54).

Typically, the TMDs comprise the putative α-helices of 12 transmembrane segments (TMSs), and the NBDs have α-helices and β-sheets arranged to form a Rossmann fold (54, 73). Although it appears that several TMSs associate with each other to form the substrate-binding site(s), this association alone is not sufficient for substrate transport across the membrane bilayer. Vectorial transport of these substrates requires energy from the hydrolysis of ATP by the NBDs. Given the variety of substrates that Pdr proteins transport, it is hardly surprising that despite the overall conservation of the domain architecture of TMDs, their primary sequences are significantly different. In contrast, the NBDs of all ABC transporters are highly conserved in terms of both primary structure and domain architecture (**Figure 1***a,b*). Two important points emerge from sequence analysis of different members of the Pdr protein subfamily. First, the N- and C-terminal NBDs of Pdr proteins segregate independently among the different fungal species, implying that full transporters, though arising from the duplication and fusion of half transporters, have since diverged during the course of evolution into two distinct clusters (85). Second, N-terminal NBDs of different proteins within a given Pdr protein cluster are more similar to one another than to their own corresponding C-terminal NBDs. This observation suggests that full-sized proteins within a single Pdr protein cluster share a common full-sized progenitor and also points to the possibility that the two halves are functionally distinct (11).

The 12 TMSs are interlinked with six extracellular (ECL1–6) and four intracellular (ICL1–4) loops. Although ECLs are diverse in sequence, their lengths are comparable among transporters. For example, ECL3 and ECL6 are the largest loops, whereas ECL1, ECL2, and ECL4 are shorter. The length of ECL5 is the most variable among transporters. Some of the ECLs are presumed to impart substrate specificity, promote proper protein assembly, and target the transporter protein to the cell surface (54). Conversely, ICLs are generally shorter but exhibit conserved primary sequences. ICL1 in particular is a communication link between NBDs and TMDs during drug

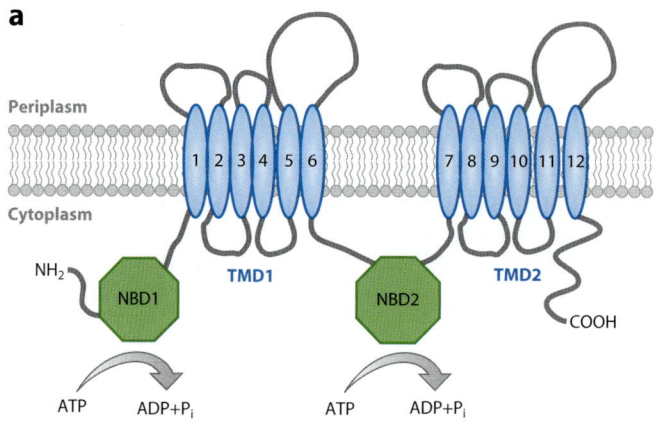

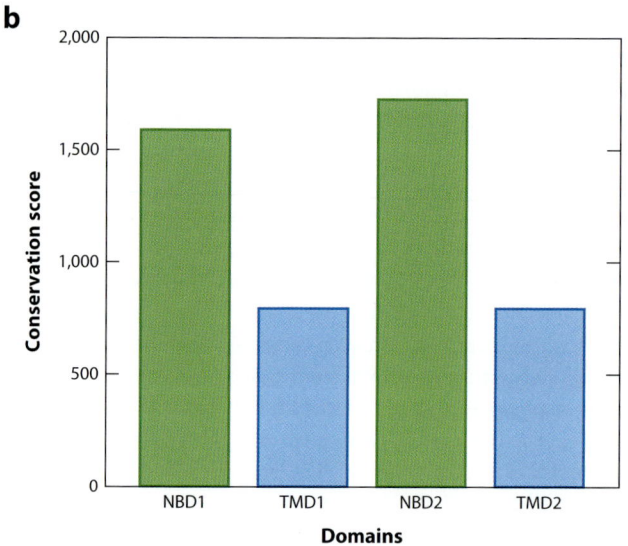

Figure 1

The predicted topology of ATP-binding cassette (ABC) transporters belonging to the pleiotropic drug resistance subfamily. (*a*) These transporters have 12 transmembrane segments (TMSs) and two nucleotide-binding domains (NBDs), organized in reverse (NBD-TMS$_6$)$_2$ topology. The two NBDs (N- and C-terminal NBDs) are shown as octagons. Transmembrane domains (TMD1 and TMD2) are indicated. Each TMD consists of six TMSs. The amino and carboxyl terminals of protein are indicated. (*b*) Relative conservation of NBDs and TMDs based on a conservation score obtained by Jalview 2.4.0.b2 (**http://www.jalview.org**).

transport (54). Hydrolysis of ATP is thought to interact with the TMDs via cytoplasmic extension of the transmembrane helices (8, 54, 58).

Nucleotide-Binding Domains

The characteristic feature of yeast Pdr proteins is that they utilize the energy of ATP hydrolysis to transport a variety of substrates across the plasma membrane. The conserved NBDs located at the cytoplasmic periphery are the hub of such activity. The N- and C-terminal NBDs of all ABC transporters, irrespective of their origin and transport substrate, share extensive amino acid sequence identity and five typical motifs. For example, the NBDs of all ABC transporters have a β-sheet subdomain containing the typical Walker A and Walker B motifs as an essential feature of all ATP-requiring enzymes (39, 40), as well as a signaling subdomain composed of α-helices that possess the ABC signature sequence (C-loop) and the Q-loops. While the inventory of *C. albicans* ABC transporters was being made, a new sequence motif characteristic of each ABC protein subfamily was identified. In the Pdr, Mdr, Mrp, and Ald protein families, the new motif occurs in the region between the Walker A and Walker B motifs, whereas in the nontransporter subfamilies (e.g., EF3 and RLI), it occurs closer to the Walker B motif. Although this new motif can be used to identify sequences from the corresponding subfamily in other organisms, its role remains to be assessed (30). Several sequence motifs and invariant residues have also been recently identified in the ECL and ICL domains as well as in all TMSs (except TM10, which is highly variable) of the fungal Pdr proteins (54).

NBD sequences possess certain conserved amino acid motifs that are critical for the domain's functionality. These motifs include the Walker A motif, with a consensus sequence GxxGxGKS/T (where "x" represents any amino acid); the Walker B motif hhhhD (where "h" represents any aliphatic residue); and the ABC signature LSGGQQ/R/KQR. Binding of a single ATP molecule requires two NBDs; one provides the Walker A and Walker B motifs and the other provides the ABC signature. Structural and biochemical analyses of NBDs show that the lysine residue of the Walker A motif binds to the β- and γ-phosphates of ribonucleotides and plays a critical role in ATP hydrolysis. Mutations of this lysine residue reduce or abolish ATP hydrolysis activity and in some cases impair nucleotide binding (39, 40). A close comparison of the primary sequences of NBDs in fungal Pdr proteins reveals a well-conserved asymmetry between the two NBDs. Interestingly, the N-terminal NBD (N-NBD) motifs display sequence degeneracy in their Walker A (GRPGAGCST) and Walker B (IQCWDN) motifs, whereas the ABC signature sequence (VSGGERKRVSIA) remains conserved. In contrast, the Walker A (GASGAGKS) and Walker B (LLFLD) motifs of the C-terminal NBD (C-NBD) are well conserved and its ABC signature motif (LNVEQRKRLTIGV) is degenerated. The NBD sequence diversity in fungal Pdr proteins is unique, and its evolution remains to be clarified (24, 53, 54) (**Figure 2**).

Biochemical analysis of a purified functional N-NBD of Cdr1p from *C. albicans* revealed that the divergent Cys193 of the Walker A motif is essential for ATP hydrolysis (40). It is fairly well established that in sensu stricto Pdr proteins, the two NBDs respond asymmetrically to the substitution of conserved residues in their respective Walker A motifs (39). Accordingly, Cdr1p chimeras containing either two N-terminal or two C-NBDs are nonfunctional (86). The functional nonequivalence between NBDs is a characteristic feature of Cdr1p and other fungal Pdr protein transporters. A mutation in the H loop selectively affects rhodamine transport by the yeast multidrug ABC transporter Pdr5p (24). The unusual Trp326, which precedes the well-conserved Asp327, in the Walker B motif of the N-NBD of Cdr1p appears to be important for ATP binding and for the accompanying conformational change (79, 80). Thus, the mutant Trp326Ala remains capable of ATP hydrolysis; however, it does so with a much higher K_M value, indicating that the docking of the substrate in the binding pocket has been altered by the mutation. The protein, however, appears capable of near-normal drug-transport function in cells expressing the full-length protein carrying the Trp326Ala substitution. This implies that the conformational change occurring upon ATP docking cannot by itself be responsible for the cross talk between the NBDs and the TMDs. Whereas the neighboring and highly conserved Asp327 of N-NBD is the catalytic carboxylate for other ABC transporters, in Cdr1p it does not appear to mediate catalysis via interaction with Mg^{2+}, but rather is essential for ATP hydrolysis and acts as a catalytic

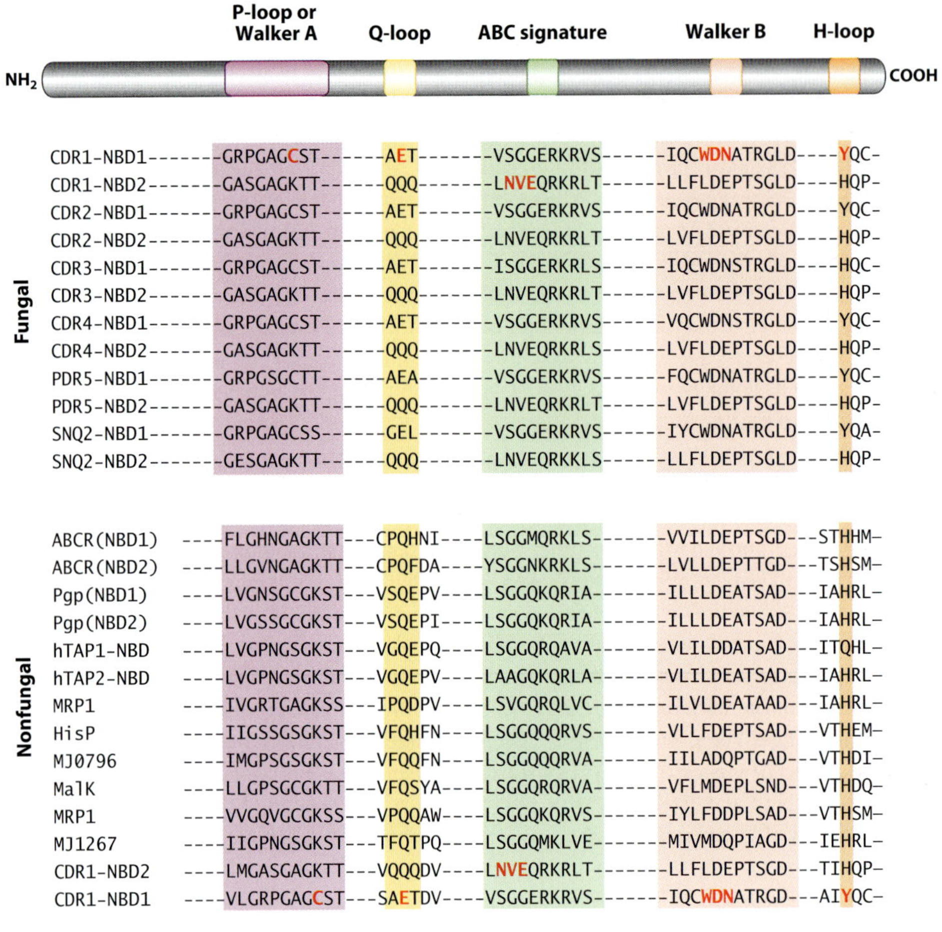

		P-loop or Walker A	Q-loop	ABC signature	Walker B	H-loop

```
              P-loop or
              Walker A      Q-loop      ABC signature       Walker B        H-loop
NH2 ━━━━━━━[         ]━━━[   ]━━━━━[        ]━━━━━━[        ]━[   ]━ COOH
```

```
Fungal
         CDR1-NBD1------GRPGAGCST-----AET------VSGGERKRVS-----IQCWDNATRGLD----YQC-
         CDR1-NBD2------GASGAGKTT-----QQQ------LNVEQRKRLT-----LLFLDEPTSGLD----HQP-
         CDR2-NBD1------GRPGAGCST-----AET------VSGGERKRVS-----IQCWDNATRGLD----YQC-
         CDR2-NBD2------GASGAGKTT-----QQQ------LNVEQRKRLT-----LVFLDEPTSGLD----HQP-
         CDR3-NBD1------GRPGAGCST-----AET------ISGGERKRLS-----IQCWDNSTRGLD----HQC-
         CDR3-NBD2------GASGAGKTT-----QQQ------LNVEQRKRLT-----LVFLDEPTSGLD----HQP-
         CDR4-NBD1------GRPGAGCST-----AET------VSGGERKRVS-----VQCWDNSTRGLD----YQC-
         CDR4-NBD2------GASGAGKTT-----QQQ------LNVEQRKRLS-----LVFLDEPTSGLD----HQP-
         PDR5-NBD1------GRPGSGCTT-----AEA------VSGGERKRVS-----FQCWDNATRGLD----YQC-
         PDR5-NBD2------GASGAGKTT-----QQQ------LNVEQRKRLT-----LVFLDEPTSGLD----HQP-
         SNQ2-NBD1------GRPGAGCSS-----GEL------VSGGERKRVS-----IYCWDNATRGLD----YQA-
         SNQ2-NBD2------GESGAGKTT-----QQQ------LNVEQRKKLS-----LLFLDEPTSGLD----HQP-

Nonfungal
         ABCR(NBD1)  ----FLGHNGAGKTT---CPQHNI----LSGGMQRKLS------VVILDEPTSGD---STHHM-
         ABCR(NBD2)  ----LLGVNGAGKTT---CPQFDA----YSGGNKRKLS------LVLLDEPTTGD---TSHSM-
         Pgp(NBD1)   ----LVGNSGCGKST---VSQEPV----LSGGQKQRIA------ILLLDEATSAD---IAHRL-
         Pgp(NBD2)   ----LVGSSGCGKST---VSQEPI----LSGGQKQRIA------ILLLDEATSAD---IAHRL-
         hTAP1-NBD   ----LVGPNGSGKST---VGQEPQ----LSGGQRQAVA------VLILDDATSAD---ITQHL-
         hTAP2-NBD   ----LVGPNGSGKST---VGQEPV----LAAGQKQRLA------VLILDEATSAD---IAHRL-
         MRP1        ----IVGRTGAGKSS---IPQDPV----LSVGQRQLVC------ILVLDEATAAD---IAHRL-
         HisP        ----IIGSSGSGKST---VFQHFN----LSGGQQQRVS------VLLFDEPTSAD---VTHEM-
         MJ0796      ----IMGPSGSGKST---VFQQFN----LSGGQQQRVA------IILADQPTGAD---VTHDI-
         MalK        ----LLGPSGCGKTT---VFQSYA----LSGGQRVA-------VFLMDEPLSND---VTHDQ-
         MRP1        ----VVGQVGCGKSS---VPQQAW----LSGGQKQRVS------IYLFDDPLSAD---VTHSM-
         MJ1267      ----IIGPNGSGKST---TFQTPQ----LSGGQMKLVE------MIVMDQPIAGD---IEHRL-
         CDR1-NBD2   ----LMGASGAGKTT---VQQQDV----LNVEQRKRLT------LLFLDEPTSGD---TIHQP-
         CDR1-NBD1   ----VLGRPGAGCST---SAETDV----VSGGERKRVS------IQCWDNATRGD---AIYQC-
```

Figure 2

Sequence alignment of conserved Walker A and B motifs and ATP-binding cassette (ABC) signature of nucleotide-binding domains of ABC transporters from fungal (*upper*) and nonfungal (*lower*) organisms. The amino acid sequences have been aligned to generate five columns of sequence conservations representing the Walker A and B motifs. The sequence degeneracy of residues in the pleiotropic drug resistance subfamily is shown in red typeface in the case of Cdr1p as an example that also exists in other similar transporters of fungal origin.

base (80). By performing a series of biochemical analyses, including fluorescence resonance energy transfer (FRET), it was demonstrated (79) that the typical amino acids such as Cys193, Trp326, and Asp327 in N-NBDs are positioned within the nucleotide-binding pocket to bind and hydrolyze ATP by a process wherein both Mg^{2+} coordination and nucleotide binding contribute to the formation of the active site. For example, FRET takes place between Trp326 and MIANS [2′-(4′-maleimidylanilino)naphthalene-6-sulfonic acid]-labeled-Cys193, indicating these two residues are within close proximity to each other in the ATP-binding pocket (79).

On the basis of the available experimental evidence, a three-step mechanism of ATP catalysis at the N-NBD of Cdr1p can be hypothesized. The first step occurs when Mg^{2+} enters the ATP-binding pocket and contacts Trp326 of the Walker B motif and Glu238 of the Q-loop. This contact induces a conformational change that drags Cys193 of the Walker A motif toward the catalytic pocket. As a result, Trp326 and Cys193 come within close proximity. Subsequently, a second conformational change occurs when ATP docks with its phosphates directed toward the pocket. Such positioning automatically brings the γ-phosphate close to the uniquely placed Asn328 residue in the extended Walker B motif. As a result of this conformational change, Cys193 comes even closer to Trp326. The sensing of the γ-phosphate and movement of the Walker A and Walker B motifs toward the substrate constitute an important event preceding ATP hydrolysis. The third step is the actual catalysis step in which the catalytic residues, Cys193 of Walker A and Asp327 of Walker B, participate. Thereafter, the protein has a more open conformation that allows ADP to leave.

P-gp: P-glycoprotein

As deduced from available data on other ATPases, the following catalytic steps lead to ATP hydrolysis at the N-NBD. Asp327, acting as a catalytic base, abstracts a proton from a water molecule present in the active site as part of the Mg–ATP complex. The hydroxyl ion thus formed attacks at the γ-phosphate, allowing the bond between the γ- and β-phosphates to weaken and in turn allowing the β-phosphate to abstract a proton from the −SH of Cys193. The consequence of this second proton abstraction is the cleavage of the phosphodiester bond between the β- and γ-phosphates, allowing the latter to leave. After ATP is hydrolyzed, the conformation of the active site relaxes back to being more open, as Asn328 cannot sense the β-phosphate of ADP. Notably, if Cys193 is replaced with Tyr or Ser in the full protein, it retains ATPase activity, implying that the −OH group of Ser or Tyr is sufficient to abstract a proton from the β-phosphate as the final step of ATP hydrolysis (79).

Because the sequence degeneracy in N-NBDs is the hallmark of all Pdr proteins among yeast ABC transporters, a mechanism of ATP catalysis similar to that observed in Cdr1p is expected to be operative in other members of the fungal Pdr protein subfamily. However, this does not seem to be the case. For example, in Pdr5p of *S. cerevisiae* and Cdr1p of *C. glabrata*, ATP hydrolysis at the N-NBD is reported to be negligible compared to hydrolysis at the C-NBD, where it is essential for drug transport. Sequence variation in NBDs of different yeast species may result in organism-specific mechanistic differences. Alternatively, the role of the N-NBD in Pdr5p of *S. cerevisiae* has been proposed to be architectural, providing a platform for interaction with the opposing NBD or having a regulatory function, as has been proposed for the N-NBD of the cystic fibrosis transmembrane conductance regulator (24, 29). Accordingly, isolated Pdr5p observed by electron microscopy was found to organize in dimeric structures of two full-sized proteins in which the two canonical C-NBDs are in physical head to foot proximity and could thus cooperate for ATP binding and hydrolysis, whereas the two degenerated N-NBDs would not contribute significantly to ATP binding and hydrolysis (26).

Uncoupled ATPase Activity

Although nucleotide hydrolysis by ABC proteins is highly conserved, some distinct features of fungal transporters have been identified. In contrast with human P-glycoprotein (P-gp), whose ATPase activity is stimulated in the presence of substrate, fungal Pdr proteins do not show drug-stimulated ATPase activity. Thus, in the strict sense, ATPase activity of all Pdr proteins tested so far remains uncoupled from drug transport activity (50). Using purified Cdr1p, Shukla et al. (96) observed that the protein can bind to the drug without ATP, confirming that ATPase activity and drug transport are uncoupled. Nonetheless, protein-ATP interactions are reported to dictate substrate selection. Thus, while both Pdr5p and Cdr1p display highly uncoupled transport systems in vitro, some cross talk between drug binding and ATP hydrolysis must occur in situ. However, the residues in NBDs that influence substrate specificity at the TMDs are not known. Sauna et al. (90) observed in Pdr5p an important interaction between TMS2 and NBD1 through ICL1.

Notably, the yeast Pdr proteins possess general nucleotidase activity wherein substrates such as ATP, UTP, GTP, and CTP are hydrolyzed. Interestingly, data suggest that the identity of the bound nucleotide can modulate which drug is transported most efficiently (15, 22, 40, 42).

The fact that the isolated N-NBD possesses intrinsic ATPase activity is interesting, particularly in the context of the coordinated function of the two NBDs that has been frequently proposed for ABC transporters. The isolated N-NBD contains all the requisites for intrinsic ATPase activity. However, it is possible that the NBDs show positive cooperativity with respect to ATP hydrolysis when the ABC signature of one NBD interacts with the other NBD in the ATP-bound state. Several reports suggest that the conserved Gln residue in the LSGGQ motif of the second C-NBD is important for the formation of the composite catalytic site with the N-NBD (92). Mutational studies in MalK, for example, indicate that this Gln is not itself a catalytic residue but that it would stimulate ATPase activity (92). It is likely that ATP docking is sensed not only by the Asn328 of Cdr1p but perhaps also by one or more residues in the LSGGQ motif of the C-NBD, as shown by a molecular dynamics simulation for MJ7096 (42). Such recognition and the subsequent conformational change might also be responsible for positive cooperativity in ATP hydrolysis. Another equally likely scenario is that the accessibility of ATP to the active site of one of the NBDs may be regulated by the other via steric considerations. However, when a drug binds to Cdr1p, conformational changes in the TMDs are sensed by the first NBD, and the signal is transduced to the second NBD via the ABC signature sequence of the former. When ATP is bound/hydrolyzed at the second NBD, it allows the domain to swing out in a conformational change that is then transduced to the first NBD via the ABC signature sequence of the second NBD. Such a sequence of events could be cooperative, and it would require concerted action by both NBDs, with neither one capable of functionally replacing the other.

Unlike N-NBDs, the C-NBDs of Cdr1p and of other fungal transporters have canonical Walker A and Walker B motifs; however, they have divergent ABC signature motifs. Whether these deviations among the ABC signature motifs simply compensate for the substitutions in the N-NBDs, or whether they are part of a new mechanism for linking ATP hydrolysis and drug efflux, is a question worth examining. Recently, the contribution of the conserved and divergent ABC signature motifs of Cdr1p to ATP hydrolysis and drug transport has been evaluated. Mutations in conserved and equipositioned residues of the N-NBD and C-NBD of Cdr1p combined with the swapping of ABC signature motifs resulted in high susceptibility to drugs, showing simultaneous abrogation of ATPase and rhodamine G efflux activities. However, some mutants displayed selective increases in drug susceptibility but displayed no differences in drug or nucleotide binding. The study concluded that, in Cdr1p, not only are the conserved Ser304, Gly306, and Glu307 residues of the N-NBD ABC signature sequence important for ATP hydrolysis and drug efflux, but also a few divergent residues (N1002 and Glu1004) of the C-NBD ABC signature motif are functionally relevant and not interchangeable (53).

The functional nonequivalence of Cdr1p and Pdr5p and of other members of the fungal Pdr protein subfamily within otherwise conserved motifs in NBDs from other ABC subfamilies is well established (39, 54). Investigations involving crystal structures of NBDs and full-sized transport proteins demonstrate that the ATP molecule is sandwiched between the Walker A and B motifs of the first NBD and the C-loop of the second NBD (42, 53). Therefore, a composite site of Cdr1p (as an example) at the N-NBD will comprise the degenerated Walker A and B motifs of the N-NBD facing the degenerated ABC signature of the C-NBD. On the other hand, the composite site at the C-NBD will display the canonical residues of its well-conserved Walker A and B motifs with the conserved ABC signature of the N-NBD. In other words, the composite site at the C-NBD represents a canonical site, whereas the N-NBD composite site is totally degenerated (**Figure 3**). The existence of structural asymmetry in NBDs became evident when an inventory of

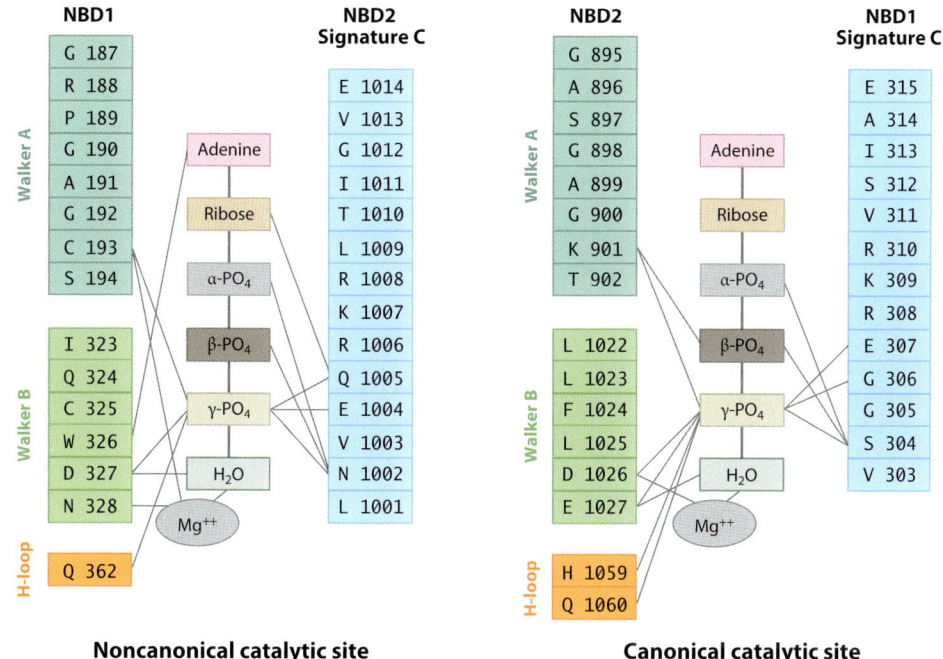

Figure 3

Schematic representation of the two composite ATP-binding sites of Cdr1p. On the basis of experimental evidence, noncanonical and canonical sites are shown highlighting the role of amino acid residues from different subdomains toward the formation of each composite site. Some residues are predicted to interact (shown as *gray lines*) with ATP, leading to the formation of a stabilized ATP-binding site before catalysis (40, 79, 80). Abbreviation: NBD, nucleotide-binding domain.

yeast Pdr proteins was undertaken by Decottignies & Goffeau (13). It was then assumed that the noncanonical ATP site of the N-NBD is probably inactive. However, overwhelming experimental evidence suggests that in Cdr1p the deviant residues at the noncanonical site have evolved to adopt newer functions (40, 53).

No NBDs from any fungal ABC transporter have yet to be crystallized. However, ample biochemical and structural data from other homologous transporters point to cooperation between the TMDs and the NBDs (58). How NBDs dimerize and orient themselves in the catalytic cycle of an ABC transporter is an important but unresolved question. An electron microscopic reconstruction of Pdr5p suggests a dimeric form wherein the four NBDs are in fixed geometric proximity, and that could allow the rotational movement of NBDs during a catalytic cycle (26). However, such a rotational mechanism is not supported by a recently deduced 3D model of Pdr5p that predicts dimerization of NBD monomers during the catalytic cycle (83). Nearly a dozen isolated NBDs from nonyeast species have been crystallized, and a great deal of structural information about NBDs is available for both bacterial and eukaryotic ABC proteins. Thus, a gradual picture is beginning to emerge about the structure, organization, and nucleotide-binding sites of NBDs. Several NBDs, such as HisP (38), MalK (19), MJ0796 (111), and Rad50 (37), crystallize as a dimeric unit, raising the probability that dimerization of NBDs is important for their in vivo functionality. Furthermore, in some NBDs such as MJ0796, ATP is sandwiched between the Walker A motif of one monomer and the ABC signature sequence of the other monomer (42). If this were indeed true in vivo, it would mean that such dimerization could be the basis for global conformational changes and drug transport by ABC transporters.

Transmembrane Domains

Delineating the architecture of the drug-binding sites of ABC transporters is not only essential for an understanding of drug-protein interactions but also helpful in designing specific inhibitors of these transporters. Owing to the known crystal structures of functional homologs from bacterial and mammalian species, and to random and site-directed mutagenesis of yeast Pdr proteins, substantial assessment of drug–protein interactions has been possible. Studies suggest that individual TMSs predominantly determine the broad substrate specificity of ABC transporters by contributing to a large, centrally located binding pocket that is accessible to either the cytosolic or the extracellular space during the reaction cycle (21, 73, 76, 97). Even though the TMS sequence is the most variable part of PDR transporters, some conservation does exist among fungal PDR transporters. A full-length multiple sequence alignment of Pdr proteins having 12 TMSs reveals that TMS2 is the most conserved and TMS10 is the least conserved (**Figure 4**). Notably, a deduced homology model of Pdr5p suggests that TMS10 faces the outside of the binding pocket, in contrast to TMS2, which aligns the binding pocket of Pdr5p (83). There are also conserved features flanking the TMDs. For example, all Pdr proteins have a stretch of amino acids (~15 residues) preceding TMS1 and TMS7 that may be important for their membrane localization. However, this hypothesis remains to be verified experimentally (54). Numerous studies addressing structure-function relationships of human P-gp suggest that nearly all the TMSs are directly or indirectly involved in drug transport. Early work defined at least two distinct substrate-binding sites, termed H (Hoechst 33342) and R (rhodamine 123). However, subsequent competition experiments suggest that P-gp contains at least seven different drug-binding sites (84). What emerges from a recent crystal structure of mouse P-gp along with biochemical data is that the protein has a large internal binding cavity that is capable of accommodating various structurally unrelated compounds of different sizes and shapes (1). However, because transporter-drug interactions are highly dynamic and because the crystal structures are only snapshots of medium resolution, the molecular details of multidrug recognition and transport remain largely unexplained.

The major ABC multidrug transporter Cdr1p of pathogenic *C. albicans* has been subjected to extensive structural and functional analysis. Iodoarylazido prazosin (IAAP, a photoaffinity analog of the P-gp substrate prazosine) and azidopine (a dihydropyridine photoaffinity analog of the P-gp modulator verapamil) specifically bind to Cdr1p. The binding of IAAP could be outcompeted by nystatin, while azidopine binding could be outcompeted only by miconazole. This raises the possibility of two different drug-binding sites for the two compounds (97). Competition experiments

Sequence alignment: a residue-by-residue relationship between protein chains (or portions of the chains)

IAAP: iodoarylazido prazosin

Figure 4

Sequence logos depicting conservation of transmembrane segment (TMS) residues of fungal pleiotropic drug resistance (PDR) transporters. Eighty-five fungal PDR transporters having 12 TMSs with (NBD-TMS$_6$)$_2$ topology were aligned for generating sequence logos (54). Each sequence logo consists of stacks of symbols, one stack for each position in the sequence. The scale indicates the certainty of occurrence of a particular amino acid at a given position and is determined by multiplying the frequency of that amino acid by the total information at that position. The residues at each position are arranged in order of predominance from top to bottom, with the highest-frequency residue at the top. The height of symbols within the stack indicates the relative frequency of each amino acid at that position. Sequence logos were generated using the WebLogo (**http://weblogo.berkeley.edu/**). The *x*-axis indicates the sequence conservation at a respective position. The *y*-axis shows the position of amino acids in the sequence from N terminus to C terminus. Abbreviation: NBD, nucleotide-binding domain.

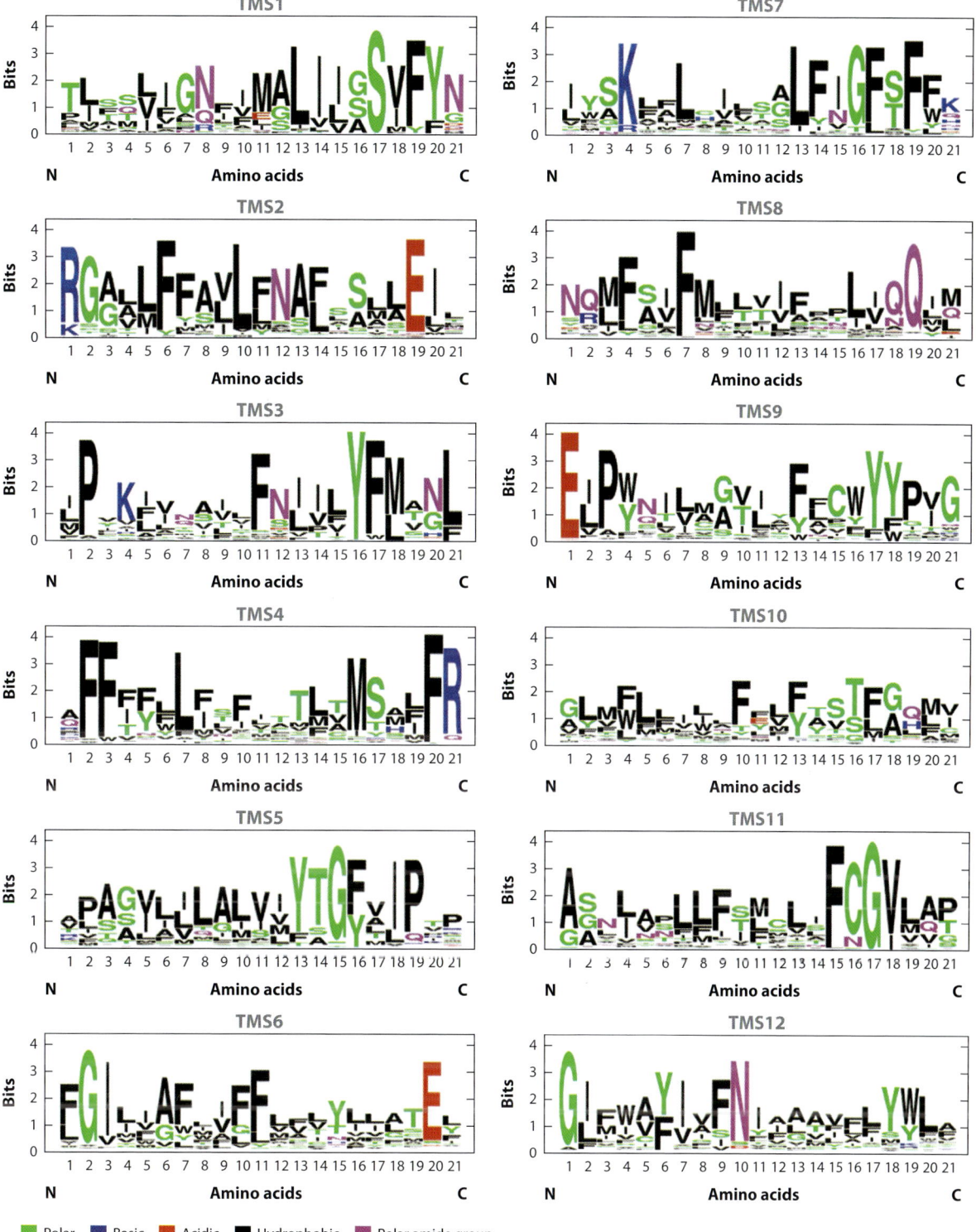

demonstrated that Cdr1p probably has at least three binding sites (76). One of the sites is probably responsible for the translocation of rhodamine 6G and azoles such as ketaconazole, miconazole, and itraconazole, and another site can only interact and transport substrate such as fluconazole. There may still be a third site where IAAP binds.

Egner et al. (21) identified several important amino acid residues of Pdr5p that are critical for drug binding and transport. Using a variety of novel substrates of Pdr5p, Golin et al. (31) reported that this transporter has at least three drug-binding sites and that some substrates might interact at more than one site. Thus, similar to Cdr1p and its mammalian P-gp homolog, Pdr5p also seems to have multiple overlapping drug-binding sites throughout the protein (31).

Although overwhelming experimental evidence from extensive mutagenesis of Cdr1p and Pdr5p suggests multiple residues scattered throughout all the TMSs that are critical for drug specificity and transport (**Figure 5**), the lack of structural information about these proteins limits the validity of the mutagenesis data. Recently, Rutledge et al. (83) presented a 3D homology model of Pdr5p and showed that Thr1213 (TMS7) and Gln1253 (TMS8), located away from the lipid-facing surface (LIPS), are critical for drug transport. Notably, the functionality of these residues was experimentally tested by alanine replacement and was validated by the homology model with regard to their predicted location and ability to form hydrogen bonds (83). On the basis of the available information, one can begin to reconstruct the critical residues that are part of the binding pockets of Cdr1p and Pdr5p, but their validation will have to wait for high-resolution crystal structures of these proteins.

SUBSTRATE PROMISCUITY OF YEAST PLEIOTROPIC DRUG RESISTANCE PROTEINS

The promiscuity in substrate specificity of some ABC transporters is well known, and yeast Pdr proteins are no exception. Thus, transport proteins belonging to the PDR, MDR, or multidrug resistance protein (MRP) subfamilies export a host of substrates with unrelated hydrophobic structures. Numerous studies of P-gp (MDR) have implicated hydrogen bond acceptor groups, aromatic rings, and hydrophobicity in transporter-substrate interactions (69, 93). One report suggests that although the topological polar surface area of a substrate is the main determinant for transport for the mammalian multidrug transporter MRP1, the presence of a pharmacophore with high affinity for the transporter could be detrimental for many substrates (25). At the same time, the transition of nonsubstrates to substrates in MRP1 could not be clearly defined by the topological polar surface area because molecular geometry is also believed to interfere with binding site recognition, but could be critical for transport of some compounds (25). Golin's (31) work on Pdr5p of *S. cerevisiae* concluded that hydrophobicity and anion makeup were relatively unimportant factors for determining whether a compound is a substrate. However, dissociation of the compound and its molecular size appeared to be important. A subsequent study correlated the molecular volume of the substrate with its efficacy for transport, and it proposed an important role for hydrogen bond acceptor groups within the substrate (31, 32). By employing the structure activity relationship (SAR), researchers have shown that high aromaticity, molecular branching, and occurrence of an atom-centered fragment (ReCHeR) are generally required for the substrates of Cdr1p. The higher aromaticity requirement of Cdr1p substrates reflects the presence of a large number of aromatic residues at its active site that probably participate in stacking interactions with the drug molecule. In addition, higher molecular branching, which is also a preferred feature for substrates of Cdr1p, probably results in easier diffusion of molecules inside the lipophilic membrane and enhanced propensity for interaction with the drug (78).

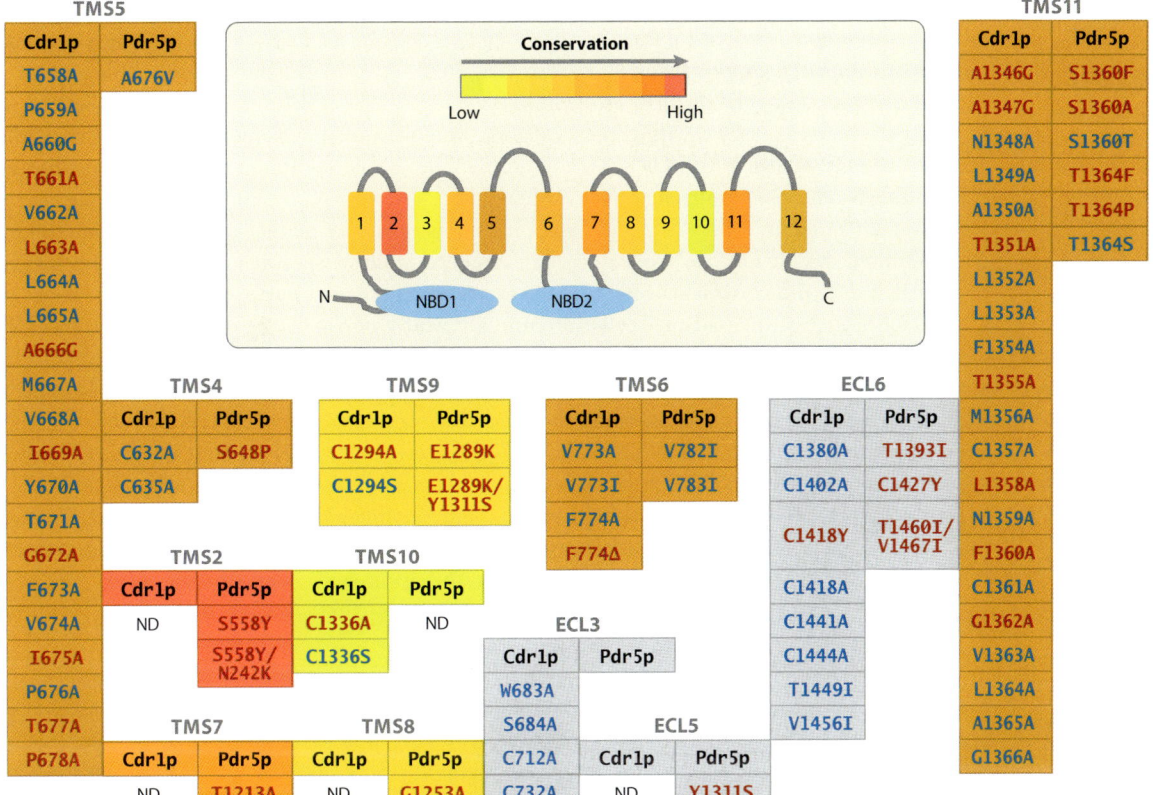

Figure 5

A list of residues of Cdr1p and Pdr5p that have been subjected to mutational analyses. Those box residues that upon substitution gave a phenotype are shown in red; the rest of the residues that upon replacement gave no phenotype are shown in black. Color gradient at top of the figure and in individual TMSs depicts a relative conservation score obtained through Jalview 2.4.0.b2 (**http://www.jalview.org**). The red color indicates highest conservation and yellow indicates the lowest score. Abbreviations: ECL, extracellular loop; NBD, nucleotide-binding domain; ND, not determined; TMS, transmembrane segment.

PHYSIOLOGICAL ROLES OF YEAST PLEIOTROPIC DRUG RESISTANCE PROTEINS

While the role of MDR transporters in removing toxic substances from cells is highly conserved throughout evolution, there is ample evidence to suggest that such a large family of proteins performs many physiological functions that may derive both from their level of expression and from their profile of exported drugs (41, 91). The tissue-specific expression of human P-gp supports this contention (98). The following sections examine some of these emerging physiological roles of yeast Pdr proteins.

Protection Against Cellular Toxins/Metabolites

Several properties of Pdr protein transporters, particularly their broad substrate specificity and occurrence, point to their role in the protection of cells against cytotoxic agents. Interestingly, the temporally regulated expression of *Pdr5p*, *Pdr15p*, and *Cdr1p* can be linked to their role in exporting toxic metabolites that accumulate during growth (47, 48, 51, 82, 110). This role is further corroborated by the fact that a double deletion of *PDR5* and *SNQ2* in *S. cerevisiae* led to impaired exponential cell growth (16, 109). To gain insight into the pathogenesis of *S. cerevisiae*, a population analysis of clinical and nonclinical isolates was performed. Several genetic variants associated with clinical isolation were identified, including polymorphisms within the coding sequences of *PDR5*. This finding implies a role for Pdr5p in cellular detoxification during *S. cerevisiae* pathogenesis (65). Although the ABC transporter Pdr12p of *S. cerevisiae* shares over 37% identity with *Pdr5p* and *Snq2p*, it confers resistance neither to nitroquinoline oxide, a substrate specific for *Snq2p*, nor to cycloheximide, a substrate specific for *PDR5* (16, 109). Weak acids induce the expression of Pdr12p, and as a result, it becomes one of the most abundant membrane proteins in acid-adapted cells. Interestingly, wild-type *S. cerevisiae* cells, if cultured at the low pH of 4.5 in the absence of weak acids, do not display active efflux of the fluorescein substrate used to monitor Pdr12p-mediated efflux (71). Further experiments have shown that Pdr12p remains inactivated by a protein kinase unless there is a need for catalyzed acid efflux (36). Taken together, it appears that the physiological function of *PDR12* is to protect cells against the toxicity of weak organic acids secreted by competitor organisms that share the same niche with yeasts, while it exports acid anions and releases them into the aqueous phase of the periplasm. The involvement of an active efflux pump for weak acids could also explain why some yeast species are capable of causing food spoilage despite the addition of weak organic acids as preservatives (71).

Membrane microdomains: specialized ordered and tightly packed microdomains that compartmentalize cellular processes by serving as organizing centers

Transport of Sterols

Steroids induce pleiotropic drug resistance in hemiascomycetes, with tremendous consequences for human fungal infections. Proteins capable of binding to steroids, such as progesterone-binding proteins and estradiol-binding proteins, are found in yeasts; however, the well-known receptor-mediated signaling present in higher eukaryotic cells is absent in yeasts and fungi. Steroids trigger a general stress response in yeast and lead to activation of heat shock proteins, cell cycle regulators, and Pdr proteins. Several reports confirm that, similar to human P-gp, yeast Pdr proteins also export human steroid hormones. Thus, transporters such as Pdr5p and Snq2p (47, 48, 60) and Aus1p and Pdr11p of *S. cerevisiae* transport steroids (45).

Cdr1p in *C. albicans* also transports β-estradiol and corticosterone; however, progesterone is not transported, and it does not affect the accumulation of either β-estradiol or corticosterone (52). Interestingly, progesterone is not a substrate of the human P-gp transporter, although it can bind to P-gp (104). This similarity reveals some conservation of substrate specificity between human and yeast drug efflux pumps. As yeast Pdr proteins mediate energy-dependent transport of human steroid hormones with high affinity and specificity, it is possible that such hormones could be the physiological substrates for these proteins. Notably, yeast drug transporters such as Cdr1p also prefer membrane microdomains (rich in ergosterol and sphingolipids) for their localization and function (67). However, the role of yeast Pdr proteins in sterol homeostasis remains an open question.

Translocation of Membrane Phospholipids

Asymmetric distribution of phospholipids across the plasma membrane of numerous cell types is well known (18). In the majority of cell types, phosphatidylethanolamine (PtdEtn) and phosphatidylserine (PtdSer) are located in the inner leaflet of the plasma membrane, whereas phosphatidylcholine (PtdCho), sphingomyelin, and glycolipids are located in the outer leaflet (18). The asymmetrical distribution of membrane lipids is very specific, and its loss has been linked to various pathological consequences (18, 103).

Interestingly, P-gp is involved in maintaining the membrane lipid asymmetry by acting as a general phospholipid translocator for different phospholipids and sphingomyelins, while mouse *mdr2* appears to be rather specific for translocating PtdCho between the two leaflets of the plasma membrane (107). The ABC proteins Cdr1p, Cdr2p, and Cdr4p of *C. albicans* elicit an energy-dependent in-to-out translocation (floppase activity) of phospholipids that is severely abrogated in knockouts (20, 88). Notably, Cdr3p does not show floppase activity (20, 88, 99). Interestingly, a *S. cerevisiae* transformant expressing *Mdr1p* (a multidrug transporter of the major facilitator superfamily) from *C. albicans* was not affected in its PtdEtn distribution pattern between the two leaflets, thus suggesting that phospholipid translocation activity is a specific feature of the ABC drug transporters (20). The Pdr proteins of S. *cerevisiae* also translocate phospholipids across the plasma membrane (47). This was demonstrated by the accumulation of fluorescent PtdEtn in vivo and by the observation that cells lacking Pdr5p and Yor1p have increased PtdEtn levels in the outer leaflet of the plasma membrane (14).

Ion Transport

Human P-gp inhibits $Cl^-/HCO3^-$ exchange to regulate Cl^- conductance and, when overexpressed in *S. cerevisiae*, to act as a H^+/K^+ pump (28). It has also been suggested that P-gp is associated with volume-regulated chloride channels (106). The yeast drug efflux pumps could also participate in H^+ transport across the plasma membrane. Antifungal peptides conjugated with FMDP [N^3-(4-methoxyfumaroyl)-L-2,3-diaminopropanoic acid] are transported into *C. albicans* cells through peptide permeases (62). The accumulated conjugate is cleaved intracellularly by peptidases, and as a consequence the released FMDP inhibits the activity of glucosamine-6-phosphate synthase. This enzyme is an important enzyme for cell wall synthesis, and hence its inhibition is lethal to *C. albicans* cells. The carrier-mediated entry of FMDP conjugates into yeast cells is pH dependent, as antifungal activity is more pronounced at low external pH (62). Interestingly, *S. cerevisiae* cells expressing Cdr1p were hypersensitive to such peptide conjugates. Furthermore, Cdr1p transformants displayed efflux of protons at three times the rate of the parental strain. It was subsequently confirmed that the lowering of external pH, presumably due to ejection of protons by Cdr1p, stimulates the uptake of FMDP conjugate and thus potentiates its antifungal activity (62). In another study, a K^+-dependent sensitivity to fluconazole was demonstrated in *S. cerevisiae*, supporting the idea that ionic movements can be linked to some MDR proteins; however, this notion remains to be confirmed (100).

Other Functions

In addition to the role of Pdr proteins in removing externally added toxic compounds from the cell membrane, the Pdr5p and Snq2p transporters play important physiological roles, significantly influencing the developmental phases and physiology of yeast populations growing in liquid culture (47). They appear to be involved in population quorum sensing, which consequently influences the level of transcription factor Pdr1p via feedback regulation (35). Recently, a number of yeast Pdr proteins have also been implicated as aging determinants, since they accumulate in the outer layer of aging colonies (105). During yeast cell division, newly synthesized transporter proteins are mainly deposited into the growing bud, whereas previously synthesized Pdr proteins remain tightly associated with the mother cell cortex. A model based on the observed dynamics of Pdr protein inheritance and turnover predicted a decline in Pdr protein activity as the mother cell advances in replicative age. As Pdr proteins have crucial roles in cellular metabolism, detoxification, and stress response, their collective decline may lead to fitness loss at an advanced age (23).

CONCLUSIONS

Each fungal species comprises 2 to 20 Pdr proteins that belong to at least 10 distinct phylogenetic clusters and, upon ATP hydrolysis, transport a huge variety of substrates involved in different metabolic pathways which in turn modulate a variety of physiological functions. This flexibility of functions of these proteins is due to the combination of a large unspecific drug-binding site and a series of conformational changes that articulate the movement of drugs and lipids across the plasma membrane. Our understanding of these structural movements occurring within specific microdomains of the plasma membrane has greatly progressed in these recent years, but their complete deciphering at the atomic level remains a challenge for the future and will probably lead to new knowledge in the evolution of multifunctional membrane proteins and possibly of new catalytic principles in bioenergetics.

SUMMARY POINTS

1. Among several mechanisms that contribute to the MDR phenomenon, the overexpression of full-sized multidrug extrusion pumps belonging to the ABC superfamily is the most frequent cause of resistance to antifungals, herbicides, anticancer drugs, and other drugs.

2. Full-sized multidrug ABC transporters are often made up of two transmembrane domains, each of which comprises six predicted transmembrane segments, and two cytoplasmic nucleotide-binding domains, in which α-helices and β-sheets form a Rossmann fold.

3. A large number of multidrug ABC transporters in fungi belong to the Pdr protein subfamily, which is characterized by a topology $(TMD-NBD)_2$ that is reversed relative to the topology $(NBD-TMD)_2$ of all other full-sized ABC transporter subfamilies.

4. The fungal Pdr protein subfamily comprises at least 10 phylogenetic clusters. In most of these clusters (sensu stricto Pdr protein), such as cluster A containing Pdr5p from *S. cerevisiae* and Cdr1p from *C. albicans*, the two nucleotide-binding domains are structurally and functionally asymmetric and have subdomains that display typical sequence degeneracy.

5. Each full-sized Pdr protein transporter has one canonical and one noncanonical ATP catalysis site. All the deviant residues of the noncanonical site in Cdr1p have evolved to adopt newer roles.

6. Extensive mutagenesis of Cdr1p and Pdr5p reveals that multiple residues scattered on all the transmembrane segments are critical for drug specificity and transport.

7. The large number of PDR genes in fungal genomes indicates their essential character. All Pdr proteins transport drugs and contribute to multiple physiological functions that are probably related to differential expression of pumps of different drug specificity.

FUTURE ISSUES

1. Refined protocols for heterologous expression of Pdr proteins must be developed so that systematic large-scale crystallization screens will be possible.

2. Purification of Pdr proteins and reconstitution of transport function in proteo-vesicles must be further developed.

3. Crystallization and X-ray analysis of Pdr proteins in different conformations must be achieved.

4. The quaternary structure of Pdr proteins in the membrane and the interaction between all lipid and protein components of Pdr proteins must be studied by fluorocytochemistry and other molecular biophysical tools.

5. Exhaustive mutagenic analyses must be pursued along with the construction and phenotypic analysis of chimeric Pdr protein molecules and engineering of synthetic Pdr proteins.

6. A systematic large-scale screening of drugs with different structures should be undertaken to analyze the resistance conferred by different PDR genes from different fungal species cloned into *S. cerevisiae*. Ultimately, novel chemisensitizers of antifungals must be developed.

7. The evolution of the fungal PDR genes through duplication and selection, and its relation to the eliciting of efflux pumps of different substrate specificity according to the nutritional niche of the host, must be further analyzed when additional fungal genomes become available.

DISCLOSURE STATEMENT

The authors are not aware of any affiliations, memberships, funding, or financial holdings that might be perceived as affecting the objectivity of this review.

ACKNOWLEDGMENTS

We thank members of our laboratory, particularly A.H. Shah, M.K. Rawal, M. Sharma, N.A. Gaur, and A. Singh, for their advice, comments, and organization of the manuscript. We are indebted to S. Krishnamurthy, S. Shukla, S. Dogra, S. Jha, V. Sharma, S. Shukla, M. Gaur, N. Puri, and H. Sanwal from Jawaharlal Nehru University, and E. Balzi, M. Ghislain, S. Ulaszewski, M. Kolaczkowski, A. Kolaszkoski, A. Decottignies, M. Boutry, and P. Baret from Université Catholique de Louvain for crucial experimental contributions. We apologize to all colleagues whose work could not be properly cited owing to space limitations. Our special thanks go to S. Ambudkar (National Institutes of Health) for his comments.

LITERATURE CITED

1. Aller SG, Yu J, Ward A, Weng Y, Chittaboina S, et al. 2009. Structure of P-glycoprotein reveals a molecular basis for poly-specific drug binding. *Science* 323:1718–22
2. Balan I, Alarco AM, Raymond M. 1997. The *Candida albicans CDR3* gene codes for an opaque-phase ABC transporter. *J. Bacteriol.* 179:7210–18
3. Balzi E, Chen W, Ulaszewski S, Capieaux E, Goffeau A. 1987. The multidrug resistance gene *PDR1* from *S. cerevisiae*. *J. Biol. Chem.* 262:16871–79
4. Balzi E, Goffeau A. 1995. Yeast multidrug resistance: the PDR network. *J. Bioenerg. Biomembr.* 27:71–76

> 5. The first report to describe the full sequence of a multidrug ABC transporter of *S. cerevisiae*.

5. **Balzi E, Wang M, Leterme S, Van Dyck L, Goffeau A. 1994. PDR5, a novel yeast multidrug resistance conferring transporter controlled by the transcription regulator PDR1. *J. Biol. Chem.* 269:2206–14**
6. Bissinger PH, Kuchler K. 1994. Molecular cloning and expression of *Saccharomyces cerevisiae STS1* gene product, a yeast ABC transporter conferring mycotoxin resistance. *J. Biol. Chem.* 269:4180–86
7. Bocer T, Zarubica A, Roussel A, Flis K, Trombik T, et al. 2012. The mammalian ABC transporter ABCA1 induces lipid-dependent drug sensitivity in yeast. *Biochim. Biophys. Acta* 1821:373–80
8. Cannon RD, Lamping E, Holmes AR, Niimi K, Baret PV, et al. 2009. Efflux-mediated antifungal drug resistance. *Clin. Microbiol. Rev.* 22:291–321
9. Capieaux E, Ulaszewski S, Balzi E, Goffeau A. 1991. Physical, transcriptional and genetical mapping of a 24 kb DNA fragment located between the PMA1 and ATE1 loci on chromosome VII from *Saccharomyces cerevisiae*. *Yeast* 7:275–80
10. Carvajal E, Van Den Hazel HB, Cybularz-Kolaczkowska A, Balzi E, Goffeau A. 1997. Molecular and phenotypic characterization of yeast *PDR1* mutants that show hyperactive transcription of various ABC multidrug transporter genes. *Mol. Gen. Genet.* 256:406–15
11. Daumke O, Knittler MR. 2001. Functional asymmetry of the ATP-binding-cassettes of the ABC transporter TAP is determined by intrinsic properties of the nucleotide binding domains. *Eur. J. Biochem.* 268:4776–86
12. Dean M, Andrey R, Rando A. 2001. The human ATP-binding cassette (ABC) transporter superfamily. *Genome Res.* 11:1156–66

> 13. The first analysis of the yeast genome to predict the existence of 29 proteins belonging to the ubiquitous ABC superfamily.

13. **Decottignies A, Goffeau A. 1997. Complete inventory of the yeast ABC proteins. *Nat. Genet.* 15:137–45**
14. Decottignies A, Grant AM, Nichols JW, De Wet H, McIntosh DB, Goffeau A. 1998. ATPase and multidrug transport activities of the overexpressed yeast ABC protein Yor1p. *J. Biol. Chem.* 273:12612–22
15. Decottignies A, Kolaczkowski M, Balzi E, Goffeau A. 1994. Solubilisation and characterization of the overexpressed PDR5 multidrug resistance nucleotide triphosphatase of yeast. *J. Biol. Chem.* 269:12797–803
16. Decottignies A, Lambert L, Catty P, Degand H, Eppings EA, et al. 1995. Identification and characterisation of SNQ2, a new multidrug ATP binding cassette transporter of the yeast plasma membrane. *J. Biol. Chem.* 270:18150–57
17. Delaveau T, Delahodde A, Carvajal E, Subik J, Jacq C. 1994. *PDR3*, a new yeast regulatory gene, is homologous to *PDR1* and controls the multidrug resistance phenomenon. *Mol. Gen. Genet.* 244:501–11
18. Diaz C, Schroit AJ. 1996. Role of translocases in the generation of phosphatidylserine asymmetry. *J. Membr. Biol.* 151:1–9
19. Diederichs K, Diez J, Greller G, Muller C, Breed J, et al. 2000. Crystal structure of MalK, the ATPase subunit of the trehalose/maltose ABC transporter of the archaeon *Thermococcus litoralis*. *EMBO J.* 19:5951–61
20. Dogra S, Krishnamurthy S, Gupta V, Dixit BL, Gupta CM, et al. 1999. Asymmetric distribution of phosphatidylethanolamine in *C. albicans*: possible mediation by *CDR1*, a multidrug transporter belonging to ATP binding cassette (ABC) superfamily. *Yeast* 15:111–21
21. Egner R, Bauer BE, Kuchler K. 2000. The transmembrane domain 10 of the yeast Pdr5p ABC antifungal efflux pump determines the substrate specificity and inhibitor susceptibility. *Mol. Microbiol.* 35:1255–63
22. Egner R, Mahe Y, Pandjaitan R, Kuchler K. 1995. Endocytosis and vacuolar degradation of the plasma membrane localized Pdr5 ATP binding cassette multidrug transporter in *Saccharomyces cerevisiae*. *Mol. Cell. Biol.* 15:5879–87
23. Eldakak A, Rancati G, Rubinstein B, Paul P, Conaway V, Li R. 2010. Asymmetrically inherited multidrug resistance transporters are recessive determinants in cellular replicative ageing. *Nat. Cell Biol.* 12:799–805
24. Ernst R, Kueppers P, Klein CM, Schwarzmueller T, Kuchler K, Schmitt L. 2008. A mutation of the H-loop selectively affects rhodamine transport by the yeast multidrug ABC transporter Pdr5. *Proc. Natl. Acad. Sci. USA* 105:5069–74
25. Fernandes J, Gattass CR. 2009. Topological polar surface area defines substrate transport by multidrug resistance associated protein 1 (MRP1/ABCC1). *J. Med. Chem.* 52:1214–18

26. Ferreira-Pereira A, Marco S, Decottignies A, Nader J, Goffeau A, Rigaud JL. 2003. Three-dimensional reconstruction of the *Saccharomyces cerevisiae* multidrug resistance protein Pdr5p. *J. Biol. Chem.* 278:11995–99
27. Franz R, Michel S, Morschhauser J. 1998. A fourth gene from the *Candida albicans* CDR family of ABC transporters. *Gene* 220:91–98
28. Fritz F, Howard EM, Hoffman MM, Roepe PD. 1999. Evidence for altered ion transport in *Saccharomyces cerevisiae* overexpressing human MDR1 protein. *Biochemistry* 38:4214–26
29. Gadsby DC, Vergani P, Csanady L. 2006. The ABC protein turned chloride channel whose failure causes cystic fibrosis. *Nature* 440:477–83
30. **Gaur M, Choudhury D, Prasad R. 2005. Complete inventory of ABC proteins in human pathogenic yeast, *Candida albicans*. *J. Mol. Microbiol. Biotechnol.* 9:3–15**
31. Golin J, Ambudkar SV, Gottesman MM, Habib AD, Sczepanski J, et al. 2003. Studies with novel Pdr5p substrates demonstrate a strong size dependence for xenobiotic efflux. *J. Biol. Chem.* 278:5963–69
32. Hanson L, May L, Tuma P, Keeven J, Mehl P, et al. 2005. The role of hydrogen bond acceptor groups in the interaction of substrates with Pdr5p, a major yeast drug transporter. *Biochemistry* 44:9703–13
33. Higgins CF. 2001. ABC transporters: physiology, structure and mechanism—an overview. *Res. Microbiol.* 152:205–10
34. Hirata D, Yano K, Miyahara K, Miyakawa T. 1994. *Saccharomyces cerevisiae YDR1*, which encodes a member of the ATP-binding cassette (ABC) superfamily, is required for multidrug resistance. *Curr. Genet.* 26:285–94
35. Hlavacek O, Kucerova H, Harant K, Palkova Z, Vachova L. 2009. Putative role for ABC multidrug exporters in yeast quorum sensing. *FEBS Lett.* 583:1107–13
36. Holyoak CD, Thompson S, Calderon CO, Hatzixanthis K, Bauer B, et al. 2000. Loss of Cmk1 Ca^{2+}-calmodulin-dependent protein kinase in yeasts results in constitutive weak organic acid resistance, associated with a post-transcriptional activation of the Pdr12 ATP-binding cassette transporter. *Mol. Microbiol.* 37:595–605
37. Hopfner KP, Karcher A, Shin DS, Craig L, Arthur LM, et al. 2000. Structural biology of Rad50 ATPase: ATP-driven conformational control in DNA double-strand break repair and the ABC-ATPase superfamily. *Cell* 101:789–800
38. Hung LW, Wang IX, Nikaido K, Liu PQ, Ames GF, Kim SH. 1998. Crystal structure of the ATP-binding subunit of an ABC transporter. *Nature* 396:703–7
39. Jha S, Dabas N, Karnani N, Saini P, Prasad R. 2004. ABC multidrug transporter Cdr1p of *Candida albicans* has divergent nucleotide-binding domains which display functional asymmetry. *FEMS Yeast Res.* 5:63–72
40. **Jha S, Karnani N, Dhar SK, Mukhopadhyay K, Shukla S, et al. 2003. Purification and characterization of N-terminal nucleotide binding domain of an ABC drug transporter of *Candida albicans*: uncommon cysteine 193 of Walker A is critical for ATP hydrolysis. *Biochemistry* 42:10822–32**
41. Johnstone RW, Ruefli AA, Smyth MJ. 2000. Multiple physiological functions for multidrug transporter P-glycoprotein. *Trends Biochem. Sci.* 25:1–6
42. Jones PM, George AM. 2007. Nucleotide-dependent allostery within the ABC transporter ATP-binding cassette: a computational study of the MJ0796 dimer. *J. Biol. Chem.* 282:22793–803
43. **Katzmann DJ, Burnett PE, Golin J, Mahe Y, Moye-Rowley WS. 1994. Transcriptional control of the yeast *PDR5* gene by the *PDR3* gene product. *Mol. Cell. Biol.* 14:4653–61**
44. Katzmann DJ, Hallstrom TC, Mahe Y, Moye-Rowley WS. 1996. Multiple Pdr1p/Pdr3p binding sites are essential for normal expression of the ATP binding cassette transporter protein-encoding gene *PDR5*. *J. Biol. Chem.* 271:23049–54
45. Kohut P, Wustner D, Hronska L, Kuchler K, Hapala I, Valachovic M. 2011. The role of ABC proteins Aus1p and Pdr11p in the uptake of external sterols in yeast: dehydroergosterol fluorescence study. *Biochem. Biophys. Res. Commun.* 404:233–38
46. Kolaczkowska A, Goffeau A. 1999. Regulation of pleiotropic drug resistance in yeast. *Drug Resist. Updates* 2:403–14

---

30. Describes the complete inventory of ABC transporters of a pathogenic yeast.

---

40. Demonstrates the relevance of sequence degeneracy of N-NBD of Cdr1p.

---

43. Documents the indispensability of transcriptional control by PDR1/PDR3 for drug transporter PDR5.

47. Kolaczkowski M, Kolaczkowska A, Luczynski J, Witek S, Goffeau A. 1998. In vivo characterization of the drug resistance profile of the major ABC transporters and other components of the yeast pleiotropic drug resistance network. *Microbial Drug Resist.* 4:143–58
48. Kolaczkowski M, van der Rest ME, Cybularz-Kolaczkowska A, Soumillion J-P, Konings WN, Goffeau A. 1996. Anticancer drugs, ionophoric peptides, and steroids as substrates of the yeast multidrug transporter Pdr5p. *J. Biol. Chem.* 271:31543–48
49. Kovalchuk A, Driessen AJ. 2010. Phylogenetic analysis of fungal ABC transporters. *BMC Genomics* 11:177
50. Krishnamurthy S, Chatterjee U, Gupta V, Prasad R, Das P, et al. 1998. Deletion of transmembrane domain 12 of *CDR1*, a multidrug transporter from *Candida albicans*, leads to altered drug specificity: expression of a yeast multidrug transporter in baculovirus expression system. *Yeast* 14:535–50
51. Krishnamurthy S, Gupta V, Prasad R, Panwar SL, Prasad R. 1998. Expression of *CDR1*, a multidrug resistance gene of *Candida albicans*: in vitro transcriptional activation by heat shock, drugs and human steroid hormones. *FEMS Microbiol. Lett.* 160:191–97
52. Krishnamurthy S, Gupta V, Snehlata P, Prasad R. 1998. Characterisation of human steroid hormone transport mediated by Cdr1p, multidrug transporter of *Candida albicans*, belonging to the ATP binding cassette super family. *FEMS Microbiol. Lett.* 158:69–74
53. Kumar A, Shukla S, Mandal A, Shukla S, Ambudkar SV, Prasad R. 2010. Divergent signature motifs of nucleotide binding domains of ABC multidrug transporter, CaCdr1p of pathogenic *Candida albicans*, are functionally asymmetric and noninterchangeable. *Biochim. Biophys. Acta* 1798:1757–66
54. Lamping E, Baret PV, Holmes AR, Monk BC, Goffeau A, Cannon RD. 2010. Fungal PDR transporters: phylogeny, topology, motifs and function. *Fungal Genet. Biol.* 47:127–42
55. **Lamping E, Monk BC, Niimi K, Holmes AR, Tsao S, et al. 2007. Characterization of three classes of membrane proteins involved in fungal azole resistance by functional hyperexpression in *Saccharomyces cerevisiae*. *Eukaryot. Cell* 6:1150–65**
56. Le Crom S, Devaux PF, Marc P, Zhang X, Moye-Rowley WS, Jacq C. 2002. New insights into the pleiotropic drug resistance network from genome-wide characterization of the YRR1 transcription factor regulation system. *Mol. Cell. Biol.* 22:2642–49
57. Leppert G, McDevitt R, Falco SC, Van Dyk TK, Ficke MB, Golin J. 1990. Cloning by gene amplification of two loci conferring multiple drug resistance in *Saccharomyces*. *Genetics* 125:13–20
58. Loo TW, Bartlett MC, Clarke DM. 2003. Drug binding in human P-glycoprotein causes conformational changes in both nucleotide-binding domains. *J. Biol. Chem.* 278:1575–78
59. MacPherson S, Larochelle M, Turcotte B. 2006. A fungal family of transcriptional regulators: the zinc cluster proteins. *Microbiol. Mol. Biol. Rev.* 70:583–604
60. Mahe Y, Lemoine Y, Kuchler K. 1996. The ATP binding cassette transporters Pdr5 and Snq2 of *Saccharomyces cerevisiae* can mediate transport of steroids in vivo. *J. Biol. Chem.* 271:26167–72
61. Meyers S, Schauer W, Balzi E, Wagner M, Goffeau A, Golin J. 1992. Interaction of the yeast pleiotropic drug resistance genes *PDR1* and *PDR5*. *Curr. Genet.* 21:431–36
62. Milewski S, Mignini F, Prasad R, Borowski E. 2001. Unusual susceptibility of a multidrug-resistance yeast strain to peptidic antifungals. *Antimicrob. Agents Chemother.* 45:223–28
63. Miranda MN, Masuda CA, Ferreira-Pereira A, Carvajal E, Ghislain M, Montero-Lomeli M. 2010. The serine/threonine protein phosphatase Sit4p activates multidrug resistance in *Saccharomyces cerevisiae*. *FEMS Yeast Res.* 10:674–86
64. Morschhauser J. 2010. Regulation of multidrug resistance in pathogenic fungi. *Fungal Genet. Biol.* 47:94–106
65. Muller LA, Lucas JE, Georgianna DR, McCusker JH. 2011. Genome-wide association analysis of clinical versus nonclinical origin provides insights into *Saccharomyces cerevisiae* pathogenesis. *Mol. Ecol.* 20:4085–97
66. Nakamura K, Niimi M, Niimi K, Holmes AR, Yates JE, et al. 2001. Functional expression of *Candida albicans* drug efflux pump Cdr1p in a *Saccharomyces cerevisiae* strain deficient in membrane transporters. *Antimicrob. Agents Chemother.* 45:3366–74
67. Pasrija R, Panwar SL, Prasad R. 2008. Multidrug transporters CaCdr1p and CaMdr1p of *Candida albicans* display different lipid specificities: both ergosterol and sphingolipids are essential for targeting of CaCdr1p to membrane rafts. *Antimicrob. Agents Chemother.* 52:694–704

---

55. Documents the heterologous expression system of membrane proteins, particularly of multidrug transporters of yeast.

68. Paumi CM, Chuk M, Snider J, Stagljar I, Michaelis S. 2009. ABC transporters in *Saccharomyces cerevisiae* and their interactors: new technology advances the biology of the ABCC (MRP) subfamily. *Microbiol. Mol. Biol. Rev.* 73:577–93
69. Penzotti JE, Lamb ML, Evensen E, Grootenhuis PD. 2002. A computational ensemble pharmacophore model for identifying substrates of P-glycoprotein. *J. Med. Chem.* 45:1737–40
70. Pety de Thozée C, Ghislain M. 2006. ER-associated degradation of membrane proteins in yeast. *Sci. World J.* 6:967–83
71. Piper P, Mahe Y, Thompson S, Pandjaitan R, Holyoak C, et al. 1998. The Pdr12 ABC transporter is required for the development of weak organic acid resistance in yeast. *EMBO J.* 17:4257–65
72. Plemper RK, Egner R, Kuchler K, Wolf DH. 1998. Endoplasmic reticulum degradation of a mutated ATP-binding cassette transporter Pdr5 proceeds in a concerted action of Sec61 and the proteasome. *J. Biol. Chem.* 273:32848–56
73. Prasad R, Sharma M, Rawal MK. 2011. Functionally relevant residues of Cdr1p: a multidrug ABC transporter of human pathogenic *Candida albicans*. *J. Amino Acids* 2011:531412
74. Prasad R, Snehlata P, Smriti. 2002. Drug resistance in yeasts—an emerging scenario. *Adv. Microb. Physiol.* 46:155–201
75. Prasad R, Worgifosse PD, Goffeau A, Balzi E. 1995. Molecular cloning and characterization of a novel gene of *C. albicans*, *CDR1*, conferring multiple resistance to drugs and antifungals. *Curr. Genet.* 27:320–29
76. Puri N, Gaur M, Sharma M, Shukla S, Ambudkar SV, Prasad R. 2009. The amino acid residues of transmembrane helix 5 of multidrug resistance protein CaCdr1p of *Candida albicans* are involved in substrate specificity and drug transport. *Biochim. Biophys. Acta* 1788:1752–61
77. Puri N, Manoharlal R, Sharma M, Sanglard D, Prasad R. 2011. Overcoming the heterologous bias: an in vivo functional analysis of multidrug efflux transporter, CgCdr1p in matched pair clinical isolates of *Candida glabrata*. *Biochem. Biophys. Res. Commun.* 404:357–63
78. Puri N, Prakash O, Manoharlal R, Sharma M, Ghosh I, Prasad R. 2010. Analysis of physico-chemical properties of substrates of ABC and MFS multidrug transporters of pathogenic *Candida albicans*. *Eur. J. Med. Chem.* 45:4813–26
79. Rai V, Gaur M, Kumar A, Shukla S, Komath SS, Prasad R. 2008. A novel catalytic mechanism for ATP hydrolysis employed by the N-terminal nucleotide-binding domain of Cdr1p, a multidrug ABC transporter of *Candida albicans*. *Biochim. Biophys. Acta* 1778:2143–53
80. Rai V, Gaur M, Shukla S, Shukla S, Ambudkar SV, et al. 2006. Conserved Asp327 of Walker B motif in the N-terminal nucleotide binding domain (NBD-1) of Cdr1p of *Candida albicans* has acquired a new role in ATP hydrolysis. *Biochemistry* 45:14726–39
81. Rank GH, Bech-Hansen NT. 1973. Single nuclear gene inherited cross resistance and collateral sensitivity to 17 inhibitors of mitochondrial function in *S. cerevisiae*. *Mol. Gen. Genet.* 126:93–102
82. Rogers B, Decottignies A, Kolaczkowski M, Carvajal E, Balzi E, Goffeau A. 2001. The pleiotropic drug ABC transporters from *Saccharomyces cerevisiae*. *J. Mol. Microbiol. Biotechnol.* 3:207–14
83. Rutledge RM, Esser L, Ma J, Xia D. 2011. Toward understanding the mechanism of action of the yeast multidrug resistance transporter Pdr5p: a molecular modeling study. *J. Struct. Biol.* 173:333–44
84. Safa AR. 2004. Identification and characterization of the binding sites of P-glycoprotein for multidrug resistance-related drugs and modulators. *Curr. Med. Chem. Anticancer Agents* 4:1–17
85. Saier MH Jr. 1994. Computer-aided analyses of transport protein sequences: gleaning evidence concerning function, structure, biogenesis, and evolution. *Microbiol. Rev.* 58:71–93
86. Saini P, Gaur NA, Prasad R. 2006. Chimeras of the ABC drug transporter Cdr1p reveal functional indispensability of transmembrane domains and nucleotide-binding domains, but transmembrane segment 12 is replaceable with the corresponding homologous region of the nondrug transporter Cdr3p. *Microbiology* 152:1559–73
87. Sanglard D, Ischer F, Monod M, Bille J. 1997. Cloning of *Candida albicans* genes conferring resistance to azole antifungal agents: characterisation of *CDR2*, a new multidrug ABC transporter gene. *Microbiology* 143:405–16

75. Documents the characterization of the first homolog of an ABC transporter in a pathogenic yeast.

81. Describes the first point mutant of *S. cerevisiae* resistant to drugs.

83. Describes the first 3D-deduced model of a yeast multidrug transporter.

87. Documents the characterization of a second major multidrug transporter of *C. albicans*.

88. Sanglard D, Ischer F, Monod M, Dogra S, Prasad R, Bille J. 1999. Analysis of the ATP binding cassette (ABC)-transporter gene *CDR4* from *Candida albicans*. In *ASM Conf. Candida Candidiasis*, Charleston, SC, 1–4 March, pp. 56.
89. Sanglard D, Odds FC. 2002. Resistance of *Candida* species to antifungal agents: molecular mechanisms and clinical consequences. *Lancet Infect. Dis.* 2:73–85
90. Sauna ZE, Bohn SS, Rutledge R, Dougherty MP, Cronin S, et al. 2008. Mutations define cross-talk between the N-terminal nucleotide-binding domain and transmembrane helix-2 of the yeast multidrug transporter Pdr5: possible conservation of a signaling interface for coupling ATP hydrolysis to drug transport. *J. Biol. Chem.* 283:35010–22
91. Schinkel AH. 1997. The physiological function of drug-transporting P-glycoproteins. *Sem. Cancer Biol.* 8:161–70
92. Schmees G, Stein A, Hunke S, Landmesser H, Schneider E. 1999. Functional consequences of mutations in the conserved 'signature sequence' of the ATP-binding-cassette protein MalK. *Eur. J. Biochem.* 266:420–30
93. Seelig A. 2006. Unraveling membrane-mediated substrate-transporter interactions. *Biophys. J.* 90:3825–26
94. Seret ML, Diffels JF, Goffeau A, Baret PV. 2009. Combined phylogeny and neighborhood analysis of the evolution of the ABC transporters conferring multiple drug resistance in hemiascomycete yeasts. *BMC Genomics* 10:459
95. Shapiro RS, Robbins N, Cowen LE. 2011. Regulatory circuitry governing fungal development, drug resistance, and disease. *Microbiol. Mol. Biol. Rev.* 75:213–67
96. Shukla S, Rai V, Banerjee D, Prasad R. 2006. Characterization of Cdr1p, a major multidrug efflux protein of *Candida albicans*: Purified protein is amenable to intrinsic fluorescence analysis. *Biochemistry* 45:2425–35
97. Shukla S, Saini P, Smriti, Jha S, Ambudkar SV, Prasad R. 2003. Functional characterization of *Candida albicans* ABC transporter Cdr1p. *Eukaryot. Cell* 2:1361–75
98. Shukla S, Wu CP, Ambudkar SV. 2008. Development of inhibitors of ATP-binding cassette drug transporters: present status and challenges. *Expert. Opin. Drug Metab. Toxicol.* 4:205–23
99. Smriti, Krishnamurthy S, Dixit BL, Gupta CM, Milewski S, Prasad R. 2002. ABC transporters Cdr1p, Cdr2p and Cdr3p of a human pathogen *Candida albicans* are general phospholipid translocators. *Yeast* 19:303–18
100. Stella CA, Burgos HI. 2001. Effect of potassium on *Saccharomyces cerevisiae* resistance to fluconazole. *Antimicrob. Agents Chemother.* 45:1–2
101. Subik J, Ulaszewski S, Goffeau A. 1986. Genetic mapping of nuclear mucidin resistance mutations in *Saccharomyces cerevisiae*. *Curr. Genet.* 10:665–70
102. Supply P, Wach P, Thines-Sempoux D, Goffeau A. 1993. Proliferation of intracellular structures upon overexpression of the PMA2 ATPase in *Saccharomyces cerevisiae*. *J. Biol. Chem.* 268:19744–52
103. Toti F, Schindler V, Riou J-F, Lombard-Platet G, Fressinaud E, et al. 1997. Another link between phospholipid transmembrane migration and ABC transporter gene family, inferred from a rare inherited disorder of phosphatidylserine externalization. *Biochem. Biophys. Res. Commun.* 241:548–52
104. Ueda K, Okamura N, Hirai M, Tanigawara Y, Saeki T, et al. 1992. Human P-glycoprotein transports cortisol, aldosterone and dexamethasone, but not progesterone. *J. Biol. Chem.* 267:24248–52
105. Vachova L, Palkova Z. 2011. Aging and longevity of yeast colony populations: metabolic adaptation and differentiation. *Biochem. Soc. Trans.* 39:1471–75
106. Valverde MA, Diaz M, Sepulveda FV. 1992. Volume-regulated chloride channels associated with the human multidrug-resistance P-glycoprotein. *Nature* 355:830–33
107. Van Helvoort A, Smith AJ, Sprong H, Fritzche I, Schinkel AH, et al. 1996. MDR1 P-glycoprotein is a lipid translocase of broad specificity, while MDR3 P-glycoprotein specifically translocates phosphatidylcholine. *Cell* 87:507–17
108. Verrier PJ, Bird D, Burla B, Dassa E, Forestier C, et al. 2008. Plant ABC proteins—a unified nomenclature and updated inventory. *Trends Plant Sci.* 13:151–59

109. Watanabe M, Mizoguchi H, Nishimura A. 2000. Disruption of the ABC transporter genes PDR5, YOR1, and SNQ2, and their participation in improved fermentative activity of a sake yeast mutant showing pleiotropic drug resistance. *J. Biosci. Bioeng.* 89:569–76
110. Wolfger H, Mamnun YM, Kuchler K. 2003. The yeast Pdr15p ATP-binding cassette (ABC) protein is a general stress response factor implicated in cellular detoxification. *J. Biol. Chem.* 279:11593–99
111. Yuan YR, Blecker S, Martsinkevich O, Millen L, Thomas PJ, Hunt JF. 2001. The crystal structure of the MJ0796 ATP-binding cassette. Implications for the structural consequences of ATP hydrolysis in the active site of an ABC transporter. *J. Biol. Chem.* 276:32313–21

'Gestalt,' Composition and Function of the *Trypanosoma brucei* Editosome

H. Ulrich Göringer

Department of Genetics, Darmstadt University of Technology, 64287 Darmstadt, Germany;
email: goringer@hrzpub.tu-darmstadt.de

Keywords

RNA editing, mitochondrial gene expression, guide RNA, African trypanosome

Abstract

RNA editing describes a chemically diverse set of biomolecular reactions in which the nucleotide sequence of RNA molecules is altered. Editing reactions have been identified in many organisms and frequently contribute to the maturation of organellar transcripts. A special editing reaction has evolved within the mitochondria of the kinetoplastid protozoa. The process is characterized by the insertion and deletion of uridine nucleotides into otherwise nontranslatable messenger RNAs. Kinetoplastid RNA editing involves an exclusive class of small, noncoding RNAs known as guide RNAs. Furthermore, a unique molecular machinery, the editosome, catalyzes the process. Editosomes are megadalton multienzyme assemblies that provide a catalytic surface for the individual steps of the reaction cycle. Here I review the current mechanistic understanding and molecular inventory of kinetoplastid RNA editing and the editosome machinery. Special emphasis is placed on the molecular morphology of the editing complex in order to correlate structural features with functional characteristics.

Contents

INTRODUCTION	66
MITOCHONDRIAL PREmRNAs AND THE EXTENT OF EDITING	67
EDITING COMPLEXES AND THEIR PROTEIN COMPOSITION	67
GROSS MORPHOLOGY OF THE 20S EDITOSOME	69
GROSS MORPHOLOGY OF THE 35-40S EDITOSOME	70
THE 20S TO 35-40S CONVERSION	71
VARIATIONS IN THE PROTEIN INVENTORY	72
A MULTISTEP REACTION CYCLE	74
A DUAL-MODE REACTION CENTER	74
CONCLUSIONS AND OUTLOOK	77

Kinetoplast: unique structural organization of mitochondrial DNA in kinetoplastid protozoa; visible as a basophilic granule at the base of the flagellum

African trypanosome: protozoan parasite; the causative agent of sleeping sickness in Africa

Kinetoplastid DNA (kDNA): network of topologically interlocked minicircle and maxicircle DNA elements within the kinetoplast

DNA maxicircle: large circular DNA elements encoding mitochondrial genes ($\geq$20 kbp)

premRNA: pre-edited messenger RNA

RNA editing: a diverse set of biochemical reactions in which the nucleotide sequence of RNA molecules is altered

INTRODUCTION

RNA editing is a characteristic trait of the order Kinetoplastida (*Excavata, Euglenozoa*). The order includes *Trypanosoma* and *Leishmania* species as well as other pathogens and some nonpathogenic species. Editing represents a unique posttranscriptional modification reaction that is directly linked to the abnormal mitochondrial DNA in these unicellular organisms (7). Unlike other eukarya, the mitochondrial genome is organized in a very large disc-shaped molecular assembly known as the kinetoplast (k). It is located within the mitochondrial matrix and consists of two classes of double-stranded circular DNA molecules, so-called maxicircles and minicircles. African trypanosomes contain about 25 to 50 presumably identical 22-kbp maxicircles and roughly 10^4 minicircles with an average size of 1 kbp. Both kDNA elements are intertwined by concatenation, thereby forming a massive DNA network that exhibits a discrete cell cycle DNA synthesis stage (reviewed in Reference 54) and accounts for up to 20% of the total cellular DNA content. Similar to other eukaryotes, DNA maxicircles encode ribosomal RNA genes and subunits of the oxidative phosphorylation system. However, unlike in other organisms, more than 50% of these genes are "incompletely" encoded: Substantial sequence information is missing—in some cases more than 50%—in addition to the presence of frameshifts and nonexisting translational start and stop codons. As a consequence, pre-edited messenger RNAs (premRNAs) of these genes require RNA editing in order to be converted to functional mRNAs for protein synthesis.

Biochemically, the reaction is characterized by the insertion and/or deletion of exclusively U nucleotides. In African trypanosomes, more than 3,000 U residues are inserted and about 300 U residues are deleted at hundreds of editing sites in 12 different premRNAs (for a recent review see 29). The process is site specific and relies on small, noncoding RNAs known as guide RNAs (gRNAs) (9, 10, 34, 83). In the majority of cases gRNAs are DNA minicircle transcripts and act as "quasi-templates" in the process. They initiate the editing reaction by forming a hybrid gRNA/premRNA molecule that adopts a three-way helix junction topology (46, 74, 94). The structure includes a short anchoring duplex that borders the sequence to be edited. Unpaired gRNA nucleotides next to the anchor region specify U-insertion events with free UTP as a substrate; non-base-paired uridylates in the premRNA become deleted (23, 38, 87, 88). By utilizing different gRNAs, alternative editing events can occur that contribute to generate protein diversity within the mitochondria of the parasites (63–65).

Early experimental evidence already suggested that RNA editing involves mitochondrial protein components and that the reaction is likely catalyzed by high-molecular-mass protein complexes (18, 28, 42, 72, 73, 77). In analogy to ribosomes and spliceosomes the complexes have been termed editosomes, and they function as a reaction platform for the catalysis of the processing reaction in a series of enzyme-driven steps (9, 23, 79). In the past years, the protein inventory of the editosome has been studied in detail and polypeptide candidates for every step of the minimal reaction cycle have been identified (reviewed in References 15 and 29).

Here I summarize the current knowledge on the individual reaction steps of the RNA-editing cycle and on the molecular components of the editing machinery, including accessory factors that contribute to the process. The discussion is limited to African trypanosomes (*Trypanosoma brucei*). A comparison to other kinetoplastid organisms has been published by Lukeš et al. (49). Special emphasis is placed on the recently published three-dimensional (3D) structures of *T. brucei* RNA-editing complexes derived from electron microscopy (EM)-based single-particle studies (26). I discuss the prominent structural landmarks of the different complexes and analyze their interconversion and functionality in the context of published biochemical and genetic data.

Guide RNAs (gRNAs): small, noncoding RNAs that act as templates in the U-nucleotide-specific RNA editing reaction

DNA minicircle: small circular, gRNA-encoding mitochondrial DNA molecules (~1 kbp in size)

Editosome: a mitochondrial high-molecular-mass complex that catalyzes the individual steps of the kinetoplastid RNA editing cycle

TAP: tandem affinity purification

MITOCHONDRIAL PREmRNAs AND THE EXTENT OF EDITING

RNA editing creates translatable mRNAs (35) by three basal modalities: (*a*) by correcting frameshifts, (*b*) by generating translational start and/or stop codons, and (*c*) by creating entire open reading frames (ORFs). Edited mRNAs encode subunits of the oxidative phosphorylation system such as the NADH-ubiquinone oxidoreductase (complex I), cytochrome bc_1 (complex III), cytochrome oxidase (complex IV), and ATP synthase (complex V). Of the 20 mitochondrial genes in *T. brucei* 12 premRNAs are substrates of the processing reaction (29 and references therein); however, the degree of editing varies considerably. The most modest form of editing exists in the case of COII (cytochrome oxidase II): only four uridylate residues are inserted. Other examples of moderate editing are Cyb (34 U insertions) and MURF2 (maxicircle unidentified reading frame 2), with 26 U residues inserted and 4 U residues deleted. COIII, on the other hand, represents an extensively edited premRNA (a phenomenon also known as pan-editing), with 547 U residues inserted and 41 U residues deleted. This affects approximately 60% of the nucleotide content of the final mRNA. Other pan-edited premRNAs are ND7 (553 U insertions/89 U deletions), A6 (447/28), ND9 (345/20), and CR4 (325/40) (summarized in **Table 1**).

EDITING COMPLEXES AND THEIR PROTEIN COMPOSITION

Native editosomes of African trypanosomes have been enriched from the endogenous steady-state pool of mitochondrial editing complexes by a variety of biochemical protocols (for a review see Reference 69). Starting materials are usually insect-stage trypanosomes, which rely on fully developed mitochondria for their energy consumption and thus are expected to have maximal RNA-editing activity. Nonsynchronized cells are harvested at mid-log growth conditions and are used to isolate mitochondrial vesicles by various cell disruption and subcellular fractionation methods (28, 32, 77). Mitochondrial vesicle preparations are converted to low-salt detergent lysates by using non-ionic detergents followed by isokinetic density gradient centrifugation techniques or column purification schemes. Following these protocols, active editing complexes harboring as few as 7 (77), 13 (3, 26), or as many as 20 polypeptides (68) have been described. However, all enrichment protocols generate low yields of complexes, suggesting either low steady-state concentrations or low kinetic/thermodynamic and/or chemical (i.e., redox) stabilities of the complexes. Therefore, more recent purification schemes have applied near-native enrichment conditions mainly following the tandem affinity purification (TAP) protocol as developed by Rigaut et al. (75). The procedure relies on transgenic trypanosomes that conditionally express TAP-tagged versions of

Table 1 Extent of RNA editing in *Trypanosoma brucei*

Mitochondrial transcript	Respiratory complex/ function	No. of U insertions/ U deletions	Length of edited mRNA (nt)
ND1	Complex I	Not edited	–
ND3	Complex I	210/13	452
ND4	Complex I	Not edited	–
ND5	Complex I	Not edited	–
ND7	Complex I	553/89	1,238
ND8	Complex I	259/46	574
ND9	Complex I	345/20	649
Cyb	Complex III	34/none	1,151
COI	Complex IV	Not edited	–
COII	Complex IV	4/none	663
COIII	Complex IV	547/41	969
A6	Complex V	447/28	811
S12	Ribosomal protein S12	132/28	325
MURF1	Unknown function	Not edited	–
MURF2	Unknown function	26/4	1,111
MURF5	Unknown function	Not edited	–
CR3	Unknown function	148/13	299
CR4	Unknown function	325/40	567
9S rRNA	SSU ribosomal RNA	3′-oligo uridylation	–
12S rRNA	LSU ribosomal RNA	3′-oligo uridylation	–

Abbreviations: ND, NADH-ubiquinone oxidoreductase subunits 1–9; Cyb, apocytochrome *b*; CO, cytochrome oxidase subunits I–III; A6, ATP synthase subunit 6; S12, small subunit ribosomal protein 12; MURF, maxicircle unidentified reading frame; CR, G- versus C-strand biased genes 3 and 4; SSU, small subunit; LSU, large subunit.

different editosomal proteins (reviewed in Reference 69). The TAP-tag contains protein A and calmodulin-binding domains separated by a tobacco etch virus protease cleavage site, which allows for chemically moderate, i.e., "native-like," chromatographic separation and elution conditions.

TAP-tagged editosome preparations have been visualized by transmission electron microscopy (TEM) and cryo-EM (26, 40, 47). Raw EM images of *T. brucei* editosomes display monodisperse populations of two classes of high-molecular-mass assemblies in addition to some high-molecular-mass aggregates (**Figure 1**). The two classes consist of large, asymmetric complexes up to 26 nm in diameter and smaller, elongated complexes with a dimension of 21 × 26 nm. Both types of complexes are characterized by well-defined, compact shapes with distinct structural features including surface areas of different electron density (**Figure 1**). In line with previous experimental data (18, 73), sedimentation analysis characterized the two particle classes as high-molecular-mass assemblies with apparent S-values (Svedberg sedimentation coefficient) of 20S and 35-40S (26). Importantly, 35-40S complexes are associated with endogenous RNA, including premRNA and gRNA (18, 26, 73), and thus likely embody the steady-state population of editing complexes actively engaged in the processing reaction. 20S complexes are protein-only assemblies (26, 77) that consist of 13 polypeptides: TbMP100, TbMP99, TbMP90, TbMP67, TbMP63, TbMP61, TbMP57, TbMP52, TbMP48, TbMP46, TbMP44, TbMP42, and TbMP24 (summarized in **Table 2**). This includes all required core activities of the editing reaction cycle (see below), and thus, 20S complexes are competent to faithfully edit synthetic insertion and deletion substrate

TEM: transmission electron microscopy

S-value: Svedberg sedimentation coefficient

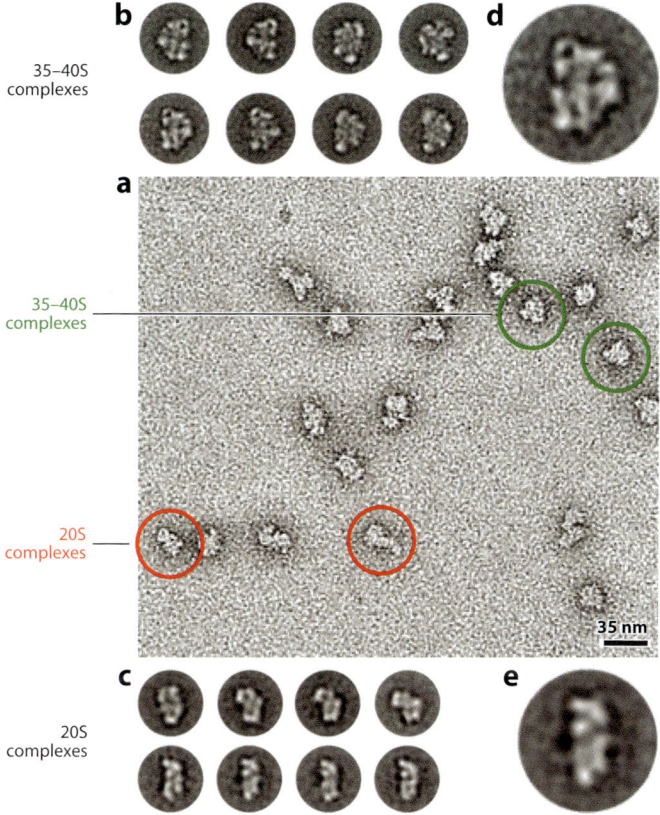

Figure 1
Single-particle electron microscopy (EM) of *Trypanosoma brucei* editosome preparations (26). (*a*) Raw EM image of negatively stained steady-state isolates of editing complexes after mild chemical fixation. The image demonstrates the presence of high-molecular-mass assemblies with a near monodisperse particle distribution. Examples of individual complexes are encircled. (*b*) Representative two-dimensional (2D) class averages of the 35-40S complex. (*c*) Representative 2D class averages of the 20S complex. (*d*) Enlarged picture of a single 35-40S 2D class average. (*e*) Enlarged picture of a single 20S 2D class average.

mRNAs in a gRNA-dependent fashion (14, 36, 37). By contrast, isolated 35-40S complexes are inactive in binding and processing synthetic premRNA molecules likely because their RNA-binding site(s) is occupied with endogenous RNA (26).

GROSS MORPHOLOGY OF THE 20S EDITOSOME

Raw EM images of purified 20S editosomes show a monodisperse population of elongated slightly bent particles with dimensions up to 26 nm. The consensus structure of the complex is characterized by an elongated, slightly bent appearance (**Figure 2**), which results in a concave/convex shape, displaying one concave contour and one convex contour on opposite sides. The particle is composed of two quasi-globular domains roughly equal in size. Both subdomains interact extensively in an interface region where a protruding arm extends on one side and a triangular protrusion emerges from the opposite side. The two subdomains differ in their structural details, indicating that 20S editosomes are not homodimers. The surface representation of the 20S particle was

Table 2 *Trypanosoma brucei* proteins involved in RNA editing[a]

Protein designation	Conserved structural motif(s)	Identified or proposed function
MP18	OB-fold	gRNA binding
MP19	OB-fold	Interaction
MP24	OB-fold	gRNA binding
MP41	U1-like	Interaction
MP42	Zn finger (2×), OB-fold	Endonuclease/exonuclease (in vitro)
MP44	U1-like, RNaseIII, Pum	Editosome integrity
MP46	U1-like, RNaseIII, Pum	Editosome integrity
MP47	U1-like	Interaction
MP48/REL2	Ligase, tau, K	RNA ligase
MP49	U1-like	Interaction
MP52/REL1	Ligase, tau, K	RNA ligase
MP57/RET2	PAP cat/assoc.	TUTase
MP61/REN2	U1-like, RNaseIII, dsRBM	U-insertion-specific endonuclease
mHel61p	DEAD-box RNA helicase	RNA helicase, RNPase
MP63	Zn finger (2×), OB-fold	Interaction
MP67/REN3	U1-like, RNaseIII, dsRBM	COXII-specific endonuclease
MP81	Zn finger (2×), OB-fold	Interaction
MP90/REN1	U1-like, RNaseIII, dsRBM	U-deletion-specific endonuclease
MP99/REX2	5′–3′ Exo, EEP	Nuclease/nucleotidyl phosphatase
MP100/REX1	5′–3′ Exo, EEP	Nuclease/nucleotidyl phosphatase
Accessory proteins		
gBP21	R-rich, Whirly-fold	Matchmaking-type RNA annealing
gBP25	R-rich, Whirly-fold	Matchmaking-type RNA annealing
RBP16	CSD	Interaction
TbRGG1	RGG-repeats	mRNA stabilization
KRET1	Zn finger, PAP cat/assoc.	TUTase

[a] Proteins are abbreviated following the nomenclature of Panigrahi et al. (68). Alternative names and acronyms have been suggested by Stuart et al. (90) and Simpson et al. (89).
Abbreviations: OB-fold, oligosaccharide/oligonucleotide-binding fold; U1-like, U1-like Zn-finger domain; Pum, Pumilio domain; tau, microtubule-associated tau motif; K, kinesin light-chain domain; PAP, poly(A) polymerase catalytic and associated (cat/assoc.) domains; dsRBM, double-stranded RNA-binding motif; 5′–3′ Exo, 5′–3′ exoribonuclease domain; Zn finger, C_2H_2-type zinc-finger domain; EEP, endonuclease-exonuclease-phosphatase domain; CSD, cold shock domain; RGG, RGG RNA-binding repeats; Whirly-fold, Whirly transcription factor-fold; TUTase, terminal uridylyltransferase.

calculated to enclose a molecular mass of 800 ± 80 kDa. The value is consistent with the sum of all proteins identified in the biochemical analysis (790 kDa), assuming that every protein is present in a single copy (26). A theoretical sedimentation coefficient was calculated to be 21–26S, in line with the experimentally determined apparent value of 20–24S (18, 73).

GROSS MORPHOLOGY OF THE 35–40S EDITOSOME

The consensus 3D structure of the 35–40S complex (**Figure 3**) is similarly characterized by defined landmarks: An elongated, straight to slightly convex platform is packed against a semispherical element of variable size. The platform extends on two sides into small head-like protrusions. One

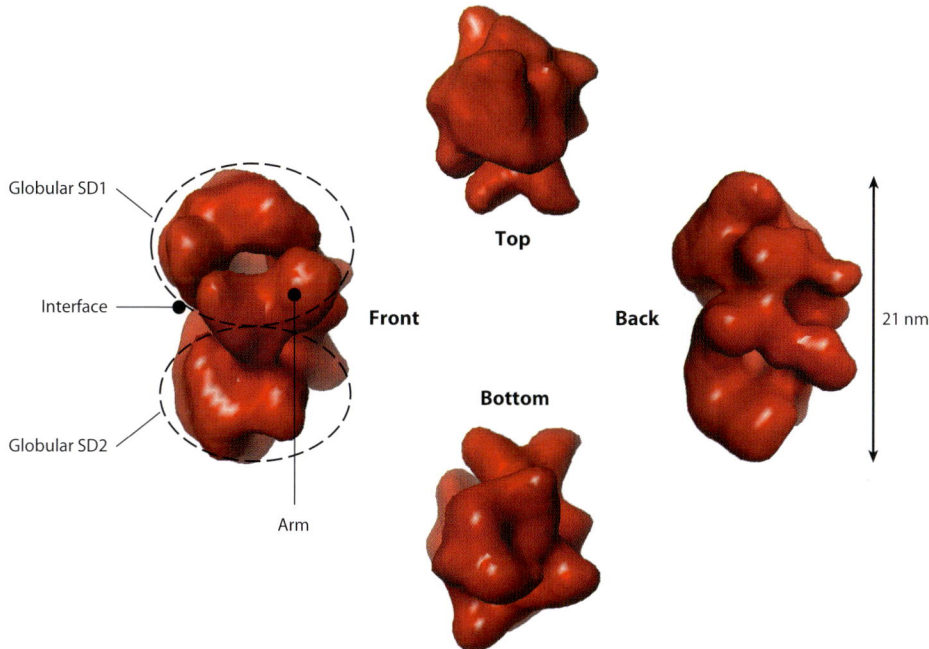

Figure 2
Consensus 3D structure of the *Trypanosoma brucei* 20S RNA-editing complex derived from cryo-EM-based single-particle studies (26). Front, back, top, and bottom representations of the consensus structure are shown (EMDB-1595). Prominent structural features are indicated. The globular subdomains 1 and 2 are highlighted by dashed circles. The model is based on the analysis of 31,641 individual complexes (609 2D class averages and 58 3D maps). The resolution is approximately 2.1 nm with a surface representation of 800 ± 80 kDa. Map data and image files can be retrieved at **http://www.pdbj.org/emnavi/** [EM Navigator, Protein Data Bank Japan (PDBj)]. Abbreviations: 2D, two-dimensional; 3D, three-dimensional; EM, electron microscopy; SD, subdomain; EMBD, Electron Microscopy Data Base.

side is oriented to the top, the other side to the bottom, forming a larger foot-like extension. The semispherical back is packed against the platform as a tight network, and the interface between both elements is marked by incisions in the upper and lower parts. The semispherical back element is asymmetric in its appearance, displaying on one side a protruding shoulder-type element and on the other side an inclination or proclivity. A structure refinement of the 35-40S complex by cryo-negative staining EM resulted in a resolution of 13 to 19 nm, and the surface of the complex was calculated to enclose a molecular mass of 1.45 ± 0.15 MDa (26). The data have been used to calculate a theoretical sedimentation coefficient of 35-41S (25), which is in agreement with the experimentally derived apparent sedimentation behavior observed in isokinetic glycerol gradients (72, 73).

THE 20S TO 35-40S CONVERSION

The structural relationship between 20S and 35-40S editosomes has been investigated by different 3D docking and alignment approaches (26). The data suggest that 20S complexes represent a significant part of the platform density of the 35-40S complexes. 20S complexes can be integrated into the head domain, the upper part of the platform density, and part of the back domain of the 35-40S complex. By contrast, the foot domain, the lower part of the platform density, and the

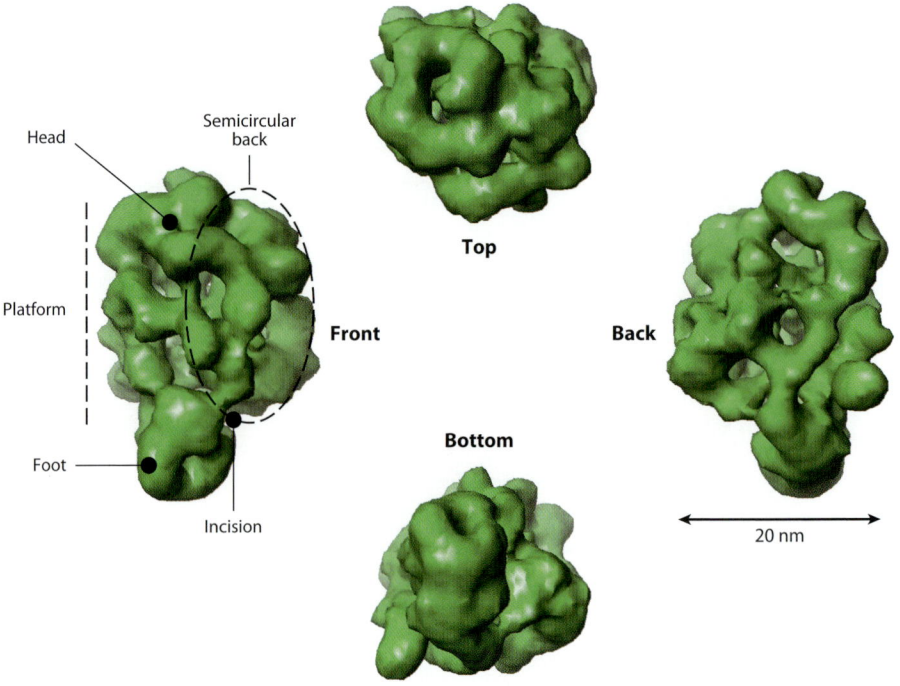

Figure 3

Consensus 3D structure of the *Trypanosoma brucei* 35-40S editosome derived from cryo-EM-based single-particle studies (26). Front, back, top, and bottom representations of the consensus structure are shown (EMDB-1594). Prominent structural features such as the substrate RNA-containing semicircular back domain (*dashed circle*) are indicated. The model is based on the analysis of 25,747 individual complexes (214 2D class averages and 175 3D maps). The resolution is approximately 1.6 nm with a surface representation of 1.45 ± 0.15 MDa. Map data and image files can be retrieved at **http://www.pdbj.org/emnavi/** [EM Navigator, Protein Data Bank Japan (PDBj)]. Abbreviations: 2D, two-dimensional; 3D, three-dimensional; EM, electron microscopy; EMBD, Electron Microscopy Data Base.

majority of the semispherical back appears to be composed of components that have no structural correlates in the 20S particles (**Figure 4**). Because the main biochemical difference between 20S and 35-40S complexes is their RNA content, it is tempting to speculate that the semispherical back of the 35-40S complexes represents gRNA and/or premRNA molecules only—although the presence of a small number of additional proteins cannot be excluded. Experimental confirmation of the suggested scenario was gained from interconversion RNase digestion experiments. Treatment of 20S and 35-40S complexes with single-strand- and double-strand-specific RNases resulted in a concentration-dependent decrease in the amount of 35-40S editosomes and an increase in 20S complexes. Conversely, incubation of isolated 20S editosomes with *T. brucei* mitochondrial RNA, which contains pre- and partially edited mRNAs as well as gRNAs, generated 35-40S complexes in a concentration-dependent fashion (26). The substrate RNA/20S interaction is of high affinity, with equilibrium dissociation constants (K_d) in the nanomolar concentration range (26).

VARIATIONS IN THE PROTEIN INVENTORY

Evidence for functionally and compositionally distinct 20S complexes has been derived from the analysis of editing complexes isolated from transgenic trypanosome strains that conditionally

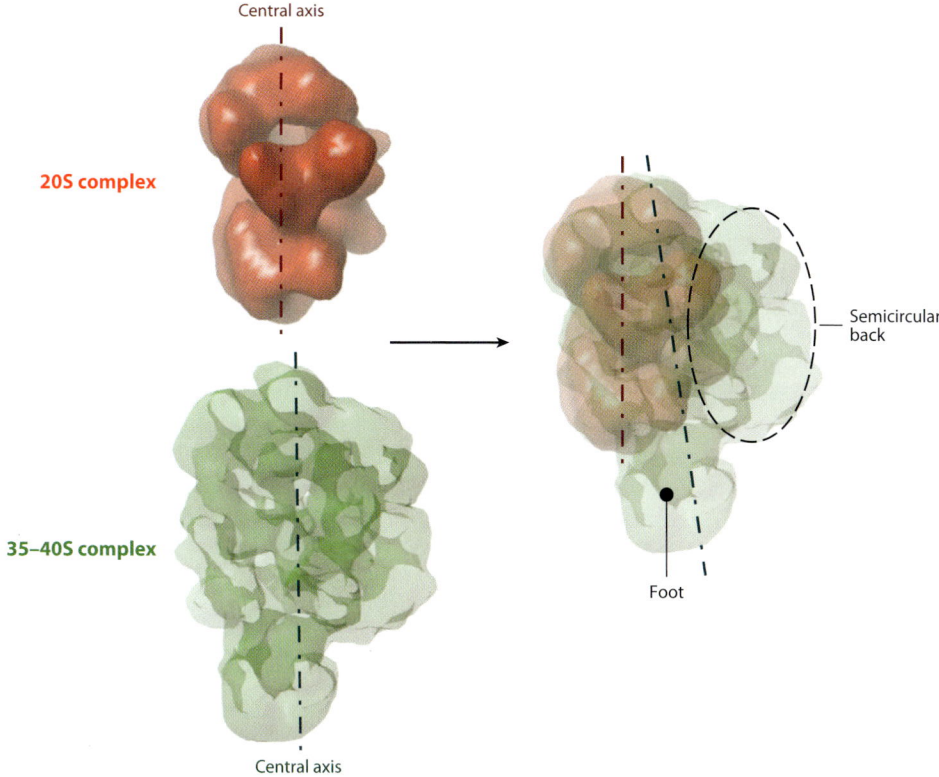

Figure 4
Three-dimensional docking/alignment of the 20S editosome into the 35-40S complex (26). The 20S particle can be integrated into the head domain, the upper part of the platform, and a part of the back domain of the 35-40S complex. The foot domain and the majority of the semispherical back (*dashed circle*) have no structural correlates in the 20S particles. The semispherical back contacts the two globular 20S subdomains and reaches into the interface region. Dashed lines indicate the central axes of the two complexes. The data suggest a structural and enzymatically active (20S) protein core with a large peripheral RNA-binding domain (semispherical back).

express TAP-tagged variants of different editosomal proteins. TAP-purified editosomes possess in some cases only a subcollection of the 20S protein inventory, and these compositional differences are mirrored in functional differences (67). For instance, TbMP90-TAP editosomes execute only deletional RNA editing, whereas TbMP61-TAP complexes process only insertional editing substrate RNAs. Protein TbMP100 is present only in TbMP90-TAP editing complexes, whereas TbMP81, TbMP63, TbMP42, TbMP24, TbMP18, TbMP46, TbMP44, TbMP57, TbMP99, TbMP52, and TbMP48 seem to be ubiquitous. On the contrary, proteins TbMP99, TbMP61, and TbMP67 have been shown to be mutually exclusive in certain complexes. On the basis of these data, Carnes et al. (17) suggested that at least three compositionally distinct editosomes exist. Whether these partially assembled machineries contribute to the endogenous, steady-state ensemble of editosomes in genetically unaltered, i.e., wild-type, trypanosomes is not known. However, the data at least demonstrate that multiple assembly pathways exist for the processing machinery and that heterogeneous populations of 20S complexes are tolerated in in vivo isolates. Furthermore, it cannot be excluded that the composition of editosomes may be in a dynamic equilibrium.

Individual components such as some of the accessory factors described below (27) may at different stages of the assembly pathway shuttle in and out of the complex, thereby further expanding the structural landscape of the particle.

> **TUTase:** terminal uridylyltransferase

A MULTISTEP REACTION CYCLE

A minimal RNA-editing reaction cycle can be formally divided into three basal reaction steps: premRNA cleavage, U-nucleotide deletion or U-nucleotide insertion, and re-ligation of the initially generated premRNA fragments. Although it is not known whether the overall reaction is distributive or processive, available in vitro RNA-editing data suggest that the reaction is carried out within a multifunctional reaction center that mediates the three partial reaction steps consecutively (**Figure 5**).

The endonucleolytic cleavage of the premRNA occurs at the first unpaired nucleotide upstream within the premRNA/gRNA duplex substrate, which adopts a three-way junction geometry (45, 46, 74, 94) (for a model see **Figure 6**). The related proteins TbMP90, TbMP63, and TbMP67, which all contain U1-like zinc-finger motifs, an RNaseIII domain, and a double-stranded RNA-binding motif sequence, possess endonuclease activity. TbMP90 is specific for the cleavage at deletion sites, while TbMP63 cleaves at insertion sites (16, 91). The function of TbMP67 is not yet fully understood. At a deletion site the endonucleolytic cleavage is followed by the removal of unpaired U nucleotides. U-nucleotide-specific ribonuclease (exoUase) activity was identified for TbMP99 and TbMP100, two related proteins with N-terminal 5'/3' exonuclease and C-terminal endo/exo/phosphatase motifs (39, 76). In addition, these proteins possess nucleotidyl phosphatase activity (62). TbMP42, a protein with two C_2H_2-Zn-finger domains and a putative oligonucleotide/oligosaccharide-binding fold, executes endo- and exoribonuclease activity in vitro, likely following a two-metal-ion reaction mechanism (13, 61). Following endonucleolytic cleavage at an insertion site, U nucleotides are added to the 3' end of the 5'-cleavage product of the mRNA in a gRNA-dependent fashion. This step of the reaction is catalyzed by TbMP57, a protein with a catalytic poly(A) polymerase domain that executes terminal uridylyltransferase (TUTase) activity (22). At the end of the processing cycle the two editing-specific RNA ligases, TbMP48 and TbMP52 (**Figure 6**), rejoin the two processed mRNA fragments (20, 24, 55, 78).

In addition to these core activities that catalyze the main steps of the reaction cycle, several accessory factors are required (reviewed in References 15 and 27). These proteins presumably bind only temporarily to the editosome to contribute additional functionality. This includes the matchmaking-type RNA/RNA annealing factors gBP21 and gBP25 (1, 4, 8, 41, 59, 60, 86) (**Figure 6**), as well as the proteins RBP16 (33, 56, 70, 71), TbRGG1 (31, 92), REAP1 (30, 50, 51), and RBP38 (81). It also includes the 3'-end-specific TUTase KRET1 (2, 5, 6) and the mitochondrial DExH/D protein mHel61p (27, 48, 57, 58) (**Figure 6**). Although the presence of accessory editing factors is unquestioned, no knowledge with respect to the dynamic interplay between these proteins and the editosome exists.

A DUAL-MODE REACTION CENTER

In addition to unraveling the protein inventory of the 20S complex, attempts have been published to describe the interconnectivity and general organization of individual proteins within the particle (84, 85). Although the precise architecture is far from being understood, it is undisputed that functional 20S particles can be assembled in the absence of pre-edited mRNA and gRNA (21). Furthermore, biochemical data suggest that insertion and deletion RNA editing may be catalyzed

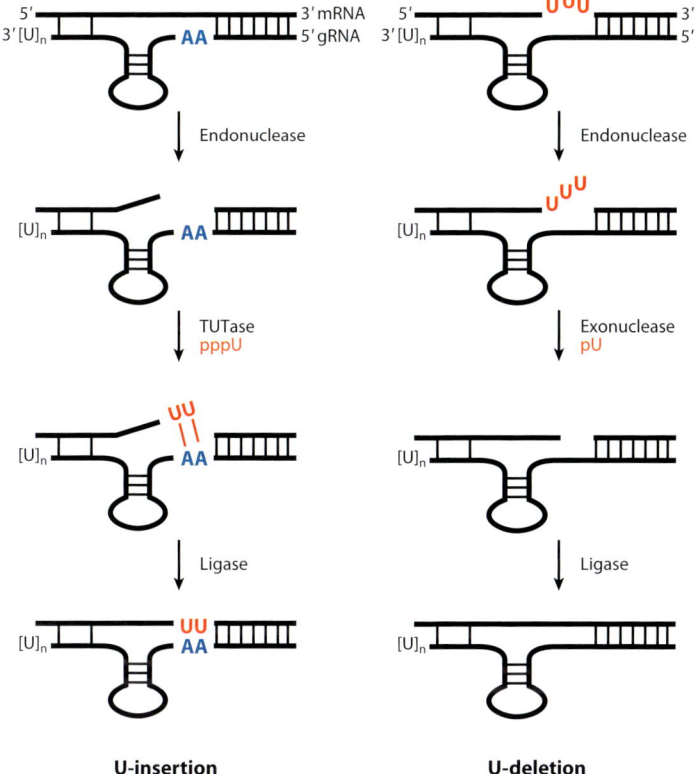

Figure 5

Mechanistic outline of the basal reaction steps of U-insertion (*left*) and U-deletion (*right*) RNA editing in *Trypanosoma brucei* (adapted from 66). gRNA molecules interact with cognate pre-edited mRNAs via antiparallel base-pairing, thereby forming a three-helix junction RNA hybrid (for a three-dimensional model see **Figure 6d**). The interaction relies on Watson-Crick-type base-pairing as well as noncanonical G:U base pairs. Endonucleolytic cleavage of the premRNA occurs at the first non-base-paired nucleotide upstream of the so-called anchor duplex. U-insertion editing continues by adding U nucleotides (from UTP) to the 3′ end of the 5′-premRNA cleavage fragment. The reaction is catalyzed by a terminal uridylyltransferase (TUTase). During U-deletion editing, U nucleotides are removed from the 3′ end of the 5′-cleavage fragment, using the activity of a 3′ exonuclease (exoUase). In both cases, the number of U nucleotides (inserted or deleted) is specified by the guiding sequence of the gRNA. Resultant premRNA cleavage fragments are finally ligated by an RNA ligase. Several editing cycles must occur until all editing sites specified by one gRNA are processed. Complete editing of a premRNA requires in most cases multiple gRNAs. The reaction proceeds with a 3′ to 5′ directionality on the premRNA. Measurements concerning the fidelity of the reaction have not been published. However, substantial misediting (19, 53) and aberrant ligation events have been identified in vivo and in vitro (11, 12, 19, 43, 53, 82). The involvement of nucleotidyl phosphatase activity has been suggested by Niemann et al. (62). Abbreviations: gRNA, guide RNA; mRNA, messenger RNA; premRNA, pre-edited messenger RNA.

by separate editing subcomplexes or subdomains (for a review, see 15). This finding is seemingly supported by the fact that the editing core activities are present in protein pairs or even higher arrangements (29, 90). Yeast two-hybrid and coimmunoprecipitation experiments identified possible protein-protein interaction partners among different editosomal proteins, which together suggest that 20S complexes consist of structurally separate insertion and deletion subdomains (84). The insertion domain is assembled around a trimeric protein core involving the terminal

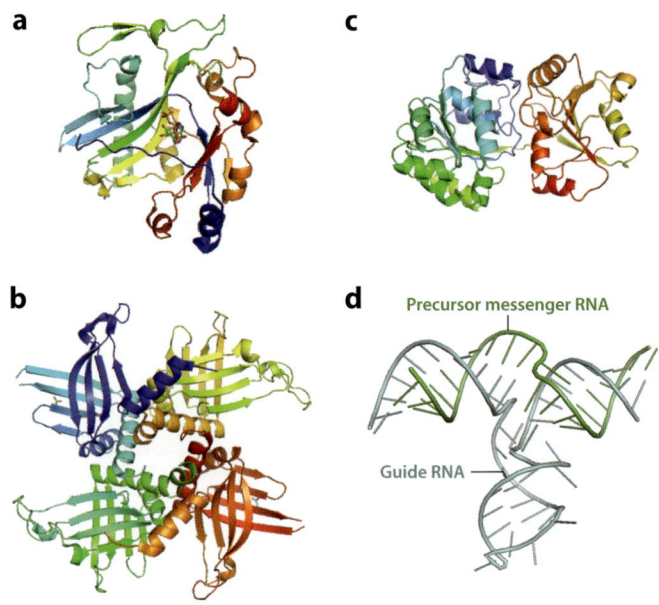

Figure 6

Representative high-resolution structures and molecular models of individual components of the *Trypanosoma brucei* editosome and the editing reaction cycle. (*a*) RNA-editing ligase TbMP52/REL1 (20). (*b*) Heteromultimeric matchmaking-type RNA/RNA annealing factor gBP21/gBP25 (86). (*c*) Molecular model of the DEAD-box RNA helicase mHel61p (27). (*d*) Three-dimensional model of the three-helix-junction architecture of a guide RNA/precursor messenger RNA hybrid molecule.

TUTase TbMP57, the RNA-editing ligase TbMP48, and the protein TbMP81. The deletion subdomain relies on the heterotrimeric interaction of the exoUase TbMP99, the RNA-editing ligase TbMP52, and the protein TbMP63. Other polypeptides such as TbMP44 and TbMP24 (80, 93) associate with both subdomains. TbMP18 likely represents a protein that is located at the interface between both subdomains (44). Some of these results have been confirmed using TAP-tagged editosome preparations (reviewed in Reference 15) and have provided insight into the structure/function correlation of various proteins in the 20S complex.

It is tempting to speculate that the compositionally different deletion and insertion subdomains correspond to the bipartite substructures of the 20S 3D consensus model (see **Figure 2**). The structure displays two nonidentical, nearly globular subdomains that are connected through an interface region in line with the above-described biochemical data. Whether both subdomains have independent RNA-binding sites is not known to date. However, on the basis of the structural details of the 35-40S complexes, in which the semispherical back of the complex was identified as bound substrate RNA, it is more likely that the 20S complex has only one RNA interaction site. The binding site covers a substantial part on one side of the 20S surface and reaches partly into the interface region between the two globular subdomains. Thus, a single catalytic core for both editing reactions might be positioned at the interface between the deletion and insertion subdomains, representing a dual-mode reaction center that is triggered by the chemical nature of the bound substrate RNA. Deletional editing substrates activate the deletion subdomain, and insertion-type substrate RNAs become processed by the insertion subdomain.

CONCLUSIONS AND OUTLOOK

Taken together, the described experimental data can be summarized to derive a first picture that correlates some of the prominent structural features of the 20S and 35-40S complexes with functional attributes of the RNA-editing reaction cycle. Steady-state editosome preparations in African trypanosomes consist of a mixture of two classes of high-molecular-mass complexes: 35-40S complexes represent the editing machinery that is actively engaged in the processing reaction, whereas 20S complexes are preassembled precursor complexes that consist of proteins only. The binding of substrate RNAs converts 20S editosomes to 35-40S complexes, and depending on the type of RNA, differently shaped 35-40S complexes are formed. Altogether 12 pre-edited mRNAs and literally hundreds of gRNAs and partially edited mRNA molecules are present in African trypanosomes. gRNAs have molecular masses of approximately 20 kDa, while the involved mRNAs vary between 60 (unedited CR3) and 450 kDa (edited ND7). As a consequence, their hydrodynamic radius must vary, and at steady-state conditions, this results in a broad structural landscape of multiple 35-40S complexes with different shapes. However, these particles differ predominantly in only one structural feature: the semispherical back element. This part of the complex represents the RNA interaction site of the editosome, which implies that 20S editosomes have only one RNA-binding domain that interacts with a large structural ensemble of differently sized and folded substrate RNAs. The RNA-binding site covers a large surface area on one side of the 20S particle and contacts the interface between the two nonidentical, globular subdomains of the 20S editosome.

The protein-only 20S editosome is characterized by a bipartite appearance with two prominent globular subdomains. Both subdomains, however, differ in their structural details, indicating that the particle is not a homodimer. In conjunction with the described biochemical data, these substructures likely represent separate insertion and deletion subdomains and they are assembled around different protein core elements in line with the different enzyme requirements that catalyze the two types of editing reactions. The two subdomains connect at an interface region, which is linked to the RNA-binding domain of the complex. This suggests a working model in which the editosome reaction center is located at the interface of the insertional and deletional subdomains, thereby presenting a catalytic core of bifunctional quality. As a result, editosome-bound substrate RNA can be in physical contact with both catalytic machineries, and depending on the RNA-editing domain, both U insertions and U deletions can be executed on the same premRNA. This is further supported by the fact that single gRNAs can mediate insertion as well as deletion editing (9, 52). Understanding the conformational dynamic and adaptive recognition at the catalytic center of the editosome is the next experimental and perhaps conceptual challenge.

SUMMARY POINTS

1. The U-nucleotide-specific RNA-editing reaction cycle within the mitochondria of kinetoplastid protozoa is catalyzed by a high-molecular-mass complex.

2. Steady-state isolates of *T. brucei* editosomes contain ensembles of two types of high-molecular-mass assemblies with apparent hydrodynamic sizes of 20S (0.8 MDa) and 35-40S (approximately 1.5 MDa).

3. 20S editosomes are protein-only complexes (≤ 20 proteins) that hold all key enzyme activities for the processing reaction.

4. Substrate RNA binding to 20S editosomes creates a structural landscape of differently sized 35-40S complexes.

5. The editing reaction follows a cascade of enzyme-catalyzed steps.

6. Editosomes have a bimodal reaction center with compositionally different subdomains for the U-deletion and U-insertion reactions.

7. Editosomes interact with accessory protein factors/complexes to modulate the reaction.

FUTURE ISSUES

1. High-resolution structures of the editing complexes, of the individual protein components, and of the substrate RNAs are needed.

2. The interplay between substrate RNAs and proteins in aligning the reactive groups of the premRNA/gRNA hybrid needs to be studied.

3. Accuracy/fidelity issues of the processing reaction should be addressed.

4. The molecular assembly pathway(s) of the editosome should be uncovered.

5. The possible interplay between the editosome and other complexes/biochemical pathways (e.g., ribosome, RNA degradosome) should be studied.

DISCLOSURE STATEMENT

The author is not aware of any affiliations, memberships, funding, or financial holdings that might be perceived as affecting the objectivity of this review.

ACKNOWLEDGMENTS

I thank all past and present members of my laboratory for their contributions. H. Stark, M.M. Golas, and B. Sander are thanked for a rewarding collaboration, and C.D. Specter, M. Brecht, and P. Morais are thanked for discussions and critical reading of the manuscript. Funding by the German Research Foundation, the Howard Hughes Medical Institute, and the Dr. Illing Foundation (http://dr.illing-stiftungen.de/) is acknowledged.

LITERATURE CITED

1. Allen TE, Heidmann S, Reed R, Myler PJ, Göringer HU, Stuart KD. 1998. Association of guide RNA binding protein gBP21 with active RNA editing complexes in *Trypanosoma brucei*. *Mol. Cell. Biol.* 18:6014–22

2. Aphasizheva I, Aphasizhev R. 2010. RET1-catalyzed uridylylation shapes the mitochondrial transcriptome in *Trypanosoma brucei*. *Mol. Cell. Biol.* 30:1555–67

3. Aphasizhev R, Aphasizheva I, Nelson RE, Gao G, Simpson AM, et al. 2003. Isolation of a U-insertion/deletion editing complex from *Leishmania tarentolae* mitochondria. *EMBO J.* 22:913–24

4. Aphasizhev R, Aphasizheva I, Nelson RE, Simpson L. 2003. A 100-kD complex of two RNA-binding proteins from mitochondria of *Leishmania tarentolae* catalyzes RNA annealing and interacts with several RNA editing components. *RNA* 9:62–76

5. Aphasizhev R, Aphasizheva I, Simpson L. 2003. A tale of two TUTases. *Proc. Natl. Acad. Sci. USA* 100:10617–22
6. Aphasizhev R, Sbicego S, Peris M, Jang SH, Aphasizheva I, et al. 2002. Trypanosome mitochondrial 3′ terminal uridylyl transferase (TUTase): the key enzyme in U-insertion/deletion RNA editing. *Cell* 108:637–48
7. **Benne R, Van Den Burg J, Brakenhoff JP, Sloof P, Van Boom JH, Tromp MC. 1986. Major transcript of the frameshifted *coxII* gene from trypanosome mitochondria contains four nucleotides that are not encoded in the DNA. *Cell* 46:819–26**

7. Gives first report of RNA editing.

8. Blom D, van den Burg J, Breek CK, Speijer D, Muijsers AO, Benne R. 2001. Cloning and characterization of two guide RNA-binding proteins from mitochondria of *Crithidia fasciculata*: gBP27, a novel protein, and gBP29, the orthologue of *Trypanosoma brucei* gBP21. *Nucleic Acids Res.* 29:2950–62
9. **Blum B, Bakalara N, Simpson L. 1990. A model for RNA editing in kinetoplastid mitochondria: "Guide" RNA molecules transcribed from maxicircle DNA provide the edited information. *Cell* 60:189–98**

9. Identifies gRNAs and gRNA-dependent RNA editing mechanism.

10. Blum B, Simpson L. 1990. Guide RNAs in kinetoplastid mitochondria have a nonencoded 3′ oligo(U) tail involved in recognition of the preedited region. *Cell* 62:391–97
11. Blum B, Simpson L. 1992. Formation of guide RNA/messenger RNA chimeric molecules in vitro, the initial step of RNA editing, is dependent on an anchor sequence. *Proc. Natl. Acad. Sci. USA* 89:11944–48
12. Blum B, Sturm NR, Simpson AM, Simpson L. 1991. Chimeric gRNA-mRNA molecules with oligo(U) tails covalently linked at sites of RNA editing suggest that U addition occurs by transesterification. *Cell* 65:543–50
13. Brecht M, Niemann M, Schlüter E, Müller UF, Stuart K, Göringer HU. 2005. TbMP42, a protein component of the RNA editing complex in African trypanosomes, has endo-exoribonuclease activity. *Mol. Cell* 17:621–30
14. Carnes J, Stuart K. 2007. Uridine insertion/deletion editing activities. *Methods Enzymol.* 424:25–54
15. Carnes J, Stuart K. 2008. Working together: the RNA editing machinery in *Trypanosoma brucei*. In *RNA Editing, Nucleic Acids and Molecular Biology*, ed. HU Göringer, 20:143–64. Heidelberg, Ger.: Springer
16. Carnes J, Trotter JR, Ernst NL, Steinberg A, Stuart K. 2005. An essential RNase III insertion editing endonuclease in *Trypanosoma brucei*. *Proc. Natl. Acad. Sci. USA* 102:16614–19
17. Carnes J, Trotter JR, Peltan A, Fleck M, Stuart K. 2008. RNA editing in *Trypanosoma brucei* requires three different editosomes. *Mol. Cell. Biol.* 28:122–30
18. Corell RA, Read LK, Riley GR, Nellissery JK, Allen TE, et al. 1996. Complexes from *Trypanosoma brucei* that exhibit deletion editing and other editing-associated properties. *Mol. Cell. Biol.* 16:1410–18
19. Decker CJ, Sollner-Webb B. 1990. RNA editing involves indiscriminate U changes throughout precisely defined editing domains. *Cell* 61:1001–11
20. **Deng J, Schnaufer A, Salavati R, Stuart KD, Hol WG. 2004. High resolution crystal structure of a key editosome enzyme from *Trypanosoma brucei*: RNA editing ligase 1. *J. Mol. Biol.* 343:601–13**

20. Provides the first high-resolution structure of a key editosome enzyme.

21. Domingo GJ, Palazzo SS, Wang B, Pannicucci B, Salavati R, Stuart KD. 2003. Dyskinetoplastic *Trypanosoma brucei* contains functional editing complexes. *Eukaryot. Cell* 2:569–77
22. Ernst NL, Panicucci B, Igo RP Jr, Panigrahi AK, Salavati R, Stuart K. 2003. TbMP57 is a 3′ terminal uridylyl transferase (TUTase) of the *Trypanosoma brucei* editosome. *Mol. Cell* 11:1525–36
23. **Frech GC, Simpson L. 1996. Uridine insertion into preedited mRNA by a mitochondrial extract from *Leishmania tarentolae*: stereochemical evidence for the enzyme cascade model. *Mol. Cell. Biol.* 16:4584–89**

23. Verifies stereochemically the enzyme cascade mechanism for RNA editing.

24. Gao G, Simpson L. 2003. Is the *Trypanosoma brucei* REL1 RNA ligase specific for U-deletion RNA editing and is the REL2 RNA ligase specific for U-insertion editing? *J. Biol. Chem.* 278:27570–74
25. García de la Torre J, Llorca O, Carrascosa JL, Valpuesta JM. 2001. HYDROMIC: prediction of hydrodynamic properties of rigid macromolecular structures obtained from electron microscopy images. *Eur. Biophys. J.* 30:457–62
26. **Golas MM, Böhm C, Sander B, Effenberger K, Brecht M, et al. 2009. Snapshots of the RNA editing machine in trypanosomes captured at different assembly stages in vivo. *EMBO J.* 28:766–78**

26. First electron microscopic visualization of editosomes.

27. Göringer HU, Brecht M, Böhm C, Kruse E. 2008. RNA editing accessory factors—the example of mHel61p. In *RNA Editing, Nucleic Acids and Molecular Biology*, ed. HU Göringer, 20:165–79. Heidelberg, Ger.: Springer
28. Göringer HU, Koslowsky DJ, Morales TH, Stuart K. 1994. The formation of mitochondrial ribonucleoprotein complexes involving gRNA molecules in Trypanosoma brucei. *Proc. Natl. Acad. Sci. USA* 91:1776–80
29. Hajduk S, Ochsenreiter T. 2010. RNA editing in kinetoplastids. *RNA Biol.* 7:229–36
30. Hans J, Hajduk SL, Madison-Antenucci S. 2007. RNA-editing-associated protein 1 null mutant reveals link to mitochondrial RNA stability. *RNA* 13:881–89
31. Hashimi H, Zíková A, Panigrahi AK, Stuart K, Lukeš J. 2008. TbRGG1, an essential protein involved in kinetoplastid RNA metabolism that is associated with a novel multiprotein complex. *RNA* 14:970–80
32. Hauser R, Pypaert M, Häusler T, Horn EK, Schneider A. 1996. In vitro import of proteins into mitochondria of Trypanosoma brucei and Leishmania tarentolae. *J. Cell Sci.* 109:517–23
33. Hayman ML, Read LK. 1999. Trypanosoma brucei RBP16 is a mitochondrial Y-box family protein with guide RNA binding activity. *J. Biol. Chem.* 274:12067–74
34. Hermann T, Schmid B, Heumann H, Göringer HU. 1997. A three-dimensional working model for a guide RNA from Trypanosoma brucei. *Nucleic Acids Res.* 25:2311–18
35. **Horváth A, Berry EA, Maslov DA. 2000. Translation of the edited mRNA for cytochrome *b* in trypanosome mitochondria. *Science* 287:1639–40**

35. Provides an experimental confirmation of kinetoplastid RNA editing on the protein level.

36. Igo RP Jr, Palazzo SS, Burgess ML, Panigrahi AK, Stuart K. 2000. Uridylate addition and RNA ligation contribute to the specificity of kinetoplastid insertion RNA editing. *Mol. Cell. Biol.* 20:8447–57
37. Igo RP Jr, Weston DS, Ernst NL, Panigrahi AK, Salavati R, Stuart K. 2002. Role of uridylate-specific exoribonuclease activity in Trypanosoma brucei RNA editing. *Eukaryot. Cell* 1:112–18
38. Kable ML, Seiwert SD, Heidmann S, Stuart K. 1996. RNA editing: a mechanism for gRNA-specified uridylate insertion into precursor mRNA. *Science* 273:1189–95
39. Kang X, Rogers K, Gao G, Falick AM, Zhou S, Simpson L. 2005. Reconstitution of uridine-deletion precleaved RNA editing with two recombinant enzymes. *Proc. Natl. Acad. Sci. USA* 102:1017–22
40. Kastner B, Fischer N, Golas MM, Sander B, Dube P, et al. 2008. GraFix: sample preparation for single-particle electron cryomicroscopy. *Nat. Methods* 5:53–55
41. Köller J, Müller UF, Schmid B, Missel A, Kruft V, et al. 1997. Trypanosoma brucei gBP21: an arginine-rich mitochondrial protein that binds to guide RNA with high affinity. *J. Biol. Chem.* 272:3749–57
42. Köller J, Nörskau G, Paul AS, Stuart K, Göringer HU. 1994. Different Trypanosoma brucei guide RNA molecules associate with an identical complement of mitochondrial proteins in vitro. *Nucleic Acids Res.* 22:1988–95
43. Koslowsky DJ, Göringer HU, Morales TH, Stuart K. 1992. In vitro guide RNA/mRNA chimaera formation in Trypanosoma brucei RNA editing. *Nature* 356:807–9
44. Law JA, O'Hearn S, Sollner-Webb B. 2007. In Trypanosoma brucei RNA editing, TbMP18 (band VII) is critical for editosome integrity and for both insertional and deletional cleavages. *Mol. Cell. Biol.* 27:777–87
45. Lescoute A, Westhof E. 2006. Topology of three-way junctions in folded RNAs. *RNA* 12:83–93
46. Leung SS, Koslowsky DJ. 2001. Interactions of mRNAs and gRNAs involved in trypanosome mitochondrial RNA editing: structure probing of an mRNA bound to its cognate gRNA. *RNA* 7:1803–16
47. Li F, Ge P, Hui WH, Atanasov I, Rogers K, et al. 2009. Structure of the core editing complex (L-complex) involved in uridine insertion/deletion RNA editing in trypanosomatid mitochondria. *Proc. Natl. Acad. Sci. USA* 106:12306–10
48. Li F, Herrera J, Zhou S, Maslov DA, Simpson L. 2011. Trypanosome REH1 is an RNA helicase involved with the 3′-5′ polarity of multiple gRNA-guided uridine insertion/deletion RNA editing. *Proc. Natl. Acad. Sci. USA* 108:3542–47
49. Lukeš J, Hashimi H, Zikova A. 2005. Unexplained complexity of the mitochondrial genome and transcriptome in kinetoplastid flagellates. *Curr. Genet.* 48:277–99
50. Madison-Antenucci S, Hajduk SL. 2001. RNA editing-associated protein 1 is an RNA binding protein with specificity for preedited mRNA. *Mol. Cell* 7:879–86
51. Madison-Antenucci S, Sabatini RS, Pollard VW, Hajduk SL. 1998. Kinetoplastid RNA-editing-associated protein 1 (REAP-1): a novel editing complex protein with repetitive domains. *EMBO J.* 17:6368–76

52. Maslov DA, Simpson L. 1992. The polarity of editing within a multiple gRNA-mediated domain is due to formation of anchors for upstream gRNAs by downstream editing. *Cell* 70:459–67
53. Maslov DA, Thiemann O, Simpson L. 1994. Editing and misediting of transcripts of the kinetoplast maxicircle G5 (ND3) cryptogene in an old laboratory strain of *Leishmania tarentolae*. *Mol. Biochem. Parasitol.* 68:155–59
54. McKean PG. 2003. Coordination of cell cycle and cytokinesis in *Trypanosoma brucei*. *Curr. Opin. Microbiol.* 6:600–7
55. McManus MT, Shimamura M, Grams J, Hajduk SL. 2001. Identification of candidate mitochondrial RNA editing ligases from *Trypanosoma brucei*. *RNA* 7:167–75
56. Miller MM, Read LK. 2003. *Trypanosoma brucei*: functions of RBP16 cold shock and RGG domains in macromolecular interactions. *Exp. Parasitol.* 105:140–48
57. Missel A, Göringer HU. 1994. *Trypanosoma brucei* mitochondria contain RNA helicase activity. *Nucleic Acids Res.* 22:4050–56
58. Missel A, Souza AE, Nörskau G, Göringer HU. 1997. Gene disruption of a mitochondrial DEAD box protein in *Trypanosoma brucei* affects edited mRNAs. *Mol. Cell. Biol.* 17:4895–903
59. Müller UF, Göringer HU. 2002. Mechanism of the gBP21-mediated RNA/RNA annealing reaction: matchmaking and charge reduction. *Nucleic Acids Res.* 30:447–55
60. Müller UF, Lambert L, Göringer HU. 2001. Annealing of RNA editing substrates facilitated by guide RNA-binding protein gBP21. *EMBO J.* 20:1394–404
61. Niemann M, Brecht M, Schlüter E, Weitzel K, Zacharias M, Göringer HU. 2008. TbMP42 is a structure-sensitive ribonuclease that likely follows a metal ion catalysis mechanism. *Nucleic Acids Res.* 36:4465–73
62. Niemann M, Kaibel H, Schlüter E, Weitzel K, Brecht M, Göringer HU. 2009. Kinetoplastid RNA editing involves a 3′ nucleotidyl phosphatase activity. *Nucleic Acids Res.* 37:1897–906
63. Ochsenreiter T, Anderson S, Wood ZA, Hajduk S. 2008. Alternative RNA editing produces a novel protein involved in mitochondrial DNA maintenance in trypanosomes. *Mol. Cell. Biol.* 28:5595–604
64. Ochsenreiter T, Cipriano M, Hajduk SL. 2008. Alternative mRNA editing in trypanosomes is extensive and may contribute to mitochondrial protein diversity. *PLoS ONE* 7:e1566
65. **Ochsenreiter T, Hajduk SL. 2006. Alternative editing of cytochrome *c* oxidase III mRNA in trypanosome mitochondria generates protein diversity. *EMBO Rep.* 7:1128–33**
66. Ochsenreiter T, Hajduk SL. 2008. The function of RNA editing in trypanosomes. In *RNA Editing, Nucleic Acids and Molecular Biology*, ed. HU Göringer, 20:181–97. Heidelberg, Ger.: Springer
67. Panigrahi AK, Ernst NL, Domingo GJ, Fleck M, Salavati R, Stuart KD. 2006. Compositionally and functionally distinct editosomes in *Trypanosoma brucei*. *RNA* 12:1038–49
68. Panigrahi AK, Schnaufer A, Carmean N, Igo RP Jr, Gygi SP, et al. 2001. Four related proteins of the *Trypanosoma brucei* RNA editing complex. *Mol. Cell. Biol.* 21:6833–40
69. Panigrahi AK, Schnaufer A, Stuart KD. 2007. Isolation and compositional analysis of trypanosomatid editosomes. *Methods Enzymol.* 424:3–24
70. Pelletier M, Miller MM, Read LK. 2000. RNA-binding properties of the mitochondrial Y-box protein RBP16. *Nucleic Acids Res.* 28:1266–75
71. Pelletier M, Read LK. 2003. RBP16 is a multifunctional gene regulatory protein involved in editing and stabilization of specific mitochondrial mRNAs in *Trypanosoma brucei*. *RNA* 9:457–68
72. Peris M, Simpson AM, Grunstein J, Liliental JE, Frech GC, Simpson L. 1997. Native gel analysis of ribonucleoprotein complexes from a *Leishmania tarentolae* mitochondrial extract. *Mol. Biochem. Parasitol.* 85:9–24
73. **Pollard VW, Harris ME, Hajduk SL. 1992. Native mRNA editing complexes from *Trypanosoma brucei* mitochondria. *EMBO J.* 11:4429–38**
74. Reifur L, Koslowsky DJ. 2008. *Trypanosoma brucei* ATPase subunit 6 mRNA bound to gA6-14 forms a conserved three-helical structure. *RNA* 14:2195–211
75. Rigaut G, Shevchenko A, Rutz B, Wilm M, Mann M, Séraphin B. 1999. A generic protein purification method for protein complex characterization and proteome exploration. *Nat. Biotechnol.* 17:1030–32
76. Rogers K, Gao G, Simpson L. 2007. Uridylate-specific 3′–5′-exoribonucleases involved in uridylate-deletion RNA editing in trypanosomatid mitochondria. *J. Biol. Chem.* 282:29073–80

65. Provides experimental evidence for alternative RNA editing.

73. Identifies high-molecular-mass RNA editing complexes as the catalytic machinery.

77. Reviews the biochemical enrichment and protein inventory of active RNA editing complexes.

77. Rusché LN, Cruz-Reyes J, Piller KJ, Sollner-Webb B. 1997. Purification of a functional enzymatic editing complex from *Trypanosoma brucei* mitochondria. *EMBO J.* 16:4069–81
78. Rusché LN, Huang CE, Piller KJ, Hemann M, Wirtz E, Sollner-Webb B. 2001. The two RNA ligases of the *Trypanosoma brucei* RNA editing complex: cloning the essential band IV gene and identifying the band V gene. *Mol. Cell. Biol.* 21:979–89
79. Sabatini R, Hajduk SL. 1995. RNA ligase and its involvement in guide RNA/mRNA chimera formation. Evidence for a cleavage-ligation mechanism of *Trypanosoma brucei* mRNA editing. *J. Biol. Chem.* 270:7233–40
80. Salavati R, Ernst NL, O'Rear J, Gilliam T, Tarun S Jr, Stuart K. 2006. KREPA4, an RNA binding protein essential for editosome integrity and survival of *Trypanosoma brucei*. *RNA* 12:819–31
81. Sbicego S, Alfonzo JD, Estevez AM, Rubio MA, Kang X, et al. 2003. RBP38, a novel RNA-binding protein from trypanosomatid mitochondria, modulates RNA stability. *Eukaryot. Cell* 2:560–68
82. Schmid B, Read LK, Stuart K, Göringer HU. 1996. Experimental verification of the secondary structures of guide RNA-pre-mRNA chimaeric molecules in *Trypanosoma brucei*. *Eur. J. Biochem.* 240:721–31
83. Schmid B, Riley GR, Stuart K, Göringer HU. 1995. The secondary structure of guide RNA molecules from *Trypanosoma brucei*. *Nucleic Acids Res.* 23:3093–102
84. Schnaufer A, Ernst NL, Palazzo SS, O'Rear J, Salavati R, Stuart K. 2003. Separate insertion and deletion subcomplexes of the *Trypanosoma brucei* RNA editing complex. *Mol. Cell* 12:307–19
85. Schnaufer A, Wu M, Park YJ, Nakai T, Deng J, et al. 2010. A protein-protein interaction map of trypanosome ~20S editosomes. *J. Biol. Chem.* 285:5282–95
86. Schumacher MA, Karamooz E, Zikova A, Trantirek L, Lukeš J. 2006. Crystal structures of *T. brucei* MRP1/MRP2 guide-RNA binding complex reveal RNA matchmaking mechanism. *Cell* 126:701–11
87. Seiwert SD, Heidmann S, Stuart K. 1996. Direct visualization of uridylate deletion in vitro suggests a mechanism for kinetoplastid RNA editing. *Cell* 84:831–41

88. Provides first experimental in vitro system to study RNA editing.

88. Seiwert SD, Stuart K. 1994. RNA editing: transfer of genetic information from gRNA to precursor mRNA in vitro. *Science* 266:114–17
89. Simpson L, Aphasizhev R, Lukeš J, Cruz-Reyes J. 2010. Guide to the nomenclature of kinetoplastid RNA editing: a proposal. *Protist* 161:2–6
90. Stuart KD, Schnaufer A, Ernst NL, Panigrahi AK. 2005. Complex management: RNA editing in trypanosomes. *Trends Biochem. Sci.* 30:97–105
91. Trotter JR, Ernst NL, Carnes J, Panicucci B, Stuart K. 2005. A deletion site editing endonuclease in *Trypanosoma brucei*. *Mol. Cell* 20:403–12
92. Vanhamme L, Perez-Morga D, Marchal C, Speijer D, Lambert L, et al. 1998. *Trypanosoma brucei* TBRGG1, a mitochondrial oligo(U)-binding protein that co-localizes with an in vitro RNA editing activity. *J. Biol. Chem.* 273:21825–33
93. Wang B, Ernst NL, Palazzo SS, Panigrahi AK, Salavati R, Stuart K. 2003. TbMP44 is essential for RNA editing and structural integrity of the editosome in *Trypanosoma brucei*. *Eukaryot. Cell* 2:578–87
94. Yu LE, Koslowsky DJ. 2006. Interactions of mRNAs and gRNAs involved in trypanosome mitochondrial RNA editing: structure probing of a gRNA bound to its cognate mRNA. *RNA* 12:1050–60

RELATED RESOURCES

Jensen RE, Englund, PT. 2012. Network news: the replication of kinetoplast DNA. *Annu. Rev. Microbiol.* 66:473–91

Katz LA. 2012. Origin and diversification of eukaryotes. *Annu. Rev. Microbiol.* 66:411–27

Physiology and Diversity of Ammonia-Oxidizing Archaea

David A. Stahl[1] and José R. de la Torre[2]

[1]Department of Civil and Environmental Engineering and Department of Microbiology, University of Washington, Seattle, Washington 98195-2700; email: dastahl@u.washington.edu

[2]Department of Biology, San Francisco State University, San Francisco, California 94132-1722; email: jdelator@sfsu.edu

Keywords

Thaumarchaeota, ammonia oxidation, nitrification, nitrogen cycle

Abstract

The discovery of ammonia-oxidizing archaea (AOA), now generally recognized to exert primary control over ammonia oxidation in terrestrial, marine, and geothermal habitats, necessitates a reassessment of the nitrogen cycle. In particular, the unusually high affinity of marine and terrestrial AOA for ammonia indicates that this group may determine the oxidation state of nitrogen available to associated micro- and macrobiota, altering our current understanding of trophic interactions. Initial comparative genomics and physiological studies have revealed a novel, and as yet unresolved, primarily copper-based pathway for ammonia oxidation and respiration distinct from that of known ammonia-oxidizing bacteria and possibly relevant to the production of atmospherically active nitrogen oxides. Comparative studies also provide compelling evidence that the lineage of *Archaea* with which the AOA affiliate is sufficiently divergent to justify the creation of a novel phylum, the *Thaumarchaeota*.

Contents

INTRODUCTION	84
A BRIEF HISTORY OF DISCOVERY	85
PHYLOGENETIC DIVERSITY	86
THAUMARCHAEOTA: A NEW DIVISION WITHIN THE *ARCHAEA*	87
HABITAT RANGE	87
PHYSIOLOGICAL PROPERTIES	88
Kinetics and Stoichiometry of Ammonia Oxidation	88
Evidence for Autotrophic, Mixotrophic, and Heterotrophic Growth	90
COMPARATIVE GENOMICS POINTS TO UNUSUAL BIOCHEMISTRY AND CELL BIOLOGY	90
The Biochemistry of Archaeal Ammonia Oxidation	91
Cell Cycle and the Machinery of Cell Division	93
POTENTIAL EFFECTS ON ATMOSPHERIC CHEMISTRY	95
FUTURE RESEARCH	96

INTRODUCTION

Our understanding of the microbiological underpinnings of the global nitrogen cycle was significantly altered in both 1999 and 2005. The year 1999 marked the first formal description of an organism mediating anaerobic ammonia oxidation (72), and in 2005 the first description of an ammonia-oxidizing archaeon (AOA) was published (36). Anaerobic decomposition of ammonia to dinitrogen gas via the anammox reaction is now recognized to be a major, if not dominant, process in many natural systems such as the Black Sea and certain marine oxygen-minimum zones (38, 39). The more recent revision in our understanding of the nitrogen cycle followed the isolation of the first AOA, *Nitrosopumilus maritimus* [*nitrosus* (Latin): nitrous; *pumilus* (Latin): dwarf; *maritimus* (Latin): of the sea], a small marine organism closely related to an abundant population of planktonic marine archaea first identified in the 1990s by Furhman et al. (21) and DeLong (17) by molecular typing of 16S rRNA genes.

However, the biogeochemical significance of marine archaea, comprising 20–40% of marine bacterioplankton (34), had remained mysterious until data from marine metagenome sequencing reported an archaeal open-reading frame coding for a protein distantly related to monooxygenases of ammonia-oxidizing bacteria (AOB) and methanotrophs (77). In these bacteria the monooxygenase activates either ammonia or methane for further metabolic transformation by inserting one atom from molecular oxygen into the substrate, yielding hydroxylamine or methanol. A homologous gene was subsequently described in an archaeal fosmid clone derived from a soil metagenome study (75). And, more recently, a related but functionally distinct monooxygenase, functioning in the hydroxylation of butane, was described for a bacterial butane oxidizer (64). The presumptive annotation of the archaeal homologs served for speculation that some marine archaea may function as ammonia oxidizers (77).

Interestingly, the alternative possibility that these divergent monooxygenases from marine and soil archaea served an alternative catabolic function was surprisingly not formally considered. The bacterial particulate methane monooxygenase (pMMO) and bacterial ammonia monooxygenase (AMO) share an amino acid similarity of approximately 74% (64). In contrast, the putative archaeal AMO was only distantly related to the bacterial pMMO and AMO. Thus, a functional assignment

AOA: ammonia-oxidizing archaea

Monooxygenase: an enzyme that incorporates one atom of molecular oxygen into its substrate as a hydroxyl group

AOB: ammonia-oxidizing bacteria

MMO: methane monooxygenase

AMO: ammonia monooxygenase

to ammonia oxidation was based on only ~40% amino acid similarity. For perspective, the BmoA subunit of the recently described putative butane monooxygenase from a gram-positive bacterium (*Nocardioides* sp. strain CF8) is more closely related to proteobacterial PmoA and AmoA sequences (37–38% average amino acid identity) than to the putative archaeal AmoA sequence (20% average amino acid identity) (64). However, given the importance of ammonia oxidation in contributing to both positive and negative environmental impacts, and the possibility that a major player in the nitrogen cycle had been overlooked since the pioneering work of Sergei Winogradsky in the late nineteenth century (81), the metagenome annotation did engender some attention. Nonetheless, it was only the isolation of a representative microorganism that conclusively demonstrated the physiological capacity of ammonia oxidation and that fully caught the attention of oceanographers, limnologists, and soil scientists (36). Numerous studies and reviews published in the past five years now provide persuasive data in support of the general dominance of archaea in controlling the fate of ammonia in both soil and aquatic systems (42, 57, 76, 84).

AmoA, AmoB, AmoC: three protein subunits composing the ammonia monooxygenase

A BRIEF HISTORY OF DISCOVERY

The paths to discovery are rarely direct, and the backstories generally not published nor attributable to an individual effort. The discovery of AOA is no exception, and we relate events that led to the isolation of *N. maritimus*. Our initial indication of a novel group of ammonia oxidizers came from our studies in the late 1990s as part of a census of bacterial diversity in the nitrifying reactor systems used to treat water in the large saltwater aquaria at the Shedd Aquarium (Chicago, Illinois). PCR amplification of bacterial 16S rRNA genes from DNA recovered from the saltwater aquaria failed to yield sequences closely related to known bacterial ammonia oxidizers (J.L. Flax & D.A. Stahl, unpublished observations). Because PCR amplification has its associated biases, the results were put aside to be revisited.

A few years later, similar observations were made in a study of the microbiology of nitrogen processing in Plum Island Sound (Massachusetts) estuary sediments (36). Again, general and specific primers for bacterial ammonia oxidizers failed to amplify the expected sequence types from nitrifying enrichments developed in collaboration with investigators at the Woods Hole Oceanographic Institution. However, by expanding the PCR-based screen to encompass archaeal 16S rRNA gene sequences from a clade of marine archaea then known as marine Group 1 crenarchaeota, related sequences were found to be abundant in the estuary sediment enrichments and in the nitrifying filtration systems at the Shedd Aquarium (36). Those results stimulated a follow-up study of nitrifying filtration systems at the Seattle Aquarium, which also revealed high representation of the same sequence type (36). Thus, an archaeal population closely related to the abundant Group 1 crenarchaeota was clearly implicated in ammonia oxidation. Subsequent development of enrichment cultures using the saltwater aquaria material as inoculant, and a medium containing a much lower concentration of ammonia than typically used to enrich bacterial ammonia oxidizers, yielded an active ammonia-oxidizing culture highly enriched in an archaeal population affiliated with the Group 1 crenarchaeota. Following another year or so of painstaking end-point dilutions, combined with selection by size and antibiotic resistance, the first pure culture of an AOA, *N. maritimus* strain SCM1, was described (36).

In a research environment where metagenomics and extensive environmental gene sequence surveys are increasingly common, this story offers a counterpoint. An organism relevant to a major environmental process was tracked down using culture-independent methods to first associate the process (ammonia oxidation) with a phylotype (Group 1 crenarchaeota) as a prelude to investing the significant effort necessary for isolation in culture. Since the isolation of *N. maritimus* in 2005,

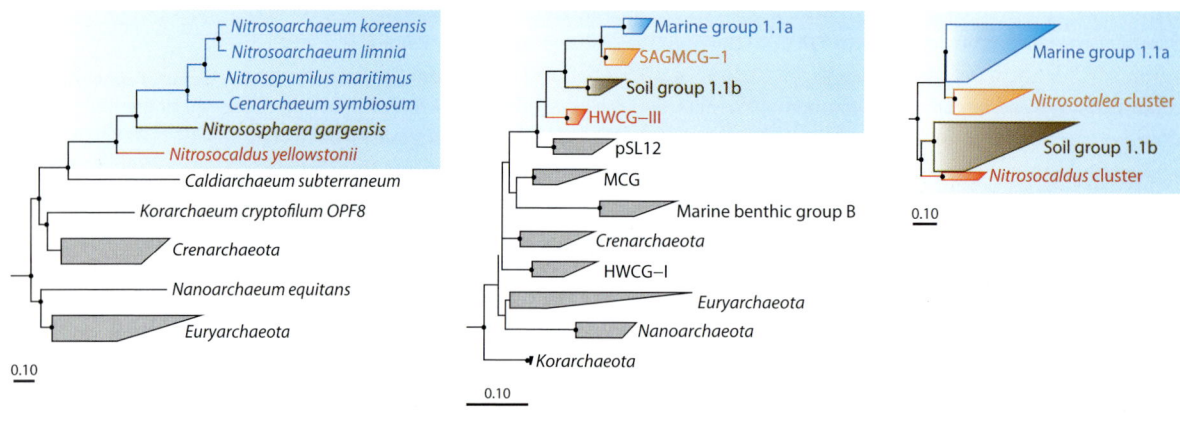

Figure 1

Phylogeny of ammonia-oxidizing archaea. (*a*) Maximum-likelihood analysis of 53 concatenated conserved ribosomal proteins constructed using 6,138 amino acid positions. Eukaryotic sequences were used as an outgroup. The tree was constructed using PhyML with an LG model with gamma correction (four site categories and using estimated proportion of invariable sites). (*b*) Maximum-likelihood tree of archaeal 16S ribosomal rRNA sequences, calculated using 594 positions and a Hasegawa-Kishino-Yano model with gamma correction (four site categories and using estimated proportion of invariable sites). Eukaryotic sequences were used as an outgroup. (*c*) Maximum-likelihood tree of archaeal *amoA* gene sequences, calculated using 1,265 positions and a Hasegawa-Kishino-Yano model with gamma correction (four site categories and using estimated proportion of invariable sites). Bacterial *amoA* and *pmoA* sequences were used as an outgroup. In all trees, nodes supported by bootstrap values greater than 80% (100 replicates) are indicated with a block dot. Blue background boxes indicate lineages belonging to the *Thaumarchaeota*. Abbreviations: SAGMCG-1, South African gold mine crenarchaeotic group 1; HWCG-I, hot water crenarchaeotic group I; HWCG-III, hot water crenarchaeotic group III; MCG, miscellaneous crenarchaeotic group.

the same general approach to enrichment and isolation has been used successfully to culture new AOA from soils and geothermal environments (5, 16, 24, 30, 41, 74).

PHYLOGENETIC DIVERSITY

Three major lineages of marine archaea were identified in the late 1990s, using 16S rRNA gene sequences as a proxy for resolving genetically distinct populations (phylotypes) (17, 21, 48). One lineage (Group 1) was proposed to be affiliated with the kingdom *Crenarchaeota* and the remaining two groups affiliated with the kingdom *Euryarchaeota*. Subsequent extensive 16S rRNA gene surveys of both soil and aquatic systems revealed that Group 1 archaea were widely distributed in moderate habitats. They were abundant throughout the marine water column (34), representing the dominant deep marine microbial population, and common in estuaries (3, 4, 14), sediments (26, 44, 65, 79), and soils (31). As additional sequences were compiled, additional divisions related to the Group 1 crenarchaeota were defined (**Figure 1**). The majority of soil and marine sequences could be assigned to two clades, defined as Group 1.1a (marine) and Group 1.1b (soil). Because no organism was available in culture, insights into the biogeochemical significance of these globally distributed and abundant microorganisms were initially inferred from in situ measurements of natural isotopic composition of diagnostic ether lipids (27, 55) and from incorporation of radiolabeled organic and inorganic substrates (25, 53, 82). These analyses suggested a significant inorganic carbon source of cellular carbon by the deep-water marine Group 1 population but also established a capacity to assimilate organic material. The physiological basis for these assimilation properties remained unknown until the isolation of the AOA and the analysis of their genomes (23, 80).

THAUMARCHAEOTA: A NEW DIVISION WITHIN THE ARCHAEA

Since the description of *N. maritimus*, the phylogenetic diversity of AOA has been under constant revision as new organisms are described. The current census reveals an assemblage that spans tremendous depth within the *Archaea*, affiliated with marine and soil groups initially assigned to the *Crenarchaeota* on the basis of comparative 16S rRNA sequence comparisons (**Figure 1**), and more divergent clades derived from geothermal environments. Phylogenetic inference based on larger sets of phylogenetically informative genes (including concatenated ribosomal protein sequence alignments) now made available by complete genome sequences has provided support for the creation of a new division within the *Archaea*, the *Thaumarchaeota*, with which all characterized AOA affiliate (7, 71).

The current understanding of phylogenetic and physiological diversity within the *Thaumarchaeota* Group 1 marine and soil archaea is based primarily on 16S rRNA sequence diversity. The few cultures of AOA can be used to physiologically anchor some of the clades initially inferred from environmentally derived 16S rRNA sequences. However, because all described members of this new kingdom have been brought into culture owing to an ability to grow via ammonia oxidation, we are left with the unusual impression that ammonia oxidation is the defining characteristic of this newly defined division, as is methanogenesis a defining characteristic of the *Euryarchaeota*. The possibility that a deeply diverging group within the *Archaea* split from the main line following development of the capacity to grow by extracting energy and electrons from ammonia would raise fundamental questions concerning both the origins of the *Archaea* and the age of the contemporary nitrogen cycle. Deeply diverging clades within the *Archaea* are dominated by anaerobic physiology, and early development of respiratory systems using oxygen as a terminal electron acceptor would seem to require an early nonphototrophic source of O_2 (11).

> *Thaumarchaeota*: a recently defined division (phylum) within the *Archaea* encompassing all known ammonia-oxidizing archaea, including organisms previously classified as Group 1 crenarchaeota
>
> **Crenarchaeol:** a glycerol dialkyl glycerol tetraether containing four cyclopentane moieties plus an additional cyclohexane moiety, associated specifically with *Thaumarchaeota*

HABITAT RANGE

The other remarkable picture emerging from the coalescence of culture-independent and culture-dependent techniques is the extraordinary range of habitats occupied by the AOA, far exceeding that of cultured bacterial ammonia oxidizers. Abundance and diversity patterns have been inferred primarily from sequencing and quantification of the gene coding for the putative archaeal *amoA* (2, 20), a measure generally well correlated with the abundance of a unique glycerol dialkyl glycerol tetraether, crenarchaeol, first associated with marine Group 1 archaea and now recognized to be made by both mesophilic and thermophilic AOA (15, 16, 42, 54). AOA appear to be the dominant archaeal clade in soils (generally comprising 1–5% of all prokaryotes) (40, 51), are a dominant marine group (comprising 20–40% of all marine bacterioplankton) (12, 34), and appear to be a major ammonia-oxidizing population in geothermal habitats (16, 19, 60, 83). Thus, their habitat range far exceeds that of their known bacterial counterparts.

Those AOA now available in culture grow at temperatures as high as 74°C (*Nitrosocaldus yellowstonii*) (16) and at pH values as low as 4 (*Nitrosotalea devanaterra*) (41). The existence of thermophilic representatives of AOA was initially indicated by the recovery of the diagnostic glycerol dialkyl glycerol tetraether lipid (crenarchaeol) previously associated only with the marine Group 1.1a and by the amplification of the archaeal homolog of *amoA* from hot springs (54). The moderate thermophile *Nitrososphaera gargensis* [*nitrosus* (Latin): nitrous; *sphaera* (Latin): spherical;

gargensis (Latin): from Garga spring], enriched from a Siberian hot spring, has an optimum growth temperature of about 46°C (24).

Continued surveys of geothermal habitats based on both nitrification rate measurements and selective amplification of the archaeal *amoA* and gene transcripts suggest their distribution includes hot springs with temperatures as high as 97°C and pH values as low as 2.5, but with greater diversity and abundance trending to springs with temperatures below 75°C (29, 60, 85). This broad habitat range is also reflected by the tremendous phylogenetic diversity of major archaeal lineages defined by 16S rRNA sequence types recognized to have ammonia-oxidizing affiliates and the corresponding diversity of putative archaeal *amoA* sequences (**Figure 1**).

> **Half-saturation constant (K_m or K_s):** the concentration of a limiting nutrient supporting one-half the maximum growth rate of an organism

PHYSIOLOGICAL PROPERTIES

Kinetics and Stoichiometry of Ammonia Oxidation

Initial physiological studies of *N. maritimus* strain SCM1, the only available marine isolate, have provided important insights into the basis for the ecological success of AOA. Studies of reaction stoichiometry using microrespirometry to measure oxygen and ammonia consumption relative to nitrite production showed that the overall stoichiometry of ammonia oxidation by *N. maritimus* is indistinguishable from that of AOB (46):

$$1\,NH_3 + 1.5\,O_2 \Rightarrow 1\,NO_2^- + H_2O + H^+.$$

Studies of ammonia oxidation kinetics subsequently demonstrated that this marine isolate is an extreme oligotroph, having an apparent half-saturation constant (K_m) for total ammonia (ammonium plus ammonia) of 132 nM (~3 nM NH$_3$ at near-neutral pH) (46). More than 50% of maximum activity could be elicited by the addition of 200 nM ammonium to resting cells. This is equivalent to the addition of one teaspoon of concentrated ammonium hydroxide to an Olympic size swimming pool. Furthermore, *N. maritimus* cells did not tolerate ammonia at concentrations significantly above 1 mM. Similar observations have been made with the moderately thermophilic AOA *N. gargensis* (24). In contrast, characterized AOB have K_m values that are more than 200-fold higher than the K_m value of *N. maritimus* (46). The extremely low apparent K_m and high maximum activity of *N. maritimus* contribute to a specific affinity for ammonia ($V_{max} \cdot K_m^{-1}$) of approximately 69,000 liter g [wet weight]$^{-1}$ h^{-1}. This is among the highest substrate affinity reported for any microorganism, not only for organic substrates but also for assimilation of ammonia by bacterial heterotrophs and marine phytoplankton for cellular growth (10, 46). As extreme oligophiles, the marine populations are plausibly responsible for maintaining oceanic concentrations of ammonia in the low nanomolar range, effectively outcompeting the bacterial ammonia oxidizers in accessing ammonia (46). In contrast to the very low ammonium concentrations required for growth, *N. maritimus* has an affinity for oxygen that is more typical of aerobic microorganisms (~4 μM) and is unable to grow anaerobically under culture conditions so far evaluated (D.A. Stahl & J.R. de la Torre, unpublished observations). Because AOA have been implicated in providing the nitrite used by anaerobic ammonia oxidizers in oxygen-minimum zones, such organisms may have significantly higher oxygen affinities than *N. maritimus*.

The soil AOA have been only more recently brought into culture and there is less direct physiological data available. *Nitrososphaera viennensis* [*nitrosus* (Latin): nitrous (nitrite producer); *sphaera* (Latin): spherically shaped; *viennensis* (Latin): from Vienna], isolated from garden soil in Vienna, Austria, tolerates higher concentrations of ammonia than does *N. maritimus*, demonstrating complete conversion of ammonia at initial ammonia concentrations as high as 3 mM (74). Growth can be initiated at higher concentrations (greater than 10 mM ammonia) but stopped when nitrite originating from ammonia oxidation exceeds ~3 mM (74). Nitrite alone is not associated with growth inhibition, as growth can be initiated in fresh culture medium supplemented with 10-mM nitrite. Thus, an unknown metabolite or intermediate was invoked as inhibitory (74). These culture-based observations are generally consistent with in situ studies showing that AOA populations grow over a wide range of ammonia concentrations in contrast to AOB that require significantly higher ammonia concentrations to initiate growth (57, 78, 84). A more recently described AOA enriched from low pH soils, *Nitrosotalea devanaterra* [*nitrosus* (Latin): nitrous (nitrite producer); *talea* (Latin): slender rod; *devana* (Latin): Aberdeen; *terra* (Latin): soil], has an optimum pH between 4 and 5 (41). This marked the first description of an acidophilic ammonia oxidizer and provided a clear microbiological explanation for significant nitrification rates measured in acidic soils, even though no such capacity was known for AOB (6). The ability of *N. devanaterra* to grow at extremely low pH values is also suggestive that ammonium, rather than un-ionized ammonia, is the substrate for growth. One explanation for the failure of AOB to grow at pH values significantly below pH 7 is their requirement for the un-ionized form as growth substrate (1, 8). Because the concentration of the un-ionized ammonia form decreases 10-fold for every 1-unit reduction in pH, the AOB would become rapidly substrate limited as pH is lowered. For example, in a system containing a total concentration of ammonia/ammonium of 100 mg total N liter^{-1} (28°C), there is 7 mg NH_3-N liter^{-1} at pH 8 and only 0.0007 mg NH_3-N liter^{-1} at pH 4. A preference, or requirement, for ammonium may also be true of the marine populations, because at a half-saturation value of 100 nM for *N. maritimus*, the concentration of un-ionized ammonia near pH 7 is only approximately 1–3 nM. We therefore suspect that the AOA use a mechanism for the collection of ammonia for oxidation (catabolism) entirely different from that used for assimilation. If not so, a high-affinity catabolic pathway would impoverish an anabolic pathway used for cellular synthesis. These data point to a novel physiology and supporting biochemistry.

In soils, studies linking nitrification rates to the nitrifier abundance and activity (using molecular proxies such as *amoA* gene and transcript abundances) have suggested that high affinity for ammonia also appears to provide soil AOA with a competitive advantage over AOB. In general, AOB tend to dominate in systems receiving high direct additions of inorganic ammonia (28, 57), whereas systems sustained by mineralization (ammonification) of organic material select for AOA (18, 52). Thus, with representatives that function as extreme oligotrophs, these chemoautotrophic organisms presumably function as a key valve in nitrification by controlling the rate of ammonia oxidation, generally considered to be the rate-limiting step in nitrification. This has significant implications for the global nitrogen cycle and for trophic interactions in both terrestrial and marine environments. Nitrification serves a key function in the nitrogen cycle by providing the oxidized species of nitrogen (nitrite and nitrate) that are essential substrates for both denitrification and anaerobic ammonia oxidation, activities that serve to return fixed nitrogen to the relatively inert atmospheric form of molecular nitrogen. The recognition of archaeal groups adapted to extremely low ammonia concentrations, acidic pH, and high temperatures now points to a fully functional nitrogen cycle in a much greater range of habitats than previously recognized. In turn, if AOA effectively compete with phytoplankton, heterotrophs, and other autotrophs in the ocean, or with plants and other microorganisms in soils, this may mean that assimilation of ammonia is a minor pathway for the metabolism of ammonia released through mineralization of organic material. Thus, the AOA may be ammonia thieves and conceivably force other groups to invest reducing power in the reduction of nitrate/nitrite to ammonia for biosynthesis. Resolution of the major operative pathway, assimilation versus oxidation of ammonia released through mineralization, is then of great significance to developing a better understanding of microbial controls of nitrogen form and availability to associated micro- and macrobiota in both marine and terrestrial systems.

Evidence for Autotrophic, Mixotrophic, and Heterotrophic Growth

Available data suggest that the AOA are generally capable of fixing CO_2. In addition to the clear demonstration of autotrophic growth by *N. maritimus*, CO_2 fixation was previously inferred from environmental studies of the isotopic composition of signature lipids and microautoradiographic imaging of natural samples incubated with labeled organic or inorganic carbon (37, 53, 55). These data also suggested some ability for the incorporation of organic carbon (27). The latter is consistent with genome sequence annotation of *N. maritimus*, pointing to the presence of a number of transporters for organic molecules. Recent studies in our laboratories have demonstrated some stimulation of *N. maritimus* growth by the addition of central metabolism intermediates (e.g., pyruvate and α-ketoglutarate) (47). A larger set of test substrates (including amino acids and fatty acids) did not stimulate growth. *N. viennensis* is also capable of chemolithoautotrophic growth. Although growth was obligately coupled to ammonia oxidation, it could be significantly enhanced by including pyruvate in the growth medium (74).

In addition to a capacity for mixotrophic growth by the marine AOA, a recent publication from the Wagner and Head laboratories (50) suggests that some *Thaumarchaeota*, although expressing an *amoA* gene closely related to that of the Group 1.1b soil clade, lack the capacity to either fix CO_2 or oxidize ammonia. This study examined a population of presumptive AOA in a petroleum refinery wastewater treatment system. Initially identified by combined 16S rRNA gene and *amoA* sequence characterization, stimulation of the ammonia oxidizers in this reactor by ammonia addition failed to stimulate incorporation of radiolabeled CO_2 by the AOA, even though coresident AOB did incorporate label. Modeling studies suggested that the AOB enumerated in the same treatment system could account for all ammonia oxidation. Thus, although active transcription of the archaeal *amoA* was confirmed, there was no evidence of a contribution to ammonia oxidation. These authors suggested that, as has been recently reported for an organism harboring a butane monooxygenase closely related to bacterial AMO and pMMO (64), the AMO-like monooxygenase of this archaeal population functions in the biodegradation of organic material, a well-characterized function of hydroxylases of petroleum-degrading bacteria. These data further emphasize the difficulty in assigning function solely on the basis of homology, even when the gene codes for a closely related and functionally well-characterized enzyme. More generally, this raises the question of whether the abundance of *amoA* genes or gene transcripts in marine or terrestrial systems can or even should be directly associated with nitrification.

COMPARATIVE GENOMICS POINTS TO UNUSUAL BIOCHEMISTRY AND CELL BIOLOGY

The completion of the genome sequence (1.64 Mbp circular chromosome) of the first isolated AOA, *N. maritimus*, revealed three major deviations from the canonical bacterial system of ammonia oxidation and carbon fixation: (*a*) a role for copper (rather than iron) as the major redox active metal in electron transfer reactions, (*b*) the absence of any homolog to the bacterial oxidoreductase (hydroxylamine oxidoreductase, HAO) responsible for the oxidation of hydroxylamine to nitrite, and (*c*) a variant of the 3-hydroxypropionate/4-hydroxybutyrate cycle for CO_2 fixation (as opposed to fixation by the ribulose bisphosphate carboxylase/oxygenase of the Calvin-Bassham-Benson cycle employed by characterized bacterial ammonia oxidizers) (80). Thus, the current picture of carbon metabolism suggests that these representatives of the marine and soil AOA derive energy and electrons primarily from the oxidation of ammonia but can supplement carbon from CO_2 fixation using a limited set of simple compounds that feed directly into central metabolism. The combined properties of having selective organic material uptake, CO_2 fixation by the 3-hydroxypropionate/ 4-hydroxybutyrate pathway, and the presence of a branched (incomplete) tricarboxylic acid (TCA) cycle now point to a relatively simple metabolic architecture of carbon assimilation and anabolism. Other distinctly novel features indicated by the genome sequences included an unusual system of cell division and the capacity to synthesize novel phosphonate compounds (80).

Phosphonate: organic compound containing a direct C-P bond

The Biochemistry of Archaeal Ammonia Oxidation

The most glaring problem presented by annotation of available genome sequences of AOA is the absence of the canonical bacterial pathway for ammonia oxidation (**Figure 2**). The available sequence information indicates that the AOA are missing all elements of the bacterial pathway other than genes coding for the presumptive AMO. *Nitrosopumilus* lacks a homolog of the bacterial HAO and the capacity for synthesis of c-type cytochromes (23, 80). Bacterial c-type cytochromes compose the redox-active centers of the HAO and mediate respiratory transfer of electrons from HAO to the terminal oxidase (**Figure 2**). These distinctive features had been previously suggested from the annotation of a metagenomic sequence assembled from a Group 1 symbiont of a marine sponge, *Cenarchaeum symbiosum* (23). Although the physiology of that uncultured archaeon is unknown, the general genome features are similar to that of *N. maritimus* and suggestive of a capacity for ammonia oxidation that may function in detoxification of sponge nitrogenous waste.

Nitroxyl (HNO): the one electron reduced and protonated congener of nitric oxide is a highly reactive nitrogen species implicated in mammalian cell signaling and as a possible intermediate in archaeal oxidation of ammonia

As yet there is no evidence that the product of ammonia oxidation by the archaeal AMO is hydroxylamine. An alternative pathway proposed by Klotz, Arp, and colleagues (80) suggested that nitroxyl (HNO) could be the product of the archaeal AMO (66) (**Figure 2**). Some clarification of the archaeal pathway for ammonia oxidation is anticipated to derive from genome comparisons of evolutionarily divergent species. Assuming a novel core pathway for ammonia oxidation is conserved among AOA (such as a novel Cu-based HAO), such features should be conserved across all lineages. *Nitrosocaldus yellowstonii* is now the most divergent representative of the AOA, distantly related to the marine and soil types, and as such a good candidate for such comparisons. As shown in **Figure 3**, the genes coding for the presumptive AMO remain the most diagnostic feature of an ammonia oxidation pathway common to all available genome sequences. However, initial comparative annotation of the *Nitrosocaldus* and *Nitrosopumilus* genomes has not yet served to further constrain the biochemistry of archaeal ammonia oxidation. In addition, only two small plastocyanin-like proteins are shared by all AOA. These redox-active copper proteins may participate in electron transfer from the unknown product of ammonia oxidation (e.g., hydroxylamine or nitroxyl) to a membrane-bound electron transfer chain (**Figure 2**).

Although the archaeal pathway for ammonia oxidation has not been resolved by comparative genomics, recent studies using nitric oxide (NO) sensitive microelectrodes are suggestive that NO may function in the biochemistry (47). Measurable amounts of NO are produced during ammonia oxidation. This has led to speculation that NO may be an intermediate or function as a redox shuttle, for instance, delivering electrons to the AMO (**Figure 2**). In contrast, the AOB draw electrons required by the monooxygenase from the membrane-associated quinone pool. Either the formation of nitroxyl as the first product of ammonia oxidation or the use of NO as an electron redox shuttle for hydroxylamine generation would eliminate the need to draw electrons directly from the quinone pool, either by obviating a requirement for reductant through formation of nitroxyl or by drawing electrons from a lower potential donor in the reduction of nitrite to NO (**Figure 2**). Equations 1 and 2 show possible recycling of an NO redox shuttle. The associated thermodynamic calculations assume that electrons for nitrite reduction originate from a donor species with an electrical potential (230 mV) approximately that of a c_1-type cytochrome and in the known range of plastocyanins.

$$2NO_2^- + 2e^- + 4H^+ \to 2NO + 2H_2O \qquad 1.$$

$$2NO + O_2 + NH_3 + H_2O \to NH_2OH + 2NO_2^- + 2H^+ \qquad 2.$$

$$\text{Rx2 } \Delta G^{o'} = -103.4 \frac{KJ}{mole\ NH_3}$$

$$\text{Rx1 } \Delta G^{o'} = -12.5 \frac{KJ}{mole\ NO_2^-}.$$

Because the formation of hydroxylamine as the immediate product of the presumptive AMO has not been demonstrated yet, the archaeal pathway for ammonia oxidation must be considered unresolved at this time.

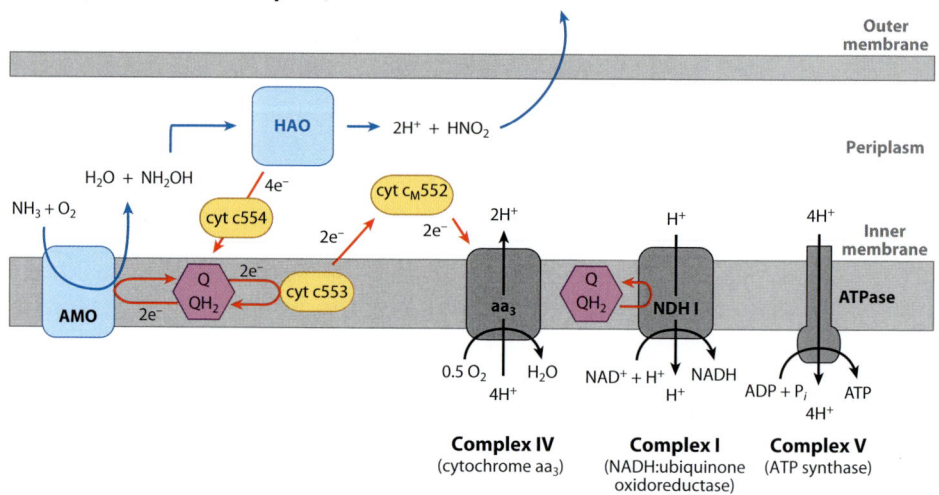

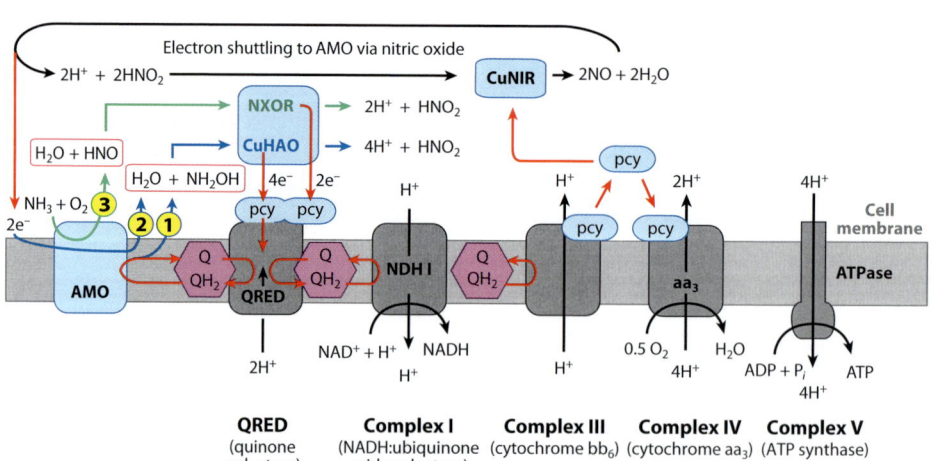

Cell Cycle and the Machinery of Cell Division

The AOA are distinctive in having genes coding for two alternative systems of cell division, the CdvABC-based and FtsZ-based division systems (43, 61). The recently described Cdv division system used by certain members of the *Crenarchaeota* is composed of three proteins, two of which (CdvB and CdvC) are homologs of the eukaryotic ESCRT-III–like sorting complex involved in vesicular sorting and cytokinesis (43, 62). The FtsZ-based division system, mediated by FtsZ protein filaments that form a constricting ring structure (45), is more widely distributed and is found in most major groups of bacteria and in the euryarchaeal and korarchaeotal branches of the *Archaea*. All the AOA genomes examined to date share the unusual characteristic of having genes diagnostic for both systems of cell division (5, 23, 35, 71, 80). Recent studies in collaboration with the Rolf Bernander group at the University of Uppsala used flow cytometry and immunofluorescence microscopy to examine the cell cycle and division system in *N. maritimus* (56). Fluorescence microscopy combined with cell staining using antibodies against the CdvA, CdvB, and CdvC proteins established that their expression was associated with cell division. Centrally positioned banding patterns of CdvA and CdvC were correlated with the presence of segregated nucleoids. Expression of two of three CdvB paralogs also correlated with segregated nucleoids, but neither formed distinct banding patterns. In contrast, the FtsZ protein was neither spatially nor temporally correlated with nucleoid segregation and strong FtsZ staining was observed in a majority of cells regardless of cell cycle state. Together, these results provided strong support for a Cdv system (ESCRT-III–like) in *N. maritimus*. As yet we can only speculate about possible function(s) of the FtsZ homolog. Hypothesized functions include a role in chromosome segregation or cell wall growth (9, 56).

Figure 2

Proposed respiratory pathways for ammonia oxidation in AOB and AOA. (*a*) Proposed pathway for ammonia oxidation in the AOB *Nitrosomonas europaea*. Ammonia is oxidized to NH_2OH by the membrane enzyme complex AMO. Subsequently, hydroxylamine is oxidized to nitrite in the periplasm by HAO. Four electrons from this oxidation are transferred to the quinone pool by cytochrome c554. Two electrons from the reduced quinone pool return to AMO and are required to initiate ammonia oxidation. The remaining two electrons enter the electron transport chain via cytochrome c553 and cytochrome $c_M 552$ to generate the proton motive force necessary for ATP synthesis. (*b*) Proposed pathway for ammonia oxidation in the AOA *Nitrosopumilus maritimus*. Three alternative pathways are indicated in this speculative diagram. In pathways 1 and 2, the immediate product of ammonia oxidation by the archaeal AMO would be hydroxylamine. However, these two pathways differ in the origin of electrons required to initiate ammonia oxidation by the monooxygenase. Pathway 1 is of the bacterial type, in which electrons produced by the oxidation of hydroxylamine to nitrite by a presumed CuHAO are transferred to pcy electron carriers and then to the quinone pool by a membrane-associated QRED. Two electrons would be recycled to AMO and the remaining two electrons would be transferred to the electron transport chain. Pathway 2 speculates that NO, produced by the reduction of nitrite by a proposed CuNIR, is the source of electrons for AMO. The possibility that HNO is the product of the archaeal AMO is shown by pathway 3. This pathway would eliminate the requirement for electron recycling during the initial oxidation of ammonia. Subsequently, HNO would be oxidized to nitrite by a presumed NXOR. The two electrons extracted during this oxidation would be transferred to QRED and the electron transport chain as indicated above. Red arrows indicate electron flow. Blue shading denotes copper-containing proteins. Hexagons containing Q and QH_2 represent the oxidized and reduced quinone pool, respectively. Abbreviations: AOA, ammonia-oxidizing archaea; AOB, ammonia-oxidizing bacteria; AMO, ammonia monooxygenase; HAO, hydroxylamine oxidoreductase; NO, nitric oxide; HNO, nitroxyl; CuHAO, copper hydroxylamine oxidoreductase; CuNIR, copper-dependent nitrite reductase; NXOR, putative nitroxyl oxidoreductase; pcy, plastocyanins; NDH, NAD(P)H:quinone oxidoreductase; NH_2OH, hydroxylamine; QRED, quinone reductase. Figure adapted with permission from Reference 80.

Group		*Nitrosopumilus maritimus*	*Cenarchaeum symbiosum*	*Nitrosoarchaeum limnia*	*Nitrosocaldus yellowstonii*	*Caldiarchaeum subterraneum*
AMO	amoA	●	●	●	●	
AMO	amoB	●	●	●	●	
AMO	amoC	●	●	●	●	
Complex I: NAD reductase	Chain N	●	●	●	●	●
Complex I: NAD reductase	Chain L	●		●	●	●
Complex I: NAD reductase	Chain M	●		●	●	●
Complex I: NAD reductase	Chain 4L	●	●	●	●	●
Complex I: NAD reductase	Chain 6	●	●	●	●	●
Complex I: NAD reductase	4 Fe-4S	●	●	●	●	●
Complex I: NAD reductase	NADH dehydrogenase	●	●	●	●	●
Complex I: NAD reductase	NADH dehydrogenase	●	●	●	●	●
Complex I: NAD reductase	NADH dehydrogenase	●	●	●	●	●
Complex I: NAD reductase	Reductase B unit	●	●	●	●	●
Complex I: NAD reductase	Chain 3	●	●	●	●	
Complex III	Rieske domain	●	●	●	●	●
Complex III	Cytb/b6 domain	●	●	●	●	●
Complex III	Blue copper	●	●	●	●	●
Complex IV (Cyt aa$_3$)	Blue copper	●	●	●	●	●
Complex IV (Cyt aa$_3$)	Heme-Cu	●	●		●	●
Complex IV (Cyt aa$_3$)	Heme-Cu	●	●	●	●	●
Complex IV (Cyt aa$_3$)	Hypothetical	●	●	●	●	●
Plastocyanins	Soluble	●			●	
Plastocyanins	Periplasmic	●	●	●	●	
Plastocyanins	Soluble	●				
Plastocyanins	Soluble	●				●
Plastocyanins	Soluble	●				
Plastocyanins	Soluble	●				
Plastocyanins	Periplasmic	●				●
Plastocyanins	Periplasmic	●				
Plastocyanins	Periplasmic	●				●
Plastocyanins	Cytoplasmic	●				●
Plastocyanins	Soluble	●		●		
Plastocyanins	Periplasmic	●				
MCO/NirK cluster	Repressor	●				●
MCO/NirK cluster	Transporter	●				●
MCO/NirK cluster	2d MCO	●				●
MCO/NirK cluster	Blue copper	●				●
MCO/NirK cluster	Oxidase	●				●
MCO/NirK cluster	3d MCO NirK	●		●		
MCO/NirK cluster	Regulator	●				●
Putative nitrogen oxide-processing cluster	MCO/blue copper fusion	●				●
Putative nitrogen oxide-processing cluster	Conserved hypothetical	●				
Putative nitrogen oxide-processing cluster	Nitroreductase	●				
Putative nitrogen oxide-processing cluster	Regulator	●				
Putative nitrogen oxide-processing cluster	Hypothetical	●				
Putative nitrogen oxide-processing cluster	Flavodoxin synthase	●				

Flow-cytometry-based analysis of the cell cycle also indicated a distinctive cell biology (56). The timing of the cell cycle of *N. maritimus* differs substantially from characterized hyperthermophilic crenarchaea, having a much longer prereplication phase (G_1) and a shorter postreplication phase (G_2, mitosis, and cell division). Replication of the small 1.64-Mbp genome required 15–18 h and tended to arrest if ammonia was depleted before replication was completed. We suspect slow replication and arrest may relate to adaptation to extreme nutrient limitation. Although ammonia is available only at generally low nanomolar concentrations in the open ocean, there is nonetheless a continuous supply of ammonia through mineralization of organic material. It is unlikely that oceanic populations of this organism ever experience complete ammonia depletion as occurs at the termination of growth in batch culture.

PEP: phosphoenolpyruvate

POTENTIAL EFFECTS ON ATMOSPHERIC CHEMISTRY

The ubiquitous and globally abundant AOA have recently been implicated as a direct or indirect source of the atmospherically reactive gasses methane (CH_4) and nitrous oxide (N_2O). The possibility that these organisms generate significant amounts of nitrous oxide was suspected by analogy with the activities of AOB known to produce N_2O in association with ammonia oxidation, or for some species to fully reduce nitrite to N_2O via a poorly characterized partial denitrification pathway (70). A recent study comparing the natural isotopic signature of bulk N_2O ($\delta^{15}N$ and $\delta^{18}O$ values) produced in cultures by AOA or AOB served to associate most oceanic production of N_2O with the AOA (63). Because the oceans are a source for as much as 30% of global N_2O inputs to the atmosphere, this observation provides additional impetus to resolve the biochemistry of archaeal ammonia oxidation.

An unsuspected association with atmospheric chemistry was the outcome of studies designed to identify phosphonate compounds predicted by the genome sequence to be synthesized by *N. maritimus*. Phosphonates are organic compounds containing a direct C-P bond and thus are distinct from the more common esterified form of phosphate. The family of biologically produced phosphonates includes antibiotics, modified extracellular polysaccharides, lipids, and phosphorus storage compounds (49). Although poorly characterized structurally, phosphonates comprise 20–30% of organic phosphorus in the oceans (13) and thus are an important source of phosphorus for organisms expressing the C-P lyase necessary to cleave the bond and release the phosphate (32, 58). A gene in the *N. maritimus* genome annotated as coding for phosphoenolpyruvate (PEP) mutase, the enzyme catalyzing the first step in a variety of pathways for phosphonate biosynthesis (conversion of PEP to phosphonopyruvate) (68, 69), was associated with a gene cluster implicated in production of both a novel phosphonate and extracellular polysaccharides. Collaborative studies between the Metcalf and van der Donk groups have since established the in vitro production of methylphosphonic acid by enzymes coded by genes in this cluster and confirmed the presence of methylphosphonate in whole-cell extracts by NMR (W.W. Metcalf, B.M. Griffin, R.M. Cicchillo,

Figure 3

Distribution of genes thought to play key roles in ammonia oxidation and energy generation in genome sequences of representative AOA and of the related archaeon "*Candidatus* Caldiarchaeum subterraneum" (of the proposed candidate phylum *Aigarchaeota*). Colored boxes indicate the presence of orthologs in each genome. Identification and clustering of orthologs were computed using OrthoMCL v1.4 and an E-value cutoff of 1E-10. Abbreviations: AOA, ammonia-oxidizing archaea; AMO, ammonia monooxygenase; MCO, multicopper oxidase; NAD+/NADH, nicotinamide adenine dinucleotide (oxidized and reduced forms, respectively).

J. Gao, S.C. Janga, H.A. Cooke, B.T. Circello, B.S. Evans, W. Martens-Habbena, D.A. Stahl & W.A. van der Donk, manuscript submitted).

Discovery of a biological source of methylphosphonate may provide a partial explanation for the long-standing ocean methane paradox (59, 67, 73). This paradox originates from the observation that the aerobic surface ocean is supersaturated in CH_4 with respect to the atmosphere, contributing as much 4% of the global methane budget (59). Because a contribution by the obligately anaerobic methanogens remains controversial, there has been no generally accepted explanation of origin. An intriguing hypothesis was put forward recently by Karl et al. (33), who suggested a new model in which methane would be produced when marine microorganisms use methylphosphonic acid as a source of phosphorus. The model was supported empirically by addition of commercially available methylphosphonate to seawater (33). The genes encoding C-P lyase are common among marine microorganisms, and those organisms were present in sufficient numbers to release methane from the added methylphosphonate and to use the released phosphate for cellular synthesis. The only significant difficulty with the model was the absence of a known biological source of methylphosphonate, which has now been provided. In addition to our discovery of the pathway in *N. maritimus*, a screen of available marine metagenome sequences for homologs of the *N. maritimus* diagnostic enzyme (methylphosphonate synthase) has shown the pathway to be present in other abundant marine clades, including representative *Prochlorococcus* and *Pelagibacter* species. Given the high abundance of these marine groups, they can produce the substantial amounts of methylphosphonate necessary to provide the missing link in the ocean methane paradox. However, apart from providing an explanation for an unknown methane source, there remains the equally significant question: What is the biological function of a phosphonate-modified outer cell wall?

FUTURE RESEARCH

These are exciting times in studies of the global nitrogen cycle. In a rapidly changing world that has more than doubled global inputs of fixed nitrogen into the biosphere since the pre-Industrial period (22), primarily through agricultural practice, it is imperative that the microbiological engines that drive the major biogeochemical cycles be more fully resolved. Ammonia-oxidizing microorganisms are essential to the proper functioning of the nitrogen cycle. As part of this critical nutrient cycle they produce the oxidized nitrogen species used by both anammox and denitrifying organisms to convert the generally biological available forms of nitrogen (ammonia, nitrate, nitrite) to the large and relatively inert atmospheric reservoir of molecular nitrogen. If the AOA do control rates of ammonia oxidation in most natural systems, as available data now indicate, it is necessary that their ecology, physiology, and underlying biochemistry are explored more fully. Such investigations are now being facilitated by the isolation of new AOA in culture and the power of comparative biology made possible by rapid genome sequencing. We suspect these studies will not only reveal novel biochemistry, but ultimately also offer a new understanding of the evolutionary origins of the *Archaea*.

SUMMARY POINTS

1. AOA are now thought to be the predominant ammonia-oxidizing population in most natural environments in which ammonia is present at very low concentrations. The naturally low concentration is attributable in part to the high affinity of the AOA for ammonia.

2. The biochemistry of archaeal ammonia oxidation is unique and as yet unresolved, sharing only genes distantly related to those coding for the AMO of characterized bacterial ammonia oxidizers.

3. Characterized AOA are chemolithoautotrophs; they use a variant of the 3-hydroxypropionate/4-hydroxybutyrate pathway for CO_2 fixation and are now believed to have only a limited capacity to assimilate different forms of fixed carbon for cellular synthesis.

4. Copper, as opposed to iron, appears to be the primary metal used in the respiratory redox chemistry of AOA.

5. NO, and possibly nitroxyl, may be important in the biochemistry of archaeal ammonia oxidation.

6. The AOA affiliate with a novel phylum, the *Thaumarchaeota*, recently recognized to constitute an early radiation within the *Archaea* on the basis of divergent features of genome sequence.

7. The AOA may contribute significantly to atmospherically active gases, including nitric and nitrous oxides and methane.

8. The habitat range of AOA, which includes hot springs and acidic soils, is significantly broader than that of characterized AOB.

DISCLOSURE STATEMENT

The authors are not aware of any affiliations, memberships, funding, or financial holdings that might be perceived as affecting the objectivity of this review.

ACKNOWLEDGMENTS

We are grateful to grants from NSF Oceanography (OCE-0623174), Molecular and Cellular Biosciences (MCB-06044482 & MCB-0949807), and the Dimensions of Biodiversity program in Biological Oceanography (OCE-1046017) for partial support of the studies reported in this article.

LITERATURE CITED

1. Allison SM, Prosser JI. 1991. Urease activity in neutrophilic autotrophic ammonia-oxidizing bacteria isolated from acid soils. *Soil Biol. Biochem.* 23:45–51
2. Beman JM, Popp BN, Francis CA. 2008. Molecular and biogeochemical evidence for ammonia oxidation by marine Crenarchaeota in the Gulf of California. *ISME J.* 2:429–41
3. Bernhard AE, Bollmann A. 2010. Estuarine nitrifiers: new players, patterns and processes. *Estuar. Coast. Shelf Sci.* 88:1–11
4. Bernhard AE, Landry ZC, Blevins A, de la Torre JR, Giblin AE, Stahl DA. 2010. Abundance of ammonia-oxidizing archaea and bacteria along an estuarine salinity gradient in relation to potential nitrification rates. *Appl. Environ. Microbiol.* 76:1285–89
5. Blainey PC, Mosier AC, Potanina A, Francis CA, Quake SR. 2011. Genome of a low-salinity ammonia-oxidizing archaeon determined by single-cell and metagenomic analysis. *PLoS One* 6:e16626

6. Booth MS, Stark JM, Rastetter E. 2005. Controls on nitrogen cycling in terrestrial ecosystems: a synthetic analysis of literature data. *Ecol. Monogr.* 75:139–57

7. Brochier-Armanet C, Boussau B, Gribaldo S, Forterre P. 2008. Mesophilic Crenarchaeota: proposal for a third archaeal phylum, the Thaumarchaeota. *Nat. Rev. Microbiol.* 6:245–52 [7. Reviews evidence supporting the creation of a new archaeal phylum.]

8. Burton SAQ, Prosser JI. 2001. Autotrophic ammonia oxidation at low pH through urea hydrolysis. *Appl. Environ. Microbiol.* 67:2952–57

9. Busiek KK, Margolin W. 2011. Split decision: A thaumarchaeon encoding both FtsZ and Cdv cell division proteins chooses Cdv for cytokinesis. *Mol. Microbiol.* 82:535–38

10. Button DK, Robertson BR, Lepp PW, Schmidt TM. 1998. A small, dilute-cytoplasm, high-affinity, novel bacterium isolated by extinction culture and having kinetic constants compatible with growth at ambient concentrations of dissolved nutrients in seawater. *Appl. Environ. Microbiol.* 64:4467–76

11. Canfield DE. 2005. The early history of atmospheric oxygen: homage to Robert M. Garrels. *Annu. Rev. Earth Planet. Sci.* 33:1–36

12. Church MJ, DeLong EF, Ducklow HW, Karner MB, Preston CM, Karl DM. 2003. Abundance and distribution of planktonic Archaea and Bacteria in the waters west of the Antarctic Peninsula. *Limnol. Oceanogr.* 48:1893–902

13. Clark LL, Ingall ED, Benner R. 1999. Marine organic phosphorus cycling: novel insights from nuclear magnetic resonance. *Am. J. Sci.* 299:724–37

14. Crump BC, Baross JA. 2000. Archaeaplankton in the Columbia River, its estuary and the adjacent coastal ocean, USA. *FEMS Microbiol. Ecol.* 31:231–39

15. Damste JSS, Schouten S, Hopmans EC, van Duin ACT, Geenevasen JAJ. 2002. Crenarchaeol: the characteristic core glycerol dibiphytanyl glycerol tetraether membrane lipid of cosmopolitan pelagic Crenarchaeota. *J. Lipid Res.* 43:1641–51

16. de la Torre JR, Walker CB, Ingalls AE, Konneke M, Stahl DA. 2008. Cultivation of a thermophilic ammonia oxidizing archaeon synthesizing crenarchaeol. *Environ. Microbiol.* 10:810–18 [16. First cultivation of a thermophilic ammonia-oxidizing microorganism growing at temperatures greater than 70°C.]

17. DeLong EF. 1992. Archaea in coastal marine environments. *Proc. Natl. Acad. Sci. USA* 89:5685–89

18. Di HJ, Cameron KC, Shen JP, Winefield CS, O'Callaghan M, et al. 2010. Ammonia-oxidizing bacteria and archaea grow under contrasting soil nitrogen conditions. *FEMS Microbiol. Ecol.* 72:386–94

19. Dodsworth JA, Hungate BA, Hedlund BP. 2011. Ammonia oxidation, denitrification and dissimilatory nitrate reduction to ammonium in two US Great Basin hot springs with abundant ammonia-oxidizing archaea. *Environ. Microbiol.* 13:2371–86

20. Francis CA, Roberts KJ, Beman JM, Santoro AE, Oakley BB. 2005. Ubiquity and diversity of ammonia-oxidizing archaea in water columns and sediments of the ocean. *Proc. Natl. Acad. Sci. USA* 102:14683–88

21. Fuhrman JA, McCallum K, Davis AA. 1992. Novel major archaebacterial group from marine plankton. *Nature* 356:148–49 [21. First description of mesophilic archaea present in the marine water column.]

22. Galloway JN, Townsend AR, Erisman JW, Bekunda M, Cai ZC, et al. 2008. Transformation of the nitrogen cycle: recent trends, questions, and potential solutions. *Science* 320:889–92

23. Hallam SJ, Konstantinidis KT, Putnam N, Schleper C, Watanabe Y, et al. 2006. Genomic analysis of the uncultivated marine crenarchaeote *Cenarchaeum symbiosum*. *Proc. Natl. Acad. Sci. USA* 103:18296–301

24. Hatzenpichler R, Lebecleva EV, Spieck E, Stoecker K, Richter A, et al. 2008. A moderately thermophilic ammonia-oxidizing crenarchaeote from a hot spring. *Proc. Natl. Acad. Sci. USA* 105:2134–39 [24. Demonstration of CO_2 assimilation at low added ammonia concentration by an enriched culture of the soil group (Group 1.1b) of AOA.]

25. Herndl GJ, Reinthaler T, Teira E, van Aken H, Veth C, et al. 2005. Contribution of Archaea to total prokaryotic production in the deep Atlantic Ocean. *Appl. Environ. Microbiol.* 71:2303–9

26. Hershberger KL, Barns SM, Reysenbach AL, Dawson SC, Pace NR. 1996. Wide diversity of Crenarchaeota. *Nature* 384:420

27. Ingalls AE, Shah SR, Hansman RL, Aluwihare LI, Santos GM, et al. 2006. Quantifying archaeal community autotrophy in the mesopelagic ocean using natural radiocarbon. *Proc. Natl. Acad. Sci. USA* 103:6442–47

28. Jia ZJ, Conrad R. 2009. Bacteria rather than Archaea dominate microbial ammonia oxidation in an agricultural soil. *Environ. Microbiol.* 11:1658–71

29. Jiang HC, Huang QY, Dong HL, Wang P, Wang FP, et al. 2010. RNA-based investigation of ammonia-oxidizing archaea in hot springs of Yunnan Province, China. *Appl. Environ. Microbiol.* 76:4538–41

30. Jung MY, Park SJ, Min D, Kim JS, Rijpstra WIC, et al. 2011. Enrichment and characterization of an autotrophic ammonia-oxidizing archaeon of mesophilic crenarchaeal Group I.1a from an agricultural soil. *Appl. Environ. Microbiol.* 77:8635–47
31. Jurgens G, Lindstrom K, Saano A. 1997. Novel group within the kingdom Crenarchaeota from boreal forest soil. *Appl. Environ. Microbiol.* 63:803–5
32. Kamat SS, Williams HJ, Raushel FM. 2011. Intermediates in the transformation of phosphonates to phosphate by bacteria. *Nature* 480:570–73
33. Karl DM, Beversdorf L, Bjorkman KM, Church MJ, Martinez A, DeLong EF. 2008. Aerobic production of methane in the sea. *Nat. Geosci.* 1:473–78
34. **Karner MB, DeLong EF, Karl DM. 2001. Archaeal dominance in the mesopelagic zone of the Pacific Ocean. *Nature* 409:507–10**
35. Kim BK, Jung MY, Yu DS, Park SJ, Oh TK, et al. 2011. Genome sequence of an ammonia-oxidizing soil archaeon, "*Candidatus* Nitrosoarchaeum koreensis" MY1. *J. Bacteriol.* 193:5539–40
36. **Könneke M, Bernhard AE, de la Torre JR, Walker CB, Waterbury JB, Stahl DA. 2005. Isolation of an autotrophic ammonia-oxidizing marine archaeon. *Nature* 437:543–66**
37. Kuypers MMM, Blokker P, Erbacher J, Kinkel H, Pancost RD, et al. 2001. Massive expansion of marine archaea during a mid-Cretaceous oceanic anoxic event. *Science* 293:92–94
38. Kuypers MMM, Lavik G, Woebken D, Schmid M, Fuchs BM, et al. 2005. Massive nitrogen loss from the Benguela upwelling system through anaerobic ammonium oxidation. *Proc. Natl. Acad. Sci. USA* 102:6478–83
39. Kuypers MMM, Sliekers AO, Lavik G, Schmid M, Jorgensen BB, et al. 2003. Anaerobic ammonium oxidation by anammox bacteria in the Black Sea. *Nature* 422:608–11
40. Lehtovirta LE, Prosser JI, Nicol GW. 2009. Soil pH regulates the abundance and diversity of Group 1.1c Crenarchaeota. *FEMS Microbiol. Ecol.* 70:367–76
41. Lehtovirta-Morley LE, Stoecker K, Vilcinskas A, Prosser JI, Nicol GW. 2011. Cultivation of an obligate acidophilic ammonia oxidizer from a nitrifying acid soil. *Proc. Natl. Acad. Sci. USA* 108:15892–97
42. **Leininger S, Urich T, Schloter M, Schwark L, Qi J, et al. 2006. Archaea predominate among ammonia-oxidizing prokaryotes in soils. *Nature* 442:806–9**
43. Lindas A-C, Karlsson EA, Lindgren MT, Ettema TJG, Bernander R. 2008. A unique cell division machinery in the Archaea. *Proc. Natl. Acad. Sci. USA* 105:18942–46
44. MacGregor BJ, Moser DP, Baker BJ, Alm EW, Maurer M, et al. 2001. Seasonal and spatial variability in Lake Michigan sediment small-subunit rRNA concentrations. *Appl. Environ. Microbiol.* 67:3908–22
45. Margolin W. 2005. FtsZ and the division of prokaryotic cells and organelles. *Nat. Rev. Mol. Cell Biol.* 6:862–72
46. **Martens-Habbena W, Berube PM, Urakawa H, de la Torre JR, Stahl DA. 2009. Ammonia oxidation kinetics determine niche separation of nitrifying Archaea and Bacteria. *Nature* 461:976–81**
47. Martens-Habbena W, Urakawa H, Costa KC, Gee AM, Stahl DA. 2009. Autotrophy-mixotrophy-heterotrophy: clues about the physiology of mesophilic ammonia-oxidizing crenarchaeota. Presented at Am. Soc. Microbiol. Gen. Meet., Philadelphia, PA
48. Massana R, DeLong EF, Pedros-Alio C. 2000. A few cosmopolitan phylotypes dominate planktonic archaeal assemblages in widely different oceanic provinces. *Appl. Environ. Microbiol.* 66:1777–87
49. Metcalf WW, van der Donk WA. 2009. Biosynthesis of phosphonic and phosphinic acid natural products. *Annu. Rev. Biochem.* 78:65–94
50. Mussmann M, Brito I, Pitcher A, Damste JSS, Hatzenpichler R, et al. 2011. Thaumarchaeotes abundant in refinery nitrifying sludges express amoA but are not obligate autotrophic ammonia oxidizers. *Proc. Natl. Acad. Sci. USA* 108:16771–76
51. Ochsenreiter T, Selezi D, Quaiser A, Bonch-Osmolovskaya L, Schleper C. 2003. Diversity and abundance of Crenarchaeota in terrestrial habitats studied by 16S RNA surveys and real time PCR. *Environ. Microbiol.* 5:787–97
52. Offre P, Prosser JI, Nicol GW. 2009. Growth of ammonia-oxidizing archaea in soil microcosms is inhibited by acetylene. *FEMS Microbiol. Ecol.* 70:99–108
53. Ouverney CC, Fuhrman JA. 2000. Marine planktonic archaea take up amino acids. *Appl. Environ. Microbiol.* 66:4829–33

34. Demonstrates the numerical significance of marine Group 1.1a archaea, showing increasing relative abundance with increasing depth in the water column.

36. Documents a capacity for chemolithoautotrophy among members of the ubiquitous group of marine archaea.

42. Demonstrates the numerical predominance of AOA in soils, generally present in numbers greatly exceeding AOB.

46. Suggests that an exceptionally high affinity for ammonia accounts for AOA predominance relative to AOB in ammonia-depleted environments.

54. Pearson A, Huang Z, Ingalls AE, Romanek CS, Wiegel J, et al. 2004. Nonmarine crenarchaeol in Nevada hot springs. *Appl. Environ. Microbiol.* 70:5229–37
55. Pearson A, McNichol AP, Benitez-Nelson BC, Hayes JM, Eglinton TI. 2001. Origins of lipid biomarkers in Santa Monica Basin surface sediment: a case study using compound-specific δC^{14} analysis. *Geochim. Cosmochim. Acta* 65:3123–37
56. Pelve EA, Lindas AC, Martens-Habbena W, de la Torre JR, Stahl DA, Bernander R. 2011. Cdv-based cell division and cell cycle organization in the thaumarchaeon *Nitrosopumilus maritimus*. *Mol. Microbiol.* 82:555–66
57. Pratscher J, Dumont MG, Conrad R. 2011. Ammonia oxidation coupled to CO_2 fixation by archaea and bacteria in an agricultural soil. *Proc. Natl. Acad. Sci. USA* 108:4170–75
58. Quinn JP, Kulakova AN, Cooley NA, McGrath JW. 2007. New ways to break an old bond: the bacterial carbon-phosphorus hydrolases and their role in biogeochemical phosphorus cycling. *Environ. Microbiol.* 9:2392–400
59. Reeburgh WS. 2007. Oceanic methane biogeochemistry. *Chem. Rev.* 107:486–513
60. Reigstad LJ, Richter A, Daims H, Urich T, Schwark L, Schleper C. 2008. Nitrification in terrestrial hot springs of Iceland and Kamchatka. *FEMS Microbiol. Ecol.* 64:167–74
61. Samson RY, Bell SD. 2011. Cell cycles and cell division in the archaea. *Curr. Opin. Biotechnol.* 14:1–7
62. Samson RY, Obita T, Freund SM, Williams RL, Bell SD. 2008. A role for the ESCRT system in cell division in archaea. *Science* 322:1710–13
63. Santoro AE, Buchwald C, McIlvin MR, Casciotti KL. 2011. Isotopic signature of N_2O produced by marine ammonia-oxidizing archaea. *Science* 333:1282–85
64. Sayavedra-Soto LA, Hamamura N, Liu CW, Kimbrel JA, Chang JH, Arp DJ. 2011. The membrane-associated monooxygenase in the butane-oxidizing gram-positive bacterium *Nocardioides* sp. strain CF8 is a novel member of the AMO/PMO family. *Environ. Microbiol. Rep.* 3:390–96
65. Schleper C, Holben W, Klenk HP. 1997. Recovery of crenarchaeotal ribosomal DNA sequences from freshwater-lake sediments. *Appl. Environ. Microbiol.* 63:321–23
66. Schleper C, Nicol GW. 2010. Ammonia-oxidising archaea: physiology, ecology and evolution. *Adv. Microb. Physiol.* 57:1–41
67. Scranton MI, Brewer PG. 1977. Occurrence of methane in near-surface waters of western subtropical North-Atlantic. *Deep-Sea Res.* 24:127–38
68. Seidel HM, Freeman S, Seto H, Knowles JR. 1988. Phosphonate biosynthesis: isolation of the enzyme responsible for the formation of a carbon phosphorus bond. *Nature* 335:457–58
69. Shao ZY, Blodgett JAV, Circello BT, Eliot AC, Woodyer R, et al. 2008. Biosynthesis of 2-hydroxyethylphosphonate, an unexpected intermediate common to multiple phosphonate biosynthetic pathways. *J. Biol. Chem.* 283:23161–68
70. Shaw LJ, Nicol GW, Smith Z, Fear J, Prosser JI, Baggs EM. 2006. *Nitrosospira* spp. can produce nitrous oxide via a nitrifier denitrification pathway. *Environ. Microbiol.* 8:214–22
71. Spang A, Hatzenpichler R, Brochier-Armanet C, Rattei T, Tischler P, et al. 2010. Distinct gene set in two different lineages of ammonia-oxidizing archaea supports the phylum Thaumarchaeota. *Trends Microbiol.* 18:331–40
72. Strous M, Fuerst JA, Kramer EHM, Logemann S, Muyzer G, et al. 1999. Missing lithotroph identified as new planctomycete. *Nature* 400:446–49
73. Tilbrook BD, Karl DM. 1995. Methane sources, distributions and sinks from California coastal waters to the oligotrophic North Pacific gyre. *Mar. Chem.* 49:51–64
74. **Tourna M, Stieglmeier M, Spang A, Konneke M, Schintlmeister A, et al. 2011. Nitrososphaera viennensis, an ammonia oxidizing archaeon from soil. *Proc. Natl. Acad. Sci. USA* 108:8420–25**

    74. Isolation of the first AOA from soil.
75. Treusch AH, Leininger S, Kletzin A, Schuster SC, Klenk HP, Schleper C. 2005. Novel genes for nitrite reductase and Amo-related proteins indicate a role of uncultivated mesophilic Crenarchaeota in nitrogen cycling. *Environ. Microbiol.* 7:1985–95
76. Urakawa H, Martens-Habbena W, Stahl DA. 2011. Physiology and genomics of ammonia-oxidizing archaea. In *Nitrification*, ed. BB Ward, MG Klotz, DJ Arp, pp. 117–55. Washington, DC: ASM Press
77. Venter JC, Remington K, Heidelberg JF, Halpern AL, Rusch D, et al. 2004. Environmental genome shotgun sequencing of the Sargasso Sea. *Science* 304:66–74

78. Verhamme DT, Prosser JI, Nicol GW. 2011. Ammonia concentration determines differential growth of ammonia-oxidising archaea and bacteria in soil microcosms. *ISME J.* 5:1067–71
79. Vetriani C, Reysenbach AL, Dore J. 1998. Recovery and phylogenetic analysis of archaeal rRNA sequences from continental shelf sediments. *FEMS Microbiol. Ecol.* 161:83–88
80. Walker CB, de la Torre JR, Klotz MG, Urakawa H, Pinel N, et al. 2010. *Nitrosopumilus maritimus* genome reveals unique mechanisms for nitrification and autotrophy in globally distributed marine crenarchaea. *Proc. Natl. Acad. Sci. USA* 107:8818–23
81. Winogradsky S. 1890. Recherches sur les organismes de la nitrification. *Ann. Inst. Pasteur* 4:257–75
82. Wuchter C, Schouten S, Boschker HT, Sinninghe Damste JS. 2003. Bicarbonate uptake by marine Crenarchaeota. *FEMS Microbiol. Lett.* 219:203–7
83. Zhang CL, Ye Q, Huang ZY, Li WJ, Chen JQ, et al. 2008. Global occurrence of archaeal *amoA* genes in terrestrial hot springs. *Appl. Environ. Microbiol.* 74:6417–26
84. **Zhang LM, Offre PR, He JZ, Verhamme DT, Nicol GW, Prosser JI. 2010. Autotrophic ammonia oxidation by soil thaumarchaea.** ***Proc. Natl. Acad. Sci. USA*** **107:17240–45**
85. Zhao WD, Song ZQ, Jiang HC, Li WJ, Mou XZ, et al. 2011. Ammonia-oxidizing Archaea in Kamchatka Hot Springs. *Geomicrobiology J.* 28:149–59

84. Demonstrates autotrophic growth and functional significance of soil AOA growing under conditions of ammonia release through mineralization of soil organic matter.

Bacterial Persistence and Toxin-Antitoxin Loci

Kenn Gerdes and Etienne Maisonneuve

Center for Bacterial Cell Biology, Institute for Cell and Molecular Biosciences, Newcastle University, Newcastle NE2 4HH, United Kingdom; email: Kenn.Gerdes@newcastle.ac.uk

Keywords

HipA, HipB, RelE, RelB, Lon, polyphosphate, (p)ppGpp

Abstract

Bacterial persistence is caused by the presence of rare, slowly growing bacteria among populations of rapidly growing cells. The slowly growing bacteria are tolerant of antibiotics and other environmental insults, whereas their isogenic, rapidly growing siblings are sensitive. Recent research has shown that persistence of the model organism *Escherichia coli* depends on toxin-antitoxin (TA) loci. Deletion of type II TA loci reduces the level of persistence significantly. Lon protease but no other known ATP-dependent proteases is required for persistence. Polyphosphate and (p)ppGpp also are required for persistence. These observations led to the proposal of a simple and testable model that explains the persistence of *E. coli*. It is now important to challenge this model and to test whether the persistence of pathogenic bacteria also depends on TA loci.

Contents

INTRODUCTION	104
BACTERIAL PERSISTENCE	105
Persistence and Stochasticity	105
Discovery of Persistence Genes	107
PROKARYOTIC TOXIN-ANTITOXIN LOCI	108
hipBA Locus of *E. coli*	110
relBE Gene Family	111
mazEF Gene Family	112
vapBC Gene Family	112
TOXIN-ANTITOXIN LOCI AND PERSISTENCE OF *ESCHERICHIA COLI*	114
A Testable Model Explaining Induction and Maintenance of Persistence	114
Importance and Consequence of the Model	116
Resuscitation from Persistence	117
PHYLOGENY OF TOXIN-ANTITOXIN LOCI	117
CONCLUDING REMARKS	118

INTRODUCTION

The discovery of penicillin and other efficient antibiotics and their introduction as magic bullets against infectious diseases in the middle of the twentieth century heralded the antibiotic era. It was believed that many serious infectious diseases would be readily cured or even eliminated, and this belief was so prevalent that, for several decades, efforts to develop new antibiotics received relatively modest attention. However, it soon became apparent that antibiotics often fail. The most obvious reason is that bacteria rapidly develop drug resistance, and it is now known that bacteria develop resistance toward most if not all antibiotics that are used clinically.

Although antibiotic resistance is a major culprit, antibiotics fail for less obvious reasons. For example, bacteria form cellular capsules and multicellular biofilms that may shield them from antibiotics. Moreover, uneven distribution of antibiotics in target organisms can reduce the beneficial effects of antibiotics. However, there are even more sophisticated mechanisms by which bacteria manage to evade killing by antibiotics. This is because clonal populations of bacteria that are sensitive to a given antibiotic almost always contain slowly growing or nongrowing cells that are tolerant of the drug (63). Penicillin causes cell lysis by corrupting cell wall synthesis and therefore kills growing bacteria efficiently (27). However, the killing efficacy of penicillin depends on the physiological state of the bacterium. Broadly speaking, slowly growing bacteria are killed less efficiently than rapidly growing bacteria, and nongrowing bacteria are usually not affected by penicillin (101). Although there are rare exceptions, similar effects have been observed with most other antibiotics. Thus, bacteria that are nongrowing owing to exhaustion of their habitat are in a state of drug indifference at the population level (75). This observation is consistent with the finding that pathogenic bacteria are usually highly sensitive to antibiotics during the initial stages of animal model infections. Experiments with such models have shown that as an infection develops, the antibiotic treatment becomes less and less efficient. One important contributing factor to the developing recalcitrance is the accumulation of slowly growing bacteria (30, 101).

However, even among populations of rapidly growing bacteria there are rare cells that are insensitive to antibiotics. This phenomenon, called bacterial persistence, was discovered by Joseph

Persistence: a phenomenon that causes bacterial cells to tolerate multiple antibiotics and other environmental insults

Bigger (9), who explored how cells of *Staphylococcus aureus* were killed by penicillin. Bigger found that penicillin often failed to sterilize cultures of exponentially growing cells. Bigger coined the surviving bacteria persisters and, as described below, correctly assumed that persisters evaded killing by the antibiotic because they were in a slowly growing or dormant state, that is, they were in a state of multidrug tolerance (MDT). Almost all bacteria that have been tested exhibit the persistence phenotype (64). Here, the concept of drug indifference is different from drug tolerance: Drug indifference is caused by the cessation of bacterial growth due to exhaustion of the environment, whereas drug tolerance, as explained below, results from a phenotypic switch into persistence. Mutations that change a bacteriocidal drug into a bacteriostatic drug are also said to confer drug tolerance (63).

After Bigger's initial discovery, the concept of bacterial persistence was almost forgotten by the scientific community, in part because the novel magic bullets had become widely available. Moreover, in the eclipse period from Bigger first recognizing the phenomenon in 1944 to the beginning of the millennium, there was a lack of sophisticated and sensitive methods to efficiently study the phenomenon. Many pathogenic bacteria, such as *Mycobacterium tuberculosis*, *Streptococcus pyogenes*, *Pseudomonas aeruginosa*, and *Salmonella enterica*, generate persistent or recurrent infections that are impossible to cure even though the responsible bacteria isolated from patients are often sensitive to antibiotics. The general topic of persistence and bacterial MDT has been reviewed in detail elsewhere (5, 30, 63, 64, 99).

> **Multidrug tolerance (MDT):** a metastable, epigenetic or a stable, genetic state that renders cells tolerant to antibiotics and other environmental challenges
>
> **Persister cell:** a differentiated cell that results in persistence

BACTERIAL PERSISTENCE

As defined here, persisters are drug-tolerant subpopulations of cells that are present in populations of exponentially growing bacteria. A typical persistence experiment is shown in **Figure 1**. At time zero, a bacteriocidal antibiotic is added to an exponentially growing, balanced bacterial culture and the killing kinetics is determined by colony counts on solid medium. The response is almost always log-linear and biphasic, with an initial rapid killing followed by a phase with much slower inactivation. The two phases, both exhibiting first-order inactivation kinetics, uncover the presence of two distinct cell populations with different but uniform drug sensitivity (64). The initial rapid killing kinetics shows that the bulk of the population is sensitive to the drug, and the much slower inactivation in the second phase reflects the presence of a subpopulation that initially survived the drug treatment. When this experiment is repeated with cells derived from the surviving colonies of an inactivation experiment, a similar response is obtained (50). This simple but crucial experiment shows that the descendants of persister cells are as drug sensitive as the ancestors of the persisters. Therefore, the persistence phenotype occurs independently of stable genetic changes, that is, persistence is a noninherited, epigenetic trait.

Persistence and Stochasticity

It is essential to understand whether the persistent state is induced by the antibiotic itself or whether it is generated by drug-independent, stochastic mechanisms. The persistence phenomenon has been analyzed at the single-cell level by fluorescence microscopy. The problem was attacked by observing fluorescent *Escherichia coli* cells growing in a microfluidic chamber (7). These investigations suggested that persister cells were present in the culture before the addition of the antibiotic, and indicated that these persisters were generated by antibiotic-independent stochastic switching from a rapidly growing, sensitive state to a slowly growing, insensitive state. Thus, according to this interpretation, the colonies that were counted as persisters in **Figure 1** originated from cells that, owing to their persistent state, survived the antibiotic treatment and later switched back to

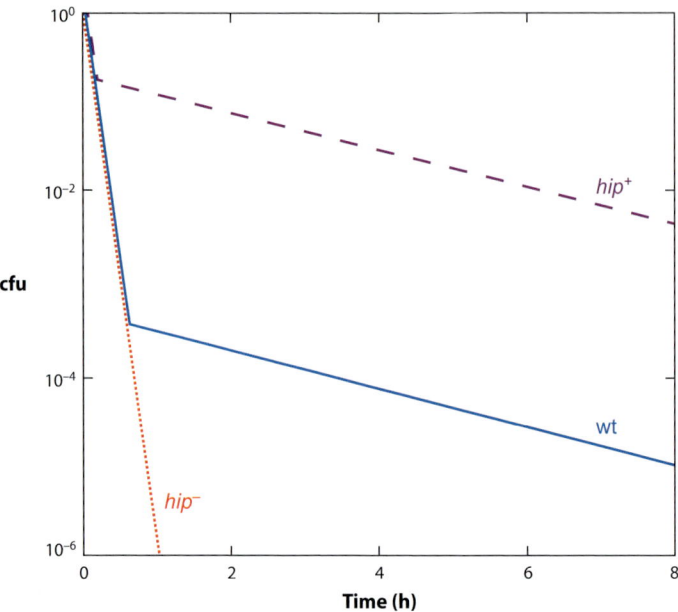

Figure 1
Schematic presentation of a typical persistence experiment. The experiment reveals a characteristic biphasic killing curve. Cells of *Escherichia coli* K-12 are grown exponentially and antibiotic is added at time zero. The killing curves for wild type (wt) (*solid line*) and a high persister (hip^+) mutant strain producing a highly elevated number of persisters ($\approx$1,000-fold; *dashed line*) are shown. The killing curve for a hypothetical strain lacking persister genes (hip^-) is also shown (*dotted line*). Initially, the killing curves for the three strains have identical slopes. The biphasic (or sometimes multiphasic) killing curves in the case of wt and hip^+ strains reveal two (or more) subpopulations of bacterial cells with increased drug tolerance. The slope of the first phase reveals the susceptibility of the bulk of the populations; the slope of the second phase reveals the resuscitation rate of persisters. Abbreviation: cfu, colony forming units.

rapid, detectable growth. At the population level, the rate of switching back to the nonpersistent, growing phenotype can be determined from the slope of the second logarithmic phase in **Figure 1**. However, the mechanisms that control phenotypic switching between the persistent and nonpersistent states, as well as those required to keep the cells in the persistent state, are unknown. The stochastic switching observed by Balaban et al. (7) was obtained with a high persistence mutant of *E. coli* (*hipA7*; see below), and the reproduction of similar studies with wild-type bacteria is highly warranted.

Balaban and colleagues (5, 7) suggest that persisters are generated mainly by conditions that lead to cessation of growth, such as amino acid starvation, stationary phase, and biofilm formation. In stationary phase, the bacterial population develops drug indifference because it exhausts the environmental resources and therefore enters a slowly growing or dormant state. In contrast, persisters generated by exponentially growing cells may have stochastically switched into slow growth. Thus, the physiological states of the two types of slowly growing cells are probably different. However, Balaban and colleagues suggest that cessation of growth leads to persister cells because there are cells that restart growth very slowly after conditions reverse to again support growth. If stationary phase is the main source of the switch to persistence, then exponential growth for many generations should dilute the persisters to low or even undetectable levels. However, maintaining cells of *E. coli* K-12 strain MG1655 in exponential growth using repeated dilutions in fresh media resulted in only a modest decrease of persisters (two- to fourfold) compared with the level observed with a short-term exponentially growing population (E. Maisonneuve & K. Gerdes, unpublished data). The validity of this observation is supported by long-term growth experiments in chemostats (97). Still, it is possible that delayed resuscitation of persisters formed in stationary phase contributes significantly to the persistence level that is actively measured (68).

A large body of literature describes the stress responses of bacteria in stationary phase and during nutrient limitations, whereas persistence by phenotypic switching is a relatively unexplored field. Here, we focus on the description of bacterial persister cells generated by exponentially growing cells. Exponentially growing cells are clinically relevant, in particular because bacteria are growing exponentially during the initial, antibiotic-susceptible phases of infections (63, 72, 73, 75). Our definition of persistence is narrower than the general phenomenon of MDT that is caused by any environmental condition that generates slow bacterial growth. However, the definition is useful because it renders the phenomenon readily approachable by experimentation and does not preclude that our discussion may apply to more realistic environmental or clinical scenarios.

Antitoxin: protein or RNA encoded by a TA locus that inhibits the activity or expression of a cognate toxin

Toxin: protein encoded by TA loci that inhibits global cellular translation either by mRNA cleavage or by inhibiting EF-Tu

TA: toxin-antitoxin

Discovery of Persistence Genes

It has been proposed that bacterial persistence reflects a fortuitous deterioration toward cell death (81). Another view conceives that persistence is a programmed, epigenetic phenomenon with a genetic basis that has evolved to allow prokaryotic organisms to survive changing environments (59). Experimental support for the latter view is steadily accumulating. The first evidence that bacterial persistence has a genetic basis came from the isolation of mutations in *E. coli* that increased the levels of persisters (**Figure 1**). After chemical mutagenesis, exponentially growing cultures of *E. coli* K-12 were exposed to ampicillin until lysis was complete; the drug was then removed and cell growth allowed to resume. Repetition of this procedure resulted in the isolation of so-called high persistence (*hip*) mutants (77). The *hip* mutants did not significantly change the minimal inhibitory concentration of ampicillin and had no growth defects. However, they exhibited a vastly elevated survival rate (1,000- to 10,000-fold) after treatment with ampicillin (**Figure 1**). Others have reported a less dramatic 10- to 100-fold increase in persistence by the *hip* mutations (55). However, these authors found that the persistence level increased ≈20-fold when cell growth approached stationary phase. Interestingly, the *hip* mutants also exhibited a highly elevated survival rate after treatment with other inhibitors of cell wall synthesis, such as phosphomycin, cycloserine with norfloxacin (an inhibitor of DNA gyrase), and after thymine starvation (77, 91, 107).

Most of the *hip* mutations were mapped to the *hipA* locus (77). HipA is encoded by the type II *hipBA* toxin-antitoxin (TA) locus. Interestingly, direct connections between the persistence phenomenon and TA loci were indicated by several other studies. Two studies showed that persister cells contained elevated levels of TA mRNAs (51, 94), and two other studies showed that deletion of TA loci reduced persistence in biofilms and during induction of the SOS response (31, 45). Moreover, we showed that ectopic overproduction of toxins encoded by TA loci induces a persistence-like state from which the cells could be resuscitated by the induction of antitoxin gene transcription (85). Thus, TA loci are excellent candidates for genes generally involved in persistence. In this connection it should be mentioned that the term toxin is semantically slightly misleading because although the inhibitors encoded by type II TA loci are efficient repressors of cell growth, they do not kill the bacteria. Even ectopic overproduction of type II toxins does not kill the cells, and in our view, it is therefore unlikely that the biological function of the toxins is generally related to cell killing. This view is supported by the fact that overproduction of type II toxins enhances rather than reduces persistence (51, 52, 69, 102).

PROKARYOTIC TOXIN-ANTITOXIN LOCI

Conditional cooperativity: the phenomenon of conditional repression or derepression of TA operon transcription by the antitoxin contingent on the [A]:[T] ratio

TA loci are highly abundant in free-living bacteria and archaea (47, 84). Usually, TA loci code for two components, a toxin that inhibits cell growth and an antitoxin that regulates toxin activity; TA modules consisting of three components are less common (18, 96) although several such loci have been discovered recently (43, 13). Three types of TA loci have been identified. Type I and type III TA loci encode small RNAs that counteract the toxins at the translational and posttranslational levels, respectively (12, 36). Toxins encoded by type II TA loci are counteracted by protein antitoxins (35). Owing to sequence conservation of the toxins, type II TA loci have been divided into families that are broadly conserved in bacteria, or in some cases both bacteria and archaea. Thus, members of the *relBE*, *vapBC*, and *hicAB* families are abundant in both domains, a modest number of *mazEF* and *phd/doc* homologs have also been identified in archaea, and $\epsilon\zeta$ loci are present in both gram-negative and gram-positive bacteria (47, 70, 79). In contrast, the *ccdAB* family is so far confined to *Gammaproteobacteria*. In the remainder of this review, TA loci refer to type II TA loci. A schematic of the general genetic organization and regulatory loops of bona fide TA loci is shown in **Figure 2a**. TA loci have the following features in common:

- The antitoxin regulates toxin activity by forming a tight complex with the toxin.
- The antitoxin usually contains a DNA-binding motif and autoregulates transcription of the TA operon.
- The antitoxin is degraded by cellular proteases. Thus, the interplay between the protease and the antitoxin determines simultaneously the activity of the corresponding toxin and the transcription rate of the TA operon.
- Usually, the toxin functions as a corepressor of TA operon transcription when in complex with the antitoxin.
- In all cases investigated so far, antitoxins are produced at a higher rate than toxins. This makes physiological sense because antitoxins are unstable and therefore must be generated at a higher rate to keep toxins inactive.
- In all cases investigated, transcription of TA operons is regulated by conditional cooperativity, a term that refers to the ability of the toxin to function both as a corepressor and a derepressor of TA operon transcription (1, 33, 83, 106). The switch between these two states of the toxin is controlled by the [A]:[T] ratio. This peculiar type of transcriptional control has evolved independently in different TA gene families.
- Most TA gene families have members on both chromosomes and mobile genetic elements. The striking similarities between these evolutionarily independent TA gene families raise the possibility that they have a common biological function.

Figure 2

Type II toxin-antitoxin (TA) loci of *Escherichia coli* K-12. (*a*) Diagram showing the genes and control loops of a prototypical type II TA locus. The rightward-pointing red arrow indicates the TA operon promoter. The promoter is repressed by the antitoxin and, in particular, by the TA complex that binds tightly to the promoter region during rapid growth when the concentration of free toxin is low. In contrast, promoter activity is derepressed by free toxin, a regulatory phenomenon called conditional cooperativity. (*b*) The chromosomal locations of 12 type II TA loci of *E. coli* K-12. RNase-encoding TA loci are shown in blue. *rnlAB* is a recently discovered TA locus encoding RNase L (54). Its role in persistence has not been investigated. It was suggested that *E. coli* K-12 has several additional type II TA loci but their role in persistence has also not been investigated (108). (*c*) The structure of the HipA-HipB-operator complex (93). Panel *c* was reproduced with permission from Maria A. Schumacher.

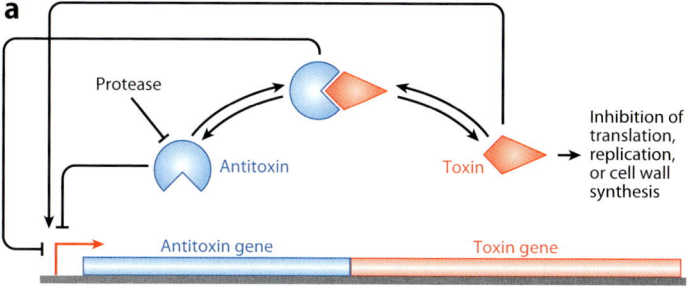
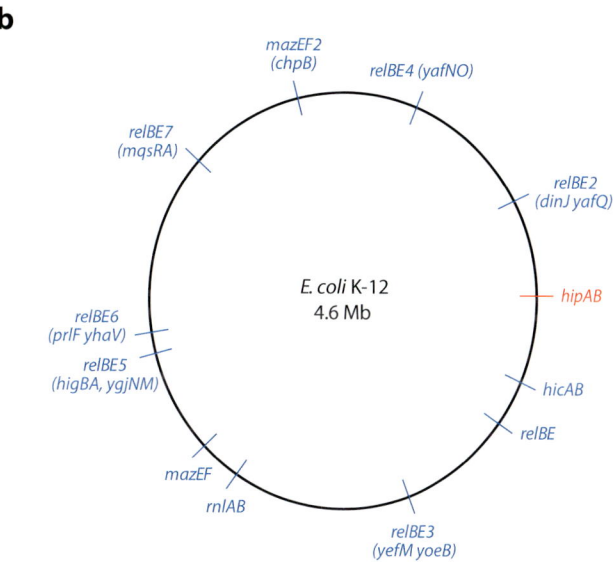
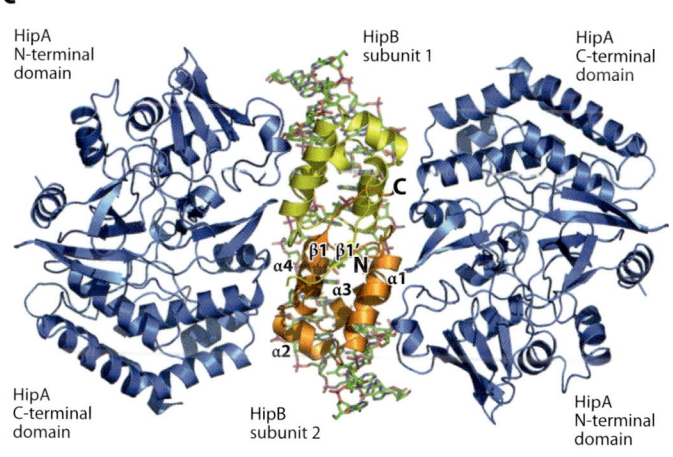

hipBA Locus of *E. coli*

Lon: a ubiquitous ATP-dependent protease

Most of the *hip* mutations described above were mapped to *hipA*, a gene encoding a 440-amino-acid protein, located at 34.3′ in the terminus region of the *E. coli* chromosome (11, 78) (**Figure 2b**). *hipA* is preceded by *hipB* (88 codons), which encodes an autorepressor of *hipBA* transcription (10, 11). HipB is a dimer that binds cooperatively to four operators in the *hipBA* promoter region via its helix-turn-helix domains and thereby represses transcription (10). Overproduction of HipA inhibits cell growth, and HipB interacts directly with HipA and inhibits HipA activity (10, 93). These observations led to the suggestion that *hipBA* constitutes a bona fide TA locus (32, 55). Mild overproduction of HipA at levels that did not impair cell growth resulted in an increased level of persisters (32), comparable to observations of ectopic expression of other toxins encoded by TA loci (69, 102). A quantitative, single-cell-based analysis showed that a threshold level of HipA was required to induce dormancy and that the amount by which this threshold was exceeded determined the duration of HipA-induced dormancy (89).

It is important to understand how mutations in *hipA* result in an elevated level of persister cells. HipA is a serine kinase that is partially phosphorylated in vivo and is autophosphorylated in vitro (28). The kinase activity of HipA is required for both the inhibition of cell growth and the stimulation of persister cell formation (28). On the basis of the crystal structure, Schumacher et al. (93) suggested that HipA defines a novel branch of eukaryote-like Ser/Thr kinases. In particular, the conserved and putative catalytic residue Asp^{309} was required for persistence, strongly suggesting that kinase function is essential to HipA-mediated drug tolerance. By inference, Schumacher et al. (93) suggested that HipA phosphorylates one or more cellular targets to inhibit cell growth and thereby induce persistence. The essential translation factor EF-Tu (elongation factor Tu) has been proposed as a putative HipA target: HipA interacts strongly in vitro with full-length EF-Tu and with a decapeptide encompassing Thr^{382} of EF-Tu. Moreover, HipA can phosphorylate EF-Tu in vitro (93). In *E. coli*, EF-Tu is phosphorylated in vivo at Thr^{382} at a low level ($\approx 5\%$) and phosphorylated EF-Tu is inactive in translation (3, 67). Thus, it is possible that HipA has the potential to inhibit translation by increasing the level of phosphorylated EF-Tu. However, it has yet to be shown that EF-Tu is the cellular target of HipA.

The binding of HipB to the operators in the *hipBA* promoter region is stimulated by HipA (11). The crystal structure of the HipA-HipB-operator complex shows that HipB dimers contact DNA in the major groove and induce a large 70° bend in operator DNA that may function to increase cooperativity of the binding (93). The *hipBA* promoter region contains an integration host factor–binding site that could further increase bending of the DNA (10). Two HipA molecules sandwich the HipB-DNA complex by contacting the sides of the HipB dimer, and noncontiguous regions of both HipA and HipB form the complex (93) (**Figure 2c**). The HipA N-terminal domain interacts with one HipB subunit, whereas the HipA C-terminal domain interacts primarily with the other HipB subunit. Interestingly, the interaction surfaces of HipB are located far from the HipA active site, and it has been suggested that HipB neutralizes HipA ATPase by locking it in an open, inactive configuration (93).

For HipA to become active, it must be released from HipB. Whether the activation occurs passively, that is, by the simple law-of-mass-action equilibrium, or whether an active mechanism is involved, is not known. Antitoxins of *E. coli*, such as RelB, YefM, and MazE, are all degraded by Lon (22, 23, 24), and all three antitoxins have intrinsically unstructured C-terminal domains that become structured upon binding to the cognate toxin (19, 48, 49, 60, 61, 65). Free HipB is also degraded by Lon (44). As seen in **Figure 2c**, the DNA-bound form of HipB forms a compact dimer in which the last 16 amino acid residues are disordered, raising the possibility that this disordered part of HipB functions as a recognition site for Lon (93). The unstructured C-terminal tail of HipB is also present in the HipBA complex, further raising the possibility that Lon can invade the complex and specifically degrade HipB and thereby activate HipA.

The structure of the HipA-HipB complex suggests an explanation for the mechanism of increased persistence of the *hipBA7* allele. HipA7, which contains two amino acid changes (G22S and D291A), mediates the high persistence phenotype in the presence of wild-type HipB (56). The D291A amino acid change alone was sufficient for the Hip$^+$ phenotype. Schumacher et al. (93) suggest that the phenotype could result from a weakened HipA-HipB interaction that would hyperactivate HipA kinase activity. In turn, a more active HipA would lead to a reduced growth rate of some cells, thus leading to an increased level of persistence. If this inference is correct, then mutations in *hipB* that reduce the interaction between HipB and HipA should also be able to confer the high persistence phenotype.

Database mining has revealed that *hipBA* loci are abundant in proteobacteria, *Firmicutes*, actinomycetes, and photosynthetic bacteria but are absent from archaea (K. Gerdes, unpublished data). As with most other bona fide TA gene families, *hipBA* loci are present on both plasmids and chromosomes.

relBE Gene Family

The *relBE* locus of *E. coli* encodes the antitoxin RelB and the toxin RelE. RelE belongs to a well-described toxin superfamily with many homologs in both bacteria and archaea (34). RelE is a riboendonuclease that cleaves mRNA positioned at the ribosomal A-site, between the second and third base of the A-site codon (20, 80, 86). Consistent with its enzymatic activity, ectopic production of RelE rapidly shuts down global cellular translation and consequently halts cell growth (20). The tertiary structure of RelE is similar to that of a family of RNases including RNase T1 (80). However, purified RelE does not cleave mRNA in the absence of ribosomes because it lacks certain conserved active-site residues found in homologous RNases (80, 86). Structural analysis of RelE in complex with the 70S ribosome confirms that RelE indeed binds to the ribosomal A-site and cleaves A-site codons (80). RelE interacts with several highly conserved regions of 16S rRNA, consistent with the broadly conserved activity of heterologous RelEs across different phyla.

RelB autoregulates transcription of the *relBE* operon by binding to two operator sequences in the *relBE* promoter region (41, 82). Lon protease degrades RelB (23, 82), and factors that inhibit translation induce strong transcription of the *relBE* operon owing to the Lon-mediated degradation of RelB (21, 23). In the absence of RelE, the RelB dimer (RelB$_2$) binds noncooperatively to two operators located in the *relBE* promoter region. In contrast, the RelB$_2$-RelE complex binds avidly and cooperatively to the operators. Interestingly, the addition of RelE to the RelB$_2$-RelE-operator complex disrupts the complex in vitro and, consistently, induces strong transcription of the *relBE* operon in vivo. Thus, the binding of RelB$_2$ to operator DNA is controlled by the [RelB]:[RelE] ratio, a phenomenon that we have coined conditional cooperativity (83). All other TA operons that have been investigated, including *vapBC*, *ccdAB*, *phd/doc*, and *mazEF*, are controlled by conditional cooperativity (1, 33, 83, 106). As argued later, it is possible that conditional cooperativity functions in the maintenance and/or resuscitation of persister cells.

As shown in **Figure 2b**, *E. coli* K-12 encodes six *relBE*-like loci (15, 25, 42, 76, 92, 109). These loci have different primary sequences, and perhaps unsurprisingly, simple BLAST queries with one RelE homolog are unable to identify any of the other homologs. Thus, structural information or functional analyses have been used to classify the genes as belonging to the *relBE* family. Consequently, some of the *relE*-like genes (e.g., *mqsR* and *yoeB*) have been used to define *relBE* subfamilies. As described below, several independent studies have implicated these loci in persistence, in particular *mqsRA* and *yafQ* (40, 45). Interestingly, the antitoxin MqsA modulates the general stress response by repressing the *rpoS* promoter during growth under nonstressed conditions. Because MqsA is degraded by Lon, environmental stresses lead to a reduced level of MqsA, resulting in increased RpoS levels and therefore increased biofilm formation and persistence (103).

mazEF Gene Family

The first member of the *mazEF* family to be discovered, *pemIK* (*parD*), was encoded by plasmid R100 (R1) that is stabilized by the locus (14, 100). It was proposed (but not shown) that type II TA loci, similar to type I TA loci, stabilize plasmids by killing the plasmid-free segregants (37, 38). The *mazEF* locus encoded by the *E. coli* K-12 chromosome was discovered by Ohtsubo and colleagues (74), who used Southern hybridization with a *pemIK* probe. Early on, it was proposed that *mazEF* mediates programmed cell death by a mechanism that depends on (p)ppGpp (2). However, a later analysis revealed that ectopic overproduction of MazF induces a bacteriostatic condition from which the cells can be resuscitated by later induction of *mazE* transcription (85). Moreover, induction of *mazEF* transcription appears to be independent of (p)ppGpp (24). Consistently, a physiological analysis of the *mazEF* homolog of plasmid R1/R100 (*pemIK*) shows that the toxin does not kill plasmid-free segregants but rather slows the growth of such cells (46).

The *mazEF* promoter is regulated in a manner similar to that of the *relBE* promoter: The unrepressed promoter is highly active, and MazE binds to the promoter region and represses transcription, with the simultaneous presence of MazE and MazF conferring an even stronger level of repression (71). Consistent with this mode of regulation, MazF enhances MazE binding to the promoter region in vitro (71, 112). As in the case of *relBE*, *mazEF* transcription is strongly induced during amino acid starvation (24) and is also controlled by conditional cooperativity (K.S. Winther & K. Gerdes, unpublished data).

MazF is a potent inhibitor of translation, both in vivo (85) and in vitro (113). Like RelE, MazF inhibits translation by cleaving mRNA (24, 113). However, MazF cleaves mRNA site specifically at ACA base motifs and does so in a ribosome-independent manner (113). This difference in mechanism of action and the different tertiary folds of the RNases (48, 66, 80) indicate that the two corresponding gene families must have evolved independently.

vapBC Gene Family

The *vapBC* family is the most abundant of the TA gene families, and some organisms have accumulated extensive cohorts of *vapBC* loci. For example, the highly persistent human pathogen *Mycobacterium tuberculosis* has at least 45 *vapBC* loci (**Figure 3a**). Similar to RelE and MazF, ectopic production of VapC inhibits translation and thereby induces dormancy from which the cells can later be resuscitated by the induction of *vapB* transcription (104). As seen with other TA loci, *vapBC* transcription in *Salmonella enterica* is repressed by VapB and corepressed by VapC. Moreover, ectopic overproduction of a nontoxic variant of VapC stimulates *vapBC* transcription in vivo and excess VapC destabilizes the binding of the VapBC complex to the promoter region in vitro (106). Thus, transcription of the *vapBC* operon is also regulated by conditional cooperativity.

Although VapC was predicted to be a ribonuclease, on the basis of its homology to eukaryotic PIN (PilT N terminus) domains, its cellular target was particularly difficult to identify. First, we showed that enteric VapC proteins do not cleave model mRNAs and other RNAs (104). Then, by employing an almost exhaustive search, we discovered that VapC of enterics (*S. enterica* and *Shigella flexneri*) are site-specific riboendonucleases that cleave initiator tRNA^{fMet} in the anticodon loop (105). Thus, like some eukaryotic PIN domain proteins, enteric VapC proteins are site-specific RNases. The identification of the first specific VapC target has led to the analysis of *vapBC* loci of other, major pathogenic bacteria.

(p)ppGpp: guanosine tetra- and pentaphosphate; intracellular signaling molecules that reprogram cellular metabolism during nutritional stresses

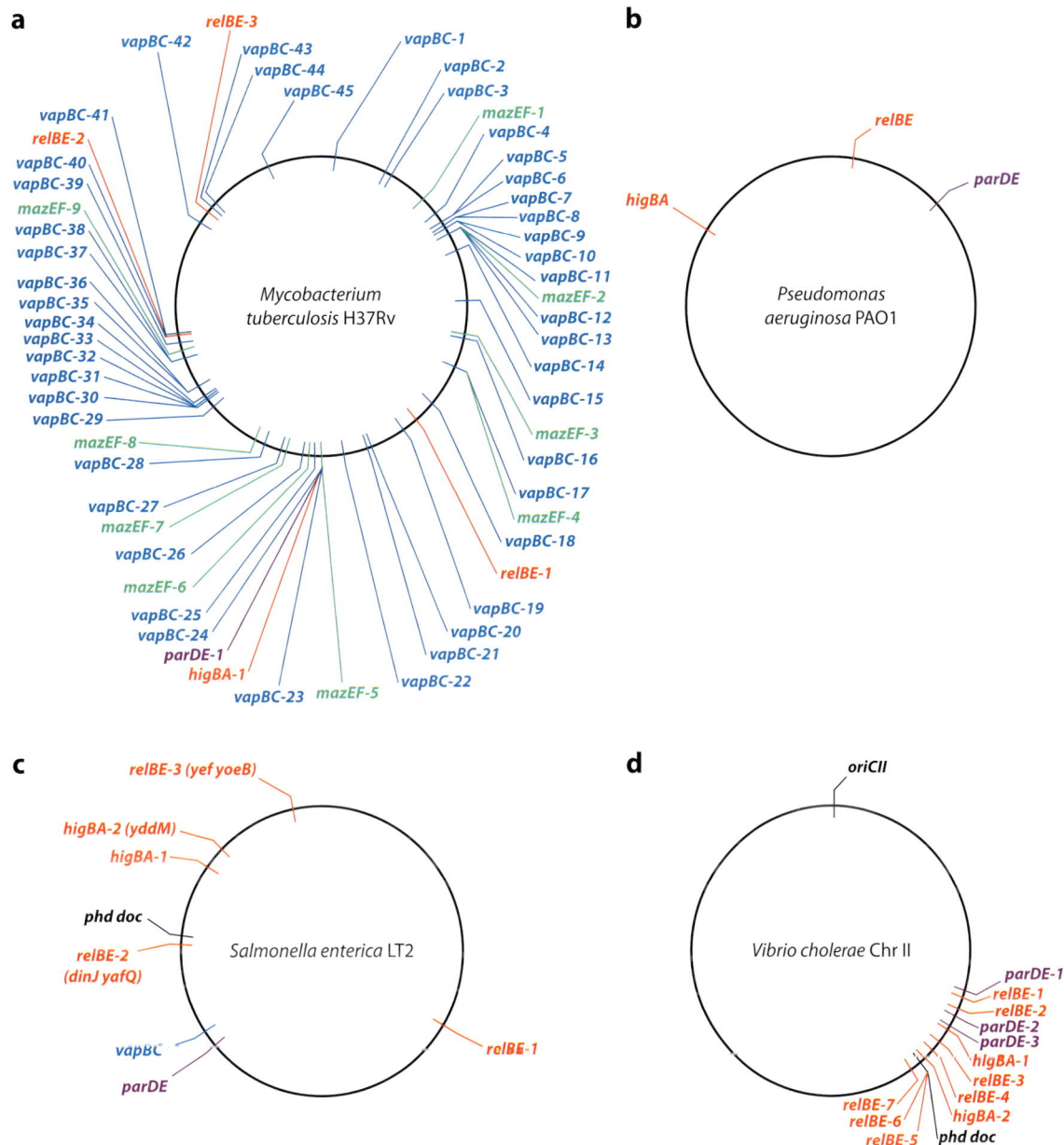

Figure 3

Type II toxin-antitoxin (TA) loci of major pathogenic bacteria. The locations of (*a*) 59 of the 88 known type II TA loci of *Mycobacterium tuberculosis* H37Rv, (*b*) *Pseudomonas aeruginosa* PAO1, (*c*) *Salmonella enterica* LT2, and (*d*) *Vibrio cholerae* Chr II. TA loci encoding toxins that belong to the same superfamily are shown in identical colors. The data were taken from References 47 and 88.

TOXIN-ANTITOXIN LOCI AND PERSISTENCE OF *ESCHERICHIA COLI*

In many cases, work done with the model organism *E. coli* K-12 has been pivotal to gaining the first mechanistic insight into biological problems and this has also been the case with bacterial persistence. Indeed, as described above, the first persister gene was *hipBA* of K-12 (77). *E. coli* K-12 has 10 TA loci, in addition to *hipBA* [and several recently identified type II TA loci (110)], that encode mRNases (**Figure 2b**). Some of these loci have been individually implicated in persistence (40, 45). However, to further understand the overall role of TA loci in the persistence phenomenon, we used a systematic genetic approach (69). Deletion of any single TA locus had no effect on the level of persistence, whereas deletion of all 10 mRNase-encoding loci reduced persistence by ≈200-fold (**Figure 4a,b**). This effect was not due to the deletion of a particular TA locus, because the progressive deletion of more and more TA loci caused a gradual decline in the persistence level (**Figure 4a**). A similar pattern was observed when the TA loci were deleted in the reverse order, supporting the notion that most if not all TA loci contribute to persistence. Because all antitoxins of K-12 that have been analyzed are degraded by Lon, a *lon* deletion strain should exhibit a phenotype resembling the multiple TA deletion strain with respect to persistence. We tested this simple prediction by measuring the persistence level of a *lon* strain. Indeed, as shown in **Figure 4b**, deletion of *lon* dramatically reduced persistence toward both ampicillin and ciprofloxacin. However, deletion of the genes encoding the other major ATP-dependent proteases of *E. coli* K-12 had no such effect (69). Further support of a link between Lon, TA loci, and persistence came from the observation that mild overproduction of Lon, such that it did not significantly reduce the growth rate, stimulated persistence in a wild-type *E. coli* K-12 strain but not in a strain that lacked the 10 TA loci (69).

A Testable Model Explaining Induction and Maintenance of Persistence

It is not known how persistence is regulated. Is it a purely random process that occurs by an accidental, stochastic activation of the TAs, or is it somehow integrated into the metabolic status of the cell? An interesting observation by Balaban's group suggests that TA-binding affinities and the degradation rates of antitoxins may be crucial in determining the persistence level (89). Our

---

Figure 4

TA loci and Lon are required for persistence of *Escherichia coli* K-12. (*a*) Effects of multiple TA gene deletions on persistence. Survival rates in persister assays using ciprofloxacin (1 μg ml^{-1}) or ampicillin (100 μg ml^{-1}). Numbers on the *x*-axis refer to the number of TA loci that were deleted in the strains tested. (*b*) Effect of deleting the gene encoding Lon (69). Deletion of *lon* but not other protease-encoding genes reduced persistence, indicating a specific role of Lon in the phenomenon. The percentage of survival after 5 h was compared with that of the wt strain (log scale). (*c*) A hypothetical working model explaining TA-dependent persistence of *E. coli* K-12. The model illuminates the proposed mechanistic roles of TA loci, Lon, PolyP, and (p)ppGpp in the formation of persister cells. The question mark indicates the possible existence of an inhibitor of Lon that is required to terminate the persistent state of a cell by reducing degradation of the antitoxins. (*d*) Schematic that explains the effect of (p)ppGpp heterogeneity on the bacterial population. (*Left*) The cellular level of (p)ppGpp fluctuates such that a minor fraction of the cells (≈10^{-4} of the population) have a high level of (p)ppGpp (*red cells*) and therefore have entered the persistent state. (*Middle*) Treatment with an antibiotic kills off cells with a low (normal) level of (p)ppGpp (*gray cells*), leaving a few survivors (persisters). (*Right*) Upon regrowth (resuscitation), the persister cells give rise to a new heterogeneous population of cells that is indistinguishable from the initial culture (and exhibits the same antibiotic sensitivity). Abbreviations: TA, toxin-antitoxin; Lon, ATP-dependent protease; PolyP, polyphosphate; (p)ppGpp, guanosine tetra- and pentaphosphate; PPK, polyphosphate kinase; PPX, exopolyphosphatase; RelA, ppGpp synthetase I; SpoT, ppGpp synthetase II; wt, wild type.

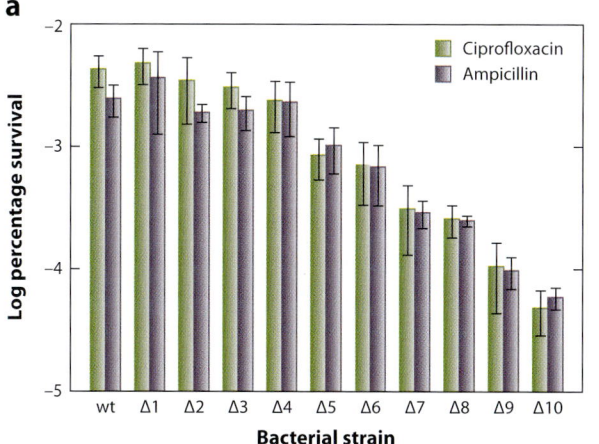
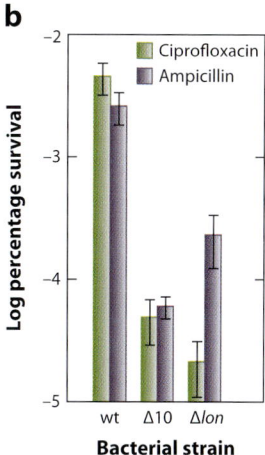
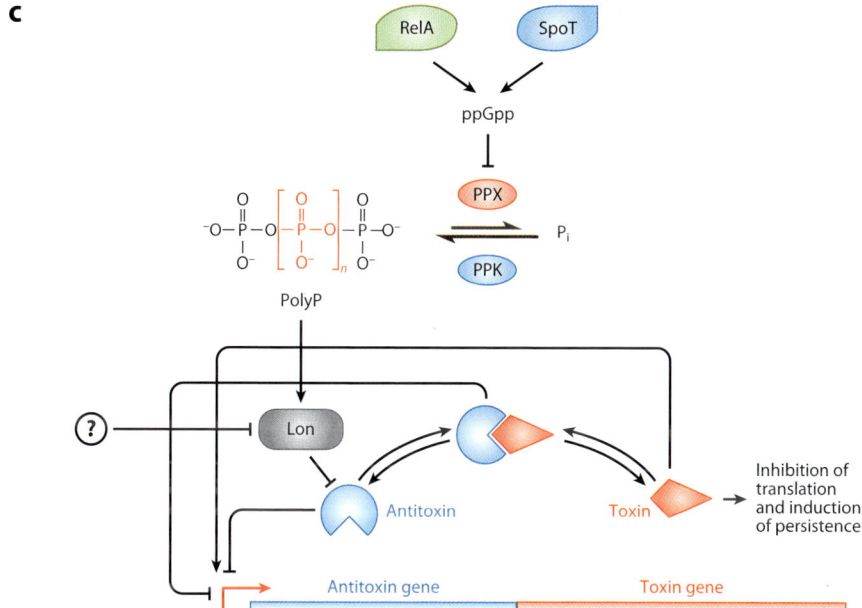
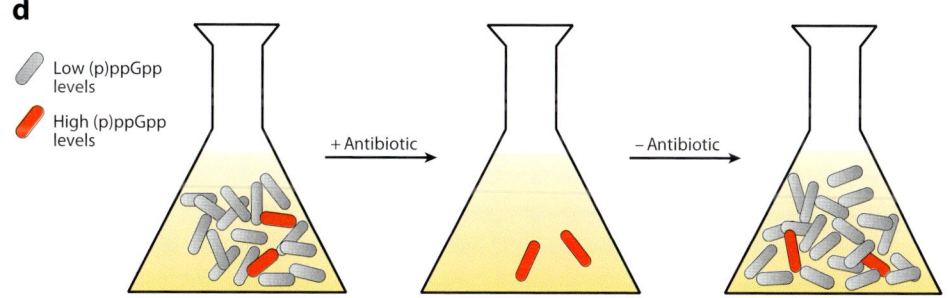

Polyphosphate (PolyP): a polymer consisting of covalently linked orthophosphates that is synthesized by PPK

Polyphosphate kinase (PPK): a ubiquitous enzyme that catalyzes the synthesis of PolyP through the transfer of a phosphoryl group of ATP to a PolyP polymer

Exopolyphosphatase (PPX): a common enzyme that degrades inorganic PolyP

RelA: (p)ppGpp synthetase I

finding that Lon is a major player in the phenomenon raises the possibility that Lon is stochastically switched on at a low frequency to generate persisters. This toggling could be due to controlled switching by a signaling pathway that regulates Lon activity or to stochastic activation of Lon by an unknown mechanism. The most straightforward explanation is that Lon activity varies in single cells due to fluctuations in the number of Lon molecules and/or molecules that regulate Lon activity (molecular noise). However, Lon can be activated by polyphosphate (PolyP), a linear polymer of orthophosphate residues (58). In *E. coli*, PolyP is synthesized by polyphosphate kinase (PPK) and degraded by exopolyphosphatase (PPX). Thus, it is possible that PolyP programs Lon to degrade the antitoxins.

Interestingly, in *E. coli*, PolyP functions as an intracellular signaling molecule controlled by the stringent response. The low-molecular-weight signaling molecule (p)ppGpp, the central mediator of the stringent response that reprograms the cells to survive nutritional limitations, competitively inhibits PPX and thereby controls the cellular level of PolyP (57). Korch et al. (55) reported that *E. coli* K-12 cells deficient in (p)ppGpp synthesis display a reduced level of persisters. These considerations led us to propose the model shown in **Figure 4c**. The model proposes that (p)ppGpp induces persistence by activating TA loci via PolyP and Lon. The phenotypic consequences of the model are visualized in **Figure 4d**. In this model, the main regulator of persistence in *E. coli* is (p)ppGpp. It is well known that the cellular level of (p)ppGpp varies inversely with growth rate (8, 90, 111). The persistence level of *E. coli* also varies inversely with growth rate (97). Thus, the model provides a simple link between the generation of persisters and the growth rate, and for the first time yields a testable explanation for the common observation that slowly but exponentially growing bacteria generate higher levels of persisters than rapidly growing ones do (30, 101). In support of this model, inactivation of the stringent response in *P. aeruginosa* reduced antibiotic tolerance in biofilms (114). Interestingly, RelA [(p)ppGpp synthetase I] is required for the long-term survival and persistence of *M. tuberculosis* in mice (29, 87). Moreover, the stringent response exhibits bistability in *Mycobacterium smegmatis*, thus offering a possible explanation for the heterogeneity of growth rates seen with mycobacterial populations (39, 98).

Importance and Consequence of the Model

Intuitively, multiple paths may lead to slow growth and thus to persistence. This is because individual cells of growing cultures of any bacterium will inevitably exhibit a range of growth rates distributed around a mean. Without assuming any programmed events at all, such a distribution could generate persisters at an appreciable rate that would depend on the broadness of the distribution. In principle, the bimodal distributions for wild-type *E. coli* observed by Lewis and colleagues (94) and for the *hipA7* mutant strain observed by Balaban, Leibler, and colleagues (7, 89) can be explained simply by a passive model stating that persistence is a consequence of random (i.e., nonprogrammed) variations in the growth rates of individual cells in a culture (4). According to this passive model, the increased persistence seen with slower but exponentially growing cells (97) is a passive consequence of the increased tailing of the growth rate distribution. Another passive model published in a theoretical paper asserts that persistence is due to slowly growing, aging cells in the bacterial culture (53).

If the passive models are valid, then the plethora of conditions and random fluctuations that lead to slow growth would lead to increased persistence and the persistence phenomenon would therefore be very difficult if not impossible to analyze systematically. We would like to argue that our recent observations challenge the passive models. The simultaneous deletion of 10 TA loci from the chromosome of *E. coli* K-12 reduced the persistence level at least 100-fold (**Figure 4a,b**). Moreover, most if not all of the 10 TA loci contributed to persistence. These robust observations indicate that the formation of at least 99 of 100 persister cells of an exponentially growing culture of *E. coli* K-12 depends on the TA loci. In other words, our results indicate that less than 1% of the persisters are generated by slow growth arising from random fluctuations in the growth rates of single cells. We look forward to learning whether TA genes also contribute to the persistence of pathogenic bacteria.

Resuscitation from Persistence

The rate of resuscitation of persisters can be calculated from the slope of the curve in the second phase of the typical persister experiment shown in **Figure 1**. This is because those persisters that start to grow in the presence of the antibiotic are killed. We do not know how phenotypic switching back to rapid growth is regulated. Therefore, a better understanding of the resuscitation mechanism may lead to the design of strategies to reduce the rate of resuscitation and thereby reduce persistence. The constant rate of resuscitation is consistent with a stochastic event (the same fraction of cells at a given time interval switches back to growth) (89). As described above, we found that transcription of the common TA loci, such as *vapBC*, *relBE*, and *mazEF*, is regulated by conditional cooperativity. In the persistent state, the cellular translation rate is low. The low rate of global translation is assumed to favor a low [A]:[T] ratio because the antitoxin is unstable. In turn, due to conditional cooperativity, the low [A]:[T] ratio derepresses transcription of TA loci and thereby secures a high rate of TA transcription and toxin production in persisters. Thus, conditional cooperativity readily explains the high levels of TA mRNAs observed in persisters (51, 94). When the Lon-mediated degradation rate of antitoxins returns to a low level (by an as yet unknown mechanism), the high transcription rates of the TA operons ensure rapid production and accumulation of the antitoxins, quench toxin activity, and resuscitate cell growth. This inference was supported by mathematical modeling of *relBE* operon transcription (17).

PHYLOGENY OF TOXIN-ANTITOXIN LOCI

The phylogeny of TA loci is unusually complex, with members of the *vapBC*, *relBE*, and *hicAB* families found in both bacteria and archaea, whereas the other families are restricted primarily to bacteria (62, 84). When analyzing the phylogenetic patterns of TA loci in ≈200 prokaryotic genomes, we found that free-living prokaryotes in particular contained many TA loci whereas most obligate intracellular organisms had few or none (47, 84). For example, *M. tuberculosis* has at least 88 TA loci (**Figure 3a**) whereas *M. leprae* has none (84, 88). This observation is striking because *M. leprae* evolved from *M. tuberculosis* by massive reductive evolution (26). *M. leprae* is an obligate intracellular parasite; *M. tuberculosis* has both an extra- and an intracellular lifestyle, consistent with the proposal that TA loci are particularly beneficial to organisms that encounter changing environments. Similarly, obligate intracellular organisms such as *Rickettsia* and *Buchnera* spp. have very few or no TA loci (84), even though rare exceptions have been discovered (62). On the basis of theoretical calculations, it has been argued that the persistence phenotype is particularly advantageous for organisms living in changing environments (59). Thus, the phylogeny of TA loci is largely consistent with the proposal that they play a role in persistence.

S. enterica LT2 and *P. aeruginosa* PAO1 have 15 and 3 known TA loci, respectively, and *Vibrio cholerae* has 13 (84) (**Figure 3b–d**). Strikingly, all 13 TA loci of *V. cholerae* are located in the superintegron of chromosome II (**Figure 3d**) and are closely linked to *attC* sites, indicating that they are bona fide integron elements that can be mobilized as integron cassettes (16). A bioinformatic survey identified 10,753 type II TA gene pairs within 1,240 prokaryotic genomes (95). An interactive database by which the cohort of TA loci present in individual genomes can be rapidly retrieved was also generated by that study and is accessible at the Toxin-Antitoxin Database (**http://bioinfo-mml.sjtu.edu.cn/TADB/**).

CONCLUDING REMARKS

We have described here a novel, simple mechanism that can explain how growing cells of the model organism *E. coli* differentiate into persisters. The abundant presence of TA loci in many important bacterial pathogens raises the possibility that these genes contribute to pathogenicity by increasing persistence and resistance to environmental and nutritional stresses. Several research groups are now engaged in investigating this hypothesis and important results are in prospect. A better understanding of the mechanisms behind bacterial persistence may lead to the design of drugs that can combat chronic and recurrent infections.

SUMMARY POINTS

1. Bacterial persistence is, at least in some cases, a programmed phenomenon that helps bacteria survive environmental insults.
2. TA loci and Lon protease are required for persistence of the model organism *E. coli*.
3. We propose here a molecular model that explains bacterial persistence.
4. According to the model, (p)ppGpp functions as a signal that determines whether a single cell differentiates into a persister cell.
5. In the model, TA loci function as effectors that are activated by (p)ppGpp due to accumulation of PolyP followed by Lon activation.
6. Fluctuations of [(p)ppGpp] in single cells determine the persistence level at a given growth condition.
7. The model is consistent with the general observation that the persistence level increases when the growth rate of a bacterial culture decreases.
8. The understanding of the molecular mechanisms behind persistence may lead to the development of strategies to combat persistent and recurrent infections.

DISCLOSURE STATEMENT

The authors are not aware of any affiliations, memberships, funding, or financial holdings that might be perceived as affecting the objectivity of this review.

ACKNOWLEDGMENTS

This work was supported by the Wellcome Trust. We thank David W. Adams for critical reading of the manuscript and the people at the Center for Bacterial Cell Biology for stimulating discussions.

LITERATURE CITED

1. Afif H, Allali N, Couturier M, Van Melderen L. 2001. The ratio between CcdA and CcdB modulates the transcriptional repression of the *ccd* poison-antidote system. *Mol. Microbiol.* 41:73–82
2. Aizenman E, Engelberg-Kulka H, Glaser G. 1996. An *Escherichia coli* chromosomal "addiction module" regulated by guanosine $3',5'$-bispyrophosphate: a model for programmed bacterial cell death. *Proc. Natl. Acad. Sci. USA* 93:6059–63

3. Alexander C, Bilgin N, Lindschau C, Mesters JR, Kraal B, et al. 1995. Phosphorylation of elongation factor Tu prevents ternary complex formation. *J. Biol. Chem.* 270:14541–47
4. Allison KR, Brynildsen MP, Collins JJ. 2011. Heterogeneous bacterial persisters and engineering approaches to eliminate them. *Curr. Opin. Microbiol.* 14:593–98
5. Balaban NQ. 2011. Persistence: mechanisms for triggering and enhancing phenotypic variability. *Curr. Opin. Genet. Dev.* 21:168–75
6. Deleted in proof
7. **Balaban NQ, Merrin J, Chait R, Kowalik L, Leibler S. 2004. Bacterial persistence as a phenotypic switch.** *Science* **305:1622–25**
8. Baracchini E, Bremer H. 1988. Stringent and growth control of rRNA synthesis in *Escherichia coli* are both mediated by ppGpp. *J. Biol. Chem.* 263:2597–602
9. Bigger JW. 1944. Treatment of staphylococcal infections with penicillin by intermittent sterilisation. *Lancet* 294:497–500
10. Black DS, Irwin B, Moyed HS. 1994. Autoregulation of hip, an operon that affects lethality due to inhibition of peptidoglycan or DNA synthesis. *J. Bacteriol.* 176:4081–91
11. Black DS, Kelly AJ, Mardis MJ, Moyed HS. 1991. Structure and organization of hip, an operon that affects lethality due to inhibition of peptidoglycan or DNA synthesis. *J. Bacteriol.* 173:5732–39
12. Blower TR, Salmond GP, Luisi BF. 2011. Balancing at survival's edge: the structure and adaptive benefits of prokaryotic toxin-antitoxin partners. *Curr. Opin. Struct. Biol.* 21:109–18
13. Bordes P, Cirinesi AM, Ummels R, Sala A, Sakr S, et al. 2011. SecB-like chaperone controls a toxin-antitoxin stress-responsive system in *Mycobacterium tuberculosis*. *Proc. Natl. Acad. Sci. USA* 108:8438–43
14. Bravo A, de Torrontegui G, Díaz R. 1987. Identification of components of a new stability system of plasmid R1, ParD, that is close to the origin of replication of this plasmid. *Mol. Gen. Genet.* 210:101–10
15. Brown BL, Grigoriu S, Kim Y, Arruda JM, Davenport A, et al. 2009. Three dimensional structure of the MqsR:MqsA complex: a novel TA pair comprised of a toxin homologous to RelE and an antitoxin with unique properties. *PLoS Pathog.* 5:e1000706
16. Cambray G, Guerout AM, Mazel D. 2010. Integrons. *Annu. Rev. Genet.* 44:141–66
17. Cataudella I, Trusina A, Sneppen K, Gerdes K, Mitarai N. 2012. Conditional cooperativity in toxin-antitoxin regulation prevents random toxin activation and promotes fast translational recovery. *Nucleic Acids Res.* doi: 10.1093/nar/gks297
18. Ceglowski P, Boitsov A, Karamyan N, Chai S, Alonso JC. 1993. Characterization of the effectors required for stable inheritance of *Streptococcus pyogenes* pSM19035-derived plasmids in *Bacillus subtilis*. *Mol. Gen. Genet.* 241:579–85
19. Cherny I, Rockah L, Gazit E. 2005. The YoeB toxin is a folded protein that forms a physical complex with the unfolded YefM antitoxin. *J. Biol. Chem.* 280:30063–72
20. Christensen SK, Gerdes K. 2003. RelE toxins from bacteria and archaea cleave mRNAs on translating ribosomes, which are rescued by tmRNA. *Mol. Microbiol.* 48:1389–400
21. Christensen SK, Gerdes K. 2004. Delayed-relaxed response explained by hyperactivation of RelE. *Mol. Microbiol.* 53:587–97
22. Christensen SK, Maenhaut-Michel G, Mine N, Gottesman S, Gerdes K, Van Melderen L. 2004. Overproduction of the Lon protease triggers inhibition of translation in *Escherichia coli*: involvement of the *yefM-yoeB* toxin-antitoxin system. *Mol. Microbiol.* 51:1705–17
23. Christensen SK, Mikkelsen M, Pedersen K, Gerdes K. 2001. RelE, a global inhibitor of translation, is activated during nutritional stress. *Proc. Natl. Acad. Sci. USA* 98:14328–33
24. Christensen SK, Pedersen K, Hansen FG, Gerdes K. 2003. Toxin-antitoxin loci as stress-response-elements: ChpAK/MazF and ChpBK cleave translated RNAs and are counteracted by tmRNA. *J. Mol. Biol.* 332:809–19
25. Christensen-Dalsgaard M, Jørgensen MG, Gerdes K. 2010. Three new RelE-homologous mRNA interferases of *Escherichia coli* differentially induced by environmental stresses. *Mol. Microbiol.* 75:333–48
26. Cole ST, Eiglmeier K, Parkhill J, James KD, Thomson NR, et al. 2001. Massive gene decay in the leprosy bacillus. *Nature* 409:1007–11

---

7. Proposes that persistence is caused by a stochastic switching from rapid growth to slow growth or dormancy.

27. Collins JF, Richmond MH. 1962. A structural similarity between *N*-acetylmuramic acid and penicillin as a basis for antibiotic action. *Nature* 195:142–43
28. Correia FF, D'Onofrio A, Rejtar T, Li L, Karger BL, et al. 2006. Kinase activity of overexpressed HipA is required for growth arrest and multidrug tolerance in *Escherichia coli*. *J. Bacteriol.* 188:8360–67
29. Dahl JL, Kraus CN, Boshoff HI, Doan B, Foley K, et al. 2003. The role of RelMtb-mediated adaptation to stationary phase in long-term persistence of *Mycobacterium tuberculosis* in mice. *Proc. Natl. Acad. Sci. USA* 100:10026–31
30. Dhar N, McKinney JD. 2007. Microbial phenotypic heterogeneity and antibiotic tolerance. *Curr. Opin. Microbiol.* 10:30–38
31. Dörr T, Vulić M, Lewis K. 2010. Ciprofloxacin causes persister formation by inducing the TisB toxin in *Escherichia coli*. *PLoS Biol.* 8:e1000317
32. Falla TJ, Chopra I. 1998. Joint tolerance to β-lactam and fluoroquinolone antibiotics in *Escherichia coli* results from overexpression of *hipA*. *Antimicrob. Agents Chemother.* 42:3282–84
33. Garcia-Pino A, Balasubramanian S, Wyns L, Gazit E, De GH, et al. 2010. Allostery and intrinsic disorder mediate transcription regulation by conditional cooperativity. *Cell* 142:101–11
34. Gerdes K. 2000. Toxin-antitoxin modules may regulate synthesis of macromolecules during nutritional stress. *J. Bacteriol.* 182:561–72
35. Gerdes K, Christensen SK, Løbner-Olesen A. 2005. Prokaryotic toxin-antitoxin stress response loci. *Nat. Rev. Microbiol.* 3:371–82
36. Gerdes K, Gultyaev AP, Franch T, Pedersen K, Mikkelsen ND. 1997. Antisense RNA-regulated programmed cell death. *Annu. Rev. Genet.* 31:1–31
37. Gerdes K, Helin K, Christensen OW, Lobner-Olesen A. 1988. Translational control and differential RNA decay are key elements regulating postsegregational expression of the killer protein encoded by the *parB* locus of plasmid R1. *J. Mol. Biol.* 203:119–29
38. Gerdes K, Rasmussen PB, Molin S. 1986. Unique type of plasmid maintenance function: postsegregational killing of plasmid-free cells. *Proc. Natl. Acad. Sci. USA* 83:3116–20
39. Ghosh S, Sureka K, Ghosh B, Bose I, Basu J, Kundu M. 2011. Phenotypic heterogeneity in mycobacterial stringent response. *BMC. Syst. Biol.* 5:18
40. Gonzalez Barrios AF, Zuo R, Hashimoto Y, Yang L, Bentley WE, Wood TK. 2006. Autoinducer 2 controls biofilm formation in *Escherichia coli* through a novel motility quorum-sensing regulator (MqsR, B3022). *J. Bacteriol.* 188:305–16
41. Gotfredsen M, Gerdes K. 1998. The *Escherichia coli relBE* genes belong to a new toxin-antitoxin gene family. *Mol. Microbiol.* 29:1065–76
42. Grady R, Hayes F. 2003. Axe-Txe, a broad-spectrum proteic toxin-antitoxin system specified by a multidrug-resistant, clinical isolate of *Enterococcus faecium*. *Mol. Microbiol.* 47:1419–32
43. Hallez R, Geeraerts D, Sterckx Y, Mine N, Loris R, Van ML. 2010. New toxins homologous to ParE belonging to three-component toxin-antitoxin systems in *Escherichia coli* O157:H7. *Mol. Microbiol.* 76:719–32
44. Hansen S, Vulić M, Min J, Yen T, Schumacher MA, et al. 2012. Regulation of the *Escherichia coli* HipBA toxin-antitoxin system by proteolysis. *PLoS One*. In press
45. Harrison JJ, Wade WD, Akierman S, Vacchi-Suzzi C, Stremick CA, et al. 2009. The chromosomal toxin gene *yafQ* is a determinant of multidrug tolerance for *Escherichia coli* growing in a biofilm. *Antimicrob. Agents Chemother.* 53:2253–58
46. Jensen RB, Grohmann E, Schwab H, Diazorejas R, Gerdes K. 1995. Comparison of *ccd* of F, *parDE* of Rp4, and *parD* of R1 using a novel conditional replication control system of plasmid R1. *Mol. Microbiol.* 17:211–20
47. Jørgensen MG, Pandey DP, Jaskolska M, Gerdes K. 2009. HicA of *Escherichia coli* defines a novel family of translation-independent mRNA interferases in bacteria and archaea. *J. Bacteriol.* 191:1191–99
48. Kamada K, Hanaoka F. 2005. Conformational change in the catalytic site of the ribonuclease YoeB toxin by YefM antitoxin. *Mol. Cell* 19:497–509
49. Kamada K, Hanaoka F, Burley SK. 2003. Crystal structure of the MazE/MazF complex: Molecular bases of antidote-toxin recognition. *Mol. Cell* 11:875–84

50. Keren I, Kaldalu N, Spoering A, Wang YP, Lewis K. 2004. Persister cells and tolerance to antimicrobials. *FEMS Microbiol. Lett.* 230:13–18
51. **Keren I, Shah D, Spoering A, Kaldalu N, Lewis K. 2004. Specialized persister cells and the mechanism of multidrug tolerance in *Escherichia coli*. *J. Bacteriol.* 186:8172–80**
52. Kim Y, Wood TK. 2010. Toxins Hha and CspD and small RNA regulator Hfq are involved in persister cell formation through MqsR in *Escherichia coli*. *Biochem. Biophys. Res. Commun.* 391:209–13
53. Klapper I, Gilbert P, Ayati BP, Dockery J, Stewart PS. 2007. Senescence can explain microbial persistence. *Microbiology* 153:3623–30
54. Koga M, Otsuka Y, Lemire S, Yonesaki T. 2011. *Escherichia coli rnlA* and *rnlB* compose a novel toxin-antitoxin system. *Genetics* 187:123–30
55. Korch SB, Henderson TA, Hill TM. 2003. Characterization of the *hipA7* allele of *Escherichia coli* and evidence that high persistence is governed by (p)ppGpp synthesis. *Mol. Microbiol.* 50:1199–213
56. Korch SB, Hill TM. 2006. Ectopic overexpression of wild-type and mutant *hipA* genes in *Escherichia coli*: effects on macromolecular synthesis and persister formation. *J. Bacteriol.* 188:3826–36
57. Kuroda A, Murphy H, Cashel M, Kornberg A. 1997. Guanosine tetra- and pentaphosphate promote accumulation of inorganic polyphosphate in *Escherichia coli*. *J. Biol. Chem.* 272:21240–43
58. Kuroda A, Nomura K, Ohtomo R, Kato J, Ikeda T, et al. 2001. Role of inorganic polyphosphate in promoting ribosomal protein degradation by the Lon protease in *E. coli*. *Science* 293:705–8
59. **Kussell E, Leibler S. 2005. Phenotypic diversity, population growth, and information in fluctuating environments. *Science* 309:2075–78**
60. Lah J, Marianovsky I, Glaser G, Engelberg-Kulka H, Kinne J, et al. 2003. Recognition of the intrinsically flexible addiction antidote MazE by a dromedary single domain antibody fragment. Structure, thermodynamics of binding, stability, and influence on interactions with DNA. *J. Biol. Chem.* 278:14101–11
61. Lah J, Simic M, Vesnaver G, Marianovsky I, Glaser G, et al. 2005. Energetics of structural transitions of the addiction antitoxin MazE: Is a programmed bacterial cell death dependent on the intrinsically flexible nature of the antitoxins? *J. Biol. Chem.* 280:17397–407
62. Leplae R, Geeraerts D, Hallez R, Guglielmini J, Dreze P, Van ML. 2011. Diversity of bacterial type II toxin-antitoxin systems: a comprehensive search and functional analysis of novel families. *Nucleic Acids Res.* 39:5513–25
63. Levin BR, Rozen DE. 2006. Non-inherited antibiotic resistance. *Nat. Rev. Microbiol.* 4:556–62
64. Lewis K. 2010. Persister cells. *Annu. Rev. Microbiol.* 64:357–72
65. Li GY, Zhang Y, Inouye M, Ikura M. 2008. Structural mechanism of transcriptional autorepression of the *Escherichia coli* RelB/RelE antitoxin/toxin module. *J. Mol. Biol.* 380:107–19
66. Li GY, Zhang Y, Inouye M, Ikura M. 2009. Inhibitory mechanism of *Escherichia coli* RelE-RelB toxin-antitoxin module involves a helix displacement near an mRNA interferase active site. *J. Biol. Chem.* 284:14628–36
67. Lippmann C, Lindschau C, Vijgenboom E, Schroder W, Bosch L, Erdmann VA. 1993. Prokaryotic elongation factor Tu is phosphorylated in vivo. *J. Biol. Chem.* 268:601–7
68. Luidalepp H, Joers A, Kaldalu N, Tenson T. 2011. Age of inoculum strongly influences persister frequency and can mask effects of mutations implicated in altered persistence. *J. Bacteriol.* 193:3598–605
69. **Maisonneuve E, Shakespeare LJ, Jørgensen MG, Gerdes K. 2011. Bacterial persistence by RNA endonucleases. *Proc. Natl. Acad. Sci. USA* 108:13206–11**
70. Makarova KS, Wolf YI, Koonin EV. 2009. Comprehensive comparative-genomic analysis of type 2 toxin-antitoxin systems and related mobile stress response systems in prokaryotes. *Biol. Direct* 4:19
71. Marianovsky I, Aizenman E, Engelberg-Kulka H, Glaser G. 2001. The regulation of the *Escherichia coli mazEF* promoter involves an unusual alternating palindrome. *J. Biol. Chem.* 276:5975–84
72. Mastroeni P, Arena A, Costa GB, Liberto MC, Bonina L, Hormaeche CE. 1991. Serum TNFα in mouse typhoid and enhancement of a salmonella infection by anti-TNFα antibodies. *Microb. Pathog.* 11:33–38
73. Mastroeni P, Skepper JN, Hormaeche CE. 1995. Effect of anti-tumor necrosis factor alpha antibodies on histopathology of primary Salmonella infections. *Infect. Immun.* 63:3674–82
74. Masuda YJ, Miyakawa K, Nishimura Y, Ohtsubo E. 1993. *chpa* and *chpb*, *Escherichia coli* chromosomal homologs of the *pem* locus responsible for stable maintenance of plasmid R100. *J. Bacteriol.* 175:6850–56

---

51. Shows that cellular fractions enriched for persisters have increased levels of TA mRNAs.

59. Argues that persistence due to stochastic switching can be favored over sensing when the environment changes infrequently.

69. Shows that the persistence of *E. coli* depends on TA loci and Lon.

75. McDermott W. 1958. Microbial persistence. *Yale J. Biol. Med.* 30:257–91
76. Motiejunaite R, Armalyte J, Markuckas A, Suziedeliene E. 2007. *Escherichia coli dinJ-yafQ* genes act as a toxin-antitoxin module. *FEMS Microbiol. Lett.* 268:112–19
77. Moyed HS, Bertrand KP. 1983. *hipA*, a newly recognized gene of *Escherichia coli* K-12 that affects frequency of persistence after inhibition of murein synthesis. *J. Bacteriol.* 155:768–75
78. Moyed HS, Broderick SH. 1986. Molecular cloning and expression of *hipA*, a gene of Escherichia coli K-12 that affects frequency of persistence after inhibition of murein synthesis. *J. Bacteriol.* 166:399–403
79. Mutschler H, Meinhart A. 2011. ε/ζ systems: their role in resistance, virulence, and their potential for antibiotic development. *J. Mol. Med.* 89:1183–94
80. Neubauer C, Gao YG, Andersen KR, Dunham CM, Kelley AC, et al. 2009. The structural basis for mRNA recognition and cleavage by the ribosome-dependent endonuclease RelE. *Cell* 139:1084–95
81. Nyström T. 2003. Conditional senescence in bacteria: death of the immortals. *Mol. Microbiol.* 48:17–23
82. Overgaard M, Borch J, Gerdes K. 2009. RelB and RelE of *Escherichia coli* form a tight complex that represses transcription via the ribbon-helix-helix motif in RelB. *J. Mol. Biol.* 394:183–96
83. **Overgaard M, Borch J, Jorgensen MG, Gerdes K. 2008. Messenger RNA interferase RelE controls *relBE* transcription by conditional cooperativity. *Mol. Microbiol.* 69:841–57**

83. Shows that *relBE* operon transcription is regulated by conditional cooperativity.

84. Pandey DP, Gerdes K. 2005. Toxin-antitoxin loci are highly abundant in free-living but lost from host-associated prokaryotes. *Nucleic Acids Res.* 33:966–76
85. Pedersen K, Christensen SK, Gerdes K. 2002. Rapid induction and reversal of a bacteriostatic condition by controlled expression of toxins and antitoxins. *Mol. Microbiol.* 45:501–10
86. Pedersen K, Zavialov AV, Pavlov MY, Elf J, Gerdes K, Ehrenberg M. 2003. The bacterial toxin RelE displays codon-specific cleavage of mRNAs in the ribosomal A site. *Cell* 112:131–40
87. Primm TP, Andersen SJ, Mizrahi V, Avarbock D, Rubin H, Barry CE III. 2000. The stringent response of *Mycobacterium tuberculosis* is required for long-term survival. *J. Bacteriol.* 182:4889–98
88. Ramage HR, Connolly LE, Cox JS. 2009. Comprehensive functional analysis of *Mycobacterium tuberculosis* toxin-antitoxin systems: implications for pathogenesis, stress responses, and evolution. *PLoS Genet.* 5:e1000767
89. Rotem E, Loinger A, Ronin I, Levin-Reisman I, Gabay C, et al. 2010. Regulation of phenotypic variability by a threshold-based mechanism underlies bacterial persistence. *Proc. Natl. Acad. Sci. USA* 107:12541–46
90. Sarubbi E, Rudd KE, Cashel M. 1988. Basal ppGpp level adjustment shown by new *spoT* mutants affect steady-state growth rates and rrnA ribosomal promoter regulation in *Escherichia coli*. *Mol. Gen. Genet.* 213:214–22
91. Scherrer R, Moyed HS. 1988. Conditional impairment of cell division and altered lethality in *hipA* mutants of *Escherichia coli* K-12. *J. Bacteriol.* 170:3321–26
92. Schmidt O, Schuenemann VJ, Hand NJ, Silhavy TJ, Martin J, et al. 2007. *prlF* and *yhaV* encode a new toxin-antitoxin system in *Escherichia coli*. *J. Mol. Biol.* 372:894–905
93. **Schumacher MA, Piro KM, Xu W, Hansen S, Lewis K, Brennan RG. 2009. Molecular mechanisms of HipA-mediated multidrug tolerance and its neutralization by HipB. *Science* 323:396–401**

93. Presents the structure of the HipA-HipB-operator DNA complex.

94. **Shah D, Zhang ZG, Khodursky A, Kaldalu N, Kurg K, Lewis K. 2006. Persisters: a distinct physiological state of E. coli. *BMC Microbiol.* 6:53**

94. Analyzes subpopulations of wild-type cells enriched for persisters, and presents evidence that cellular fractions enriched for persisters generated by a wild-type strain have increased levels of TA mRNAs.

95. Shao Y, Harrison EM, Bi D, Tai C, He X, et al. 2011. TADB: a web-based resource for Type 2 toxin-antitoxin loci in bacteria and archaea. *Nucleic Acids Res.* 39:D606–11
96. Smith ASG, Rawlings DE. 1998. Autoregulation of the pTF-FC2 proteic poison-antidote plasmid addiction system (*pas*) is essential for plasmid stabilization. *J. Bacteriol.* 180:5463–65
97. Sufya N, Allison DG, Gilbert P. 2003. Clonal variation in maximum specific growth rate and susceptibility towards antimicrobials. *J. Appl. Microbiol.* 95:1261–67
98. Sureka K, Ghosh B, Dasgupta A, Basu J, Kundu M, Bose I. 2008. Positive feedback and noise activate the stringent response regulator *rel* in mycobacteria. *PLoS One* 3:e1771
99. Tischler AD, McKinney JD. 2010. Contrasting persistence strategies in *Salmonella* and *Mycobacterium*. *Curr. Opin. Microbiol.* 13:93–99
100. Tsuchimoto S, Ohtsubo H, Ohtsubo E. 1988. Two genes, *pemK* and *pemI*, responsible for stable maintenance of resistance plasmid R100. *J. Bacteriol.* 170:1461–66

101. Tuomanen E, Cozens R, Tosch W, Zak O, Tomasz A. 1986. The rate of killing of *Escherichia coli* by β-lactam antibiotics is strictly proportional to the rate of bacterial growth. *J. Gen. Microbiol.* 132:1297–304
102. Vazquez-Laslop N, Lee H, Neyfakh AA. 2006. Increased persistence in *Escherichia coli* caused by controlled expression of toxins or other unrelated proteins. *J. Bacteriol.* 188:3494–97
103. Wang X, Kim Y, Hong SH, Ma Q, Brown BL, et al. 2011. Antitoxin MqsA helps mediate the bacterial general stress response. *Nat. Chem. Biol.* 7:359–66
104. Winther KS, Gerdes K. 2009. Ectopic production of VapCs from Enterobacteria inhibits translation and trans-activates YoeB mRNA interferase. *Mol. Microbiol.* 72:918–30
105. **Winther KS, Gerdes K. 2011. Enteric virulence associated protein VapC inhibits translation by cleavage of initiator tRNA. *Proc. Natl. Acad. Sci. USA* 108:7403–7**
106. Winther KS, Gerdes K. 2012. Regulation of enteric *vapBC* transcription: induction by VapC toxin dimer-breaking. *Nucleic Acids Res.* 40:4347–57
107. Wolfson JS, Hooper DC, McHugh GL, Bozza MA, Swartz MN. 1990. Mutants of *Escherichia coli* K-12 exhibiting reduced killing by both quinolone and β-lactam antimicrobial agents. *Antimicrob. Agents Chemother.* 34:1938–43
108. Yamaguchi Y, Inouye M. 2011. Regulation of growth and death in *Escherichia coli* by toxin-antitoxin systems. *Nat. Rev. Microbiol.* 9:779–90
109. Yamaguchi Y, Park JH, Inouye M. 2009. MqsR, a crucial regulator for quorum sensing and biofilm formation, is a GCU-specific mRNA interferase in *Escherichia coli*. *J. Biol. Chem.* 284:28746–53
110. Yamaguchi Y, Park JH, Inouye M. 2011. Toxin-antitoxin systems in bacteria and archaea. *Annu. Rev. Genet.* 45:61–79
111. Zacharias M, Goringer HU, Wagner R. 1992. Analysis of the Fis-dependent and Fis-independent transcription activation mechanisms of the *Escherichia coli* ribosomal RNA P1 promoter. *Biochemistry* 31:2621–28
112. Zhang JJ, Zhang YL, Inouye M. 2003. Characterization of the interactions within the *mazEF* addiction module of *Escherichia coli*. *J. Biol. Chem.* 278:32300–6
113. Zhang YL, Zhang JJ, Hoeflich KP, Ikura M, Qing GL, Inouye M. 2003. MazF cleaves cellular mRNAs specifically at ACA to block protein synthesis in *Escherichia coli*. *Mol. Cell* 12:913–23
114. Nguyen D, Joshi-Datar A, Lepine F, Bauerle E, Olakanmi O, et al. 2011. Active starvation responses mediate antibiotic tolerance in biofilms and nutrient-limited bacteria. *Science* 334:982–86

105. Shows that enteric VapC toxins inhibit translation by site-specific cleavage of initiator tRNA.

Activating Transcription in Bacteria

David J. Lee, Stephen D. Minchin, and Stephen J.W. Busby

School of Biosciences, University of Birmingham, Birmingham B15 2TT, United Kingdom; email: d.lee@bham.ac.uk, s.d.minchin@bham.ac.uk, s.j.w.busby@bham.ac.uk

Keywords

promoters, RNA polymerase, sigma factors, transcription factors, recruitment, appropriation, integration

Abstract

Bacteria use a variety of mechanisms to direct RNA polymerase to specific promoters in order to activate transcription in response to growth signals or environmental cues. Activation can be due to factors that interact at specific promoters, thereby increasing transcription directed by these promoters. We examine the range of architectures found at activator-dependent promoters and outline the mechanisms by which input from different factors is integrated. Alternatively, activation can be due to factors that interact with RNA polymerase and change its preferences for target promoters. We summarize the different mechanistic options for activation that are focused directly on RNA polymerase.

Contents

PREAMBLE 126
BACTERIAL TRANSCRIPTION APPARATUS AND ECONOMICS 127
TRANSCRIPT INITIATION AND THE OPTIONS FOR ACTIVATION 127
SIMPLE TRANSCRIPTION ACTIVATION 129
   Activation by Altering Promoter Conformation 130
   Activation by Direct Contact with RNA Polymerase 131
ACTIVATOR–RNA POLYMERASE INTERACTIONS 133
   Activation by Bacteriophage λ cI Protein at the λ P_{RM} Promoter 133
   Domain 4 of σ Factors as a Target for Transcription Activators 134
   Activation by CRP at the *lac* Promoter 135
   αCTD as a Target for Transcription Activators 136
   Organization of Class II CRP- and FNR-Dependent Promoters 137
   Promoters with Tandem DNA Sites for CRP or FNR 139
PROMOTERS DEPENDENT ON σ^{54}: THE SECOND
   ACTIVATION PARADIGM 140
RNA POLYMERASE APPROPRIATION 141
   Alternative σ Factors 141
   Appropriation and Pre-Recruitment 141
INDIRECT ACTIVATION 143
INTEGRATION OF REGULATORY SIGNALS AT PROMOTERS 143
CONCLUDING COMMENTS 146

PREAMBLE

In his autobiography (62), Francois Jacob recalls the Sunday afternoon, late in July 1958, when he suddenly realized that a host of experimental observations concerning bacteriophage development and enzyme induction in *Escherichia coli* could be accounted for simply by postulating the existence of repressors that "acted on the DNA." This delightful eureka moment marks the start of the scientific age of transcription factors, and most readers will know how it led immediately to the discovery of the lactose operon repressor; the discovery of the bacteriophage λ cI repressor; and the subsequent establishment of transcription factors as key players in just about every area of biological regulation, adaptation, and development. However, many readers will be unaware that, at the time, most people thought that all this regulation would be done by repressors, and little credence was given to the existence of transcription activators (123). In their seminal 1984 review (106), Raibaud & Schwartz recall two decades of meticulous research, based mainly on genetics with the *E. coli* arabinose, maltose, and cyclic AMP systems, that overturned the prejudice and placed transcription activators on an even footing with transcription repressors.

   Right from the start, it was clear how transcription repressors could work, simply by behaving as blocking agents at their targets. In contrast, up to the 1990s, most transcription activators were regarded as mysterious agents that could somehow, as if by magic, turn on the expression of their target genes. Although there are still a few areas where the magic remains, we now understand the mechanism by which many transcription activators do their job. This is due mainly to big advances in our knowledge of how the bacterial transcription apparatus is organized, and one of the major tasks of this review is to explain the modus operandi of some of these activators and how they interface with the transcription apparatus. Later sections are concerned with the variety of mechanisms by which transcription can be activated, and how promoters can act as integrators of different activator signals.

BACTERIAL TRANSCRIPTION APPARATUS AND ECONOMICS

All the transcription activation mechanisms discussed in this chapter are concerned with changing the distribution of RNA polymerase molecules between the different transcription units in a bacterium. Transcription in all bacteria is due to the multisubunit DNA-dependent RNA polymerase, originally characterized in the 1960s by Stevens, Burgess, and others (17, 18). The core enzyme subunit composition is $\beta\beta\alpha_2\omega$, and this form of polymerase is capable of DNA-dependent RNA synthesis but is unable to locate promoters and direct specific transcript initiation. The key factor for specific transcript initiation at promoters is the σ subunit, which carries the major determinants for promoter recognition, and the $\beta\beta\alpha_2\omega\sigma$ form of polymerase is known as the holoenzyme and is competent for transcription initiation (19). Many of the functions of RNA polymerase core and holoenzyme are now well understood at the structural level following high-resolution structural studies using polymerase from thermophilic bacteria (97, 98). These functions have been reviewed in detail elsewhere, but some aspects of promoter recognition and transcription initiation are summarized schematically in **Figure 1**.

Concerning *E. coli*, the principal σ factor is the 613-amino-acid σ^{70}, the founder member of the largest group of σ factors (102). σ^{70} is often called the housekeeping sigma factor because it is responsible for most transcription, including all essential genes in *E. coli* (60). Like all members of this family, σ^{70} contains four distinct domains, and determinants in these conserved domains recognize the different promoter elements (50, 96). Many groups have attempted to quantify the levels of σ^{70} and RNA polymerase core enzymes in vivo and there is a wide variation in the measurements reported (49, 60, 105). Most recent reports, including a comprehensive analysis from Carol Gross's laboratory (49), show that several thousand core enzyme molecules are present per cell (the exact number depends upon growth conditions) with 1.3- to 1.6-fold more σ^{70} molecules. One consequence of this is that the activities of alternative σ factors are limited by competition, and this is especially critical in *E. coli*, where σ^{70} has a higher affinity for core RNA polymerase than do any of the six alternative σ factors (86). Another important consideration is that the number of RNA polymerase molecules per cell exceeds the number of growing RNA chains, and so cells contain a pool of unemployed polymerase. Most of this polymerase is probably sequestered either at random chromosomal targets or by RNA, and this acts as a reserve to supply polymerase during transcription activation (49).

TRANSCRIPT INITIATION AND THE OPTIONS FOR ACTIVATION

The first step in transcription initiation is promoter recognition in which holo RNA polymerase recognizes promoter elements located upstream of a transcript start point. This is followed by open complex formation in which a short segment of promoter DNA including the start point is unwound and the template strand is inserted into the active site of the polymerase (96). These open complexes are competent for transcript formation provided that appropriate nucleoside triphosphates are present. As illustrated in **Figure 1**, the main promoter elements that facilitate specific transcript initiation by RNA polymerase are the UP (upstream) element, the −35 element, the extended −10 element, and the −10 element, with other elements located in the spacer region between the −10 and −35 elements and the discriminator element from position −4 to −6. Determinants in conserved domains 4, 3, 2, and 1 of the σ factor are responsible for recognition of

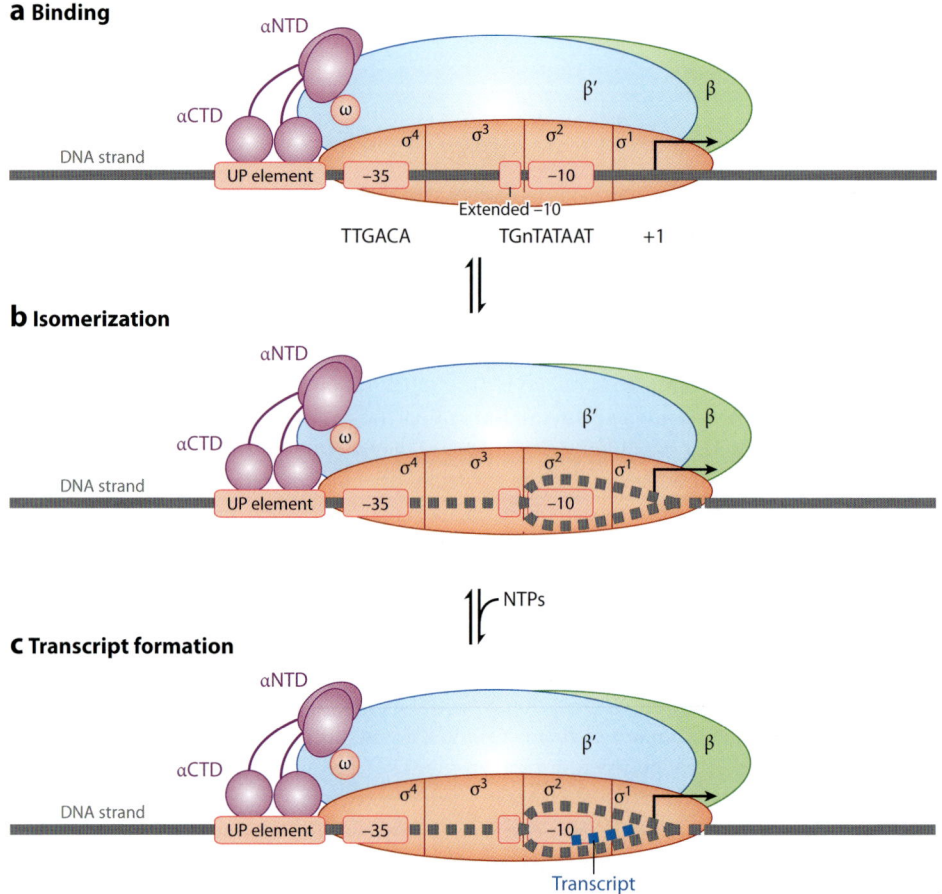

Figure 1

RNA polymerase interactions at a promoter and the steps to transcription initiation. (*a*) A schematic representation of RNA polymerase bound to a promoter. The location of the transcription start site is indicated by the bent arrow. (*b*) Isomerization to the open complex is signified by unwinding of the duplex DNA around the transcription start site. (*c*) Formation of transcripts (*blue dashed line*) ensues in the presence of NTPs. Abbreviations: NTD, N-terminal domain; CTD, C-terminal domain; UP, upstream; NTP, nucleoside triphosphate.

the −35 element, certain bases in the spacer, the extended −10 element, the −10 element, and the discriminator element, respectively. Additional promoter recognition determinants are provided by the C-terminal domains of the two RNA polymerase α subunits (αCTDs) that interact with UP elements, located upstream of the −35 element (39, 44).

An important aspect of αCTD is that it is separated from the α subunit N-terminal domain (αNTD) by an unstructured linker, and this confers a degree of flexibility in its positioning at promoters. The two αCTDs can be thought of as antennae for RNA polymerase, because they are peripheral to the core enzyme and play no role in its assembly or in the catalysis of RNA synthesis. However, at many promoters, a big contribution is made by the promoter-proximal αCTD, which binds to the minor groove of promoter DNA at position −41 and interacts directly with Domain 4 of the σ subunit bound to the −35 element (44, 113). The contribution of the different elements differs greatly from one promoter to another, and bacterial promoters appear to have evolved by a pick-and-mix mechanism that has resulted in hierarchies of promoter activities that permit a 1000-fold dynamic range in transcript initiation frequencies (55, 92). Although to date the time course of recognition of the different elements has been studied at very few promoters, it seems likely that for many cases the initial encounter of polymerase involves the promoter UP element and −35 element, with interactions with the downstream elements coming later (16, 30, 116, 124). In particular, recent structural studies from Feklistov & Darst (40) and Zhang & Ebright (personal communication) show that interactions between σ and the −10 element take place only after promoter melting.

Irrespective of differences in the functional elements and the pathway to the open complex, the overall organization at the heart of most final open RNA polymerase holoenzyme-promoter complexes appears to be very similar. At many promoters, these complexes, which are competent for transcript initiation, can form without help from any other factor. A key point, established by early studies with the *lac* operon promoter, is that both activator-dependent and activator-independent promoters use the same overall pathway to transcript initiation (1, 130). Hence, the different determinants for promoter recognition and open complex formation outlined above are the same at activator-dependent promoters. Thus, it is useful to consider activator-dependent promoters as activator-independent promoters that have become handicapped in some way. For example, a promoter may be activator dependent because it carries one or more defective elements so that the initial binding of polymerase is reduced, and it is easy to see how the degree of dependence on an activator could be set by the precise nature of these elements (115). Put simply, reducing the intrinsic activity of a promoter leads to the possibility of increasing its activity by another factor, thereby introducing a step at which promoter activity can be regulated by some environmental factor. Of course, some promoters are so defective that they exhibit no factor-independent activity.

Bacteria use two distinct sets of mechanisms for activation of transcript initiation and these are focused either on the promoter or on the RNA polymerase. In promoter-centric mechanisms, factors interact with the promoter to improve its ability to guide RNA polymerase to initiate transcription, either by providing additional functional determinants to the promoter or by reversing the action of a repressor. In RNA polymerase–centric mechanisms, factors interact with RNA polymerase to alter its promoter preference. The simplest scenario is when the housekeeping sigma factor of a subpopulation of polymerase molecules is replaced by an alternative sigma factor (50, 60). In other cases, holoenzyme containing the housekeeping factor is altered in order to direct it to certain promoters, and this is often referred to as RNA polymerase appropriation (52).

SIMPLE TRANSCRIPTION ACTIVATION

In this section we consider the simplest scenario of transcription activation in which a single activator molecule binding at a bacterial target promoter is essential for it to be served by holo-RNA polymerase containing σ^{70}. Most simple activators function either by stabilizing the initial polymerase-promoter complex or by accelerating the transition to the open complex. An increase in the rate of polymerase passage through the initiation pathway will result if the free energy of the initial polymerase-promoter complex is reduced, or if the energy barrier for the transition to the open complex is reduced (115). In principle, there are two ways in which a transcription factor could directly influence these parameters. The activator might alter the conformation of the promoter DNA to improve the promoter quality, or the activator could interact directly with RNA polymerase to compensate for the defects in a promoter.

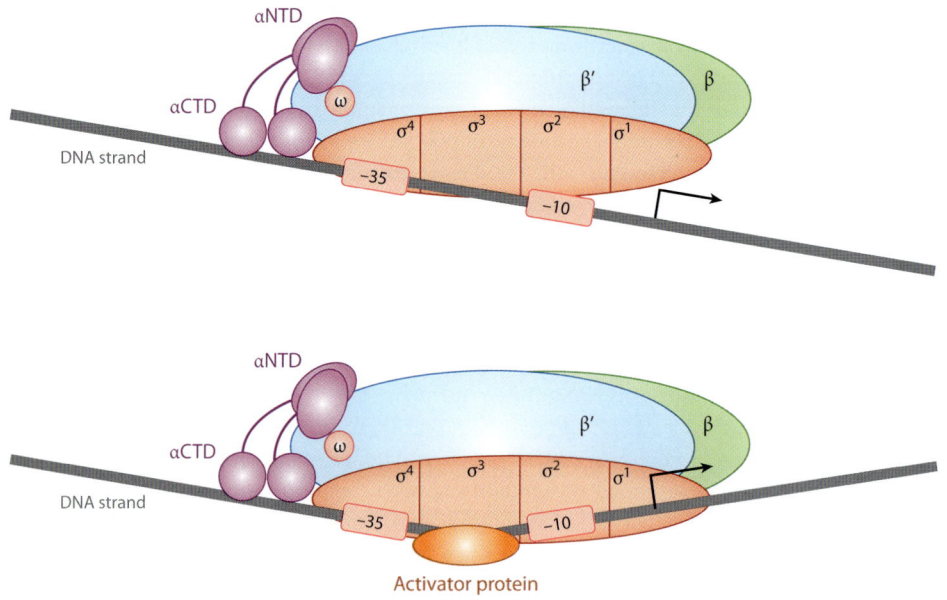

Figure 2

Simple activation by conformational change. The activator protein binds between the −10 and −35 elements and induces a conformational change in the DNA, allowing RNA polymerase to fully engage with the promoter. Abbreviations: CTD, C-terminal domain; NTD, N-terminal domain.

Activation by Altering Promoter Conformation

A few well-characterized bacterial regulatory factors work by inducing a conformation change in promoter DNA. The best-understood cases involve members of the MerR family of transcription factors that mostly bind between the −10 and −35 elements at target promoters (11, 107). Examples include the Tn*901* MerR protein at the *merP* promoter, which controls expression of a mercury resistance determinant, and BmtR, a MerR family member found in *Bacillus subtilis* that controls expression of efflux pumps and whose activity is triggered by xenometabolites. Target promoters for these and other related activators are defective because of nonoptimal spacing between the −35 and −10 elements. Thus, after initial binding of RNA polymerase to the promoter UP element and −35 element, the −10 element is misplaced and promoter melting, unwinding, and interaction with Domain 2 of the σ subunit are hindered (**Figure 2**). This hindrance is overcome by activated MerR or BmtR, which causes a twist in the spacer that results in the promoter −10 element being brought into register with the −35 element, thereby triggering transcription activation. For many years, it was believed that the twist was a smooth change in the winding of the spacer element, but high-resolution structural studies from Brennan and colleagues (51, 99) have now shown how MerR family activators cause a specific localized distortion that aligns the −10 element with the −35 element. This mechanism was originally thought to be restricted to the MerR family, but a recent report has suggested that other unrelated activators may use a similar mechanism (61).

The fact that a promoter can be activated by altering its conformation suggests the possibility of regulation by supercoiling without the direct intervention of transcription factors. In *E. coli* and related bacteria, it appears that supercoiling plays a global role in setting hierarchies of promoter activity, but it is rarely responsible for activation at specific promoters in response to specific signals (33). In contrast, in *Mycoplasma* species that contain very few transcription factors, changes in DNA topology may be responsible for the osmoregulation of certain genes, although it is unclear how promoter specificity is set (147). One possibility is that specific sequences located near a promoter transcript start point make it more or less sensitive to modulation by supercoiling. In this context, results from Hatfield and colleagues (101, 128) and Travers and colleagues (2, 103, 136) have shown how specific binding of certain *E. coli* nucleoid-associated proteins adjacent to a promoter can induce local changes in topology that can modulate promoter activity. Recall that the bacterial nucleoid-associated proteins are a small group of abundant proteins whose main functions are thought to be the folding and compaction of bacterial chromosomes, but many of these proteins also play key roles in transcriptional regulation (14). Hence, working with the *E. coli ilvG* promoter, Hatfield and colleagues (128) showed that upstream-bound IHF (integration host factor; a nucleoid-associated protein) influences the conformation of the neighboring promoter sequences, and defined the specific DNA base sequence that rendered the promoter susceptible to this activation. Here, IHF binding sets the conformation of neighboring sequences, thereby triggering promoter activation. Similarly, at the *tyrT* promoter, Travers and colleagues (2) found that upstream binding of another nucleoid-associated protein, FIS (factor for inversion stimulation), constrains DNA supercoiling, thereby activating the formation of transcriptionally competent complexes.

Activation by Direct Contact with RNA Polymerase

Genetic and biochemical studies at many different bacterial promoters have shown that a direct activator–RNA polymerase contact is essential for activation. At the majority of promoters that are dependent on a single activator, the activator binds to a DNA target either upstream of or overlapping the promoter −35 element (**Figure 3a,b**), so the bound activator can make a direct interaction with RNA polymerase when it engages with the promoter (20, 53, 104). Remarkably, irrespective of whether polymerase binding is dependent on or independent of an activator, the final organization of the open complex appears to be similar. Thus, most activator-dependent promoters are simply defective activator-independent promoters, with activator protein–polymerase interactions replacing promoter DNA–polymerase interactions (12, 53). The key evidence for this comes from the existence of positive control mutations that result in single amino acid substitutions that reduce or destroy the ability of an activator to interact with RNA polymerase without affecting its other functions such as DNA binding or regulation. Positive control mutations reduce or stop transcription activation and have been used to identify the amino acid side chains in activators that are essential for the direct interaction with polymerase. Analysis of positive control mutations in several bacterial transcription factors has shown that these side chains are often clustered and form small surface-exposed patches known as activating regions (20, 53, 80).

The discovery of direct contacts between bacterial transcription activators and RNA polymerase that are essential for promoter activation leads to interesting mechanistic questions. Are the contacts essential for the initial recruitment of polymerase to the promoter, required for isomerization to the open complex, or needed for promoter escape? Alternatively, could the contact push some activatory button in polymerase that, for example, somehow tweaks the active site? Although examples of all these possibilities are mentioned below, two important strands of evidence suggest that many (if not most) activators intervene early in the pathway to transcription initiation at most target promoters. First, in the two cases for which we have a high-resolution structure of a ternary complex between promoter DNA and an activator making contact with its cognate target in RNA polymerase (see below), it is clear that the structure of the polymerase target is unaltered by interaction with the activator (7, 64). Thus, the interaction appears to

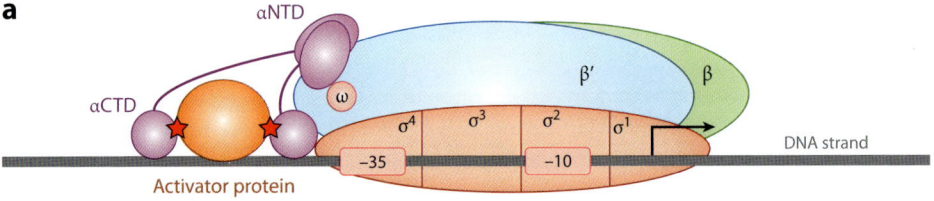

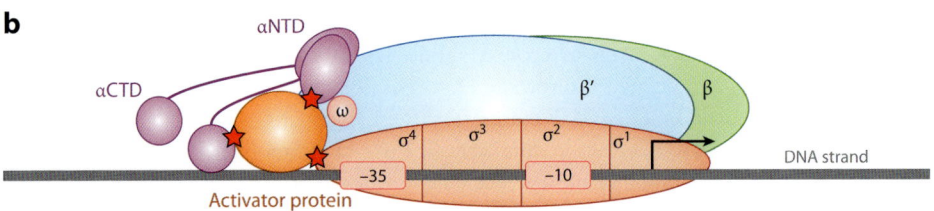

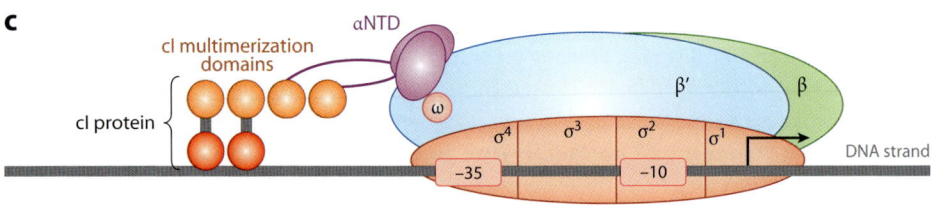

Figure 3
Simple activation by recruitment. (*a*) The activator protein binds upstream of RNA polymerase and contacts (*red stars*) one or both α subunit C-terminal domains (CTDs). (*b*) The activator binds close to RNA polymerase and interacts with different polymerase subunits. (*c*) Artificial activation in which the DNA-bound cI (bacteriophage λ repressor) protein activates transcription by interacting with the cI multimerization domains fused to the RNA polymerase α subunit N-terminal domain (NTD).

function merely as molecular velcro. Second, Hochschild and colleagues (37) have demonstrated that RNA polymerase can be recruited to promoters by nonnative contacts, and deduced that this recruitment is sufficient for promoter activation. The Hochschild experiments work by engineering *E. coli* to contain polymerase with a supplementary protein domain that can then be used to target the polymerase to specific promoters. One set of experiments exploit the fact that bacteriophage λ repressor (cI) consists of separate DNA-binding and multimerization domains, and use a synthetic test promoter with an upstream DNA site for cI located at a position where cI alone has no effect on transcription with normal polymerase (35, 37). However, if the cI multimerization domain is attached to the RNA polymerase α or ω subunits, then the test promoter can be activated by cI because the cI multimerization domain associates with the multimerization domain attached to the RNA polymerase α or ω subunits (**Figure 3c**). Hence, an arbitrary nonnative interaction can activate transcription. Another study from Hochschild and colleagues (46) exploits the bacteriophage T4 AsiA protein, which normally inactivates RNA polymerase holoenzyme by interacting with Domain 4 of σ^{70} and changing its conformation. AsiA was fused to a DNA-binding module and the hybrid protein was able to direct inactivated polymerase to promoters that carried the cognate target sequence for the DNA-binding module, resulting in activation at these promoters. Taken together, these experiments define the minimum requirement for transcription activation as any contact that recruits RNA polymerase to a target promoter, and they show that complex mechanisms involving activatory buttons and extensive conformational changes may often be unnecessary.

ACTIVATOR–RNA POLYMERASE INTERACTIONS

In this section, we describe the best-understood examples of simple transcription activation that involve direct activator–RNA polymerase interactions. Some of the examples cited have been studied for over 30 years and, rightly or wrongly, are taken as the paradigms for our understanding of less-well-established systems.

Activation by Bacteriophage λ cI Protein at the λ P_{RM} Promoter

Although bacteriophage λ cI protein is primarily a repressor, it also activates its own expression at the λ rightward promoter for the maintenance of lysogeny (P_{RM}). Activation results from the binding of a single cI dimer to a DNA site that abuts the P_{RM} −35 hexamer element. Activation is due to direct interactions between λ cI residues E34 and D38, which are immediately adjacent to the cI DNA recognition helix, and σ^{70} residues R588 and R596, located in Domain 4, just next to the determinant that contacts the −35 element. Residues E34 and D38 cI were originally found as the sites of positive control substitutions, and the cognate target in σ^{70} Domain 4 was identified from suppression genetic experiments in which allele-specific suppressors of positive control mutations in cI were found and characterized (82). These interactions have been confirmed by high-resolution structural studies (64) of the ternary complex between the DNA-binding domain of cI, a segment of DNA carrying the cI operator sequence and the P_{RM} −35 element, and Domain 4 of the housekeeping σ factor from *Thermus aquaticus* (which was used in preference to its *E. coli* counterpart to facilitate the structural analysis). From these studies, it is clear that the key activator–RNA polymerase contact takes place when both partners are DNA bound and is mediated by interactions between small clusters of amino acid side chains on cI and σ Domain 4 that are positioned immediately adjacent to the DNA. In addition, the interaction causes little or no conformation change in either protein partner, and this is a theme that is repeated in other systems (see below).

The action of cI at P_{RM} is of special interest, not only because it is a key component of the first genetic switch to be understood in detail, but also because, against all expectations, cI at P_{RM} has no effect on the initial binding of RNA polymerase but rather accelerates the formation of the open complex. Because the structural analysis shows that the cI–σ Domain 4 interaction has no effect on the conformation of σ Domain 4, the conclusion must be that the cI-activating region makes contact with its cognate target only after both cI and RNA polymerase have bound to the promoter, thereby driving RNA polymerase forward through the pathway to the transcriptionally competent open complex (64). Interestingly, with RNA polymerase containing σ^{70} with the RH596 suppressor substitution in Domain 4, cI-dependent activation at P_{RM} takes place via accelerated formation of the initial binding of RNA polymerase (81). Hence, an interaction between the same two surfaces can have different kinetic consequences, depending on the stage in the initiation pathway at which the interaction comes into play.

Key lessons of the λ cI story are that molecular velcro is sufficient for activation and that there is no need for activator-induced conformational changes in RNA polymerase, even to drive open complex formation. These lessons were underscored by another series of elegant experiments by Hochschild and colleagues (36) showing that the cI–σ Domain 4 contact in a different context could be used to drive RNA polymerase recruitment to a promoter. These experiments used the synthetic promoter described above, with a DNA site for cI located upstream of the promoter −35 element. At this promoter, the activating region of cI is unable to contact its target in σ Domain 4, and hence cI cannot activate transcription with normal RNA polymerase. However, with RNA polymerase carrying an ectopic σ Domain 4 fused to one of its α subunits, cI can activate, because the flexible α subunit linker permits formation of the cI–σ Domain 4 interaction that then recruits the polymerase to the promoter.

Another key lesson from the λ cI story is that, even though cI is a homodimer, only one of the two subunits makes contact with σ Domain 4 during activation. This is unsurprising because each RNA polymerase carries only one σ factor, but it underscores that, although a single protein-protein interaction is sufficient for transcription activation at a promoter, binding specificity of a transcription factor is dependent on reading 8–10 base pairs. This requires two recognition helix determinants that, in cI and in many other homodimeric transcription factors, are carried by the two subunits. During activation at P_{RM}, cI binds to the same face of the DNA helix as σ Domain 4, and hence it is the downstream subunit of the cI dimer that makes the activatory contact. This was shown by Kim & Hu (72), who engineered an asymmetric version of the cI dimer that bound in a specific orientation to an asymmetric DNA site for cI. Positive control substitutions could then be targeted to either cI subunit to identify which subunit makes the key contact with σ Domain 4.

Domain 4 of σ Factors as a Target for Transcription Activators

The positively charged residues in σ^{70} Domain 4 that interact with λ cI during activation at the λ P_{RM} promoter are located on one face of an α-helix that is immediately adjacent to the interface between Domain 4 and the promoter −35 element. As might be expected, many other transcription activators have evolved to exploit this target (34, 84). Examples include bacteriophage Mu Mor protein, E. coli FNR (regulator of fumarate and nitrate reduction) protein, many response regulators, and most members of the AraC family of transcription activators. In order to make contact with this target, activators have to be precisely positioned and bind to the same face of the promoter as σ Domain 4. This is because the positioning of Domain 4 with respect to the rest of RNA polymerase at promoters is critical, with little room for flexibility (96). Thus, for many AraC family members, transcription activation at target promoters is contingent on binding to a target that overlaps the upstream end of the −35 hexamer, and this binding is triggered by an activatory ligand (90). Examples include the action of arabinose with AraC (121), melibiose with MelR (45), and rhamnose with RhaS and RhaR (142). As with all AraC family members, these factors all contain a ∼100-amino-acid module that carries two helix-turn-helix DNA-binding determinants that insert into two adjacent major grooves. The bound module is oriented by this interaction. Genetic analyses with RhaR, RhaS, and MelR have shown that the C-terminal helix-turn-helix binds adjacent to σ Domain 4 bound at target promoter −35 elements and identified a conserved negatively charged amino acid residue (D261 in MelR and D241 in RhaS) that interacts with σ^{70} residues R588 and R596 (8, 45). A similar set of interactions appears to be involved with MarA, SoxS, and AraC (and doubtless dozens of other AraC family members). Hence, here we have a rerun of activation by cI at P_{RM}, and it is interesting to note that AraC promotes both the initial binding of RNA polymerase at the *araBAD* promoter and passage to the open complex (148). This should not be surprising because we know that σ Domain 4 engagement with −35 elements is central to promoter recognition and that evolution uses the same mechanisms over and over again.

The key point is that for an activator to contact σ^{70} residues such as R588 and R596, it has to be carefully positioned. This is neatly illustrated by the Ada transcription factor, whose activity is triggered by methylation damage. Hence, at the *aidB* and *ada* promoters, the principal Ada-activating region interacts with a target in σ Domain 4 that overlaps the target for λ cI and AraC family members (76). However, at the *alkA* promoter, the Ada-binding target is offset upstream by a few base pairs, and thus completely different residues in Ada and σ supply the activatory contact (77). Note that some activators contact DNA targets that overlap target promoters' −35 element, but are unable to contact σ Domain 4 (discussed below).

Activation by CRP at the *lac* Promoter

Transcription initiation at the *E. coli lac* operon promoter is highly dependent on the cyclic AMP receptor protein (CRP), a global regulator whose activity is dependent on cyclic AMP (23, 129, 149). Thus, even when the *lac* repressor is induced, the *lac* promoter requires activation by the binding of homodimeric CRP to a 22-bp DNA target centered between base pairs 61 and 62 upstream from the transcript start point (i.e., position −61.5). When bound at this target it is clear that direct interaction with σ^{70} Domain 4 would be impossible unless there were a massive conformation change in the promoter DNA (38). The solution to the puzzle came from Ishihama and colleagues (58, 59), who worked with holo-RNA polymerase reconstituted with α subunits lacking the C-terminal domain (residues 235–329). They found that, whereas this form of RNA polymerase was competent for transcription initiation at many promoters, it was defective for CRP-dependent activation at the *lac* promoter. The subsequent discovery that the RNA polymerase αCTD could fold into an autonomous DNA-binding unit, and that it was joined to the αNTD by a flexible linker (9, 65, 112), prompted the suggestion that CRP activated the *lac* promoter by interacting with αCTD. This has been proved by work from Ebright and colleagues (78) that culminated with the structure of the ternary CRP–holo-RNA polymerase–promoter complex, deduced from electron microscopy, exploiting high-resolution X-ray structures of the component parts (56). The structure shows one αCTD sandwiched between the bound CRP dimer and σ Domain 4. Strikingly, αCTD, σ Domain 4, and the other parts of RNA polymerase bind to the promoter almost exactly as if it were a factor-independent promoter.

CRP provides a supplemental functional determinant to the *lac* promoter that stabilizes the formation of the initial RNA polymerase–promoter complex (22, 78). Genetic studies defined this activatory determinant as a small surface-exposed patch (known as activating region 1, AR1), comprising amino acid side chains in a β-turn in the DNA-binding domain, immediately adjacent to the DNA recognition. High-resolution structural analysis (7) of the ternary CRP-αCTD-DNA complex shows that this determinant interacts directly with specific residues of αCTD near position 287 (known as the 287 determinant). Thus, αCTD in the ternary CRP–holo-RNA polymerase–promoter complex makes direct contact with AR1 of CRP via the 287 determinant, with the *lac* promoter UP element via the 265 determinant (residues near R265), and with σ Domain 4 via the 261 determinant (residues near D261) (27, 56, 119). Significantly, the interaction between CRP and αCTD causes no detectable change in the conformation of either CRP or αCTD, and again, this argues that CRP, like cI, activates by a simple recruitment mechanism involving molecular velcro.

An interesting aspect of CRP-dependent activation at the *lac* promoter is that, although each subunit of the CRP homodimer contains AR1 and RNA polymerase contains two αCTDs, activation requires only one AR1-αCTD interaction. Experiments with CRP dimers containing one wild-type AR1 and one mutated nonfunctional AR1 show that only the AR1 in the downstream subunit is required for activation (150), and this AR1 contacts the αCTD that interacts with σ Domain 4 (56, 78). In these experiments, CRP heterodimers were oriented by using derivatives carrying substitutions in the DNA recognition helix and asymmetric 22-bp DNA sites for CRP. The location of the second αCTD during *lac* promoter activation is less well defined, but it appears to bind upstream and to make no functional interaction with AR1 in the upstream subunit of the CRP dimer. This result was confirmed using RNA polymerase reconstituted with one full-length α subunit and one α subunit lacking αCTD (83). This one-armed RNA polymerase is fully functional for CRP-dependent activation at promoters with an architecture similar to that of the *lac* promoter. Hence, as with cI, although the binding specificity of CRP requires two subunits, the activating region in only one subunit is required for the activation function.

αCTD as a Target for Transcription Activators

The RNA polymerase αCTD acts as an antenna for RNA polymerase, by interacting either with UP elements or with transcription activators such as CRP, and although the number of proven and completely characterized cases is still quite small, it is likely to be a target for scores of activators in *E. coli* and other bacteria (**Figure 3a**). Because αCTD is joined to αNTD, and hence the rest of RNA polymerase, by a nonstructured linker, there is considerable flexibility with respect to where an activator can be positioned for productive activation (39). An interesting aspect of the *lac* promoter is that, if the DNA site for CRP is moved upstream by 11 base pairs from position -61.5 to -72.5, CRP still can make an activatory interaction with RNA polymerase (133). Systematic studies showed that CRP could activate transcription from a single site at positions -72.5 and -82.5, and some activation was found from position -93.5 (42, 137). Because the action of CRP from all these locations is stopped by substitutions that render AR1 nonfunctional, activation from each of these locations must depend on AR1 interacting with αCTD.

A similar set of observations was reported for FNR (145), which is the global transcription regulator for adaptation to anaerobic conditions and is structurally and functionally related to CRP (see sidebar CRP AND FNR). The helical periodicity for activation in these cases must be due to the need for the activator to be lined up on the same face of promoter DNA as RNA polymerase, and it is now clear that this causes wrapping of upstream promoter sequences. Note, however, that it is possible for factors to make activatory interactions with one or more αCTDs without binding to the same face of the DNA as the rest of RNA polymerase. Such activators interact with a different surface of αCTD. This is likely to be the case for some response regulators [e.g., NarL at the *E. coli ogt* promoter (131) or BvgA at the *Bordetella pertussis fim3* promoter (31)], but the best-understood case is the activation of the bacteriophage λ P_{RE} promoter by the cII protein.

CRP AND FNR

The cyclic AMP receptor protein (also known as CAP, catabolite gene activator protein) was originally discovered as an essential activator of the *E. coli lac* promoter but was subsequently found to modulate transcription at hundreds of promoters (129, 149). In 1983, Guest and colleagues (127) reported homologies between CRP and FNR, the global transcription activator that controls fumarate and nitrate reduction and other genes induced during anaerobic growth. Since then, the CRP family has continued to grow. CRP was the first transcription activator to be purified and characterized structurally and, together with FNR, has proved to be a useful paradigm for understanding transcription activation (78). Members of the CRP family of transcription factors play diverse roles in many different bacteria.

CLASS I AND CLASS II ACTIVATION

The terms Class I and Class II were introduced in the 1990s, following studies on CRP, to describe activation by upstream-bound CRP that interacts with the C-terminal domain of the RNA polymerase α subunit (Class I), or activation by CRP bound to a target that overlaps the promoter −35 element, where αCTD is dispensable (Class II) (58, 137) (as in **Figure 4a,b**). These terms have proven useful and are often applied loosely to describe the actions of other bacterial transcription activators. However, it is now clear that many activator-dependent promoters do not fit with such a simple classification. The term Class III is often used to describe promoters in which two transcription factors make independent contacts with polymerase (as in **Figure 4c,d** and **Figure 7a①,②**).

At λ P_{RE}, cII binds as a tetramer that recognizes two four-base elements that flank the promoter −35 hexamer element. Although it appears at first sight that cII might interact directly with σ Domain 4, model building, based on high-resolution structures of cII bound to its target, argues for a contact between the cII subunit bound to the upstream four-base element and the αCTD bound at position −41 (63), and this is supported by a vast amount of mutational analysis (70). The significance of this is that even transcription factors that bind to targets overlapping with promoter −35 elements can function by interacting with αCTD. Note too that the αCTD that contacts cII is located exactly where it would be at an activator-independent promoter, and again, it is unnecessary to postulate any activator-induced conformational changes in RNA polymerase.

Organization of Class II CRP- and FNR-Dependent Promoters

Early in the study of transcription activation by CRP and FNR, it became apparent that the majority of their target promoters did not follow the *lac* promoter model (58). Such promoters are often referred to as Class II promoters, and these are defined by their target site for CRP or FNR overlapping the promoter −35 element (21) (in contrast to Class I promoters such as *lac* for which the activator binding site is upstream of the −35 element) (see sidebar Class I and Class II Activation; **Figure 4a,b**). In fact, at these Class II promoters, the location of bound CRP or FNR is similar to that of cI protein at λ P_{RM}. Extensive genetic and biochemical studies have shown that bound CRP (or FNR) prevents αCTD from contacting σ Domain 4, that αCTD is displaced to bind just upstream of the activator, and that the displaced αCTD makes a direct interaction with the upstream subunit of the activator homodimer (5, 144, 151). For CRP, this interaction involves AR1 and the 287 determinant of αCTD, and FNR contains an equivalent to AR1 (79, 120). Although these interactions contribute to activation at most Class II promoters, they are not indispensable (21). Hence, holo-RNA polymerase reconstituted with α subunits lacking the C-terminal domain is competent for Class II activation (58). For CRP-dependent Class II promoters, the major activatory determinant is a second activating region, AR2, that is a small, positively charged surface, distinct from AR1, that interacts with a target on the surface of the αNTD (100). Kinetic analysis shows that, while the role of the AR1-αCTD interaction promotes RNA polymerase recruitment, the AR2-αNTD interaction promotes transition to the open complex (100, 110, 134).

Wild-type CRP appears not to make any direct activatory interaction with σ at Class II promoters, but substitution of CRP residue K52 unmasks a third activating region, AR3, that interacts with σ Domain 4 (108, 109). Hence, CRP can easily be converted from activating by interaction with one part of RNA polymerase (αNTD) to activating by interaction with another part (σ Domain 4). Experiments with oriented heterodimers of CRP have shown that AR2 and AR3

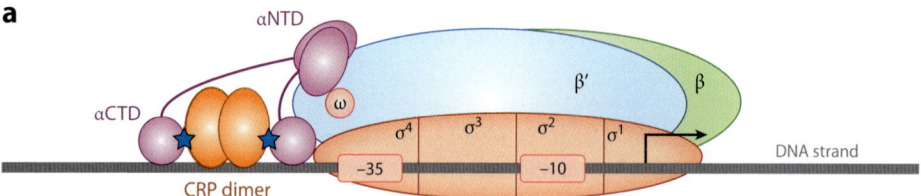
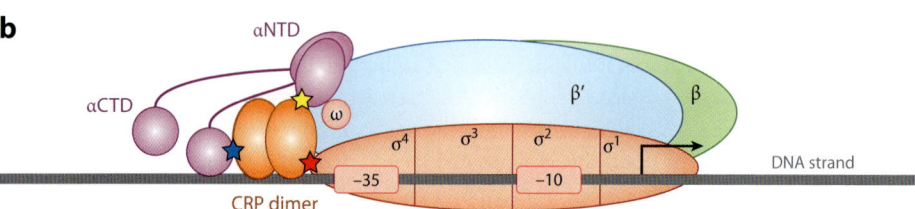
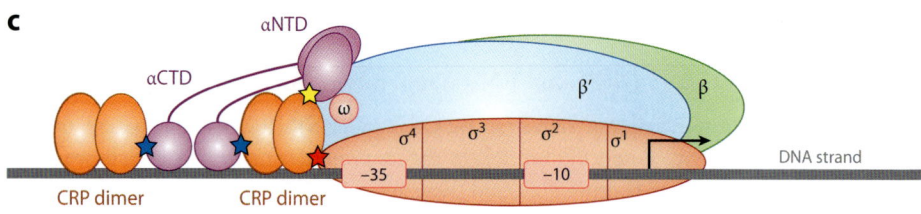
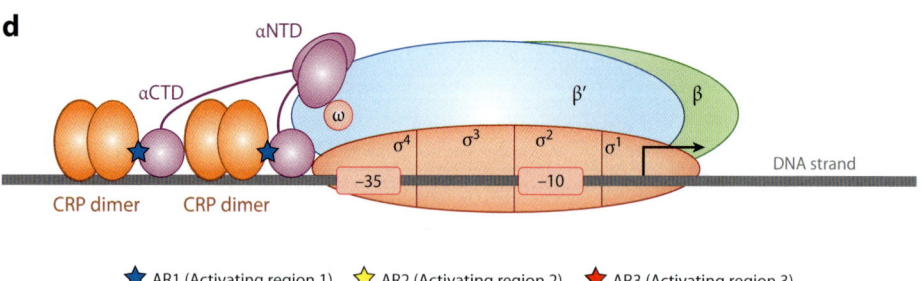

Figure 4

Activation of transcription by the cyclic AMP receptor protein (CRP). (*a*) The CRP dimer binds upstream of RNA polymerase and contacts one or both α subunit C-terminal domains (CTDs) via AR1 (activating region 1) (*blue star*). (*b*) CRP binds adjacent to RNA polymerase and interacts with different polymerase subunits: AR1 interacts with αCTD, AR2 interacts with an α subunit N-terminal domain (NTD), and AR3 interacts with σ^4. (*c*) Activation by tandem-bound CRP dimers where the downstream CRP is in a Class II position (as in panel *b*). (*d*) Activation by tandem-bound CRP dimers where the downstream CRP is in a Class I position (as in panel *a*).

are functional in the downstream subunit of the homodimer (100, 143). Hence, at Class II promoters, both subunits of the CRP dimer make activatory interactions with RNA polymerase. For FNR-dependent Class II promoters, the major activatory determinant in FNR corresponds to AR3 of CRP and AR2 plays a much smaller role (80). Thus, for members of the CRP-FNR family, different combinations of functional activating regions must have evolved, thereby enabling activation at promoters with differing architectures. It is easy to imagine how similar evolution has occurred in other transcription factor families.

Promoters with Tandem DNA Sites for CRP or FNR

Figure 4 illustrates some of the diverse organizations found at CRP-dependent promoters. Many CRP-regulated promoters carry two or more DNA sites for CRP (129, 149). At some promoters, full activation can be achieved with just one bound CRP and thus the other target sites are redundant. In other cases, two bound CRPs are essential for optimal activation, and at many of these promoters, the two sites have different binding affinities for CRP. This can result in a nonlinear output from the promoter during growth transitions when CRP is activated or deactivated. It is presumed that this complexity has evolved to tailor the output of certain promoters during such transitions (23, 73).

At promoters that are dependent on CRP binding to tandem targets, each bound CRP makes an independent activatory contact with one of the two RNA polymerase αCTDs (6, 69, 95). This mechanism works because, at simple CRP-dependent promoters, there is always a spare αCTD available for interaction with a correctly positioned second CRP (83). The most commonly found organization is for one CRP site to overlap the promoter -35 element and to activate by the Class II mechanism described above (**Figure 4c**). Further activation is observed when a second CRP binds upstream near positions -74, -82, -93, or -102, and extensive biochemical studies have shown that AR1 in the upstream-bound CRP recruits the spare αCTD to the promoter (6, 95). At this type of promoter, often referred to as a Class III promoter, the two bound CRPs together exploit the two RNA polymerase αCTDs to generate enough velcro for optimal RNA polymerase recruitment. Hence, one-armed RNA polymerase reconstituted with one full-length α subunit and one α subunit lacking αCTD, though competent for CRP-mediated activation at a promoter dependent on one bound CRP, is unable to activate at a promoter dependent on tandem-bound CRP (83). Similar synergy between tandem-bound CRPs can be generated when both CRPs bind upstream of the promoter -35 element (**Figure 4d**) (4, 69, 135). The simplest situation is when the promoter-proximal CRP, located at position -61.5, as at the *lac* promoter, activates together with upstream-bound CRP near position -92 or -103. Again, both αCTDs are essential for such activation.

Because FNR is so similar to CRP, it was expected that some promoters would be codependent on tandem-bound FNR for optimal activation. However, surprisingly, FNR-dependent promoters cannot be substantially further activated by upstream-bound FNR, and indeed, when located at certain positions, upstream-bound FNR is inhibitory (3). The clearest example of this is the *E. coli yfiD* promoter, which is dependent on FNR binding to a single target centered at position -40.5 (87). Binding of FNR to a second weaker upstream site at position -93.5 causes a repression of *yfiD* promoter activity, and this appears to be due to a direct interaction involving AR1 of FNR. This complexity provides a mechanism for microanaerobic induction of gene expression. Thus, when FNR becomes activated in response to oxygen depletion, it occupies the site at position -40.5 and activates *yfiD* expression. Subsequently, as FNR becomes fully activated, the upstream site at -93.5 is filled, resulting in downregulation of expression. Hence, the interplay of FNR binding at two targets with different regulatory outcomes results in transient promoter activation when the environment switches from aerobic to fully anaerobic conditions.

PROMOTERS DEPENDENT ON σ^{54}: THE SECOND ACTIVATION PARADIGM

Unlike holo-RNA polymerase containing σ^{70} (and most alternative σ factors), polymerase containing σ^{54} is unable to serve target promoters independently of a transcription activator. Genes encoding σ^{54}-like proteins are found in ~60% of bacteria but the origins of σ^{54} are unclear because it is structurally distinct from σ factors related to σ^{70} and does not share their characteristic domains (91).

RNA polymerase containing σ^{54} serves a variety of target promoters, and the key sequence elements at these promoters are located near positions −12 and −24. RNA polymerase can recognize these promoters, but it is unable to form transcriptionally competent open complexes without the intervention of a specialized transcription activator that contains an AAA+ activator domain that interacts directly with σ^{54} (43). These activators, often known as enhancer-binding proteins, typically contain three domains, a regulatory domain responsive to a particular metabolic signal, the AAA+ activator domain, and a DNA-binding domain, and they self-assemble into hexamers that bind to DNA targets at σ^{54}-dependent promoters. These targets are mostly found 100–150 base pairs upstream of the transcript start, but they can function up to 1,000 base pairs upstream and a DNA loop has to form in order for the activator domain to contact σ^{54}. DNA-bending proteins such as IHF can help in the formation of these loops (**Figure 5**), and a careful study of the requirements for positioning IHF sites led to the conclusion that the activator domain was presented to RNA polymerase from the opposite side of the target DNA (57). Cryo-electron microscopy of complexes between the PspF activator domain and RNA polymerase σ^{54} holoenzyme showed this directly, and it was possible to visualize protrusions from the activator domain that correspond to surface-exposed loops that make direct contact with σ^{54} in the holoenzyme (10).

AAA+ activator domains couple ATP hydrolysis to movement in a wide variety of systems in all kingdoms of life. Thus, the current model for σ^{54}-dependent activation is that ATP hydrolysis causes motion of the surface-exposed loops in the activator domain so that they interact with σ^{54} to overcome the blockage between the closed and open complexes (67). This has been referred to as the second paradigm for transcription activation because the activator domain is targeted directly to push an essential button needed to remove the obstacle to open complex formation

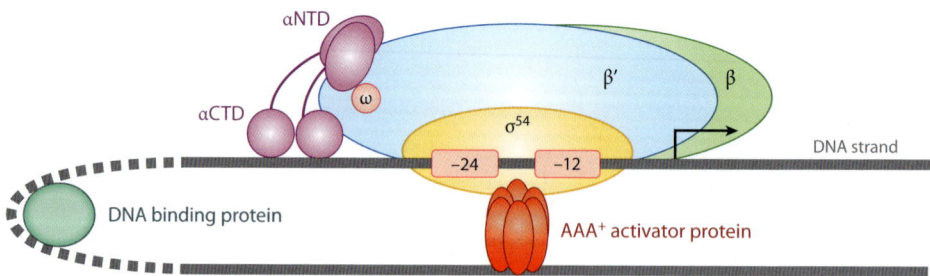

Figure 5

Transcription activation at a σ^{54}-dependent promoter. RNA polymerase containing σ^{54} binds to the −12 and −24 promoter elements. Interaction with the AAA+ activator protein is often facilitated by DNA looping caused by a DNA-bending protein. Abbreviations: CTD, C-terminal domain; NTD, N-terminal domain.

(15). This is in contrast to most of the mechanisms discussed above, where the activator's strategy is to recruit or guide RNA polymerase holoenzyme, but then let it get on with transcript formation as an activator-independent promoter.

At least in the cases that have been studied thus far, there are no big fluctuations in σ^{54} levels during bacterial growth or adaptation (66). Thus, regulation is imposed by the different AAA+ activator proteins whose activities are modulated by diverse mechanisms in response to different metabolic or environmental conditions. This is in sharp contrast with most other alternative σ factors whose activity is regulated either by their level or their availability, without the intervention of transcription factors (see below).

RNA POLYMERASE APPROPRIATION

Alternative σ Factors

RNA polymerase–centric activation mechanisms involve proteins that reprogram the promoter preferences of the polymerase holoenzyme (as opposed to improving the attractiveness of promoters for RNA polymerase); some of these mechanisms are illustrated in **Figure 6**. The best example is when, in response to specific signals, the housekeeping factor is replaced with an alternative σ factor, thereby changing the promoter preferences of a proportion of the cell's polymerase (50). In some cases, this reprogramming mediates global responses to a general stress, while in others, the alternative σ factor participates in driving a developmental pathway such as sporulation. The activity of most alternative σ factors is controlled by their availability in the cell, and therefore, target promoters tend not to require transcription activators and contain recognition elements that resemble the consensus (74). However, there are some exceptions. For example, the σ^{28}-dependent *E. coli aer* promoter is activated by CRP (54). The consequence of this is that, in certain conditions, when σ^{28} levels are rising or falling, the *aer* promoter has an advantage over other σ^{28}-dependent promoters. Although it was previously thought that regulons controlled by different σ factors contain distinct sets of genes, genome-wide studies have now shown that many genes can be served by RNA polymerases carrying different σ factors (140).

Appropriation and Pre-Recruitment

It can be argued that mechanisms that focus on RNA polymerase rather than promoters are the regulatory method of choice when bacteria need to respond efficiently to ensure their survival. Hence, a sudden heat shock demands instant action to avoid death, and this can be done by σ factors appropriating RNA polymerase, whereas the choice to metabolize arabinose or lactose is not so pressing and is resolved by transcription factors that change the competition between different promoters (50). Examples of this include the *E. coli* SoxS and MarA transcription activators, each of which induces expression of large stress regulons by binding to a common DNA target known as the Mar box (88). Both factors are AraC family members that contain just the DNA binding/activation domain and their activity is controlled by their level. They were thought to bind to Mar boxes at target promoters and activate by the Class I or Class II mechanisms outlined above. However, there is now some evidence that SoxS and MarA bind directly to RNA polymerase away from target promoters, and this has been dubbed pre-recruitment (47, 48, 89). Genetic and biochemical analyses argue that the binding target in RNA polymerase is the determinant in αCTD that interacts with UP elements (29, 126). Hence, upon binding to RNA polymerase, SoxS and MarA switch the αCTD DNA-binding determinant from recognizing UP elements to recognizing Mar boxes. Because UP elements play a big role in directing large amounts of

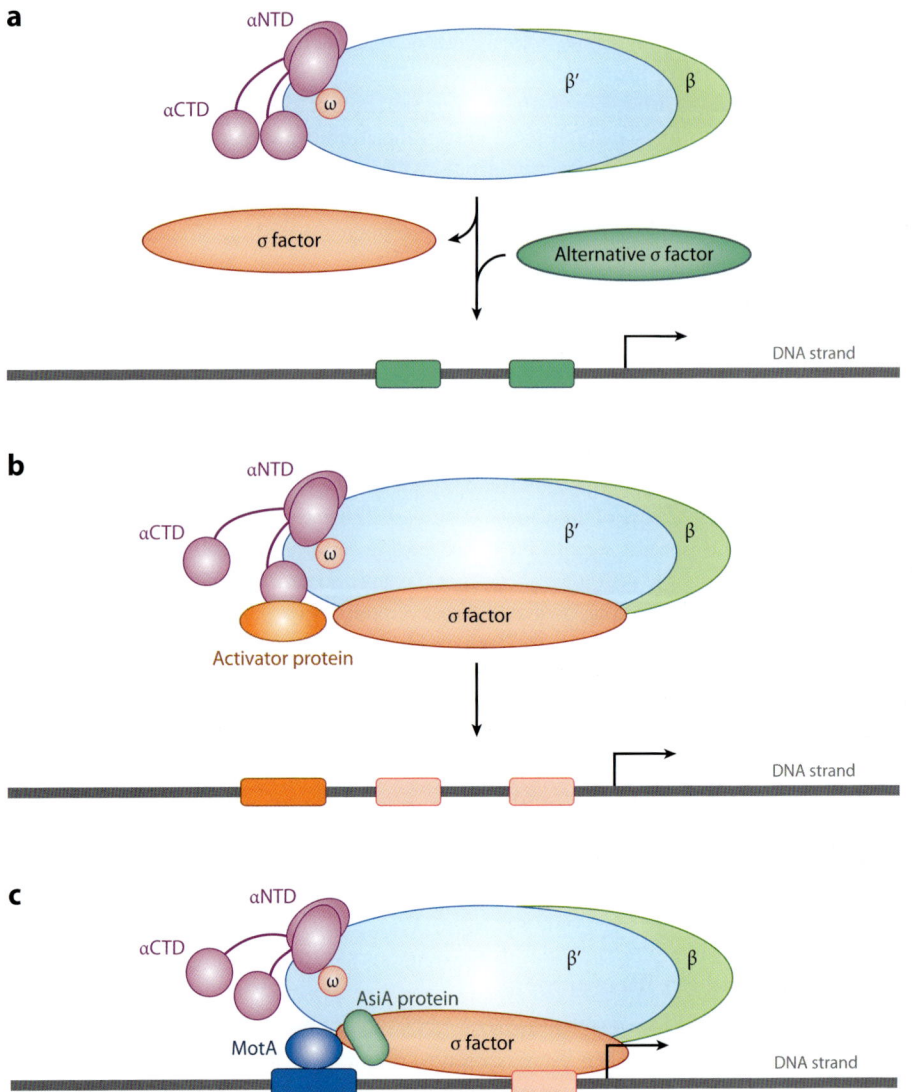

Figure 6

Activation by appropriation and pre-recruitment. (*a*) σ factor appropriation. RNA polymerase is reprogrammed by an alternative σ factor, thus altering the promoter preference. (*b*) Pre-recruitment by an activator protein directs RNA polymerase to a specific set of promoters. (*c*) Appropriation of RNA polymerase by the T4 phage AsiA protein. AsiA remodels σ^{70}, enabling MotA to interact and direct transcription from middle phase T4 promoters. Abbreviations: CTD, C-terminal domain; NTD, N-terminal domain.

RNA polymerase to transcription units involved in growth, it is easy to see how the SoxS/MarA pre-recruitment can result in a population of RNA polymerase molecules tasked to deal with an urgent stress rather than with growth (141, 146).

As expected, bacterial viruses are the masters of appropriation. The best example is perhaps the AsiA gene product of bacteriophage T4, which is dependent on the host multisubunit RNA polymerase for its development because its genome does not encode its own bespoke polymerase. AsiA protein is expressed early during infection, binds tightly to σ^{70}, and was originally thought to be just an anti sigma, thereby shutting down host transcription (28). However, it was subsequently shown that AsiA binds to and remodels Domain 4 of σ^{70} in the host RNA polymerase holoenzyme, and that this remodeling is essential for the function of the T4-encoded MotA transcription activator that is required for the middle phase of T4 transcription (24, 52, 75). Thus, AsiA can be thought of as an appropriator that remodels part of σ^{70} to prevent recognition of host -35 elements while facilitating MotA action and the development of T4 infection.

INDIRECT ACTIVATION

Alongside the direct transcription activation mechanisms discussed above, bacteria use other ingenious indirect ways to control transcription positively. For example, upregulation of gene expression can be due to inversion switching of a DNA segment carrying a promoter, or it can result from the insertion of a transposon that carries a full promoter or a promoter element (118). Alternatively, in some bacteria, clonal variation of promoter sequences is responsible for up- or downregulation of gene expression (94).

The most frequently found mechanism for indirect activation of transcription in bacteria is the relief of repression. In one scenario *dam*-mediated DNA methylation prevents repressor binding, and at some promoters this is an important mechanism for toggling promoters between on and off states (85). However, in most cases, relief of repression is due to the binding of a protein factor that then disrupts repression caused either by a specific repressor (41) or by nucleoid-associated proteins (32). Such protein factors, which appear to function as activators, are perhaps more aptly termed antirepressors. Good examples of antirepressors are found in many pathogenic bacteria where the expression of genes involved in virulence is repressed by the H-NS nucleoid-associated protein. Hence, Ler protein in enteropathogenic *E. coli* and VirB in *Shigella flexneri* are unable to activate transcription directly at promoters, but they trigger the expression of virulence genes by disrupting repression mediated by H-NS (132). Such repression can be disrupted in many different ways, including by a bona fide repressor. Thus, in a clever piece of synthetic biology, Caramel & Schnetz (25) showed that the Lac repressor could break H-NS-dependent repression of the *E. coli bgl* promoter, thereby demonstrating that the best-known transcription repressor could also become involved in transcription activation.

INTEGRATION OF REGULATORY SIGNALS AT PROMOTERS

Bacterial transcription factors have evolved to couple environmental signals to transcription output; thus, each factor's activity is directly or indirectly controlled by a specific signal. Because bacteria have to sense different combinations of signals, it is unsurprising to find that most promoters are regulated by multiple transcription factors and that the role of the promoter is to integrate the different signals into one output. Many promoters are codependent on two or more transcription activators, and the most common scenario is for each of the activators to bind independently and make independent contacts with RNA polymerase. Thus, just as some CRP-dependent promoters depend on CRP binding at two upstream sites, a promoter can be dependent on two different upstream-bound activators. The first reported cases of this were an artificial promoter dependent on activation by CRP and phage λ cI protein (68) and the *E. coli ansB* promoter (125), but subsequently, scores of examples have been found and they appear to fall into two groups (71). In the first group, the promoter-proximal activator overlaps the target promoter -35 element and functions as a Class II activator (**Figure 7a**①). In the second group, the promoter-proximal activator binds

a Independent contacts

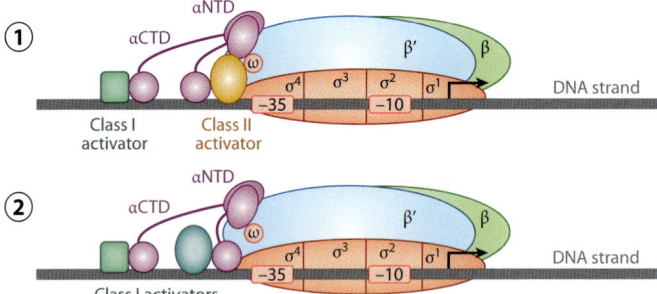

b Cooperative binding

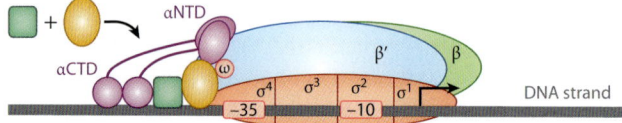

c Antirepression

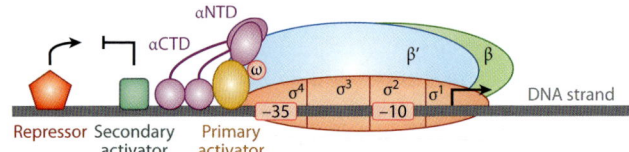

d Repositioning

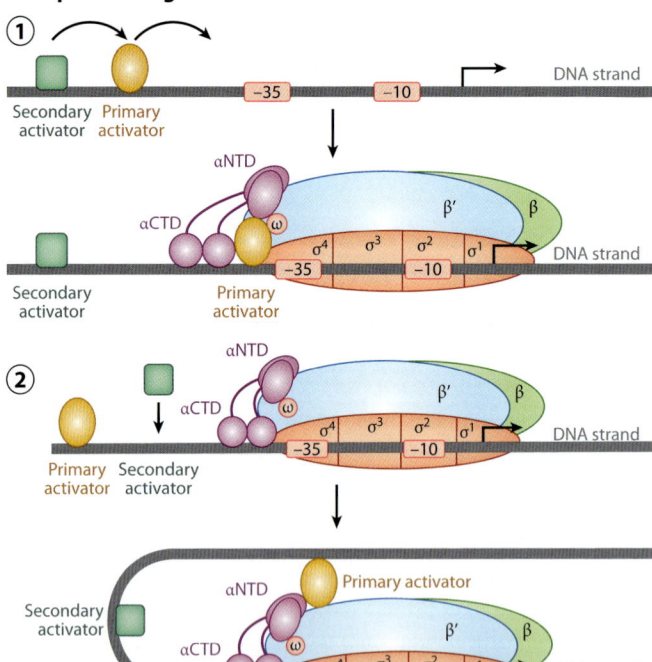

upstream of the −35 element and functions as a Class I activator (**Figure 7a**②). In both groups, the promoter-distal activator contacts the spare αCTD. At such promoters, the flexible linker connecting the α subunit domains allows considerable variability in the way the activator targets can be organized. Note too that for a promoter to be codependent on two activators, it must evolve such that each individual activator is unable to do the job by itself. This is often fixed by weaker binding sites, suboptimal −35 elements or UP elements, or suboptimal spacing. For example, in the pathogenic enteroaggregative *E. coli* strain 042, the promoter-controlling expression of the plasmid-encoded toxin depends on both upstream-bound Fis and CRP binding to a target that overlaps the −35 element. CRP appears unable to activate the promoter alone because its DNA site is too close to the −10 element (114). Hence, insertion of a single base between the promoter −10 and −35 elements relieves the dependence on Fis.

At most bacterial promoters where more than one different activator is required, these bind independently. A likely explanation for this is that bacteria rely on rapid evolution for their survival, and one element of this involves mixing and matching transcription activators to activate synergically at promoters. This is facilitated if direct interactions between the different factors are not required. However, there are a few examples in which different activators bind cooperatively (**Figure 7b**), such as MelR and CRP at the *E. coli melAB* promoter (139); possibly ToxR and TcpP at the *Vibrio cholerae* promoter (93); and GadE and RcsB at the *E. coli gadA* promoter, where a GadE/RcsB heterodimer forms (26).

At some bacterial promoters that are codependent on two activators, the primary activator has the potential to fully activate the target promoter, but its activity is suppressed by repressors (**Figure 7c**). An example is the *E. coli nir* promoter, which can be fully activated by FNR, but activation is suppressed by upstream binding of two nucleoid-associated proteins, IHF and Fis (13). This suppression can be released by the binding of either of the nitrate-activated factors, NarL and NarP, that displace IHF from one of its targets.

At other codependent promoters, the primary activator is unable to make an activatory contact with RNA polymerase without being repositioned by a secondary activator (**Figure 7d**). For example, at the *E. coli malE* promoter, CRP repositions MalT from a location where it is unable to activate transcription to one where it can activate (111). Other examples are found at promoters where an essential activator binds at a far-upstream target and requires a DNA-bending protein such as IHF to bring it closer to the core promoter elements. This appears to be the case for the *E. coli narG* promoter (122), and this is also often found at σ^{54}-dependent promoters where IHF is needed to position the enhancer-binding protein so that it can interact with σ^{54} in RNA polymerase holoenzyme (138). This is done by inducing a sharp bend that brings the enhancer binding protein close to the promoter −12 region.

Figure 7

Integration of multiple regulatory signals. (*a*) Independent contacts by different activator proteins. In ①, transcription is dependent on independent contacts from a Class II activator and a Class I activator. In ②, transcription is dependent on independent contacts from two Class I activators. (*b*) Cooperative binding. The binding of the first activator requires binding of the second activator. (*c*) Antirepression. Maximum activation by the primary activator is facilitated by the second activator, which overcomes the inhibitory action of the repressor (*red*). (*d*) Repositioning. In ①, the secondary activator repositions the primary activator, enabling interactions with RNA polymerase at the promoter. In ②, the secondary activator alters the DNA conformation, thereby presenting the primary activator for interaction with RNA polymerase. Abbreviations: CTD, C-terminal domain; NTD, N-terminal domain.

CONCLUDING COMMENTS

Although transcription activation at bacterial promoters has been studied directly for nearly 40 years, the number of cases for which the details are fully understood is still small. Fortunately, the models that have been developed over this period appear to be applicable to most newly sequenced genomes, but it is unthinkable that new paradigms won't emerge from some of these genomes over the next few years. One big area of current weakness is a quantitative understanding of transcription initiation, and the recent discovery of cases in which promoter escape is limiting and regulated adds an extra dimension of complexity. Another area in need of progress is our understanding of how the folding of bacterial chromosomes affects transcription. It has recently been suggested that location within these structures may be important (117). Perhaps the most pressing need is to find a simple method by which to monitor directly the distribution of RNA polymerase throughout an operon in a time-resolved manner. Methods based on chromatin immunoprecipitation have potential but single-cell approaches are needed. Finally, the challenge is to exploit our newfound knowledge and devise antibacterial therapies that target signaling pathways and transcription factors.

DISCLOSURE STATEMENT

The authors are not aware of any affiliations, memberships, funding, or financial holdings that might be perceived as affecting the objectivity of this review.

ACKNOWLEDGMENTS

The authors' research has been supported by a Wellcome Trust Program grant.

LITERATURE CITED

1. Amouyal M, Buc H. 1987. Topological unwinding of strong and weak promoters by RNA polymerase. A comparison between the *lac* wild-type and the UV5 sites of *Escherichia coli*. *J. Mol. Biol.* 195:795–808
2. Auner H, Buckle M, Deufel A, Kutateladze T, Lazarus L, et al. 2003. Mechanism of transcriptional activation by FIS: role of core promoter structure and DNA topology. *J. Mol. Biol.* 331:331–44
3. Barnard AM, Green J, Busby SJ. 2003. Transcription regulation by tandem-bound FNR at *Escherichia coli* promoters. *J. Bacteriol.* 185:5993–6004
4. Beatty CM, Browning DF, Busby SJ, Wolfe AJ. 2003. Cyclic AMP receptor protein-dependent activation of the *Escherichia coli acsP2* promoter by a synergistic class III mechanism. *J. Bacteriol.* 185:5148–57
5. Belyaeva TA, Bown JA, Fujita N, Ishihama A, Busby SJ. 1996. Location of the C-terminal domain of the RNA polymerase alpha subunit in different open complexes at the *Escherichia coli* galactose operon regulatory region. *Nucleic Acids Res.* 24:2242–51
6. Belyaeva TA, Rhodius VA, Webster CL, Busby SJ. 1998. Transcription activation at promoters carrying tandem DNA sites for the *Escherichia coli* cyclic AMP receptor protein: organisation of the RNA polymerase α subunits. *J. Mol. Biol.* 277:789–804
7. Benoff B, Yang H, Lawson CL, Parkinson G, Liu J, et al. 2002. Structural basis of transcription activation: the CAP-α CTD-DNA complex. *Science* 297:1562–66
8. Bhende PM, Egan SM. 2000. Genetic evidence that transcription activation by RhaS involves specific amino acid contacts with sigma 70. *J. Bacteriol.* 182:4959–69
9. Blatter EE, Ross W, Tang H, Gourse RL, Ebright RH. 1994. Domain organization of RNA polymerase α subunit: C-terminal 85 amino acids constitute a domain capable of dimerization and DNA binding. *Cell* 78:889–96
10. Bose D, Pape T, Burrows PC, Rappas M, Wigneshweraraj SR, et al. 2008. Organization of an activator-bound RNA polymerase holoenzyme. *Mol. Cell* 32:337–46

11. Brown NL, Stoyanov JV, Kidd SP, Hobman JL. 2003. The MerR family of transcriptional regulators. *FEMS Microbiol. Rev.* 27:145–63
12. Browning DF, Busby SJ. 2004. The regulation of bacterial transcription initiation. *Nat. Rev. Microbiol.* 2:57–65
13. Browning DF, Cole JA, Busby SJ. 2004. Transcription activation by remodelling of a nucleoprotein assembly: the role of NarL at the FNR-dependent *Escherichia coli nir* promoter. *Mol. Microbiol.* 53:203–15
14. Browning DF, Grainger DC, Busby SJ. 2010. Effects of nucleoid-associated proteins on bacterial chromosome structure and gene expression. *Curr. Opin. Microbiol.* 13:773–80
15. Buck M, Bose D, Burrows P, Cannon W, Joly N, et al. 2006. A second paradigm for gene activation in bacteria. *Biochem. Soc. Trans.* 34:1067–71
16. Buckle M, Pemberton IK, Jacquet MA, Buc H. 1999. The kinetics of sigma subunit directed promoter recognition by *E. coli* RNA polymerase. *J. Mol. Biol.* 285:955–64
17. Burgess RR. 1969. Separation and characterization of the subunits of ribonucleic acid polymerase. *J. Biol. Chem.* 244:6168–76
18. Burgess RR. 1971. RNA polymerase. *Annu. Rev. Biochem.* 40:711–40
19. Burgess RR, Travers AA, Dunn JJ, Bautz EK. 1969. Factor stimulating transcription by RNA polymerase. *Nature* 221:43–46
20. Busby S, Ebright RH. 1994. Promoter structure, promoter recognition, and transcription activation in prokaryotes. *Cell* 79:743–46
21. Busby S, Ebright RH. 1997. Transcription activation at class II CAP-dependent promoters. *Mol. Microbiol.* 23:853–59
22. Busby S, Ebright RH. 1999. Transcription activation by catabolite activator protein (CAP). *J. Mol. Biol.* 293:199–213
23. Busby S, Kolb A. 1996. The CAP modulon. In *Regulation of Gene Expression in Escherichia coli*, ed. ECC Lin, AS Lynch, pp. 255–79. Austin, TX: Landes
24. Campbell EA, Westblade LF, Darst SA. 2008. Regulation of bacterial RNA polymerase σ factor activity: a structural perspective. *Curr. Opin. Microbiol.* 11:121–27
25. Caramel A, Schnetz K. 1998. Lac and λ repressors relieve silencing of the *Escherichia coli bgl* promoter. Activation by alteration of a repressing nucleoprotein complex. *J. Mol. Biol.* 284:875–83
26. Castanié-Cornet MP, Cam K, Bastiat B, Cros A, Bordes P, Gutierrez C. 2010. Acid stress response in *Escherichia coli*: mechanism of regulation of *gadA* transcription by RcsB and GadE. *Nucleic Acids Res.* 38:3546–54
27. Chen H, Tang H, Ebright RH. 2003. Functional interaction between RNA polymerase α subunit C-terminal domain and σ^{70} in UP-element- and activator-dependent transcription. *Mol. Cell* 11:1621–33
28. Colland F, Orsini G, Brody EN, Buc H, Kolb A. 1998. The bacteriophage T4 AsiA protein: a molecular switch for sigma 70-dependent promoters. *Mol. Microbiol.* 27:819–29
29. Dangi B, Gronenborn AM, Rosner JL, Martin RG. 2004. Versatility of the carboxy-terminal domain of the α subunit of RNA polymerase in transcriptional activation: use of the DNA contact site as a protein contact site for MarA. *Mol. Microbiol.* 54:45–59
30. Davis CA, Bingman CA, Landick R, Record MT Jr, Saecker RM. 2007. Real-time footprinting of DNA in the first kinetically significant intermediate in open complex formation by *Escherichia coli* RNA polymerase. *Proc. Natl. Acad. Sci. USA* 104:7833–38
31. Decker KB, Chen Q, Hsieh ML, Boucher P, Stibitz S, Hinton DM. 2011. Different requirements for σ region 4 in BvgA activation of the *Bordetella pertussis* promoters P_{fim3} and P_{fhaB}. *J. Mol. Biol.* 409:692–709
32. Dillon SC, Dorman CJ. 2010. Bacterial nucleoid-associated proteins, nucleoid structure and gene expression. *Nat. Rev. Microbiol.* 8:185–95
33. Dorman CJ. 2006. DNA supercoiling and bacterial gene expression. *Sci. Prog.* 89:151–66
34. Dove SL, Darst SA, Hochschild A. 2003. Region 4 of σ as a target for transcription regulation. *Mol. Microbiol.* 48:863–74
35. Dove SL, Hochschild A. 1998. Conversion of the ω subunit of *Escherichia coli* RNA polymerase into a transcriptional activator or an activation target. *Genes Dev.* 12:745–54

36. Dove SL, Huang FW, Hochschild A. 2000. Mechanism for a transcriptional activator that works at the isomerization step. *Proc. Natl. Acad. Sci. USA* 97:13215–20
37. Dove SL, Joung JK, Hochschild A. 1997. Activation of prokaryotic transcription through arbitrary protein-protein contacts. *Nature* 386:627–30
38. Ebright RH. 1993. Transcription activation at Class I CAP-dependent promoters. *Mol. Microbiol.* 8:797–802
39. Ebright RH, Busby S. 1995. The *Escherichia coli* RNA polymerase α subunit: structure and function. *Curr. Opin. Genet. Dev.* 5:197–203
40. Feklistov A, Darst SA. 2011. Structural basis for promoter -10 element recognition by the bacterial RNA polymerase σ subunit. *Cell* 147:1257–69
41. Frederix M, Edwards A, McAnulla C, Downie JA. 2011. Co-ordination of quorum-sensing regulation in *Rhizobium leguminosarum* by induction of an anti-repressor. *Mol. Microbiol.* 81:994–1007
42. Gaston K, Bell A, Kolb A, Buc H, Busby S. 1990. Stringent spacing requirements for transcription activation by CRP. *Cell* 62:733–43
43. Ghosh T, Bose D, Zhang X. 2010. Mechanisms for activating bacterial RNA polymerase. *FEMS Microbiol. Rev.* 34:611–27
44. Gourse RL, Ross W, Gaal T. 2000. UPs and downs in bacterial transcription initiation: the role of the alpha subunit of RNA polymerase in promoter recognition. *Mol. Microbiol.* 37:687–95
45. Grainger DC, Webster CL, Belyaeva TA, Hyde EI, Busby SJ. 2004. Transcription activation at the *Escherichia coli melAB* promoter: interactions of MelR with its DNA target site and with domain 4 of the RNA polymerase σ subunit. *Mol. Microbiol.* 51:1297–309
46. Gregory BD, Deighan P, Hochschild A. 2005. An artificial activator that contacts a normally occluded surface of the RNA polymerase holoenzyme. *J. Mol. Biol.* 353:497–506
47. Griffith KL, Shah IM, Myers TE, O'Neill MC, Wolf RE Jr. 2002. Evidence for "pre-recruitment" as a new mechanism of transcription activation in *Escherichia coli*: the large excess of SoxS binding sites per cell relative to the number of SoxS molecules per cell. *Biochem. Biophys. Res. Commun.* 291:979–86
48. Griffith KL, Wolf RE Jr. 2004. Genetic evidence for pre-recruitment as the mechanism of transcription activation by SoxS of *Escherichia coli*: the dominance of DNA binding mutations of SoxS. *J. Mol. Biol.* 344:1–10
49. Grigorova IL, Phleger NJ, Mutalik VK, Gross CA. 2006. Insights into transcriptional regulation and σ competition from an equilibrium model of RNA polymerase binding to DNA. *Proc. Natl. Acad. Sci. USA* 103:5332–37
50. Gruber TM, Gross CA. 2003. Multiple sigma subunits and the partitioning of bacterial transcription space. *Annu. Rev. Microbiol.* 57:441–66
51. Heldwein EE, Brennan RG. 2001. Crystal structure of the transcription activator BmrR bound to DNA and a drug. *Nature* 409:378–82
52. Hinton DM, Pande S, Wais N, Johnson XB, Vuthoori M, et al. 2005. Transcriptional takeover by σ appropriation: remodelling of the σ^{70} subunit of *Escherichia coli* RNA polymerase by the bacteriophage T4 activator MotA and co-activator AsiA. *Microbiology* 151:1729–40
53. Hochschild A, Dove SL. 1998. Protein-protein contacts that activate and repress prokaryotic transcription. *Cell* 92:597–600
54. Hollands K, Lee DJ, Lloyd GS, Busby SJ. 2010. Activation of σ^{28}-dependent transcription in *Escherichia coli* by the cyclic AMP receptor protein requires an unusual promoter organization. *Mol. Microbiol.* 75:1098–111
55. Hook-Barnard IG, Hinton DM. 2007. Transcription initiation by mix and match elements: flexibility for polymerase binding to bacterial promoters. *Gene Regul. Syst. Bio.* 1:275–93
56. Hudson BP, Quispe J, Lara-Gonzalez S, Kim Y, Berman HM, et al. 2009. Three-dimensional EM structure of an intact activator-dependent transcription initiation complex. *Proc. Natl. Acad. Sci. USA* 106:19830–35
57. Huo YX, Zhang YT, Xiao Y, Zhang X, Buck M, et al. 2009. IHF-binding sites inhibit DNA loop formation and transcription initiation. *Nucleic Acids Res.* 37:3878–86

58. Igarashi K, Hanamura A, Makino K, Aiba H, Mizuno T, et al. 1991. Functional map of the α subunit of *Escherichia coli* RNA polymerase: two modes of transcription activation by positive factors. *Proc. Natl. Acad. Sci. USA* 88:8958–62

59. Igarashi K, Ishihama A. 1991. Bipartite functional map of the *E. coli* RNA polymerase α subunit: involvement of the C-terminal region in transcription activation by cAMP-CRP. *Cell* 65:1015–22

60. Ishihama A. 2000. Functional modulation of *Escherichia coli* RNA polymerase. *Annu. Rev. Microbiol.* 54:499–518

61. Islam MS, Bingle LE, Pallen MJ, Busby SJ. 2011. Organization of the *LEE1* operon regulatory region of enterohaemorrhagic *Escherichia coli* O157:H7 and activation by GrlA. *Mol. Microbiol.* 79:468–83

62. Jacob F. 1988. *The Statue Within: An Autobiography*. New York: Basic Books

63. Jain D, Kim Y, Maxwell KL, Beasley S, Zhang R, et al. 2005. Crystal structure of bacteriophage λcII and its DNA complex. *Mol. Cell* 19:259–69

64. Jain D, Nickels BE, Sun L, Hochschild A, Darst SA. 2004. Structure of a ternary transcription activation complex. *Mol. Cell* 13:45–53

65. Jeon YH, Negishi T, Shirakawa M, Yamazaki T, Fujita N, et al. 1995. Solution structure of the activator contact domain of the RNA polymerase α subunit. *Science* 270:1495–97

66. Jishage M, Iwata A, Ueda S, Ishihama A. 1996. Regulation of RNA polymerase sigma subunit synthesis in *Escherichia coli*: intracellular levels of four species of sigma subunit under various growth conditions. *J. Bacteriol.* 178:5447–51

67. Joly N, Zhang N, Buck M, Zhang X. 2012. Coupling AAA protein function to regulated gene expression. *Biochim. Biophys. Acta* 1823:108–16

68. Joung JK, Koepp DM, Hochschild A. 1994. Synergistic activation of transcription by bacteriophage λ cI protein and *E. coli* cAMP receptor protein. *Science* 265:1863–66

69. Joung JK, Le LU, Hochschild A. 1993. Synergistic activation of transcription by *Escherichia coli* cAMP receptor protein. *Proc. Natl. Acad. Sci. USA* 90:3083–87

70. Kedzierska B, Lee DJ, Wegrzyn G, Busby SJ, Thomas MS. 2004. Role of the RNA polymerase α subunits in cII-dependent activation of the bacteriophage λ p_E promoter: identification of important residues and positioning of the α C-terminal domains. *Nucleic Acids Res.* 32:834–41

71. Keseler IM, Collado-Vides J, Santos-Zavaleta A, Peralta-Gil M, Gama-Castro S, et al. 2011. EcoCyc: a comprehensive database of *Escherichia coli* biology. *Nucleic Acids Res.* 39:D583–90

72. Kim YI, Hu JC. 1997. Oriented DNA binding by one-armed λ repressor heterodimers and contacts between repressor and RNA polymerase at P_{RM}. *Mol. Microbiol.* 25:311–18

73. Kolb A, Busby S, Buc H, Garges S, Adhya S. 1993. Transcriptional regulation by cAMP and its receptor protein. *Annu. Rev. Biochem.* 62:749–95

74. Koo BM, Rhodius VA, Nonaka G, deHaseth PL, Gross CA. 2009. Reduced capacity of alternative σs to melt promoters ensures stringent promoter recognition. *Genes Dev.* 23:2426–36

75. Lambert LJ, Wei Y, Schirf V, Demeler B, Werner MH. 2004. T4 AsiA blocks DNA recognition by remodeling σ^{70} region 4. *EMBO J.* 23:2952–62

76. Landini P, Bown JA, Volkert MR, Busby SJ. 1998. Ada protein-RNA polymerase σ subunit interaction and α subunit-promoter DNA interaction are necessary at different steps in transcription initiation at the *Escherichia coli ada* and *aidB* promoters. *J. Biol. Chem.* 273:13307–12

77. Landini P, Busby SJ. 1999. The *Escherichia coli* Ada protein can interact with two distinct determinants in the σ^{70} subunit of RNA polymerase according to promoter architecture: identification of the target of Ada activation at the *alkA* promoter. *J. Bacteriol.* 181:1524–29

78. Lawson CL, Swigon D, Murakami KS, Darst SA, Berman HM, Ebright RH. 2004. Catabolite activator protein: DNA binding and transcription activation. *Curr. Opin. Struct. Biol.* 14:10–20

79. Lee DJ, Wing HJ, Savery NJ, Busby SJ. 2000. Analysis of interactions between activating region 1 of *Escherichia coli* FNR protein and the C-terminal domain of the RNA polymerase α subunit: use of alanine scanning and suppression genetics. *Mol. Microbiol.* 37:1032–40

80. Li B, Wing H, Lee D, Wu HC, Busby S. 1998. Transcription activation by *Escherichia coli* FNR protein: similarities to, and differences from, the CRP paradigm. *Nucleic Acids Res.* 26:2075–81

81. Li M, McClure WR, Susskind MM. 1997. Changing the mechanism of transcriptional activation by phage λ repressor. *Proc. Natl. Acad. Sci. USA* 94:3691–96

82. Li M, Moyle H, Susskind MM. 1994. Target of the transcriptional activation function of phage λ cI protein. *Science* 263:75–77
83. Lloyd GS, Niu W, Tebbutt J, Ebright RH, Busby SJ. 2002. Requirement for two copies of RNA polymerase α subunit C-terminal domain for synergistic transcription activation at complex bacterial promoters. *Genes Dev.* 16:2557–65
84. Lonetto MA, Rhodius V, Lamberg K, Kiley P, Busby S, Gross C. 1998. Identification of a contact site for different transcription activators in region 4 of the *Escherichia coli* RNA polymerase σ^{70} subunit. *J. Mol. Biol.* 284:1353–65
85. Low DA, Casadesus J. 2008. Clocks and switches: bacterial gene regulation by DNA adenine methylation. *Curr. Opin. Microbiol.* 11:106–12
86. Maeda H, Fujita N, Ishihama A. 2000. Competition among seven *Escherichia coli* σ subunits: relative binding affinities to the core RNA polymerase. *Nucleic Acids Res.* 28:3497–503
87. Marshall FA, Messenger SL, Wyborn NR, Guest JR, Wing H, et al. 2001. A novel promoter architecture for microaerobic activation by the anaerobic transcription factor FNR. *Mol. Microbiol.* 39:747–53
88. Martin RG, Bartlett ES, Rosner JL, Wall ME. 2008. Activation of the *Escherichia coli marA/soxS/rob* regulon in response to transcriptional activator concentration. *J. Mol. Biol.* 380:278–84
89. Martin RG, Gillette WK, Martin NI, Rosner JL. 2002. Complex formation between activator and RNA polymerase as the basis for transcriptional activation by MarA and SoxS in *Escherichia coli*. *Mol. Microbiol.* 43:355–70
90. Martin RG, Rosner JL. 2001. The AraC transcriptional activators. *Curr. Opin. Microbiol.* 4:132–37
91. Merrick MJ. 1993. In a class of its own—the RNA polymerase sigma factor σ^{54} (σ^{N}). *Mol. Microbiol.* 10:903–9
92. Miroslavova NS, Busby SJ. 2006. Investigations of the modular structure of bacterial promoters. *Biochem. Soc. Symp.* 73:1–10
93. Morgan SJ, Felek S, Gadwal S, Koropatkin NM, Perry JW, et al. 2011. The two faces of ToxR: activator of *ompU*, co-regulator of *toxT* in *Vibrio cholerae*. *Mol. Microbiol.* 81:113–28
94. Moxon ER. 2009. Bacterial variation, virulence and vaccines. *Microbiology* 155:997–1003
95. Murakami K, Owens JT, Belyaeva TA, Meares CF, Busby SJ, Ishihama A. 1997. Positioning of two alpha subunit carboxy-terminal domains of RNA polymerase at promoters by two transcription factors. *Proc. Natl. Acad. Sci. USA* 94:11274–78
96. Murakami KS, Darst SA. 2003. Bacterial RNA polymerases: the wholo story. *Curr. Opin. Struct. Biol.* 13:31–39
97. Murakami KS, Masuda S, Campbell EA, Muzzin O, Darst SA. 2002. Structural basis of transcription initiation: an RNA polymerase holoenzyme-DNA complex. *Science* 296:1285–90
98. Murakami KS, Masuda S, Darst SA. 2002. Structural basis of transcription initiation: RNA polymerase holoenzyme at 4 Å resolution. *Science* 296:1280–84
99. Newberry KJ, Brennan RG. 2004. The structural mechanism for transcription activation by MerR family member multidrug transporter activation, N terminus. *J. Biol. Chem.* 279:20356–62
100. Niu W, Kim Y, Tau G, Heyduk T, Ebright RH. 1996. Transcription activation at class II CAP-dependent promoters: two interactions between CAP and RNA polymerase. *Cell* 87:1123–34
101. Opel ML, Aeling KA, Holmes WM, Johnson RC, Benham CJ, Hatfield GW. 2004. Activation of transcription initiation from a stable RNA promoter by a Fis protein-mediated DNA structural transmission mechanism. *Mol. Microbiol.* 53:665–74
102. Paget MS, Helmann JD. 2003. The σ^{70} family of sigma factors. *Genome Biol.* 4:203
103. Pemberton IK, Muskhelishvili G, Travers AA, Buckle M. 2002. FIS modulates the kinetics of successive interactions of RNA polymerase with the core and upstream regions of the *tyrT* promoter. *J. Mol. Biol.* 318:651–63
104. Pérez-Rueda E, Gralla JD, Collado-Vides J. 1998. Genomic position analyses and the transcription machinery. *J. Mol. Biol.* 275:165–70
105. Piper SE, Mitchell JE, Lee DJ, Busby SJ. 2009. A global view of *Escherichia coli* Rsd protein and its interactions. *Mol. Biosyst.* 5:1943–47
106. Raibaud O, Schwartz M. 1984. Positive control of transcription initiation in bacteria. *Annu. Rev. Genet.* 18:173–206

107. Reyes-Caballero H, Campanello GC, Giedroc DP. 2011. Metalloregulatory proteins: metal selectivity and allosteric switching. *Biophys. Chem.* 156:103–14
108. Rhodius VA, Busby SJ. 2000. Interactions between activating region 3 of the *Escherichia coli* cyclic AMP receptor protein and region 4 of the RNA polymerase σ^{70} subunit: application of suppression genetics. *J. Mol. Biol.* 299:311–24
109. Rhodius VA, Busby SJ. 2000. Transcription activation by the *Escherichia coli* cyclic AMP receptor protein: determinants within activating region 3. *J. Mol. Biol.* 299:295–310
110. Rhodius VA, West DM, Webster CL, Busby SJ, Savery NJ. 1997. Transcription activation at class II CRP-dependent promoters: the role of different activating regions. *Nucleic Acids Res.* 25:326–32
111. Richet E, Vidal-Ingigliardi D, Raibaud O. 1991. A new mechanism for coactivation of transcription initiation: repositioning of an activator triggered by the binding of a second activator. *Cell* 66:1185–95
112. Ross W, Gosink KK, Salomon J, Igarashi K, Zou C, et al. 1993. A third recognition element in bacterial promoters: DNA binding by the α subunit of RNA polymerase. *Science* 262:1407–13
113. Ross W, Schneider DA, Paul BJ, Mertens A, Gourse RL. 2003. An intersubunit contact stimulating transcription initiation by *E. coli* RNA polymerase: interaction of the α C-terminal domain and σ region 4. *Genes Dev.* 17:1293–307
114. Rossiter AE, Browning DF, Leyton DL, Johnson MD, Godfrey RE, et al. 2011. Transcription of the plasmid-encoded toxin gene from enteroaggregative *Escherichia coli* is regulated by a novel co-activation mechanism involving CRP and Fis. *Mol. Microbiol.* 81:179–91
115. Roy S, Garges S, Adhya S. 1998. Activation and repression of transcription by differential contact: two sides of a coin. *J. Biol. Chem.* 273:14059–62
116. Saecker RM, Record MT Jr, Dehaseth PL. 2011. Mechanism of bacterial transcription initiation: RNA polymerase - promoter binding, isomerization to initiation-competent open complexes, and initiation of RNA synthesis. *J. Mol. Biol.* 412:754–71
117. Sánchez-Romero MA, Lee DJ, Sánchez-Morán E, Busby SJ. 2012. Location and dynamics of an active promoter in *Escherichia coli* K-12. *Biochem. J.* 441:481–85
118. Saunders JR. 1999. Switch systems. In *Prokaryotic Gene Expression*, ed. S Baumberg, pp. 229–52. Oxford, UK: Oxford Univ. Press
119. Savery NJ, Lloyd GS, Busby SJ, Thomas MS, Ebright RH, Gourse RL. 2002. Determinants of the C-terminal domain of the *Escherichia coli* RNA polymerase α subunit important for transcription at class I cyclic AMP receptor protein-dependent promoters. *J. Bacteriol.* 184:2273–80
120. Savery NJ, Lloyd GS, Kainz M, Gaal T, Ross W, et al. 1998. Transcription activation at Class II CRP-dependent promoters: identification of determinants in the C-terminal domain of the RNA polymerase α subunit. *EMBO J.* 17:3439–47
121. Schleif R. 2010. AraC protein, regulation of the L-arabinose operon in *Escherichia coli*, and the light switch mechanism of AraC action. *FEMS Microbiol. Rev.* 34:779–96
122. Schroder I, Darie S, Gunsalus RP. 1993. Activation of the *Escherichia coli* nitrate reductase (*narGHJI*) operon by NarL and Fnr requires integration host factor. *J. Biol. Chem.* 268:771–74
123. Schwartz M. 2003. Another route. In *Origins of Molecular Biology: A Tribute to Jacques Monod*, ed. A Ullmann, pp. 207–16. Washington, DC: ASM Press
124. Sclavi B, Zaychikov E, Rogozina A, Walther F, Buckle M, Heumann H. 2005. Real-time characterization of intermediates in the pathway to open complex formation by *Escherichia coli* RNA polymerase at the T7A1 promoter. *Proc. Natl. Acad. Sci. USA* 102:4706–11
125. Scott S, Busby S, Beacham I. 1995. Transcriptional co-activation at the *ansB* promoters: involvement of the activating regions of CRP and FNR when bound in tandem. *Mol. Microbiol.* 18:521–31
126. Shah IM, Wolf RE Jr. 2004. Novel protein-protein interaction between *Escherichia coli* SoxS and the DNA binding determinant of the RNA polymerase α subunit: SoxS functions as a co-sigma factor and redeploys RNA polymerase from UP-element-containing promoters to SoxS-dependent promoters during oxidative stress. *J. Mol. Biol.* 343:513–32
127. Shaw DJ, Rice DW, Guest JR. 1983. Homology between CAP and Fnr, a regulator of anaerobic respiration in *Escherichia coli*. *J. Mol. Biol.* 166:241–47
128. Sheridan SD, Opel ML, Hatfield GW. 2001. Activation and repression of transcription initiation by a distant DNA structural transition. *Mol. Microbiol.* 40:684–90

129. Shimada T, Fujita N, Yamamoto K, Ishihama A. 2011. Novel roles of cAMP receptor protein (CRP) in regulation of transport and metabolism of carbon sources. *PLoS One* 6:e20081
130. Spassky A, Busby S, Buc H. 1984. On the action of the cyclic AMP-cyclic AMP receptor protein complex at the *Escherichia coli* lactose and galactose promoter regions. *EMBO J.* 3:43–50
131. Squire DJ, Xu M, Cole JA, Busby SJ, Browning DF. 2009. Competition between NarL-dependent activation and Fis-dependent repression controls expression from the *Escherichia coli yeaR* and *ogt* promoters. *Biochem. J.* 420:249–57
132. Stoebel DM, Free A, Dorman CJ. 2008. Anti-silencing: overcoming H-NS-mediated repression of transcription in gram-negative enteric bacteria. *Microbiology* 154:2533–45
133. Straney DC, Straney SB, Crothers DM. 1989. Synergy between *Escherichia coli* CAP protein and RNA polymerase in the *lac* promoter open complex. *J. Mol. Biol.* 206:41–57
134. Sun L, Dove SL, Panaghie G, deHaseth PL, Hochschild A. 2004. An RNA polymerase mutant deficient in DNA melting facilitates study of activation mechanism: application to an artificial activator of transcription. *J. Mol. Biol.* 343:1171–82
135. Tebbutt J, Rhodius VA, Webster CL, Busby SJ. 2002. Architectural requirements for optimal activation by tandem CRP molecules at a class I CRP-dependent promoter. *FEMS Microbiol. Lett.* 210:55–60
136. Travers A, Muskhelishvili G. 1998. DNA microloops and microdomains: a general mechanism for transcription activation by torsional transmission. *J. Mol. Biol.* 279:1027–43
137. Ushida C, Aiba H. 1990. Helical phase dependent action of CRP: effect of the distance between the CRP site and the −35 region on promoter activity. *Nucleic Acids Res.* 18:6325–30
138. Valls M, Silva-Rocha R, Cases I, Muñoz A, de Lorenzo V. 2011. Functional analysis of the integration host factor site of the σ^{54} *Pu* promoter of *Pseudomonas putida* by in vivo UV imprinting. *Mol. Microbiol.* 82:591–601
139. Wade JT, Belyaeva TA, Hyde EI, Busby SJ. 2001. A simple mechanism for co-dependence on two activators at an *Escherichia coli* promoter. *EMBO J.* 20:7160–67
140. Wade JT, Roa DC, Grainger DC, Hurd D, Busby SJ, et al. 2006. Extensive functional overlap between sigma factors in *Escherichia coli*. *Nat. Struct. Mol. Biol.* 13:806–14
141. Wall ME, Markowitz DA, Rosner JL, Martin RG. 2009. Model of transcriptional activation by MarA in *Escherichia coli*. *PLoS Comp. Biol.* 5:e1000614
142. Wickstrum JR, Egan SM. 2004. Amino acid contacts between sigma 70 domain 4 and the transcription activators RhaS and RhaR. *J. Bacteriol.* 186:6277–85
143. Williams RM, Rhodius VA, Bell AI, Kolb A, Busby SJ. 1996. Orientation of functional activating regions in the *Escherichia coli* CRP protein during transcription activation at class II promoters. *Nucleic Acids Res.* 24:1112–18
144. Wing HJ, Green J, Guest JR, Busby SJ. 2000. Role of activating region 1 of *Escherichia coli* FNR protein in transcription activation at class II promoters. *J. Biol. Chem.* 275:29061–65
145. Wing HJ, Williams SM, Busby SJ. 1995. Spacing requirements for transcription activation by *Escherichia coli* FNR protein. *J. Bacteriol.* 177:6704–10
146. Zafar MA, Shah IM, Wolf RE Jr. 2010. Protein-protein interactions between σ^{70} region 4 of RNA polymerase and *Escherichia coli* SoxS, a transcription activator that functions by the prerecruitment mechanism: evidence for "off-DNA" and "on-DNA" interactions. *J. Mol. Biol.* 401:13–32
147. Zhang W, Baseman JB. 2011. Transcriptional regulation of MG_149, an osmoinducible lipoprotein gene from *Mycoplasma genitalium*. *Mol. Microbiol.* 81:327–39
148. Zhang X, Reeder T, Schleif R. 1996. Transcription activation parameters at *ara* p_{BAD}. *J. Mol. Biol.* 258:14–24
149. Zheng D, Constantinidou C, Hobman JL, Minchin SD. 2004. Identification of the CRP regulon using in vitro and in vivo transcriptional profiling. *Nucleic Acids Res.* 32:5874–93
150. Zhou Y, Busby S, Ebright RH. 1993. Identification of the functional subunit of a dimeric transcription activator protein by use of oriented heterodimers. *Cell* 73:375–79
151. Zhou Y, Pendergrast PS, Bell A, Williams R, Busby S, Ebright RH. 1994. The functional subunit of a dimeric transcription activator protein depends on promoter architecture. *EMBO J.* 13:4549–57

Herpesvirus Transport to the Nervous System and Back Again

Gregory Smith

Department of Microbiology-Immunology, Northwestern University Feinberg School of Medicine, Chicago, Illinois 60611; email: g-smith3@northwestern.edu

Keywords

neuroinvasion, neurotropism, neurovirulence, axon, neuron, sensory ganglion

Abstract

Herpes simplex virus, varicella zoster virus, and pseudorabies virus are neurotropic pathogens of the *Alphaherpesvirinae* subfamily of the *Herpesviridae*. These viruses efficiently invade the peripheral nervous system and establish lifelong latency in neurons resident in peripheral ganglia. Primary and recurrent infections cycle virus particles between neurons and the peripheral tissues they innervate. This remarkable cycle of infection is the topic of this review. In addition, some of the distinguishing hallmarks of the infections caused by these viruses are evaluated in terms of their underlying similarities.

Contents

INTRODUCTION 154
TRANSMISSION FROM THE SITE OF EXPOSURE TO NERVE ENDINGS .. 155
DELIVERY OF VIRAL PARTICLES FROM AXON TERMINALS
    TO PERIPHERAL GANGLIA 156
        Entering the Nerve Ending 156
        Tegument Disassociation 157
        Cortical Actin 159
        Retrograde Axon Transport 159
        From Microtubule Organizing Center to Nuclear Pore Complexes 160
        Nuclear Injection of the Viral Genome 160
SELECTIVE LOSS OF VP16 DURING NEURONAL INFECTION 161
POSTREPLICATIVE SPREAD 161
    Transport to the Cytoplasmic Envelopment Site 162
AXON TARGETING 163
    Anterograde Axon Transport 164
    Release from the Axon Terminal 166
    Spread Between Neurons and to Distant Sites 166
CONCLUSIONS 167

INTRODUCTION

One of the defining characteristics of viruses belonging to the *Herpesviridae* family is the establishment of lifelong infections by means of a latency program. During latency the infection persists in a dormant state during which the viral genome is stably maintained but no viral particles are assembled. Neurons of the peripheral nervous system (PNS) host the latent infection for a subset of viruses belonging to the *Alphaherpesvirinae*, a subfamily of the *Herpesviridae* (**Figure 1**). The first demonstration of the neurotropic properties of these viruses came from a study of the veterinary pathogen, pseudorabies virus (PRV), published by Dr. Albert Sabin (145). Since that time, it is widely recognized that instillation of neurotropic herpesviruses such as herpes simplex virus (HSV-1, HSV-2) or PRV into rodents typically results in transmission of infection from the PNS to the central nervous system (CNS), producing a lethal encephalitis. Transmission from the PNS to the CNS occurs across synapse-linked neurons, and the resulting self-amplifying circuit-specific spread has been exploited to trace neural connections in the vertebrate nervous system (47, 158). How these neuroinvasive and neurovirulent properties relate to the more typically benign infections of natural hosts remains an open question. HSV-1 only rarely causes severe encephalitis in humans.

Productive infections consist of a cycle of transmission between latently infected neurons in ganglia of the PNS and the somatic cells to which they project. Virus particles never cross a synapse in recurrent infections of the natural host, and yet these viruses are exquisitely capable of doing just that. Why do these viruses maintain this devastating potential, and how is this property suppressed in the natural host, where disease is most often mild or nonexistent? This sets the neurotropic herpesviruses apart from other agents such as poliovirus, rabies virus, measles virus, and Japanese encephalitis virus, which are virulent upon entering the nervous system (175). Nevertheless, because of the high prevalence of these viruses in ourselves and our livestock, the

PNS: peripheral nervous system

PRV: pseudorabies virus

HSV: herpes simplex virus

CNS: central nervous system

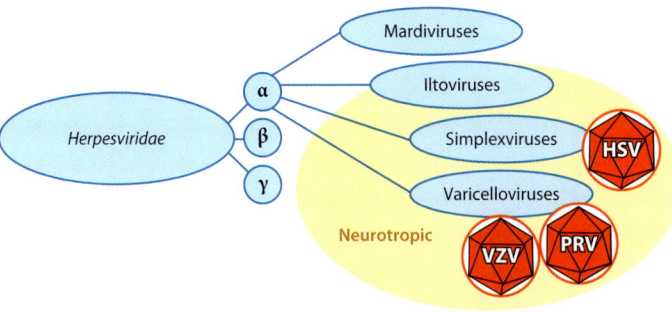

Figure 1

The *Alphaherpesvirinae* subfamily consists of four genera. Two of the genera, the *Simplexvirus* and *Varicellovirus*, consist of human and veterinary pathogens that establish latency in peripheral neurons. Herpes simplex virus (HSV) is a simplexvirus, while varicella zoster virus (VZV) and pseudorabies virus (PRV) are varicelloviruses. At least one member of the *Iltovirus* genus is also proposed to establish latent infections in peripheral neurons (177), but the *Mardivirus* genus lacks neurotropism.

low incidence of severe forms of disease takes a large toll. For example, HSV-1 pathogenesis is generally limited to cold sores but the virus is also the leading cause of infectious blindness in the United States (92). Reactivation of varicella zoster virus (VZV) produces shingles, which can transform into postherpetic neuralgia. And neonatal transmission of HSV-2 results in a high incidence of CNS and disseminated infections in newborns (72).

There are many outstanding questions regarding the neurotropic herpesvirus infectious cycle. "It must be explained, for example, why a virus causing no more than a cold sore in one person can produce fatal encephalitis in another." This problem, which was articulated by Richard Johnson and Cedric Mims in 1968, remains relevant today (67). I review the current understanding of the neurotropic infectious cycle that has resulted from more than half a century of study.

TRANSMISSION FROM THE SITE OF EXPOSURE TO NERVE ENDINGS

Nerve endings are generally not exposed to the outside world. As such, the neurotropic herpesvirus infectious cycle exhibits dual tropism: replication in somatic cells such as epithelia, followed by transmission into neurons (**Figure 2**). How these viruses transmit between cell types is poorly understood. While *Betaherpesvirinae* and *Gammaherpesvirinae* can effect changes in cell tropism by modulating the composition of the infectious viral particle, there is currently no evidence that neuroinvasive herpesviruses employ a similar strategy (16, 149).

Despite the apparent lack of a neuron-specific tropism switch, viral effectors are required for neuroinvasion. The ICP34.5 protein of HSV-1 was described as a neurovirulence determinant due to its selective requirement for infections in animals (27). However, since its discovery, ICP34.5 has been recognized for its role in evading the host immune response (91, 101, 137). Although ICP34.5 is required for spread to neurons from the mouse cornea, for example, this is in large part due to loss of viral replication in the cornea itself (172). One factor that appears to specifically govern viral transmission into the nervous system without loss of viral replication at sites of peripheral inoculation is the PRV deubiquitinase activity housed within the amino terminus of the pUL36 (VP1/2) tegument protein (17, 89). All herpesviruses consist of a membrane envelope that contains an icosahedral capsid and a collection of additional proteins collectively referred to as the tegument, of which VP1/2 is one component. The capsid and tegument are deposited into cells upon fusion-mediated entry (**Figure 3**). How the deubiquitinase activity of VP1/2 contributes to

VZV: varicella zoster virus

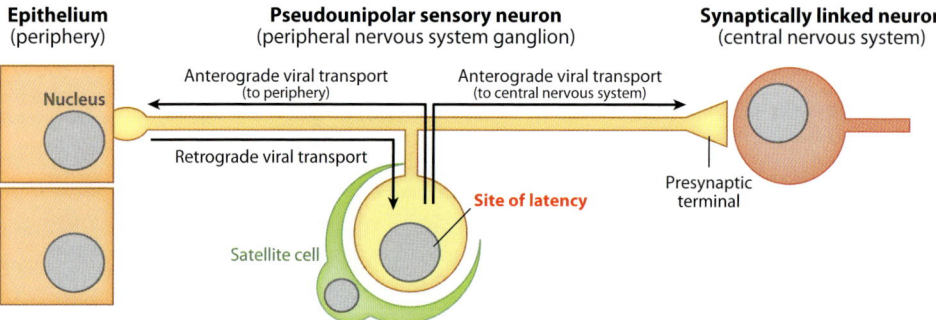

Figure 2

Fundamentals of a neurotropic herpesvirus niche. Although the herpesviruses discussed in this article are referred to as neurotropic, these agents are more accurately described as dual tropic or multitropic. Primary infections begin in exposed tissues such as mucosal epithelia (*left*). Subsequent spread is normally restricted to innervating neurons resident in peripheral ganglia, where latent infections are maintained, following a single round of retrograde axon transport (*middle*). Reactivation results in the production of new viruses and the return to peripheral tissues (anterograde axon transport to epithelia). Severe disease associated with invasion of second-order neurons (anterograde axon transport to CNS) occurs only rarely in the natural host but may be frequent in secondary hosts (*right*). Anterograde and retrograde spread is dictated by the orientation of axonal microtubules and should not be confused with the direction of action potential propagation in the pseudounipolar neuron. Neurotropic herpesviruses can also infect cells of the circulatory system to varying degrees (not illustrated).

the transmission of infection to the nervous system is currently unknown, but the phenotype of the mutant virus indicates that an innate barrier exists between epithelial tissues and innervating neurons that prevents viral transmission into the PNS. In this context, it is perhaps noteworthy that recent studies of poliovirus demonstrate that neuroinvasion is inefficient unless damage has been inflicted on the innervated tissue (86, 130). Whereas poliovirus is an enteric pathogen that infrequently invades the nervous system, the infectious cycles of HSV and PRV efficiently breach the epithelia-neuron barrier. Identifying the relevant substrates of the herpesvirus deubiquitinase will be required to better define the innate barrier function and how these viruses overcome it.

DELIVERY OF VIRAL PARTICLES FROM AXON TERMINALS TO PERIPHERAL GANGLIA

Entering the Nerve Ending

Upon cell contact, three events are triggered in the HSV-1 virion: There is an internal restructuring of the tegument, the virion orients so that the dense pole of the particle faces away from the cell surface, and the fusion apparatus is triggered (54, 104, 109). Virion restructuring results in redistribution of tegument proteins in the virion that were initially symmetrically proportioned around the capsid to an asymmetric distribution, and its significance is unknown. Similar morphologic changes in HSV-1 virions can occur simply by allowing the particles to age (123). While the meaning of these findings can only be speculated on, they are suggestive of an internal virion trigger mechanism that is required at the moment of entry. Consistent with this, binding of HSV-1 virions to cells induces the disassociation of the pUL16 tegument protein from the capsid (109). This event is required for infection (108). Whether release of pUL16 from the capsid contributes to the larger morphological changes visualized in the tegument is currently unknown but seems likely.

How HSV-1 enters cells, including neurons, by membrane fusion is extensively reviewed elsewhere (31, 62, 70, 157). It should be noted here, however, that the principal HSV-1 envelope protein that triggers entry into cells is glycoprotein D (gD), which binds several cell membrane receptors including Nectin-1, HVEM (herpesvirus entry mediator), and 3-O-sulfated heparan sulfate (55, 114, 152). HSV-1 is dependent upon gD to enter cells, with Nectin-1 serving as the primary entry receptor on neurons and HVEM playing a supporting role (69, 77, 93, 164). However, the importance of gD in the neurotropic herpesviruses is not conserved. The prototypic varicellovirus, VZV, does not encode a gD homolog and instead may enter neurons through a different receptor interaction (160). In PRV, gD is dispensable for cell-cell spread and neuroinvasion, and gD mutant viruses spread independently of Nectin-1 and HVEM (25, 61, 134, 135, 140). Because PRV cannot produce extracellular plaque-forming units in the absence of gD, it can be inferred that PRV spread in the nervous system does not require the release of cell-free virions from infected cells.

Although all herpesviruses enter cells by membrane fusion, the fusion event can occur at either the plasma membrane or an endosomal membrane (111, 126). Several lines of evidence indicate that HSV-1 and PRV enter axon nerve endings by fusion at the plasma membrane: (*a*) entry is pH independent, (*b*) entry at the plasma membrane has been observed by transmission electron microscopy, and (*c*) capsids are not associated with fluid phase markers following entry (98, 125, 154).

Tegument Disassociation

Following fusion-mediated entry into cells, the herpesvirus envelope is lost and the tegument and capsid are deposited into the cytosol (**Figure 3**). A recent analysis of tegument protein associations with capsids based on resistance to detergent and salt extraction classified seven proteins as components of the inner tegument: pUL14, pUL16, pUL21, pUL36 (VP1/2), pUL37, pUs3, and ICP0 (139). The inner tegument consists of proteins in close juxtaposition to the capsid that likely are acquired prior to the final budding event that produces the enveloped virion. Although this study was not a comprehensive examination of all tegument proteins, it seems a reasonable first approximation that the identified proteins may remain bound to capsids upon entry as part of a retrograde transport complex. This is the case for at least three of these proteins. VP1/2 and pUL37 cotransport with capsids upon entry into sensory neurons according to fluorescence time-lapse imaging of HSV-1 and PRV infections (9, 96). The retention of VP1/2 and pUL37 on capsids postfusion is also evident by imaging fixed cell lines by electron microscopy and fluorescence methods (33, 58). The same approaches have confirmed that the pUs3 protein kinase is also cotransported with capsids following entry into neurons and nonneuronal cells infected with PRV (30, 58). Because tegument proteins that remain capsid bound may modulate retrograde transport, determining which proteins are retained on the capsid is of great interest. Consistent with this, a truncated version of VP1/2 that is poorly retained on HSV-1 capsids following entry fails to deliver genomes to nuclei (148). Of the seven recognized inner tegument proteins, four have been implicated in capsid delivery to the nucleus (**Table 1**).

HSV-1 and PRV tegument disassociation appears to be similar: both retain VP1/2 and pUL37 and remove pUL46 (VP11/12), pUL47 (VP13/14), pUL48 (VP16), and pUL49 (VP22) (9, 96). However, whereas removal of the last four tegument proteins appears to be efficient for PRV, HSV-1 retains small amounts of VP11/12 and VP16 on capsids postfusion (9). Although this seems to be a subtle difference between the two viruses, it may have relevance to the infectious cycle, as discussed below.

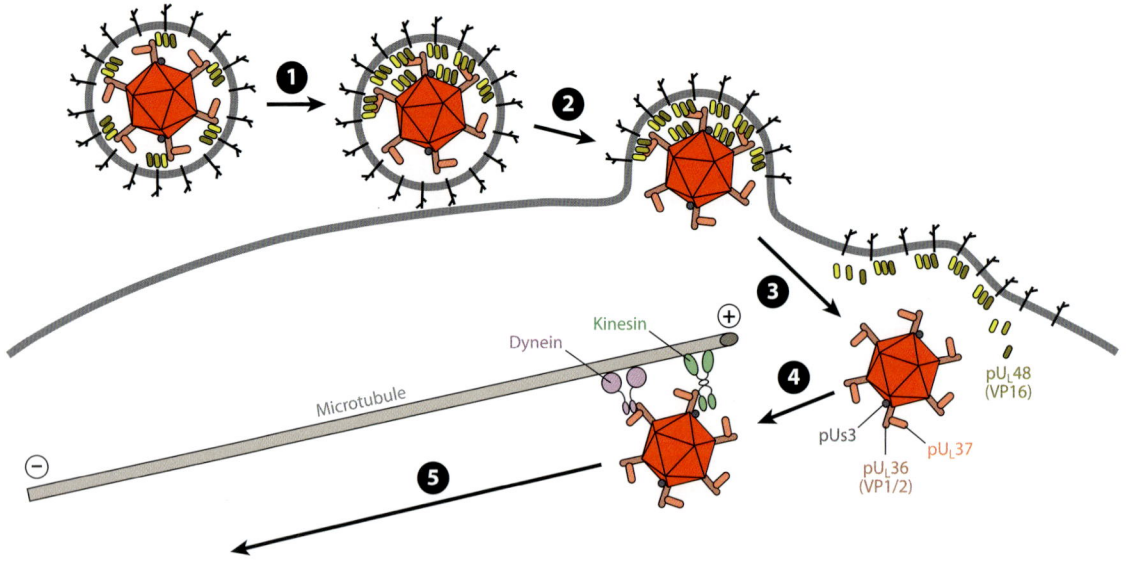

Figure 3

Early events in herpesvirus infection. ❶ Virion contacts plasma membrane of a somatic cell or the axon membrane of a neuron. Tegument proteins redistribute and the virion orients with the bulk of its mass away from the cell. ❷ Fusion between the virion envelope and the cell membrane deposits the capsid and tegument proteins into the cytosol. ❸ The capsid releases from the majority of tegument and envelope proteins (including VP16), but a subset of inner tegument proteins (including pUs3, VP1/2, pUL37) remain capsid bound and together compose the retrograde transport complex. VP16, which enters the nucleus in somatic cells and promotes productive infection, may be lost upon entering a neuron owing to an inability to efficiently participate in retrograde axon transport. ❹ The retrograde transport complex traverses cortical actin (not illustrated) and associates with dynein and kinesin motors that in turn bind microtubules. Microtubules in axons are almost uniformly oriented, with plus-ends facing the axon terminals. ❺ Dynein dominates over kinesin activity, resulting in transport toward the minus-ends of microtubules and trafficking to the neural soma (retrograde transport).

Table 1 Factors governing capsid delivery to the nucleus

Step	Viral proteins	Cellular proteins	References
Passage through cortical actin?	gD	ROCK1 FAK PI3K Rho GTPases	26, 28, 36, 52, 127, 129
Retrograde microtubule transport	pUL35 (VP26)?	Dynein Dynactin	7, 44–46
Nuclear delivery[a]	pUL14 pUL36 (VP1/2) pUL37 ICP0	Proteasome	33, 38, 39, 78, 143, 148, 181
Nuclear docking and genome release	pUL6 (portal) pUL25 pUL36 (VP1/2)	Importin-β Ran Nup214 Nup358hCG1	1, 2, 4, 14, 33, 68, 75, 122, 128, 131, 132, 138, 144

[a]Factors listed are required for efficient delivery of capsids to the nucleus, and maybe involved in passage through cortical actin or microtubule transport.

Cortical Actin

Binding of herpesvirus to cells triggers Rho GTPases that can in turn induce rearrangements in the actin cytoskeleton (28, 36, 129). There is increasing evidence that these signaling events may increase cell susceptibility to infection (reviewed in References 171 and 183). Whether these events pertain to infection at axon terminals has not been addressed. Applying cryo-electron tomography to image HSV-1 capsids deposited in the cytosol after fusion into synaptosomes, which are axon terminals severed from neurons, reveals capsids underneath the plasma membrane surrounded by dense meshworks of actin filaments (104). Although fusion-mediated entry occurred in less than one minute, cytosolic capsids remained near the plasma membrane for more than one hour. This suggests actin may be a barrier to initial infection.

Evidence that virus-induced signal cascades assist capsid translocation through cortical actin can be inferred from time-lapse imaging. Following contact with a terminal of an intact sensory axon, fluorescently tagged PRV capsids display motion that is biphasic (154). Initially, capsids move retrograde with slow kinetics as they enter the axon shaft. Fast retrograde transport consistent with dynein motion (see below) engages shortly thereafter. Capturing recordings of entry events remains challenging and has so far precluded an in-depth analysis of initial intracellular transport dynamics. Nevertheless, the available observations are consistent with navigation of capsids from the plasma membrane to microtubules via a directed process that may be promoted by alterations in axonal actin. In natural infections, nerve endings are generally not freely accessible to extracellular virions as they are in culture models of infection. Although highly speculative, the transmission of infection across epithelia-neuron contacts may promote virus-induced rearrangements of axon actin in a process analogous to the virological synapses reported for HIV and HSV (12, 115).

Retrograde Axon Transport

Time-lapse imaging of fluorescently tagged capsids of HSV-1 and PRV in cultured neurons has revealed that retrograde transport is a robust and sustained process that efficiently delivers capsids to the distant nucleus (9, 50, 94, 154). Transport occurs along microtubules at rates in excess of 1 $\mu m\ s^{-1}$ (9, 80, 168). The fast retrograde axonal transport exhibited by HSV-1 and PRV capsids can be mediated only by active microtubule transport by the dynein motor complex (reviewed in References 71 and 170). Retrograde capsid motion is not continuous; brief reversals in transport direction occur in the axon (154). This transient anterograde motion implies the presence of a kinesin motor on the capsid in addition to the dynein motor complex. The presence of opposing microtubule motors is consistent with endogenous cellular cargoes as well as other viruses in axons (59, 147). Although not intuitive, the opposing pull of the kinesin motor may act to promote dynein retrograde transport of the capsid (6).

Purified extracellular virions stripped of their envelopes bind dynein and kinesin motors, and this binding is enhanced by extracting outer tegument proteins from the particles (139, 180). These findings implicate either the capsid or tegument proteins closely juxtaposed to the capsid surface as microtubule motor binding sites. Several genes encoding tegument proteins can be deleted without impeding retrograde axon transport, but the inner tegument proteins have critical roles in virion assembly, making the study of mutant viruses during the initial stages of neuron infection difficult (7). Among these, VP1/2 is considered a top candidate for recruiting microtubule motors to capsids. Antibodies directed against VP1/2 or truncation of the VP1/2 carboxyl-terminus interferes with capsid delivery to the nucleus and VP1/2 is required for microtubule transport during late infection in cell lines (33, 97, 148). In addition, VP1/2 carrying an amino acid change that confers temperature sensitivity dramatically accumulates in aggresomes at the nonpermissive temperature (1, 3). While this phenotype may simply result from an abundance of misfolded VP1/2

MTOC: microtubule organizing center

NPC: nuclear pore complex

proteins inside cells, a recent model of virus retrograde transport posits that viruses may mimic protein aggregates to recruit dynein and effect their transport to the microtubule organizing center (MTOC) (176).

Viruses lacking the inner tegument proteins ICP0, pUL14, or pUL37 suffer from delayed delivery of capsids to the nucleus in cell lines (**Table 1**). Whether these proteins contribute to dynein recruitment and retrograde transport or another step in capsid delivery is also of great interest. The pUL37 tegument protein was examined in a nuclear delivery assay between nuclei in syncytia and in this context was dispensable (143). This finding indicates that pUL37 does not function at nuclear pore complexes (NPCs), but rather at an earlier step in infection: perhaps overcoming the cortical actin barrier or microtubule transport.

Dynein recruitment does not have to be mediated by a tegument protein. The pUL35 (VP26) capsid protein can bind dynein and was implicated in capsid retrograde transport (46). However, in the absence of VP26, PRV capsids transport at wild-type velocity, and transport of HSV-1 is unimpeded both in culture and in a mouse model of infection (7, 40, 44). Therefore, if VP26 binding to dynein is biologically relevant, it must be redundant with another dynein recruitment mechanism. Furthermore, de novo–assembled capsids isolated from the nucleus of infected cells do not bind to dynein in vitro, arguing that capsid proteins may not be involved in dynein recruitment; however, the possibility of posttranslational modifications to capsids in the cytosol cannot be ruled out (139, 180). While studies of HSV-1 and PRV have propelled our understanding of retrograde axon transport, a new neuron chamber infection model has the promise of extending these studies to VZV (103).

From Microtubule Organizing Center to Nuclear Pore Complexes

Minus-end-directed transport along microtubules is expected to end at the MTOC. However, HSV-1 capsids only transiently accumulate at the MTOC prior to moving to NPCs in the nuclear membrane (156). The MTOC is typically located adjacent to the nucleus. Whether the MTOC is close enough to allow subsequent passive diffusion of capsids to NPCs is debated. For adenovirus, capsid translocation from the MTOC to NPCs is dependent upon the nuclear export factor CRM1 (159). Whether CRM1 or a factor exported from the nucleus by CRM1 is necessary for proper adenovirus targeting is unknown, but the dependence on CRM1 argues for a facilitated process. Evidence of a plus-end kinesin motor activity associated with retrogradely trafficking HSV-1 and PRV capsids has led to the suggestion that movement from the MTOC to NPCs may be mediated by plus-end transport along perinuclear microtubules (45).

Nuclear Injection of the Viral Genome

By transmission electron microscopy, HSV-1 capsids are observed docked at NPCs following entry into cell lines (14, 156). The interactions between capsids and NPC proteins that mediate this coupling are not fully defined. Capsid docking is dependent on importin-β and Ran GTPase in an in vitro reconstituted model, and the Nup358 component of the NPC is required for capsid docking in intact cells (33, 131). Tegument components of the retrograde transport complex remain capsid bound once docked at nuclear pores, where the genome is released into the nucleus. Once docking occurs, the VP1/2 tegument protein is proteolytically processed, and this cleavage is required for subsequent release of the DNA genome from the capsid into the nuclear pore (68). A recent study determined that pUL37, and to a lesser extent VP1/2, is lost from docked capsids by 4 h postinfection (5). Whether this loss is of functional significance is not clear. Although the proteins may simply turn over after genome injection, their loss could be a functional consequence of VP1/2 proteolytic processing and genome release from the capsid. HSV-1 encoding a temperature-sensitive mutation in VP1/2 is not processed at the nonpermissive temperature and fails to release its genome upon NPC docking (14, 68, 75). VP1/2 is not expected to maintain the encapsidated genome, so its cleavage likely triggers a conformational change that allows escape of the DNA through the portal vertex of the capsid (122). VP1/2 is bound directly to the capsid surface through an interaction with the pUL25 capsid protein and additionally makes contact with the pUL17 and VP5 capsid proteins (23, 29, 132). Notably, pUL25 is required for stable genome encapsidation during assembly and packaging in the nucleus (107). Although deletion of UL25 prevents production of infectious virions, HSV-1-encoding UL25 mutations that are defective for genome injection after NPC docking have been isolated (138). In addition to binding VP1/2, pUL25 binds two NPC components: Nup214 and hCG1 (132). VP1/2 encodes a nuclear localization signal that is essential for productive infection (2). It will be of great interest to learn whether capsid-bound VP1/2 binds importin-β via the essential nuclear localization signal, and how this pertains to capsid docking and genome injection.

SELECTIVE LOSS OF VP16 DURING NEURONAL INFECTION

Once in the nucleus, a single viral genome can establish a replication compartment as an early step to viral amplification (76). In neurons of the natural host, this potential is suppressed and latency is instead established. The decision to replicate or enter latency is tied to the fate of the VP16 tegument protein. VP16 is a transactivator of immediate early gene expression that enters the nucleus of somatic cells in a complex with host cell factor C1 (HCF-C1) to promote productive infection (84, 85). Upon entering a cell, VP16 is removed from the capsid (**Figure 3**) (58). In the context of a neuron, this disassociation eliminates an obvious means to deliver VP16 to the neural soma upon entry at the distal axon (9, 96). Input VP16 may be otherwise unable to reach neuronal nuclei, as retrograde axon transport of cytosolic proteins targeted for nuclear import can be blocked in healthy neurons (60). These findings suggest that the neuron polar architecture is predisposed to favor latency establishment. However, VP16 expression in neural soma is not sufficient to trigger productive infection, indicating that there is a second barrier to VP16 nuclear delivery in neurons (150). This second block appears to be at the level of nuclear import. HCF-C1 is sequestered from the nucleus in neurons, which is expected to prevent VP16 nuclear import (81). These two properties functioning in tandem may serve as the primary determinant of latency establishment in neurons. In accordance with this model, VP16 is also a critical determinant for reactivation of HSV-1 from latency (165).

In the case of HSV-1, the presence of a small amount of capsid cotransported VP16 may help explain why, upon initial seeding of the nervous system, some neurons establish an acute productive infection that sends progeny virions back to peripheral innervated tissues, whereas other neurons become latently infected (15, 166). The difference in neuron fate could be attributed to small doses of VP16 delivered to the neural soma by variable numbers of retrogradely moving capsids. Consistent with such a model, VP16 was absent from capsids once docked at the nuclear membrane, indicating that this tegument protein is released from capsids following axon transport (9). Whether capsid-dependent trafficking of VP16 to nuclear pores requires HCF-C1 for subsequent nuclear import is unknown.

POSTREPLICATIVE SPREAD

Herpesvirus replication in the nucleus is followed by capsid assembly and genome encapsidation. There is a long-standing debate on how nucleocapsids egress to the cytosol, but current consensus

TGN: *trans*-Golgi network

supports a budding event through the inner nuclear membrane followed by a fusion event at the outer nuclear membrane (reviewed in References 66 and 110). Prior to envelopment at the *trans*-Golgi network (TGN), PRV capsids acquire inner tegument proteins, including VP1/2, pUL37, and pUs3 (57, 74). In fact, the composition of PRV capsids newly deposited into the cytosol, whether from extracellular virions or newly replicated from the nucleus, is remarkably similar. How capsids are differentially targeted to the nucleus early in infection and to the TGN postreplication is an open issue. Because the complement of tegument proteins that are capsid bound in the cytosol is not fully known during either stage of infection, differences in the composition of the capsid-tegument complexes may exist.

For HSV-1, this conundrum could be explained by observations that a subset of outer tegument proteins are acquired on capsids in the nucleus prior to reaching the cytosol; however, these observations are generally inconsistent. VP16 was detected by immunogold electron microscopy on capsids that had budded into the inner nuclear membrane (118). This reactivity was quite weak, however, and in several independent studies VP16 was not detected on purified nuclear capsids (139, 141, 180, 182). The pUL41 outer tegument protein (virion host shutoff, Vhs) also has mixed reports regarding whether it is present on nuclear capsids (139, 141, 180). Similar to reports on these outer tegument proteins, reports vary regarding the site of acquisition for HSV-1 inner tegument proteins, including VP1/2 and pUL37 (20, 124, 139, 169, 180).

In contrast to the varying reports for HSV-1, tegument proteins have generally not been detected on PRV nuclear capsids (57). However, the PRV VP1/2 tegument protein is expressed as multiple isoforms, one of which is a carboxyl-terminal fragment that associates with capsids in the nucleus and expedites nuclear egress (90). Although this new finding has not yet been examined in the context of HSV-1 infection, there is evidence of a carboxyl-terminal species of VP1/2 associated with nuclear HSV-1 capsids (139). Given the technical challenges inherent to these studies, a case can be argued for the need of side-by-side experiments comparing HSV-1 and PRV tegument acquisition to determine whether divergence in assembly pathways of these neurotropic herpesviruses truly exists. A report that HSV-1 uses gB and gH to egress from the nucleus, whereas a separate report found no role for these proteins in PRV nuclear egress, further emphasizes the need for direct comparative analysis to define the assembly and egress pathways used by these viruses (49, 73). This is especially important given that HSV-1 and PRV assembly and egress are essentially indistinguishable when viewed by transmission electron microscopy (56).

Transport to the Cytoplasmic Envelopment Site

While the carboxyl-terminal species of VP1/2 plays an accessory role in the egress of capsids from the nucleus, full-length VP1/2 is essential for cytoplasmic envelopment and egress (41, 53). In addition, time-lapse imaging has shown that UL36-null PRV capsids fail to transport along microtubules. UL37-null PRV envelopes in the cytoplasm poorly but retains vestigial intracellular transport (97). Because UL37-null PRV acquires capsid-bound VP1/2, the directed motion exhibited by this mutant supports a role for VP1/2 in microtubule transport that is enhanced by pUL37 (74). As discussed above, VP1/2 and pUL37 also contribute to capsid delivery to the nucleus, very likely by recruiting the dynein motor to capsids. Implicit in these observations is a puzzle. Capsid-VP1/2-pUL37-pUs3 complexes are present in the cytosol following both entry and postreplication (**Figures 3** and **4**), with both potentially recruiting the dynein motor, and yet the trafficking of these particles is quite different.

Two broad possibilities that could explain how capsids target to the nucleus during early infection and to the cytoplasmic envelopment site during egress require further examination. First, capsid-associated kinesin activities could be specifically enhanced during the egress stage of

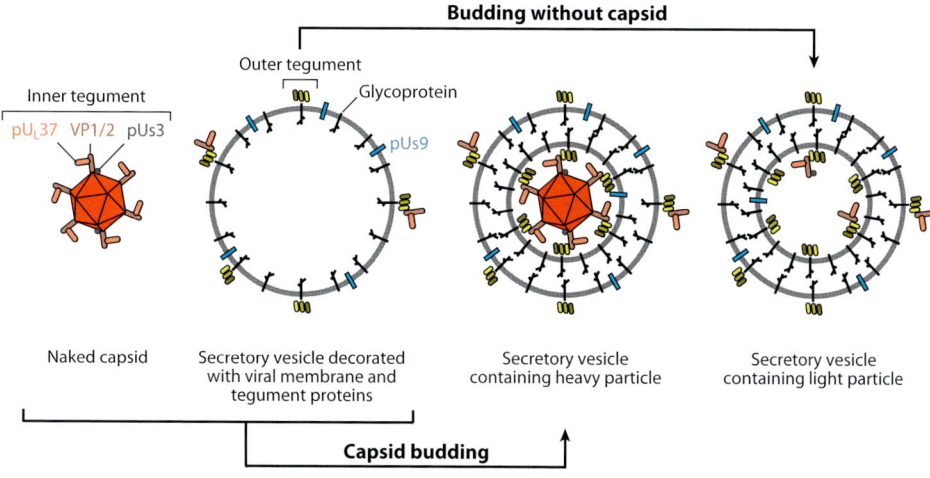

Figure 4

Four types of cytoplasmic viral particles present during the egress stage of infection. Unenveloped (naked) capsids newly egressed from the nucleus acquire inner tegument proteins beginning with VP1/2, which binds capsids directly. The composition of these particles and the retrograde transport complex that is released into the cytosol following entry are similar, if not equivalent (see **Figure 3**). Outer tegument proteins associate primarily with viral membrane proteins resident in the biosynthetic pathway (secretory vesicle), which may also acquire some inner tegument proteins in the absence of capsids. Membrane proteins consist of glycoproteins and nonglycosylated membrane-associated proteins that include pUs9. Infectious virions (heavy particles) form when capsids bud into membranes decorated with viral membrane and outer tegument proteins and subsequently exocytose from the cell. Light particles are noninfectious secreted viral particles that consist of a viral envelope and tegument proteins but lack a capsid. The three membrane-associated intracellular particles expose membrane protein tails and tegument proteins to the cytosolic surface. Among the exposed proteins, pUs9 is enriched in the vesicle membrane surface and may direct trafficking of particles to the distal axon.

infection. Capsid-tegument complexes bind both dynein and kinesins (139). During retrograde axon transport, kinesin-based motion is weak and the predominating dynein-based motion results in capsid delivery to the neural soma (154). The presence of kinesin activity during retrograde axon transport is not intuitive but may be explained if kinesin and dynein recruitment also occurs during egress. In this scenario, capsid-associated kinesin activities could be enhanced to traffic the capsids to the site of envelopment. Second, capsids may recruit the dynein motor and move retrograde at both stages of infection. The presence of viral tegument and glycoproteins in the biosynthetic pathway during late infection could redirect capsid trafficking to the TGN for envelopment, as capsids move by dynein-based motion toward the MTOC. Similar to that observed in nonneuronal cell lines, capsids bud into the TGN in neurons and are released from the neural soma by exocytosis (113).

AXON TARGETING

In the absence of either gE, gI, or the pUs9 envelope proteins, PRV cannot transport through neural circuits in the anterograde direction (13, 18, 19, 21, 82, 83, 116, 117, 174). In cultured neurons the phenotype of the deletion mutants is not absolute, but PRV or HSV-1 lacking any one of these proteins is dramatically reduced for anterograde transmission in cultured neurons (24, 100, 155). When the genes for all three proteins are simultaneously deleted, anterograde transmission is eliminated (24). The gE, gI, and pUs9 viral envelope proteins coordinate the delivery of cytoplasmic viral particles from the neural soma into the axon. However, the requirement for these proteins in cultured axons is present only in neurons that have been cultured for prolonged periods: two weeks is typically used. In short-term cultures, anterograde axon transport of PRV is unimpeded in the absence of pUs9 (G. Smith & L.W. Enquist, unpublished data). Therefore, the barrier overcome by these proteins is slow to manifest following axon outgrowth and is consistent with the maturation of the axon initial segment (178).

The transmembrane domain and cytosolic tail of gE are dispensable for anterograde trafficking (167). This last point is remarkable for two reasons. First, gE is not structurally incorporated into virions in the absence of the transmembrane anchor and cytosolic tail. Second, the truncated protein is secreted. Taken together, these findings argue that gE is facilitating the delivery of cytoplasmic viral particles by interacting with extracellular proteins away from the viral particle.

Although the mechanisms by which these proteins function need to be further addressed, pUs9 appears to function in a way that is distinct from gE and gI. PRV lacking pUs9 is far more attenuated for anterograde transmission than are gE or gI mutants, yet the pUs9-null virus has no defect in plaque size, whereas gE-null and gI-null viruses do (24). The function of pUs9 is discussed further below.

Another puzzling facet to these proteins is that they are all selected against during serial passage in culture. PRV spontaneously loses gE expression during serial passage, and an extensive passage regime that was used to make an attenuated vaccine strain of PRV resulted in the deletion of all three genes. Similarly, isolates of HSV-1 with spontaneous mutations in gI or pUs9 have been reported (121, 161).

Anterograde Axon Transport

Trafficking of viral particles to the distal axon occurs by fast microtubule-based transport (153). In a remarkable case report of a nine-year-old boy that experienced herpetic pain from the sciatic nerve and zoster lesions six days later, a crude estimate of herpes anterograde spread in its natural setting was found to be consistent with the measured transport velocities of individual PRV capsids in culture neurons (162). This observation provides a unique validation of the use of neuron culture models in the study of herpesvirus infection. Unfortunately, studies of infection in neural culture models have resulted in a debate regarding the mechanics of anterograde axon transport. Two types of viral particle could potentially enter axons and engage in anterograde transport: cytosolic capsids that emerge from the nucleus (referred to as naked capsids owing to the absence of a membrane envelope) and enveloped virions in secretory vesicles resulting from the budding of naked capsids into the TGN. In both instances tegument proteins would be positioned to recruit motors and effect transport of the particle, whether it is a naked capsid or an enveloped virion (**Figure 4**). Prior to budding, naked capsids become decorated with inner tegument proteins, which together have the capacity to bind dynein and kinesin motors (139, 180). Upregulation of kinesin-bound motors would provide a hypothetical mechanism by which the distal axon could be targeted. On the other hand, enveloped virions are readily detected in neuronal cell bodies (22, 37, 113). Because enveloped virions reside in a membrane vesicle derived from the host biosynthetic pathway, these vesicles could intrinsically be targeted to the distal axon without the need for viral effectors. Alternatively, viral proteins, including tegument proteins and the cytosolic tails of envelope proteins, could serve as a platform to recruit microtubule motors to the cytosolic surface of the vesicle. Support for the latter comes from the observation that envelope and tegument proteins transport anterograde in axons in the absence of capsids (8, 37, 50, 112). In fact, tegument proteins transport to distal axons in advance of capsid-containing particles (96).

Therefore, membrane-bound envelope-tegument complexes have the capacity to transport efficiently in axons. In further support of enveloped virion transport, TGN-derived vesicles containing viral particles isolated from HSV-1-infected cells exhibit kinesin-based motion in a reconstituted in vitro microtubule transport model (88).

Enveloped virions are observed by electron microscopy in axons in animal and culture models of infection, for both HSV-1 (32, 64, 79, 87, 99, 120) and PRV (24, 50, 51, 102), adding further support to the vesicle model of axon transport. However, this model was challenged by a study of HSV-1 in cultured human neurons that reported that all viral particles seen throughout the entire length of axons were naked capsids; enveloped virions were universally restricted to the neuronal cell body (136). This dramatic result had the explicit implication that capsids, or capsid-tegument complexes, directly recruited a kinesin motor independent of the neuronal secretory pathway. However, a subsequent reassessment of this finding concluded that the majority of capsids in axons were in fact enveloped but were restricted to axon terminals and varicosities along the axon (146). Therefore, the debate evolved into a discussion of why naked and enveloped viral particles were coresident in axons and which of these particles represented the species that transports anterograde in axons. Interested readers are directed to the many reports describing the debate that have presented differing viewpoints and have also provided descriptions of additional studies that contribute to aspects of the debate that go beyond the current overview (for example, see References 34 and 43).

Recently, work from several labs, including new electron microscopy studies and time-lapse imaging of viral particle composition during active anterograde transport (10, 65, 120, 179), has provided compelling data that enveloped capsids of HSV-1 are transported anterograde in axons. The only remaining question is whether naked capsids also transport anterograde in axons. Because the tegument and envelope of herpes virions are heterogeneous, with the copy number of protein species varying widely from one particle to the next, this question will be difficult to conclusively answer (37). Viral particles containing low amounts of a tagged envelope protein will often be scored as lacking an envelope, particularly with the added challenge of imaging the particles while they are moving at speeds in excess of $1~\mu m~s^{-1}$. Although use of sensitive electron-multiplying charge-coupled device cameras coupled with bright light sources has allowed detection of dim emissions from rapidly moving viral particles in axons, detecting all viral particles by means of a heterogeneous antigen such as an envelope protein will likely require total internal reflection microscopy or a new technique of equivalent sensitivity to determine whether all capsids actively moving in axons are associated with envelope components (10). Several different investigations have provided additional insight into the nature of viral particle association with the transport vesicle. HSV-1-containing vesicles isolated from infected cell cytoplasm frequently do not contain a fully budded virion within the lumen of the vesicle, but instead have a capsid that is partially wrapped in the vesicle membrane (88). Similar structures are seen in axons of neurons infected with HSV-1 (146). Close inspection of transmission electron micrographs of PRV-containing vesicles in axons further hints at these structures (24). There is also functional evidence for axon transport of partially budded PRV (30). Although more studies are necessary to confirm the presence of stable budded intermediates serving as cargoes of kinesin-based anterograde transport, these preliminary observations may have implications for the spread of virions from axons to neighboring cells (see below). Moving forward, additional characterization of the transport vesicle hijacked by HSV-1 and PRV to reach the distal axon is needed. The cellular Vamp2 protein, which functions at presynaptic axon terminals, is often cotransported with HSV-1 and PRV transport vesicles and may indicate that virus-containing vesicles are predisposed to target the axon terminal (10). VP1/2 may be one protein that contributes to this targeting (151). In addition, the pUs9 protein is an intriguing candidate as a membrane-bound effector of anterograde axon transport (163).

A GFP-pUs9 fusion protein expressed by a recombinant of PRV cotransports with capsids to the distal axon, consistent with PRV anterograde transport occurring in a vesicle. Although this finding in itself is not unexpected, it is more notable that GFP-pUs9 was significantly diminished in extracellular viral particles. The reduction in GFP-pUs9 incorporation in extracellular particles relative to viral particles actively undergoing anterograde axon transport can easily be explained only if the GFP-pUs9 protein is enriched in the transport vesicle membrane that surrounds the viral particle (**Figure 4**). A mutation in GFP-pUs9 that prevents anterograde spread of PRV remained competent to move anterograde in axons, but capsids were no longer cotransported with the GFP signal. These findings provide a compelling case for pUs9 as an anterograde transport effector in neurons. Because pUs9 is dispensable for anterograde capsid transport in short-term neuron cultures, its contribution may be to overcome the barrier function of a mature axon initial segment. Alternatively, if pUs9 recruits a kinesin motor to effect anterograde transport, then redundant mechanisms of kinesin recruitment must be available in short-term neural cultures that become restricted as neurons mature.

Release from the Axon Terminal

Using chambered neuron culture models, investigators cannot detect release of free HSV-1 and PRV virions from axons of infected neurons (24, 106). This is due, not to a lack of anterograde transport in axons, but to virions remaining tethered to the axon surface after emerging from the axon (35). Infection by the surface-bound virions requires axon-cell contact and is dependent upon gB in the virions.

As PRV infection in cultured neurons progresses, the neurons display changes in electrophysiology (105). Action potentials fire at increased frequency and eventually neurons display synchronous firing. This electrical coupling of infected neurons is promoted by axon-axon fusion pores that form in part by the action of gB. Currently, no evidence indicates that the activity of gB that forms axon-axon fusion pores is the result of the population of gB present in virions at the axon surface. But if cell surface virions promote electrical coupling, it may hint at the presence of a pore between the axon and surface virion that would allow for subsequent connectivity upon fusion to a neighboring axon. In this context, the previously discussed observations that viral particles may not fully bud into transport vesicles could be relevant to electrical coupling if these structures are preserved after exocytosis. Changes in electrophysiology are predicted to cause the caustic itching (pruritis) exhibited during PRV infection, and a related phenomenon could underlie the severe pain (postherpetic neuralgia) often suffered following VZV reactivation.

Spread Between Neurons and to Distant Sites

As previously noted, herpes virions can egress from the cell bodies of neurons as well as from axons. The biological consequence of virion release from the neuron cell body in peripheral ganglia has broad implications in the pathogenesis of the neurotropic herpesviruses. In the simplest scenario, HSV or PRV would only transmit between neurons and the cells they innervate at the periphery by repeated cycles of anterograde and retrograde axonal transport (**Figure 2**). The tendency for HSV-1 lesions (herpes labialis, or cold sores) to creep from one reactivation episode to the next would be explained by lateral spread of infection in the mucosal epithelium, allowing for seeding of new neurons that innervate cells adjacent to the active lesion. This view is consistent with the presentation of cold sore lesions, which tend to be focal and exhibit limited dissemination in the innervated dermatome. This is in contrast to reactivated VZV infections, which are discussed below.

Spread of HSV-1 within sensory ganglia may be limited by satellite cells that surround each neuron. Although these support cells are susceptible to PRV and HSV-1, they generally do not appear to be productively infected (22, 32, 63, 87). Perhaps the best evidence for the absence of lateral spread within peripheral ganglia comes from an in vivo study of PRV infection (158). PRV inoculated into the anterior chamber of the rat eye transits by retrograde axon transport to neuron cell bodies in the superior cervical ganglion. Subsequent labeling of the subset of superior cervical ganglion neurons that project to the anterior chamber by retrograde labeling with wheat germ agglutinin conjugated to horseradish peroxidase demonstrated that PRV remained confined to the neurons projecting axons to the anterior chamber of the eye and did not spread to neighboring neurons projecting to other sites. These results provided compelling evidence that viral dissemination occurred only by circuit-specific transmission. However, the PRV strain used in this seminal study, PRV Bartha, is an attenuated vaccine strain. Whether these results hold true for wild-type isolates of PRV or HSV-1 has not, to the best of my knowledge, been examined. But available evidence argues that wild-type virus strains are not always subject to this restriction. HSV inoculated on the mouse flank results in a zosteriform spread throughout the innervated dermatome, which suggests HSV spreads between neurons in dorsal root sensory ganglia (DRG) before returning back to the skin (173). Applying PRV to the flank model does not produce zoster-like lesions in the skin, but nevertheless the virus transmits through the dermatome, indicating that PRV, like HSV, disseminates laterally within the peripheral ganglia (19). In contrast, the attenuated Bartha strain remains confined to the inoculation site in the skin and does not spread through the dermatome.

In contrast to HSV and PRV, VZV reactivation often produces infection throughout a dermatome, which is the prototypic shingles manifestation (herpes zoster). Observations of human DRG infected with VZV either from human patients or in a mouse xenograft model are consistent with productive infections of neurons and satellite cells (48, 142). Furthermore, satellite cell and neural syncytia in human DRG are also evident. Lateral spread in sensory ganglia would likely result in infection of sensory neurons that project to internal organs. These general visceral afferent neurons are similar to somatosensory pseudounipolar neurons and are coresident in sensory ganglia. Because visceral infections are not associated with cold sore eruptions, this would argue that HSV-1 does not randomly spread between neurons in sensory ganglia in adult humans. However, neonatal transmission of HSV-2 can produce life-threatening infections that involve disseminated infections in internal organs (72). VZV spread to internal organs can also occur, although this outcome is more common in immunocompromised individuals (11).

Disseminated infections can occur with HSV, PRV, and VZV by means of cell-associated viremia. While this is most prevalent in VZV infections, PRV transplacental spread occurs by monocytes that transmit the infection (119). HSV infections are generally not associated with cell-associated viremia, but in the absence of an interferon response, infection transmits efficiently to the liver and other organs (95, 133). Neonatal disseminated HSV-2 infections are also associated with cell-associated viremia (42).

CONCLUSIONS

Whereas many viruses transport retrogradely in axons of neurons, neuroinvasive herpesviruses are notable for their efficient entry into the nervous system and coordination of retrograde and anterograde trafficking at opposing stages of the infectious cycle. These properties make viruses such as HSV-1 candidates for the development of gene delivery vectors that target the nervous system, but also underlie the potential for these viruses to become highly virulent following dissemination into the central nervous system. Ongoing research will resolve how herpesvirus assembly and egress are coupled to neuroinvasion and pathogenesis, and may yield the tools to produce both novel gene vectors and antiviral treatments to combat severe forms of neuroinvasive disease.

> **SUMMARY POINTS**
>
> 1. Herpesviruses overcome multiple barriers to establish latent infections in the nervous system.
> 2. Tegument proteins retained on capsids likely recruit microtubule motors necessary for axon transport.
> 3. Several viral and cellular factors have been identified that contribute to capsid delivery to the nucleus and to our understanding of how and when these functions delineate steps in neuronal infection.
> 4. Removal of VP16 from capsids upon axon entry may promote establishment of latency in the infected neuron.
> 5. HSV and PRV effectively transmit between synapse-linked neurons, but productive recurrent infections in natural hosts do not involve synaptic transmission.
> 6. Neurologic responses such as pain and pruritis may indicate productive infection and spread within ganglia.
> 7. In addition to their characteristic neurotropism, HSV, PRV, and VZV share broad-tissue tropism, which accounts for replication in the mucosa, visceral organs, and cell-associated viremia.

DISCLOSURE STATEMENT

The author is not aware of any affiliations, memberships, funding, or financial holdings that might be perceived as affecting the objectivity of this review.

ACKNOWLEDGMENTS

Research in the author's laboratory was funded by the National Institutes of Health. Many thanks go to Lynn W. Enquist and Gary Pickard for their helpful discussions. I apologize to those many investigators whose interesting work could not be cited because of page limits.

LITERATURE CITED

1. Abaitua F, Daikoku T, Crump CM, Bolstad M, O'Hare P. 2011. A single mutation responsible for temperature-sensitive entry and assembly defects in the VP1-2 protein of herpes simplex virus. *J. Virol.* 85:2024–36
2. Abaitua F, O'Hare P. 2008. Identification of a highly conserved, functional nuclear localization signal within the N-terminal region of herpes simplex virus type 1 VP1-2 tegument protein. *J. Virol.* 82:5234–44
3. Abaitua F, Souto RN, Browne H, Daikoku T, O'Hare P. 2009. Characterization of the herpes simplex virus (HSV)-1 tegument protein VP1-2 during infection with the HSV temperature-sensitive mutant tsB7. *J. Gen. Virol.* 90:2353–63
4. Addison C, Rixon FJ, Palfreyman JW, O'Hara M, Preston VG. 1984. Characterisation of a herpes simplex virus type 1 mutant which has a temperature-sensitive defect in penetration of cells and assembly of capsids. *Virology* 138:246–59

5. Aggarwal A, Miranda-Saksena M, Boadle RA, Kelly BJ, Diefenbach RJ, et al. 2012. Ultrastructural visualization of individual tegument protein dissociation during entry of herpes simplex virus 1 into human and rat dorsal root ganglion neurons. *J. Virol.* 86:6123–37
6. Ally S, Larson AG, Barlan K, Rice SE, Gelfand VI. 2009. Opposite-polarity motors activate one another to trigger cargo transport in live cells. *J. Cell Biol.* 187:1071–82
7. Antinone SE, Shubeita GT, Coller KE, Lee JI, Haverlock-Moyns S, et al. 2006. The herpesvirus capsid surface protein, VP26, and the majority of the tegument proteins are dispensable for capsid transport toward the nucleus. *J. Virol.* 80:5494–98
8. Antinone SE, Smith GA. 2006. Two modes of herpesvirus trafficking in neurons: Membrane acquisition directs motion. *J. Virol.* 80:11235–40
9. Antinone SE, Smith GA. 2010. Retrograde axon transport of herpes simplex virus and pseudorabies virus: a live-cell comparative analysis. *J. Virol.* 84:1504–12
10. Antinone SE, Zaichick SV, Smith GA. 2010. Resolving the assembly state of herpes simplex virus during axon transport by live-cell imaging. *J. Virol.* 84:13019–30
11. Arvin AM. 1996. Varicella-zoster virus. *Clin. Microbiol. Rev.* 9:361–81
12. Aubert M, Yoon M, Sloan DD, Spear PG, Jerome KR. 2009. The virological synapse facilitates herpes simplex virus entry into T cells. *J. Virol.* 83:6171–83
13. Babic N, Klupp B, Brack A, Mettenleiter TC, Ugolini G, Flamand A. 1996. Deletion of glycoprotein gE reduces the propagation of pseudorabies virus in the nervous system of mice after intranasal inoculation. *Virology* 219:279–84
14. Batterson W, Furlong D, Roizman B. 1983. Molecular genetics of herpes simplex virus. VIII. Further characterization of a temperature-sensitive mutant defective in release of viral DNA and in other stages of the viral reproductive cycle. *J. Virol.* 45:397–407
15. Blyth WA, Harbour DA, Hill TJ. 1984. Pathogenesis of zosteriform spread of herpes simplex virus in the mouse. *J. Gen. Virol.* 65(Pt. 9):1477–86
16. Borza CM, Hutt-Fletcher LM. 2002. Alternate replication in B cells and epithelial cells switches tropism of Epstein-Barr virus. *Nat. Med.* 8:594–99
17. Böttcher S, Maresch C, Granzow H, Klupp BG, Teifke JP, Mettenleiter TC. 2008. Mutagenesis of the active-site cysteine in the ubiquitin-specific protease contained in large tegument protein pUL36 of pseudorabies virus impairs viral replication in vitro and neuroinvasion in vivo. *J. Virol.* 82:6009–16
18. Brideau AD, Card JP, Enquist LW. 2000. Role of pseudorabies virus Us9, a type II membrane protein, in infection of tissue culture cells and the rat nervous system. *J. Virol.* 74:834–45
19. Brittle EE, Reynolds AE, Enquist LW. 2004. Two modes of pseudorabies virus neuroinvasion and lethality in mice. *J. Virol.* 78:12951–63
20. Bucks MA, O'Regan KJ, Murphy MA, Wills JW, Courtney RJ. 2007. Herpes simplex virus type 1 tegument proteins VP1/2 and UL37 are associated with intranuclear capsids. *Virology* 361:316–24
21. Card JP, Levitt P, Enquist LW. 1998. Different patterns of neuronal injection after intracerebral infection of two strains of pseudorabies virus. *J. Virol.* 72:4434–41
22. Card JP, Rinaman L, Lynn RB, Lee BH, Meade RP, et al. 1993. Pseudorabies virus infection of the rat central nervous system: ultrastructural characterization of viral replication, transport, and pathogenesis. *J. Neurosci.* 13:2515–39
23. Cardone G, Newcomb WW, Cheng N, Wingfield PT, Trus BL, et al. 2012. The UL36 tegument protein of herpes simplex virus 1 has a composite binding site at the capsid vertices. *J. Virol.* 86:4058–64
24. Ch'ng TH, Enquist LW. 2005. Neuron-to-cell spread of pseudorabies virus in a compartmented neuronal culture system. *J. Virol.* 79:10875–89
25. Ch'ng TH, Spear PG, Struyf F, Enquist LW. 2007. Glycoprotein D-independent spread of pseudorabies virus infection in cultured peripheral nervous system neurons in a compartmented system. *J. Virol.* 81:10742–57
26. Cheshenko N, Liu W, Satlin LM, Herold BC. 2005. Focal adhesion kinase plays a pivotal role in herpes simplex virus entry. *J. Biol. Chem.* 280:31116–25
27. Chou J, Kern ER, Whitley RJ, Roizman B. 1990. Mapping of herpes simplex virus-1 neurovirulence to gamma 134.5, a gene nonessential for growth in culture. *Science* 250:1262–66

28. Clement C, Tiwari V, Scanlan PM, Valyi-Nagy T, Yue BY, Shukla D. 2006. A novel role for phagocytosis-like uptake in herpes simplex virus entry. *J. Cell Biol.* 174:1009–21
29. Coller KE, Lee JI, Ueda A, Smith GA. 2007. The capsid and tegument of the alphaherpesviruses are linked by an interaction between the UL25 and VP1/2 proteins. *J. Virol.* 81:11790–97
30. Coller KE, Smith GA. 2008. Two viral kinases are required for sustained long distance axon transport of a neuroinvasive herpesvirus. *Traffic* 9:1458–70
31. Connolly SA, Jackson JO, Jardetzky TS, Longnecker R. 2011. Fusing structure and function: a structural view of the herpesvirus entry machinery. *Nat. Rev. Microbiol.* 9:369–81
32. Cook ML, Stevens JG. 1973. Pathogenesis of herpetic neuritis and ganglionitis in mice: evidence for intra-axonal transport of infection. *Infect. Immun.* 7:272–88
33. Copeland AM, Newcomb WW, Brown JC. 2009. Herpes simplex virus replication: roles of viral proteins and nucleoporins in capsid-nucleus attachment. *J. Virol.* 83:1660–68
34. Curanovic D, Enquist L. 2009. Directional transneuronal spread of α-herpesvirus infection. *Future Virol.* 4:591
35. Curanovic D, Enquist LW. 2009. Virion-incorporated glycoprotein B mediates transneuronal spread of pseudorabies virus. *J. Virol.* 83:7796–804
36. De Regge N, Nauwynck HJ, Geenen K, Krummenacher C, Cohen GH, et al. 2006. α-Herpesvirus glycoprotein D interaction with sensory neurons triggers formation of varicosities that serve as virus exit sites. *J. Cell Biol.* 174:267–75
37. del Rio T, Ch'ng TH, Flood EA, Gross SP, Enquist LW. 2005. Heterogeneity of a fluorescent tegument component in single pseudorabies virus virions and enveloped axonal assemblies. *J. Virol.* 79:3903–19
38. Delboy MG, Nicola AV. 2011. A pre-immediate-early role for tegument ICP0 in the proteasome-dependent entry of herpes simplex virus. *J. Virol.* 85:5910–18
39. Delboy MG, Roller DG, Nicola AV. 2008. Cellular proteasome activity facilitates herpes simplex virus entry at a postpenetration step. *J. Virol.* 82:3381–90
40. Desai P, DeLuca NA, Person S. 1998. Herpes simplex virus type 1 VP26 is not essential for replication in cell culture but influences production of infectious virus in the nervous system of infected mice. *Virology* 247:115–24
41. Desai PJ. 2000. A null mutation in the UL36 gene of herpes simplex virus type 1 results in accumulation of unenveloped DNA-filled capsids in the cytoplasm of infected cells. *J. Virol.* 74:11608–18
42. Diamond C, Mohan K, Hobson A, Frenkel L, Corey L. 1999. Viremia in neonatal herpes simplex virus infections. *Pediatr. Infect. Dis. J.* 18:487–89
43. Diefenbach RJ, Miranda-Saksena M, Douglas MW, Cunningham AL. 2008. Transport and egress of herpes simplex virus in neurons. *Rev. Med. Virol.* 18:35–51
44. Dohner K, Radtke K, Schmidt S, Sodeik B. 2006. Eclipse phase of herpes simplex virus type 1 infection: efficient dynein-mediated capsid transport without the small capsid protein VP26. *J. Virol.* 80:8211–24
45. Dohner K, Wolfstein A, Prank U, Echeverri C, Dujardin D, et al. 2002. Function of dynein and dynactin in herpes simplex virus capsid transport. *Mol. Biol. Cell* 13:2795–809
46. Douglas MW, Diefenbach RJ, Homa FL, Miranda-Saksena M, Rixon FJ, et al. 2004. Herpes simplex virus type 1 capsid protein VP26 interacts with dynein light chains RP3 and Tctex1 and plays a role in retrograde cellular transport. *J. Biol. Chem.* 279:28522–30
47. Enquist LW, Husak PJ, Banfield BW, Smith GA. 1998. Infection and spread of alphaherpesviruses in the nervous system. *Adv. Virus Res.* 51:237–347
48. Esiri MM, Tomlinson AH. 1972. Herpes Zoster. Demonstration of virus in trigeminal nerve and ganglion by immunofluorescence and electron microscopy. *J. Neurol. Sci.* 15:35–48
49. Farnsworth A, Wisner TW, Webb M, Roller R, Cohen G, et al. 2007. Herpes simplex virus glycoproteins gB and gH function in fusion between the virion envelope and the outer nuclear membrane. *Proc. Natl. Acad. Sci. USA* 104:10187–92
50. Feierbach B, Bisher M, Goodhouse J, Enquist LW. 2007. In vitro analysis of transneuronal spread of an alphaherpesvirus infection in peripheral nervous system neurons. *J. Virol.* 81:6846–57
51. Field HJ, Hill TJ. 1974. The pathogenesis of pseudorabies in mice following peripheral inoculation. *J. Gen. Virol.* 23:145–57

52. Frampton AR Jr, Uchida H, von Einem J, Goins WF, Grandi P, et al. 2010. Equine herpesvirus type 1 (EHV-1) utilizes microtubules, dynein, and ROCK1 to productively infect cells. *Vet. Microbiol.* 141:12–21

53. Fuchs W, Klupp BG, Granzow H, Mettenleiter TC. 2004. Essential function of the pseudorabies virus UL36 gene product is independent of its interaction with the UL37 protein. *J. Virol.* 78:11879–89

54. Fuller AO, Santos RE, Spear PG. 1989. Neutralizing antibodies specific for glycoprotein H of herpes simplex virus permit viral attachment to cells but prevent penetration. *J. Virol.* 63:3435–43

55. Geraghty RJ, Krummenacher C, Cohen GH, Eisenberg RJ, Spear PG. 1998. Entry of alphaherpesviruses mediated by poliovirus receptor-related protein 1 and poliovirus receptor. *Science* 280:1618–20

56. Granzow H, Klupp BG, Fuchs W, Veits J, Osterrieder N, Mettenleiter TC. 2001. Egress of alphaherpesviruses: comparative ultrastructural study. *J. Virol.* 75:3675–84

57. Granzow H, Klupp BG, Mettenleiter TC. 2004. The pseudorabies virus US3 protein is a component of primary and of mature virions. *J. Virol.* 78:1314–23

58. Granzow H, Klupp BG, Mettenleiter TC. 2005. Entry of pseudorabies virus: an immunogold-labeling study. *J. Virol.* 79:3200–5

59. Gross SP. 2003. Dynactin: coordinating motors with opposite inclinations. *Curr. Biol.* 13:R320–22

60. Hanz S, Perlson E, Willis D, Zheng JQ, Massarwa R, et al. 2003. Axoplasmic importins enable retrograde injury signaling in lesioned nerve. *Neuron* 40:1095–104

61. Heffner S, Kovacs F, Klupp BG, Mettenleiter TC. 1993. Glycoprotein gp50-negative pseudorabies virus: a novel approach toward a nonspreading live herpesvirus vaccine. *J. Virol.* 67:1529–37

62. Heldwein EE, Krummenacher C. 2008. Entry of herpesviruses into mammalian cells. *Cell. Mol. Life Sci.* 65:1653–68

63. Hill TJ, Field HJ. 1973. The interaction of herpes simplex virus with cultures of peripheral nervous tissue: an electron microscopic study. *J. Gen. Virol.* 21:123–33

64. Hill TJ, Field HJ, Roome AP. 1972. Intra-axonal location of herpes simplex virus particles. *J. Gen. Virol.* 15:233–35

65. Huang J, Lazear HM, Friedman HM. 2011. Completely assembled virus particles detected by transmission electron microscopy in proximal and mid-axons of neurons infected with herpes simplex virus type 1, herpes simplex virus type 2 and pseudorabies virus. *Virology* 409:12–16

66. Johnson DC, Baines JD. 2011. Herpesviruses remodel host membranes for virus egress. *Nat. Rev. Microbiol.* 9:382–94

67. Johnson RT, Mims CA. 1968. Pathogenesis of viral infections of the nervous system. *N. Engl. J. Med.* 278:23–30

68. Jovasevic V, Liang L, Roizman B. 2008. Proteolytic cleavage of VP1-2 is required for release of herpes simplex virus 1 DNA into the nucleus. *J. Virol.* 82:3311–19

69. Karaba AH, Kopp SJ, Longnecker R. 2011. Herpesvirus entry mediator and nectin-1 mediate herpes simplex virus 1 infection of the murine cornea. *J. Virol.* 85:10041–47

70. Karasneh GA, Shukla D. 2011. Herpes simplex virus infects most cell types in vitro: clues to its success. *Virol. J.* 8:481

71. Kardon JR, Vale RD. 2009. Regulators of the cytoplasmic dynein motor. *Nat. Rev. Mol. Cell Biol.* 10:854–65

72. Kimberlin DW, Whitley RJ. 2005. Neonatal herpes: What have we learned. *Semin. Pediatr. Infect. Dis.* 16:7–16

73. Klupp B, Altenschmidt J, Granzow H, Fuchs W, Mettenleiter TC. 2008. Glycoproteins required for entry are not necessary for egress of pseudorabies virus. *J. Virol.* 82:6299–309

74. Klupp BG, Fuchs W, Granzow H, Nixdorf R, Mettenleiter TC. 2002. Pseudorabies virus UL36 tegument protein physically interacts with the UL37 protein. *J. Virol.* 76:3065–71

75. Knipe DM, Batterson W, Nosal C, Roizman B, Buchan A. 1981. Molecular genetics of herpes simplex virus. VI. Characterization of a temperature-sensitive mutant defective in the expression of all early viral gene products. *J. Virol.* 38:539–47

76. Kobiler O, Brodersen P, Taylor MP, Ludmir EB, Enquist LW. 2011. Herpesvirus replication compartments originate with single incoming viral genomes. *mBio* 2(6):e00278–11

77. Kopp SJ, Banisadr G, Glajch K, Maurer UE, Grunewald K, et al. 2009. Infection of neurons and encephalitis after intracranial inoculation of herpes simplex virus requires the entry receptor nectin-1. *Proc. Natl. Acad. Sci. USA* 106:17916–20
78. Krautwald M, Fuchs W, Klupp BG, Mettenleiter TC. 2009. Translocation of incoming pseudorabies virus capsids to the cell nucleus is delayed in the absence of tegument protein pUL37. *J. Virol.* 83:3389–96
79. Kristensson K, Ghetti B, Wisniewski HM. 1974. Study on the propagation of herpes simplex virus (type 2) into the brain after intraocular injection. *Brain Res.* 69:189–201
80. Kristensson K, Lycke E, Röyttä M, Svennerholm B, Vahlne A. 1986. Neuritic transport of herpes simplex virus in rat sensory neurons in vitro. Effects of substances interacting with microtubular function and axonal flow [nocodazole, taxol and erythro-9-3-(2-hydroxynonyl)adenine]. *J. Gen. Virol.* 67:2023–28
81. Kristie TM, Vogel JL, Sears AE. 1999. Nuclear localization of the C1 factor (host cell factor) in sensory neurons correlates with reactivation of herpes simplex virus from latency. *Proc. Natl. Acad. Sci. USA* 96:1229–33
82. Kritas SK, Nauwynck HJ, Pensaert MB. 1995. Dissemination of wild-type and gC-, gE- and gI-deleted mutants of Aujeszky's disease virus in the maxillary nerve and trigeminal ganglion of pigs after intranasal inoculation. *J. Gen. Virol.* 76:2063–66
83. Kritas SK, Pensaert MB, Mettenleiter TC. 1994. Role of envelope glycoproteins gI, gp63 and gIII in the invasion and spread of Aujeszky's disease virus in the olfactory nervous pathway of the pig. *J. Gen. Virol.* 75:2319–27
84. La Boissière S, Hughes T, O'Hare P. 1999. HCF-dependent nuclear import of VP16. *EMBO J.* 18:480–89
85. La Boissière S, O'Hare P. 2000. Analysis of HCF, the cellular cofactor of VP16, in herpes simplex virus-infected cells. *J. Virol.* 74:99–109
86. Lancaster KZ, Pfeiffer JK. 2010. Limited trafficking of a neurotropic virus through inefficient retrograde axonal transport and the type I interferon response. *PLoS Pathog.* 6:e1000791
87. LaVail JH, Topp KS, Giblin PA, Garner JA. 1997. Factors that contribute to the transneuronal spread of herpes simplex virus. *J. Neurosci. Res.* 49:485–96
88. Lee GE, Murray JW, Wolkoff AW, Wilson DW. 2006. Reconstitution of herpes simplex virus microtubule-dependent trafficking in vitro. *J. Virol.* 80:4264–75
89. Lee JI, Sollars PJ, Baver SB, Pickard GE, Leelawong M, Smith GA. 2009. A herpesvirus encoded deubiquitinase is a novel neuroinvasive determinant. *PLoS Pathog.* 5:e1000387
90. Leelawong M, Lee JI, Smith GA. 2012. Nuclear egress of pseudorabies virus capsids is enhanced by a subspecies of the large tegument protein that is lost upon cytoplasmic maturation. *J. Virol.* 86:6303–14
91. Leib DA, Machalek MA, Williams BR, Silverman RH, Virgin HW. 2000. Specific phenotypic restoration of an attenuated virus by knockout of a host resistance gene. *Proc. Natl. Acad. Sci. USA* 97:6097–101
92. Liesegang TJ. 2001. Herpes simplex virus epidemiology and ocular importance. *Cornea* 20:1–13
93. Ligas MW, Johnson DC. 1988. A herpes simplex virus mutant in which glycoprotein D sequences are replaced by beta-galactosidase sequences binds to but is unable to penetrate into cells. *J. Virol.* 62:1486–94
94. Liu WW, Goodhouse J, Jeon NL, Enquist LW. 2008. A microfluidic chamber for analysis of neuron-to-cell spread and axonal transport of an alpha-herpesvirus. *PLoS ONE* 3:e2382
95. Luker GD, Prior JL, Song J, Pica CM, Leib DA. 2003. Bioluminescence imaging reveals systemic dissemination of herpes simplex virus type 1 in the absence of interferon receptors. *J. Virol.* 77:11082–93
96. Luxton GW, Haverlock S, Coller KE, Antinone SE, Pincetic A, Smith GA. 2005. Targeting of herpesvirus capsid transport in axons is coupled to association with specific sets of tegument proteins. *Proc. Natl. Acad. Sci. USA* 102:5832–37
97. Luxton GW, Lee JI, Haverlock-Moyns S, Schober JM, Smith GA. 2006. The pseudorabies virus VP1/2 tegument protein is required for intracellular capsid transport. *J. Virol.* 80:201–9
98. Lycke E, Hamark B, Johansson M, Krotochwil A, Lycke J, Svennerholm B. 1988. Herpes simplex virus infection of the human sensory neuron. An electron microscopy study. *Arch. Virol.* 101:87–104
99. Lycke E, Kristensson K, Svennerholm B, Vahlne A, Ziegler R. 1984. Uptake and transport of herpes simplex virus in neurites of rat dorsal root ganglia cells in culture. *J. Gen. Virol.* 65:55–64
100. Lyman MG, Feierbach B, Curanovic D, Bisher M, Enquist LW. 2007. PRV Us9 directs axonal sorting of viral capsids. *J. Virol.* 81:11363–71

101. Ma Y, Jin H, Valyi-Nagy T, Cao Y, Yan Z, He B. 2011. Inhibition of TANK binding kinase 1 by herpes simplex virus 1 facilitates productive infection. *J. Virol.* 86:2188–96
102. Maresch C, Granzow H, Negatsch A, Klupp BG, Fuchs W, et al. 2010. Ultrastructural analysis of virion formation and anterograde intraaxonal transport of the alphaherpesvirus pseudorabies virus in primary neurons. *J. Virol.* 84:5528–39
103. Markus A, Grigoryan S, Sloutskin A, Yee MB, Zhu H, et al. 2011. Varicella-zoster virus (VZV) infection of neurons derived from human embryonic stem cells: direct demonstration of axonal infection, transport of VZV, and productive neuronal infection. *J. Virol.* 85:6220–33
104. Maurer UE, Sodeik B, Grunewald K. 2008. Native 3D intermediates of membrane fusion in herpes simplex virus 1 entry. *Proc. Natl. Acad. Sci. USA* 105:10559–64
105. McCarthy KM, Tank DW, Enquist LW. 2009. Pseudorabies virus infection alters neuronal activity and connectivity in vitro. *PLoS Pathog.* 5:e1000640
106. McGraw HM, Awasthi S, Wojcechowskyj JA, Friedman HM. 2009. Anterograde spread of herpes simplex virus type 1 requires glycoprotein E and glycoprotein I but not Us9. *J. Virol.* 83:8315–26
107. McNab AR, Desai P, Person S, Roof LL, Thomsen DR, et al. 1998. The product of the herpes simplex virus type 1 UL25 gene is required for encapsidation but not for cleavage of replicated viral DNA. *J. Virol.* 72:1060–70
108. Meckes DG Jr, Wills JW. 2007. Dynamic interactions of the UL16 tegument protein with the capsid of herpes simplex virus. *J. Virol.* 81:13028–36
109. Meckes DG Jr, Wills JW. 2008. Structural rearrangement within an enveloped virus upon binding to the host cell. *J. Virol.* 82:10429–35
110. Mettenleiter TC, Klupp BG, Granzow H. 2009. Herpesvirus assembly: an update. *Virus Res.* 143:222–34
111. Milne RS, Nicola AV, Whitbeck JC, Eisenberg RJ, Cohen GH. 2005. Glycoprotein D receptor-dependent, low-pH-independent endocytic entry of herpes simplex virus type 1. *J. Virol.* 79:6655–63
112. Miranda-Saksena M, Boadle RA, Aggarwal A, Tijono B, Rixon FJ, et al. 2009. Herpes simplex virus utilizes the large secretory vesicle pathway for anterograde transport of tegument and envelope proteins and for viral exocytosis from growth cones of human fetal axons. *J. Virol.* 83:3187–99
113. Miranda-Saksena M, Boadle RA, Armati P, Cunningham AL. 2002. In rat dorsal root ganglion neurons, herpes simplex virus type 1 tegument forms in the cytoplasm of the cell body. *J. Virol.* 76:9934–51
114. Montgomery RI, Warner MS, Lum BJ, Spear PG. 1996. Herpes simplex virus-1 entry into cells mediated by a novel member of the TNF/NGF receptor family. *Cell* 87:427–36
115. Mothes W, Sherer NM, Jin J, Zhong P. 2010. Virus cell-to-cell transmission. *J. Virol.* 84:8360–68
116. Mulder W, Pol J, Kimman T, Kok G, Priem J, Peeters B. 1996. Glycoprotein D-negative pseudorabies virus can spread transneuronally via direct neuron-to-neuron transmission in its natural host, the pig, but not after additional inactivation of gE and gI. *J. Virol.* 70:2191–200
117. Mulder WAM, Jacobs L, Priem J, Kok GL, Wagenaar F, et al. 1994. Glycoprotein gE-negative pseudorabies virus has a reduced capability to infect second- and third-order neurons of the olfactory and trigeminal routes in the porcine central nervous system. *J. Gen. Virol.* 75:3095–106
118. Naldinho-Souto R, Browne H, Minson T. 2006. Herpes simplex virus tegument protein VP16 is a component of primary enveloped virions. *J. Virol.* 80:2582–84
119. Nauwynck HJ, Pensaert MB. 1992. Abortion induced by cell-associated pseudorabies virus in vaccinated sows. *Am. J. Vet. Res.* 53:489–93
120. Negatsch A, Granzow H, Maresch C, Klupp BG, Fuchs W, et al. 2010. Ultrastructural analysis of virion formation and intraaxonal transport of herpes simplex virus type 1 in primary rat neurons. *J. Virol.* 84:13031–35
121. Negatsch A, Mettenleiter TC, Fuchs W. 2011. Herpes simplex virus type 1 strain KOS carries a defective US9 and a mutated US8A gene. *J. Gen. Virol.* 92:167–72
122. Newcomb WW, Booy FP, Brown JC. 2007. Uncoating the herpes simplex virus genome. *J. Mol. Biol.* 370:633–42
123. Newcomb WW, Brown JC. 2009. Time-dependent transformation of the herpesvirus tegument. *J. Virol.* 83:8082–89
124. Newcomb WW, Brown JC. 2010. Structure and capsid association of the herpesvirus large tegument protein UL36. *J. Virol.* 84:9408–14

125. Nicola AV, Hou J, Major EO, Straus SE. 2005. Herpes simplex virus type 1 enters human epidermal keratinocytes, but not neurons, via a pH-dependent endocytic pathway. *J. Virol.* 79:7609–16
126. Nicola AV, McEvoy AM, Straus SE. 2003. Roles for endocytosis and low pH in herpes simplex virus entry into HeLa and Chinese hamster ovary cells. *J. Virol.* 77:5324–32
127. Nicola AV, Straus SE. 2004. Cellular and viral requirements for rapid endocytic entry of herpes simplex virus. *J. Virol.* 78:7508–17
128. O'Hara M, Rixon FJ, Stow ND, Murray J, Murphy M, Preston VG. 2010. Mutational analysis of the herpes simplex virus type 1 UL25 DNA packaging protein reveals regions that are important after the viral DNA has been packaged. *J. Virol.* 84:4252–63
129. Oh MJ, Akhtar J, Desai P, Shukla D. 2010. A role for heparan sulfate in viral surfing. *Biochem. Biophys. Res. Commun.* 391:176–81
130. Ohka S, Matsuda N, Tohyama K, Oda T, Morikawa M, et al. 2004. Receptor (CD155)-dependent endocytosis of poliovirus and retrograde axonal transport of the endosome. *J. Virol.* 78:7186–98
131. Ojala PM, Sodeik B, Ebersold MW, Kutay U, Helenius A. 2000. Herpes simplex virus type 1 entry into host cells: reconstitution of capsid binding and uncoating at the nuclear pore complex in vitro. *Mol. Cell. Biol.* 20:4922–31
132. Pasdeloup D, Blondel D, Isidro AL, Rixon FJ. 2009. Herpesvirus capsid association to the nuclear pore complex and viral DNA release involve the nucleoporin CAN/Nup214 and the capsid protein pUL25. *J. Virol.* 83:6610–23
133. Pasieka TJ, Collins L, O'Connor MA, Chen Y, Parker ZM, et al. 2011. Bioluminescent imaging reveals divergent viral pathogenesis in two strains of Stat1-deficient mice, and in $\alpha\beta\gamma$ interferon receptor-deficient mice. *PLoS ONE* 6:e24018
134. Peeters B, de Wind N, Hooisma M, Wagenaar F, Gielkens A, Moormann R. 1992. Pseudorabies virus envelope glycoproteins gp50 and gII are essential for virus penetration, but only gII is involved in membrane fusion. *J. Virol.* 66:894–905
135. Peeters B, Pol J, Gielkens A, Moormann R. 1993. Envelope glycoprotein gp50 of pseudorabies virus is essential for virus entry but is not required for viral spread in mice. *J. Virol.* 67:170–77
136. Penfold MET, Armati P, Cunningham AL. 1994. Axonal transport of herpes simplex virions to epidermal cells: evidence for a specialized mode of virus transport and assembly. *Proc. Natl. Acad. Sci. USA* 91:6529–33
137. Poon AP, Roizman B. 1997. Differentiation of the shutoff of protein synthesis by virion host shutoff and mutant $\gamma_1 34.5$ genes of herpes simplex virus 1. *Virology* 229:98–105
138. Preston VG, Murray J, Preston CM, McDougall IM, Stow ND. 2008. The UL25 gene product of herpes simplex virus type 1 is involved in uncoating of the viral genome. *J. Virol.* 82:6654–66
139. Radtke K, Kieneke D, Wolfstein A, Michael K, Steffen W, et al. 2010. Plus- and minus-end directed microtubule motors bind simultaneously to herpes simplex virus capsids using different inner tegument structures. *PLoS Pathog.* 6:e1000991
140. Rauh I, Mettenleiter TC. 1991. Pseudorabies virus glycoproteins gII and gp50 are essential for virus penetration. *J. Virol.* 65:5348–56
141. Read GS, Patterson M. 2007. Packaging of the virion host shutoff (Vhs) protein of herpes simplex virus: two forms of the Vhs polypeptide are associated with intranuclear B and C capsids, but only one is associated with enveloped virions. *J. Virol.* 81:1148–61
142. Reichelt M, Zerboni L, Arvin AM. 2008. Mechanisms of varicella-zoster virus neuropathogenesis in human dorsal root ganglia. *J. Virol.* 82:3971–83
143. Roberts AP, Abaitua F, O'Hare P, McNab D, Rixon FJ, Pasdeloup D. 2009. Differing roles of inner tegument proteins pUL36 and pUL37 during entry of herpes simplex virus type 1. *J. Virol.* 83:105–16
144. Rode K, Dohner K, Binz A, Glass M, Strive T, et al. 2011. Uncoupling uncoating of herpes simplex virus genomes from their nuclear import and gene expression. *J. Virol.* 85:4271–83
145. Sabin AB. 1938. Progression of different nasally instilled viruses along different nervous pathways in the same host. *Proc. Soc. Exp. Biol.* 38:270–75
146. Saksena MM, Wakisaka H, Tijono B, Boadle RA, Rixon F, et al. 2006. Herpes simplex virus type 1 accumulation, envelopment, and exit in growth cones and varicosities in mid-distal regions of axons. *J. Virol.* 80:3592–606

147. Salinas S, Bilsland LG, Henaff D, Weston AE, Keriel A, et al. 2009. CAR-associated vesicular transport of an adenovirus in motor neuron axons. *PLoS Pathog.* 5:e1000442
148. Schipke J, Pohlmann A, Diestel R, Binz A, Rudolph K, et al. 2012. The C-terminus of the large tegument protein pUL36 contains multiple capsid binding sites that function differently during assembly and cell entry of herpes simplex virus. *J. Virol.* 86:3682–700
149. Scrivano L, Sinzger C, Nitschko H, Koszinowski UH, Adler B. 2011. HCMV spread and cell tropism are determined by distinct virus populations. *PLoS Pathog.* 7(1):e1001256
150. Sears AE, Hukkanen V, Labow MA, Levine AJ, Roizman B. 1991. Expression of the herpes simplex virus 1 alpha transinducing factor (VP16) does not induce reactivation of latent virus or prevent the establishment of latency in mice. *J. Virol.* 65:2929–35
151. Shanda SK, Wilson DW. 2008. UL36p is required for efficient transport of membrane-associated herpes simplex virus type 1 along microtubules. *J. Virol.* 82:7388–94
152. Shukla D, Liu J, Blaiklock P, Shworak NW, Bai X, et al. 1999. A novel role for 3-O-sulfated heparan sulfate in herpes simplex virus 1 entry. *Cell* 99:13–22
153. Smith GA, Gross SP, Enquist LW. 2001. Herpesviruses use bidirectional fast-axonal transport to spread in sensory neurons. *Proc. Natl. Acad. Sci. USA* 98:3466–70
154. Smith GA, Pomeranz L, Gross SP, Enquist LW. 2004. Local modulation of plus-end transport targets herpesvirus entry and egress in sensory axons. *Proc. Natl. Acad. Sci. USA* 101:16034–39
155. Snyder A, Polcicova K, Johnson DC. 2008. Herpes simplex virus gE/gI and US9 proteins promote transport of both capsids and virion glycoproteins in neuronal axons. *J. Virol.* 82:10613–24
156. Sodeik B, Ebersold MW, Helenius A. 1997. Microtubule-mediated transport of incoming herpes simplex virus 1 capsids to the nucleus. *J. Cell Biol.* 136:1007–21
157. Spear PG, Manoj S, Yoon M, Jogger CR, Zago A, Myscofski D. 2006. Different receptors binding to distinct interfaces on herpes simplex virus gD can trigger events leading to cell fusion and viral entry. *Virology* 344:17–24
158. Strack AM, Loewy AD. 1990. Pseudorabies virus: a highly specific transneuronal cell body marker in the sympathetic nervous system. *J. Neurosci.* 10:2139–47
159. Strunze S, Trotman LC, Boucke K, Greber UF. 2005. Nuclear targeting of adenovirus type 2 requires CRM1-mediated nuclear export. *Mol. Biol. Cell* 16:2999–3009
160. Suenaga T, Satoh T, Somboonthum P, Kawaguchi Y, Mori Y, Arase H. 2010. Myelin-associated glycoprotein mediates membrane fusion and entry of neurotropic herpesviruses. *Proc. Natl. Acad. Sci. USA* 107:866–71
161. Szpara ML, Parsons L, Enquist LW. 2010. Sequence variability in clinical and laboratory isolates of herpes simplex virus 1 reveals new mutations. *J. Virol.* 84:5303–13
162. Tannous R, Grose C. 2011. Calculation of the anterograde velocity of varicella-zoster virions in a human sciatic nerve during shingles. *J. Infect. Dis.* 203:324–26
163. Taylor MP, Kramer T, Lyman MG, Kratchmarov R, Enquist LW. 2012. Visualization of an alphaherpesvirus membrane protein that is essential for anterograde axonal spread of infection in neurons. *mBio* 3(2):e00063–12
164. Taylor JM, Lin E, Susmarski N, Yoon M, Zago A, et al. 2007. Alternative entry receptors for herpes simplex virus and their roles in disease. *Cell Host Microbe* 2:19–28
165. Thompson RL, Preston CM, Sawtell NM. 2009. De novo synthesis of VP16 coordinates the exit from HSV latency in vivo. *PLoS Pathog.* 5:e1000352
166. Thompson RL, Stevens JG. 1983. Biological characterization of a herpes simplex virus intertypic recombinant which is completely and specifically non-neurovirulent. *Virology* 131:171–79
167. Tirabassi RS, Townley RA, Eldridge MG, Enquist LW. 1997. Characterization of pseudorabies virus mutants expressing carboxy-terminal truncations of gE: evidence for envelope incorporation, virulence, and neurotropism domains. *J. Virol.* 71:6455–64
168. Topp KS, Meade LB, LaVail JH. 1994. Microtubule polarity in the peripheral processes of trigeminal ganglion cells: relevance for the retrograde transport of herpes simplex virus. *J. Neurosci.* 14:318–25
169. Trus BL, Newcomb WW, Cheng N, Cardone G, Marekov L, et al. 2007. Allosteric signaling and a nuclear exit strategy: binding of UL25/UL17 heterodimers to DNA-filled HSV-1 capsids. *Mol. Cell* 26:479–89

170. Vallee RB, Williams JC, Varma D, Barnhart LE. 2004. Dynein: an ancient motor protein involved in multiple modes of transport. *J. Neurobiol.* 58:189–200
171. Van den Broeke C, Favoreel HW. 2011. Actin' up: herpesvirus interactions with Rho GTPase signaling. *Viruses* 3:278–92
172. Verpooten D, Feng Z, Valyi-Nagy T, Ma Y, Jin H, et al. 2009. Dephosphorylation of eIF2α mediated by the γ134.5 protein of herpes simplex virus 1 facilitates viral neuroinvasion. *J. Virol.* 83:12626–30
173. Weeks BS, Ramchandran RS, Hopkins JJ, Friedman HM. 2000. Herpes simplex virus type-1 and -2 pathogenesis is restricted by the epidermal basement membrane. *Arch. Virol.* 145:385–96
174. Whealy ME, Card JP, Robbins AK, Dubin JR, Rziha H-J, Enquist LW. 1993. Specific pseudorabies virus infection of the rat visual system requires both gI and gp63 glycoproteins. *J. Virol.* 67:3786–97
175. Whitley RJ, Gnann JW. 2002. Viral encephalitis: familiar infections and emerging pathogens. *Lancet* 359:507–13
176. Wileman T. 2007. Aggresomes and pericentriolar sites of virus assembly: cellular defense or viral design? *Annu. Rev. Microbiol.* 61:149–67
177. Williams RA, Bennett M, Bradbury JM, Gaskell RM, Jones RC, Jordan FT. 1992. Demonstration of sites of latency of infectious laryngotracheitis virus using the polymerase chain reaction. *J. Gen. Virol.* 73:2415–20
178. Winckler B, Mellman I. 1999. Neuronal polarity: controlling the sorting and diffusion of membrane components. *Neuron* 23:637–40
179. Wisner TW, Sugimoto K, Howard PW, Kawaguchi Y, Johnson DC. 2011. Anterograde transport of herpes simplex virus capsids in neurons by both separate and married mechanisms. *J. Virol.* 85:5919–28
180. Wolfstein A, Nagel CH, Radtke K, Dohner K, Allan VJ, Sodeik B. 2006. The inner tegument promotes herpes simplex virus capsid motility along microtubules in vitro. *Traffic* 7:227–37
181. Yamauchi Y, Kiriyama K, Kubota N, Kimura H, Usukura J, Nishiyama Y. 2008. The UL14 tegument protein of herpes simplex virus type 1 is required for efficient nuclear transport of the alpha transinducing factor VP16 and viral capsids. *J. Virol.* 82:1094–106
182. Yao F, Courtney RJ. 1992. Association of ICP0 but not ICP27 with purified virions of herpes simplex virus type 1. *J. Virol.* 66:2709–16
183. Zaichick SV, Bohannon KP, Smith GA. 2011. Alphaherpesviruses and the cytoskeleton in neuronal infections. *Viruses* 3:941–81

A Virological View of Innate Immune Recognition

Akiko Iwasaki

Department of Immunobiology, Yale University School of Medicine, New Haven, Connecticut 06520; email: akiko.iwasaki@yale.edu

Keywords

Toll-like receptor, RIG-I-like receptor, NOD-like receptor, inflammasome, interferons

Abstract

The innate immune system uses multiple strategies to detect viral infections. Because all viruses rely on host cells for their synthesis and propagation, the molecular features used to detect viral infections must be unique to viruses and absent from host cells. Research in the past decade has advanced our understanding of various cell-intrinsic and cell-extrinsic modes of virus recognition. This review examines the innate recognition from the point of view of virus invasion and replication strategies, and places innate sensors in the context of detecting viral genome, replication intermediate, transcriptional by-product, and other viral invasion strategies. On the basis of other unique features common to viral infections, undiscovered areas of virus detection are discussed.

Contents

INTRODUCTION	178
PATHWAYS ENGAGED FOLLOWING INNATE VIRUS RECOGNITION	178
NUCLEIC ACID–BASED VIRAL RECOGNITION	180
Endosomal Recognition: Virion-Associated Viral Genomes and Viral Replication Intermediates	181
Cytosolic Recognition	181
NUCLEIC ACID–INDEPENDENT VIRAL RECOGNITION	189
Recognition of Viral Structural Proteins	189
Innate Recognition of Viral Invasion Activities	189
UNEXPLORED AREAS OF INNATE VIRUS RECOGNITION	190

INTRODUCTION

Viruses are the most abundant life form on earth, inhabiting nearly every ecosystem, including animals, plants, and bacteria. Research over the past decade has provided enormous insights into the mechanism by which viruses are detected by infected cells. It is clear that the innate immune system is equipped with multiple sensors that detect different molecular signatures of a viral infection. Some sensors are expressed in specialized cell types, whereas others are virtually ubiquitous. The principle of innate virus recognition falls largely into two categories: recognition of pathogen-associated molecular patterns (PAMPs) via pattern recognition receptors (PRRs) and detection of pathogen-inflicted damage or stress. Viral PAMPs often carry distinct molecular or subcellular signatures not found in host cells, such as unique molecular features of the viral genome or viral replication intermediates. On the other hand, stress or damage inflicted by viral infection is recognized through pathways that are shared with other stress-sensing pathways. The reader is referred to many excellent reviews on molecular descriptions of sensors, signaling pathways, and antiviral effectors elicited by innate viral recognition (2, 70, 92, 98). This review attempts to describe innate virus recognition from a virological perspective. I describe recent developments in our understanding of innate virus recognition by focusing on viral replication and invasion strategies, and highlight unexplored features of viral infections that might serve as signatures recognized by the innate immune system.

PAMP: pathogen-associated molecular pattern

PRR: pattern recognition receptor

Inflammasome: a high-order cytosolic protein complex that forms in response to activation of the NLR or ALR proteins

PATHWAYS ENGAGED FOLLOWING INNATE VIRUS RECOGNITION

Innate sensors of viruses induce two distinct outcomes (**Figure 1**). The first outcome is that the engagement of PRRs induces signals, resulting in the transcriptional activation of cytokines and type I interferon (IFN) genes. Most cytokines are downstream of the transcription factor NF-κB, while the IFN genes are regulated by interferon regulatory factors 3 and 7 (IRF3 and IRF7) (39). The second outcome of PRR engagement is the activation of caspase-1 through the formation of inflammasomes. The inflammasomes enable proteolytic activation of caspase-1, which in turn can cleave multiple substrates including pro-interleukin (IL)-1β and pro-IL-18 (62). The posttranslational modification (caspase-1 cleavage) of these cytokines is required for their extracellular release and activity (**Figure 1**). Both PRR-induced transcriptional and inflammasome pathways can also engage programmed cell death through apoptosis and pyroptosis, respectively,

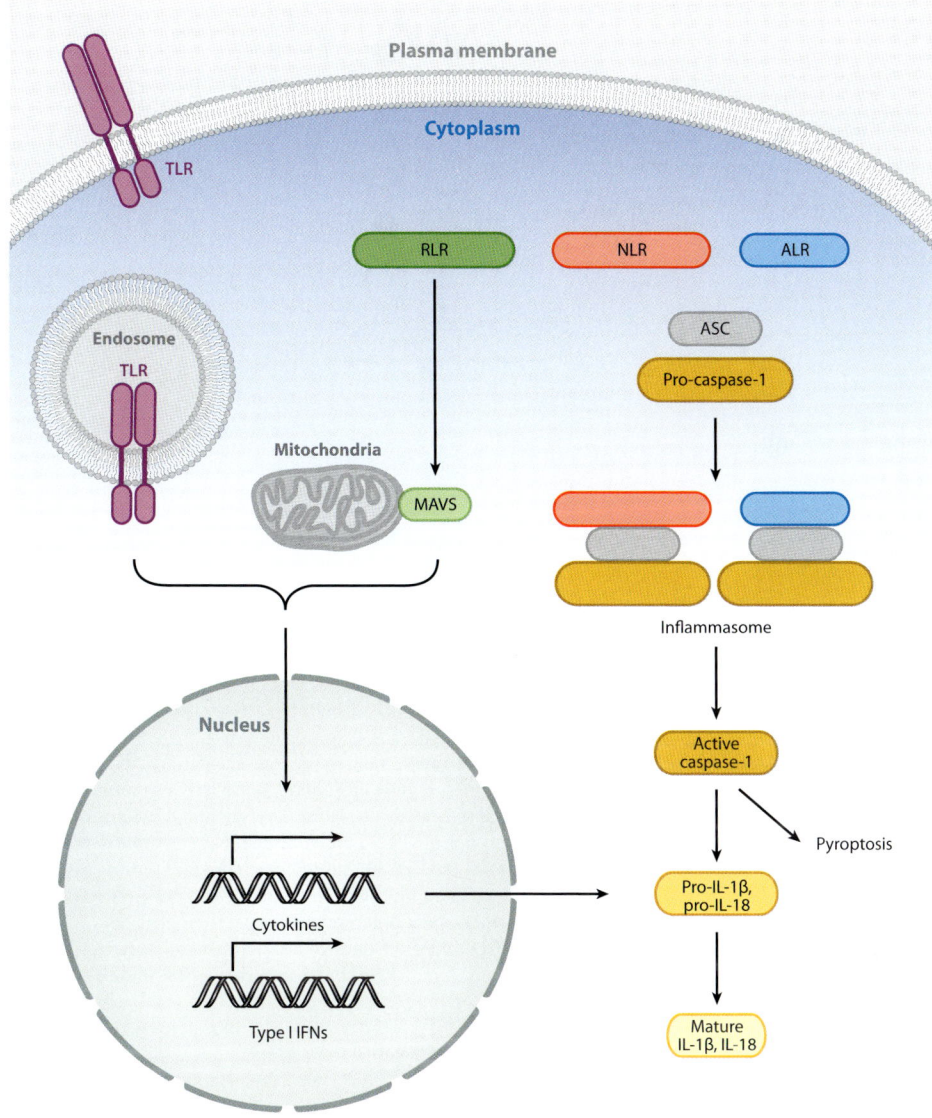

Figure 1

Pathways engaged following activation of innate viral sensors. TLRs reside either on the cell surface or in the endosomes, the latter requires cleavage for signaling. RLRs are present in the cytosol. Upon engagement of TLRs and RLRs by viruses, the receptor transmits signals that lead to the transcriptional activation of hundreds of genes including cytokines and type I IFNs. NLR and ALR proteins are localized in the cytosol. Certain virus infection leads to the activation of these receptors to form inflammasome, a large multimeric complex consisting of a subset of NLR/ALR, ASC, and pro-caspase-1. Caspase-1 becomes activated and cleaves its substrates including pro-IL-1β and pro-IL-18 for extracellular release. Cross talks between these pathways and exceptions are discussed throughout this review. Abbreviations: TLR, Toll-like receptor; RLR, RIG-I-like receptor; ALR, AIM2-like receptor; ASC, apoptosis-associated speck-like protein containing a caspase recruitment domain; MAVS, mitochondria antiviral signaling protein; NLR, Nod-like receptor; IL, interleukin; IFN, interferon.

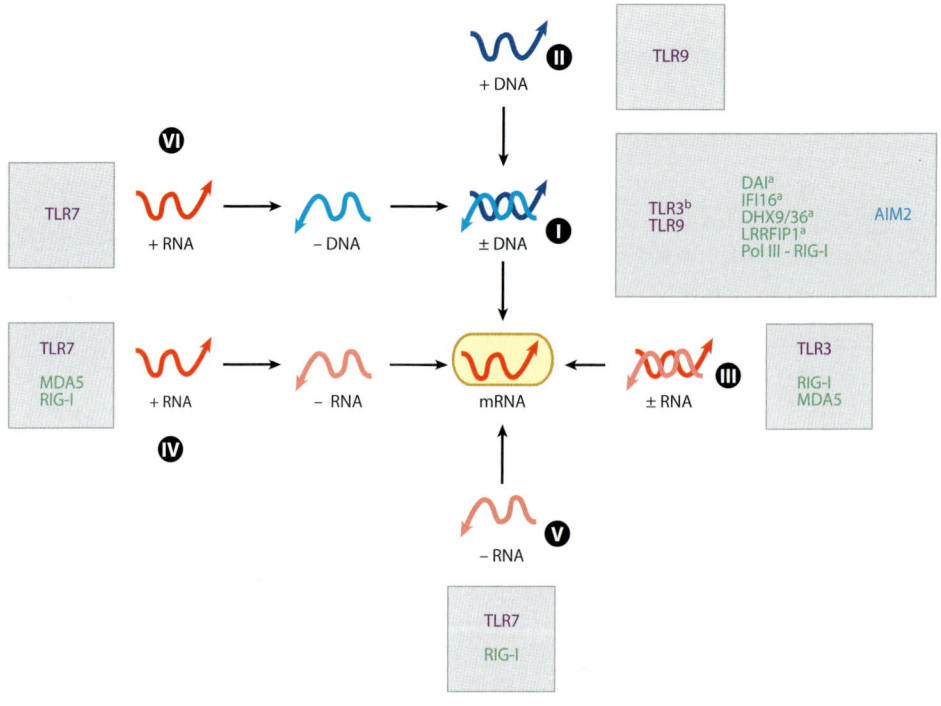

Figure 2

Innate sensors and Baltimore classification of viruses. All viruses fall into one of seven groups depending on a combination of their genomes (DNA, RNA), strandedness (single or double), sense (sense or antisense), and mode of replication. This classification enables innate sensors to be placed into functional categories. TLRs, RLRs, and other sensors that recognize respective groups of viruses are indicated. Superscript a denotes sensors that have been identified by genetic knockdown studies, and superscript b denotes sensors associated with virus-induced diseases in humans. Abbreviations: TLR, Toll-like receptor; RLR, RIG-I-like receptor; mRNA, messenger RNA; Pol, polymerase; MDA5, melanoma differentiation-associated gene 5; RIG-I, retinoic-acid-inducible gene I; DAI, DNA-dependent activator of interferon-regulatory factors; IFI, interferon-inducible protein; DHX, DEAH box protein; LRRFIP, leucine-rich repeat flightless-interacting protein.

in an effort to prevent pathogen replication and spread. Here, we consider natural viral ligands for PRRs that engage these two types of biological outcomes.

NUCLEIC ACID–BASED VIRAL RECOGNITION

There are an estimated 10^{31} viruses on earth (14). In 1971, David Baltimore proposed a classification of viruses based on the mechanism of mRNA production (7). According to the Baltimore classification, all viruses fall into one of seven groups depending on a combination of their genomes (DNA, RNA), strandedness (single or double), sense (sense or antisense), and mode of replication (**Figure 2**). I utilize this classification throughout this review, as it is particularly useful for understanding distinct modes of innate viral recognition strategies. The virus genome and the viral strategy used to generate mRNA from its genome provide a suitable framework to classify host innate sensors. The best-characterized mode of innate viral recognition is the detection of viral nucleic acids. Nucleic acid–based recognition can sense either virion-associated viral genomes (replication independent) or replication products, including the whole genome, replication intermediates, or viral transcripts.

Endosomal Recognition: Virion-Associated Viral Genomes and Viral Replication Intermediates

Toll-like receptors (TLRs) are innate sensors that detect PAMPs from a variety of pathogens (2). Many TLRs are expressed on the cell surface, but some are expressed in the endosomes, dedicated to recognizing viral genomes associated with virions (**Figure 1**). Upon endocytosis of viruses, endosomal TLRs sense viral genomes presumably after the envelopes and capsids are uncoated by the degradative enzymes therein, and trigger cytokine and type I IFN transcription. TLR7 and TLR9 recruit MyD88 and IRF7 to stimulate cytokine and type I IFN genes from the endosome. Signaling downstream of TLR7 and TLR9 is studied most extensively in a specialized cell type, plasmacytoid dendritic cells (pDCs), which use these receptors exclusively to recognize a wide array of viruses and produce copious amounts of type I IFNs (31). These TLRs require proteolytic processing for signaling (26, 68). TLR9 recognizes double-stranded DNA (dsDNA) viral genomes of Group I viruses in the endosome (**Figures 2** and **3**). Recognition via TLR9 does not require viral replication nor sequence-specific motifs (33). Group II viruses contain single-stranded DNA (ssDNA) genomes, and a member of this group of viruses (adeno-associated virus) stimulates TLR9 (109). TLR7 senses ssRNA viral genomes of Group IV, V, and VI viruses in the endosome. In humans, TLR8 is expressed by myeloid dendritic cells (DCs) and is similarly capable of recognizing ssRNA in the endosome (36). Uridine and ribose, the defining signatures of RNA, are both necessary and sufficient for TLR7 stimulation (23). Whereas influenza virus (6, 22) and retrovirus genomes (12, 46) are recognized via TLR7 in a replication-independent manner, other ssRNA viruses, such as vesicular stomatitis virus (VSV) and paramyxoviruses, require replication (55) and autophagy for recognition by TLR7 (55, 61). Why some RNA viruses require replication and autophagy for TLR7 recognition while others do not is unclear. Autophagy-dependent recognition may be required for viruses that fuse at the plasma membrane or escape endosomes before they progress to the late maturation stage needed for TLR signaling (10). In this case, cytosolic viral RNA is delivered to the TLR-containing endolysosomes via autophagy. Another intriguing correlation is that the viruses that are recognized independently of autophagy replicate in the nucleus, whereas those dependent on autophagy for recognition replicate in the cytosol.

TLR3 was originally identified as a sensor of dsRNA viruses, as TLR3-deficient splenocytes failed to upregulate CD69 upon stimulation with an isolated reovirus genome (3). However, TLR3 is not required for innate or adaptive immune responses against lymphocytic choriomeningitis virus (LCMV), VSV, murine cytomegalovirus (MCMV), and reovirus (24). Instead, TLR3 may be important in detecting virally infected cells when they are phagocytosed by DCs for cross-priming (89). In humans, genetic deficiencies in TLR3 and its signaling pathway have been associated with herpes simplex encephalitis (107), indicating that TLR3 may play a key role in protecting the central nervous system against herpes simplex virus 1 (HSV-1) infection (**Figure 2**). Recent data indicate that TLR3-deficient mice succumb to central nervous system infection upon vaginal HSV-2 challenge due to a lack of type I IFN production by astrocytes (80), revealing the importance of innate recognition of HSV by TLR3 in the brain.

Cytosolic Recognition

Several classes of sensors detect viral infection in the cytosol. Cytosolic sensors can be divided in terms of structurally related family members: RIG-I-like receptors (RLRs), which are sensors of RNA; AIM2-like receptors (ALRs), which are sensors of DNA; and nucleotide-binding

TLR: Toll-like receptor

Autophagy: a catabolic process whereby cytosolic cellular components are degraded through the lysosomal machinery

RLR: RIG-I-like receptor

ALR: AIM2-like receptor

oligomerization domain (NOD)-like receptors (NLRs), which are sensors of viral PAMPs and virus-inflicted cellular stress (discussed in detail below) (**Figure 1**). In general, RLRs signal via an adaptor molecule called the mitochondria antiviral signaling protein (MAVS) (90) [also known as IPS-1 (48), Cardif (64), or VISA (102)], which is found on the mitochondrial membrane (90). AIM2 (absent in melanoma 2) and NLRs form the inflammasome complex via an adaptor protein ASC (apoptosis-associated speck-like protein containing a caspase recruitment domain) within the cytosol. Engagement of RLRs results in MAVS-dependent transcription of cytokines and type I IFNs; engagement of NLRs results in inflammasome activation (**Figure 1**), with some exceptions as described below. ALR stimulation results in either cytokine/IFN induction or inflammasome activation.

NLR: NOD-like receptor

MAVS: mitochondria antiviral signaling protein

Innate recognition of replicated viral genomes in the cytosol. Retinoic-acid-inducible gene I (RIG-I) recognizes 5′-ppp RNA upon infection with Group V viruses (41, 71) (**Figure 2**). Recognition through RIG-I requires replication, indicating that virion-associated genomes are not sufficient to serve as the viral ligand. A recent study showed that the majority of RIG-I-stimulating activity is associated with whole viral genomic RNA (vRNA) that is generated upon replication (**Figure 3**), but not with the complementary positive-strand RNA (cRNA) (79). Preferential recognition of vRNA to cRNA may be due to their ratio in infected cells. This study demonstrated that nongenomic viral transcripts, short replication intermediates, and cleaved self-RNA do not contribute substantially to IFN induction in cells infected with these negative-strand RNA viruses (79). However, RIG-I may also recognize partial replication intermediates containing 5′ ppp as described below.

Innate recognition of viral replication intermediates in the cytosol. In addition to the whole genome, viral replication intermediates serve as PAMPs for cytosolic viral sensors. Unique RNA and DNA intermediates generated in the course of virus infection are recognized by different groups of innate sensors and effector molecules.

RNA sensors. Evidence indicates that virus infection generates distinct species of RNA that are recognized by host sensors.

    RIG-I. In an effort to obtain unbiased information on RIG-I ligand during viral infection, Baum et al. (11) performed deep sequencing of RIG-I-bound RNA from cells infected with the Sendai virus (Cantell strain) and influenza (strain delta NS1). This study revealed that RIG-I

Figure 3

Known and putative viral PAMPs. Innate sensors can detect viral genomes in the endosomes (*purple boxes*) or in the cytosol inside infected cells (*yellow boxes*). Green letters denote cytosolic sensors, purple letters denote endosomal sensors, and blue letters denote antiviral effector ISGs. Host counterparts, where appropriate, are depicted at the bottom. Viral signatures predicted to serve as PAMPs are indicated by the pink boxes. These include sfRNA (Group IV), cap 0 structure of Sindbis virus mRNA, ssDNA in the nucleus (Group II), circular DNA (Group I, II), ssDNA and dsDNA in the cytosol (Group VI), and leader RNA (Group V). Abbreviations: PAMP, pathogen-associated molecular pattern; ISG, interferon-stimulated gene; sfRNA, subgenomic *Flavivirus* RNA; ssDNA, single-stranded DNA; dsDNA, double-stranded DNA; PKR, double-stranded RNA-activated protein kinase; OAS, 2′,5′-oligoadenylate synthase; RIG-I, retinoic-acid-inducible gene I; IFIT, interferon-induced tetratricopeptide repeat protein; MDA5, melanoma differentiation-associated protein 5; TREX1, three prime repair exonuclease 1; TLR, Toll-like receptor; DI, defective interfering; EBV, Epstein-Barr virus; EBER, EBV-encoded RNA; DHX, DEAH box protein; ppp, triphosphate; pA, poly(A) tail; Vpg, viral protein genome-linked; mRNA, messenger RNA; DAI, DNA-dependent activator of interferon-regulatory factors; AIM2, absent in melanoma 2; KSHV, Kaposi's sarcoma-associated herpesvirus; IRES, internal ribosomal entry site; tRNA, transfer RNA.

sfRNA: subgenomic *Flavivirus* RNA

specifically bound the genome of the defective interfering (DI) particle and did not bind the full-length virus genome or any other viral RNAs in Sendai virus–infected cells (**Figure 3**). In influenza-infected cells RIG-I preferentially bound shorter genomic segments as well as subgenomic DI particles. Other reports demonstrated that RIG-I detects negative-strand RNA viruses that contain blunt, short, double-stranded 5′-triphosphate RNA (88) in the panhandle region of their single-stranded genome (87). These reports collectively indicate that a major viral RIG-I ligand is the short dsRNA with a 5′-ppp end, likely a DI particle (**Figure 3**). In addition to Group V viruses, RIG-I is the primary sensor for hepatitis C virus (HCV) (Group IV) (30). Of note, the HCV genome has 5′ ppp (**Figure 3**). When various regions of the HCV genome were tested, the polyuridine motif of the 3′-untranslated region and its replication intermediate were identified to be the PAMP substrate of RIG-I (84). Flaviviruses (genus *Flavivirus*), but not other genera of the *Flaviviridae* family (such as HCV), produce a unique, small, noncoding RNA (∼0.5 kb) derived from the 3′-untranslated region of the genomic RNA, which is required for their cytopathicity and pathogenicity (73) as well as replication (27). This subgenomic *Flavivirus* RNA (sfRNA) is a product of incomplete degradation of vRNA by cellular ribonucleases (73) (**Figure 3**). It is tempting to speculate that sfRNA may bind to an RNA sensor, such as RIG-I, to inhibit innate signaling.

MDA5. MDA5 belongs to the RLR family. Unlike RIG-I, MDA5 does not recognize 5′-ppp RNA. Instead, replication intermediates generated upon infection with Group IV viruses are recognized. Some flaviviruses (Group IV) are recognized by both RIG-I and MDA5 (dengue, West Nile Virus), where others are recognized only by RIG-I (Japanese encephalitis virus, HCV) (57). Reovirus (Group III) infection is also recognized by both RIG-I and MDA5 (57). The precise target of MDA5 is still unclear. The DI particle generated in Sendai virus–infected cells can activate MDA5 and induce IRF3-dependent genes (**Figure 3**). DI particles can overcome viral immune evasion via the Sendai virus V protein (106). Another study reported that highly structured RNA in infected cells is recognized by MDA5 (72). Interestingly, Sendai virus infection, which is known to trigger RIG-I but not MDA5 in vitro (47), revealed the importance of MDA5 in antiviral defense in vivo (32). Type I IFNs and cytokine responses were intact in MDA5-deficient mice until day 5 postinfection, suggesting that MDA5 may induce an IFN-independent antiviral program that is not entirely countered by the Sendai virus V protein.

Other sensors. NOD2 is a member of the NLR. A report (82) indicated that NOD2 binds to MAVS and induces IRF3 activation upon Group V ssRNA viral infections (respiratory syncytial virus, VSV, parainfluenza virus) but not following vaccinia virus (VV) (Group I, DNA) infection. Ultra-violet treatment of viruses diminished IRF3-dependent type I IFN response, suggesting that NOD2 is stimulated by RNA generated after Group V viral infection (82). In another report, in human peripheral blood mononuclear cells or mouse bone-marrow-derived DCs, VSV was shown to activate the RIG-I/ASC/caspase-1 inflammasome, independent of Nod-like receptor protein 3 (NLRP3) (75). In this case, RIG-I senses viral RNA and activates caspase-1 instead of activating type I IFN synthesis downstream of MAVS. This study showed that synthetic 5′-ppp RNA also triggers inflammasomes in a RIG-I-dependent manner. It remains unclear under what circumstances RIG-I induces cytokine/IFN transcription via MAVS versus inflammasome activation via ASC.

RNA-sensing executors. In addition to RNA sensors that trigger the synthesis of IFNs and cytokines, signatures of viral RNA are sensed by the effector molecules that carry out antiviral functions. These effector molecules are themselves IFN induced and require the presence of unique RNA structures synthesized during viral infection for their activity. Both double-stranded RNA-activated protein kinase (PKR) and 2′,5′-oligoadenylate synthase bind to dsRNA to become active enzymes (83). Most viral infections result in dsRNA synthesis during their replication cycle (**Figure 3**). A recent study has identified that interferon-stimulated genes (ISGs), IFIT1, IFIT2, and IFIT3 (interferon-induced tetratricopeptide repeat protein 1/2/3), bind 5′-ppp RNA in order to restrict replication of Group V viruses, including Rift Valley fever virus, VSV, and influenza A virus (69). Thus, in addition to RIG-I, IFIT proteins utilize 5′-ppp RNA to execute their viral restriction effector function. In addition, as discussed below, IFIT1 and IFIT2 also sense viral mRNAs that lack 2′ O-methylation (20, 111) in order to restrict viral replication.

PKR: double-stranded RNA-activated protein kinase

DNA sensors. Multiple sensors recognize DNA in the cytosol from various sources including DNA from viruses, bacteria, and apoptotic cells. Synthetic B-form DNA (poly dA:dT) and IFN stimulatory DNA (ISD) have been used to probe distinct pathways of cytosolic DNA recognition. ISDs are dsDNA that contain oligonucleotides of at least 25 base pairs, which in a sequence-independent manner trigger the stimulation of type I IFNs downstream of TBK1 and IRF3 but not MAVS (93). Interestingly, ISD does not engage NF-κB or MAPK pathways, thus activating only IRF3-induced pathways. The ISD recognition pathway exists only in primary cells and is lost from transformed cells (93). The search for DNA sensors of B-form DNA or ISD leading to IFN production is ongoing. DNA-dependent activator of IRFs (DAI) (also known as DLM-1/ZBP1) was identified as a candidate intracellular DNA-sensing molecule (96). In vivo, DAI deficiency can be compensated for by other DNA-sensing receptors (44).

IFI16/p204. IFI16 (interferon-inducible protein 16) and its mouse ortholog, p204, are members of the PYHIN (pyrin and HIN domain-containing protein) protein family, which contains a pyrin domain and two DNA-binding HIN (hemopoietic expression, interferon-inducibility, nuclear localization) domains. Stimulation of IFI16 by HSV-1 infection triggers NF-κB and IRF3 activation in bone-marrow-derived macrophages (99). IFI16 is recruited to synthetic ds-DNA (60- or 70-mer) and STING (stimulator of interferon genes) in the cytosolic compartment. Requirement for IFI16/p204 for inhibiting viral replication appears to be restricted to MCMV and human cytomegalovirus (HCMV), as expression of p204 mutants had no effect on the replication of HSV-1, ectromelia virus, or VSV. In contrast to the role of IFI16 in IFN induction, IFI16 forms ASC-dependent inflammasomes after Kaposi's sarcoma-associated herpesvirus infection in primary human endothelial cells (50), presumably upon recognition of nuclear viral DNA. The mechanism that determines whether IFI16 induces IFN or forms the inflammasome complex upon DNA viral infection is unknown.

AIM2. AIM2 is an IFN-inducible protein that contains the N-terminal pyrin domain and the C-terminal HIN-200 domain. The HIN-200 domain binds dsDNA in the cytosol. AIM2 recognizes certain dsDNA viruses (MCMV and VV, but not HSV-1) upon infection in primed macrophages (28, 40) (**Table 1**). AIM2 is also triggered by synthetic dsDNA poly dA:dT in phorbol myristate acetate–differentiated THP-1 cells (15) or in mouse bone-marrow-derived macrophages (81). In macrophages primed with TLR ligands, cytosolic dsDNA binds to AIM2, leading to formation of an inflammasome consisting of AIM2/ASC/pro-caspase-1. Although not formally tested, pyroptosis following AIM2 inflammasome activation likely has an antiviral role by eliminating infected cells.

DHX9 AND DHX36. DHX9 and DHX36 are aspartate-glutamate-any amino acid-aspartate/histidine (DExD/H)-box helicase (DHX) proteins that localize in the cytosol. In pDCs,

Table 1 Inflammasome activation by viruses

Virus	Genome (Baltimore Group)	Inflammasome	Trigger	Use of NLRP/ALR knockout	Virus replication in knockout	Reference(s)
Influenza	−ssRNA (V)	NLRP3/ASC	M2 ion channel	+	NT (43) High (42)	42, 43
			Viral RNA?	+	High (4) No change (97)	4, 97
Adenovirus	dsDNA (I)	NLRP3/ASC	?	+	NT	65
			Lysosomal penetration	−	NT	9
Vaccinia virus	dsDNA (I)	AIM2/ASC	dsDNA	+	NT	28, 40
Myxomavirus (M013KO)	dsDNA (I)	NLRP3/ASC Not AIM2	Lysosomal cathepsin B	Knockdown	NT	77
MCMV	dsDNA (I)	AIM2/ASC	dsDNA	+	NT	40
HSV-1	dsDNA (I)	Not AIM2	?	+	NT	40
Varicella-zoster	dsDNA (I)	NLRP3 (complex formation)	?	−	NT	67
KSHV	dsDNA (I)	IFI16/ASC	Viral DNA in nucleus	Knockdown	NT	50
EMCV	+ssRNA (IV)	NLRP3/ASC	?	+	NT	75, 78
Modified Vaccinia Ankara	dsDNA (I)	NLRP3/ASC	Replication independent, endocytosis dependent	+	NT	21
HIV-1	+ssRNA (VI)	NLRP3 (expression)	?	−	NT	76
VSV	−ssRNA (V)	RIG-I/ASC	ssRNA	+	NT	75
		NLRP3/ASC	?	+	NT	78

Abbreviations: ssRNA, single-stranded RNA; dsDNA, double-stranded DNA; VSV, vesicular stomatitis virus; MCMV, murine cytomegalovirus; KSHV, Kaposi's sarcoma-associated herpesvirus; EMCV, encephalomyocarditis virus; NLRP, Nod-like receptor protein; ALR, AIM2-like receptor; ASC, apoptosis-associated speck-like protein containing a caspase recruitment domain; RIG-I, retinoic-acid-inducible gene 1; IFI, interferon-inducible protein; AIM2, absent in melanoma 2. Question marks indicate that the exact trigger for a given pathway has not been identified. NT, not tested.

DHX9 and DHX36 bind to synthetic oligodeoxynucleotides, CpG-A and CpG-B, respectively, and induce MyD88-dependent, TLR9-independent IFN production (51). The two forms of CpG-motif-containing oligodeoxynucleotides, CpG-A and CpG-B, have distinct molecular signatures and immunological phenotype in pDCs (49). Knocking down DHX9 or DHX36 reduces the cytokine responses of pDCs to HSV-1 but has no effect on the cytokine responses to influenza virus. Both DHX9 and DHX36 are localized within the cytosol and both are directly bound to the TIR domain of MyD88 via their helicase-associated domain 2 and DUF domains. These molecules provide pDCs with a TLR-independent cytosolic DNA recognition mechanism consistent with residual cytokine secretion in TLR9-deficient pDCs infected with certain DNA viruses [HSV-1 and MCMV (37, 38) but not HSV-2 (59)] (**Figure 3**). Whether these molecules have a role in viral detection in non-pDC cell types is unknown.

LRRFIP1. An siRNA screening for molecules required for IFN-β production following *Listeria monocytogenes* infection identified LRRFIP1. LRRFIP1 binds to both B-form dsDNA and

Z-form dsDNA (poly dG:dC), and it enhances the expression of IFN-β. Of note, LRRFIP1 enhances IFN-β production by both dsRNA (poly I:C) and dsDNA. LRRFIP1 promotes the activation of β-catenin, which increases IFN-β expression by binding to the C-terminal domain of the transcription factor IRF3 and recruiting the acetyltransferase p300 to the IFN-β enhanceosome via IRF3. Therefore, LRRFIP1 and its downstream partner β-catenin constitute a coactivator pathway for IRF3-mediated production of type I IFN (105).

DNA choppers. Presence of DNA in the cytosol is a hallmark of viral infection and can trigger a direct antiviral response. TREX1 (three prime repair exonuclease 1), a cytosolic exonuclease, is a negative regulator of the ISD pathway by inhibiting excess accumulation of DNA products from endogenous retroelements (91). This DNA exonuclease activity is important to prevent the autoimmune disease Aicardi-Goutieres syndrome in humans. However, TREX1 is also involved in degrading nonproductive reverse transcriptase products generated during HIV-1 infection, enabling HIV-1 to remain undetected by the putative DNA sensor (104) (**Figure 3**).

Innate recognition of DNA viral transcripts by host RNA polymerase III. In 2009, two independent studies (1, 18) demonstrated that RIG-I recognizes DNA viral infections. In infected cells, host RNA polymerase III (Pol III) transcribes a certain region of the DNA virus genome to produce 5'-ppp RNA, which becomes a target of recognition by RIG-I. Pol III-dependent generation of RIG-I ligands can be mimicked by introducing poly dA:dT into the cytosol of cells by transfection. RNA Pol III normally transcribes cellular pre-tRNA, 5S rRNA, and U6 snRNA, all of which have conserved promoter sequences (**Figure 3**). Therefore, Pol III transcription is highly regulated and is sequence specific. What is the natural physiological viral ligand recognized by the Pol III–RIG-I pathway? Certain classes of viruses utilize Pol III to transcribe small, noncoding RNAs to counter host innate defense mechanisms, specifically PKR (**Figure 4**). These viruses include adenovirus and Epstein-Barr virus (EBV). Therefore, the Pol III–RIG-I pathway of viral recognition represents a counter-countermeasure by the host to guarantee type I IFN production even in the face of viral evasion. However, as described below, the virus wins in the end by ensuring that the Pol III transcripts block the pathway downstream of type I IFN signaling.

Adenovirus uses Pol III to generate small, noncoding viral associated RNAs (VA RNAs). VA RNAs are small, highly structurally conserved noncoding RNAs (∼160 nucleotides in length) synthesized at high levels (10^8 copies per cell) during adenovirus replication (63). Adenovirus lacking VA RNAs induces minimal IFNs or cytokines in infected cells, indicating that VA RNAs are the primary target of recognition (103). VA RNAs are likely recognized by both RIG-I and MDA5, because mouse embryonic fibroblasts deficient in either of these molecules still respond robustly to adenovirus infection, whereas MAVS-deficient mouse embryonic fibroblasts are incapable of inducing IFN (103). VA RNAs are synthesized by the virus to counteract two host cell defense mechanisms: the PKR (52, 66, 100) and the Dicer/RNA-induced silencing complex (5) (**Figure 4**). VA RNAs bind to the dsRNA-binding domain of PKR and inactivate it. Similarly, VA RNAs bind to Dicer and act as a competitive inhibitor. In adenovirus mutants lacking one of the VA RNAs, PKR and Dicer are activated and viral replication is severely attenuated (52). Thus, even though RLRs can recognize VA RNAs to induce IFNs, adenovirus circumvents this by enabling VA RNAs to inactivate the downstream antiviral effectors.

EBV-encoded RNAs (EBERs) are structurally similar to the adenovirus VA RNAs, which are similarly small (∼160–170 nucleotides) untranslated RNAs transcribed by Pol III. EBERs are expressed by EBV-infected cells during latency and are associated with resistance to apoptosis in Burkitt's lymphoma. EBERs are recognized by RIG-I (1, 85). Like VA RNAs, EBERs also bind PKR, inhibit its phosphorylation, and thereby prevent type-I IFN-mediated apoptosis in infected

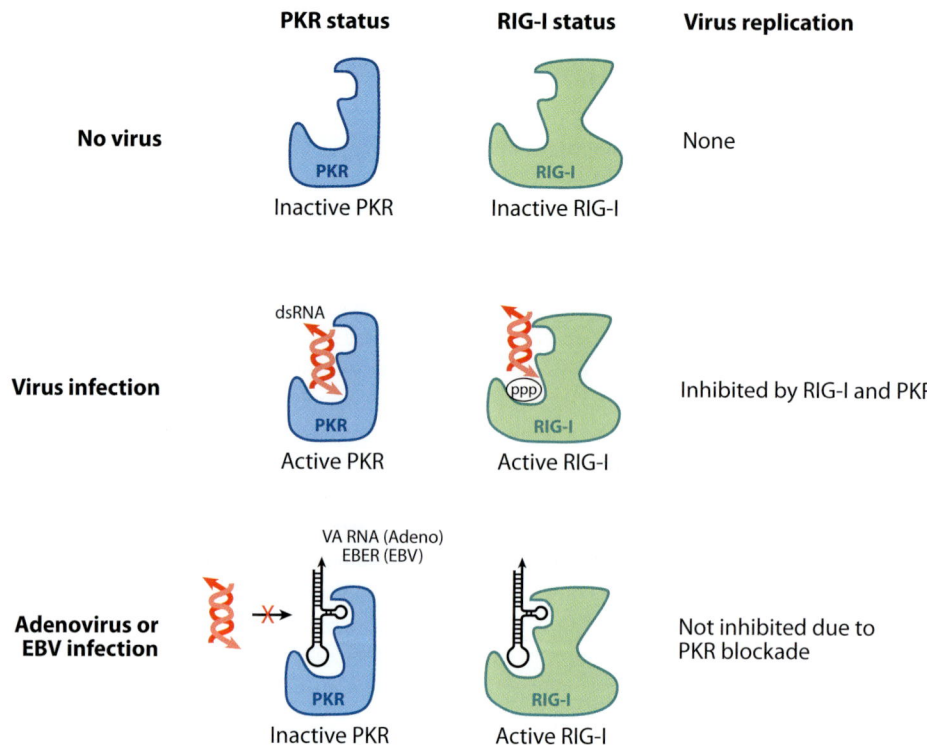

Figure 4

Pol III viral transcripts activate RIG-I but inhibit PKR. Pol III transcripts generated during adenoviral infection (VA RNA) and EBV infection (EBER) are small, noncoding RNAs that bind to and block PKR activation. In the absence of viral infection, neither PKR nor RIG-I is activated. In cells infected with a virus (excluding adenovirus and EBV), dsRNA structure generated in the cytosol triggers the activation of PKR and 5′-ppp RNA triggers RIG-I activation, resulting in an antiviral state. In cells infected with adenovirus or EBV, noncoding RNA Pol III transcripts bind to RIG-I and stimulate IFN synthesis. However, Pol III transcripts bind to PKR and block its activity by disabling binding of stimulatory dsRNA. Abbreviations: Pol, polymerase; VA RNA, viral associated RNA; EBV, Epstein-Barr virus; EBER, EBV-encoded RNA; RIG-I, retinoic-acid-inducible gene 1; PKR, double-stranded RNA-activated protein kinase; dsRNA, double-stranded RNA; IFN, interferon.

cells. Interestingly, unlike the ISD pathway, the Pol III–RIG-I pathway is preserved in transformed cells (18), enabling EBV-transformed B cells to stimulate such a pathway. However, even though EBERs induce type I IFNs, as in the case of VA RNA, EBERs provide EBV a replicative advantage by blocking PKR and enabling translation of viral proteins (**Figure 4**).

Innate recognition of viral mRNA cap structures. All eukaryotic mRNA are modified at the 5′ end with a highly conserved cap structure shortly after the start of transcription. m^{7}GpppN (referred to as cap 0) is further modified by cap-specific 2′-O RNA methyltransferases in the nucleus and cytoplasm that add a methyl group to ribose 2′-hydroxyl positions of the first and second nucleotides, giving rise to m^{7}GpppNm (cap 1) and m^{7}GpppNmNm (cap 2) structures, respectively. Viruses are equipped with various mechanisms to add 5′ cap to mimic the host mRNA. For example, 2′ O-methyl transferases are encoded by coronaviruses, VV, and flaviviruses to disguise their mRNA as "self." The absence of 2′ O-methyl group is recognized by MDA5 (111), IFIT1, and IFIT2 (20, 111).

Sindbis virus, an alphavirus, generates cap 0 structure, which is rarely found in mammalian mRNA in the cytosol (35). The 5′-terminal nucleotide of the Sindbis virus is modified such that the sequence is m^{7}GpppApUpGp. This cap structure lacks a 2′ O-methyl group on both the first and the second nucleotides and represents a likely target of innate recognition by either PRRs (e.g., MDA5) or ISGs (e.g., IFIT1). To this end, the Sindbis virus uses its macromolecular host shutoff mechanism to ensure that the viral infection does not trigger IFN synthesis mediated by MDA5 (16). Whether this virus also blocks the functions of IFIT proteins would be interesting to explore.

NUCLEIC ACID–INDEPENDENT VIRAL RECOGNITION

Recognition of Viral Structural Proteins

In addition to viral nucleic acids, other signatures of viruses or viral infection processes are detected by the host cells to trigger innate defenses. Of these defenses, stimulation of PRRs by viral structural proteins has been the target of intense research (29). Surface TLRs (**Figure 1**), including TLR2 and TLR4, are stimulated mainly through surface glycoproteins of a variety of viruses. TLR2 is stimulated by a variety of viruses including HCMV (19), MCMV (95), HSV-1 and HSV-2 (53, 86), HCV (17), LCMV (108), measles virus (13), VZV (101), and VV (110). TLR4 activation is triggered by respiratory syncytial virus (54) and mouse mammary tumor virus (MMTV) (45). Macrophages and DCs that express TLR2 and TLR4 are stimulated by these respective viruses and secrete a variety of cytokines. However, inflammatory monocytes appear to be the predominant producer of type I IFNs upon recognition of viruses such as VV and MCMV by TLR2 (8). Interestingly, monocyte recognition of viruses by TLR2 requires endocytosis, whereas recognition of bacterial TLR2 ligands does not.

In contrast to the usage of TLRs to induce the antiviral state by the host, certain viruses rely upon TLR signaling as a survival mechanism (13, 45). MMTV persists indefinitely in wild-type mice but is rapidly cleared by the cytotoxic T cell response in mice deficient in TLR4 (45). MMTV stimulates interleukin (IL)-10 production by B cells through DC and macrophage activation mediated by TLR4 signaling. IL-10, which is an immunosuppressive cytokine, protects the MMTV-infected B cells from removal by cytotoxic T cells. Another example in which TLR-virus interaction benefits the virus is the measles virus. The hemagglutinin (HA) protein of measles virus induces cytokine secretion in a TLR2-dependent manner. Interestingly, this interaction results in upregulated expression of the measles virus receptor, CD150, suggesting that HA-TLR2 interactions in fact benefit the virus at the expense of the host (13).

Innate Recognition of Viral Invasion Activities

In addition to recognizing unique features of viral nucleic acids or viral structural proteins, macrophages and DCs sense damage inflicted by viral invasion. This type of recognition is best characterized for the NLR protein NLRP3 following infection by viruses. NLRP3 forms a complex with ASC and caspase-1 to form inflammasomes (**Figure 1**). Two important functions of NLR/ALR-inflammasome activation are (*a*) to release caspase-1-dependent cytokines such as IL-1β and IL-18, which act on multiple cell types to induce activation of innate leukocytes and lymphocytes, and (*b*) to induce pyroptosis in order to kill infected cells. The formation of the NLRP3 inflammasome requires two signals. The first signal is often transduced by TLRs, leading to the expression of NLRP3; the second signal enables the assembly of a multiprotein complex

Internal ribosomal entry site: RNA sequence that allows ribosomal entry for translation initiation independent of the 5′ mRNA cap

Unfolded protein response: a sequence of reactions that restores normal function of the cell through translation arrest and production of molecular chaperones involved in protein folding

involving NLRP3, ASC, and caspase-1 (62). Although many distinct stimuli act as the second signal for the NLRP3 inflammasome, many of them have in common the ability to perturb membranes (94). Most viral infections impose by necessity some form of membrane stress or damage during entry, fusion, replication, and budding. Not surprisingly, emerging evidence indicates that host cells sense viral infections through NLRP3 and other innate sensors (**Table 1**). **Table 1** summarizes the current evidence for how each of these viruses activates the inflammasome; in many cases neither the trigger for activation nor the antiviral relevance of inflammasomes is known. One exception is influenza virus, for which we understand the molecular trigger and the immune consequences of inflammasome activation. Influenza virus infection triggers NLRP3 inflammasome activation through the action of the M2 ion channel (43). The M2 ion channel transports H^+ ions across lysosomal and Golgi membranes so that the virus can enter the host cells and maintain HA protein in a nonfusogenic conformation while transiting through the acidic *trans*-Golgi network, respectively (74). The latter activity is recognized by the influenza-infected macrophages and DCs via NLRP3 (43). Influenza virus is capable of inducing both signal 1 (via TLR7) and signal 2 (via M2) within the infected phagocytes. Adenovirus, on the other hand, activates the NLRP3 inflammasomes in phorbol myristate acetate–differentiated Pam3CysK (TLR2 agonist)-primed THP-1 cells through disruption of the lysosomal membrane (9). Activation of NLRP3 by adenovirus (9) and myxomavirus lacking an ortholog of PYRIN-domain-containing protein, M013 (77), depends on reactive oxygen species and cathepsin B activity. Similarly, modified vaccinia Ankara (MVA) activates NLRP3 in a replication-independent, endocytosis-dependent manner (21). Although the precise mechanisms are unclear, other viruses including encephalomyocarditis virus (EMCV) (75, 78) and VSV (78) trigger the NLRP3 inflammasome in Pam3CysK-primed THP-1 cells. Unlike the influenza viruses that activate both signal 1 and signal 2, priming of THP-1 cells through TLR activation is required to activate the NLRP3 inflammasomes upon adenovirus, MVA, EMCV, and VSV infection. HIV-1 infection has been reported to induce expression of NLRP3 and stimulates IL-1β release from healthy human peripheral blood mononuclear cells (76). Recognition of virus-inflicted damage through NLRP3 leads to the activation of caspase-1 and release of its substrates including IL-1β and IL-18, which activate adaptive immune responses to influenza virus (42). IL-1R is expressed by DCs and can signal to activate their migration, antigen presentation, and costimulation (58). Thus, these cytokines may be particularly important in stimulating noninfected DCs to prime immune responses against viruses that incapacitate directly infected antigen-presenting cells.

UNEXPLORED AREAS OF INNATE VIRUS RECOGNITION

Although progress in the past decade has brought us enormous insights into the molecular and cellular mechanisms of innate viral sensing and the ensuing effector functions, many areas of host viral recognition remain unaddressed. In particular, PRR-independent sensing of viral infections, likely through engaging cellular stress responses, remains largely undefined. Lytic virus infection almost always results in the shutdown of host transcription and translation, converting the infected cells into a virus factory. This conversion is accompanied by a sudden drop in cellular mRNA in the cytosol and a blockade of synthesis of cellular proteins in general. Instead, cellular translation machinery is taken over by the virus to produce large amounts of viral proteins. For some viruses, translation is mediated exclusively by the internal ribosomal entry site (**Figure 3**). I hypothesize that host cells must be equipped to sense drastic changes in cellular mRNA and protein levels through unknown mechanisms and to induce an antiviral program, likely through the induction of apoptosis. Most viruses encode factors to block apoptosis at multiple levels in order to maximize viral production. Similarly, the unfolded protein response induced by viral protein synthesis may induce an antiviral program (34), although studies indicate that the unfolded protein response may also block antiviral responses (56).

In addition to virus recognition systems through these general features of viral infections, there may also be location-specific recognition systems. For instance, most DNA viruses replicate in the nucleus. Nuclear domain 10 (ND10, or nuclear bodies), small nuclear substructures defined by the presence of promyelocytic leukemia protein (PML), becomes the early replication site of many DNA viruses (25). A number of proteins involved in DNA repair reside in or are recruited to ND10, and they may be utilized by DNA viruses for efficient replication. Structures such as the PML body may contain viral sensors and/or effectors of antiviral defense. In fact, PML and several other ND10 proteins are upregulated by type I IFNs. Consistent with this idea, ND10 structures are disrupted by viral proteins. Viruses deficient in genes capable of ND10 disruption are repressed for viral gene expression or DNA replication. Another possible location-based recognition system might exist to recognize viral replication that occurs on membranous structures. Many Group IV viruses (positive-strand ssRNA) generate unique 70- to 100-nm membrane vesicles that wrap around the active replicating viral RNA, providing a microenvironment optimal for viral replication (60). Such membrane structures are usually generated by a viral nonstructural protein, using full complement of cellular machinery. It is tempting to speculate whether host cells might be capable of sensing viral infection by detecting unusual membrane formation or pre-existing antiviral defense mechanisms in sites such as the PML bodies.

PML: promyelocytic leukemia protein

SUMMARY POINTS

1. Virus-associated molecular patterns are detected by innate immune cells and infected cells through pattern recognition receptors, whereas virus-inflicted damage is recognized by stress sensors and NLRP3.

2. Viral genomic nucleic acids serve as viral PAMPs for endosomal TLRs. Certain RNA viruses are recognized by TLR7 after entry and replication in the cytosol via autophagic delivery of replication intermediates to the endosome.

3. Viral replication intermediates and by-products are recognized by cytosolic viral sensors, RLRs, NLRs, and ALRs. Engagement of these sensors can lead either to transcriptional activation of type I IFNs and cytokines or to the assembly of inflammasomes and secretion of IL-1β and IL-18.

4. Adenovirus and EBV contain sequences transcribed by RNA Pol III to generate small, noncoding RNA. This RNA serves to block PKR activation but is recognized by RIG-I.

5. Viruses encode enzymes to modify their mRNA cap to resemble host structure. The unmodified viral mRNA cap is recognized by IFIT1 and IFIT2, which block replication.

DISCLOSURE STATEMENT

The author is not aware of any affiliations, memberships, funding, or financial holdings that might be perceived as affecting the objectivity of this review.

ACKNOWLEDGMENTS

I am grateful to Drs. Ruslan Medzhitov, Daniel DiMaio, Stacy Horner, and Brett Lindenbach for their helpful comments and discussions on this review. The National Institutes of Health grants AI 054359, AI 081884, AI 062428, and AI 064705 and the Midwest Center of Excellence in Biodefense and Emerging Infectious Disease Research support the work performed in my lab.

LITERATURE CITED

1. **Ablasser A, Bauernfeind F, Hartmann G, Latz E, Fitzgerald KA, Hornung V. 2009. RIG-I-dependent sensing of poly(dA:dT) through the induction of an RNA polymerase III-transcribed RNA intermediate.** *Nat. Immunol.* **10:1065–72**

    1. Shows that Pol III transcripts from dsDNA viruses are recognized by RIG-I (see also Reference 18).

2. Akira S, Uematsu S, Takeuchi O. 2006. Pathogen recognition and innate immunity. *Cell* 124:783–801
3. Alexopoulou L, Holt AC, Medzhitov R, Flavell RA. 2001. Recognition of double-stranded RNA and activation of NF-κB by Toll-like receptor 3. *Nature* 413:732–38
4. Allen IC, Scull MA, Moore CB, Holl EK, McElvania-TeKippe E, et al. 2009. The NLRP3 inflammasome mediates in vivo innate immunity to influenza A virus through recognition of viral RNA. *Immunity* 30:556–65
5. Andersson MG, Haasnoot PC, Xu N, Berenjian S, Berkhout B, Akusjarvi G. 2005. Suppression of RNA interference by adenovirus virus-associated RNA. *J. Virol.* 79:9556–65
6. Asselin-Paturel C, Boonstra A, Dalod M, Durand I, Yessaad N, et al. 2001. Mouse type I IFN-producing cells are immature APCs with plasmacytoid morphology. *Nat. Immunol.* 2:1144–50
7. **Baltimore D. 1971. Expression of animal virus genomes.** *Bacteriol. Rev.* **35:235–41**

    7. Describes a classification system in which all viruses can be classified into seven different groups on the basis of their paths from genome to mRNA production.

8. Barbalat R, Lau L, Locksley RM, Barton GM. 2009. Toll-like receptor 2 on inflammatory monocytes induces type I interferon in response to viral but not bacterial ligands. *Nat. Immunol.* 10:1200–7
9. Barlan AU, Griffin TM, McGuire KA, Wiethoff CM. 2011. Adenovirus membrane penetration activates the NLRP3 inflammasome. *J. Virol.* 85:146–55
10. Barton GM, Kagan JC. 2009. A cell biological view of Toll-like receptor function: regulation through compartmentalization. *Nat. Rev. Immunol.* 9:535–42
11. **Baum A, Sachidanandam R, Garcia-Sastre A. 2010. Preference of RIG-I for short viral RNA molecules in infected cells revealed by next-generation sequencing.** *Proc. Natl. Acad. Sci. USA* **107:16303–8**

    11. Uses deep sequencing to identify that DI particles are the natural ligand for RIG-I.

12. Beignon AS, McKenna K, Skoberne M, Manches O, Dasilva I, et al. 2005. Endocytosis of HIV-1 activates plasmacytoid dendritic cells via Toll-like receptor-viral RNA interactions. *J. Clin. Invest.* 115:3265–75
13. Bieback K, Lien E, Klagge IM, Avota E, Schneider-Schaulies J, et al. 2002. Hemagglutinin protein of wild-type measles virus activates Toll-like receptor 2 signaling. *J. Virol.* 76:8729–36
14. Breitbart M, Rohwer F. 2005. Here a virus, there a virus, everywhere the same virus? *Trends Microbiol.* 13:278–84
15. Bürckstümmer T, Baumann C, Blüml S, Dixit E, Dürnberger G, et al. 2009. An orthogonal proteomic-genomic screen identifies AIM2 as a cytoplasmic DNA sensor for the inflammasome. *Nat. Immunol.* 10:266–72
16. Burke CW, Gardner CL, Steffan JJ, Ryman KD, Klimstra WB. 2009. Characteristics of alpha/beta interferon induction after infection of murine fibroblasts with wild-type and mutant alphaviruses. *Virology* 395:121–32
17. Chang S, Dolganiuc A, Szabo G. 2007. Toll-like receptors 1 and 6 are involved in TLR2-mediated macrophage activation by hepatitis C virus core and NS3 proteins. *J. Leukoc. Biol.* 82:479–87
18. Chiu YH, Macmillan JB, Chen ZJ. 2009. RNA polymerase III detects cytosolic DNA and induces type I interferons through the RIG-I pathway. *Cell* 138:576–91
19. Compton T, Kurt-Jones EA, Boehme KW, Belko J, Latz E, et al. 2003. Human cytomegalovirus activates inflammatory cytokine responses via CD14 and Toll-like receptor 2. *J. Virol.* 77:4588–96
20. **Daffis S, Szretter KJ, Schriewer J, Li J, Youn S, et al. 2010. 2′-O methylation of the viral mRNA cap evades host restriction by IFIT family members.** *Nature* **468:452–56**

    20. Reveals that 2′ O-methylases are encoded by viruses to disguise their mRNA as to prevent detection by IFIT proteins (see also Reference 111).

21. Delaloye J, Roger T, Steiner-Tardivel QG, Le Roy D, Knaup Reymond M, et al. 2009. Innate immune sensing of modified vaccinia virus Ankara (MVA) is mediated by TLR2-TLR6, MDA-5 and the NALP3 inflammasome. *PLoS Pathog.* 5:e1000480

22. Diebold SS, Kaisho T, Hemmi H, Akira S, Reis e Sousa C. 2004. Innate antiviral responses by means of TLR7-mediated recognition of single-stranded RNA. *Science* 303:1529–31
23. Diebold SS, Massacrier C, Akira S, Paturel C, Morel Y, Reis e Sousa C. 2006. Nucleic acid agonists for Toll-like receptor 7 are defined by the presence of uridine ribonucleotides. *Eur. J. Immunol.* 36:3256–67
24. Edelmann KH, Richardson-Burns S, Alexopoulou L, Tyler KL, Flavell RA, Oldstone MB. 2004. Does Toll-like receptor 3 play a biological role in virus infections? *Virology* 322:231–38
25. Everett RD. 2006. Interactions between DNA viruses, ND10 and the DNA damage response. *Cell Microbiol.* 8:365–74
26. Ewald SE, Lee BL, Lau L, Wickliffe KE, Shi GP, et al. 2008. The ectodomain of Toll-like receptor 9 is cleaved to generate a functional receptor. *Nature* 456:658–62
27. Fan YH, Nadar M, Chen CC, Weng CC, Lin YT, Chang RY. 2011. Small noncoding RNA modulates Japanese encephalitis virus replication and translation in trans. *Virol. J.* 8:492
28. Fernandes-Alnemri T, Yu JW, Datta P, Wu J, Alnemri ES. 2009. AIM2 activates the inflammasome and cell death in response to cytoplasmic DNA. *Nature* 458:509–13
29. Finberg RW, Wang JP, Kurt-Jones EA. 2007. Toll-like receptors and viruses. *Rev. Med. Virol.* 17:35–43
30. Foy E, Li K, Sumpter R Jr, Loo YM, Johnson CL, et al. 2005. Control of antiviral defenses through hepatitis C virus disruption of retinoic acid-inducible gene-I signaling. *Proc. Natl. Acad. Sci. USA* 102:2986–91
31. Gilliet M, Cao W, Liu YJ. 2008. Plasmacytoid dendritic cells: sensing nucleic acids in viral infection and autoimmune diseases. *Nat. Rev. Immunol.* 8:594–606
32. Gitlin L, Benoit L, Song C, Cella M, Gilfillan S, et al. 2010. Melanoma differentiation-associated gene 5 (MDA5) is involved in the innate immune response to *Paramyxoviridae* infection in vivo. *PLoS Pathog.* 6:e1000734
33. Haas T, Metzger J, Schmitz F, Heit A, Muller T, et al. 2008. The DNA sugar backbone 2′ deoxyribose determines Toll-like receptor 9 activation. *Immunity* 28:315–23
34. He B. 2006. Viruses, endoplasmic reticulum stress, and interferon responses. *Cell Death Differ.* 13:393–403
35. Hefti E, Bishop DH, Dubin DT, Stollar V. 1975. 5′ nucleotide sequence of sindbis viral RNA. *J. Virol.* 17:149–59
36. Heil F, Hemmi H, Hochrein H, Ampenberger F, Kirschning C, et al. 2004. Species-specific recognition of single-stranded RNA via Toll-like receptor 7 and 8. *Science* 303:1526–29
37. Hochrein H, Schlatter B, O'Keeffe M, Wagner C, Schmitz F, et al. 2004. Herpes simplex virus type-1 induces IFN-α production via Toll-like receptor 9-dependent and -independent pathways. *Proc. Natl. Acad. Sci. USA* 101:11416–21
38. Hokeness-Antonelli KL, Crane MJ, Dragoi AM, Chu WM, Salazar-Mather TP. 2007. IFN-$\alpha\beta$-mediated inflammatory responses and antiviral defense in liver is TLR9-independent but MyD88-dependent during murine cytomegalovirus infection. *J. Immunol.* 179:6176–83
39. Honda K, Takaoka A, Taniguchi T. 2006. Type I interferon gene induction by the interferon regulatory factor family of transcription factors. *Immunity* 25:349–60
40. Hornung V, Ablasser A, Charrel-Dennis M, Bauernfeind F, Horvath G, et al. 2009. AIM2 recognizes cytosolic dsDNA and forms a caspase-1-activating inflammasome with ASC. *Nature* 458:514–18
41. Hornung V, Ellegast J, Kim S, Brzozka K, Jung A, et al. 2006. 5′-Triphosphate RNA is the ligand for RIG-I. *Science* 314:994–97
42. Ichinohe T, Lee HK, Ogura Y, Flavell R, Iwasaki A. 2009. Inflammasome recognition of influenza virus is essential for adaptive immune responses. *J. Exp. Med.* 206:79–87
43. Ichinohe T, Pang IK, Iwasaki A. 2010. Influenza virus activates inflammasomes via its intracellular M2 ion channel. *Nat. Immunol.* 11:404–10
44. Ishii KJ, Kawagoe T, Koyama S, Matsui K, Kumar H, et al. 2008. TANK-binding kinase-1 delineates innate and adaptive immune responses to DNA vaccines. *Nature* 451:725–29
45. Jude BA, Pobezinskaya Y, Bishop J, Parke S, Medzhitov RM, et al. 2003. Subversion of the innate immune system by a retrovirus. *Nat. Immunol.* 4:573–78
46. Kane M, Case LK, Wang C, Yurkovetskiy L, Dikiy S, Golovkina TV. 2011. Innate immune sensing of retroviral infection via Toll-like receptor 7 occurs upon viral entry. *Immunity* 35:135–45

47. Kato H, Takeuchi O, Sato S, Yoneyama M, Yamamoto M, et al. 2006. Differential roles of MDA5 and RIG-I helicases in the recognition of RNA viruses. *Nature* 441:101–5
48. Kawai T, Takahashi K, Sato S, Coban C, Kumar H, et al. 2005. IPS-1, an adaptor triggering RIG-I- and Mda5-mediated type I interferon induction. *Nat. Immunol.* 6:981–88
49. Kerkmann M, Rothenfusser S, Hornung V, Towarowski A, Wagner M, et al. 2003. Activation with CpG-A and CpG-B oligonucleotides reveals two distinct regulatory pathways of type I IFN synthesis in human plasmacytoid dendritic cells. *J. Immunol.* 170:4465–74
50. Kerur N, Veettil MV, Sharma-Walia N, Bottero V, Sadagopan S, et al. 2011. IFI16 acts as a nuclear pathogen sensor to induce the inflammasome in response to Kaposi sarcoma-associated herpesvirus infection. *Cell Host Microbe* 9:363–75
51. Kim T, Pazhoor S, Bao M, Zhang Z, Hanabuchi S, et al. 2010. Aspartate-glutamate-alanine-histidine box motif (DEAH)/RNA helicase A helicases sense microbial DNA in human plasmacytoid dendritic cells. *Proc. Natl. Acad. Sci. USA* 107:15181–86

52. Demonstrates the role of adenovirus Pol III transcript (VA RNA) in blocking PKR activation.

52. **Kitajewski J, Schneider RJ, Safer B, Munemitsu SM, Samuel CE, et al. 1986. Adenovirus VAI RNA antagonizes the antiviral action of interferon by preventing activation of the interferon-induced eIF-2α kinase. *Cell* 45:195–200**
53. Kurt-Jones EA, Chan M, Zhou S, Wang J, Reed G, et al. 2004. Herpes simplex virus 1 interaction with Toll-like receptor 2 contributes to lethal encephalitis. *Proc. Natl. Acad. Sci. USA* 101:1315–20
54. Kurt-Jones EA, Popova L, Kwinn L, Haynes LM, Jones LP, et al. 2000. Pattern recognition receptors TLR4 and CD14 mediate response to respiratory syncytial virus. *Nat. Immunol.* 1:398–401
55. Lee HK, Lund JM, Ramanathan B, Mizushima N, Iwasaki A. 2007. Autophagy-dependent viral recognition by plasmacytoid dendritic cells. *Science* 315:1398–401
56. Liu J, HuangFu WC, Kumar KG, Qian J, Casey JP, et al. 2009. Virus-induced unfolded protein response attenuates antiviral defenses via phosphorylation-dependent degradation of the type I interferon receptor. *Cell Host Microbe* 5:72–83

57. Provides a systematic analysis of RNA virus recognition by RIG-I and MDA5.

57. **Loo YM, Fornek J, Crochet N, Bajwa G, Perwitasari O, et al. 2008. Distinct RIG-I and MDA5 signaling by RNA viruses in innate immunity. *J. Virol.* 82:335–45**
58. Luft T, Jefford M, Luetjens P, Hochrein H, Masterman K-A, et al. 2002. IL-1β enhances CD40 ligand-mediated cytokine secretion by human dendritic cells (DC): a mechanism for T cell-independent DC activation. *J. Immunol.* 168:713–22
59. Lund J, Sato A, Akira S, Medzhitov R, Iwasaki A. 2003. Toll-like receptor 9-mediated recognition of herpes simplex virus-2 by plasmacytoid dendritic cells. *J. Exp. Med.* 198:513–20
60. Mackenzie J. 2005. Wrapping things up about virus RNA replication. *Traffic* 6:967–77
61. Manuse MJ, Briggs CM, Parks GD. 2010. Replication-independent activation of human plasmacytoid dendritic cells by the paramyxovirus SV5 requires TLR7 and autophagy pathways. *Virology* 405:383–89
62. Martinon F, Mayor A, Tschopp J. 2009. The inflammasomes: guardians of the body. *Annu. Rev. Immunol.* 27:229–65
63. Mathews MB, Shenk T. 1991. Adenovirus virus-associated RNA and translation control. *J. Virol.* 65:5657–62
64. Meylan E, Curran J, Hofmann K, Moradpour D, Binder M, et al. 2005. Cardif is an adaptor protein in the RIG-I antiviral pathway and is targeted by hepatitis C virus. *Nature* 437:1167–72
65. Muruve DA, Petrilli V, Zaiss AK, White LR, Clark SA, et al. 2008. The inflammasome recognizes cytosolic microbial and host DNA and triggers an innate immune response. *Nature* 452:103–7
66. Nayak R, Pintel DJ. 2007. Adeno-associated viruses can induce phosphorylation of eIF2α via PKR activation, which can be overcome by helper adenovirus type 5 virus-associated RNA. *J. Virol.* 81:11908–16
67. Nour AM, Reichelt M, Ku CC, Ho MY, Heineman TC, Arvin AM. 2011. Varicella-zoster virus infection triggers formation of an interleukin-1β (IL-1β)-processing inflammasome complex. *J. Biol. Chem.* 286:17921–33
68. Park B, Brinkmann MM, Spooner E, Lee CC, Kim YM, Ploegh HL. 2008. Proteolytic cleavage in an endolysosomal compartment is required for activation of Toll-like receptor 9. *Nat. Immunol.* 9:1407–14
69. Pichlmair A, Lassnig C, Eberle CA, Gorna MW, Baumann CL, et al. 2011. IFIT1 is an antiviral protein that recognizes 5′-triphosphate RNA. *Nat. Immunol.* 12:624–30

70. Pichlmair A, Reis e Sousa C. 2007. Innate recognition of viruses. *Immunity* 27:370–83
71. Pichlmair A, Schulz O, Tan CP, Naslund TI, Liljestrom P, et al. 2006. RIG-I-mediated antiviral responses to single-stranded RNA bearing 5′-phosphates. *Science* 314:997–1001
72. Pichlmair A, Schulz O, Tan CP, Rehwinkel J, Kato H, et al. 2009. Activation of MDA5 requires higher-order RNA structures generated during virus infection. *J. Virol.* 83:10761–69
73. Pijlman GP, Funk A, Kondratieva N, Leung J, Torres S, et al. 2008. A highly structured, nuclease-resistant, noncoding RNA produced by flaviviruses is required for pathogenicity. *Cell Host Microbe* 4:579–91
74. Pinto LH, Lamb RA. 2006. The M2 proton channels of influenza A and B viruses. *J. Biol. Chem.* 281:8997–9000
75. Poeck H, Bscheider M, Gross O, Finger K, Roth S, et al. 2010. Recognition of RNA virus by RIG-I results in activation of CARD9 and inflammasome signaling for interleukin 1β production. *Nat. Immunol.* 11:63–69
76. Pontillo A, Silva LT, Oshiro TM, Finazzo C, Crovella S, Duarte AJ. 2012. HIV-1 induces NALP3-inflammasome expression and interleukin-1β secretion in dendritic cells from healthy individuals but not from HIV-positive patients. *AIDS* 26:11–18
77. Rahman MM, McFadden G. 2011. Myxoma virus lacking the pyrin-like protein M013 is sensed in human myeloid cells by both NLRP3 and multiple Toll-like receptors, which independently activate the inflammasome and NF-κB innate response pathways. *J. Virol.* 85:12505–17
78. Rajan JV, Rodriguez D, Miao EA, Aderem A. 2011. The NLRP3 inflammasome detects encephalomyocarditis virus and vesicular stomatitis virus infection. *J. Virol.* 85:4167–72
79. **Rehwinkel J, Tan CP, Goubau D, Schulz O, Pichlmair A, et al. 2010. RIG-I detects viral genomic RNA during negative-strand RNA virus infection. *Cell* 140:397–408**
80. Reinert LS, Harder L, Holm CK, Iversen MB, Horan KA, et al. 2012. TLR3 deficiency renders astrocytes permissive to herpes simplex virus infection and facilitates establishment of CNS infection in mice. *J. Clin. Invest.* 122:1368–76
81. Roberts TL, Idris A, Dunn JA, Kelly GM, Burnton CM, et al. 2009. HIN-200 proteins regulate caspase activation in response to foreign cytoplasmic DNA. *Science* 323:1057–60
82. Sabbah A, Chang TH, Harnack R, Frohlich V, Tominaga K, et al. 2009. Activation of innate immune antiviral responses by Nod2. *Nat. Immunol.* 10:1073–80
83. Sadler AJ, Williams BR. 2008. Interferon-inducible antiviral effectors. *Nat. Rev. Immunol.* 8:559–68
84. Saito T, Owen DM, Jiang F, Marcotrigiano J, Gale M Jr. 2008. Innate immunity induced by composition-dependent RIG-I recognition of hepatitis C virus RNA. *Nature* 454:523–27
85. Samanta M, Iwakiri D, Kanda T, Imaizumi T, Takada K. 2006. EB virus-encoded RNAs are recognized by RIG-I and activate signaling to induce type I IFN. *EMBO J.* 25:4207–14
86. Sato A, Linehan MM, Iwasaki A. 2006. Dual recognition of herpes simplex viruses by TLR2 and TLR9 in dendritic cells. *Proc. Natl. Acad. Sci. USA* 103:17343–48
87. Schlee M, Roth A, Hornung V, Hagmann CA, Wimmenauer V, et al. 2009. Recognition of 5′ triphosphate by RIG-I helicase requires short blunt double-stranded RNA as contained in panhandle of negative-strand virus. *Immunity* 31:25–34
88. Schmidt A, Schwerd T, Hamm W, Hellmuth JC, Cui S, et al. 2009. 5′-triphosphate RNA requires base-paired structures to activate antiviral signaling via RIG-I. *Proc. Natl. Acad. Sci. USA* 106:12067–72
89. Schulz O, Diebold SS, Chen M, Näslund TI, Nolte MA, et al. 2005. Toll-like receptor 3 promotes cross-priming to virus-infected cells. *Nature* 433:887–92
90. Seth RB, Sun L, Ea CK, Chen ZJ. 2005. Identification and characterization of MAVS, a mitochondrial antiviral signaling protein that activates NF-κB and IRF 3. *Cell* 122:669–82
91. Stetson DB, Ko JS, Heidmann T, Medzhitov R. 2008. Trex1 prevents cell-intrinsic initiation of autoimmunity. *Cell* 134:587–98
92. Stetson DB, Medzhitov R. 2006. Antiviral defense: interferons and beyond. *J. Exp. Med.* 203:1837–41
93. Stetson DB, Medzhitov R. 2006. Recognition of cytosolic DNA activates an IRF3-dependent innate immune response. *Immunity* 24:93–103
94. Sutterwala FS, Ogura Y, Flavell RA. 2007. The inflammasome in pathogen recognition and inflammation. *J. Leukoc. Biol.* 82:259–64

79. Identifies major viral ligand for RIG-I as the intact negative-strand RNA genome.

95. Szomolanyi-Tsuda E, Liang X, Welsh RM, Kurt-Jones EA, Finberg RW. 2006. Role for TLR2 in NK cell-mediated control of murine cytomegalovirus in vivo. *J. Virol.* 80:4286–91
96. Takaoka A. 2007. DAI (DLM-1/ZBP1) is a cytosolic DNA sensor and an activator of innate immune response. *Nature* 448:501–5
97. Thomas PG, Dash P, Aldridge JR Jr, Ellebedy AH, Reynolds C, et al. 2009. The intracellular sensor NLRP3 mediates key innate and healing responses to influenza A virus via the regulation of caspase-1. *Immunity* 30:566–75
98. Thompson MR, Kaminski JJ, Kurt-Jones EA, Fitzgerald KA. 2011. Pattern recognition receptors and the innate immune response to viral infection. *Viruses* 3:920–40
99. Unterholzner L, Keating SE, Baran M, Horan KA, Jensen SB, et al. 2010. IFI16 is an innate immune sensor for intracellular DNA. *Nat. Immunol.* 11:997–1004
100. Wahid AM, Coventry VK, Conn GL. 2008. Systematic deletion of the adenovirus-associated RNAI terminal stem reveals a surprisingly active RNA inhibitor of double-stranded RNA-activated protein kinase. *J. Biol. Chem.* 283:17485–93
101. Wang JP, Kurt-Jones EA, Shin OS, Manchak MD, Levin MJ, Finberg RW. 2005. Varicella-zoster virus activates inflammatory cytokines in human monocytes and macrophages via Toll-like receptor 2. *J. Virol.* 79:12658–66
102. Xu LG, Wang YY, Han KJ, Li LY, Zhai Z, Shu HB. 2005. VISA is an adapter protein required for virus-triggered IFN-β signaling. *Mol. Cell* 19:727–40
103. Yamaguchi T, Kawabata K, Kouyama E, Ishii KJ, Katayama K, et al. 2010. Induction of type I interferon by adenovirus-encoded small RNAs. *Proc. Natl. Acad. Sci. USA* 107:17286–91
104. Yan N, Regalado-Magdos AD, Stiggelbout B, Lee-Kirsch MA, Lieberman J. 2010. The cytosolic exonuclease TREX1 inhibits the innate immune response to human immunodeficiency virus type 1. *Nat. Immunol.* 11:1005–13
105. Yang P, An H, Liu X, Wen M, Zheng Y, et al. 2010. The cytosolic nucleic acid sensor LRRFIP1 mediates the production of type I interferon via a β-catenin-dependent pathway. *Nat. Immunol.* 11:487–94

> 106. Shows that DI particles are a major ligand for MDA5.

106. **Yount JS, Gitlin L, Moran TM, López CB. 2008. MDA5 participates in the detection of paramyxovirus infection and is essential for the early activation of dendritic cells in response to Sendai virus defective interfering particles. *J. Immunol.* 180:4910–18**
107. Zhang SY, Jouanguy E, Ugolini S, Smahi A, Elain G, et al. 2007. TLR3 deficiency in patients with herpes simplex encephalitis. *Science* 317:1522–27
108. Zhou S, Kurt-Jones EA, Mandell L, Cerny A, Chan M, et al. 2005. MyD88 is critical for the development of innate and adaptive immunity during acute lymphocytic choriomeningitis virus infection. *Eur. J. Immunol.* 35:822–30
109. Zhu J, Huang X, Yang Y. 2009. The TLR9-MyD88 pathway is critical for adaptive immune responses to adeno-associated virus gene therapy vectors in mice. *J. Clin. Invest.* 119:2388–98
110. Zhu J, Martinez J, Huang X, Yang Y. 2007. Innate immunity against vaccinia virus is mediated by TLR2 and requires TLR-independent production of IFN-β. *Blood* 109:619–25
111. Züst R, Cervantes-Barragan L, Habjan M, Maier R, Neuman BW, et al. 2011. Ribose 2′-O-methylation provides a molecular signature for the distinction of self and non-self mRNA dependent on the RNA sensor Mda5. *Nat. Immunol.* 12:137–43

DNA Replication and Genomic Architecture of Very Large Bacteria

Esther R. Angert

Department of Microbiology, Cornell University, Ithaca, New York 14853;
email: era23@cornell.edu

Keywords

nucleoid, *Epulopiscium*, homopolymeric tract, polyploid, bacterial development, sporulation

Abstract

Large cell size is not restricted to a particular bacterial lifestyle, dispersal method, or cell envelope type. What is conserved among the very large bacteria are the quantity and arrangement of their genomic resources. All large bacteria described to date appear to be highly polyploid. This review focuses on *Epulopiscium* sp. type B, which maintains tens of thousands of genome copies throughout its life cycle. Only a tiny proportion of mother cell DNA is inherited by intracellular offspring, but surprisingly DNA replication takes place in the terminally differentiated mother cell as offspring grow. Massive polyploidy supports the acquisition of unstable genetic elements normally not seen in essential genes. Further studies of how large bacteria manage their genomic resources will provide insight into how simple cellular modifications can support unusual lifestyles and exceptional cell forms.

Contents

INTRODUCTION ... 198
FACTORS THAT INFLUENCE BACTERIAL CYTOARCHITECTURE AND
  DNA DYNAMICS ... 199
  Reliance on Diffusive Transport Impacts Bacterial Cell Size and Form ... 199
  The Quantity and Arrangement of DNA in Bacteria ... 200
UNUSUAL REPRODUCTIVE PROCESSES INFLUENCE
  REPLICATION AND INHERITANCE ... 200
  Reproduction in *Epulopiscium* Affects the Location of DNA ... 200
  Multiple Endospore Formation Requires Regulatory Changes
    in DNA Replication ... 202
  DNA Inheritance and Replication Dynamics in *Epulopiscium* ... 204
VERY LARGE BACTERIA ARE POLYPLOID ... 204
  Arrangement of DNA in *Epulopiscium* sp. Type B ... 204
  Estimates of Genome Size and Copy Number in *Epulopiscium* sp. Type B ... 204
  Arrangement of DNA in *Thiomargarita* ... 205
  Polyploidy in Bacteria ... 206
  Effect of Extreme Polyploidy on the *Epulopiscium* Genome ... 207
CONCLUSIONS ... 208

INTRODUCTION

Very large bacteria (>200 μm long or wide, and visible to the unaided human eye) come in a variety of shapes and sizes. Described early in the nineteenth century, *Spirochaeta plicatilis* (spirilla 80–250 μm long) and sulfur-oxidizing *Beggiatoa* spp. (conspicuous filaments composed of stacks of disc-shaped cells) are among the first on record (8, 64). With the application of molecular phylogeny, and the trained eyes of those who work with sulfur-oxidizing *Gammaproteobacteria*, our appreciation of the global distribution and diversity of members of this group has greatly expanded in recent years. The family *Beggiatoaceae* includes the spherical, 800-μm-diameter *Thiomargarita namibiensis* (which means "sulfur pearl of Namibia") (66, 67), as well as stalked and budding forms of *Thiomargarita*-like bacteria that attach to a variety of substrates and may be found riding on gastropods (7), in addition to the more classic filamentous forms (64). The physiology and microbial ecology of giant sulfur-oxidizing bacteria have been extensively studied. Large cell size allows bacteria better control over their position in physicochemical gradients and the ability to carry substantial intracellular chemical reserves to support respiration (68). In the *Firmicutes*, the largest known heterotrophic bacteria form a phylogenetically diverse group of 70- to 600-μm-long, cigar-shaped cells that represent at least two distinct but related clades collectively called *Epulopiscium fishelsoni* and the 100- to 300-μm-long *Epulopiscium* sp. type B (5). These too are members of a morphologically diverse group of bacteria that inhabit the intestinal tract of surgeonfish (2, 12). The size of these motile, submarine-shaped bacteria may allow them to maintain their position in the gut while helping them to avoid predation by ciliates that cohabitate the surgeonfish intestine (43). Big bacteria are not restricted to a particular phylum, lifestyle, dispersal method, cell envelope type, or mode of carbon or energy acquisition—features that may be considered the fundamental defining characters of a bacterial group. What is conserved among these massive microbes are the quantity and arrangement of their genomic resources. All large

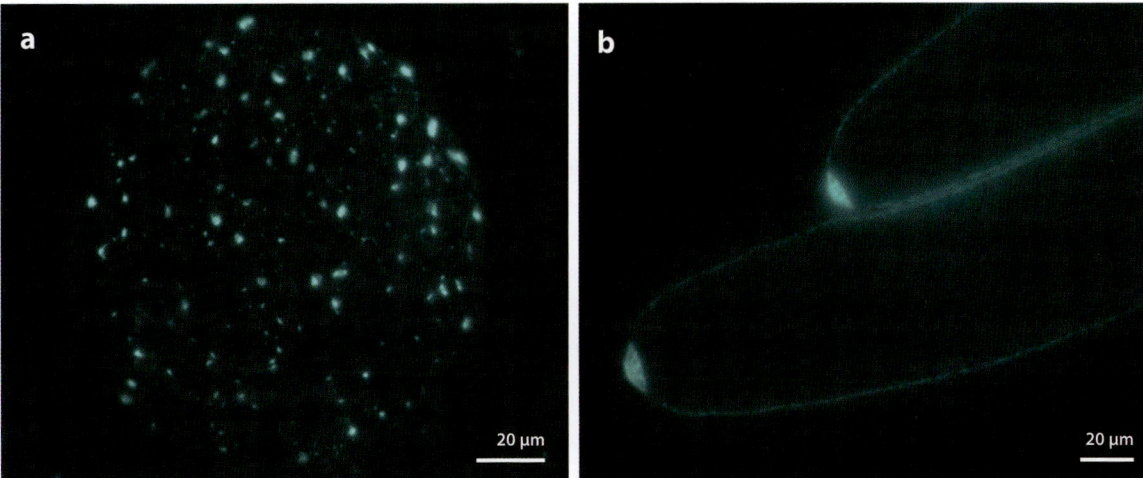

Figure 1

Large bacteria, stained with DAPI, display a similar arrangement of their cellular DNA. (*a*) Numerous nucleoids found in the peripheral, active cytoplasm of a spherical *Thiomargarita namibiensis* cell are associated with the plasma membrane. A surface layer focal plane of a small *T. namibiensis* cell is shown. (*b*) Half of a large *Epulopiscium* sp. type B mother cell. This medial section illustrates the peripheral DNA layers of two large internal offspring. Granddaughter cell primordia are seen at the tips of the offspring. At this stage in development, the mother cell DNA has disassembled and is no longer visible. The image of *T. namibiensis* was kindly provided by Verena Salman. Abbreviation: DAPI, 4′,6-diamidino-2-phenylindole.

bacteria described to date appear to be proportionately polyploid. Genomic amplification may be a unifying theme in overcoming the constraints that seem to keep most bacteria small.

This review focuses on recent developments in our understanding of the genomic structure and functional arrangement of DNA in very large bacteria (**Figure 1**), including *Epulopiscium* sp. type B, which are the most extensively studied bacterial behemoths in terms of cellular biology and DNA metabolism. A few tantalizing images of the poly-nucleoid *T. namibiensis* have been published (66), and these are a prelude to future characterization of the genomic composition and cell biology of these exceptional microbes. Although technically challenging, in part because none of these giants can be grown readily in pure culture, studies of their cell and developmental biology are providing insight into cellular modifications that test the limits of bacterial cell size.

FACTORS THAT INFLUENCE BACTERIAL CYTOARCHITECTURE AND DNA DYNAMICS

Reliance on Diffusive Transport Impacts Bacterial Cell Size and Form

Most bacterial and archaeal cells are small and appear ultrastructurally simple, especially when compared to microbial eukaryotes. The compartmentalization of cellular functions, the motor-protein-facilitated trafficking of vesicles and organelles over the cytoskeletal network, the expansion of genomic resources, and the acquisition of endosymbionts that became energy-generating organelles have all been credited for the advancement of the size and complexity of the eukaryotic cell (34, 68). In stark contrast, the cytoplasm of most bacteria appears homogenous with its scattered ribosomes and patches of nucleic acids (42, 60). Until recently, the bacterial cell was considered a disorganized membrane-bound compartment (38, 70). The reliance on diffusion to encounter nutrients and to move intracellular biomolecules was deemed to be the overriding principle dictating the limits of cell size. To acquire nutrients at an adequate rate to support

Polyploid: describes a cell that possesses two or more complete copies of its chromosome(s)

Nucleoid: a defined structure in the bacterial (or archaeal) cell that consists of a chromosome(s) and associated proteins

Binary fission:
process of cellular reproduction, used by most bacteria, which entails replication and segregation of genetic material and division of the cell to form two equivalent offspring

metabolic processes, bacteria maintain a high surface area relative to their total cytoplasmic volume. Even most large bacteria fit this model; they are long and thin, or they contain large intracellular inclusions, thereby limiting their cytoplasmic volume or thickness relative to their surface. *Epulopiscium* spp. are exceptions to this rule. With their complex intracellular membrane system, copious DNA, and unusual reproduction, it is easy to see why *Epulopiscium* spp. were first described as microbial eukaryotes (20, 48). The impacts of the unusual morphological and developmental features of *Epulopiscium* spp. on cellular function are still under investigation. However, some traction has been gained from work on characterizing the genomic content of these cells.

The Quantity and Arrangement of DNA in Bacteria

Bacteria contain a tremendous amount of DNA relative to their cell volume (28). The DNA and its associated proteins form an ordered yet dynamic nucleoprotein mass referred to as the nucleoid (14, 39, 60). When viewed in live bacteria, the nucleoid appears to fill much of the cytoplasm (41). High-resolution imaging of the nucleoid reveals a compact center with loops of presumably transcriptionally active DNA emanating from this core (9, 14). Treatment with translation inhibitors leads to nucleoid collapse in live cells, as DNA withdraws from the peripheral cytoplasm (26, 90). The association of DNA with the cell membrane appears vital in cells that are active, and this arrangement may be a consequence of transertion, the linked processes of transcription, translation, and placement of membrane proteins (42, 53). In contrast, treatments with antibiotics that halt transcription make the nucleoid relax and fill the cytoplasm. Fine-scale mapping of regions of the chromosome has been accomplished in several model systems using two complementary methods (52, 82, 87). Fluorescent in situ hybridization of sequence-specific DNA probes has been applied to fixed cells. In addition, dynamic changes in localization have been recorded in live cells using fluorescently tagged DNA-binding proteins (such as Lac repressor, LacI) expressed in genetically modified cells in which a tandem array of cognate, high-affinity binding sequences (such as Lac operator, *lacO*) have been inserted into specific places in the chromosome. Both approaches have provided evidence that regions of the chromosomal DNA of the nucleoid maintain a particular cell-cycle-dependent arrangement within a cell (82). A recent report suggests that localized transcription of genes in bacteria, and limited diffusion of mRNA, may be one mechanism by which transcripts and thus proteins are assigned to discrete cellular positions (47). The arrangement of chromosomal DNA and its association with the cell membrane may be essential for efficient placement of transport proteins and metabolic processing, particularly in large heterotrophic bacteria.

In model organisms such as *Bacillus subtilis*, *Escherichia coli*, and *Caulobacter crescentus*, DNA replication and segregation of nucleoids prior to division have been characterized in great detail (19, 78). DNA replication origins are the first segments of the chromosome to separate, and eventually two nucleoid masses part from one another as a growing cell progresses toward division. Binary fission is the primary means of reproduction for most bacteria, although notable exceptions have been described (1). *Epulopiscium* spp., for example, reproduce by the formation of multiple intracellular offspring.

UNUSUAL REPRODUCTIVE PROCESSES INFLUENCE REPLICATION AND INHERITANCE

Reproduction in *Epulopiscium* Affects the Location of DNA

Thus far, the largest *Epulopiscium* spp. have been found in the intestinal tract of only a few species of surgeonfish (Acanthuridae), including *Acanthurus nigrofuscus*, *A. lineatus*, *A. triostegus*, and *Naso*

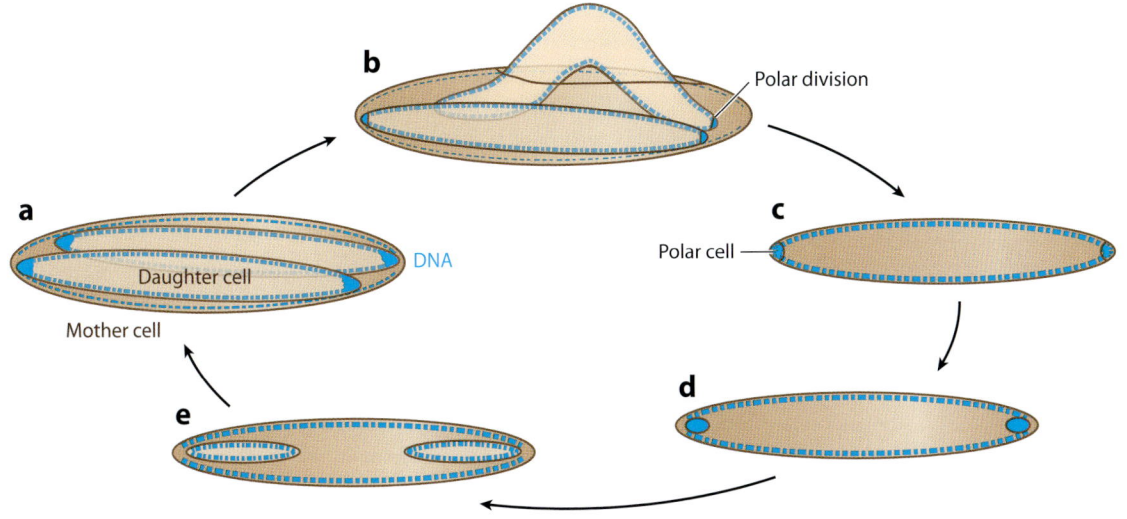

Figure 2

The *Epulopiscium* sp. type B life cycle. No evidence of binary fission has been seen in *Epulopiscium* sp. type B cells; instead they produce multiple intracellular offspring. (*a*) In *Epulopiscium* sp. type B cells, the initiation of granddaughter cells happens in intracellular daughter cells prior to their release from the mother cell. DNA accumulates at the tip of the daughter cells just prior to polar division. (*b*) Bipolar division takes place, usually before daughter cells emerge from the dying mother cell. (*c*,*d*) The polar cells are engulfed and (*e*) these new offspring begin to grow. In this diagram of cell cross-sections, the cell envelope and division septa are shown as brown lines and DNA is in blue.

tonganus (5, 12, 20). Smaller but related intestinal bacteria (or epulos) inhabit these and other surgeonfish hosts (22). *Epulopiscium* spp. and their relatives tend to be found in herbivorous, omnivorous, and detritivorous species of surgeonfish and likely contribute to host digestive processes and nutrition (12).

On the basis of 16S rRNA gene surveys, the largest of these symbionts (cigar-shaped cells ∼100–600 μm in length) represent at least three distinct phylogenetic clusters (2, 5). *Epulopiscium fishelsoni* comprises two clades of bacteria and these groups share ∼91% 16S rRNA gene sequence identity. *Epulopiscium* sp. type B populations are genetically much more homogeneous and typically show 99–100% 16S rRNA gene identity. All *Epulopiscium* spp. share an unusual mode of reproduction in which a cell produces multiple internal offspring. The daily production and release of active intracellular daughter cells, at the expense of the mother cell, is a process referred to as vivipary or viviparity to distinguish it from binary fission and the formation of multiple dormant endospores (1, 85). Vivipary follows a fairly predictable 24-h cycle leading to naturally synchronized cells with respect to offspring development (4, 48). Reproductive progression dictates the location of cellular DNA and influences inheritance of genetic material.

Thus far, no *Epulopiscium* has been maintained for more than a day outside its fish host. Therefore, the reproductive process has been pieced together from observations of microbial communities harvested from different individual fish. This has led to some conflicting reports regarding the reliance on binary fission (4, 48), but all accounts agree on the fundamentals of vivipary. Descriptions of offspring development or cell biology reported here focus on *Epulopiscium* sp. type B.

Intracellular offspring formation in *Epulopiscium* sp. type B (**Figure 2**) begins with a shift in the distribution of cellular DNA (4). Offspring (granddaughter cells) begin to form in daughter cells while still contained within their mother cell. During much of its growth phase, DNA is evenly distributed around the peripheral cytoplasm, but as a cell transitions toward reproduction,

Vivipary (or viviparity): development of offspring within the mother until the offspring can survive independently, at which point the active offspring emerge from the mother

DAPI: 4′,6-diamidino-2-phenylindole

DNA accumulates at the cell poles. The cell divides at both poles, extremely close to the tips. As with other bacteria, the division site is determined by the localization of FtsZ rings (4). Polar DNA is trapped within each primordial offspring cell, and with time, the pole-associated DNA that was not captured during division is often redistributed; either it is translocated into the newly formed offspring or it dissipates or relaxes away from the offspring (4). Other times pole-associated mother cell DNA remains and encircles the growing offspring (85). Each polar cell is engulfed (10), and within a membrane-bound compartment, the offspring grows. Engulfment leaves conspicuous polar voids in the mother cell DNA network. The inherited DNA is packaged in a highly ordered and condensed form that is easily visualized with fluorescent DNA dyes (such as DAPI, 4′,6-diamidino-2-phenylindole) and classic light microscopy methods such as Feulgen or Giemsa staining (59). Only a small proportion of mother cell DNA is passed on to offspring; most remains within the mother cell. As the offspring grow, the bias of DNA toward the cell periphery begins to emerge (see **Supplemental Movie 1**; follow the **Supplemental Material link** from the Annual Reviews home page at **http://www.annualreviews.org**) (4). Eventually the offspring cells nearly fill the mother cell cytoplasm. Late in development, the mother cell appears to undergo programmed cell death, which may provide a mechanism for recycling some of the resources accumulated in the mother cell (85). Most evident are the diminution and eventual disappearance of mother cell DNA. Mature offspring emerge through a split in the mother cell envelope, which marks the end of the mother cell. Any bits of cytoplasm or envelope remaining are quickly broken down and assimilated by the host fish or devoured by other microbes.

Multiple Endospore Formation Requires Regulatory Changes in DNA Replication

Intracellular offspring development in *Epulopiscium* spp. likely evolved from endospore formation (3, 46). For most of the best-characterized spore-forming *Firmicutes*, an endospore is produced as a last resort to enhance survival and dispersal of a bacterium when nutrients have been depleted or the bacterium senses perilously high concentrations of fermentation products, e.g., fatty acids (17, 18, 27, 56). In these bacteria, only binary fission is used for reproduction. When an endospore is produced, resources are limited and DNA replication is tightly regulated (83). In *B. subtilis*, one round of DNA replication may occur at the start of sporulation but throughout the rest of the process replication initiation is repressed (11, 23). One chromosome copy is packaged into the developing spore and the other remains in the mother cell. Both copies are essential to direct the discrete yet coordinated genetic programs of each cell type to ensure the formation of a fully mature and resilient endospore.

The closest relatives of *Epulopiscium* spp. are endospore-forming bacteria such as *Metabacterium polyspora* (3), a bacterium with the noteworthy ability to produce multiple endospores, as many as eight or nine per mother cell (61). This 12- to 35-μm-long bacterium makes its living cycling in and out of the gastrointestinal tract of guinea pigs. Endospore production is a principal means of propagation for *M. polyspora* when it is associated with its host, although cells may also undergo binary fission (6). The close phylogenetic relationship between *M. polyspora* and *Epulopiscium* led to the hypothesis that vivipary in *Epulopiscium* spp. arose from endospore formation (3). Further support has come from shared morphological transitions common to both processes (1) and from the discovery that some smaller epulos reproduce solely by the formation of multiple phase-bright endospores (22).

The reproductive processes of *M. polyspora* serve as a counterpoint to those of *Epulopiscium* spp. by providing an intermediate strategy between the more typical endospore formation and active internal offspring production seen in *Epulopiscium*. The relationship between *M. polyspora* and its guinea pig host may be driving the reliance on sporulation for reproduction (6). Guinea pigs are coprophagous, like many rodents that rely on hindgut fermentation. *Metabacterium* and relatives take advantage of the coprophagous nature of their host to cycle through its gastrointestinal tract. As a strict anaerobe, the *M. polyspora* endospore survives the external and oxygen-rich environment. Unlike actively growing cells, the spore survives mastication, exposure to lytic enzymes, and passage through the stomach. The endospore germinates in the small intestine. The emerging cell often exhibits signs of spore formation and has already divided at one or both poles. If binary fission occurs, it happens just after germination. The *M. polyspora* cells pass from the small intestine into the cecum, which is the predominant organ of the intestinal tract of a guinea pig. Here, forespores continue to develop, and most of the *M. polyspora* cells show advanced stages of sporulation. Cells then pass from the cecum and can be found in the feces.

For *M. polyspora*, endospore formation may be hardwired and an essential part of its life cycle (6, 84). To accommodate this novel reproductive strategy, DNA replication management and the patterns of division during sporulation differ from those seen in *B. subtilis*. To form multiple forespores, *M. polyspora* divides at both poles and, for a short time after engulfment, each forespore may undergo binary fission (6). The location and timing of DNA replication during developmental progression have been investigated in *M. polyspora* by labeling cells with the nucleotide analog bromodeoxyuridine (BrdU) and by immunolocalization to reveal sites of incorporation into DNA (84). Localization patterns in cells that were pulse-labeled or continuously exposed to BrdU suggest that DNA replication continues throughout sporulation in *M. polyspora*, with the strongest localization signal found inside forespores (**Figure 3***a*). Because sporulation commences immediately after germination, there is little time for the cell to replicate its chromosome. If DNA replication was immediately suppressed just after initiation of sporulation, the reproductive potential of these cells could be compromised, as offspring number would be predetermined. It seems more likely that DNA replication occurs during sporulation within developing forespores, when the cells are in the nutrient-rich cecal environment. This would allow each spore to be

Coprophagous: feeding on feces

BrdU: bromodeoxyuridine

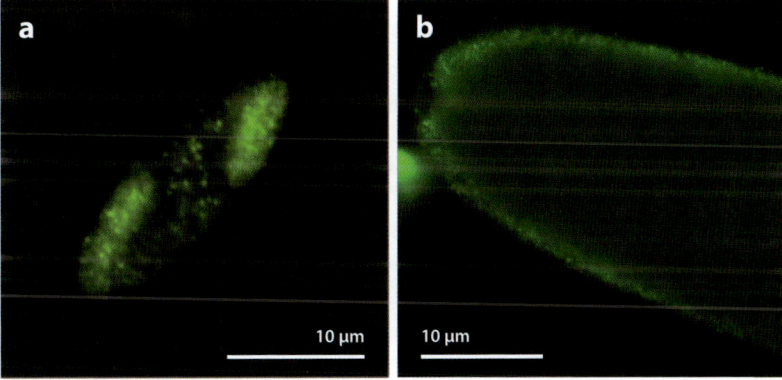

Figure 3

Patterns of replication during development in *Metabacterium polyspora* and *Epulopiscium* sp. type B. Cells within the gut of the host were pulse-labeled with the nucleotide analog BrdU. Gut contents were fixed and sites of replication were viewed by immunolocalization. (*a*) During sporulation, BrdU accumulates in the developing *M. polyspora* forespores (large polar bodies) and the mother cell. (*b*) In *Epulopiscium*, BrdU is incorporated into DNA within offspring but also displays a striking pattern of integration into DNA in the terminally differentiated mother cell. Shown here is a cross-section through the tip of a mother cell. Panel *a* is reprinted from Reference 84 with permission from the publisher. Abbreviation: BrdU, bromodeoxyuridine.

loaded with multiple copies of the genome and for cells to accumulate resources throughout their reproductive cycle.

DNA Inheritance and Replication Dynamics in *Epulopiscium*

For *Epulopiscium* sp. type B, reproductive progression resembles that seen in *M. polyspora* in that a large mother cell produces two or more internal offspring. Unlike *M. polyspora*, *Epulopiscium* sp. type B is maintained in the intestinal tract of *Naso tonganus* for extended periods of time—*Epulopiscium* cells do not cycle out and back into the gastrointestinal tract after each reproductive cycle. This change may have reduced pressure to produce an internal offspring that is as tough as an endospore. There are two other important differences that bear mentioning. *Epulopiscium* sp. type B does not have the option to undergo binary fission. Also, a much smaller proportion of the total cellular DNA in *Epulopiscium* is partitioned to offspring. We estimated that less than 1% of mother cell DNA is directly inherited (4). BrdU incorporation into live *Epulopiscium* sp. type B cells was used to follow the accumulation, segregation, and perhaps recycling of DNA (85). As expected, DNA replication occurred within the growing offspring. However, cells midway through development had an even more striking pattern of replication (**Figure 3b**). Well past a stage where mother cell DNA could be directly inherited (illustrated in **Figure 2e**), DNA within the mother cell continued to replicate (85). DNA replication in a cell destined to die indicated that the genetic material harbored in the mother cell must be performing a vital function. We suspect that additional genome copies are required to support the metabolic needs of the mother cell and its growing offspring. This unusual reproductive strategy and large cell size have given rise to nucleoids with separate fates; some nucleoids are inherited by offspring and could be characterized as germline nucleoids, whereas others are not and these represent somatic nucleoids that proliferate.

VERY LARGE BACTERIA ARE POLYPLOID

Arrangement of DNA in *Epulopiscium* sp. Type B

From the first descriptions of *E. fishelsoni* (20, 48), it was apparent that *Epulopiscium* spp. contain a considerable amount of DNA. Combined with its unusual arrangement, it was challenging to determine by flow cytometry or microscopy the quantity of DNA contained in individuals. No good comparable microbe with a characterized genomic content was available to serve as a meter. Instead, the average DNA content in *Epulopiscium* sp. type B cells was estimated by handpicking cells and quantifying the amount of DNA recovered from extractions of 5,000 cells selected from different populations (43). Two populations of cells were targeted for this analysis. One population consisted of large cells with two large offspring (as in **Figure 2a**) and the other consisted of newly emerged cells with small internal offspring (**Figure 2d**). By examining these extremes of developmental progression, it was determined that an *Epulopiscium* sp. type B always maintains a large quantity of DNA and on average contains approximately 85–250 pg of DNA (43). For comparison, a diploid human cell has about 6 pg of DNA, and depending on its stage of development, *B. subtilis* harbors ∼5–10 fg of DNA.

Estimates of Genome Size and Copy Number in *Epulopiscium* sp. Type B

Despite advances in genomics of uncultured bacteria, a completely assembled genome for a big bacterium has yet to be published (51). Characterized bacterial genomes range in size from ∼0.14

to 13 Mb and it is probable that the *Epulopiscium* and *Thiomargarita* genomes fall somewhere within this range. Although it seems unlikely that very large bacteria have enormous genomes, it is possible that genes essential for maintaining a large cell could be preferentially amplified.

Most of the available data on genome size and copy number come from work with *Epulopiscium* sp. type B. Classic methods for determining genome size and complexity, such as restriction fragment length estimates of the fractionated genome on a pulsed-field gel, have not been applied successfully to populations of *Epulopiscium* cells. Instead, three unlinked single-copy genes have been assayed using quantitative PCR (qPCR) in individual *Epulopiscium* (43). Copy numbers for *dnaA*, *recA*, and *ftsZ* ranged from ∼20,000 to over 400,000. The results suggested that individuals contained tens of thousands to hundreds of thousands of genome copies. With this 20-fold gene copy variance in *Epulopiscium* sp. type B populations, it was impossible to tell if the selected single-copy genes were equally represented using single-cell qPCR assays.

As an alternative approach to determine whether these genes are equally represented in *Epulopiscium*, qPCR assays were performed on purified DNA (43). The results showed that *dnaA*, *recA*, and *ftsZ* genes were equally represented in DNA extracted from a population of large *Epulopiscium* cells with mature offspring. In DNA extracted from small cells, gene copy numbers were equivalent, except the numbers of *dnaA* copies were significantly higher than the numbers of *recA* and *ftsZ* genes. These small cells were taken at an early stage of offspring development, presumably when DNA replication occurs. The *dnaA* gene is linked to the chromosomal origin of replication (*oriC*) in many bacterial genomes (54). The higher number of *dnaA* genes in DNA from small cells is consistent with idea that these were growing cells and that some chromosomes were caught in the process of replication (43). In all, approximately 40,900 copies of each single-copy gene were found in 156 pg of genomic DNA. Although preferential amplification of small genomic regions cannot yet be ruled out, these data are consistent with *Epulopiscium* sp. type B having a genome of approximately 3.8 Mb that is present in tens of thousands of copies per cell.

Bacteria regulate their DNA content in response to environmental conditions. During exponential growth, *B. subtilis* maintains a consistent DNA-to-cytoplasmic volume ratio even across conditions that lead to different growth rates (71). In *E. coli*, DNA replication status is linked to cell growth (15, 16). Likewise, *Epulopiscium* sp. type B appears to have some mechanism that regulates genome copy number and coordinates it with cell size (43). Individual *Epulopiscium* sp. type B cells, taken at different stages of development, were photographed and then assayed for *ftsZ* gene copy number using qPCR. From each photograph, cell length and width measurements were used to estimate cell volume. Plotting gene copy number against cell volume revealed a clear linear relationship. On average, *Epulopiscium* sp. type B cells maintain one genome copy for every 1.9 μm^3 of cell over a size range of 28,200 to 436,000 μm^3 (43). For comparison, during exponential growth *B. subtilis* maintains ∼0.7 μm^3 of cell volume per chromosome. While *Epulopiscium* sp. type B cells may maintain a larger ratio of cell volume to genome copy compared with *B. subtilis*, it appears that for *Epulopiscium* cell growth and replication are linked.

Arrangement of DNA in *Thiomargarita*

Less is known about the quantity and content of DNA in *T. namibiensis*. However, it appears that some of the general features described for *Epulopiscium* spp. hold true for these exceptionally large proteobacteria. *T. namibiensis* cells can reach 800 μm in diameter, but like *Epulopiscium*, extremely large cells are rare and most cells in a population range from 100 to 300 μm (66). The cell contains a centrally located, fluid-filled vacuole that takes up about 98% of the cell volume. The active cytoplasm is pressed into a 0.5- to 2-μm layer just under the cytoplasmic membrane. Even with this arrangement, the cytoplasmic volume of *T. namibiensis* exceeds that of *Epulopiscium*

CYANOBACTERIA

The maintenance of multiple genome copies is a common feature of cyanobacteria. A recent survey cataloged examples of cyanobacteria that appear to hold a single genome copy, but most maintain several to hundreds of complete genomes (25). Some strains of *Synechocystis* PCC 6803 exhibit dramatic growth-rate-dependent shifts in copy number, from 40 to >200 copies per cell. Published images of large spherical cyanobacteria show nucleoids localized to the cell periphery (79) in an arrangement similar to that of *T. namibiensis*. In pleurocapsalean cyanobacteria that reproduce by reductive division, nucleoids are distributed evenly throughout the cytoplasm of spherical cells entering their reproductive phase (1, 86). Polyploidy is persistent even in genetic models, such as *Anabaena* sp. PCC7120 and *Synechocystis* sp. PCC 6803 (Kazusa strain), which maintain about 8–12 copies of their chromosome (29, 33). It is noteworthy in this context to consider chloroplasts, which maintain a highly reduced genome present in tens to hundreds of copies (58, 63). Polyploidy in cyanobacteria may be important for repairing DNA efficiently, for replacing mutated genes (25) and for meeting the demands of producing light-harvesting and photocenter apparati. The arrangement and number of nucleoids vary according to the dictates of reproductive biology and life histories.

sp. type B (68). DAPI staining of large *Thiomargarita* reveals numerous distinct nucleoids that appear to be closely associated with the plasma membrane (66). The density of these nucleoids appears lower than estimates of genome copy numbers in *Epulopiscium* sp. type B. This may be explained by the possibility that each nucleoid of *T. namibiensis* holds multiple genome copies. Alternatively, different ratios may reflect differences in metabolism. These alternatives remain to be addressed.

Polyploidy in Bacteria

Polyploidy is still a bit of an oddity for many. Consider for example these recent announcements of evidence of modest polyploidy (45, 55). Indeed, polyploidy appears fairly common and phylogenetically widespread. Polyploidy is well known in cyanobacteria (25) and may be a regular feature of proteobacteria as well (57). When certain rhizobia establish occupancy in galegoid legume nodules and transform to symbiotic bacteroids, they undergo rounds of DNA replication without cell division (44). The tiny predatory proteobacterium *Bdellovibrio bacteriovorus* becomes polyploid during its elongation phase as it grows within the periplasm of its prey (73). The exceptional DNA repair systems of *Deinococcus radiodurans*, which is capable of stitching together a shattered genome with thousands of double-strand breaks, relies in part on the presence of multiple complete genome copies per cell (89). Polyploidy in *Borrelia hermsii* may be important for antigenic variation (31). Some *Neisseria* spp. are reported to be polyploid but this feature may be limited to pathogens (77). High genome copy numbers (100–350 copies per cell) have been observed in diverse bacterial endosymbionts including *Buchnera aphidicola*, the obligate proteobacterial symbionts of the pea aphid (32), and the flavobacterial endosymbionts of the cockroach *Blatta orientalis* (37). Both substrate and aerial mycelia of streptomycetes and other filamentous actinobacteria are polyploid. The sporogenic cell of *Streptomyces coelicolor* may contain 50 or more unsegregated genome copies that condense into individual nucleoids. Each nucleoid is then packaged into a single spore at the very final stages of sporulation (21). Although this is not an exhaustive list, the above examples represent six of the most studied bacterial phyla.

For bacteria that may be under constant pressure to conserve resources and energy use, ploidy appears to be a target for selection. In other situations, polyploidy provides advantages that outweigh the cost of carrying multiple genome copies, especially in bacteria that have predictably good carbon and energy supplies. Efficient DNA repair, large cell size, and enhanced metabolic cooperativity in symbiotic associations, as well as the ability to grow and divide more quickly than chromosome replication rates would allow, are all advantages supported by polyploidy.

Effect of Extreme Polyploidy on the *Epulopiscium* Genome

The ploidy level of *Epulopiscium* spp. and the apportionment of a number of chromosomes to offspring suggest that *Epulopiscium* may tolerate genome copy diversification (43). Genome duplication and the successive divergence of alleles have been an important mechanism of genome evolution in eukaryotes (40, 69, 80). While gene duplication has been important in the evolution of microbial genomes, genome streamlining appears to be an overriding evolutionary force. As an initial assessment of genome variation in individual *Epulopiscium* cells, clone libraries were developed from several PCR-amplified single-copy genes (43). The sequences of clones generated from a single amplicon were determined and compared. By this approach no evidence was found for genetic divergence within any of the three genes surveyed. Although this is a limited genetic survey, it suggests that variants of essential genes are not maintained for long periods in *Epulopiscium* sp. type B. Another potential source of genetic variation would be if genome copies, especially somatic copies, recombined during development. This is seen in other terminally differentiated bacterial cells such as the mother cell of sporulating *B. subtilis* (74) or cyanobacterial heterocysts (24), although these are not universal regulatory mechanisms widely distributed in endospore or heterocyst-forming lineages.

The *dnaA* gene was one of three genes targeted for gene copy divergence analysis in *Epulopiscium* sp. type B (43). DnaA is a highly conserved mediator of DNA replication that also acts as a transcription factor (35, 65). In its role of replication initiator, DnaA monomers bind to sites near *oriC*. Chromosome binding and hydrolysis of ATP determine both structural changes and function of DnaA (49). The nucleoprotein complex formed by oligomerized DnaA and *oriC* facilitates double-strand DNA melting and loading of the replication complex. In addition, DnaA serves as a liaison between replication and cell division by binding to and modulating expression of cell division proteins such as FtsL and in some cases FtsZ (65). While the proper timing and regulation of initiation are essential for survival, DnaA levels and activity have been adapted for particular cell types and serve key regulatory functions in developmental transitions in *Caulobacter crescentus* (13, 30, 76), *Myxococcus xanthus* (62), *Streptomyces coelicolor* (72), and sporulating *B. subtilis* (81). DnaA-mediated activities in *Epulopiscium* spp. will likely show genus-specific fine-tuning to facilitate their unusual replication patterns and reproduction.

In *Epulopiscium* sp. type B, the *dnaA* gene appears usual in most respects: It is located next to the putative *oriC* (**Figure 4**), which contains all features of a typical bacterial replication origin (54). The predicted DnaA protein has the conserved structural features of DnaA proteins from

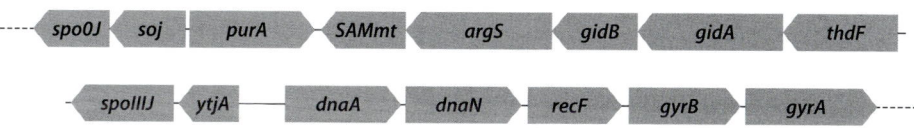

Figure 4

Diagrammatic representation of the putative *Epulopiscium* sp. type B. chromosomal replication origin (*oriC*). This region was identified by the presence of origin-linked genes normally found in *Firmicutes*: *dnaA*, *dnaN*, *recF*, *gyrA*, *gyrB*, *gidA*, *gidB*, and the *parA/B* homologs *soj* and *spo0J*. Twelve canonical origin-proximal DnaA boxes (TTATCCACA) have been identified (not shown).

characterized model systems. One unique feature of the *dnaA* gene in *Epulopiscium* sp. type B is that it contains a long mononucleotide repeat of 10 adenines located near the middle of the gene (43). Mononucleotide repeats of this length are rare in bacterial and archaeal genes, presumably owing to the high frequency of polymerase slippage in mononucleotide runs of this length (50). When present, these motifs tend to be found near either the 3′ end or the 5′ end of genes that encode variable surface proteins, and slippage within the homopolymeric sequence facilitates phase variation (36). Exceptions are found within A+T-rich endosymbiont genomes that often maintain long poly(A) tracts within genes (88). In *Blochmannia* and *Buchnera*, slippage by RNA polymerase in genes with homopolymeric repeats can restore the original gene reading frame and produce full-length, functional proteins (75). Although *Epulopiscium* is likely subject to purifying selection during each round of reproduction, perhaps its extreme polyploidy tolerates unstable genomic elements such as poly(A) tracts in essential genes. For *Epulopiscium* sp. type B, polymerase slippage in the poly(A) tract of *dnaA* would produce a frameshift leading to a premature stop codon. The functional significance of this *dnaA* adenine-deletion variant and the frequency of its expression has yet to be determined. If produced, the truncated protein would contain an oligomerization domain and an ATP-binding site but it would be devoid of its DNA-binding domain. It is possible that the truncated form is unstable and targeted for immediate proteolytic recycling. Or it may play some unforeseen regulatory role that alters replication initiation.

CONCLUSIONS

Exceptionally large bacteria are pushing the envelope of our concept of the bacterial cell. From observations thus far, it appears that big bacteria share one salient cytoarchitectural element that helps support a large volume of cytoplasm. Known large bacteria are invariably polyploid and the distribution of nucleoids appears biased toward the plasma membrane. Polyploidy is widespread in the *Bacteria* and, although not novel to very large bacteria, this feature may be taken to extremes to support the volume of cytoplasm and metabolic demands of these cells. The peripheral arrangement of genomic resources may allow large cells to produce macromolecular machinery where it is needed in a cell—often near the cytoplasmic membrane. Such a factor may be particularly influential in heterotrophic cells such as *Epulopiscium* sp. type B that may require excessive transport flux to support the daily production of offspring. It may be less of an issue for *Thiomargarita* spp. that are less metabolically constrained by reliance on obtaining organic carbon from environmental resources. Polyploidy underpins the metabolic demands of large cell size but it may also support reproductive flexibility, unstable genetic elements that allow unusual forms of regulation, and advantages normally seen only in eukaryotic cells.

SUMMARY POINTS

1. Very large bacteria have greatly expanded genomic resources.
2. Ploidy level in bacteria is subject to selection. The advantages of maintaining multiple genome copies include rapid growth, efficient repair of DNA lesions, large cell size, and enhanced metabolic cooperativity in symbiotic associations. These advantages often outweigh the cost of carrying multiple genome copies, especially in bacteria that have predictably good carbon and energy supplies.
3. Individual *Epulopiscium* sp. type B harbor tens of thousands of copies of their genome throughout their life cycle, and copy number varies with cytoplasmic volume.

4. Extreme polyploidy and the unusual reproductive strategy of *Epulopiscium* sp. type B have given rise to nucleoids with disparate fates; some nucleoids are inherited by offspring and could be characterized as germline nucleoids, whereas others are not and these represent somatic nucleoids that proliferate.

FUTURE ISSUES

1. Do large bacteria with different cell morphologies exhibit similar cytoplasm-to-genome-copy ratios?
2. For other lineages, does DNA content vary with size?
3. How do large bacteria manage and segregate multiple chromosomes?
4. Is tolerance of unstable elements greater in large bacteria than in obligate endosymbionts?

DISCLOSURE STATEMENT

The author is not aware of any affiliations, memberships, funding, or financial holdings that might be perceived as affecting the objectivity of this review.

ACKNOWLEDGMENTS

I am grateful to my colleagues for their support and insights into many of the topics covered in this review. Research in my laboratory is supported by the National Science Foundation grant 0721583.

LITERATURE CITED

1. Angert ER. 2005. Alternatives to binary fission in bacteria. *Nat. Rev. Microbiol.* 3:214–24
2. Angert ER. 2006. The enigmatic cytoarchitecture of *Epulopiscium* spp. In *Microbiology Monographs*, ed. JM Shively, pp. 285–301. Berlin/Heidelberg: Springer-Verlag
3. Angert ER, Brooks AE, Pace NR. 1996. Phylogenetic analysis of *Metabacterium polyspora*: clues to the evolutionary origin of daughter cell production in *Epulopiscium* species, the largest bacteria. *J. Bacteriol.* 178:1451–56
4. Angert ER, Clements KD. 2004. Initiation of intracellular offspring in *Epulopiscium*. *Mol. Microbiol.* 51:827–35
5. Angert ER, Clements KD, Pace NR. 1993. The largest bacterium. *Nature* 362:239–41
6. Angert ER, Losick RM. 1998. Propagation by sporulation in the guinea pig symbiont *Metabacterium polyspora*. *Proc. Natl. Acad. Sci. USA* 95:10218–23
7. **Bailey JV, Salman V, Rouse GW, Schulz-Vogt HN, Levin LA, Orphan VJ. 2011. Dimorphism in methane seep-dwelling ecotypes of the largest known bacteria. *ISME J.* 5:1926–35**
8. Blakemore RP, Canale-Parola E. 1973. Morphological and ecological characteristics of *Spirochaeta plicatilis*. *Arch. Mikrobiol.* 89:273–89
9. Bohrmann B, Villiger W, Johansen R, Kellenberger E. 1991. Coralline shape of the bacterial nucleoid after cryofixation. *J. Bacteriol.* 173:3149–58
10. Bresler V, Montgomery WL, Fishelson L, Pollak PE. 1998. Gigantism in a bacterium, *Epulopiscium fishelsoni*, correlates with complex patterns in arrangement, quantity, and segregation of DNA. *J. Bacteriol.* 180:5601–11

7. Includes eye-catching images of new morphologies and reproductive modes observed in enormous sulfur-oxidizing bacteria.

11. Castilla-Llorente V, Munoz-Espin D, Villar L, Salas M, Meijer WJ. 2006. Spo0A, the key transcriptional regulator for entrance into sporulation, is an inhibitor of DNA replication. *EMBO J.* 25:3890–99
12. Clements KD, Sutton DC, Choat JH. 1989. Occurrence and characteristics of unusual protistan symbionts from surgeonfishes (Acanthuridae) of the Great Barrier Reef Australia. *Marine Biol.* 102:403–12
13. Collier J, Murray SR, Shapiro L. 2006. DnaA couples DNA replication and the expression of two cell cycle master regulators. *EMBO J.* 25:346–56
14. Dillon SC, Dorman CJ. 2010. Bacterial nucleoid-associated proteins, nucleoid structure and gene expression. *Nat. Rev. Microbiol.* 8:185–95
15. Donachie WD. 1968. Relationship between cell size and time of initiation of DNA replication. *Nature* 219:1077–79
16. Donachie WD, Blakely GW. 2003. Coupling the initiation of chromosome replication to cell size in *Escherichia coli. Curr. Opin. Microbiol.* 6:146–50
17. Dürre P. 2011. Ancestral sporulation initiation. *Mol. Microbiol.* 80:584–87
18. Errington J. 2003. Regulation of endospore formation in *Bacillus subtilis. Nat. Rev. Microbiol.* 1:117–26
19. Errington J, Murray H, Wu LJ. 2005. Diversity and redundancy in bacterial chromosome segregation mechanisms. *Philos. Trans. R. Soc. Lond. B* 360:497–505
20. Fishelson L, Montgomery WL, Myrberg AA. 1985. A unique symbiosis in the gut of a tropical herbivorous surgeonfish (Acanthuridae: Teleostei) from the Red Sea. *Science* 229:49–51
21. Flardh K, Buttner MJ. 2009. *Streptomyces* morphogenetics: dissecting differentiation in a filamentous bacterium. *Nat. Rev. Microbiol.* 7:36–49
22. Flint JF, Drzymalski D, Montgomery WL, Southam G, Angert ER. 2005. Nocturnal production of endospores in natural populations of *Epulopiscium*-like surgeonfish symbionts. *J. Bacteriol.* 187:7460–70
23. Fujita M, Losick R. 2005. Evidence that entry into sporulation in *Bacillus subtilis* is governed by a gradual increase in the level and activity of the master regulator Spo0A. *Genes Dev.* 19:2236–44
24. Golden JW, Wiest DR. 1988. Genome rearrangement and nitrogen fixation in *Anabaena* blocked by inactivation of *xisA* gene. *Science* 242:1421–23
25. Griese M, Lange C, Soppa J. 2011. Ploidy in cyanobacteria. *FEMS Microbiol. Lett.* 323:124–31
26. Hashimoto M, Ichimura T, Mizoguchi H, Tanaka K, Fujimitsu K, et al. 2005. Cell size and nucleoid organization of engineered *Escherichia coli* cells with a reduced genome. *Mol. Microbiol.* 55:137–49
27. Hilbert DW, Piggot PJ. 2004. Compartmentalization of gene expression during *Bacillus subtilis* spore formation. *Microbiol. Mol. Biol. Rev.* 68:234–62
28. Holmes VF, Cozzarelli NR. 2000. Closing the ring: links between SMC proteins and chromosome partitioning, condensation, and supercoiling. *Proc. Natl. Acad. Sci. USA* 97:1322–24
29. Hu B, Yang G, Zhao W, Zhang Y, Zhao J. 2007. MreB is important for cell shape but not for chromosome segregation of the filamentous cyanobacterium *Anabaena* sp. PCC 7120. *Mol. Microbiol.* 63:1640–52
30. Jonas K, Chen YE, Laub MT. 2011. Modularity of the bacterial cell cycle enables independent spatial and temporal control of DNA replication. *Curr. Biol.* 21:1092–101
31. Kitten T, Barbour AG. 1992. The relapsing fever agent *Borrelia hermsii* has multiple copies of its chromosome and linear plasmids. *Genetics* 132:311–24
32. Komaki K, Ishikawa H. 2000. Genomic copy number of intracellular bacterial symbionts of aphids varies in response to developmental stage and morph of their host. *Insect Biochem. Mol. Biol.* 30:253–58
33. Labarre J, Chauvat F, Thuriaux P. 1989. Insertional mutagenesis by random cloning of antibiotic resistance genes into the genome of the cyanobacterium *Synechocystis* strain PCC 6803. *J. Bacteriol.* 171:3449–57

34. Discusses the links between genome complexity, the acquisition of energy-producing organelles, and the potential for morphological complexity.

34. **Lane N, Martin W. 2010. The energetics of genome complexity. *Nature* 467:929–34**
35. Leonard AC, Grimwade JE. 2011. Regulation of DnaA assembly and activity: taking directions from the genome. *Annu. Rev. Microbiol.* 65:19–35
36. Lin WH, Kussell E. 2012. Evolutionary pressures on simple sequence repeats in prokaryotic coding regions. *Nucleic Acids Res.* 40:2399–413
37. López-Sánchez MJ, Neef A, Patiño-Navarrete R, Navarro L, Jiménez R, et al. 2008. Blattabacteria, the endosymbionts of cockroaches, have small genome sizes and high genome copy numbers. *Environ. Microbiol.* 10:3417–22

38. Losick R, Shapiro L. 1999. Changing views on the nature of the bacterial cell: from biochemistry to cytology. *J. Bacteriol.* 181:4143–45
39. Luijsterburg MS, Noom MC, Wuite GJL, Dame R. 2006. The architectural role of nucleoid-associated proteins in the organization of bacterial chromatin: a molecular perspective. *J. Struct. Biol.* 156:262–72
40. Maere S, De Bodt S, Raes J, Casneuf T, Van Montagu M, et al. 2005. Modeling gene and genome duplications in eukaryotes. *Proc. Natl. Acad. Sci. USA* 102:5454–59
41. Margolin W. 2010. Imaging the bacterial nucleoid. In *Bacterial Chromatin*, ed. RT Dame, CJ Dorman, pp. 13–30. Berlin: Springer-Verlag
42. Matias VRF, Al-Amoudi A, Dubochet J, Beveridge TJ. 2003. Cryo-transmission electron microscopy of frozen-hydrated sections of *Escherichia coli* and *Pseudomonas aeruginosa*. *J. Bacteriol.* 185:6112–18
43. **Mendell JE, Clements KD, Choat JH, Angert ER. 2008. Extreme polyploidy in a large bacterium.** ***Proc. Natl. Acad. Sci. USA*** **105:6730–40**
44. Mergaert P, Uchiumi T, Alunni B, Evanno G, Cheron A, et al. 2006. Eukaryotic control on bacterial cell cycle and differentiation in the *Rhizobium*-legume symbiosis. *Proc. Natl. Acad. Sci. USA* 103:5230–35
45. Michelsen O, Hansen FG, Albrechtsen B, Jensen PR. 2010. The MG1363 and IL1403 laboratory strains of *Lactococcus lactis* and several dairy strains are diploid. *J. Bacteriol.* 192:1058–65
46. Miller DA, Choat JH, Clements KD, Angert ER. 2011. The *spoIIE* homolog of *Epulopiscium* sp. type B is expressed early in intracellular offspring development. *J. Bacteriol.* 193:2642–46
47. Montero Llopis P, Jackson AF, Sliusarenko O, Surovtsev I, Heinritz J, et al. 2010. Spatial organization of the flow of genetic information in bacteria. *Nature* 466:77–81
48. Montgomery WL, Pollak PE. 1988. *Epulopiscium fishelsoni* n.g., n.s., a protist of uncertain taxonomic affinities from the gut of an herbivorous reef fish. *J. Protozool.* 35:565–69
49. Mott ML, Berger JM. 2007. DNA replication initiation: mechanisms and regulation in bacteria. *Nat. Rev. Microbiol.* 5:343–54
50. Moxon ER, Rainey PB, Nowak MA, Lenski RE. 1994. Adaptive evolution of highly mutable loci in pathogenic bacteria. *Curr. Biol.* 4:24–33
51. Mussmann M, Hu FZ, Richter M, de Beer D, Preisler A, et al. 2007. Insights into the genome of large sulfur bacteria revealed by analysis of single filaments. *PLoS Biol.* 5:1923–37
52. Niki H, Hiraga S. 1998. Polar localization of the replication origin and terminus in *Escherichia coli* nucleoids during chromosome partitioning. *Genes Dev.* 12:1036–45
53. Norris V, Turnock G, Sigee D. 1996. The *Escherichia coli* enzoskeleton. *Mol. Microbiol.* 19:197–204
54. Ogasawara N, Yoshikawa H. 1992. Genes and their organization in the replication origin region of the bacterial chromosome. *Mol. Microbiol.* 6:629–34
55. Ohtani N, Tomita M, Itaya M. 2010. An extreme thermophile, *Thermus thermophilus*, is a polyploid bacterium. *J. Bacteriol.* 192:5499–505
56. Paredes CJ, Alsaker KV, Papoutsakis ET. 2005. A comparative genomic view of clostridial sporulation and physiology. *Nat. Rev. Microbiol.* 3:969–78
57. Pecoraro V, Zerulla K, Lange C, Soppa J. 2011. Quantification of ploidy in proteobacteria revealed the existence of monoploid, (mero-)oligoploid and polyploid species. *PLoS ONE* 6:e16392
58. Rauwolf U, Golczyk H, Greiner S, Herrmann R. 2009. Variable amounts of DNA related to the size of chloroplasts III. Biochemical determinations of DNA amounts per organelle. *Mol. Genet. Genomics* 283:35–47
59. Robinow C, Angert ER. 1998. Nucleoids and coated vesicles of "Epulopiscium" spp. *Arch. Microbiol.* 170:227–35
60. Robinow C, Kellenberger E. 1994. The bacterial nucleoid revisited. *Microbiol. Rev.* 58:211–32
61. Robinow CF. 1957. Kurzer hinweis auf *Metabacterium polyspora*. *Z. Tropenmed. Parasitol.* 8:225–27
62. Rosario CJ, Singer M. 2010. Developmental expression of *dnaA* is required for sporulation and timing of fruiting body formation in *Myxococcus xanthus*. *Mol. Microbiol.* 76:1322–33
63. Rowan BA, Bendich AJ. 2009. The loss of DNA from chloroplasts as leaves mature: fact or artefact? *J. Exp. Bot.* 60:3005–10
64. Salman V, Amann R, Girnth AC, Polerecky L, Bailey JV, et al. 2011. A single-cell sequencing approach to the classification of large, vacuolated sulfur bacteria. *Syst. Appl. Microbiol.* 34:243–59

43. Provides evidence of the exceptional number of chromosome copies held by *Epulopiscium* sp. type B and explores the impact of polyploidy on genome evolution and *Epulopiscium* biology.

65. Scholefield G, Veening JW, Murray H. 2011. DnaA and ORC: more than DNA replication initiators. *Trends Cell Biol.* 21:188–94
66. Schulz HN. 2006. The genus *Thiomargarita*. In *The Prokaryotes*, ed. M Dworkin, S Falkow, E Rosenberg, KH Schleifer, E Stackebrandt, pp. 1156–63. New York: Springer-Verlag
67. Schulz HN, Brinkhoff T, Ferdelman TG, Marine MH, Teske A, Jorgensen BB. 1999. Dense populations of a giant sulfur bacterium in Namibian shelf sediments. *Science* 284:493–95

68. **Schulz HN, Jorgensen BB. 2001. Big bacteria.** ***Annu. Rev. Microbiol.*** **55:105–37**

> 68. Analyzes the limits of bacterial cell size and bacteria that go beyond those limits.

69. Sémon M, Wolfe KH. 2007. Consequences of genome duplication. *Curr. Opin. Genet. Dev.* 17:505–12
70. Shapiro L, Losick R. 1997. Protein localization and cell fate in bacteria. *Science* 276:712–18
71. Sharpe ME, Hauser PM, Sharpe RG, Errington J. 1998. *Bacillus subtilis* cell cycle as studied by fluorescence microscopy: constancy of cell length at initiation of DNA replication and evidence for active nucleoid partitioning. *J. Bacteriol.* 180:547–55
72. Smulczyk-Krawczyszyn A, Jakimowicz D, Ruban-Ośmialowska B, Zawilak-Pawlik A, Majka J, et al. 2006. Cluster of DnaA boxes involved in regulation of *Streptomyces* chromosome replication: from in silico to in vivo studies. *J. Bacteriol.* 188:6184–94
73. Sockett RE. 2009. Predatory lifestyle of *Bdellovibrio bacteriovorus*. *Annu. Rev. Microbiol.* 63:523–39
74. Stragier P, Kunkel B, Kroos L, Losick R. 1989. Chromosomal rearrangement generating a composite gene for a developmental transcription factor. *Science* 243:507–12

75. **Tamas I, Wernegreen JJ, Nystedt B, Kauppinen SN, Darby AC, et al. 2008. Endosymbiont gene functions impaired and rescued by polymerase infidelity at poly(A) tracts.** ***Proc. Natl. Acad. Sci. USA*** **105:14934–39**

> 75. The surprising discovery that polymerase slippage in poly(A) tracts within a damaged gene can recover a full-length functional protein.

76. Taylor JA, Ouimet MC, Wargachuk R, Marczynski GT. 2011. The *Caulobacter crescentus* chromosome replication origin evolved two classes of weak DnaA binding sites. *Mol. Microbiol.* 82:312–26
77. Tobiason DM, Seifert HS. 2010. Genomic content of *Neisseria* species. *J. Bacteriol.* 192:2160–68
78. Toro E, Shapiro L. 2010. Bacterial chromosome organization and segregation. *Cold Spring Harb. Perspect. Biol.* 2:a000349
79. Tschermak-Woess E, Schöller A. 1982. Distribution and partition of the DNA-network in some *Cyanophyceae* as shown by DAPI-fluorescence. *Plant Syst. Evol.* 140:207–23
80. Van de Peer Y, Maere S, Meyer A. 2009. The evolutionary significance of ancient genome duplications. *Nat. Rev. Genet.* 10:725–32
81. Veening JW, Murray H, Errington J. 2009. A mechanism for cell cycle regulation of sporulation initiation in *Bacillus subtilis*. *Genes Dev.* 23:1959–70
82. Viollier PH, Thanbichler M, McGrath PT, West L, Meewan M, et al. 2004. Rapid and sequential movement of individual chromosomal loci to specific subcellular locations during bacterial DNA replication. *Proc. Natl. Acad. Sci. USA* 101:9257–62
83. Wagner JK, Marquis KA, Rudner DZ. 2009. SirA enforces diploidy by inhibiting the replication initiator DnaA during spore formation in *Bacillus subtilis*. *Mol. Microbiol.* 73:963–74
84. Ward RJ, Angert ER. 2008. DNA replication during endospore development in *Metabacterium polyspora*. *Mol. Microbiol.* 67:1360–70

85. **Ward RJ, Clements KD, Choat JH, Angert ER. 2009. Cytology of terminally differentiated *Epulopiscium* mother cells.** ***DNA Cell Biol.*** **28:57–64**

> 85. Provides evidence that populations of somatic nucleoids in the mother cell continue to replicate.

86. Waterbury JB, Stanier RY. 1978. Patterns of growth and development in pleurocapsalean cyanobacteria. *Microbiol. Rev.* 42:2–44
87. Webb CD, Teleman A, Gordon S, Straight A, Belmont A, et al. 1997. Bipolar localization of the replication origin regions of chromosomes in vegetative and sporulating cells of *B. subtilis*. *Cell* 88:667–74
88. Wernegreen JJ, Kauppinen SN, Degnan PH. 2010. Slip into something more functional: Selection maintains ancient frameshifts in homopolymeric sequences. *Mol. Biol. Evol.* 27:833–39
89. Zahradka K, Slade D, Bailone A, Sommer S, Averbeck D, et al. 2006. Reassembly of shattered chromosomes in *Deinococcus radiodurans*. *Nature* 443:569–73
90. Zimmerman SB. 2002. Toroidal nucleoids in *Escherichia coli* exposed to chloramphenicol. *J. Struct. Biol.* 138:199–206

Large T Antigens of Polyomaviruses: Amazing Molecular Machines

Ping An, Maria Teresa Sáenz Robles, and James M. Pipas

Department of Biological Sciences, University of Pittsburgh, Pittsburgh, Pennsylvania 15260; email: pipas@pitt.edu

Keywords

tumorigenesis, p53, Rb, DNA replication, DNA damage response

Abstract

The large tumor antigen (T antigen) encoded by simian virus 40 is an amazing molecular machine because it orchestrates viral infection by modulating multiple fundamental viral and cellular processes. T antigen is required for viral DNA replication, transcription, and virion assembly. In addition, T antigen targets multiple cellular pathways, including those that regulate cell proliferation, cell death, and the inflammatory response. Ectopic T antigen expression results in the immortalization and transformation of many cell types in culture and T antigen induces neoplasia when expressed in rodents. The analysis of the mechanisms by which T antigen carries out its many functions has proved to be a powerful way of gaining insights into cell biology. The accelerating pace at which new polyomaviruses are being discovered provides a collection of novel T antigens that, like simian virus 40, can be used to discover and study key cellular regulatory systems.

Contents

INTRODUCTION 214
    Simian Virus 40 214
    Structure of Large T Antigen 217
LARGE T ANTIGEN ORCHESTRATES VIRAL DNA REPLICATION 220
    The Helicase Activity of SV40 Large T Antigen 220
    In Vitro Replication of SV40 Viral DNA 223
    SV40 DNA Replication Under the Cellular DNA Damage Response 224
CELL TRANSFORMATION AND TUMORIGENESIS
    MEDIATED BY T ANTIGEN 225
    The pRb Pathway 225
    The p53 Pathway 228
    Additional Targets: How Important Are They? 228
    Transformation by Other Polyomaviruses 229
    Contribution of Other Viral Proteins to Transformation 229
REGULATION OF CELLULAR GENE EXPRESSION
    BY LARGE T ANTIGEN 229
CONCLUSIONS 230

INTRODUCTION

Large T antigens (T antigen) are proteins encoded by members of the *Polyomaviridae*, a family of small double-stranded DNA (dsDNA) viruses. The T antigens are multifunctional proteins expressed early in the infectious cycle that regulate viral and cellular gene expression, as well as function in viral DNA replication. Some polyomaviruses induce tumors in their natural hosts or in animal models. In these cases, T antigen is essential for the initiation and maintenance of the transformed state.

Polyomaviruses have been found in many hosts, including birds, rodents, nonhuman primates, and humans (**Table 1**). In general, polyomaviruses appear to infect growth-arrested differentiated cells. Polyomaviruses trophic for the kidney, lung, skin, and blood have been identified. With the exception of the avian polyomaviruses, which can be acutely pathogenic, polyomaviruses usually establish asymptomatic, lifelong persistent infections in their natural hosts. At present, nine human polyomaviruses have been identified. Four of these, BKV (BK polyomavirus), JCV (JC polyomavirus), TSV (Trichodysplasia spinulosa–associated polyomavirus), and MCV (Merkel cell polyomavirus), have been linked to disease. As of yet, efficient cell culture systems for the human viruses have not been developed and this has hampered the study of their infectious cycles. In contrast, simian virus 40 (SV40) replicates well in culture and encodes a potent oncoprotein, large T antigen, which has been studied intensively and has served as a model for understanding polyomavirus infection and the biology of the host cell.

Simian Virus 40

The SV40 virion is composed of three capsid proteins (VP1, VP2, and VP3) and contains a single molecule of circular dsDNA and host histones. The commonly used laboratory strain of SV40 has a genome of 5,243 base pairs that can be divided into an early region that encodes the T antigens; a late region that encodes the capsid proteins and in some viruses auxiliary proteins; and a regulatory region that contains the origin of viral DNA replication (*ori*), the early region

---

Large T antigen (T antigen): oncogene expressed from the early region of polyomaviruses and responsible for the induction of transformation in vivo and in vitro

***Polyomaviridae*:** a family of small DNA viruses with non-enveloped virions and circular double stranded DNA

Simian virus 40 (SV40): a primate-infecting polyomavirus; also called Simian vacuolating virus 40

Table 1 Representative polyomaviruses[a]

Virus	Taxonomy ID	Accession number	Host	Tropism	Large T antigen length	Middle T antigen length	Small t antigen length
LPV	12480	NC_00476	African green monkey	Blood	NP_848008	–	NP_848009
BPCV1[b]	479058	NC_010107	Western barred bandicoot	–	ND	–	YP_001595473
BPCV2[b]	500654	NC_101817	Western barred bandicoot	–	ND	–	YP_001955937
BKV	10629	NC_001538	Human	Kidney, respiratory system	YP_717940	–	YP_717941
BPyV	10627	NC_001442	Bovine	–	NP_040788	–	NP_040789
BFPyV	10625	NC_004764	Avian	–	YP_004061429	–	YP_004061430
SLPyV1	715223	NC_013796	California sea lion	–	YP_003429323	–	YP_003429324
CaPyV	881945	GU345044	Canary	–	ADM88652	–	ADM88651
Chimpanzee polyomavirus	305667	NC_014743	Chimpanzee	–	YP_004046683	–	YP_004046684
CPyV	349563	NC_007922	Avian	–	YP_529828	–	YP_529829
FPyV	349564	NC_007923	Avian	–	YP_529834	–	YP_529835
GHPyV	208491	NC_004800	Avian	–	NP_849170	–	NP_849171
GggPyV1	928214	HQ385752	Gorilla	–	ADQ54205	–	ADQ54206
HaPyV	10626	NC_001663	Hamster	–	NP_056730	NP_056731	NP_056732
HPyV9	943908	NC_015150	Human	Blood	YP_004243706	–	YP_004243707
Japanese eel endothelial cells-infection virus	712037	NC_015123	Japanese eel	–	YP_004222838	–	–
JCV	10632	NC_001699	Human	Kidney, respiratory system, brain	NP_043512 AEC15994	–	NP_043513 AEC15995
KIPyV	423445	NC_009238	Human	Lung	YP_001111259	–	YP_001111260
Mastomys polyomavirus	888441	AB588640	Rodent	–	BAJ53087	–	BAJ53088
MCV, MCPyV	493803	NC_010277	Human	Skin	AŁR38265	–	YP_001651047
K virus	229278	NC_001505	Mouse	Lung	NP_041232	–	NP_041233
MuPyV, MPyV	–	NC_001515	Mouse	–	NP_041264	NP_041265	NP_041266
Myotis polyomavirus VM-2008	563775	NC_011310	Bat	–	YP_002261489	–	YP_002261490
Orangutan polyomavirus	638786	NC_013439	Orangutan	–	YP_003264534	–	YP_003264535

(Continued)

Table 1 (Continued)

Virus	Taxonomy ID	Accession number	Host	Tropism	Large T antigen length	Middle T antigen length	Small t antigen length
Pan troglodytes verus polyomavirus 1a	928211	HQ385746	Chimpanzee	–	ADQ54175	–	ADQ54179
Pan troglodytes verus polyomavirus 1b	927576	HQ385747	Chimpanzee	–	ADQ54180	–	ADQ54181
Pan troglodytes verus polyomavirus 2a	928212	HQ385748	Chimpanzee	–	ADQ54185	–	ADQ54186
Pan troglodytes verus polyomavirus 2c	928213	HQ385749/ HQ385750/ HQ385751	Chimpanzee	–	ADQ54190 ADQ54195 ADQ54200	–	ADQ54191 ADQ54196 ADQ54201
HPyV6	746830	NC_014406	Human	Skin	YP_003848919	–	YP_003848920
HPyV7	746831	NC_014407	Human	Skin	YP_003848924	–	YP_003848925
SA12	614086	NC_007611	Monkey	–	YP_406555	–	YP_406556
Simian virus 12	46771	NC_012122	Baboon	–	YP_002635567	–	YP_002635568
SV40	10633	NC_001669	Rhesus macaque	Kidney	YP_003708382	–	YP_003708383
Squirrel monkey polyomavirus	452475	NC_009951	Squirrel monkey	–	YP_001531349	–	YP_001531350
TSV, TSPyV	862909	NC_014361	Human	Skin	YP_003800007	–	YP_003800008
WUPyV	440266	NC_009539	Human	Lung	YP_001285488	–	YP_001285489

[a]The expression of large, middle, and small tumor antigens are summarized. Gene bank accession numbers for T antigen proteins are listed when available. ND, not determined.
[b]BPCV1 and 2 are classified as members of *Papillomaviridae*.
Abbreviations: LPV, lymphotropic papovavirus, or African green monkey polyomavirus; BPCV1, Bandicoot papillomatosis carcinomatosis virus type 1; BPCV2, Bandicoot papillomatosis carcinomatosis virus type 2; BKV, BK polyomavirus; BPyV, Bovine polyomavirus; BFPyV, Budgerigar fledgling disease polyomavirus; SLPyV 1, California sea lion polyomavirus 1; CaPyV, Canary polyomavirus; CPyV, Crow polyomavirus; FPyV, Finch polyomavirus; GHPyV, Goose hemorrhagic polyomavirus; GggPyV 1, Gorilla gorilla gorilla polyomavirus 1; HaPyV, Hamster polyomavirus; HPyV 9, Human polyomavirus 9; JCV, JC polyomavirus; KIPyV, KI polyomavirus; MCV, MCPyV, Merkel cell polyomavirus; K virus, Murine pneumotropic polyomavirus; MuPyV, MPyV, Murine polyomavirus; HPyV6, Human polyomavirus 6; HPyV7, Human polyomavirus 7; SA12, Simian agent 12; SV40, Simian virus 40; TSV, TSPyV, Trichodysplasia spinulosa-associated polyomavirus; WUPyV, WU polyomavirus.

promoter, the late region promoter, and enhancer sequences. The genome structure, infectious cycle, and transforming properties of SV40 have been reviewed extensively (24, 30, 74).

Following attachment to cell receptors, SV40 virions enter the cytoplasm by the caveolae-mediated endocytic pathway. The virions are transported via the endoplasmic reticulum, where they are partially uncoated and delivered to the nuclear pores. Viral chromatin enters the nucleus, where the early promoter recruits the cellular transcription apparatus and initiates expression of the early genes. A single early primary transcript is differentially spliced to encode different early region proteins. All polyomaviruses encode two common early proteins, large T antigen and small t antigen (t antigen). Some viruses, including SV40, also express additional T antigens but their roles in infection are not yet clear. The SV40 small t antigen functions to alter the cellular environment primarily by acting on the cellular phosphatase pp2A (81), but it is dispensable for productive infection in cell culture under most conditions. However, the conservation of the small t antigen gene in all polyomaviruses suggests that it plays an essential role for the virus in the wild.

Large T antigen contributes to multiple steps in infection and thus is essential for the production of progeny virions. Early in infection, T antigen drives infected cells into S phase and induces a DNA damage response. This results in the coordinate upregulation of several hundred genes controlling the cellular DNA replication and repair apparatus as well as genes encoding enzymes responsible for producing a larger nucleotide pool. Simultaneously, T antigen binds p53 and prevents it from activating p53-dependent genes that induce cell cycle arrest and apoptosis. T antigen then functions in viral DNA replication, capturing the host's replication machinery, positioning it on the viral *ori*, and acting as the replicative DNA helicase. The viral late promoter is activated coincident with the onset of viral DNA replication, and genetic evidence indicates that T antigen is required for late transcription. Expression of the late viral messenger RNAs (mRNAs) results in the production of VP1, VP2, and VP3 and their assembly into new virions. Again, genetic and biochemical studies indicate that T antigen plays a role in virion assembly.

SV40 and murine polyomavirus grow well in culture and much of what we know about the role of T antigens in the infectious cycle is based on these two viruses. Most other polyomaviruses have not been successfully propagated in culture, because they grow poorly or establish persistent infections with very low virus yield. Therefore, it is not clear that their T antigens function like the SV40 T antigen. The fact that the key elements involved in viral DNA replication are conserved in all T antigens strongly suggests that these fundamental mechanisms are common. However, the regulation of viral or host gene expression, and the activation or inactivation of cellular proliferation and death pathways, may differ among polyomaviruses that infect different types of tissues or different hosts. The characterization of T antigens from the 25 or so relatively unstudied polyomaviruses is likely to result in the identification of novel cellular targets and mechanisms that regulate four fundamental steps in viral infection: gene expression, DNA replication, assembly, and the inactivation of antiviral defenses.

p53: protein 53 or tumor suppressor protein 53

Cellular targets: host proteins altered by viral proteins

Chaperones: proteins that assist the noncovalent folding/unfolding in molecular biology

E2F: E2F transcription factor

pRb: retinoblastoma protein

Hsc70: heat shock cognate 70-kDa protein

Structure of Large T Antigen

The structure of the full-length SV40 T antigen has been difficult to obtain, most likely because of the relatively large disordered regions in the protein. However, structures for the four well-conserved domains have been solved and structures for each of these domains bound to key cellular targets are available (**Figures 1** and **2**). These studies, coupled with an extensive genetic analysis, have given important insights into how SV40 T antigen acts on its various cellular binding partners as well as the biological consequences of these actions. In many cases, the effects of T antigen are mediated through the coordinate action of different domains. Understanding the structure and dynamics of these multiprotein complexes is one of the next challenges of polyomavirus biology.

All polyomavirus T antigens contain four conserved domains: J domain, origin-binding domain (OBD), zinc (Zn)-binding domain, and ATPase domain (**Figure 1**). The J domain is at the amino terminus of T antigen, and T antigen functions as an authentic DnaJ molecular chaperone (94). Mutations in the J domain are defective for viral DNA replication in cell culture (71). However, the J domain is dispensable for DNA replication in vitro (19). The reason for this inconsistency is not known. One possible explanation is that the chaperone activity of T antigen is needed to remove an inhibitor of replication that is present in cells but that has been eliminated from cell-free replication systems. The J domain is also required for T antigen to free E2F transcription factors from their association with retinoblastoma (Rb) proteins. In the case of the p130-E2F4 repressive complex, T antigen binds to p130 via its LXCXE motif and then energy derived from J domain–stimulated ATP hydrolysis by Hsc70 disrupts the complex. Genetic studies suggest that the J domain has additional functions (93).

The OBD is a sequence-specific DNA-binding domain that recognizes the sequence GAGGC. The viral *ori* is centered by four of these elements and this interaction is essential for the initiation of viral DNA replication. The association of the OBD with replication protein A (RPA) also is essential for replication (**Figure 1a** and **Figure 3**). The OBD has also been reported to bind several cellular transcription factors; however, the consequences of these associations have not yet been studied. The Zn-binding and ATPase domains form the enzymatic core of T antigen's DNA helicase activity. The Zn-binding domain is responsible primarily for the formation of T antigen hexamers, the active helicase form, and the ATPase domain provides the energy needed for helicase activity. Full helicase activity requires the coupled actions of the OBD, Zn-binding, and ATPase domains. Understanding how different conformations are adopted by these domains in response to their interaction with *ori*, nucleotide binding, and hydrolysis, or association with components of the cellular replication apparatus, will yield important insights into how multiprotein molecular machines act.

Computational studies suggest that all polyomavirus T antigens have a flexible, partially unstructured region between the J domain and the OBD (43 amino acids in SV40 T antigen). This region harbors the nuclear localization signal as well as several binding motifs for cellular proteins. The Rb proteins, the checkpoint kinase Bub1, and the cullin complex anchored by Cul7 associate with SV40 T antigen through motifs in this region (**Figure 1a,b**). The flexible nature of these sequences also means that the J domain can adopt multiple conformations with respect to the other T antigen domains. This predicts that the J domain–Hsc70 chaperone function can be positioned to act on different T antigen–cell protein complexes. Interestingly, the size of the flexible region varies greatly among different polyomavirus T antigens, ranging from 43 amino acids in SV40 to 151 amino acids in murine polyomavirus (MuPyV). One intriguing possibility is that polyomaviruses with a longer flexible region might have additional motifs that bind as yet unidentified cellular targets.

Figure 1

Domain structure and biological activities of SV40 large T antigen and cellular binding partners. (*a*) SV40 large T antigen consists of four well-folded domains [J domain, origin-binding domain (OBD), zinc (Zn)-binding domain, and AAA$^+$ ATPase domain; represented by *blue ovals*] and two large variable disordered regions (shown by *curves*). Boundaries of each domain are indicated by the amino acid residue numbers. The J domain binds to Hsc70 and functions as its co-chaperone. The J domain also interacts with DNA polymerase (Pol) α primase. The N-terminal disordered region immediately downstream of the J domain harbors the LXCXE motif (*diamond*). This motif is critical for the interaction between the T antigen and the pRb proteins. Additional cellular targets of this region include Bub1 and Cul7. The OBD binds to the SV40 replication origin, as well as to two host proteins, replication protein A (RPA) and Nijmegen breakage syndrome 1 (Nbs1). The Zn-binding domain mediates oligomerization of T antigen. The AAA$^+$ ATPase domain binds to and hydrolyzes ATP, which is essential for SV40 T antigen helicase to unwind its template DNA during viral DNA replication. This domain also interacts with two cellular proteins, p53 and topoisomerase I (Topo I). The C-terminal disordered region contains the host range (HR) activity and the adenovirus-helper function. The thick brown curve highlights the region critical for the HR activity. SV40 T antigen also binds the Fbw7 ubiquitin ligase through a phosphodegron motif within the HR region. The boxes next to pRb, RPA, and p53 show the crystal structures of T antigen in complex with the corresponding cellular targets. Protein Data Bank identifiers (PDB ID) are indicated. (*b*) A comparison of functional motifs within a group of polyomaviruses. SV40 large T antigen contains all motifs listed in the table. The HPDKGG motif (Hsc70 binding), the Walker A box (nucleotide binding) motif, and the Zn-binding motif are the most highly conserved motifs among the polyomaviruses shown here. The BFPyV and Myotis PyV have single amino acid variations in their LXCXE motifs (highlighted in *red*). For polyoma T antigens containing the Bub1-binding motif, the three tryptophans are invariable. The HR region and the TPPP phosphodegron motif are unique to SV40, JCV, BKV, and SA12 (simian agent 12) (not shown). +, presence of the motif; –, absence of the motif. Abbreviations: SV40, simian virus 40; JCV, JC polyomavirus; BKV, BK polyomavirus; WUPyV, WU polyomavirus; KIPyV, KI polyomavirus; MCV, Merkel cell polyomavirus; HPyV6, human polyomavirus 6; HPyV7, human polyomavirus 7; TSV, Trichodysplasia spinulosa-associated polyomavirus; HPyV9, human polyomavirus 9; LPV, lymphotropic papovavirus; MuPyV, murine polyomavirus; BPCV2, Bandicoot papillomatosis carcinomatosis virus type 2; GHPyV, goose hemorrhagic polyomavirus; BFPyV, Budgerigar fledgling disease polyomavirus; SLPyV, California sea lion polyomavirus; K virus, murine pneumotropic polyomavirus; BPyV, Bovine polyomavirus; PyV, polyomavirus.

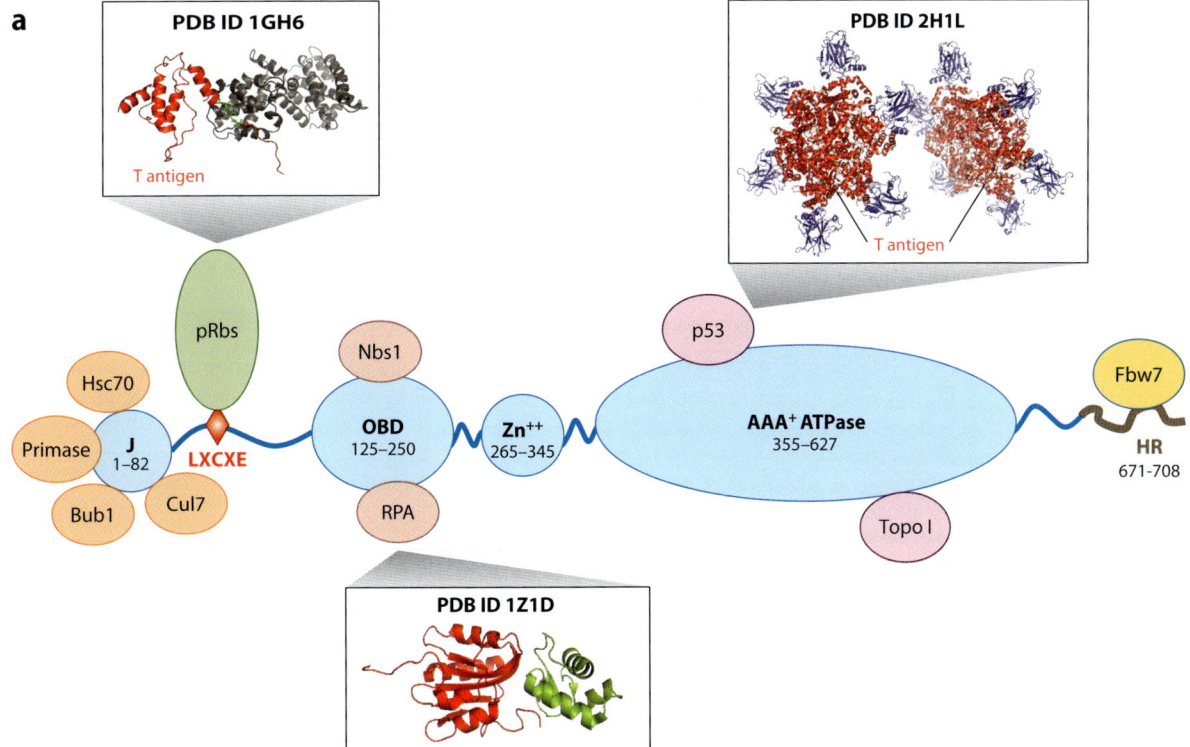

Host range region: unstructured and functionally complex region spanning the last 38 amino acids of SV40 large T antigen

Four T antigens, including SV40 T antigen, have another stretch of unstructured sequences at their carboxy terminus. In SV40, the last 38 amino acids compose the host range (HR) domain, a small but complex region. It harbors a phosphodegron centered by a threonine residue, T701, that can be phosphorylated. However, mutation of T701 to alanine has no effect on SV40 infection. This region also contains a function essential for viral replication on some cell lines but less so on others that is thought to be the same activity as the adenovirus helping function. This activity was discovered in human adenovirus–SV40 hybrids that were able to grow in monkey cells because they expressed the last 38 amino acids of T antigen. Neither the molecular basis for this activity nor its exact roles in viral infection are understood.

LARGE T ANTIGEN ORCHESTRATES VIRAL DNA REPLICATION

SV40 large T antigen is the key protein that orchestrates viral DNA replication during a productive life cycle. It recognizes and binds to the viral *ori*, and assembles into double hexamer DNA helicase to unwind template dsDNA for replication (30). SV40 large T antigen is also capable of coordinating dynamic interactions with multiple factors from the cellular DNA replication machinery to facilitate viral DNA replication.

The Helicase Activity of SV40 Large T Antigen

In addition to its interactions with viral *ori* and multiple cellular replication factors, SV40 large T antigen possesses an intrinsic ATPase activity and is able to assemble into double hexamers, function as a DNA helicase to initiate the unwinding of the dsDNA template at the origin, and continue to unwind during the elongation step.

The structure of the SV40 replication origin is illustrated in **Figure 2a** (12). The OBD of SV40 large T antigen specifically recognizes and binds to the consensus pentanucleotide sequences. Crystal structures of T antigen's OBD in complex with short DNA fragments containing the origin pentanucleotides (65) show residues from A1 loop (S147-T155) and B2

Figure 2

Simian virus 40 (SV40) large T antigen is a DNA helicase. (*a*) Two T antigen origin-binding domain (OBD)/zinc (Zn) binding/ATPase domain fragments bound to the replication origin (*ori*). The SV40 *ori* is composed of a 64-nucleotide region and can be divided into three parts: four pairs of GAGGC pentanucleotides (*red*) in the center and the AT-rich element (AT) and early palindrome site (EP) flanking each side. (*b*) Crystal structure of T antigen OBD bound to the origin DNA. Two pentanucleotides are highlighted in orange. Residues critical for DNA binding are highlighted (A1 loop in red sticks, and B2 element in blue ribbon). (*c*) Instead of a flat ring, the T antigen OBD hexamer forms an open spiral, creating an opening in the structure. Top (*left*) and side (*right*) views are shown. The spiral structure and the opening are more obvious from the side view. (*d*) The three-dimensional cryo-electron microscopy (EM) images of T antigen helicase in complex with DNA show a three-module unit, with the N-terminal J domains and OBDs from both hexamers joined closely in the center, and the helicase domains (C-ter) from the two hexamers flanking each side of the central N-terminal unit (reprinted from Reference 21 with permission). Two major conformational states, the parallel and displaced conformations, are indicated. (*e*) EM image of double-stranded DNA (dsDNA) unwinding by T antigen double hexamer helicase. (Reprinted from Reference 103 with permission. Labels for corresponding dsDNA, ssDNA, and Tag are enlarged for better visibility.) In the presence of ATP, double hexamers of SV40 large T assemble onto the SV40 origin and bidirectionally unwind dsDNA. EM studies revealed unwinding intermediates, in which two unwound single-stranded DNA (ssDNA) form the rabbit ear–like structure. (*f*) Side view of T antigen helicase with the zinc-binding domains on the top and the AAA$^+$ ATPase domains on the bottom. Spheres represent zinc ions. (*g*) Crystal structures of T antigen helicase (the zinc-binding domain and the AAA$^+$ ATPase domain) under various nucleotide (Nt)-binding states (Nt-free, left; ADP bound, center; ATP bound, right). The structures confirmed that T antigen helicase is a hexameric ring with a central channel. The hexameric T antigen helicase is either Nt-free or completely occupied with a single type of Nt (*spheres*). The binding and hydrolysis of ATP lead to conformational changes between neighboring large T monomers, and to an iris motion-like expansion/constriction of the central channel. In panels *f* and *g*, the zinc-binding domains are shown in yellow, and ATPase domains in dark gray. The red sticks highlight the tip (K512 and H513) of a β-hairpin that directly contacts DNA along the inner face of the channel. When available, Protein Data Bank identifiers (PDB ID) for the structures shown are indicated.

element (H203-N210) making extensive contacts with both the bases and the backbone of the DNA (**Figure 2b**). The GAGGC consensus pentanucleotide sequence is not present in avian polyomaviruses. It is possible that they use a different sequence for origin recognition; in which case their OBDs may adopt different structures to interact with the corresponding origins.

After binding to the origin, SV40 T antigen monomers assemble into a double hexamer in a head-to-head direction. The complete assembly of this dodecamer requires binding to origin DNA and the presence of ATP or ADP (12). Protein-protein interactions involving the OBD and Zn-binding domain among T antigen monomers also facilitate the assembly (80). The fully assembled double hexamer is capable of melting the origin without the power of its helicase activity. Current structural studies from several groups have shown that SV40 large T antigen is a highly flexible molecule capable of adopting various conformations and that the conformational changes are coupled to DNA binding and ATP/ADP binding (21, 33, 99). These results not only provided important implications regarding how conformational dynamics may contribute to melting and

Large T antigen helicase: domain within large T antigen with the capacity to distort and separate double stranded nucleic acid using energy derived from ATP hydrolysis

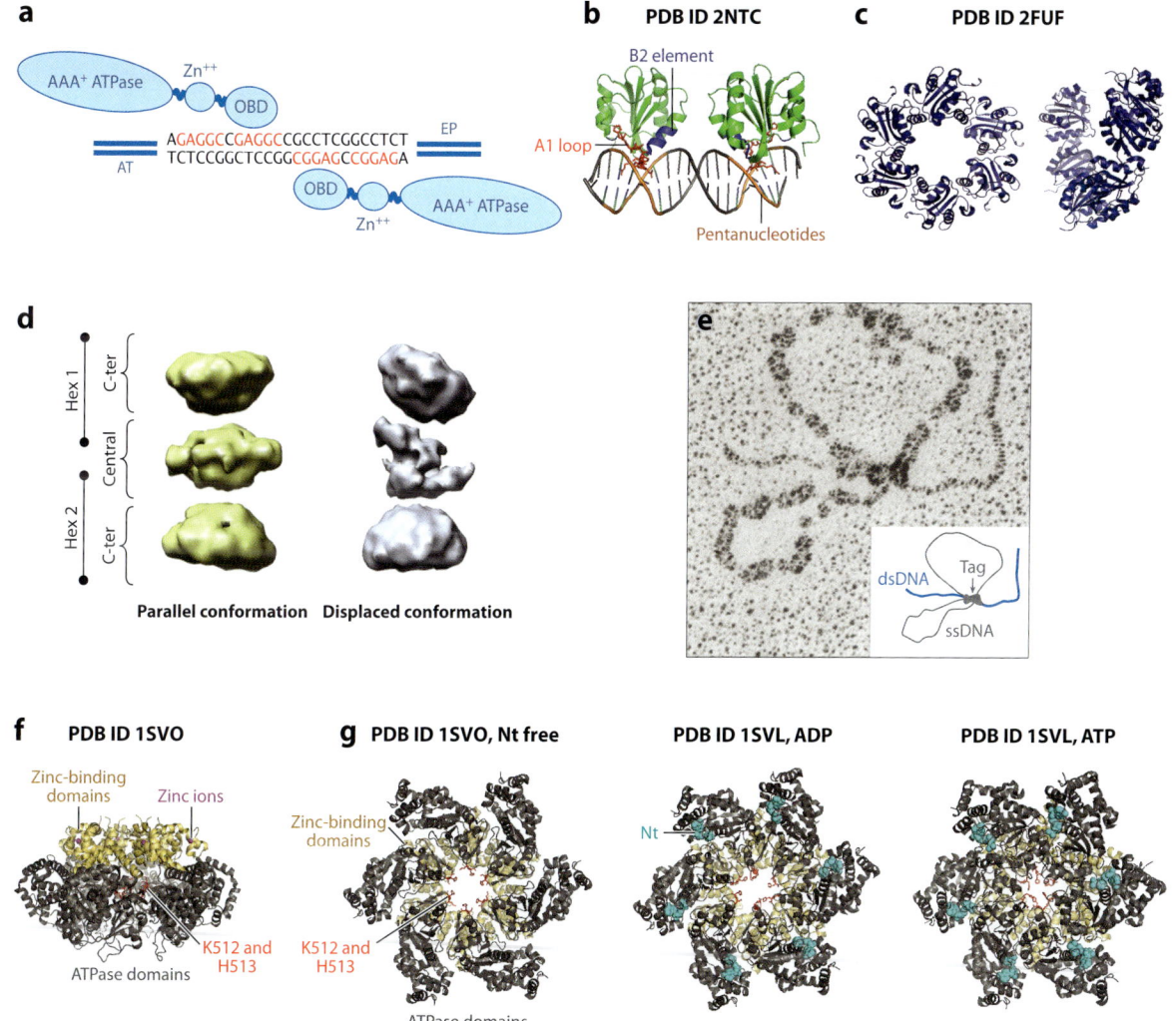

distorting the replication origin, but also shed light on mechanisms of continuous bidirectional unwinding by large T antigen helicase during viral DNA replication.

Three-dimensional cryo-electron microscopy (cryo-EM) and X-ray crystallography are among the major approaches used to solve molecular structures and thus to determine the binding of large T antigen to the *ori*, the assembly of helicase, and its local action on the origin. Recent cryo-EM data show the structure of full-length SV40 T antigen double hexamer bound to origin DNA in the presence of ADP (21) (**Figure 2d**). The fact that several atomic structures, including the assembly of OBDs as an open spiral (64) (**Figure 2c**), the SV40 J domain (37) (**Figure 1a**), and the ADP-bound hexameric helicase (33) (**Figure 2f**), can fit into the above cryo-EM structure is indeed encouraging.

Two major conformational states of the T antigen–DNA complex, the parallel and the displaced conformations, are observed (**Figure 2d**). Complexes of T antigen mutants with deletions of J domain and HR domain (a large T antigen of 108–627 aa) adopt predominantly the parallel conformation and display reduced flexibility and reduced efficiency in supporting DNA replication. The high level of flexibility might be important for helicase activity, and the displaced conformation able to provide stronger bending of DNA might be critical for efficient distortion and melting of the origin.

As a multidomain, multifunctional protein, SV40 large T antigen also belongs to a large enzyme family, namely the AAA+ (ATPase associated with diverse cellular activities) protein family. These proteins function in general as molecular motors by binding and hydrolyzing ATP and by coupling the energy from ATP hydrolysis to mechanical force through conformational changes. In the case of the SV40 DNA helicase, this force is important for melting and unwinding viral DNA to support DNA replication. SV40 large T antigen helicase shares characteristics of an AAA+ ATPase, including the conserved Walker A box (**Figure 1**) and the overall hexameric ring structure. However, the T antigen helicase is unique in its ability to initiate DNA melting and unwinding starting from the replication origin, whereas the other helicases only act on DNA premelted by corresponding initiators. Previous studies have demonstrated that in the presence of ATP, T antigen assembles into double hexamers on SV40 origin and bidirectionally unwinds dsDNA. Images of unwinding intermediates, in which the two unwound ssDNA form a rabbit ear–like structure (103) (**Figure 2e**), have been revealed by EM studies. DNA footprinting further proved that T antigen helicase contacts both the dsDNA and two unwound ssDNA (103).

The hexameric ring structure of the helicase actually appears as a two-layer ring (**Figure 2f**), with the Zn-binding domain forming the small layer on the top and the rest of the helicase domain as the large layer at the bottom (22). The hexameric ring has a central channel with a positively charged inner surface for binding DNA (**Figure 2g**). Side channels, or holes, may provide an exit path for unwound ssDNA along the side of the hexamer. The above structural features are consistent with the more recent cryo-EM findings.

Sequential ATP binding and hydrolysis, as well as the consequent conformational changes, are essential to the actions of SV40 large T antigen helicase. Crystal structures of the helicase under various nucleotide-binding states have been solved (**Figure 2g**), providing critical insights into understanding its mechanism and dynamics (33). Different from typical AAA+ ATPase, the nucleotide binding for T antigen helicase resembles an all-or-none mode. The hexameric helicase is either nucleotide free or occupied with a single type of nucleotide, regardless of the concentration of the nucleotide or mixture of ADP and ATP, a fact compatible with a concerted ATP hydrolysis mechanism. Also important, the T antigen helicase contains an additional unusual lysine finger besides the canonical arginine finger for ATP hydrolysis, which would allow the helicase to hydrolyze ATP in both association and dissociation modes.

Furthermore, comparing the structures of the hexameric helicase in different nucleotide-binding states and determining their differences suggest that the binding and hydrolysis of ATP lead to changes in relative positioning of residues between neighboring T antigen monomers (*trans*-effect) rather than within the same monomer (*cis*-effect). The *trans*-effect in the context of the hexamer results in the twisting/untwisting between the two layers of the Zn-binding domains and the AAA+ ATPase domains, as well as the expansion and constriction of the hexameric channel (**Figure 2g**), implying that these two conformational changes are coupled to melting of the origin and continuous unwinding of helicase activity.

In Vitro Replication of SV40 Viral DNA

Replication of the SV40 DNA has been reconstituted successfully in vitro, providing important insights into our understanding of the mechanisms of eukaryotic DNA replication. Multiple cellular proteins are needed to complete replication of SV40 DNA (101). The large T antigen interacts with RPA, DNA polymerase α primase, and Topo I during replication (**Figure 3**). The domains and motifs important for mediating these interactions have been mapped through mutagenesis and biochemical assays (4, 46, 47, 54). Recent NMR studies have provided more information on hRPA C-terminal-mediated assembly of the SV40 replisome (4) and on the docking site for α primase on the large T antigen helicase domain (46). Protein-protein interactions between T antigen and the cellular replication factors are critical for orchestrating the multiple steps involved in synthesizing progeny viral DNA, although the underlying mechanistic details have not been completely elucidated.

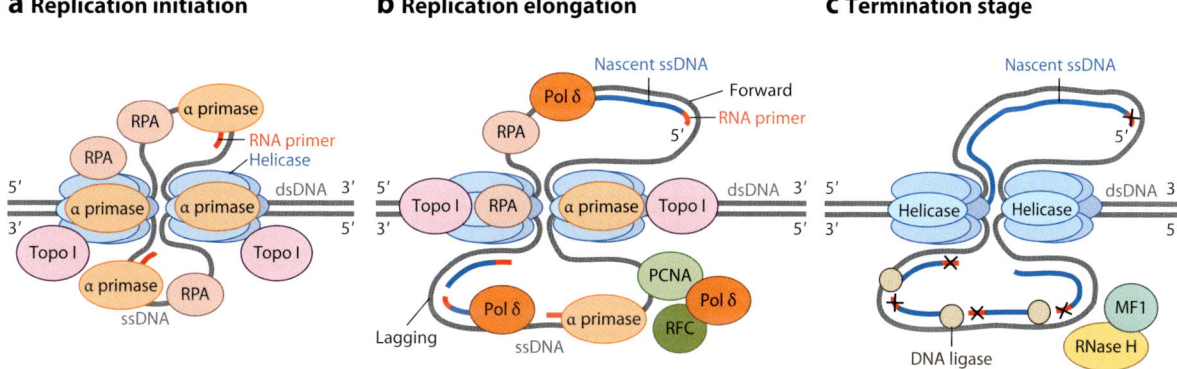

Figure 3

Simian virus 40 (SV40) large T antigen is the master molecule directing viral DNA replication. (*a*) Simplified schematic of the initiation process of SV40 viral DNA replication. T antigen double hexamer helicase (*two sets of six ovals*) initiates the distortion and melting of SV40 viral origin and subsequently unwinds the double-stranded DNA (dsDNA) template bidirectionally (*represented by two parallel gray lines*). The unwound single-stranded DNA (ssDNA) is shown by disordered gray curves. In addition to T antigen, nine cellular factors are required to reconstitute SV40 DNA replication in vitro. The cellular proteins that interact with T antigen at this stage are RPA, DNA polymerase α primase, and topoisomerase I (Topo I). Replication protein A (RPA) is a ssDNA-binding protein necessary for unwinding the double-stranded template DNA, whose C-terminal domain is required for interaction with T antigen. DNA polymerase α primase synthesizes RNA primers (*short red curves*) about 11 nucleotides in size, which serve as a starting point of DNA synthesis. Topo I and II function to resolve topological problems caused by unwinding and to establish and maintain the double helical configuration of daughter dsDNA. (*b*) Replication elongation of SV40 DNA. SV40 T antigen helicases continue to unwind template DNA and recruit RPA, α primase, and Topo I through specific interactions. More cellular replicative factors are involved in the elongation process. Replication factor C (RFC) and proliferating cell nuclear antigen (PCNA) facilitate the switch from α primase to DNA polymerase (Pol) δ, which then extends the nascent ssDNA (*blue curves*) from the primer. For synthesis of the lagging strand, the α primase has to produce primers repeatedly. (*c*) During the termination stage of viral DNA replication, RNase H and maturation factor 1 (MF1), a 5' to 3' nuclease, are required to remove the primer. Finally, DNA ligase covalently closes the gaps of the newly synthesized strands and completes the replication.

SV40 DNA Replication Under the Cellular DNA Damage Response

SV40 large T antigen is capable of driving resting cells into S phase and inducing the DNA damage response (DDR) to facilitate the replication of viral DNA. The SV40 DNA replisomes are bound to nuclear scaffolds and localize adjacent to promyelocytic leukemia protein bodies (also referred to as nuclear dots or nuclear domains) (48), which are distinct subnuclear foci involved in transcription and DDR. A similar observation has been described for BK virus (49, 51). DNA damage normally leads to the activation of cell cycle checkpoints controlled by two master kinases, ATM (ataxia telangiectasia mutated) and ATR (ATM Rad3 related), and to further inhibition of DNA replication. UV radiation has been shown to inhibit both cellular and SV40 DNA replication at both the initiation and elongation steps (66). Clearly, T-antigen-induced activation of DDR and the downstream events are different from UV radiation. There has been accumulating evidence that DDR signaling is required for SV40 DNA replication, as well as for JCV DNA replication. Functional downregulation of DDR signaling through inhibitors against ATM and ATR (108), or RNA-interference-mediated knockdown of ATM (92), Rad51 (8), or promyelocytic leukemia protein (9), results in impaired viral DNA replication.

DNA damage response (DDR): cellular pathways activated in response to DNA injuries that slow down cell cycle progression and allow the cell to repair the damage prior to replication

ATM: ataxia telangiectasia mutated

ATR: ATM Rad3 related

Nbs1: nibrin; DNA repair protein associated with Nijmegen breakage syndrome 1

Transformation: heritable change in properties that expands the proliferative and survival capacity of cells

SV40 large T antigen interacts with Bub1 and Nbs1, components of host DDR signaling, through the linker downstream of J domain and the OBD, respectively (20, 104). These interactions may provide entry points for T antigen to further deregulate cellular DDR for the benefit of viral DNA replication. T antigen association with Cul7 is also implicated in the ATM-dependent degradation of the Mre11-Rad50-Nbs1 complex (108). ATM phosphorylates large T antigen on S120 (92), and phosphorylation of S120 inhibits DNA unwinding activity of T antigen (15). However, the S120A mutant is completely deficient in viral DNA replication (89). Further investigations are needed to gain more insights into the complicated and dynamic interplay between T antigen and DDR.

SV40 large T antigen is subjected to various posttranslational modifications, especially phosphorylation, and SV40 DNA replication can be regulated by differential phosphorylation states of T antigen (89). The signaling pathways mediating these phosphorylation events and the mechanisms of how phosphorylation may affect DNA replication (through conformational change of T antigen, DNA binding, protein-protein interaction, assembly and/or activity of the helicase) remain to be fully characterized.

Large T antigens from all polyomaviruses share similar domain structures, including the OBD, Zn-binding domain, and the AAA+ ATPase domain (**Figure 1**). It seems reasonable to assume that all polyomavirus T antigens operate with similar mechanisms for viral DNA replication. However, very few studies have been conducted on viral DNA replication of the other polyomaviruses. Purified JCV large T antigen is able to support replication of a plasmid containing JCV origin in a cell-free system (68). Interestingly, SV40 large T antigen but not MuPyV large T antigen can substitute JCV T antigen in supporting initiation of DNA replication (68). This may result from the failure of MuPyV T antigen to recognize the JCV *ori*, since the same MuPyV large T antigen was able to actively replicate plasmid with MuPyV *ori*. Infections of murine cells with SV40 or BKV do not result in subsequent viral DNA replication, even though the large T antigens are expressed. Continuous expression and accumulation of SV40 T antigen in nonpermissive cells eventually lead to cellular transformation and produce tumors in lab animals such as mice and hamsters. The underlying mechanism for this cell type restriction is not clear.

Further investigations are needed to gain a complete understanding of polyomavirus DNA replication. Accumulating data suggesting that SV40 hijacks DDR enzymes rather than the normal cellular DNA replication machinery to replicate viral DNA are especially provocative. Results from these studies will likely provide mechanistic insights into eukaryotic DNA replication, as has been achieved for the SV40 DNA replication model.

CELL TRANSFORMATION AND TUMORIGENESIS MEDIATED BY T ANTIGEN

To perpetuate the integrity of organs and tissues, all multicellular systems need to balance proliferation and cell death. Alterations due to heredity, environmental causes, or extraneous agents such as viruses can disrupt these processes and lead to cancer, during which cells undergo transformation, a thorough or dramatic change in form or appearance. Viral products capable of altering the growth control of cells in which they are expressed, thus inducing tumorigenesis, have been instrumental to the study of cancer.

Tumor suppressor: a protein that negatively regulates cell growth and prevents a cell from becoming tumorigenic

Three main approaches have been used to study tumorigenesis in model systems.

- Experimental animals, mainly rodents but also higher mammals, can be infected with tumor viruses, or with viral vectors expressing viral oncoproteins. Some of these animals then develop tumors, depending on the site and method of delivery, the dosage, and the virus or vector being tested.
- Cell culture assays measure changes in proliferation and survival *in vitro*. Numerous cell types can be adapted to grow in culture, and under certain circumstances, their properties can be immortalized and/or transformed by expressing oncoproteins. Transformation is usually assessed through a series of assays such as focus formation, growth in soft agar, survival, and growth in low serum.
- Transgenic or genetically modified mice are powerful tools for analyzing the effects of oncoprotein expression in an organismal setting. Oncoproteins can be expressed ectopically in specific tissues or cell types to assess their effects on tissue homeostasis. Gene knockout technology allows for the tissue-specific removal of cellular genes that may play a role in tumorigenesis.

The polyomavirus most extensively studied by all of the above methods is SV40, with particular emphasis on its main transforming product, large T antigen. Although other proteins encoded within the SV40 genome can contribute to the transformed phenotype, T antigen is necessary and sometimes sufficient to induce cell transformation and tumorigenesis in multiple cell types from several hosts and sources (2, 17, 34, 84, 90, 100). The current understanding of the process indicates that large T antigen binds two essential tumor suppressors, the pRb proteins and p53, and by doing so drives uncontrolled cell proliferation. In fact, many tumor viruses encode proteins that disrupt the retinoblastoma and p53 tumor suppressor pathways, suggesting that these pathways are key switches in cellular growth control.

The pRb Pathway

The retinoblastoma proteins pRb, p107, and p130 negatively regulate cell proliferation. The currently accepted model is that the Rb family controls cell proliferation via binding and inhibition of the E2F family of transcription factors (**Figure 4**). The E2F family of transcription factors has been divided into two subfamilies: the repressing E2F factors, which include E2F4 and E2F5, and the activating E2F factors, which include E2F1, E2F2, and E2F3a. The promoters of repressed E2F-dependent genes are usually occupied by p130 bound to E2F4 or E2F5. The p130-E2F4/5 association anchors a large repressive complex, the DREAM complex. Growth-positive stimuli received by the cell activate signaling pathways that ultimately phosphorylate the Rb proteins, leading to their disassociation from E2Fs and dissolution of the repressive complex. In contrast, growth-negative signals maintain Rb proteins in a hypophosphorylated state, thereby stabilizing the repressive complex. Under growth-positive conditions, when the Rb proteins are hyperphosphorylated, the activating E2Fs are derepressed and bind to the promoters of E2F-dependent

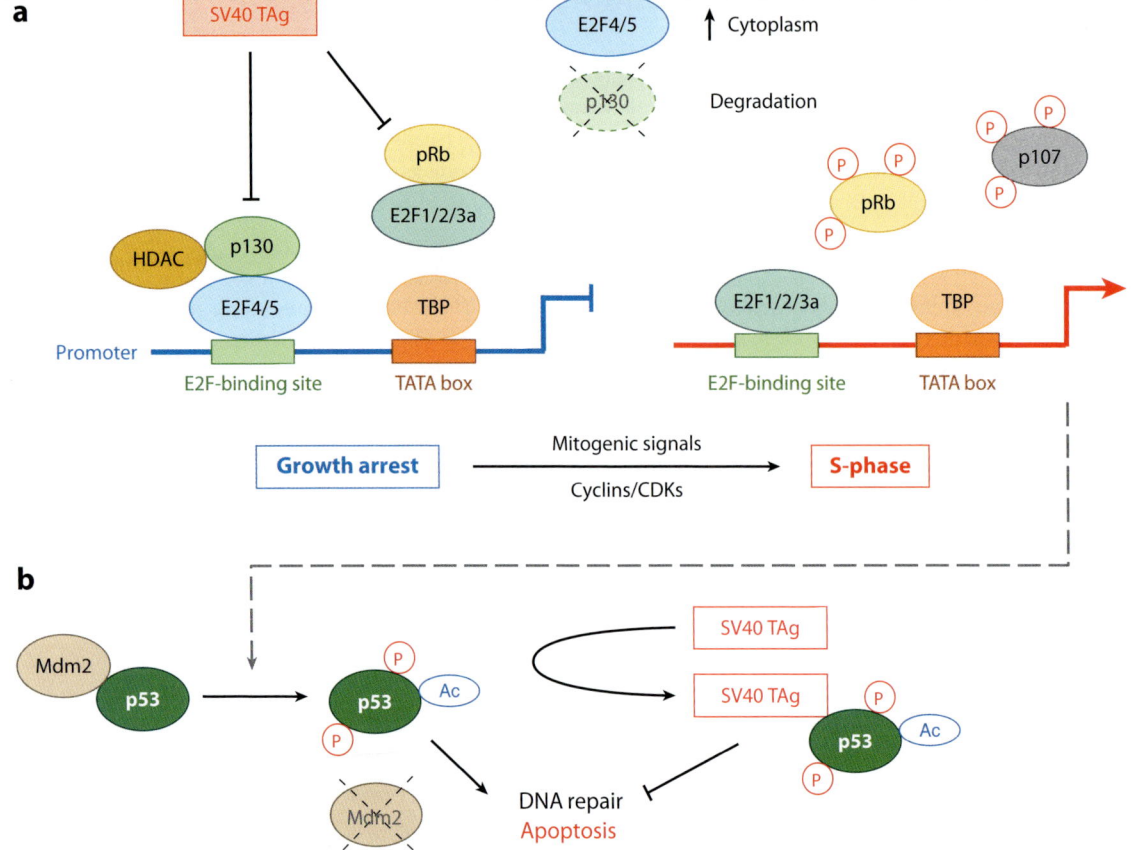

Figure 4

Manipulation of the host cell machinery by simian virus 40 (SV40) T antigen. (*a*) The pRb connection. Differentiated cells are prevented from re-entering the cycle by the retinoblastoma (Rb) family of tumor suppressors, which consists of pRb, p107, and p130. Hypophosphorylated pRb prevents replication by binding and inhibiting E2F1/2/3a, transcriptional activators that control genes required for the transition from G1 into S phase. Such genes include those controlling nucleotide synthesis, DNA repair and DNA replication, cell cycle checkpoints, and apoptosis. At the same time, p130 in complex with the E2F4-5 transcriptional repressors binds to the promoter of E2F target genes and actively blocks transcription. This complex also recruits histone deacetylase (HDAC) to chromatin, contributing to transcriptional blockage. Cells thus remain in G1 phase. Abnormal conditions and mitogenic stimuli stimulate cyclins/cyclin-dependent kinases in vitro to phosphorylate pRb and release the activators E2F1/2/3a. Similarly, phosphorylation of p130 disrupts the p130/E2F4 complex, and activator E2Fs take its place in the promoter, triggering E2F-dependent transcription and allowing S phase to proceed. Although similar to the role of pRb, the role of p107 during the cell cycle is less understood. SV40 T antigen mimics mitogenic signals by acting as a chaperone machine that first binds Rb proteins via its LXCXE motif (**Figure 1***a*) and then recruits cellular Hsc70 via its J domain. Triggering the ATPase function of Hsc70 disrupts the p130/E2F4 complex; p130 becomes ubiquitinated and is targeted for degradation (59). Thus, the brakes imposed on proliferation are released and S-phase transition ensues. (*b*) The p53 connection. Multiple cellular stresses, including abnormal upregulation of E2F targets and subsequent cell proliferation, raise the levels of p53, which are kept very low in normal cells by the E3 ubiquitin ligase Mdm2. Phosphorylation and subsequent acetylation allow p53 to become an active transcriptional regulator (reviewed in Reference 6) of genes driving apoptosis and/or senescence (13), thus restricting cellular expansion under unfavorable conditions. TAg avoids premature cell death by binding and blocking p53, and T antigen mutants unable to alter p53 are defective for transformation (72). Unlike other viral oncogenes that trigger p53 degradation, the association between SV40 T antigen and the DNA-binding domain of p53 stabilizes p53 while blocking p53-dependent gene transcription (91). In fact, this interaction is so strong that it allowed for the initial identification of p53 in complex with SV40 T antigen (58, 60). The dashed line with an arrow between panels *a* and *b* indicates that ectopic proliferation triggers the activation of p53 and subsequent apoptosis, which is prevented by SV40 T antigen interaction with p53. Abbreviations: TBP, TATA-binding protein; Ac, acetylation; P, phosphorylation; CDK, cyclin-dependent kinase; TAg, T antigen.

genes, stimulating transcription. The expression of E2F-dependent genes is thought to be necessary for S-phase entry and progression.

Large T antigen binds and inactivates all three Rb proteins, releasing the brake imposed on E2Fs, subsequently upregulating E2F targets and inducing cell re-entry into S phase. As a consequence, quiescent cells undergo ectopic proliferation and, in some cases, transformation. To elicit this response and transform certain cell types *in vitro* or *in vivo*, large T antigen requires both an intact LXCXE motif and a functional J domain (73, 84) (**Table 1**) (**Figure 4**). For instance, in both cell culture and transgenic systems, expression of N-terminal truncations of large T antigen containing the LXCXE motif and the J domain leads to increased E2F-dependent gene transcription and proliferation (14, 78, 86, 97). Large T antigen mutants lacking either a functional LXCXE motif or the J domain fail to increase E2F-dependent transcription and do not stimulate proliferation (14, 78, 79, 84). These results indicate that induction of proliferation by large T antigen is dependent on its capacity to block the ability of the Rb proteins to antagonize E2F-dependent gene expression.

This model predicts that the genetic ablation of the Rb proteins should mimic the effect of large T antigens on E2F-dependent transcription and cell proliferation. In fact, the knockout of all three Rb proteins in primary mouse embryonic fibroblasts (MEFs) results in a continuous cell cycle (23). Similar results are seen in knockout mouse models. The intestinal enterocytes are terminally differentiated, growth-arrested cells. The expression of SV40 T antigen in enterocytes results in the elevated transcription of E2F-dependent genes and re-entry into the cell cycle. The inactivation of either pRb/p130 or pRb/p107 in enterocytes results in ectopic proliferation and hyperplasia and thus partly mimics large T antigen expression (39). Nevertheless, expression of T antigen in enterocytes produces a much more robust phenotype, suggesting either that removal of all three pRb proteins is required to completely abolish control of cell proliferation or that T antigen targets additional host components required in normal homeostasis.

The large T antigens from other polyomaviruses, including JCV, BKV, SA12 (simian agent 12), LPV (lymphotropic papovavirus), MuPyV, and MCV, also form complexes with pRb (28, 45), although in some cases these interactions are detected only when large amounts of total protein are used (40). Presumably, as with SV40, this interaction results in uncontrolled S-phase transition and proliferation. In fact, BKV large T antigen, but not a mutant unable to bind the pRb proteins, stimulates the transcription of DNMT1, an E2F target gene (63). Similarly, the expression of a truncated form of MCV large T antigen and, in particular, its interference with the Rb pathway (45) seem required to induce Merkel cell carcinoma in human skin. However, much more work is needed to determine whether large T antigens from other polyomaviruses act on the Rb pathway in the same manner as SV40 large T antigen, and to assess the importance of the Rb-E2F axis in the transforming capabilities of the polyoma T antigens.

Although the current model proposes that SV40 large T antigen induces cell proliferation using the same components used by host cells during normal proliferation, recent data show that there are two types of cell cycles, distinguished by their requirement of activating E2Fs (E2F1–3) and by their response to ablation of the *retinoblastoma* gene. The T-antigen-mediated re-entry of enterocytes into the cell cycle requires activating E2F (85). However, the normal cycling of stem cells and of pluripotent progenitor cells is independent of the activating E2Fs (18, 102). In these cases, the activating E2Fs function as repressors cooperating with the Rb proteins to inhibit E2F-dependent transcription in preparation for growth arrest and terminal differentiation. One hypothesis that is consistent with all these results is that stem cell and progenitor cell proliferation is independent of the activating E2Fs. However, once cells have undergone growth arrest and differentiation, their re-entry into the cell cycle is dependent on the activating E2Fs. Thus, SV40 and perhaps the other polyomaviruses may elicit proliferation through a tissue repair or wound-healing mechanism rather than by altering the normal cell cycle.

The p53 Pathway

Cells respond to the abnormal upregulation of E2F targets and cell proliferation with defensive mechanisms, such as upregulation of ARF (alternative reading frame of CDKN2A locus) and triggering of the p53 pathway. Under normal conditions p53 levels are kept very low, but in response to stress, DNA damage or oncogenes, the half-life of p53 rises dramatically, leading to the rapid accumulation of the protein (**Figure 4**). Large T antigen binds and blocks p53, therefore avoiding growth arrest and premature cell death. This binding takes place through residues on the surface of the ATPase domain of T antigen, and mutants unable to bind p53 are defective for transformation (1, 55). T antigen blankets the DNA-binding domain of p53 and thereby blocks its ability to associate with promoters and activate the expression of p53-dependent genes. Like the Rb proteins, p53 is targeted by many different tumor viruses. T antigen can also block some p53-dependent functions without directly binding p53, suggesting that SV40 deals with this tumor suppressor in multiple ways.

Whereas most viral oncoproteins trigger p53 degradation, T antigen stabilizes p53 and SV40-transformed cells harbor very high p53 levels (58, 75, 91). In these cases T antigen is clearly blocking p53-dependent transcription, but does the T antigen-p53 complex possess some as yet undetermined activity? Several reports strongly suggest such a gain of function for p53 in SV40-transformed cells (43). Determining whether this is the case and, if it is, establishing the biological role of the T antigen-p53 complex in infection and tumorigenesis are imperatives.

Large T antigens from other polyomaviruses (e.g., BKV, LPV) can also form complexes with p53 and increase its half-life (10, 40, 52, 95), although some controversy still remains (52). However, how those complexes contribute to transformation has not been studied in detail. Thus, at this point it is unclear whether all polyomavirus T antigens possess the same transforming functions, nor is it certain that they act on the Rb and p53 pathways in the same way.

Additional Targets: How Important Are They?

Although the pRb and p53 pathways play critical roles in SV40-mediated transformation, other T antigen activities may also contribute. For example, cells coexpressing a T antigen truncation mutant that retains the ability to inactivate Rb proteins and a dominant-negative p53 do not exhibit a transformed phenotype (82). This indicates either that T antigen must act on targets in addition to the Rb proteins and p53 to induce transformation, or that an active T antigen-p53 complex is necessary. In another instance, both full-length T antigen and a truncation mutant capable of inactivating Rb proteins induce hyperplasia when expressed in murine intestinal enterocytes (56, 57, 78). However, full-length T antigen induces a much stronger transformed phenotype, including marked dysplasia. Because enterocytes do not express p53, some other T antigen target must be responsible for the stronger phenotype induced by the full-length protein (62).

Several additional binding partners of T antigen have been implicated in transformation: Bub1, a kinase involved in the mitotic spindle checkpoint; Cul7, part of an E3 ubiquitin-protein ligase complex; Nbs1, a component of the cellular response to DNA double-strand breaks; and the IGF/insulin pathway (3, 20, 25, 35, 42, 53, 98, 104). As T antigen interacts with many cellular partners, it is important to consider whether distinct subsets of T antigen molecules bind to different proteins, perhaps depending on the cell type, or whether each molecule binds and bridges several different molecules at the same time.

One particular host cell target deserves special attention, the histone acetyltransferase and molecular adaptor CBP/p300. SV40 large T antigen is capable of binding CBP/p300 (29) and this binding is required for T antigen to fully transform cells in culture (1). The interaction between T antigen and CBP/p300 requires p53 and results in acetylation of T antigen residue K697 (11). It is not clear how this modification affects T antigen activity. The adenovirus E1A protein also targets CBP/p300 and this interaction is essential for E1A-mediated transformation. E1A binds CBP/p300 directly (5, 106). T antigen complements adenovirus E1A mutants defective for CBP/p300 interaction (105). The interaction between E1A and CBP/p300 results in decreased levels of cellular histone H3 lysine 18 acetylation (44) and in a global relocalization of pRbs and CBP/p300 on promoters. This leads to a change in the histone acetylation pattern that stimulates cell cycle entry and progression while inhibiting antiviral responses and cellular differentiation (31). T antigen increases the steady-state levels of CBP/p300 in multiple systems, with a concomitant increase in the acetylation levels of specific histone residues, thus perhaps altering gene expression in a not yet fully understood manner (M.T. Sáenz Robles, C.S. Shivalila & J.M. Pipas, unpublished data). It will be important to determine whether other polyomavirus large T antigens are similarly capable of increasing CBP/p300 levels and altering the histone acetylation pattern. Furthermore, it is still not clear how the CBP/p300 interaction contributes to SV40 transformation.

CBP/p300:
CREB-binding protein/E1A-binding protein p300

Transformation by Other Polyomaviruses

Many polyomaviruses can transform cells, and at least one, MCV, has been linked to a human cancer (50). In most cases, including BKV, JCV, and LPV, transformation capabilities have been linked to the large T antigen or at least the early region (16, 36, 41, 52, 67, 70, 100). However, the mechanisms of transformation have not been explored in detail. Although the large T antigens encoded by different polyomaviruses potentially have the capacity to form complexes with several host target proteins, it is not clear whether they transform through the same targets and mechanisms used by SV40.

Contribution of Other Viral Proteins to Transformation

Many polyomaviruses encode proteins in addition to large T antigen that contribute to transformation. These include the small t and middle T antigens. A detailed review of these proteins is beyond the scope of this article. The middle T antigen encoded by MuPyV has been the subject of several excellent reviews (17, 32, 88). The only small t antigen studied in detail to date is the SV40 protein. SV40 small t antigen cooperates with large T antigen to induce transformation in several systems, including primary MEFs and human telomerase reverse transcriptase (hTERT)-immortalized human fibroblasts or epithelial cells (38, 61). In these cases, the contribution of small t antigen to transformation requires its interaction with the cellular phosphatase pp2A. Similarly, a variant of the BKV early region unable to express small t antigen shows reduced capacity for transformation (69). The SV40 small t antigen and its role in transformation have been reviewed (81).

REGULATION OF CELLULAR GENE EXPRESSION BY LARGE T ANTIGEN

Large T antigen is thought to regulate cellular gene expression by acting on transcription factors themselves or on the proteins that regulate them. Its role in antagonizing the pRb and p53 proteins, thus allowing the transcription of E2F-dependent genes and blocking p53-dependent gene expression, is well established. For instance, practically all T-antigen-regulated E2F-dependent gene expression is dependent upon binding and inactivation of the pRb proteins (14, 78). Because cells contain very low levels of p53 under normal conditions, p53-dependent genes are not expressed. SV40-transformed cells contain high steady-state levels of p53, but p53-dependent genes are not expressed because T antigen blocks p53 from binding to promoters and from stimulating transcription. In theory, the activation of p53 gene expression by these T antigen mutants should result in growth arrest and cell death. The fact that these cells proliferate rapidly with little cell death suggests that T antigens possess a mechanism(s) for blocking the effects of p53-responsive gene products.

The action of T antigens on the Rb pathway also has cell-type-specific effects on cellular gene expression. For example, the P450 detoxification pathway in murine enterocytes is downregulated by T antigen. Repression of the detoxification pathway requires T antigen binding to the pRb tumor suppressors (87). Furthermore, the genetic ablation of pRb in pluripotent progenitor cells of the intestine results in loss of expression of genes in the P450 pathway (83). This suggests that pRb positively regulates the P450 pathway and that T antigen blocks this effect.

Apart from its effects on pRbs and p53, SV40 T antigen also binds TATA-binding protein (TBP), several TBP-associated factors, and other transcription factors, interactions that could influence cellular transcription. However, the effects of these interactions on viral infection, transformation, or cellular transcription have not been determined. In addition to acting on polymerase II promoters, T antigen activates transcription of the rRNA genes, apparently through T-antigen-mediated changes in the phosphorylation state of key polymerase II transcription factors (107).

Several studies indicate that T antigen regulates many genes that are not part of the Rb-E2F or p53 axis, although the mechanisms involved have not been explored. For instance, many interferon-stimulated genes (ISGs) are highly transcribed in MEFs stably transformed by SV40 (14, 77). SV40-transformed cells contain activated STAT-1 (signal transducers and activators of transcription 1) but little or no interferon expression, indicating that T antigen might regulate the transcription of ISGs by activating the interferon pathway upstream of STAT-1. This effect is independent of the J domain but requires both the LXCXE motif and some unknown activity residing in the carboxy terminus of T antigen. Because T antigen action on pRb proteins requires the J domain in concert with the LXCXE motif, it is likely that an as yet unidentified cellular target of T antigen is responsible for elevating ISG expression. Whether other polyomavirus T antigens induce a similar inflammatory response remains to be determined.

CONCLUSIONS

We often think of viruses as agents that cause acute disease and then are eliminated from our bodies by our immune defenses. However, direct or indirect evidence has indicated the abundant presence of multiple polyomaviruses, including BKV, JCV, and LPV, in human populations (26, 27, 76, 96). Polyomaviruses are present in nearly all of us, never leave us, and can cause serious disease if we are weakened. Thus, it is pressing to understand the mechanism that polyomaviruses use to favor infection and evade host defenses.

The discovery of new polyomaviruses is accelerating, and more are likely to be discovered. SV40 is a model system, and as such, it would be interesting to know how accurately the properties and actions of SV40 T antigen reflect those of all T antigens. Furthermore, T antigens occur not only in polyomaviruses but also in other novel types of viruses (7). The domain organization of large T antigens and their replicative functions appear to be broadly conserved. However, the mechanisms by which these proteins control host functions may differ among viruses. These differences can be exploited to study cellular pathways and to provide unique opportunities for therapeutic intervention into the important emerging diseases caused by these viruses. The multiple viral and cellular systems altered by T antigen and the diverse mechanisms used to regulate these systems truly make SV40 large T antigen the most amazing molecule in the universe.

SUMMARY POINTS

1. All known polyomaviruses encode large T antigens containing four common domains: J domain, OBD, Zn-binding domain, and AAA+ ATPase domain. The similarities in the domain structures of large T antigens suggest (and in many cases have proved) conserved biological activities.

2. SV40 large T antigen is required for the viral life cycle and for cellular transformation. The multiple functions of large T antigen are achieved through interactions with various cellular targets, including pRb, p53, and Hsc70.

3. As the master protein directing viral DNA replication, SV40 large T antigen binds to the viral origin of DNA replication, assembles into double hexamers, and functions as a DNA helicase to unwind template DNA during replication. Its interactions with the cellular DNA replication machinery also play important roles during this process.

4. SV40 large T antigen is capable of driving cells into S-phase and inducing the DNA damage response to facilitate replication of viral DNA.

5. Proliferation induced by polyomavirus large T antigens depends on their capacity to block Rb proteins and trigger E2F-dependent expression.

6. Large T antigens from different polyomaviruses can form complexes with p53, though how and whether those complexes contribute to transformation are unclear.

7. In addition to deregulating the pRb/E2F and p53 pathways, SV40 T antigen affects the expression of other molecular pathways, notably detoxification and inflammatory response.

DISCLOSURE STATEMENT

The authors are not aware of any affiliations, memberships, funding, or financial holdings that might be perceived as affecting the objectivity of this review.

LITERATURE CITED

1. Ahuja D, Rathi AV, Greer AE, Chen XS, Pipas JM. 2009. A structure-guided mutational analysis of simian virus 40 large T antigen: identification of surface residues required for viral replication and transformation. *J. Virol.* 83:8781–88
2. Ahuja D, Saenz-Robles MT, Pipas JM. 2005. SV40 large T antigen targets multiple cellular pathways to elicit cellular transformation. *Oncogene* 24:7729–45
3. Ali SH, Kasper JS, Arai T, DeCaprio JA. 2004. Cul7/p185/p193 binding to simian virus 40 large T antigen has a role in cellular transformation. *J. Virol.* 78:2749–57
4. Arunkumar AI, Klimovich V, Jiang X, Ott RD, Mizoue L, et al. 2005. Insights into hRPA32 C-terminal domain–mediated assembly of the simian virus 40 replisome. *Nat. Struct. Mol. Biol.* 12:332–39

5. Barbeau D, Marcellus RC, Bacchetti S, Bayley ST, Branton PE. 1992. Quantitative analysis of regions of adenovirus E1A products involved in interactions with cellular proteins. *Biochem. Cell. Biol.* 70:1123–34
6. Beckerman R, Prives C. 2010. Transcriptional regulation by p53. *Cold Spring Harb. Perspect. Biol.* 2:a000935
7. Bennett MD, Woolford L, Stevens H, Van Ranst M, Oldfield T, et al. 2008. Genomic characterization of a novel virus found in papillomatous lesions from a southern brown bandicoot (*Isoodon obesulus*) in Western Australia. *Virology* 376:173–82
8. Boichuk S, Hu L, Hein J, Gjoerup OV. 2010. Multiple DNA damage signaling and repair pathways deregulated by simian virus 40 large T antigen. *J. Virol.* 84:8007–20
9. Boichuk S, Hu L, Makielski K, Pandolfi PP, Gjoerup OV. 2011. Functional connection between Rad51 and PML in homology-directed repair. *PLoS ONE* 6:e25814
10. Bollag B, Chuke WF, Frisque RJ. 1989. Hybrid genomes of the polyomaviruses JC virus, BK virus, and simian virus 40: identification of sequences important for efficient transformation. *J. Virol.* 63:863–72
11. Borger DR, DeCaprio JA. 2006. Targeting of p300/CREB binding protein coactivators by simian virus 40 is mediated through p53. *J. Virol.* 80:4292–303
12. Borowiec JA, Dean FB, Bullock PA, Hurwitz J. 1990. Binding and unwinding—how T antigen engages the SV40 origin of DNA replication. *Cell* 60:181–84
13. Brady CA, Jiang D, Mello SS, Johnson TM, Jarvis LA, et al. 2011. Distinct p53 transcriptional programs dictate acute DNA-damage responses and tumor suppression. *Cell* 145:571–83
14. Cantalupo PG, Saenz-Robles MT, Rathi AV, Beerman RW, Patterson WH, et al. 2009. Cell-type specific regulation of gene expression by simian virus 40 T antigens. *Virology* 386:183–91
15. Cegielska A, Moarefi I, Fanning E, Virshup DM. 1994. T-antigen kinase inhibits simian virus 40 DNA replication by phosphorylation of intact T antigen on serines 120 and 123. *J. Virol.* 68:269–75
16. Chen JD, Neilson K, Van Dyke T. 1989. Lymphotropic papovavirus early region is specifically regulated transgenic mice and efficiently induces neoplasia. *J. Virol.* 63:2204–14
17. Cheng J, DeCaprio JA, Fluck MM, Schaffhausen BS. 2009. Cellular transformation by simian virus 40 and murine polyoma virus T antigens. *Semin. Cancer Biol.* 19:218–28
18. **Chong JL, Wenzel PL, Saenz-Robles MT, Nair V, Ferrey A, et al. 2009. E2f1-3 switch from activators in progenitor cells to repressors in differentiating cells. *Nature* 462:930–34**
19. Collins BS, Pipas JM. 1995. T antigens encoded by replication-defective simian virus 40 mutants dl1135 and 5080. *J. Biol. Chem.* 270:15377–84
20. Cotsiki M, Lock RL, Cheng Y, Williams GL, Zhao J, et al. 2004. Simian virus 40 large T antigen targets the spindle assembly checkpoint protein Bub1. *Proc. Natl. Acad. Sci. USA* 101:947–52
21. **Cuesta I, Nunez-Ramirez R, Scheres SH, Gai D, Chen XS, et al. 2010. Conformational rearrangements of SV40 large T antigen during early replication events. *J. Mol. Biol.* 397:1276–86**
22. Danna K, Nathans D. 1971. Specific cleavage of simian virus 40 DNA by restriction endonuclease of *Haemophilus influenzae*. *Proc. Natl. Acad. Sci. USA* 68:2913–17
23. Dannenberg JH, van Rossum A, Schuijff L, te Riele H. 2000. Ablation of the retinoblastoma gene family deregulates G_1 control causing immortalization and increased cell turnover under growth-restricting conditions. *Genes. Dev.* 14:3051–64
24. Das D, Imperiale MJ. 2009. Transformation by polyomaviruses. In *DNA Tumor Viruses*, ed. B Damania, JM Pipas, pp. 25–52. New York: Springer
25. DeAngelis T, Chen J, Wu A, Prisco M, Baserga R. 2006. Transformation by the simian virus 40 T antigen is regulated by IGF-I receptor and IRS-1 signaling. *Oncogene* 25:32–42
26. Del Valle L, Pina-Oviedo S, Perez-Liz G, Augelli BJ, Azizi SA, et al. 2010. Bone marrow-derived mesenchymal stem cells undergo JCV T-antigen mediated transformation and generate tumors with neuroectodermal characteristics. *Cancer Biol. Ther.* 9(4)
27. Delbue S, Tremolada S, Elia F, Carloni C, Amico S, et al. 2010. Lymphotropic polyomavirus is detected in peripheral blood from immunocompromised and healthy subjects. *J. Clin. Virol.* 47:156–60
28. **Dyson N, Bernards R, Friend SH, Gooding LR, Hassell JA, et al. 1990. Large T antigens of many polyomaviruses are able to form complexes with the retinoblastoma protein. *J. Virol.* 64:1353–56**
29. Eckner R, Ludlow JW, Lill NL, Oldread E, Arany Z, et al. 1996. Association of p300 and CBP with simian virus 40 large T antigen. *Mol. Cell. Biol.* 16:3454–64

---

18. Establishes the requirement of E2f1/2/3 for cell survival, but not for cell division, and indicates that E2f1/2/3 change from repressors to activators in the absence of pRb, leading to ectopic cell divisions.

21. Shows the conformational flexibility of the double hexameric T antigens in complex with DNA and suggests its contribution to distortion and opening of the DNA double strand.

28. Shows conserved interactions between different large T antigens and Rb proteins from human or mouse origin.

30. Fanning E, Xiaorong Zhao XJ. 2009. Polyomavirus life cycle. In *DNA Tumor Viruses*, ed. B Damania, JM Pipas, pp. 1–24. New York: Springer
31. Ferrari R, Pellegrini M, Horwitz GA, Xie W, Berk AJ, Kurdistani SK. 2008. Epigenetic reprogramming by adenovirus e1a. *Science* 321:1086–88
32. Fluck MM, Schaffhausen BS. 2009. Lessons in signaling and tumorigenesis from polyomavirus middle T antigen. *Microbiol. Mol. Biol. Rev.* 73:542–63
33. **Gai D, Zhao R, Li D, Finkielstein CV, Chen XS. 2004. Mechanisms of conformational change for a replicative hexameric helicase of SV40 large tumor antigen. *Cell* 119:47–60**
34. Gill JA, Lowe L, Nguyen J, Liu PP, Blake T, et al. 2010. Enforced expression of Simian virus 40 large T-antigen leads to testicular germ cell tumors in zebrafish. *Zebrafish* 7:333–41
35. Gjoerup OV, Wu J, Chandler-Militello D, Williams GL, Zhao J, et al. 2007. Surveillance mechanism linking Bub1 loss to the p53 pathway. *Proc. Natl. Acad. Sci. USA* 104:8334–39
36. Grossi MP, Corallini A, Valieri A, Balboni PG, Poli F, et al. 1982. Transformation of hamster kidney cells by fragments of BK virus DNA. *J. Virol.* 41:319–25
37. Gum JR Jr, Hicks JW, Gillespie AM, Rius JL, Treseler PA, et al. 2001. Mouse intestinal goblet cells expressing SV40 T antigen directed by the MUC2 mucin gene promoter undergo apoptosis upon migration to the villi. *Cancer Res.* 61:3472–79
38. Hahn WC, Dessain SK, Brooks MW, King JE, Elenbaas B, et al. 2002. Enumeration of the simian virus 40 early region elements necessary for human cell transformation. *Mol. Cell. Biol.* 22:2111–23
39. Haigis K, Sage J, Glickman J, Shafer S, Jacks T. 2006. The related retinoblastoma (pRb) and p130 proteins cooperate to regulate homeostasis in the intestinal epithelium. *J. Biol. Chem.* 281:638–47
40. Harris KF, Christensen JB, Imperiale MJ. 1996. BK virus large T antigen: interactions with the retinoblastoma family of tumor suppressor proteins and effects on cellular growth control. *J. Virol.* 70:2378–86
41. Hayashi H, Endo S, Suzuki S, Tanaka S, Sawa H, et al. 2001. JC virus large T protein transforms rodent cells but is not involved in human medulloblastoma. *Neuropathology* 21:129–37
42. Hein J, Boichuk S, Wu J, Cheng Y, Freire R, et al. 2009. Simian virus 40 large T antigen disrupts genome integrity and activates a DNA damage response via Bub1 binding. *J. Virol.* 83:117–27
43. Hermannstadter A, Ziegler C, Kuhl M, Deppert W, Tolstonog GV. 2009. Wild-type p53 enhances efficiency of simian virus 40 large-T-antigen-induced cellular transformation. *J. Virol.* 83:10106–18
44. Horwitz GA, Zhang K, McBrian MA, Grunstein M, Kurdistani SK, Berk AJ. 2008. Adenovirus small e1a alters global patterns of histone modification. *Science* 321:1084–85
45. Houben R, Adam C, Baeurle A, Hesbacher S, Grimm J, et al. 2012. An intact retinoblastoma protein-binding site in Merkel cell polyomavirus large T antigen is required for promoting growth of Merkel cell carcinoma cells. *Int. J. Cancer* 130:847–56
46. Huang H, Weiner BE, Zhang H, Fuller BE, Gao Y, et al. 2010. Structure of a DNA polymerase α-primase domain that docks on the SV40 helicase and activates the viral primosome. *J. Biol. Chem.* 285:17112–22
47. Huang H, Zhao K, Arnett DR, Fanning E. 2010. A specific docking site for DNA polymerase α-primase on the SV40 helicase is required for viral primosome activity, but helicase activity is dispensable. *J. Biol. Chem.* 285:33475–84
48. Ishov AM, Maul GG. 1996. The periphery of nuclear domain 10 (ND10) as site of DNA virus deposition. *J. Cell Biol.* 134:815–26
49. Jiang M, Entezami P, Gamez M, Stamminger T, Imperiale MJ. 2011. Functional reorganization of promyelocytic leukemia nuclear bodies during BK virus infection. *MBio* 2:e00281–10
50. Johnson EM. 2010. Structural evaluation of new human polyomaviruses provides clues to pathobiology. *Trends Microbiol.* 18:215–23
51. Jul-Larsen A, Visted T, Karlsen BO, Rinaldo CH, Bjerkvig R, et al. 2004. PML-nuclear bodies accumulate DNA in response to polyomavirus BK and simian virus 40 replication. *Exp. Cell Res.* 298:58–73
52. Kang S, Folk WR. 1992. Lymphotropic papovavirus transforms hamster cells without altering the amount or stability of p53. *Virology* 191:754–64
53. Kasper JS, Kuwabara H, Arai T, Ali SH, DeCaprio JA. 2005. Simian virus 40 large T antigen's association with the CUL7 SCF complex contributes to cellular transformation. *J. Virol.* 79:11685–92

33. Provides mechanistic insights on how T antigen helicase may function through conformational changes induced by ATP binding and hydrolysis using structural biology.

54. Khopde S, Simmons DT. 2008. Simian virus 40 DNA replication is dependent on an interaction between topoisomerase I and the C-terminal end of T antigen. *J. Virol.* 82:1136–45
55. Kierstead TD, Tevethia MJ. 1993. Association of p53 binding and immortalization of primary C57BL/6 mouse embryo fibroblasts by using simian virus 40 T-antigen mutants bearing internal overlapping deletion mutations. *J. Virol.* 67:1817–29
56. Kim SH, Roth KA, Coopersmith CM, Pipas JM, Gordon JI. 1994. Expression of wild-type and mutant simian virus 40 large tumor antigens in villus-associated enterocytes of transgenic mice. *Proc. Natl. Acad. Sci. USA* 91:6914–18
57. Kim SH, Roth KA, Moser AR, Gordon JI. 1993. Transgenic mouse models that explore the multistep hypothesis of intestinal neoplasia. *J. Cell Biol.* 123:877–93
58. Lane DP, Crawford LV. 1979. T antigen is bound to a host protein in SV40-transformed cells. *Nature* 278:261–63
59. Lin JY, DeCaprio JA. 2003. SV40 large T antigen promotes dephosphorylation of p130. *J. Biol. Chem.* 278:46482–87
60. **Linzer DI, Levine AJ. 1979. Characterization of a 54K dalton cellular SV40 tumor antigen present in SV40-transformed cells and uninfected embryonal carcinoma cells.** ***Cell*** **17:43–52**

> 60. Seminal work discovering a host protein, p53, targeted by a tumor virus, SV40 (see also Reference 58).

61. Loeken MR. 1992. Simian virus 40 small t antigen trans activates the adenovirus E2A promoter by using mechanisms distinct from those used by adenovirus E1A. *J. Virol.* 66:2551–55
62. Markovics JA, Carroll PA, Robles MT, Pope H, Coopersmith CM, Pipas JM. 2005. Intestinal dysplasia induced by simian virus 40 T antigen is independent of p53. *J. Virol.* 79:7492–502
63. McCabe MT, Low JA, Imperiale MJ, Day ML. 2006. Human polyomavirus BKV transcriptionally activates DNA methyltransferase 1 through the pRb/E2F pathway. *Oncogene* 25:2727–35
64. Meinke G, Bullock PA, Bohm A. 2006. Crystal structure of the simian virus 40 large T-antigen origin-binding domain. *J. Virol.* 80:4304–12
65. Meinke G, Phelan P, Moine S, Bochkareva E, Bochkarev A, et al. 2007. The crystal structure of the SV40 T-antigen origin binding domain in complex with DNA. *PLoS Biol.* 5:e23
66. Miao H, Seiler JA, Burhans WC. 2003. Regulation of cellular and SV40 virus origins of replication by Chk1-dependent intrinsic and UVC radiation-induced checkpoints. *J. Biol. Chem.* 278:4295–304
67. Nakshatri H, Pater MM, Pater A. 1988. Functional role of BK virus tumor antigens in transformation. *J. Virol.* 62:4613–21
68. Nesper J, Smith RW, Kautz AR, Sock E, Wegner M, et al. 1997. A cell-free replication system for human polyomavirus JC DNA. *J. Virol.* 71:7421–28
69. Pagnani M, Negrini M, Reschiglian P, Corallini A, Balboni PG, et al. 1986. Molecular and biological properties of BK virus-IR, a BK virus variant isolated from a human tumor. *J. Virol.* 59:500–5
70. Pater A, Pater MM. 1986. Transformation of primary human embryonic kidney cells to anchorage independence by a combination of BK virus DNA and the Harvey-ras oncogene. *J. Virol.* 58:680–83
71. Peden KW, Pipas JM. 1992. Simian virus 40 mutants with amino acid substitutions near the amino terminus of large T antigen. *Virus Genes* 6:107–18
72. Peden KW, Srinivasan A, Vartikar JV, Pipas JM. 1998. Effects of mutations within the SV40 large T antigen ATPase/p53 binding domain on viral replication and transformation. *Virus Genes* 16:153–65
73. Pipas JM. 1992. Common and unique features of T antigens encoded by the polyomavirus group. *J. Virol.* 66:3979–85
74. Pipas JM. 2009. SV40: cell transformation and tumorigenesis. *Virology* 384:294–303
75. Pipas JM, Levine AJ. 2001. Role of T antigen interactions with p53 in tumorigenesis. *Semin. Cancer Biol.* 11:23–30
76. Ramamoorthy S, Devaraj B, Miyai K, Luo L, Liu YT, et al. 2010. John Cunningham virus T-antigen expression in anal carcinoma. *Cancer* 117:2379–85
77. Rathi AV, Cantalupo PG, Sarkar SN, Pipas JM. 2010. Induction of interferon-stimulated genes by Simian virus 40 T antigens. *Virology* 406:202–11
78. Rathi AV, Saenz Robles MT, Cantalupo PG, Whitehead RH, Pipas JM. 2009. Simian virus 40 T-antigen-mediated gene regulation in enterocytes is controlled primarily by the Rb-E2F pathway. *J. Virol.* 83:9521–31

79. Rathi AV, Saenz Robles MT, Pipas JM. 2007. Enterocyte proliferation and intestinal hyperplasia induced by simian virus 40 T antigen require a functional J domain. *J. Virol.* 81:9481–89
80. Rose PE, Schaffhausen BS. 1995. Zinc-binding and protein-protein interactions mediated by the polyomavirus large T antigen zinc finger. *J. Virol.* 69:2842–49
81. Sablina AA, Hahn WC. 2008. SV40 small T antigen and PP2A phosphatase in cell transformation. *Cancer Metastasis Rev.* 27:137–46
82. Sachsenmeier KF, Pipas JM. 2001. Inhibition of Rb and p53 is insufficient for SV40 T-antigen transformation. *Virology* 283:40–48
83. Saenz Robles MT, Case A, Chong JL, Leone G, Pipas JM. 2011. The retinoblastoma tumor suppressor regulates a xenobiotic detoxification pathway. *PLoS ONE* 6:e26019
84. Saenz Robles MT, Pipas JM. 2009. T antigen transgenic mouse models. *Semin. Cancer Biol.* 19:229–35
85. Saenz-Robles MT, Markovics JA, Chong JL, Opavsky R, Whitehead RH, et al. 2007. Intestinal hyperplasia induced by simian virus 40 large tumor antigen requires E2F2. *J. Virol.* 81:13191–99
86. Saenz Robles MT, Symonds H, Chen J, Van Dyke T. 1994. Induction versus progression of brain tumor development: differential functions for the pRB- and p53-targeting domains of simian virus 40 T antigen. *Mol. Cell. Biol.* 14:2686–98
87. **Saenz-Robles MT, Toma D, Cantalupo P, Zhou J, Gong H, et al. 2007. Repression of intestinal drug metabolizing enzymes by the SV40 large T antigen. *Oncogene* 26:5124–31**
88. Schaffhausen BS, Roberts TM. 2009. Lessons from polyoma middle T antigen on signaling and transformation: a DNA tumor virus contribution to the war on cancer. *Virology* 384:304–16
89. Schneider J, Fanning E. 1988. Mutations in the phosphorylation sites of simian virus 40 (SV40) T antigen alter its origin DNA-binding specificity for sites I or II and affect SV40 DNA replication activity. *J. Virol.* 62:1598–605
90. Shahid JM, Iwamuro M, Sasamoto H, Kubota Y, Seita M, et al. 2010. Establishment of an immortalized porcine liver cell line JSNK-1 with retroviral transduction of SV40T. *Cell Transplant* 19:849–56
91. Sheppard HM, Corneillie SI, Espiritu C, Gatti A, Liu X. 1999. New insights into the mechanism of inhibition of p53 by simian virus 40 large T antigen. *Mol. Cell. Biol.* 19:2746–53
92. Shi Y, Dodson GE, Shaikh S, Rundell K, Tibbetts RS. 2005. Ataxia-telangiectasia-mutated (ATM) is a T-antigen kinase that controls SV40 viral replication in vivo. *J. Biol. Chem.* 280:40195–200
93. Spence SL, Pipas JM. 1994. SV40 large T antigen functions at two distinct steps in virion assembly. *Virology* 204:200–9
94. **Srinivasan A, McClellan AJ, Vartikar J, Marks I, Cantalupo P, et al. 1997. The amino-terminal transforming region of simian virus 40 large T and small t antigens functions as a J domain. *Mol. Cell Biol.* 17:4761–73**
95. Symonds H, Chen JD, Van Dyke T. 1991. Complex formation between the lymphotropic papovavirus large tumor antigen and the tumor suppressor protein p53. *J. Virol.* 65:5417–24
96. Takemoto KK, Segawa K. 1983. A new monkey lymphotropic papovavirus: characterization of the virus and evidence of a related virus in humans. *Prog. Clin. Biol. Res.* 105:87–96
97. Tevethia MJ, Bonneau RH, Griffith JW, Mylin L. 1997. A simian virus 40 large T-antigen segment containing amino acids 1 to 127 and expressed under the control of the rat elastase-1 promoter produces pancreatic acinar carcinomas in transgenic mice. *J. Virol.* 71:8157–66
98. Tognon M, Corallini A, Martini F, Negrini M, Barbanti-Brodano G. 2003. Oncogenic transformation by BK virus and association with human tumors. *Oncogene* 22:5192–200
99. Valle M, Chen XS, Donate LE, Fanning E, Carazo JM. 2006. Structural basis for the cooperative assembly of large T antigen on the origin of replication. *J. Mol. Biol.* 357:1295–305
100. von Hoyningen-Huene V, Kurth M, Deppert W. 1992. Selection against large T-antigen expression in cells transformed by lymphotropic papova virus. *Virology* 190:155–67
101. **Waga S, Bauer G, Stillman B. 1994. Reconstitution of complete SV40 DNA replication with purified replication factors. *J. Biol. Chem.* 269:10923–34**
102. Wenzel PL, Chong JL, Saenz-Robles MT, Ferrey A, Hagan JP, et al. 2011. Cell proliferation in the absence of E2F1-3. *Dev. Biol.* 351:35–45

87. Presents a novel role for SV40 T antigen, linking the molecule to the regulation of detoxification pathways in a cell-specific manner.

94. Demonstrates that the amino terminus of T antigen acts as co-chaperone of Hsc70 to hydrolyze ATP.

101. Identifies cellular replication factors required for replicating SV40 DNA that facilitates our understanding of mechanisms underlying eukaryotic DNA replication.

103. Wessel R, Schweizer J, Stahl H. 1992. Simian virus 40 T-antigen DNA helicase is a hexamer which forms a binary complex during bidirectional unwinding from the viral origin of DNA replication. *J. Virol.* 66:804–15

> 103. Captures SV40 T antigen helicase in action using electron microscopy and demonstrates the bidirectional unwinding by T antigen helicase with biochemical analyses.

104. Wu X, Avni D, Chiba T, Yan F, Zhao Q, et al. 2004. SV40 T antigen interacts with Nbs1 to disrupt DNA replication control. *Genes Dev.* 18:1305–16
105. Yaciuk P, Carter MC, Pipas JM, Moran E. 1991. Simian virus 40 large-T antigen expresses a biological activity complementary to the p300-associated transforming function of the adenovirus E1A gene products. *Mol. Cell. Biol.* 11:2116–24
106. Yee SP, Branton PE. 1985. Detection of cellular proteins associated with human adenovirus type 5 early region 1A polypeptides. *Virology* 147:142–53
107. Zhai W, Tuan JA, Comai L. 1997. SV40 large T antigen binds to the TBP-TAF(I) complex SL1 and coactivates ribosomal RNA transcription. *Genes Dev.* 11:1605–17
108. Zhao X, Madden-Fuentes RJ, Lou BX, Pipas JM, Gerhardt J, et al. 2008. Ataxia telangiectasia-mutated damage-signaling kinase- and proteasome-dependent destruction of Mre11-Rad50-Nbs1 subunits in simian virus 40-infected primate cells. *J. Virol.* 82:5316–28

Peroxisome Assembly and Functional Diversity in Eukaryotic Microorganisms

Laurent Pieuchot and Gregory Jedd

Temasek Life Sciences Laboratory and Department of Biological Sciences, National University of Singapore, 117604 Singapore; email: laurent@tll.org.sg, gregory@tll.org.sg

Keywords

glycosome, Woronin body, meiocyte, biogenesis, peroxisomal targeting signal, evolution

Abstract

Peroxisomes are core eukaryotic organelles that generally function in lipid metabolism and detoxification of reactive oxygen species, but they are increasingly associated with taxa-specific metabolic, cellular, and developmental functions. Here, we present a brief overview of peroxisome assembly, followed by a discussion of their functional diversification. Matrix protein import occurs through a remarkable translocon that can accommodate folded and even oligomeric proteins. Metabolically specialized peroxisomes include glycosomes of trypanosomes, which have come to compartmentalize most of the glycolytic pathway and play a role in developmental signal transduction. The differentiation of physically distinct subcompartments also contributes to peroxisome diversification; in the clade of filamentous ascomycetes, dense-core Woronin bodies bud from peroxisomes to gate cell-to-cell channels. Here, the import of oligomeric cargo is central to the mechanism of subcompartment specification. In general, the acquisition of a tripeptide peroxisome targeting signal by nonperoxisomal proteins appears to be a recurrent step in the evolution of peroxisome diversity.

Contents

INTRODUCTION	238
PEROXISOME ASSEMBLY	238
Matrix Protein Import	238
Membrane Targeting and Biogenesis	241
Peroxisome Proliferation	242
FUNCTIONAL DIVERSIFICATION OF PEROXISOMES	245
Glycolysis in Trypanosomes	245
Other Taxa-Specific Functions	247
Woronin Body Subcompartment	248

INTRODUCTION

Peroxisomes are single-membrane-bound eukaryotic organelles involved in diverse metabolic functions. They generally function in lipid metabolism and detoxification of reactive oxygen species, but they are also involved in diverse taxa-specific functions. These include catabolism of very-long-chain, D-amino acids, and polyamines in mammals; biosynthesis of plasmalogens in mammals (181); photorespiration in leaves (132); the glyoxylate cycle in germinating seeds (90); and assimilation of methanol in some yeasts (172). In trypanosomes, peroxisomes called glycosomes compartmentalize components of the glycolytic pathway and are implicated in a signaling pathway associated with parasite development (158). Peroxisomes also play important roles in signal transduction, such as salicylic acid signaling in plants (183), and have been implicated as platforms for signaling in vertebrate innate immunity (18). In filamentous fungi, peroxisomes are involved in penicillin biosynthesis (63), plant pathogenicity (67), and sexual development (8), and they have evolved the capacity to develop a peroxisome subcompartment known as the Woronin body, which performs an adaptive function in gating fungal cell-to-cell channels. Here we review peroxisome biogenesis, function, and dynamics, and provide an overview of their remarkable functional plasticity in eukaryotic microorganisms.

Peroxisome: single-membrane-bound eukaryotic organelle generally associated with lipid metabolism

Glycosome: specialized peroxisome housing a subset of enzymes involved in glycolysis and playing a role in developmental signal transduction

Woronin body: a physically and functionally distinct peroxisome subcompartment involved in cellular wound healing in filamentous ascomycetes

PTS1: peroxisome targeting signal 1

PTS2: peroxisome targeting signal 2

PEROXISOME ASSEMBLY

Matrix Protein Import

More than 30 genes involved in the biogenesis of peroxisomes (PEX genes) have been identified. Mutations in genes coding for the components of the matrix protein import machinery lead to the formation of "ghosts," or empty peroxisome remnants, in which membrane proteins are still inserted into the lipid bilayer. This demonstrates that matrix and membrane proteins are imported by distinct pathways.

Cargo recognition. Peroxisome matrix proteins are synthesized on free polyribosomes in the cytosol and imported posttranslationally (29, 180). Their targeting to peroxisomes depends on short sequences known as peroxisomal targeting signal (PTS) type 1 and type 2. PTS1 is present in the majority of matrix proteins and consists of a C-terminal tripeptide (S/A/C)(K/R/H)L (38). PTS2 is a degenerated nonapeptide (R/K)(L/V/I)X_5(H/Q)(L/A) found near the N terminus that,

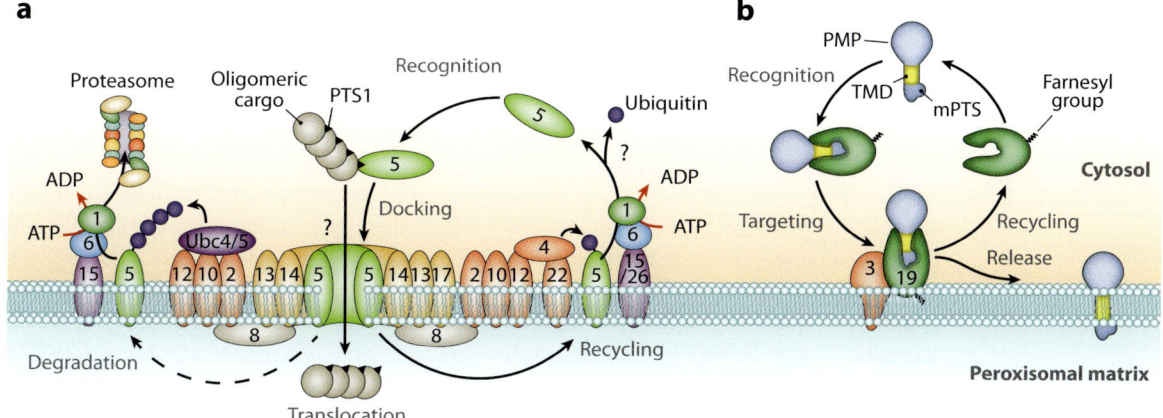

Figure 1

(*a*) PTS1 matrix import cycle. Cargo, which can be oligomeric, is recognized by Pex5 and docks with the membrane through the docking peroxins Pex13, Pex14, and Pex17. (*Right*) The cargo-receptor complex crosses the membrane via an unknown mechanism that may involve the formation of a transient pore. The Pex5 receptor is monoubiquitinated and recycled to the cytosol through the AAA ATPases Pex1 and Pex6. (*Left*) In the RADAR pathway, dysfunctional Pex5 can be polyubiquitinated and targeted to the proteasome for degradation. Adapted from Reference 145 with permission. (*b*) PMP integration. PMPs possessing a mPTS are recognized posttranslationally by Pex19p and delivered to the membrane through interaction with Pex3, where the PMP is released and integrated into the membrane. Adapted from Reference 40 with permission. Abbreviations: PTS1/2, peroxisome targeting signal 1/2; RADAR, receptor accumulation and degradation in the absence of recycling; PMP, peroxisomal membrane protein; mPTS, membrane peroxisome targeting signal; TMD, transmembrane domain.

in some species, is cleaved off after import into the peroxisomal matrix (117, 123, 157). PTS1 and PTS2 are recognized by cycling cytosolic receptors Pex5 and Pex7, respectively.

Pex5 contains a conserved C-terminal domain composed of tetratricopeptide repeats (TPRs) that directly bind the PTS1 of peroxisomal cargo (11, 33). The crystal structure of the Pex5 TPR domains shows that this protein undergoes dramatic conformational changes upon cargo binding (154). In a few cases the N-terminal half of Pex5 can also mediate the binding to cargoes lacking canonical PTS1 sequences (43, 69). This region is less conserved, disordered (154), and responsible for docking to the cytoplasmic face of the peroxisome membrane through interaction with the docking peroxins (Pex13, Pex14, and Pex17) (118, 119, 171) (**Figure 1***a*).

The targeting of PTS2-bearing proteins is mediated by Pex7 together with a coreceptor that varies from species to species: Pex18 and Pex21 in *Saccharomyces cerevisiae* (130), Pex20 in filamentous ascomycetes (82, 120, 150, 167), and Pex5L, a splicing variant of Pex5, in plants and mammals (9, 78). Thus, Pex5 is required for both PTS1 and PTS2 import in plants and mammals, whereas in fungi, Pex20 allows these two pathways to function more independently.

Surprisingly, the PTS2 pathway seems to be absent in certain organisms including *Caenorhabditis elegans* (106), diatoms (36), and probably the red alga *Cyanidioschyzon merolae*, in which no PTS2-like sequences have been identified in silico (149). Instead, in those organisms, orthologs of PTS2-containing proteins have acquired PTS1 signals (36, 106). Thus, switching of targeting signals appears to have allowed the PTS2 pathway to be lost in certain phylogenetic lineages.

Docking and translocation. Both PTS receptor systems address their cargo to a docking/translocation platform composed of Pex13, Pex14 and, in some yeast, Pex17. Pex13 and Pex14 bind to each other (144) and interact with both PTS receptors (19, 24, 119, 155) through multiple binding sites. Remarkably, whereas Pex5 behaves like a soluble protein in the cytosol,

TPR: tetratricopeptide repeat

Peroxin: protein associated with peroxisome biogenesis and proliferation

RING: Really Interesting New Gene

AAA ATPase: ATPases associated with diverse cellular activities

membrane-associated Pex5 behaves like a transmembrane protein (39). Subsequent work showed that during the import cycle, both receptor and cargo cross the peroxisome membrane, and this finding has been demonstrated for Pex5 (17), Pex7 (110), and Pex20 (82). How the cargo is released has not yet been determined. However, the intraperoxisomal peroxin Pex8 interacts with Pex5 in vitro and stimulates the release of PTS1 peptides, suggesting one possible mechanism (182).

Recently, a new docking peroxin Pex14/17 (114), also known as Pex33 (89), has been identified in filamentous ascomycetes. This peroxin interacts with Pex5 (89) and appears to be functionally equivalent to yeast Pex17. Deletion mutants significantly reduce the efficiency of matrix import, indicating that this is a bona fide new peroxin. Interestingly, in the filamentous fungus *Podospora anserina*, *PEX14* is required for matrix import in vegetative mycelium but it is dispensable at certain developmental stages in meiocytes (121). In this specialized cell type, matrix import also occurs in a *pex14*, *pex14/17* double mutant but depends on the presence of PEX13. In *Hansenula polymorpha*, overproduction of Pex5 can stimulate residual matrix import in a *pex14* mutant (140), further suggesting that matrix import can occur in the absence of Pex14. Interestingly, in *Pichia pastoris*, the PTS2-dependent import of Pex8 requires Pex14, but not Pex13 and Pex17 (86). Together, these data suggest that the composition of the docking complex can vary in a species-, development-, and cargo-dependent manner.

Receptor recycling. Pex5 receptor recycling is initiated by monoubiquitination (128) on a conserved N-terminal cysteine residue (185). This reaction is mediated by the RING-finger E3 ligase Pex12 (127) and the ubiquitination conjugating enzyme Pex4, which is soluble and is recruited to the membrane by Pex22 (71) (**Figure 1***a*). In yeast, the intraperoxisomal peroxin Pex8 functions as a bridge between the docking receptors and the RING complex to form the overall importomer (1); Pex3 might also play a role in this process (48).

Following monoubiquitination, PEX5 is recycled back to the cytosol by the AAA ATPase peroxins Pex1 and Pex6 (103, 126), which are anchored to the peroxisomal membrane through the tail-anchored peroxin, Pex15, in yeast (6) and the orthologous function of Pex26 in animals (95) and filamentous fungi (83) (**Figure 1***a*). The N-terminal region of the PTS2 coreceptor Pex20 is similar to the N terminus of the PTS1 receptor Pex5 and contains conserved residues that are essential for its recycling from the peroxisomal membrane, suggesting that Pex20 and Pex5 are recycled through a similar mechanism (81, 82). In addition to the ATP consumed by ubiquitin activation, receptor export is believed to be the only ATP-consuming step of the matrix import cycle (103), leading to the concept of an export-driven import, which posits that receptor recycling by the AAA ATPases is mechanistically coupled to cargo translocation across the membrane (145).

When components of the receptor recycling machinery are mutated, Pex5 and Pex20 cannot be recycled back to the cytosol. Under these conditions, the receptors are polyubiquitinated at N-terminal lysine residues and directed to the proteasome for degradation (61, 81, 82, 127, 184). This process, called RADAR (receptor accumulation and degradation in the absence of recycling), appears to constitute a quality-control system that prevents obstruction of the import machinery.

Import of oligomeric cargo. Some peroxisomal matrix proteins are preassembled in the cytosol prior to import and cross the membrane as oligomers (35, 80, 98, 164, 167). Matrix proteins with mutant PTS signals can be imported by their PTS-containing binding partners, and these types of experiments provided definitive evidence for piggyback import into peroxisomes (21, 35, 98, 188). The fact that oligomeric cargo can be imported suggests a translocation machinery that can expand to accommodate substrates of variable dimensions. The presence of a large and flexible translocon is further indicated by experiments showing that 4- to 9-nm PTS1-coated colloidal gold particles can be imported (180). This remarkable feature of the matrix import pathway leads to a central question concerning how these particles cross the membrane, and indicates that this mechanism is fundamentally different from the translocation of proteins across the endoplasmic reticulum (ER) and mitochondrial membrane.

Electron microscopy reveals the frequent occurrence of aggregates and crystals in the peroxisomal matrix (68, 175, 179), indicating that some matrix proteins form extremely high-order oligomers, and in general the functional significance of these proteins remains enigmatic. One clear case in which functions have been ascribed to these high-order matrix oligomers is found in the fungal HEX (Hexagonal peroxisome) protein, which uses a PTS1 signal and self-assembles to form the Woronin body crystalline core (189). Interestingly, HEX crosses the membrane as an oligomer, and mutants that disrupt self-assembly lead to dominant-negative effects on PTS1 import, but not on PTS2 import (83). This suggests that under normal conditions, HEX oligomers promote import efficiency by allowing each cycle of PEX5 action to import multiple HEX proteins. This provides one physiological context in which the import of oligomers is beneficial. HEX oligomers also appear to influence subcompartment specification (83), further suggesting that import of oligomers can influence peroxisome fate.

PMP: peroxisomal membrane protein

mPTS: peroxisomal membrane protein targeting signal

Nature of the translocon. Several models have been proposed to account for the ability to import folded and oligomeric proteins. These include a membrane invagination mechanism and the opening of a static pore (99, 156). Another model suggests the formation of a transient pore (156) formed by the import receptors themselves (25). Several lines of evidence support this model. Membrane-associated Pex5 behaves like an integral membrane protein (39), and patch-clamp experiments show that large-conductance channels are present in the membranes of mammalian peroxisomes (76, 79). Recently, a membrane-associated Pex5-Pex14 complex was purified, reconstituted into planar lipid membranes, and subjected to current recordings to obtain evidence for channel activity (101). PTS1-bearing peptides alone had no effect on these preparations. However, purified cytosolic Pex5 loaded with cargo induced conductance channel gating events consistent with a pore of 2.8 nm, and when a large oligomeric cargo was used this value increased up to 9.25 nm. The putative translocon present in these preparations was produced from a *pex8* mutant and was not associated with the RING complex. Thus, a major challenge for the future is to reconstitute the entire importomer complex and the full import cycle.

Membrane Targeting and Biogenesis

The majority of PEX genes are implicated in the process of matrix protein import or the regulation of peroxisome number and size. Mutants defective in matrix import typically accumulate empty peroxisomes that contain peroxisomal membrane proteins (PMPs). However, in yeast, loss of one of only two genes, *pex3* and *pex19*, leads to the complete absence of peroxisomal membrane remnants, suggesting that these functions are essential for membrane biogenesis (49).

Direct targeting of PMPs to peroxisomes. In the pathway for direct targeting, PMPs are believed to be synthesized on free cytosolic ribosomes and subsequently imported directly to the peroxisome membrane. This process depends on membrane peroxisome targeting signals (mPTSs) consisting of short stretches of basic amino acids associated with hydrophobic transmembrane domains (133). In this pathway, Pex19 functions as a cycling chaperone, which recognizes the mPTS of PMPs and ferries them to the membrane where PMP integration follows docking to Pex3 (28, 58, 133) (**Figure 1*b***). In vitro systems have been used to reconstitute PMP membrane integration, and these studies show that Pex19 can keep PMPs in an import-competent state before docking to the membrane through a physical interaction with Pex3 (97, 124). Consistent with this, fluorescence resonance energy transfer (FRET) experiments suggest that the interaction between Pex3 and Pex19 occurs mainly at the peroxisomal membrane (109). Interestingly, cargo-loaded Pex19 has a higher affinity for Pex3 than does free Pex19, suggesting that the mPTS or transmembrane domains of the PMP also contribute to Pex3 binding (124). The C-terminal region of Pex19p forms a globular α-helical domain that binds the mPTS of PMPs (147), and a distinct N-terminal region is responsible for binding to Pex3 (143, 146).

Pex19 also possesses a farnesylated C-terminal CaaX box. Yeast mutants in which Pex19 cannot be farnesylated produce peroxisomes, but contain significantly reduced steady-state levels of PMPs and show defects in matrix import, which are presumably a secondary consequence of defects in PMP import (136). In addition, PMP recognition is 10 times more efficient when Pex19 is farnesylated (136). Together, these observations suggest that farnesylation promotes Pex19 function, possibly by promoting its interaction with the membrane and/or hydrophobic transmembrane domains associated with the mPTS.

PMP trafficking via the ER. N- and O-linked glycosylation occurs in the lumen of the ER, and several PMPs receive these types of modification, suggesting ER–to–peroxisome trafficking. These PMPs include Pex2 and Pex16 in *Yarrowia lipolytica* (165) and Pex15 (20, 77) and Pex3 (72) in *S. cerevisiae*. In addition, the putative mPTS signals of Pex3 and Pex22 do not interact with Pex19 (47), implying that these proteins follow a different pathway to the membrane. Furthermore, mutations that interfere with the Sec61 translocon (163, 173) and ER exit (165) also interfere with PMP trafficking. Get3 is a chaperone responsible for the integration of tail-anchored proteins into the ER, and loss of function in this system leads to mistargeting of the tail-anchored peroxin Pex15 (148, 173). Further evidence for ER trafficking comes from pulse-chase experiments that follow the reappearance of peroxisomes in peroxisome-free cells. Here, a Pex3 GFP (green fluorescent protein) fusion is first seen in the ER and later in punctate structures that mature to eventually attain the capacity to import matrix proteins (50, 72, 105, 160). In *P. pastoris*, Pex30 and Pex31 also appear to traffic via the ER (187). Although there is abundant evidence to show that PMPs can transit the ER en route to the peroxisome, more work is required to identify the machinery involved in trafficking. Interestingly, mutations in Pex1 and Pex6 delay ER exit of Pex2 and Pex16 in *Y. lipolytica* (165). Pex1 and Pex6 have also been associated with fusion of pre-peroxisomal vesicles (166), suggesting that they may be involved in ER–to–peroxisome trafficking (**Figure 2**).

Peroxisome Proliferation

Peroxisome abundance can vary dramatically depending on cellular need and environmental conditions. Peroxisomes can proliferate through growth and division of preexistent peroxisomes and they can also arise de novo from the ER.

Proliferation by growth and division. Peroxisome proliferation follows a multistep process involving the Pex11-dependent tubulation of peroxisomes, followed by dynamin GTPase-dependent scission of the elongated organelles (reviewed in Reference 116) (**Figure 2**). Pex11p is the first protein identified as being involved in peroxisome proliferation in yeast (23). Here, loss of Pex11 leads to reduced peroxisome abundance and enlarged peroxisomes (23). By contrast, its overexpression produces cells crowded with elongated clusters of peroxisomes (92). Pex25 and Pex27 are two additional *S. cerevisiae* isoforms that share weak similarities with Pex11. Both are involved in peroxisome division and appear to have partially redundant functions with Pex11p (134). The activity of Pex11p appears to be tightly regulated. In yeast, *PEX11* expression is stimulated when

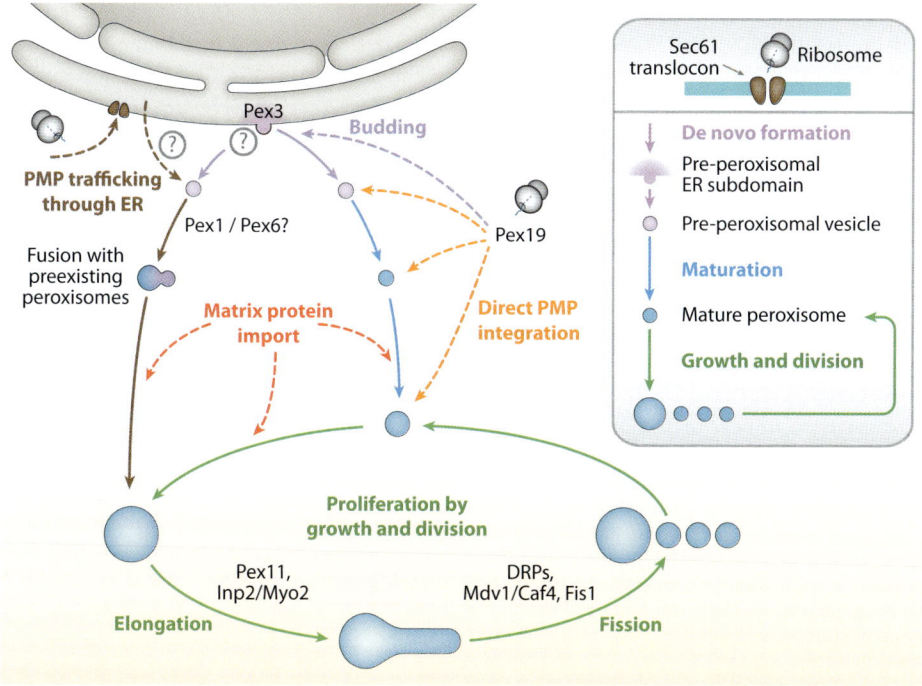

Figure 2

Peroxisome biogenesis and proliferation. Peroxisomes can multiply by growth and division or rise de novo from the ER. In the de novo pathway, Pex3 is targeted to the ER membrane and may bud in a Pex19-dependent manner to form a pre-peroxisome, which matures through the import of additional PMPs to eventually acquire the capacity to import matrix proteins. PMPs may also be inserted into the ER membrane and traffic to the peroxisome membrane. Please see text for additional information. Abbreviations: ER, endoplasmic reticulum; PMP, peroxisomal membrane protein; DRP, dynamin-related protein.

peroxisome proliferation is induced (151, 174), and the activity of Pex11 is also regulated by phosphorylation (70, 139). Interestingly, recent work shows that Pex11 has the inherent ability to sculpt membrane curvature through the insertion of a conserved N-terminal amphipathic helix into the membrane, and this provides a molecular mechanism for its mode of action (115). A number of other peroxins have been associated with peroxisome division and they display varying interactions with one another and Pex11 (168, 176, 177, 187), suggesting that the number and size of peroxisomes are controlled by a complex network of interacting peroxins. In *Y. lipolytica*, Pex16 is intraperoxisomal and functions as a negative regulator of division. As peroxisomes mature to attain a high matrix protein density, acyl-CoA oxidase (Aox) is relocalized from the matrix to the inner face of the membrane, where it inhibits Pex16 function and triggers modification of membrane lipid composition. This further promotes recruitment of the cytosolic division machinery (44, 45) and provides a means of coordinating peroxisome maturation and division. The absence of Pex16 in most other yeast suggests that this type of regulation is achieved through diverse mechanisms.

After tubulation, peroxisome division requires the function of dynamin-like protein (DLP) GTPases, which are involved in various cellular membrane fission events. DLPs oligomerize around membranes, forming ring-like structures that, after GTP hydrolysis, cause deformation and ultimately scission of the membranes (for review, see Reference 129). The DLPs Vps1

DLP: dynamin-like protein

and Dnm1 are involved in peroxisome fission in yeast (51, 75, 105, 178). Dnm1 is recruited to peroxisomes via Mdv1 and Caf4, which are linked to the peroxisomal membrane via the tail-anchored protein Fis1 (107). Fis1 is targeted to mitochondria, where it promotes fission through the same factors (for review, see Reference 113). This suggests that these two organelles share this fission machinery and that this might function to coordinate their biogenesis. Vps1 also functions in peroxisome fission, but it is not recruited to peroxisomes via Fis1 and instead seems to require Pex19 for peroxisome association (178). A recent study in *S. cerevisiae* shows that Pex34 interacts with Pex11, Pex25p, and Pex27p, as well as Fis1p, establishing in yeast a link between Pex11 proteins and the fission machinery (168).

De novo formation of peroxisomes from the ER. In yeast, *pex3* and *pex19* mutants are devoid of peroxisome membranes. However, their reintroduction can induce the de novo formation of peroxisomes. In this case, Pex3p traffics from the ER to punctate structures that gradually attain the capacity to import matrix proteins in a process that depends on Pex19 (50, 72, 105, 160) (**Figure 2**). On the basis of their known function in promoting PMP integration, Pex3 and Pex19 are believed to recruit additional PMPs to pre-peroxisomal vesicles to eventually reconstitute the matrix import pathway. Moreover, Pex3 targeted to mitochondria can induce the formation of import-competent peroxisomes in Pex3-deficient *S. cerevisiae* cells, further suggesting that Pex3 can initiate the de novo peroxisome biogenesis pathway from diverse membranous precursor (135). In *H. polymorpha*, the reintroduction of peroxisomes also requires Pex25, a relative of Pex11, and the Rho1 GTPase (142). Pex25 was previously shown to interact with Rho1 in *S. cerevisiae* (91). However, whether these represent conserved or taxa-specific functions required for the de novo pathway remains to be determined. In animal cells, *pex16* mutants also lack peroxisome remnants, and Pex16 is believed to be responsible for recruiting Pex3 to the membrane either directly (96) or via the ER (66). Pex16 is absent in most yeast but is present in *Y. lipolytica*, where it acts as a negative regulator of peroxisome division (45). Interestingly, Pex16 is present in all filamentous ascomycetes (64), and in *Neurospora crassa*, a *pex16* mutant is defective in the PTS1 matrix import pathway (84), suggesting a more central role for this peroxin in the filamentous ascomycetes, which warrants further investigation.

The degree to which de novo biogenesis and the division of preexisting organelles contribute to normal peroxisome proliferation remains an open question. The importance of these two pathways may vary significantly from species to species and depending on cellular physiology. In yeast, the de novo pathway was initially observed only in the absence of preexistent peroxisomes. Using pulse-chase and mating assays, it has been shown that the de novo pathway is slow and engaged only when peroxisomes are absent due to defects in inheritance (105). Interestingly, when a *pex19* mutant background is used to accumulate Pex3p in the ER and this strain is mated with a wild-type strain, Pex3 appears to rapidly traffic directly to preexistent peroxisomes rather than follow the slow de novo pathway. This, together with the rapid kinetics of ER–to–peroxisome trafficking in *Y. lipolytica* (165), conforms to the notion that the pathway for sorting PMPs from the ER to peroxisomes may be partially or fully independent of the de novo peroxisome assembly pathway (**Figure 2**).

In vitro budding assays. In vitro assays have recently been developed to examine the production of pre-peroxisomal vesicles from the ER. In *S. cerevisiae*, a modified form of the tail-anchored peroxin Pex15 (Pex15G) containing a lumenal glycosylation domain appears to traffic from the ER to peroxisomes in wild-type cells and accumulates in the ER of *pex19* mutant cells (77). Using microsomes from this strain, in vitro assays show that Pex15G together with Pex3 is packaged into vesicles in a reaction that requires Pex19, ATP, and additional cytosolic factors (77), but is independent of the COPII (coat protein complex II) coat. Similar results have been obtained in a complementary *P. pastoris* system (2), in which both Pex11 and Pex3 are integrated into vesicles emerging from the ER, and here too the budding mechanism is cytosol, ATP, and Pex19 dependent. Surprisingly, Pex11 budding was independent of Pex3, but these vesicles are devoid of most of the PMPs present in Pex3 vesicles and might represent peroxisomal remnants (2). Nevertheless, this raises the possibility that Pex19 can promote vesicle budding independently of Pex3. The mode of action of Pex19 in the budding process is unclear. In animal cells, Pex16 lacking its mPTS is trapped in the ER, suggesting that the mPTS is required for Pex16 trafficking to peroxisomes (66). This implies that Pex19 might act in pulling out the budding vesicle via its interaction with the mPTS of PMPs (85). Pex19 alone is not sufficient for budding; thus, the identification of essential cytosolic factors should help define the budding mechanism and its ATP requirement.

COPII: coat protein complex II

FUNCTIONAL DIVERSIFICATION OF PEROXISOMES

The set of proteins involved in peroxisome biogenesis and maintenance is highly conserved, suggesting a single evolutionary origin of this cellular compartment. However, the peroxisomal enzymatic content can vary substantially in different groups of organisms and this appears to be the result of the evolutionary acquisition of PTS targeting signals. The most pronounced example of this is the peroxisomal glycosome of trypanosomes, which has evolved to compartmentalize most of the enzymes for glycolysis. Woronin bodies of filamentous ascomycetes provide another example of functional diversification through the evolution of a complex sorting machinery that enables production of a physically distinct peroxisome subcompartment.

Glycolysis in Trypanosomes

Trypanosomatids are the causal agents of sleeping sickness, leishmaniasis, and Chagas' disease. In addition to glycolysis, glycosomes also contain the canonical peroxisome functions and a variety of other pathways. Glycolysis in other eukaryotes is essentially cytosolic; thus, its compartmentalization in these cells is unique and of great interest.

Glycosomes during the trypanosome life cycle.
Trypanosomes are transmitted between their mammalian hosts by blood-feeding insects and experience highly different environments within these hosts. The enzymatic content of glycosomes varies considerably during the life cycle. In the animal host, the parasite (long-slender form) is in a glucose-rich environment, where mitochondrial oxidative phosphorylation is repressed and all the ATP is generated through glycosomal glycolysis (10). At this point, glycolytic enzymes represent 90% of the protein content of the organelle (102). By contrast, in the insect host, the concentration of sugars is low. While still in the bloodstream, the parasite can differentiate into a short-stumpy form preadapted for life in the insect. This form of the parasite has a partially derepressed mitochondrial system, which allows the parasite to survive in the low-sugar environment of the insect midgut. Here, exposure to environmental signals (16) triggers differentiation into a procyclic form that is fully adapted to the insect host.

Differentiation of the short-stumpy form into the procyclic form is repressed by the activity of a tyrosine phosphatase, TbPTP1 (159). Upon ingestion by the insect host, TbPTP1 is inactivated and differentiation into the procyclic form occurs. The downstream component in this developmental signaling pathway is a Ser/Thr phosphatase, TbPIP39, which is inhibited by tyrosine phosphorylation and is thus negatively regulated by TbPTP1 (158). During differentiation, TbPIP39 is rapidly phosphorylated by a yet unidentified kinase and its RNAi-mediated depletion inhibits

phospho-: phosphorylated form

HXK: hexokinase

PFK: phosphofructokinase

development of the procyclic form, indicating a key role in promoting this developmental transition. Remarkably, TbPIP39 possesses a consensus PTS1 signal (−SRL, serine-arginine-lysine or Ser-Arg-Lys), which is required for function, and shows perfect colocalization with a glycosomal marker (158). TbPTP1 is presumably cytosolic, and this raises the question of how it can regulate TbPIP39. TbPIP39 protein is induced with differentiation, and this nascent TbPIP39 can be subject to phosphorylation/dephosphorylation in the cytosol before import into the peroxisome.

How phospho-TbPIP39 promotes differentiation from within the glycosome, and whether this relates to glycosomal metabolism or further signal transduction, remains an open question. Glycosomes are predicted to contain over 200 proteins (13), and in principle any of these could be subject to regulation by TbPIP39. Other members of this family possess lipid phosphatase activity (158), suggesting possible regulation through second messengers.

The importance of glycolysis compartmentation. Glycolysis is initiated by hexokinase (HXK) and phosphofructokinase (PFK), which phosphorylate glucose to produce fructose 1,6-bisphosphate (Fru1,6BP). These initial reactions consume two ATP molecules, while net ATP production only comes from downstream reactions. Because this ATP can fuel the activity of HXK and PFK beyond the capacity of downstream reactions, the overall reaction is autocatalytic and can result in the depletion of cellular ATP and the accumulation of hexose phosphate intermediates [glucose 6-phosphate (Glc6P), fructose 6-phosphate (Fru6P), and Fru1,6BP] to toxic levels. In most organisms, glycolysis is cytosolic and this consequence is avoided by tight negative-feedback regulation of HXK and PFK by hexose phosphates. This regulation ensures that upstream reactions do not exceed the capacity of downstream reactions. This type of regulation is absent in trypanosomes (15, 112). However, glycosomal reactions result in no net ATP synthesis, which only occurs in terminal cytosolic reactions, where this ATP is unavailable to glycosomal HXK and PFK. Thus, in trypanosomes, feedback regulation appears to be replaced by partial glycosomal compartmentation.

Several studies using RNAi to disrupt glycosomal function support this model. RNAi against Pex2 (42), Pex14 (30, 46, 60), Pex6, Pex10, or Pex12 (73) leads to the cytosolic mislocalization of glycosomal enzymes and cell death. Cell death can also be triggered by the addition of glucose (30), which results in the accumulation of Glc6P (46), and depletion of HXK has a protective effect on glycosome-deficient trypanosomes (60). The role of compartmentalization is further demonstrated in *Leishmania donovani*, another kinetoplastid parasite, in which the expression of a catalytically active PTS2-truncated HXK located in the cytoplasm also engenders glucose toxicity (74). Together, these data suggest that glycosomal compartmentation provides an alternative to allosteric regulation of HXK and PFK and protects trypanosomes from the autocatalytic nature of glycolysis (46).

Origins of glycosomal compartmentation. Analysis shows that the majority of glycolytic enzymes use PTS1 signals, whereas others utilize PTS2 and internal PTS (iPTS) signals (reviewed in Reference 14). Interestingly, functional analyses of these various PTS1 targeting signals show that glycosomes tolerate a greater degree of degeneration in the PTS1 sequence than do animal peroxisomes. For example, the signal -SSL is functional in trypanosomes but nonfunctional in animal cells (7). From an evolutionary perspective, the shift of glycolysis from a cytosolic to a glycosomal localization presents an interesting problem given that the PTS1 signal is acquired by random mutation and must have been acquired sequentially by these various enzymes. In this case, it is likely that early intermediates in these signals were partially functional and led to dual localization in peroxisomes and the cytosol as previously suggested (31, 41). This situation could sustain functional glycolysis until all the required enzymes had acquired rudimentary PTSs, at which point these could be refined by further mutation for full functionality. Further sampling of the peroxisomal environment may have been facilitated by the loose glycosomal PTS1 consensus (7), which is due presumably to variation in the PTS1 receptor PEX5.

Glycosomes as a target for therapeutic drugs. The emergence of drug-resistant trypanosomes is becoming a major problem, making the development of new medicines critical. The identification of several known peroxins and their associated functions in *Trypanosoma brucei* clearly supports the view that membrane biogenesis (3) and protein import (32) mechanisms are conserved between peroxisomes and glycosomes. However, the sequence conservation between human and trypanosomal peroxins and glycosomal enzymes is low (14). For example, TbPex19 possesses only 20% sequence identity to human PEX19 (3), 32% between TbPex7 and its human homolog (32), and 36% between HsPex6 and TbPex6 (73). This suggests that specific inhibitors can be designed to interfere with essential glycosomal functions without affecting human peroxisomal function.

Other Taxa-Specific Functions

Biotin biosynthesis. Among the eukaryotes, only plants and some fungi are able to synthesize biotin. Although the final steps are well characterized and occur in mitochondria, the initial events leading to the biosynthesis of biotin were unknown until recently. In the filamentous fungi *Aspergillus oryzae* and *Aspergillus nidulans*, peroxisome mutants are auxotrophic for biotin (87, 161). Moreover, the biotin biosynthetic enzyme encoded by the *bioF* gene (8-amino-7-oxononanoate synthase) has a PTS1 and its peroxisomal localization is required for biotin biosynthesis (87, 161), as is an intact β-oxidation cycle (87). Biotin auxotrophies induced by defects in β-oxidation can be complemented by pimelic acid supplementation, suggesting that the substrate for BioF is pimeloyl-CoA generated via peroxisomal β-oxidation (87). Peroxisomal targeting of the BioF ortholog also occurs in *Arabidopsis*, suggesting that peroxisomes are important for biotin biosynthesis in plants.

Secondary metabolism: penicillin biosynthesis. β-lactam antibiotics such as penicillins and related cephalosporins are produced as secondary metabolites by certain actinomycetes and filamentous fungi (e.g., *Penicillium*, *Aspergillus*, and *Acremonium* species). These compounds are of particular interest in the treatment of bacterial infections and contribute to over 40% of the total antibiotic market. The principal organism used for their production is the fungus *Penicillium chrysogenum*. Penicillin biosynthesis is initiated in the cytoplasm, where the three amino acids α-aminoadipic acid, cysteine, and valine are converted to a peptide by the nonribosomal peptide synthetase, δ-(L-aminoadipyl)-L-cysteinyl-D-valine. This peptide is then cyclized by isopenicillin N synthase to form a β-lactam to produce isopenicillin N (IPN). These first two steps occur in the cytosol and the intermediate IPN is then imported into the peroxisome through action of the PMP CefP (170). The final two reactions occur in the peroxisome: Isopenicillin-N:acyl-CoA acyltransferase (IAT) replaces the α-aminoadipyl side chain of IPN for a hydrophobic acyl group, which is provided by activity of the phenylacetyl CoA ligase (PCL) (100), and both of these enzymes possess consensus PTS1 signals (−ARL, alanine-arginine-leucine or Ala-Arg-Leu in IAT, and −AKL, alanine-lysine-leucine or Ala-Lys-Leu in PCL). The related cephalosporins (produced by *Acremonium chrysogenum*) are also produced from the IPN precursor, and two of these enzymes appear to be peroxisomal and utilize PTS1 signals (62, 93).

Several lines of evidence suggest that peroxisomal localization plays an important role in penicillin biosynthesis. A mutant in IAT lacking its PTS1 signal fails to produce penicillin (108), and a variety of mutants in *Penicillium chrysogenum* (100, 114) and *Aspergillus nidulans* (153) defective in matrix import are also defective in penicillin production. Peroxisome abundance also appears to correlate positively with penicillin production (100). Interestingly, overproduction of either PEX14/17 (114) or PEX11 (63) leads to a twofold increase in penicillin production. The latter case is probably working through increased peroxisome abundance. However, not all conditions that increase peroxisome abundance lead to increased penicillin production (100). For example, growth on fatty acids leads to increased peroxisome abundance and decreased penicillin production. Thus, physiological factors also affect the peroxisomal contribution to penicillin production. Recently, the penicillin biosynthetic pathway has been reconstituted in the yeast *H. polymorpha*, and here too strains lacking peroxisomes produce less penicillin (34). From all these data it is clear that peroxisomal localization is important for the biosynthesis of β-lactam antibiotics. However, precisely how peroxisomes support these biosynthetic pathways remains to be determined.

Secondary metabolism: AK-toxin and paxilline biosynthesis. The filamentous fungus *Alternaria alternata* includes seven pathogenic variants that produce host-specific AK (*Alternaria kikuchiana*) toxins, and causes black spot disease of Japanese pear. Three enzymes responsible for AK-toxin biosynthesis (Akt1p, Akt2p, and Akt3p) are targeted to peroxisomes through their PTS1 signals (SKI, serine-lysine-isoleucine or Ser-Lys-Ile; SKL, serine-lysine-leucine or Ser-Lys-Leu; and PKL, proline-lysine-leucine or Pro-Lys-Leu) (55). In addition, a *pex6* mutant is deficient in AK-toxin production and pathogenicity, suggesting that peroxisome localization of these enzymes is essential for AK-toxin biosynthesis. Paxilline is another secondary metabolite produced by *Penicillium paxilli*. This indole-diterpene has the ability to block calcium-activated potassium channels (141). Paxilline biosynthesis requires the activity of the geranylgeranyl-pyrophosphate (GGPP) synthase, PaxG, and this enzyme is localized in peroxisomes through a PTS1 signal, which is also required for PaxG function (137). From these two examples, it seems likely that additional links between peroxisomes and other secondary metabolic pathways will emerge in the future.

Sexual development in a filamentous ascomycete. A series of studies on the filamentous ascomycete *Podospora anserina* reveals a role for peroxisomes in sexual development (5, 8, 122). In the filamentous ascomycetes, sexual development takes place in multicellular fruiting bodies in which meiocytes (asci) differentiate to support karyogamy, meiosis, and sporulation (191). In *P. anserina*, deficiency for the RING complex peroxin PEX2 was originally shown to block meiotic commitment in meiocytes just before karyogamy (5). The two other components of the RING complex, PEX10 and PEX12, are also required for this process (122), but both matrix import receptors, PEX5 and PEX7, are dispensable (8) as are the docking peroxins PEX14 and PEX14/17 (121). However, the docking peroxin PEX13 and the PTS2 coreceptor PEX20 are required for meiocyte differentiation as are both PEX3 and PEX19. The requirement of a known cycling import receptor (PEX20), a docking peroxin (PEX13), and the RING peroxins suggests that the meiocyte-specific function of the peroxisome is associated with a modified matrix import cycle. Identifying the meiotic substrates of this import pathway remains a key challenge for the future and this should help resolve whether the meiocyte-specific function of peroxisomes is metabolic or cellular in nature. In addition, contrary to the case for *Podospora*, meiosis in *Aspergillus nidulans* does not require Pex13 (54) or Pex2 (53), and understanding this difference as well as the phylogenetic distribution of the link between peroxisomes and meiocyte development is another important question for future work.

Woronin Body Subcompartment

Filamentous fungi grow through the extension of cellular filaments (hyphae), in which individual cellular compartments are interconnected through conducting channels called septal pores.

GGPP: geranylgeranyl-pyrophosphate

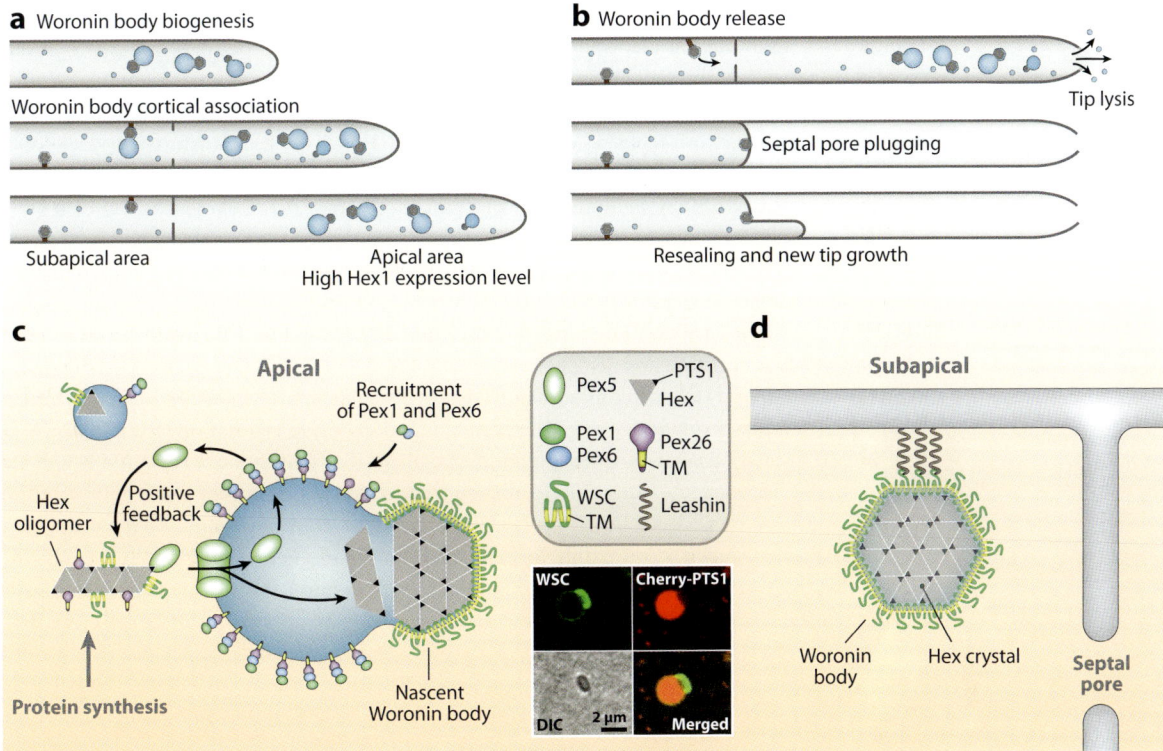

Figure 3

Woronin body biogenesis and function in *Neurospora crassa*. (*a*) Woronin body biogenesis in apical compartments. Woronin bodies are manufactured continuously in differentiated apical peroxisomes through a process determined in part by polarized *hex* gene expression. Coincident with septation, Woronin bodies attach to the cell cortex in a step that promotes their segregation into subapical compartments. (*b*) Woronin bodies in septal pore plugging, membrane resealing, and tip growth. Tip lysis triggers Woronin body release and septal pore plugging, followed by membrane resealing and the generation of new hyphal tips. Because tip lysis is likely the most common form of hyphal damage in nature, Woronin body production in apical compartments ensures that overall cell damage is minimal. (*c*) Model for differentiation of Woronin body–producing peroxisomes in *N. crassa*. Woronin bodies are produced in enlarged peroxisomes, which differentiate through the action of HEX oligomers. HEX acts by recruiting WSC (Woronin sorting complex) and PEX26 to these peroxisomes. (*Left*) HEX and PEX26 compose elements of a positive-feedback loop that promotes matrix import and the emergence of these differentiated peroxisomes. See text for additional information. (*Right*) A nascent Woronin body budding from the peroxisome. (*d*) The mature Woronin body is immobilized at the cell cortex by the tethering protein Leashin until signals from wounding trigger its release. Abbreviations: TM, transmembrane.

This syncytial multicellular organization allows the transport of cytoplasm and organelles between cells, permits cellular cooperativity, and is especially suited to support the invasive growth of saprotrophs and pathogens alike. However, this organization also carries the risk of uncontrolled cytoplasmic bleeding when hyphae are damaged. Woronin bodies evolved over 400 mya in the common ancestor of filamentous ascomycetes; they perform an adaptive function to seal septal pores in response to cellular wounding (56, 169) (**Figure 3*a,b***). Their biogenesis requires a dedicated machinery that promotes budding from the peroxisome and tethering to the cell cortex for segregation (**Figure 3*c***). Interestingly, the import of oligomeric PTS1-bearing cargo is central to this mechanism of subcompartment differentiation.

The self-assembled Woronin body dense-core. The Woronin body dense-core is composed of the PTS1-containing HEX protein, which is imported into peroxisomes (57, 88), where it self-assembles into solid micrometer-scale assemblies. The size of these structures appears to always exceed the diameter of the septal pore and this relationship relates to the pore-plugging function of the Woronin body (169). The deletion of *hex* abrogates Woronin body formation and leads to protoplasmic bleeding from septal pores in response to cell lysis (57, 94, 152). Moreover, expression of *HEX* in yeast leads to the formation of Woronin body–like dense-cores inside peroxisomes (57, 186), and recombinant HEX spontaneously crystallizes in vitro (57). Together, these data show that HEX is necessary and sufficient for Woronin body dense-core formation. The HEX crystal structure reveals three intermolecular contacts that promote formation of the HEX protein matrix. Mutants disrupting this assembly produce a soluble core, and despite attaining normal dimensions, these Woronin bodies are nonfunctional because they deform and pass through the septal pore during wound-induced plugging (189).

The overall fold of HEX is highly similar to that of eIF5a proteins, which are ancient cytoplasmic proteins that play a role in polypeptide chain elongation during translation (138). In addition, HEX and eIF5a share primary sequence homology, suggesting that they are related through ancestral gene duplication (57). Interestingly, the sequence at the C terminus of contemporary eIF5a proteins appears to be a single amino acid substitution away from attaining a PTS1 signal (56), and eIF5a has been shown to reversibly form tetramers and hexamers (12). This leads to the hypothesis that following *eIF5a* gene duplication, one copy acquired a PTS1 signal and was capable of undergoing a degree of self-assembly in the condensing environment of the peroxisome. This could have provided rudimentary Woronin body function that was improved over time through extensive mutation.

Apical biogenesis and budding. Woronin body biogenesis occurs in the growing apical compartment, where peroxisomes producing nascent Woronin bodies move in the cytoplasm in a generally tip-directed manner (**Figure 3a**). At this point HEX assemblies appear to be in the process of budding from the peroxisome; with a timing that roughly coincides with septation, these can be observed to associate with the cell cortex, where they are immobilized (104, 162). Following cortex association, these hybrid organelles undergo fission, which separates the Woronin body from its mother peroxisome, and this event depends on the action of Pex11 (26). The elastic hyphal tip is easily ruptured by hypotonic shock and this is likely to occur frequently in nature. Thus, apical biogenesis ensures that the first subapical compartment has functional Woronin bodies, and as a result the outcome of tip-lysis is usually pore-plugging at the first subapical septum (94) (**Figure 3b**). The expression of *HEX* mRNA is maximal in apical hyphal compartments and decreases rapidly in subapical regions of the colony (162). When the *HEX* expression pattern is respecified by swapping its promoter for one that is active subapically, Woronin body formation is redetermined to subapical compartments (162). Thus, the spatial pattern of *HEX* gene expression is a key determinant of the apical localization of Woronin body biogenesis. Many other transcripts are spatially regulated within the colony (59), and understanding the molecular basis for this control is an interesting area for future research.

Woronin body cores can be observed to bud from the peroxisome matrix at the level of both light and electron microscopy, and this process requires the Woronin body–specific membrane protein, WSC (Woronin sorting complex) (84). In *wsc* mutants, instead of budding, HEX assemblies vibrate randomly in the matrix of apical peroxisomes. Furthermore, these structures fail to associate with the cell cortex as they do in wild-type cells, resulting in their accumulation in the apical compartment. Moreover, this failure in segregation explains why *wsc* mutants display Woronin body loss-of-function phenotypes despite producing HEX assemblies. WSC is a four-pass transmembrane protein and in wild-type cells, it accumulates over budding HEX assemblies in nascent Woronin bodies and coats mature Woronin bodies. However, it is found at very low levels in the majority of peroxisomes, indicating that it is a Woronin body–specific membrane protein. WSC physically associates with HEX and together these data suggest that WSC is a membrane receptor for HEX assemblies. Interestingly, WSC behaves like a generic PMP in a *hex* deletion, and this was the first evidence that HEX might play a regulatory role in subcompartment specification (see below).

Overproduction of *WSC* in a *hex* deletion mutant results in the cortical association of peroxisomes that contain elevated levels of WSC. This suggests that cortex association depends on appropriate levels of WSC in the membrane. As WSC levels in nascent Woronin bodies depend on HEX, this provides a mechanism to ensure that segregation of Woronin bodies occurs only when HEX assemblies have reached appropriate dimensions. In yeast, peroxisome segregation is controlled by a balance between retention in mother cells through cortex binding and actomyosin-dependent transport into daughter cells. In this case, cortex association depends on the peroxisomal Inp1 protein, and the proportion of peroxisomes retained in mother cells can be increased by increasing levels of Inp1 (reviewed in Reference 27). Thus, protein-level-dependent cortex association may be widely used to control peroxisome segregation in fungal systems.

WSC is related to the PMP22 family of PMPs, which are also four-pass transmembrane PMPs. *PMP22* is lost in some fungal species, suggesting that it does not execute an essential function, and this conforms to the notion that *WSC* could have evolved through co-option of the PMP22 function. The nearest *PMP22* homolog of *WSC* is unable to complement a *wsc* deletion despite being targeted to peroxisomes, and this further supports the idea that WSC has new and divergent activities.

A tether for Woronin body segregation. In most of the Pezizomycotina, Woronin bodies are found in the vicinity of the septal pore. Upon being pulled away from the septum with laser tweezers (4), these recoil to their original position, suggesting attachment to the septum through an elastic tether. A mutant screen in *Neurospora crassa* identified the *leashin* locus and these large (∼500-kDa) cytosolic proteins appear to encode the Woronin body tether (111). *leashin* mutants accumulate WSC-enveloped nascent Woronin bodies in the apical compartment, indicating that Leashin is required for cortex association. An N-terminal domain of Leashin physically associates with WSC for localization to the Woronin body and the C terminus appears to be required for cortex binding. Central regions of Leashin proteins are not conserved but retain a similar character: they are acidic and enriched for the amino acids, PELS (proline–glutamic acid–leucine–serine or Pro-Glu-Leu-Ser). Similar PEVK repeats in the muscle protein titin adopt a random-coil structure that forms an elastic filament and this may also be true of Leashin. In addition, Woronin bodies are tethered to the pore at a distance of about 200 nm, and this spacing is in reasonable agreement with predicted dimensions of Leashin monomers.

Neurospora and its close relative *Sordaria* are unique within the clade Pezizomycotina in that they do not tether Woronin bodies to the septal pores, but rather localize them to the cell cortex in a dispersed pattern. Phylogenetic analysis suggests that this pattern is derived from the pore-associated pattern, and variation in Leashin appears to account for this difference. In *Neurosopora*, the *leashin* locus encodes two adjacent genes. The 5′ gene (*leashin-1*) encodes the N-terminal half of ancestral Leashin and maintains a function in Woronin body segregation, and the 3′ gene (*leashin-2*) encodes the C-terminal region and is localized to the septal pore. Remarkably, in *Neurospora*, a chromosomally encoded fusion of *leashin-1* and *leashin-2* can reproduce the ancestral pattern, suggesting that splitting of ancestral *leashin* indeed underlies the evolutionary transition in organelle localization. Interestingly, deletion of *leashin-2* does not interfere with Woronin body

NAD: nicotinamide adenine dinucleotide

FAD: flavin adenine dinucleotide

segregation but produces a defect in hyphal growth, suggesting that Leashin plays additional roles in controlling cellular development. Extensive protoplasmic streaming occurs through septal pores in both *Neurospora* (131) and *Sordaria*, and the emergence of this cellular physiology may have provided selective pressures for evolution of the delocalized pattern of cell cortex association (125).

HEX oligomers promote functional peroxisome differentiation.

HEX and WSC use consensus peroxisome targeting signals (PTS1 and mPTS, respectively), indicating that Woronin bodies are not differentiated from peroxisomes by special targeting signals. Moreover, Woronin bodies comprise a minor fraction of the total peroxisome population (84), and this leads to the question of how the abundance and composition of the subcompartment are controlled. This problem was resolved in part by experiments showing that enlarged Woronin body–producing peroxisomes are hypercompetent for matrix import and receive the majority of nascent PTS1 proteins. Thus, a disparity in import competence can account for the difference between Woronin body and peroxisome abundance (83).

Remarkably, differentiation of this subpopulation is self-organized by HEX. When HEX is absent, PTS1 trafficking becomes uniformly distributed to abundant small peroxisomes. Moreover, HEX is imported as an oligomer and mutations that abolish oligomer formation also abolish functional peroxisome differentiation. The activation of matrix import in a subset of peroxisomes suggested that HEX oligomers act by influencing the activity or localization of a key component of the matrix import pathway. Indeed, the tail-anchored peroxin PEX26, which functions to promote AAA ATPase membrane recruitment for receptor recycling (**Figure 1**), is associated with differentiation. PEX26 is found at elevated levels in the membrane of differentiated peroxisomes, and as with WSC (84), deletion of *HEX* results in uniform targeting of PEX26 to all peroxisomes. PEX26 physically interacts with HEX through its major cytoplasmic domain, suggesting that HEX might directly influence PEX26 localization. A *pex26* hypomorph that can still support matrix import but shows defects in differentiation results in increased Woronin body abundance, reduced size, and diminished organelle function. Together, these data suggest that HEX and PEX26 compose a positive-feedback loop to promote functional peroxisome differentiation and control subcompartment abundance (83) (**Figure 3c**). In this model, stochastic variations in the level of HEX import are amplified by the ability of HEX to recruit a key component of the import machinery, resulting in the import of more HEX proteins.

More work is required to validate this model and determine whether other peroxins are required for differentiation. In most cell types such as animal tissue culture cells, mature peroxisomes are uniform in size and composition (52), and this type of system can be used to reconstitute HEX-dependent peroxisome differentiation and define its minimal machinery. In the preimplex hypothesis, mutually multivalent interactions between nascent oligomeric matrix proteins and components of the import machinery were proposed to play an important role in the import process (37). Although the data concerning HEX do not suggest that its oligomerization is essential for import, they are consistent with a role in increasing import efficiency and promoting functional organelle differentiation. Examining how multivalent interactions between components of the import machinery and oligomeric cargo influence these processes will be an interesting area for future investigation.

Additional functions for Woronin bodies?

A number of diverse proteins involved in signaling and development have been associated with the Woronin body. TmpL is a multipass transmembrane protein containing N-terminal AMP-binding domain and C-terminal NAD(P)/flavin adenine nucleotide (FAD)-binding, which appears to segregate specifically to the membrane of Woronin bodies (65). TmpL is required for virulence of plant and animal pathogens; in *Alternaria brassicicola*, *tmpL* mutants are hypersensitive to oxidative stress and produce an excess of reactive oxygen species during plant infection, suggesting a role in redox homeostasis or signaling. In *Sordaria macrospora*, PRO40 is a WW domain protein required for fruiting body formation. PRO40 does not have putative transmembrane domains, but it localizes to Woronin bodies (22). Finally, in *Aspergillus nidulans*, the ApsB (anucleate primary sterigmata) protein is a component of the spindle pole body and a unique fungal septum-associated microtubule-organizing center. ApsB interacts with HEX in the yeast two-hybrid system and localizes to a subset of peroxisomes using a PTS2 signal (190). Although it remains unclear how HEX influences ApsB function, these data provide further evidence for heterogeneity of fungal peroxisomes and imply a possible role in cytoskeletal regulation.

In the case of TmpL and PRO40, the functional significance of Woronin body localization remains to be determined. A network of protein-protein interactions is required for Woronin body specification. HEX physically associates with WSC and PEX26 and Leashin interacts with WSC. These types of interactions could also account for the localization of other proteins to the Woronin body membrane or surface. Woronin bodies are physically distinct from the overall peroxisome population and localize to cell-to-cell channels near the plasma membrane. It is tempting to speculate that this unique organelle environment is suited especially to support functions in redox regulation and the control of multicellular development.

The clade Pezizomycotina is estimated to comprise 90% of ascomycetes and half of all fungi, and harbors the majority of plant and animal pathogens. On the basis of the phylogenetic distribution of *HEX*, *WSC*, and *leashin*, Woronin bodies were fully evolved at the origin of this group (56). In the future, it will be interesting to learn to what extent Woronin body function was further diversified to support additional functions and promote evolutionary radiation within this diverse group.

SUMMARY POINTS

1. The peroxisome can import oligomeric cargo. The mechanism of import may involve a transient aqueous pore composed of components of the docking complex and the cycling PTS receptors themselves.

2. PEX19 and its receptor PEX3 are believed to direct nascent PMPs directly to the peroxisome membrane, but they have also been associated with budding of pre-peroxisomes from the ER.

3. Peroxisomes proliferate by growth and division but can also form de novo from ER-derived precursors and are thus bona fide components of the endomembrane system.

4. Peroxisome function has diversified through the acquisition of new metabolic pathways or through subcompartment development, as exemplified by glycosomes and Woronin bodies, respectively.

5. A key step in the evolution of peroxisome diversity occurs when nonperoxisomal proteins attain PTS1 signals. The simplicity and degeneracy of this signal are likely to promote peroxisome evolvability.

6. Both glycosomes and Woronin bodies have been associated with signal transduction and cellular development.

7. The import of oligomeric HEX promotes both import efficiency and subcompartment differentiation. The latter depends on the ability of HEX import oligomers to influence the localization of specific PMPs.

FUTURE ISSUES

1. Pex19 and Pex3 have been associated with direct targeting of nascent PMPs to the peroxisome membrane and with budding of pre-peroxisomal vesicles from the ER. Further studies are required to understand whether these activities are distinct or mechanistically related.

2. In the yeast *Saccharomyces cerevisiae*, the major pathway for peroxisome proliferation is growth and division. However, the extent to which de novo formation contributes to peroxisome renewal in other microorganisms and during developmental transitions remains to be determined.

3. PMPs appear to traffic both directly to the peroxisome and via the ER. Understanding the intrinsic targeting signals, chaperones, and membrane integrases that differentiate these pathways will help clarify the mechanisms involved in peroxisome membrane biogenesis.

4. Peroxisomes have been associated with signal transduction and developmental decisions, but more work is required to determine their precise roles in these processes.

5. As genomes are sequenced in diverse eukaryotes, bioinformatic searches for PTS1 signals in predicted proteomes can be used to further explore peroxisome diversity.

DISCLOSURE STATEMENT

The authors are not aware of any affiliations, memberships, funding, or financial holdings that might be perceived as affecting the objectivity of this review.

ACKNOWLEDGMENTS

Work in the Jedd lab is supported by the Temasek Life Sciences Laboratory. Our apologies to those whose work was not included due to space limitations.

LITERATURE CITED

1. Agne B, Meindl NM, Niederhoff K, Einwächter H, Rehling P, et al. 2003. Pex8p: an intraperoxisomal organizer of the peroxisomal import machinery. *Mol. Cell* 11(3):635–46
2. Agrawal G, Joshi S, Subramani S. 2011. Cell-free sorting of peroxisomal membrane proteins from the endoplasmic reticulum. *Proc. Natl. Acad. Sci. USA* 108(22):9113–18
3. Banerjee SK, Kessler PS, Saveria T, Parsons M. 2005. Identification of trypanosomatid PEX19: functional characterization reveals impact on cell growth and glycosome size and number. *Mol. Biochem. Parasitol.* 142(1):47–55
4. Berns MW, Aist JR, Wright WH, Liang H. 1992. Optical trapping in animal and fungal cells using a tunable, near-infrared titanium-sapphire laser. *Exp. Cell Res.* 198(2):375–78
5. Berteaux-Lecellier V, Picard M, Thompson-Coffe C, Zickler D, Panvier-Adoutte A, Simonet JM. 1995. A nonmammalian homolog of the *PAF7* gene (Zellweger syndrome) discovered as a gene involved in caryogamy in the fungus *Podospora anserina*. *Cell* 81(7):1043–51
6. Birschmann I, Stroobants AK, van den Berg M, Schäfer A, Rosenkranz K, et al. 2003. Pex15p of *Saccharomyces cerevisiae* provides a molecular basis for recruitment of the AAA peroxin Pex6p to peroxisomal membranes. *Mol. Biol. Cell* 14(6):2226–36
7. Blattner J, Swinkels B, Dörsam H, Prospero T, Subramani S, Clayton C. 1992. Glycosome assembly in trypanosomes: variations in the acceptable degeneracy of a COOH-terminal microbody targeting signal. *J. Cell Biol.* 119(5):1129–36

8. Bonnet C, Espagne E, Zickler D, Boisnard S, Bourdais A, Berteaux-Lecellier V. 2006. The peroxisomal import proteins PEX2, PEX5 and PEX7 are differently involved in *Podospora anserina* sexual cycle. *Mol. Microbiol.* 62(1):157–69
9. Braverman N, Dodt G, Gould SJ, Valle D. 1998. An isoform of Pex5p, the human PTS1 receptor, is required for the import of PTS2 proteins into peroxisomes. *Hum. Mol. Genet.* 7(8):1195–205
10. Bringaud F, Rivière L, Coustou V. 2006. Energy metabolism of trypanosomatids: adaptation to available carbon sources. *Mol. Biochem. Parasitol.* 149(1):1–9
11. Brocard C, Kragler F, Simon MM, Schuster T, Hartig A. 1994. The tetratricopeptide repeat-domain of the PAS10 protein of *Saccharomyces cerevisiae* is essential for binding the peroxisomal targeting signal-SKL. *Biochem. Biophys. Res. Commun.* 204(3):1016–22
12. Chung SI, Park MH, Folk JE, Lewis MS. 1991. Eukaryotic initiation factor 5A: the molecular form of the hypusine-containing protein from human erythrocytes. *Biochim. Biophys. Acta* 1076(3):448–51
13. Colasante C, Ellis M, Ruppert T, Voncken F. 2006. Comparative proteomics of glycosomes from bloodstream form and procyclic culture form *Trypanosoma brucei brucei*. *Proteomics* 6(11):3275–93
14. Coley AF, Dodson HC, Morris MT, Morris JC. 2011. Glycolysis in the African trypanosome: targeting enzymes and their subcellular compartments for therapeutic development. *Mol. Biol. Int.* 2011:123702
15. Cronin CN, Tipton KF. 1987. Kinetic studies on the reaction catalysed by phosphofructokinase from *Trypanosoma brucei*. *Biochem. J.* 245(1):13–18
16. Czichos J, Nonnengaesser C, Overath P. 1986. *Trypanosoma brucei*: *cis*-aconitate and temperature reduction as triggers of synchronous transformation of bloodstream to procyclic trypomastigotes in vitro. *Exp. Parasitol.* 62(2):283–91
17. **Dammai V, Subramani S. 2001. The human peroxisomal targeting signal receptor, Pex5p, is translocated into the peroxisomal matrix and recycled to the cytosol. *Cell* 105(2):187–96**
18. Dixit E, Boulant S, Zhang Y, Lee ASY, Odendall C, et al. 2010. Peroxisomes are signaling platforms for antiviral innate immunity. *Cell* 141(4):668–81
19. Elgersma Y, Kwast L, Klein A, Voorn-Brouwer T, van Den Berg M, et al. 1996. The SH3 domain of the *Saccharomyces cerevisiae* peroxisomal membrane protein Pex13p functions as a docking site for Pex5p, a mobile receptor for the import PTS1-containing proteins. *J. Cell Biol.* 135(1):97–109
20. Elgersma Y, Kwast L, van Den Berg M, Snyder WB, Distel B, et al. 1997. Overexpression of Pex15p, a phosphorylated peroxisomal integral membrane protein required for peroxisome assembly in *S. cerevisiae*, causes proliferation of the endoplasmic reticulum membrane. *EMBO J.* 16(24):7326–41
21. Elgersma Y, Vos A, van Den Berg M, van Roermund CW, van der Sluijs P, et al. 1996. Analysis of the carboxyl-terminal peroxisomal targeting signal 1 in a homologous context in *Saccharomyces cerevisiae*. *J. Biol. Chem.* 271(42):26375–82
22. Engh I, Würtz C, Witzel-Schlömp K, Zhang HY, Hoff B, et al. 2007. The WW domain protein PRO40 is required for fungal fertility and associates with Woronin bodies. *Eukaryot. Cell* 6(5):831–43
23. Erdmann R, Blobel G. 1995. Giant peroxisomes in oleic acid-induced *Saccharomyces cerevisiae* lacking the peroxisomal membrane protein Pmp27p. *J. Cell Biol.* 128(4):509–23
24. Erdmann R, Blobel G. 1996. Identification of Pex13p a peroxisomal membrane receptor for the PTS1 recognition factor. *J. Cell Biol.* 135(1):111–21
25. Erdmann R, Schliebs W. 2005. Peroxisomal matrix protein import: the transient pore model. *Nat. Rev. Mol. Cell Biol.* 6(9):738–42
26. Escaño CS, Juvvadi PR, Jin FJ, Takahashi T, Koyama Y, et al. 2009. Disruption of the Ao*pex11-1* gene involved in peroxisome proliferation leads to impaired Woronin body formation in *Aspergillus oryzae*. *Eukaryot. Cell* 8(3):296–305
27. Fagarasanu A, Mast FD, Knoblach B, Rachubinski RA. 2010. Molecular mechanisms of organelle inheritance: lessons from peroxisomes in yeast. *Nat. Rev. Mol. Cell Biol.* 11(9):644–54
28. Fang Y, Morrell JC, Jones JM, Gould SJ. 2004. PEX3 functions as a PEX19 docking factor in the import of class I peroxisomal membrane proteins. *J. Cell Biol.* 164(6):863–75
29. Fujiki Y, Lazarow PB. 1985. Post-translational import of fatty acyl-CoA oxidase and catalase into peroxisomes of rat liver in vitro. *J. Biol. Chem.* 260(9):5603–609
30. Furuya T, Kessler P, Jardim A, Schnaufer A, Crudder C, Parsons M. 2002. Glucose is toxic to glycosome-deficient trypanosomes. *Proc. Natl. Acad. Sci. USA* 99(22):14177–82

17. The first paper to show that a matrix import receptor cycles through the peroxisome matrix.

31. Gabaldón T. 2010. Peroxisome diversity and evolution. *Philos. Trans. R. Soc. Lond. B Biol. Sci.* 365(1541):765–73
32. Galland N, Demeure F, Hannaert V, Verplaetse E, Vertommen D, et al. 2007. Characterization of the role of the receptors PEX5 and PEX7 in the import of proteins into glycosomes of *Trypanosoma brucei*. *Biochim. Biophys. Acta* 1773(4):521–35
33. Gatto GJ, Geisbrecht BV, Gould SJ, Berg JM. 2000. Peroxisomal targeting signal-1 recognition by the TPR domains of human PEX5. *Nat. Struct. Biol.* 7(12):1091–95
34. Gidijala L, Kiel JAKW, Douma RD, Seifar RM, van Gulik WM, et al. 2009. An engineered yeast efficiently secreting penicillin. *PLoS ONE* 4(12):e8317

> 35. The first paper to demonstrate piggyback import of a matrix protein.

35. **Glover JR, Andrews DW, Rachubinski RA. 1994. *Saccharomyces cerevisiae* peroxisomal thiolase is imported as a dimer. *Proc. Natl. Acad. Sci. USA* 91(22):10541–45**
36. Gonzalez NH, Felsner G, Schramm FD, Klingl A, Maier U-G, Bolte K. 2011. A single peroxisomal targeting signal mediates matrix protein import in diatoms. *PLoS ONE* 6(9):e25316
37. Gould SJ, Collins CS. 2002. Opinion: peroxisomal-protein import: is it really that complex? *Nat. Rev. Mol. Cell Biol.* 3(5):382–89
38. Gould SJ, Keller GA, Hosken N, Wilkinson J, Subramani S. 1989. A conserved tripeptide sorts proteins to peroxisomes. *J. Cell Biol.* 108(5):1657–64
39. Gouveia AM, Reguenga C, Oliveira ME, Sá-Miranda C, Azevedo JE. 2000. Characterization of peroxisomal Pex5p from rat liver. Pex5p in the Pex5p-Pex14p membrane complex is a transmembrane protein. *J. Biol. Chem.* 275(42):32444–51
40. Girzalsky W, Saffian D, Erdmann R. 2010. Peroxisomal protein translocation. *Biochim. Biophys. Acta* 1803(6):724–31
41. Gualdrón-López M, Brennand A, Hannaert V, Quiñones W, Cáceres AJ, et al. 2012. When, how and why glycolysis became compartmentalised in the Kinetoplastea. A new look at an ancient organelle. *Int. J. Parasitol.* 42(1):1–20
42. Guerra-Giraldez C, Quijada L, Clayton CE. 2002. Compartmentation of enzymes in a microbody, the glycosome, is essential in *Trypanosoma brucei*. *J. Cell. Sci.* 115(Pt. 13):2651–58
43. Gunkel K, van Dijk R, Veenhuis M, van der Klei IJ. 2004. Routing of *Hansenula polymorpha* alcohol oxidase: an alternative peroxisomal protein-sorting machinery. *Mol. Biol. Cell* 15(3):1347–55
44. Guo T, Gregg C, Boukh-Viner T, Kyryakov P, Goldberg A, et al. 2007. A signal from inside the peroxisome initiates its division by promoting the remodeling of the peroxisomal membrane. *J. Cell Biol.* 177(2):289–303
45. Guo T, Kit YY, Nicaud J-M, Le Dall M-T, Sears SK, et al. 2003. Peroxisome division in the yeast *Yarrowia lipolytica* is regulated by a signal from inside the peroxisome. *J. Cell Biol.* 162(7):1255–66
46. Haanstra JR, van Tuijl A, Kessler P, Reijnders W, Michels PAM, et al. 2008. Compartmentation prevents a lethal turbo-explosion of glycolysis in trypanosomes. *Proc. Natl. Acad. Sci. USA* 105(46):17718–23
47. Halbach A, Rucktäschel R, Rottensteiner H, Erdmann R. 2009. The N-domain of Pex22p can functionally replace the Pex3p N-domain in targeting and peroxisome formation. *J. Biol. Chem.* 284(6):3906–16
48. Hazra PP, Suriapranata I, Snyder WB, Subramani S. 2002. Peroxisome remnants in *pex3*Δ cells and the requirement of Pex3p for interactions between the peroxisomal docking and translocation subcomplexes. *Traffic* 3(8):560–74
49. Hettema EH, Girzalsky W, van Den Berg M, Erdmann R, Distel B. 2000. *Saccharomyces cerevisiae* Pex3p and Pex19p are required for proper localization and stability of peroxisomal membrane proteins. *EMBO J.* 19(2):223–33

> 50. Demonstrates the de novo production of peroxisomes from an ER-derived precursor.

50. **Hoepfner D, Schildknegt D, Braakman I, Philippsen P, Tabak HF. 2005. Contribution of the endoplasmic reticulum to peroxisome formation. *Cell* 122(1):85–95**
51. Hoepfner D, van Den Berg M, Philippsen P, Tabak HF, Hettema EH. 2001. A role for Vps1p, actin, and the Myo2p motor in peroxisome abundance and inheritance in Saccharomyces cerevisiae. *J. Cell Biol.* 155(6):979–90
52. Huybrechts SJ, Van Veldhoven PP, Brees C, Mannaerts GP, Los GV, Fransen M. 2009. Peroxisome dynamics in cultured mammalian cells. *Traffic* 10(11):1722–33
53. Hynes MJ, Murray SL, Kahn FK. 2010. Deletion of the RING-finger peroxin 2 gene in *Aspergillus nidulans* does not affect meiotic development. *FEMS Microbiol. Lett.* 306(1):67–71

54. Hynes MJ, Murray SL, Khew GS, Davis MA. 2008. Genetic analysis of the role of peroxisomes in the utilization of acetate and fatty acids in *Aspergillus nidulans*. *Genetics* 178(3):1355–69
55. Imazaki A, Tanaka A, Harimoto Y, Yamamoto M, Akimitsu K, et al. 2010. Contribution of peroxisomes to secondary metabolism and pathogenicity in the fungal plant pathogen Alternaria alternata. *Eukaryot. Cell* 9(5):682–94
56. Jedd G. 2011. Fungal evo-devo: organelles and multicellular complexity. *Trends Cell Biol.* 21(1):12–19
57. Jedd G, Chua NH. 2000. A new self-assembled peroxisomal vesicle required for efficient resealing of the plasma membrane. *Nat. Cell Biol.* 2(4):226–31
58. Jones JM, Morrell JC, Gould SJ. 2004. PEX19 is a predominantly cytosolic chaperone and import receptor for class 1 peroxisomal membrane proteins. *J. Cell Biol.* 164(1):57–67
59. Kasuga T, Glass NL. 2008. Dissecting colony development of *Neurospora crassa* using mRNA profiling and comparative genomics approaches. *Eukaryot. Cell* 7(9):1549–64
60. Kessler PS, Parsons M. 2005. Probing the role of compartmentation of glycolysis in procyclic form *Trypanosoma brucei*: RNA interference studies of PEX14, hexokinase, and phosphofructokinase. *J. Biol. Chem.* 280(10):9030–36
61. Kiel JAKW, Emmrich K, Meyer HE, Kunau W-H. 2005. Ubiquitination of the peroxisomal targeting signal type 1 receptor, Pex5p, suggests the presence of a quality control mechanism during peroxisomal matrix protein import. *J. Biol. Chem.* 280(3):1921–30
62. Kiel JAKW, van den Berg MA, Fusetti F, Poolman B, Bovenberg RAL, et al. 2009. Matching the proteome to the genome: the microbody of penicillin-producing *Penicillium chrysogenum* cells. *Funct. Integr. Genomics* 9(2):167–84
63. Kiel JAKW, van der Klei IJ, van den Berg MA, Bovenberg RAL, Veenhuis M. 2005. Overproduction of a single protein, Pc-Pex11p, results in 2-fold enhanced penicillin production by *Penicillium chrysogenum*. *Fungal Genet. Biol.* 42(2):154–64
64. Kiel JAKW, Veenhuis M, van der Klei IJ. 2006. PEX genes in fungal genomes: common, rare or redundant. *Traffic* 7(10):1291–303
65. Kim K-H, Willger SD, Park S-W, Puttikamonkul S, Grahl N, et al. 2009. TmpL, a transmembrane protein required for intracellular redox homeostasis and virulence in a plant and an animal fungal pathogen. *PLoS Pathog.* 5(11):e1000653
66. Kim PK, Mullen RT, Schumann U, Lippincott-Schwartz J. 2006. The origin and maintenance of mammalian peroxisomes involves a de novo PEX16-dependent pathway from the ER. *J. Cell Biol.* 173(4):521–32
67. Kimura A, Takano Y, Furusawa I, Okuno T. 2001. Peroxisomal metabolic function is required for appressorium-mediated plant infection by *Colletotrichum lagenarium*. *Plant Cell* 13(8):1945–57
68. Kleff S, Sander S, Mielke G, Eising R. 1997. The predominant protein in peroxisomal cores of sunflower cotyledons is a catalase that differs in primary structure from the catalase in the peroxisomal matrix. *Eur. J. Biochem.* 245(2):402–10
69. Klein ATJ, van den Berg M, Bottger G, Tabak HF, Distel B. 2002. *Saccharomyces cerevisiae* acyl-CoA oxidase follows a novel, non-PTS1, import pathway into peroxisomes that is dependent on Pex5p. *J. Biol. Chem.* 277(28):25011–19
70. Knoblach B, Rachubinski RA. 2010. Phosphorylation-dependent activation of peroxisome proliferator protein PEX11 controls peroxisome abundance. *J. Biol. Chem.* 285(9):6670–80
71. Koller A, Snyder WB, Faber KN, Wenzel TJ, Rangell L, et al. 1999. Pex22p of *Pichia pastoris*, essential for peroxisomal matrix protein import, anchors the ubiquitin-conjugating enzyme, Pex4p, on the peroxisomal membrane. *J. Cell Biol.* 146(1):99–112
72. Kragt A, Voorn-Brouwer T, van den Berg M, Distel B. 2005. Endoplasmic reticulum-directed Pex3p routes to peroxisomes and restores peroxisome formation in a *Saccharomyces cerevisiae pex3Δ* strain. *J. Biol. Chem.* 280(40):34350–57
73. Krazy H, Michels PAM. 2006. Identification and characterization of three peroxins—PEX6, PEX10 and PEX12—involved in glycosome biogenesis in *Trypanosoma brucei*. *Biochim. Biophys. Acta* 1763(1):6–17
74. Kumar R, Gupta S, Srivastava R, Sahasrabuddhe AA, Gupta CM. 2010. Expression of a PTS2-truncated hexokinase produces glucose toxicity in Leishmania donovani. *Mol. Biochem. Parasitol.* 170(1):41–44

75. Kuravi K, Nagotu S, Krikken AM, Sjollema K, Deckers M, et al. 2006. Dynamin-related proteins Vps1p and Dnm1p control peroxisome abundance in *Saccharomyces cerevisiae*. *J. Cell. Sci.* 119(Pt. 19):3994–4001
76. Labarca P, Wolff D, Soto U, Necochea C, Leighton F. 1986. Large cation-selective pores from rat liver peroxisomal membranes incorporated to planar lipid bilayers. *J. Membr. Biol.* 94(3):285–91

77. Defines an in vitro system that allows the study of PEX3- and PEX19-dependent vesicle production from the ER (also see Reference 2)

77. **Lam SK, Yoda N, Schekman R. 2010. A vesicle carrier that mediates peroxisome protein traffic from the endoplasmic reticulum.** *Proc. Natl. Acad. Sci. USA* **107(50):21523–28**
78. Lee JR, Jang HH, Park JH, Jung JH, Lee SS, et al. 2006. Cloning of two splice variants of the rice PTS1 receptor, OsPex5pL and OsPex5pS, and their functional characterization using *pex5*-deficient yeast and *Arabidopsis. Plant J.* 47(3):457–66
79. Lemmens M, Verheyden K, Van Veldhoven P, Vereecke J, Mannaerts GP, Carmeliet E. 1989. Single-channel analysis of a large conductance channel in peroxisomes from rat liver. *Biochim. Biophys. Acta* 984(3):351–59
80. Léon S, Goodman JM, Subramani S. 2006. Uniqueness of the mechanism of protein import into the peroxisome matrix: transport of folded, co-factor-bound and oligomeric proteins by shuttling receptors. *Biochim. Biophys. Acta* 1763(12):1552–64
81. Léon S, Subramani S. 2007. A conserved cysteine residue of *Pichia pastoris* Pex20p is essential for its recycling from the peroxisome to the cytosol. *J. Biol. Chem.* 282(10):7424–30
82. Léon S, Zhang L, McDonald WH, Yates J, Cregg JM, Subramani S. 2006. Dynamics of the peroxisomal import cycle of PpPex20p: ubiquitin-dependent localization and regulation. *J. Cell Biol.* 172(1):67–78

83. Shows that a piggyback-imported matrix oligomer can increase import efficiency and promote functional peroxisome differentiation.

83. **Liu F, Lu Y, Pieuchot L, Dhavale T, Jedd G. 2011. Import oligomers induce positive feedback to promote peroxisome differentiation and control organelle abundance.** *Dev. Cell* **21(3):457–68**
84. Liu F, Ng SK, Lu Y, Low W, Lai J, Jedd G. 2008. Making two organelles from one: Woronin body biogenesis by peroxisomal protein sorting. *J. Cell Biol.* 180(2):325–39
85. Ma C, Agrawal G, Subramani S. 2011. Peroxisome assembly: matrix and membrane protein biogenesis. *J. Cell Biol.* 193(1):7–16
86. Ma C, Schumann U, Rayapuram N, Subramani S. 2009. The peroxisomal matrix import of Pex8p requires only PTS receptors and Pex14p. *Mol. Biol. Cell* 20(16):3680–89
87. Magliano P, Flipphi M, Arpat BA, Delessert S, Poirier Y. 2011. Contributions of the peroxisome and the β-oxidation cycle to biotin synthesis in fungi. *J. Biol. Chem.* 286(49):42133–40
88. Managadze D, Würtz C, Sichting M, Niehaus G, Veenhuis M, Rottensteiner H. 2007. The peroxin PEX14 of *Neurospora crassa* is essential for the biogenesis of both glyoxysomes and Woronin bodies. *Traffic* 8(6):687–701
89. Managadze D, Würtz C, Wiese S, Schneider M, Girzalsky W, et al. 2010. Identification of PEX33, a novel component of the peroxisomal docking complex in the filamentous fungus *Neurospora crassa*. *Eur. J. Cell Biol.* 89(12):955–64
90. Mano S, Nishimura M. 2005. Plant peroxisomes. *Vitam. Horm.* 72:111–54
91. Marelli M, Smith JJ, Jung S, Yi E, Nesvizhskii AI, et al. 2004. Quantitative mass spectrometry reveals a role for the GTPase Rho1p in actin organization on the peroxisome membrane. *J. Cell Biol.* 167(6):1099–112
92. Marshall PA, Krimkevich YI, Lark RH, Dyer JM, Veenhuis M, Goodman JM. 1995. Pmp27 promotes peroxisomal proliferation. *J. Cell Biol.* 129(2):345–55
93. Martín JF, Ullán RV, García-Estrada C. 2010. Regulation and compartmentalization of β-lactam biosynthesis. *Microb. Biotechnol.* 3(3):285–99
94. Maruyama J-I, Juvvadi PR, Ishi K, Kitamoto K. 2005. Three-dimensional image analysis of plugging at the septal pore by Woronin body during hypotonic shock inducing hyphal tip bursting in the filamentous fungus *Aspergillus oryzae*. *Biochem. Biophys. Res. Commun.* 331(4):1081–88
95. Matsumoto N, Tamura S, Fujiki Y. 2003. The pathogenic peroxin Pex26p recruits the Pex1p-Pex6p AAA ATPase complexes to peroxisomes. *Nat. Cell Biol.* 5(5):454–60
96. Matsuzaki T, Fujiki Y. 2008. The peroxisomal membrane protein import receptor Pex3p is directly transported to peroxisomes by a novel Pex19p- and Pex16p-dependent pathway. *J. Cell Biol.* 183(7):1275–86
97. Matsuzono Y, Fujiki Y. 2006. In vitro transport of membrane proteins to peroxisomes by shuttling receptor Pex19p. *J. Biol. Chem.* 281(1):36–42

98. McNew JA, Goodman JM. 1994. An oligomeric protein is imported into peroxisomes in vivo. *J. Cell Biol.* 127(5):1245–57
99. McNew JA, Goodman JM. 1996. The targeting and assembly of peroxisomal proteins: Some old rules do not apply. *Trends Biochem. Sci.* 21(2):54–58
100. Meijer WH, Gidijala L, Fekken S, Kiel JAKW, van den Berg MA, et al. 2010. Peroxisomes are required for efficient penicillin biosynthesis in *Penicillium chrysogenum*. *Appl. Environ. Microbiol.* 76(17):5702–9
101. Meinecke M, Cizmowski C, Schliebs W, Krüger V, Beck S, et al. 2010. The peroxisomal importomer constitutes a large and highly dynamic pore. *Nat. Cell Biol.* 12(3):273–77
102. Misset O, Bos OJ, Opperdoes FR. 1986. Glycolytic enzymes of *Trypanosoma brucei*. Simultaneous purification, intraglycosomal concentrations and physical properties. *Eur. J. Biochem.* 157(2):441–53
103. Miyata N, Fujiki Y. 2005. Shuttling mechanism of peroxisome targeting signal type 1 receptor Pex5: ATP-independent import and ATP-dependent export. *Mol. Cell. Biol.* 25(24):10822–32
104. Momany M, Richardson EA, Van Sickle C, Jedd G. 2002. Mapping Woronin body position in *Aspergillus nidulans*. *Mycologia* 94(2):260–66
105. Motley AM, Hettema EH. 2007. Yeast peroxisomes multiply by growth and division. *J. Cell Biol.* 178(3):399–410
106. Motley AM, Hettema EH, Ketting R, Plasterk R, Tabak HF. 2000. *Caenorhabditis elegans* has a single pathway to target matrix proteins to peroxisomes. *EMBO Rep.* 1(1):40–46
107. Motley AM, Ward GP, Hettema EH. 2008. Dnm1p-dependent peroxisome fission requires Caf4p, Mdv1p and Fis1p. *J. Cell. Sci.* 121(Pt. 10):1633–40
108. Müller WH, Bovenberg RA, Groothuis MH, Kattevilder F, Smaal EB, et al. 1992. Involvement of microbodies in penicillin biosynthesis. *Biochim. Biophys. Acta* 1116(2):210–13
109. Muntau AC, Roscher AA, Kunau W-H, Dodt G. 2003. The interaction between human PEX3 and PEX19 characterized by fluorescence resonance energy transfer (FRET) analysis. *Eur. J. Cell Biol.* 82(7):333–42
110. Nair DM, Purdue PE, Lazarow PB. 2004. Pex7p translocates in and out of peroxisomes in *Saccharomyces cerevisiae*. *J. Cell Biol.* 167(4):599–604
111. Ng SK, Liu F, Lai J, Low W, Jedd G. 2009. A tether for Woronin body inheritance is associated with evolutionary variation in organelle positioning. *PLoS Genet.* 5(6):e1000521
112. Nwagwu M, Opperdoes FR. 1982. Regulation of glycolysis in *Trypanosoma brucei*: hexokinase and phosphofructokinase activity. *Acta Trop.* 39(1):61–72
113. Okamoto K, Shaw JM. 2005. Mitochondrial morphology and dynamics in yeast and multicellular eukaryotes. *Annu. Rev. Genet.* 39:503–36
114. Opaliński Ł, Kiel JAKW, Homan TG, Veenhuis M, van der Klei IJ. 2010. *Penicillium chrysogenum* Pex14/17p—a novel component of the peroxisomal membrane that is important for penicillin production. *FEBS J.* 277(15):3203–18
115. Opaliński Ł, Kiel JAKW, Williams C, Veenhuis M, van der Klei IJ. 2011. Membrane curvature during peroxisome fission requires Pex11. *EMBO J.* 30(1):5–16
116. Opaliński Ł, Veenhuis M, van der Klei IJ. 2011. Peroxisomes: membrane events accompanying peroxisome proliferation. *Int. J. Biochem. Cell Biol.* 43(6):847–51
117. Osumi T, Tsukamoto T, Hata S, Yokota S, Miura S, et al. 1991. Amino-terminal presequence of the precursor of peroxisomal 3-ketoacyl-CoA thiolase is a cleavable signal peptide for peroxisomal targeting. *Biochem. Biophys. Res. Commun.* 181(3):947–54
118. Otera H, Harano T, Honsho M, Ghaedi K, Mukai S, et al. 2000. The mammalian peroxin Pex5pL, the longer isoform of the mobile peroxisome targeting signal (PTS) type 1 transporter, translocates the Pex7p.PTS2 protein complex into peroxisomes via its initial docking site, Pex14p. *J. Biol. Chem.* 275(28):21703–14
119. Otera H, Setoguchi K, Hamasaki M, Kumashiro T, Shimizu N, Fujiki Y. 2002. Peroxisomal targeting signal receptor Pex5p interacts with cargoes and import machinery components in a spatiotemporally differentiated manner: Conserved Pex5p WXXXF/Y motifs are critical for matrix protein import. *Mol. Cell. Biol.* 22(6):1639–55
120. Otzen M, Wang D, Lunenborg MGJ, van der Klei IJ. 2005. *Hansenula polymorpha* Pex20p is an oligomer that binds the peroxisomal targeting signal 2 (PTS2). *J. Cell. Sci.* 118(Pt. 15):3409–18

---

101. Provides evidence supporting peroxisome matrix import through a transient pore that opens to varying degrees depending on the size of cargo.

105. Indicates that yeast peroxisomes proliferate mainly by growth and division.

115. Shows that Pex11 is a membrane morphogenic protein, providing a mechanism for tubule formation in the process of peroxisome proliferation.

121. Peraza-Reyes L, Arnaise S, Zickler D, Coppin E, Debuchy R, Berteaux-Lecellier V. 2011. The importomer peroxins are differentially required for peroxisome assembly and meiotic development in *Podospora anserina*: insights into a new peroxisome import pathway. *Mol. Microbiol.* 82(2):365–77
122. Peraza-Reyes L, Zickler D, Berteaux-Lecellier V. 2008. The peroxisome RING-finger complex is required for meiocyte formation in the fungus *Podospora anserina*. *Traffic* 9(11):1998–2009
123. Petriv OI, Tang L, Titorenko VI, Rachubinski RA. 2004. A new definition for the consensus sequence of the peroxisome targeting signal type 2. *J. Mol. Biol.* 341(1):119–34
124. Pinto MP, Grou CP, Alencastre IS, Oliveira ME, Sá-Miranda C, et al. 2006. The import competence of a peroxisomal membrane protein is determined by Pex19p before the docking step. *J. Biol. Chem.* 281(45):34492–502
125. Plamann M. 2009. Cytoplasmic streaming in *Neurospora*: disperse the plug to increase the flow? *PLoS Genet.* 5(6):e1000526
126. Platta HW, Debelyy MO, Magraoui El F, Erdmann R. 2008. The AAA peroxins Pex1p and Pex6p function as dislocases for the ubiquitinated peroxisomal import receptor Pex5p. *Biochem. Soc. Trans.* 36(Pt. 1):99–104
127. Platta HW, Magraoui El F, Bäumer BE, Schlee D, Girzalsky W, Erdmann R. 2009. Pex2 and pex12 function as protein-ubiquitin ligases in peroxisomal protein import. *Mol. Cell. Biol.* 29(20):5505–16
128. Platta HW, Magraoui El F, Schlee D, Grunau S, Girzalsky W, Erdmann R. 2007. Ubiquitination of the peroxisomal import receptor Pex5p is required for its recycling. *J. Cell Biol.* 177(2):197–204
129. Praefcke GJK, McMahon HT. 2004. The dynamin superfamily: universal membrane tubulation and fission molecules? *Nat. Rev. Mol. Cell Biol.* 5(2):133–47
130. Purdue PE, Yang X, Lazarow PB. 1998. Pex18p and Pex21p, a novel pair of related peroxins essential for peroxisomal targeting by the PTS2 pathway. *J. Cell Biol.* 143(7):1859–69
131. Ramos-García SL, Roberson RW, Freitag M, Bartnicki-García S, Mouriño-Pérez RR. 2009. Cytoplasmic bulk flow propels nuclei in mature hyphae of *Neurospora crassa*. *Eukaryot. Cell* 8(12):1880–90
132. Reumann S, Weber APM. 2006. Plant peroxisomes respire in the light: Some gaps of the photorespiratory C_2 cycle have become filled—others remain. *Biochim. Biophys. Acta* 1763(12):1496–510
133. Rottensteiner H, Kramer A, Lorenzen S, Stein K, Landgraf C, et al. 2004. Peroxisomal membrane proteins contain common Pex19p-binding sites that are an integral part of their targeting signals. *Mol. Biol. Cell* 15(7):3406–17
134. Rottensteiner H, Stein K, Sonnenhol E, Erdmann R. 2003. Conserved function of Pex11p and the novel Pex25p and Pex27p in peroxisome biogenesis. *Mol. Biol. Cell* 14(10):4316–28
135. Rucktäschel R, Halbach A, Girzalsky W, Rottensteiner H, Erdmann R. 2010. De novo synthesis of peroxisomes upon mitochondrial targeting of Pex3p. *Eur. J. Cell Biol.* 89(12):947–54
136. Rucktäschel R, Thoms S, Sidorovitch V, Halbach A, Pechlivanis M, et al. 2009. Farnesylation of Pex19p is required for its structural integrity and function in peroxisome biogenesis. *J. Biol. Chem.* 284(31):20885–96
137. Saikia S, Scott B. 2009. Functional analysis and subcellular localization of two geranylgeranyl diphosphate synthases from *Penicillium paxilli*. *Mol. Genet. Genomics* 282(3):257–71
138. Saini P, Eyler DE, Green R, Dever TE. 2009. Hypusine-containing protein eIF5A promotes translation elongation. *Nature* 459(7243):118–21
139. Saleem RA, Knoblach B, Mast FD, Smith JJ, Boyle J, et al. 2008. Genome-wide analysis of signaling networks regulating fatty acid-induced gene expression and organelle biogenesis. *J. Cell Biol.* 181(2):281–92
140. Salomons FA, Kiel JA, Faber KN, Veenhuis M, van der Klei IJ. 2000. Overproduction of Pex5p stimulates import of alcohol oxidase and dihydroxyacetone synthase in a *Hansenula polymorpha* Pex14 null mutant. *J. Biol. Chem.* 275(17):12603–11
141. Sanchez M, McManus OB. 1996. Paxilline inhibition of the alpha-subunit of the high-conductance calcium-activated potassium channel. *Neuropharmacology* 35(7):963–68
142. Saraya R, Krikken AM, Veenhuis M, van der Klei IJ. 2011. Peroxisome reintroduction in *Hansenula polymorpha* requires Pex25 and Rho1. *J. Cell Biol.* 193(5):885–900
143. Sato Y, Shibata H, Nakatsu T, Nakano H, Kashiwayama Y, et al. 2010. Structural basis for docking of peroxisomal membrane protein carrier Pex19p onto its receptor Pex3p. *EMBO J.* 29(24):4083–93

144. Schell-Steven A, Stein K, Amoros M, Landgraf C, Volkmer-Engert R, et al. 2005. Identification of a novel, intraperoxisomal Pex14-binding site in Pex13: association of Pex13 with the docking complex is essential for peroxisomal matrix protein import. *Mol. Cell. Biol.* 25(8):3007–18
145. Schliebs W, Girzalsky W, Erdmann R. 2010. Peroxisomal protein import and ERAD: variations on a common theme. *Nat. Rev. Mol. Cell Biol.* 11(12):885–90
146. Schmidt F, Treiber N, Zocher G, Bjelic S, Steinmetz MO, et al. 2010. Insights into peroxisome function from the structure of PEX3 in complex with a soluble fragment of PEX19. *J. Biol. Chem.* 285(33):25410–17
147. Schueller N, Holton SJ, Fodor K, Milewski M, Konarev P, et al. 2010. The peroxisomal receptor Pex19p forms a helical mPTS recognition domain. *EMBO J.* 29(15):2491–500
148. Schuldiner M, Metz J, Schmid V, Denic V, Rakwalska M, et al. 2008. The GET complex mediates insertion of tail-anchored proteins into the ER membrane. *Cell* 134(4):634–45
149. Shinozaki A, Sato N, Hayashi Y. 2009. Peroxisomal targeting signals in green algae. *Protoplasma* 235(1–4):57–66
150. Sichting M, Schell-Steven A, Prokisch H, Erdmann R, Rottensteiner H. 2003. Pex7p and Pex20p of *Neurospora crassa* function together in PTS2-dependent protein import into peroxisomes. *Mol. Biol. Cell* 14(2):810–21
151. Smith JJ, Marelli M, Christmas RH, Vizeacoumar FJ, Dilworth DJ, et al. 2002. Transcriptome profiling to identify genes involved in peroxisome assembly and function. *J. Cell Biol.* 158(2):259–71
152. Soundararajan S, Jedd G, Li X, Ramos-Pamploña M, Chua NH, Naqvi NI. 2004. Woronin body function in *Magnaporthe grisea* is essential for efficient pathogenesis and for survival during nitrogen starvation stress. *Plant Cell* 16(6):1564–74
153. Spröte P, Brakhage AA, Hynes MJ. 2009. Contribution of peroxisomes to penicillin biosynthesis in *Aspergillus nidulans*. *Eukaryot. Cell* 8(3):421–23
154. Stanley WA, Wilmanns M. 2006. Dynamic architecture of the peroxisomal import receptor Pex5p. *Biochim. Biophys. Acta* 1763(12):1592–98
155. Stein K, Schell-Steven A, Erdmann R, Rottensteiner H. 2002. Interactions of Pex7p and Pex18p/Pex21p with the peroxisomal docking machinery: implications for the first steps in PTS2 protein import. *Mol. Cell. Biol.* 22(17):6056–69
156. Subramani S. 2002. Hitchhiking fads en route to peroxisomes. *J. Cell Biol.* 156(3):415–17
157. Swinkels BW, Gould SJ, Bodnar AG, Rachubinski RA, Subramani S. 1991. A novel, cleavable peroxisomal targeting signal at the amino-terminus of the rat 3-ketoacyl-CoA thiolase. *EMBO J.* 10(11):3255–62
158. **Szöor B, Ruberto I, Burchmore R, Matthews KR. 2010. A novel phosphatase cascade regulates differentiation in *Trypanosoma brucei* via a glycosomal signaling pathway. *Genes Dev.* 24(12):1306–16**
159. Szöor B, Wilson J, McElhinney H, Tabernero L, Matthews KR. 2006. Protein tyrosine phosphatase TbPTP1: a molecular switch controlling life cycle differentiation in trypanosomes. *J. Cell Biol.* 175(2):293–303
160. Tam YYC, Fagarasanu A, Fagarasanu M, Rachubinski RA. 2005. Pex3p initiates the formation of a preperoxisomal compartment from a subdomain of the endoplasmic reticulum in *Saccharomyces cerevisiae*. *J. Biol. Chem.* 280(41):34933–39
161. Tanabe Y, Maruyama J-I, Yamaoka S, Yahagi D, Matsuo I, et al. 2011. Peroxisomes are involved in biotin biosynthesis in *Aspergillus* and *Arabidopsis*. *J. Biol. Chem.* 286(35):30455–61
162. Tey WK, North AJ, Reyes JL, Lu YF, Jedd G. 2005. Polarized gene expression determines Woronin body formation at the leading edge of the fungal colony. *Mol. Biol. Cell* 16(6):2651–59
163. Thoms S, Harms I, Kalies K-U, Gärtner J. 2012. Peroxisome formation requires the endoplasmic reticulum channel protein Sec61. *Traffic* 13:599–609
164. Titorenko VI, Nicaud J-M, Wang H, Chan H, Rachubinski RA. 2002. Acyl-CoA oxidase is imported as a heteropentameric, cofactor-containing complex into peroxisomes of *Yarrowia lipolytica*. *J. Cell Biol.* 156(3):481–94
165. Titorenko VI, Rachubinski RA. 1998. Mutants of the yeast *Yarrowia lipolytica* defective in protein exit from the endoplasmic reticulum are also defective in peroxisome biogenesis. *Mol. Cell. Biol.* 18(5):2789–803

158. Provides strong evidence linking glycosomes and regulation of a developmental transition.

166. Titorenko VI, Rachubinski RA. 2000. Peroxisomal membrane fusion requires two AAA family ATPases, Pex1p and Pex6p. *J. Cell Biol.* 150(4):881–86
167. Titorenko VI, Smith JJ, Szilard RK, Rachubinski RA. 1998. Pex20p of the yeast *Yarrowia lipolytica* is required for the oligomerization of thiolase in the cytosol and for its targeting to the peroxisome. *J. Cell Biol.* 142(2):403–20
168. Tower RJ, Fagarasanu A, Aitchison JD, Rachubinski RA. 2011. The peroxin Pex34p functions with the Pex11 family of peroxisomal divisional proteins to regulate the peroxisome population in yeast. *Mol. Biol. Cell* 22(10):1727–38
169. Trinci AP, Collinge AJ. 1974. Occlusion of the septal pores of damaged hyphae of *Neurospora crassa* by hexagonal crystals. *Protoplasma* 80(1):57–67
170. Ullán RV, Teijeira F, Guerra SM, Vaca I, Martín JF. 2010. Characterization of a novel peroxisome membrane protein essential for conversion of isopenicillin N into cephalosporin C. *Biochem. J.* 432(2):227–36
171. Urquhart AJ, Kennedy D, Gould SJ, Crane DI. 2000. Interaction of Pex5p, the type 1 peroxisome targeting signal receptor, with the peroxisomal membrane proteins Pex14p and Pex13p. *J. Biol. Chem.* 275(6):4127–36
172. van der Klei IJ, Veenhuis M. 1997. Yeast peroxisomes: function and biogenesis of a versatile cell organelle. *Trends Microbiol.* 5(12):502–9
173. van der Zand A, Braakman I, Tabak HF. 2010. Peroxisomal membrane proteins insert into the endoplasmic reticulum. *Mol. Biol. Cell* 21(12):2057–65
174. van Zutphen T, Baerends RJS, Susanna KA, de Jong A, Kuipers OP, et al. 2010. Adaptation of *Hansenula polymorpha* to methanol: a transcriptome analysis. *BMC Genomics* 11:1
175. Veenhuis M, Harder W, van Dijken JP, Mayer F. 1981. Substructure of crystalline peroxisomes in methanol-grown *Hansenula polymorpha*: evidence for an in vivo crystal of alcohol oxidase. *Mol. Cell. Biol.* 1(10):949–57
176. Vizeacoumar FJ, Torres-Guzman JC, Bouard D, Aitchison JD, Rachubinski RA. 2004. Pex30p, Pex31p, and Pex32p form a family of peroxisomal integral membrane proteins regulating peroxisome size and number in *Saccharomyces cerevisiae*. *Mol. Biol. Cell* 15(2):665–77
177. Vizeacoumar FJ, Torres-Guzman JC, Tam YYC, Aitchison JD, Rachubinski RA. 2003. YHR150w and YDR479c encode peroxisomal integral membrane proteins involved in the regulation of peroxisome number, size, and distribution in *Saccharomyces cerevisiae*. *J. Cell Biol.* 161(2):321–32
178. Vizeacoumar FJ, Vreden WN, Fagarasanu M, Eitzen GA, Aitchison JD, Rachubinski RA. 2006. The dynamin-like protein Vps1p of the yeast *Saccharomyces cerevisiae* associates with peroxisomes in a Pex19p-dependent manner. *J. Biol. Chem.* 281(18):12817–23
179. Völkl A, Baumgart E, Fahimi HD. 1988. Localization of urate oxidase in the crystalline cores of rat liver peroxisomes by immunocytochemistry and immunoblotting. *J. Histochem. Cytochem.* 36(4):329–36

180. Shows that 4- to 9-nm PTS1 coated gold particles can be imported to the peroxisome matrix.

180. **Walton PA, Hill PE, Subramani S. 1995. Import of stably folded proteins into peroxisomes. *Mol. Biol. Cell* 6(6):675–83**
181. Wanders RJA, Waterham HR. 2006. Biochemistry of mammalian peroxisomes revisited. *Annu. Rev. Biochem.* 75:295–332
182. Wang D, Visser NV, Veenhuis M, van der Klei IJ. 2003. Physical interactions of the peroxisomal targeting signal 1 receptor Pex5p, studied by fluorescence correlation spectroscopy. *J. Biol. Chem.* 278(44):43340–45
183. Weber H. 2002. Fatty acid-derived signals in plants. *Trends Plant Sci.* 7(5):217–24
184. Williams C, van den Berg M, Geers E, Distel B. 2008. Pex10p functions as an E3 ligase for the Ubc4p-dependent ubiquitination of Pex5p. *Biochem. Biophys. Res. Commun.* 374(4):620–24
185. Williams C, van den Berg M, Sprenger RR, Distel B. 2007. A conserved cysteine is essential for Pex4p-dependent ubiquitination of the peroxisomal import receptor Pex5p. *J. Biol. Chem.* 282(31):22534–43
186. Würtz C, Schliebs W, Erdmann R, Rottensteiner H. 2008. Dynamin-like protein-dependent formation of Woronin bodies in *Saccharomyces cerevisiae* upon heterologous expression of a single protein. *FEBS J.* 275(11):2932–41
187. Yan M, Rachubinski DA, Joshi S, Rachubinski RA, Subramani S. 2008. Dysferlin domain-containing proteins, Pex30p and Pex31p, localized to two compartments, control the number and size of oleate-induced peroxisomes in *Pichia pastoris*. *Mol. Biol. Cell* 19(3):885–98

188. Yang X, Purdue PE, Lazarow PB. 2001. Eci1p uses a PTS1 to enter peroxisomes: either its own or that of a partner, Dci1p. *Eur. J. Cell Biol.* 80(2):126–38
189. Yuan P, Jedd G, Kumaran D, Swaminathan S, Shio H, et al. 2003. A HEX-1 crystal lattice required for Woronin body function in *Neurospora crassa*. *Nat. Struct. Biol.* 10(4):264–70
190. Zekert N, Veith D, Fischer R. 2010. Interaction of the *Aspergillus nidulans* microtubule-organizing center (MTOC) component ApsB with gamma-tubulin and evidence for a role of a subclass of peroxisomes in the formation of septal MTOCs. *Eukaryot. Cell* 9(5):795–805
191. Zickler D, Arnaise S, Coppin E, Debuchy R, Picard M. 1995. Altered mating-type identity in the fungus *Podospora anserina* leads to selfish nuclei, uniparental progeny, and haploid meiosis. *Genetics* 140(2):493–503

Microbial Population and Community Dynamics on Plant Roots and Their Feedbacks on Plant Communities

James D. Bever, Thomas G. Platt, and Elise R. Morton

Department of Biology, Indiana University, Bloomington, Indiana 47405;
email: jbever@indiana.edu, tgplatt@indiana.edu, ermorton@indiana.edu

Keywords

rhizosphere, competition, trade-offs, mutualism, pathogen, specificity

Abstract

The composition of the soil microbial community can be altered dramatically due to association with individual plant species, and these effects on the microbial community can have important feedbacks on plant ecology. Negative plant-soil feedback plays primary roles in maintaining plant community diversity, whereas positive plant-soil feedback may cause community conversion. Host-specific differentiation of the microbial community results from the trade-offs associated with overcoming plant defense and the specific benefits associated with plant rewards. Accumulation of host-specific pathogens likely generates negative feedback on the plant, while changes in the density of microbial mutualists likely generate positive feedback. However, the competitive dynamics among microbes depends on the multidimensional costs of virulence and mutualism, the fine-scale spatial structure within plant roots, and active plant allocation and localized defense. Because of this, incorporating a full view of microbial dynamics is essential to explaining the dynamics of plant-soil feedbacks and therefore plant community ecology.

Contents

INTRODUCTION	266
SOIL COMMUNITY FEEDBACK AND PLANT COMMUNITY DYNAMICS	266
Evidence of Plant-Soil Feedbacks in Shaping Plant Communities	267
Microbial Agents of Plant-Soil Feedback	269
A MICROBIAL PERSPECTIVE ON PLANT-SOIL FEEDBACK	270
Microbial Specialization on Plant Roots	270
Feedback on Plant Growth from Differentiated Microbial Communities	272
Establishment of Plant-Associated Microbial Diversity	274
SIGNIFICANCE AND FUTURE DIRECTIONS	276
Microbial Drivers of Plant-Soil Feedback	276
Microbial Mediation of Feedbacks in the Plant Ecology Context	276

INTRODUCTION

Ecologists have historically focused on resource partitioning as the primary force structuring communities (83). Communities of competing species can be stabilized by strong negative intraspecific interactions relative to interspecific interactions (30). Traditionally, strong negative intraspecific interactions have been thought to result from high resource use overlap (83, 131), a framework that has been successful at explaining patterns of animal communities (28). A variant of resource partitioning theory, built around the Monod model of microbial growth (95), has been effectively used for explaining the dynamics of microbial communities (145). Resource partitioning also explains patterns of microbial communities in evolving laboratory communities (63) and in the field (124). The Monod models of resource partitioning have also been developed into an influential framework for understanding plant community dynamics (131). However, years of plant competition studies have produced only limited evidence of coexistence of competing plant species through resource partitioning (36, 91).

The limited success of resource partitioning theory in explaining the dynamics of plant communities may result from neglecting soil microorganisms, which act as drivers of terrestrial ecology, as shown by a growing body of work. The composition of soil microbial communities has large impacts on plant-plant interactions (47, 96) and consequently on plant diversity and composition (134, 135, 140). Therefore, a complete understanding of plant community structure and plant dynamics requires integrating microbial perspectives into our conceptual frameworks.

Several frameworks for integrating microbes into plant community dynamics have been developed (14, 118, 119). The framework of plant-soil community feedback (10, 13, 17) has become increasingly influential in plant ecology, as it is instrumental in explanations of plant diversity and community structure (13, 69, 85, 104, 108), plant species invasion (25, 34, 37, 69, 117, 139), and succession dynamics (62, 89, 136). Here, we review the conceptual framework of plant-soil feedback, and the evidence of its importance in structuring plant communities, and then dissect the microbial interactions that drive these feedbacks.

Feedback: a process whereby the plant alters its environment (e.g., the microbial community) in such a way that it in turn affects the plant's growth and fitness

SOIL COMMUNITY FEEDBACK AND PLANT COMMUNITY DYNAMICS

The plant-soil feedback framework builds on the well-established observation that plant species differ in their response to individual microbial species, as both the negative effects of soil pathogens and the positive effects of root symbionts are host specific. Growth rates of microbes are also host specific; components of the soil biota rapidly change in response to plant identity, and this change in microbial composition generates a feedback on plant relative performance that defines the long-term influence of soil microbes on plant species coexistence (13, 17).

Soil community feedback involves two steps: First, the density and/or composition of the soil community changes in response to the composition of the plant community, and second, the change in composition alters the relative growth rates of individual plant species (**Figure 1**). As plant-microbe interactions likely occur at a local scale, the feedbacks can be analyzed at the scale of individual plants. Changes in the microbial community due to the identity of the resident plant can not only alter a plant's growth rate, but also affect survival, reproduction, and the likelihood of being replaced by an individual of the same species. These influences may be identified as the direct feedbacks of the soil community on the fitness of the resident species, represented by α_A and β_B in **Figure 2**. The microbial change may also alter the likelihood that the resident plant species is replaced by an individual of a second species, which can be measured as indirect feedbacks represented by α_B and β_A in **Figure 2**. The net pairwise dynamics depends on the relative magnitude of the direct and indirect feedbacks (13, 17). Accumulation of soil microbes that promote their hosts' fitness better than that of neighboring competitor plants generates a positive-feedback dynamic that leads to a loss of local-scale diversity and contributes to alternative stable states (**Figure 1**). In contrast, a negative-feedback dynamic that allows for the coexistence of competing plant species results when plants promote the growth of soil microbes that antagonize their own fitness relative to that of their competitors (**Figure 1**).

Evidence of Plant-Soil Feedbacks in Shaping Plant Communities

Mounting empirical evidence suggests that soil community feedbacks are major determinants of plant species coexistence. Pot studies, which isolate the microbial effects, have demonstrated negative soil feedbacks among co-occurring plant species (10, 69, 70, 85, 108). Further, seedling performance in the field frequently declines with proximity to conspecific adults (32, 51, 85, 104, 144). This is a pattern frequently referred to as the Janzen-Connell hypothesis. Janzen originally imagined that this pattern was driven by species-specific seed predation (56); however, empirical work demonstrates that host-specific soil pathogens play a dominant role (7, 85, 104). Moreover, the strength of measured feedbacks positively correlates with relative plant species abundance (32, 69, 85), and simulation models identify that this pattern is expected only when the soil community feedbacks maintain plant diversity (85).

Soil feedbacks may also represent an important dimension of the success of invasive species. Consistent with this, several studies have shown that invading plant species benefit from escape from host-specific soil pathogens (25, 52, 117). Positive soil community feedback is also important in the success of invasive plant species (14, 100, 114, 139), potentially contributing to a self-accelerating decline in community composition known as invasion meltdown (125).

Both positive and negative feedbacks contribute to plant community change during succession (62, 89). Early successional plant species tend to have limited defenses and therefore are vulnerable to a buildup of pathogens that later successional species can resist (136). Early successional plant species also tend to have a low dependence on mycorrhizal fungi, the buildup of which generates a positive feedback that promotes the success of later successional species (55, 89).

In agricultural settings, negative soil community feedbacks drive the seasonal rotation of monocrops throughout the world (23, 68). Management of soil pathogens through chemical means is often not economical. Rather, as most soil pathogens are host specific, their population densities are managed by rotation with nonhost crops. The corn-soybean rotation that dominates much of

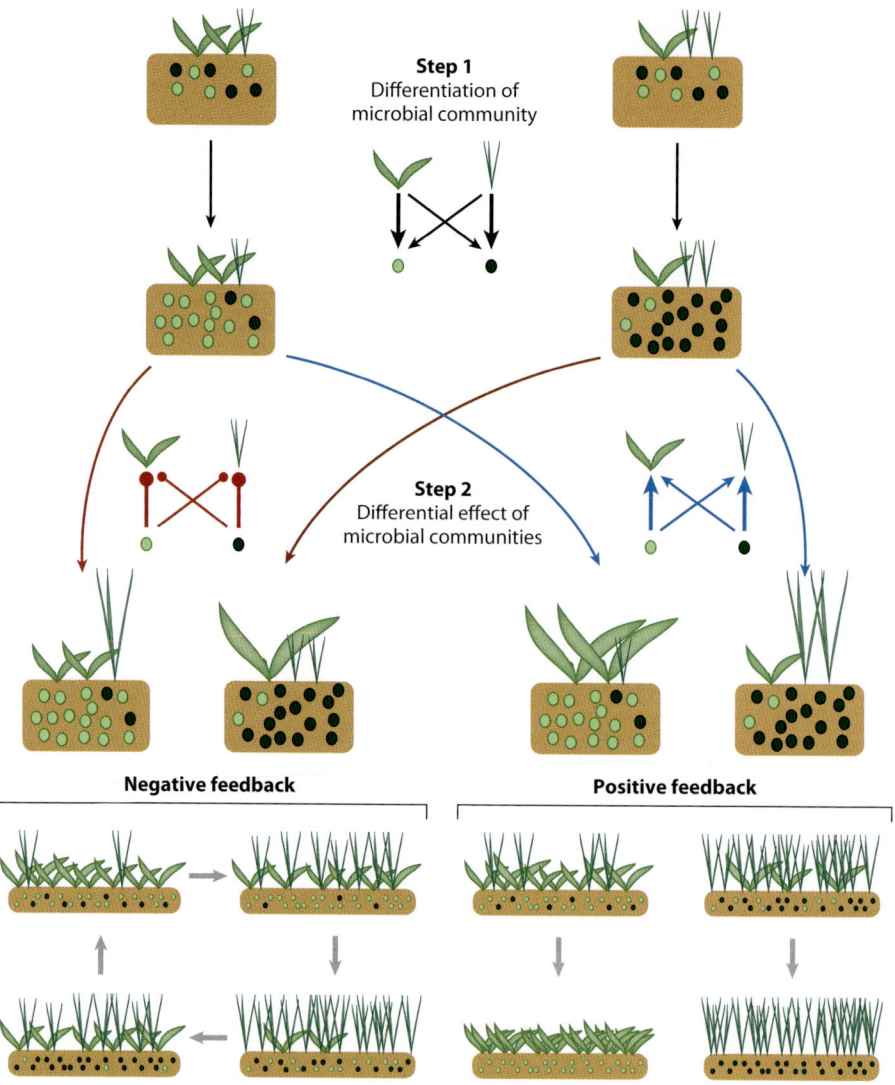

Figure 1

Soil microbial feedback involves two steps. First, the composition of the microbial community differentiates on the plants because of host-specific microbial growth rates. In this illustration, the light green microbe has higher growth rates on the broad-leaved grass and the dark green microbe has higher growth rates on the narrow-leaved grass. The relative benefit to the microbes is represented by the thickness of the arrows in the fitness diagram. The second step involves differential effects of the microbes on the plant species. (*Left*) For host-specific pathogens, the light and dark green microbes have strongest negative effects on the broad- and narrow-leaved plants, respectively, with relative virulence represented by the thickness of the red clubs. As a result, the plants that were initially most abundant have the lowest growth rates. The net consequence of this negative feedback on plant communities is illustrated at the bottom left, with both species maintained in the community over time. (*Right*) However, for host-specific mutualists, the light and dark green microbes have strongest positive effects on the broad- and narrow-leaved plants, respectively. As a result, the plants that were initially most common grow best. The net result of this positive feedback on the community is a loss of diversity on a local scale with a potential for the community to reach alternate stable states.

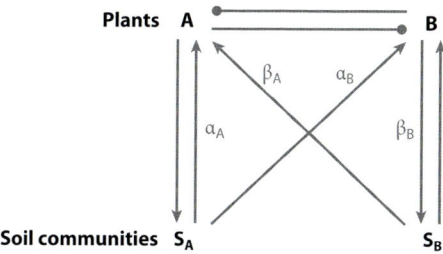

Figure 2
In this conceptual representation of soil community feedback, the presence of Plant A causes a change in the composition of the soil community, represented by S_A. This change in the soil community can directly alter the population growth rate of Species A (represented by α_A) and it can alter the growth rate (α_B) of competing plant Species B. Similarly, the presence of Plant B can cause a change in the composition of the soil community (S_B), which can directly feed back (β_B) on the population growth rate of Plant B or indirectly feed back on the growth rate of Plant B through changes in the growth rate (β_A) of competing Plant A. The exponential model predicts that the net effect of soil community dynamics on plant species coexistence is determined by the sign and magnitude of an interaction coefficient $I_S = (\alpha_A + \beta_B) - (\alpha_B + \beta_A)$, which represents the net pairwise feedback. Adapted with permission from References 13 and 17.

agriculture in the United States, for example, is motivated by the escape of host-specific pathogens (bacterial, fungal, and oomycete) and nematodes (19, 23, 98, 112).

Microbial Agents of Plant-Soil Feedback

Although there is growing conceptual clarity on, and accumulating empirical support for, the importance of microbial feedbacks on plant community dynamics, there has been less explicit focus on the microbial agents of these feedbacks in unmanaged systems. In part, this results from the challenge imposed by the stunning taxonomic and functional diversity present in soil microbial communities (115). Which members of these diverse communities are the agents driving plant-soil feedbacks?

THEORY OF PLANT-SOIL FEEDBACKS

An exponential model of plant-soil feedback (17) identified that net pairwise dynamics depend upon an interaction coefficient I_s, where I_s equals the difference in the direct feedbacks and the indirect feedbacks ($I_S = (\alpha_A + \beta_B) - (\alpha_B + \beta_A)$). When I_S is positive, the change in microbial composition increases the relative performance of the locally abundant plant species, generating a positive feedback dynamic that would lead to loss of diversity at a local scale. Conversely, competing plant species can coexist when the change in microbial composition decreases the relative performance of the locally abundant plant species, generating a net negative feedback, as reflected by a negative I_S (17).

The basic conclusions of the simple exponential model (17) are generally upheld by inclusion of greater complexities. The inclusion of negative density dependence and interspecific competition among plant species demonstrates that microbial dynamics can contribute to plant species coexistence, even of strong competitors such as plants utilizing the same resources (13). Spatially explicit models reveal that negative feedback leads to coexistence regardless of the scale of interaction, though the resulting spatial pattern can vary with the spatial scale of interaction and dispersal (94). Conversely, while positive feedback always leads to loss of diversity at the local scale, local interactions can contribute to the stability of uniform patch types and therefore to heterogeneity at the larger geographic scale (93).

AM: arbuscular mycorrhizal

EM: ectomycorrhizal

Trade-off: a negative correlation between two aspects of fitness such as growth rate on one host and growth rate on a second host

Cost of mutualism: the metabolic and fitness costs associated with delivering benefit to a second species

Virulence: the ability of a pathogenic agent to cause disease (often measured as the negative effect on host fitness)

Rhizosphere: region in the soil surrounding plant roots

Negative feedbacks result from an accumulation of host-specific bacterial and fungal pathogens (23, 68); oomycetes are particularly important in both grassland and forest systems (7, 92, 104). Negative feedbacks may also result from changes in the composition of other components of the soil community (17, 133). For example, shifts in the composition of beneficial arbuscular mycorrhizal (AM) fungi can generate negative feedback on plant growth (12, 26).

Conversely, positive feedback commonly results from changes in the density of host-specific mutualists, including AM fungi, ectomycorrhizal (EM) fungi and symbiotic nitrogen-fixing bacteria, as depicted in **Figure 1**. However, pathogen accumulation on a tolerant host may generate net positive effects on its own growth when pathogen accumulation suppresses competing plants, as has been found with invasive plant species (38, 86). Reciprocal changes in the pathogen composition of this type would generate positive feedback (13).

A MICROBIAL PERSPECTIVE ON PLANT-SOIL FEEDBACK

Given the great taxonomic and functional diversity of microbes and the potentially complex and counterintuitive ways they can generate dynamically equivalent net pairwise feedbacks, one may question the predictability of soil microbial dynamics. However, similar net feedback dynamics may be driven by a set of forces common across microbial communities. We outline a general conceptual framework for understanding the forces that structure microbial population dynamics on plant roots and how these processes generate positive and negative feedbacks on plant growth. Our intent is to find commonalities across functionally distinct categories of microbes such as pathogens and symbionts and to build toward the common factors that influence their dynamics. To do this, we first identify common features of the plant-microbe system that make feedback on plant growth likely.

- Microbes are likely to differ in their growth rates on plant roots because plants are defended both by chemicals and by specific immune responses. Trade-offs associated with overcoming specific defenses generate host-specific differentiation of microbial communities.
- Microbes have high rates of turnover, so small differences in fitness are amplified, making microbial dynamics and microbial specialization on their host fast relative to changes in the plant community.
- Soil is a viscous medium, leading to high spatial structure and patchy distributions of microbes, so changes in microbial communities are likely to feed back on the growth of that same plant type on which they accumulated.
- The sign of these feedbacks is generated by the relationship between the microbe's competitive ability on the roots and the effect of that microbe on that plant. Competitive ability of microbes is determined by the interplay of costs of mutualism and virulence, the burden of overcoming plant defenses, benefits of plant allocation, and fine-scale microbial spatial structure.

In the remaining sections, we discuss first the selective forces that drive differentiation of the microbial community on plant roots, and then the consequences of this differentiation on plants. Throughout this review, we identify essential instabilities that will likely generate turnover within the microbial community. The rate of this turnover will be a function of the rate of (re)introduction of new microbial types. Accordingly we discuss the means by which microbes can be introduced or reintroduced into a root system.

Microbial Specialization on Plant Roots

The interface of plant roots and the surrounding soil, the rhizosphere, harbors a diverse and dynamic microbial community (115). Plant roots exude a wide range of molecules into the rhizosphere, thereby altering soil chemistry and providing nutrient sources that resident microbes can utilize. However, plant roots are not passive targets, as a significant portion of plant exudates include a diversity of defensive secondary metabolites (141). The suite of molecules exuded by roots shape the rhizosphere environment and consequently also help shape the composition of microbial communities (20). Root exudates generally vary substantially between different plant species and genotypes and can depend on a variety of factors such as nutrient levels, disease, stress, and even the microbial community itself (74, 147). This variation can have dramatic effects on the composition and dynamics of microbial communities, as the microbes present must be able to tolerate or utilize the plant's exudates. Trade-offs associated with tolerating host defenses structure microbial competition on hosts, thereby determining the differentiation of microbial communities on different host species. In fact, there is extensive evidence of such host specialization, in which particular microbial communities, species, or strains associate with specific plant species or genotypes (5, 6, 20).

Apoptosis: host-initiated programmed cell death in response to stimulus such as a pathogen

PAMP: pathogen-associated molecular patterns

***R*:** resistance

Specific and sometimes reciprocal responses between plants and microbes often influence the differentiation of symbiotic microbes. The interaction between rhizobial bacteria and leguminous plants offers one of the best-characterized examples. During the mutualistic association of nitrogen-fixing bacteria and leguminous plants, the microbe produces chemical cues that elicit developmental and exudate changes within the plant, and the bacteria also respond to root exudates secreted by the plant to alter its behavior. Plant-exuded flavonoids trigger its bacterial partner to express numerous genes involved in nodulation (Nod) (60). The resulting Nod factors vary among rhizobial species and strains (107) and trigger plant host changes necessary for nodulation such as root hair curling. Further, plant and bacterial factors contribute to the subsequent maturation of the legume-rhizobia symbiosis (60). The specificity of these factors determines which plant and rhizobial species/genotypes can associate (48). AM fungi use a set of host recognition cues similar to that used by rhizobia, and preliminary work suggests that EM fungi do as well (50). The degree of specificity of these cues appears to be low in AM fungi; however, after infection, the growth rates of AM fungal species differ across host species, generating distinct AM fungal community composition on different host plants (12, 15).

Agrobacterial plant pathogens are close relatives of rhizobial mutualists (39), and the benefits provided to the infecting bacteria also depend on manipulation of plant host exudates (111). These pathogens respond to plant-dependent cues such as low pH and release of plant phenolics by expressing genes involved with the genetic transformation of plant cells (146). The resultant genetic manipulation causes the plant cell to misregulate its growth hormones, resulting in tumor production. The transformed plant cells are also induced to exude unusual metabolites called opines, which promote the fitness of the infecting agrobacterial pathogen's relatives (44, 45, 111).

Plants are also sometimes able to shunt resources to, or away from, particular symbionts. The dynamics of mutualists can be influenced by plant control of nutrient flow into the rhizosphere, which uses sanctions or preferential allocation to promote the growth of particular mutualists (16, 65, 66, 103). The two-tiered plant immune system provides plants a means of diverting resources away from pathogens by undergoing targeted apoptosis of pathogen-associated cells (31, 59). The first tier of the plant immune system depends on the recognition of slowly evolving pathogen-associated molecular patterns (PAMPs) by transmembrane pattern recognition receptors (101). The perception of PAMPs triggers an immune defense that can prevent the pathogen from establishing an infection. Pathogens may subvert host defenses by injecting effector proteins that interfere with the immune response into the plant cell. The second tier of the plant immune system depends on the recognition of the pathogen's effector proteins by R (resistance) proteins (31, 59). Detection of bacterial effectors by R proteins results in an amplified PAMP-triggered response and a hypersensitive cell death response, preventing further infection by the pathogen.

This effector-triggered immunity involves a high degree of specificity between R proteins and the cognate effectors.

The gene-for-gene interactions between the plant's R loci and the pathogen's *avr* (avirulence) loci encoding the effector proteins can generate host-specific differentiation of associated microbial pathogens (9, 73, 130). Differentiation of the pathogen community depends on trade-offs emerging from the costs associated with overcoming host defenses. Although the existence of these costs has been controversial, several studies give direct evidence of their significance (57, 130). Moreover, in the absence of costs of virulence, one would expect pathogens to evolve the ability to infect many hosts. Contrary to this, alleles conferring the ability to infect hosts are rapidly lost when those hosts are no longer available (41, 106, 137).

Feedback on Plant Growth from Differentiated Microbial Communities

The change in microbial composition with host plant feeds back onto plant dynamics when the components of the microbial community exert differential effects on plant growth. Although mutualists and pathogens can generate similar net pairwise feedbacks (17), they likely differ in their basic tendencies (e.g., **Figure 1**). We address the forces structuring these feedbacks by discussing the dynamics of mutualists and pathogens.

Dynamics of microbial mutualists.
The differential accumulation of microbial mutualists, as expected from host recognition systems, should positively affect the growth of compatible hosts, compared to incompatible hosts, thereby generating positive feedback. Positive feedback is also generated among plant hosts that differ in their degree of responsiveness to microbial symbionts when there is a positive correlation between the quality of the plant as a host and the dependence of that plant on the symbionts (133). Alternatively, changes in density of microbial mutualists could generate a negative feedback if the most responsive plant species were also a poorer host for AM fungi (133). However, although more data are needed, the few studies suggest that responsive host plants are also better hosts for their mutualists (122). Moreover, positive feedback via changes in mutualist density has been observed.

Positive feedback through changes in mutualist density is important at early stages of succession, where many colonizing species do not associate or have weak associations with mycorrhizal fungi, whereas later successional species can have strong dependence (55). Concomitantly, invasive plants with low dependence on AM fungi may decrease the density of mycorrhizal mutualists, thereby inhibiting re-establishment of native species. For example, high levels of glucosinolates produced by *Alliaria petiolata*, an invasive species of North American forests, inhibit AM fungi and thereby impede establishment of native plants (24, 74, 128). Similarly, dominance by nonnative species in California grasslands can decrease the density of mycorrhizal fungi, which may limit the growth of native species that tend to be highly dependent on these fungi (114, 139). By degrading soil microbial mutualisms, nonnative plants transform terrestrial communities. A similar transformation can occur when nonnative plants species are paired with locally novel microbial symbiosis, e.g., the invasive EM fungal association with invasive pines in South America (100) and *Frankia* association with invasive *Myrica faya* in Hawaii (138).

Host-specific changes in symbiont composition can also feed back positively or negatively on net pairwise plant dynamics if the competitively dominant symbionts improve or decrease plant growth, respectively (11). Experimental evidence suggests that plants can use rhizobial cues to preferentially associate with superior mutualists (42, 49), generating a positive correlation between plant and fungal relative fitness and positive feedback. Recognition cues involved in specificity of association may also generate positive feedback in EM fungal communities (22).

However, given that all plant root mutualists' associations may involve individual plants simultaneously associating with multiple symbionts that vary in their benefit, the overall feedback will be a function of the microbial competitive dynamics within plant roots. The most competitive symbionts may be the least beneficial because of the energetic cost of providing resources to the host (i.e., there is a substantial cost of mutualism). For example, it is energetically costly for an AM fungus to acquire phosphorus, transport it along its hyphae, and then deliver it to the host plant. Consistent with such costs, several studies have shown that the least beneficial AM fungi are the most competitive (8, 12). A negative correlation between measures of host and rhizobium fitness suggests that a similar cost operates within this interaction (48); however, tests with EM fungi have produced mixed results (64). In AM fungi, the competitive shift on hosts generates negative feedback on plant growth (12, 26). Similarly, *Bacillus mycoides* generally improves plant growth but is detrimental when interacting with the host species from which it is isolated (143). Further work is required to evaluate whether this dynamic occurs within other communities of beneficial plant microbes.

Because the spread of less beneficial "cheater" strains could degrade the mutualism, the processes maintaining mutualists have been an area of active investigation. Plant hosts have been observed to sanction ineffective rhizobia (66, 103) and preferentially allocate the most effective AM fungal mutualists (16, 65). The ability of plants to sanction and preferentially allocate likely varies among plant species (42, 49) and the quality of the mutualist (16, 49, 103). The potential of sanctions and preferential allocation to overcome the competitive advantage of the poorer mutualists may depend on the level of mixing of the symbionts within the plant roots. As an extreme example, sanctioning would be ineffective if all nodules contained mixed infections of good and bad rhizobia. In an empirical test with AM fungi, the most beneficial symbiont dominated when symbionts were spatially separated within the root system (16). However, nonbeneficial fungi dominated (16) and negative feedback was observed (12) when beneficial and nonbeneficial fungi were well mixed within the root system of their host. Nodules occupied by nitrogen-fixing bacteria and root tips associated with EM fungi represent discrete spatial structures that plants can manipulate. Whether the dominance of better mutualists enabled by sanctions and preferential allocation will generate positive feedback depends upon whether microbial dynamics are coupled with host-specific differences in growth promotion—an area that requires additional research. Several studies have found evidence that host-specific changes in the composition of AM fungal communities can generate positive feedback on plant growth (84, 149).

Dynamics of plant pathogens. The accumulation of a specialist pathogen on its host generates feedbacks that favor the growth of nonhost plant species. Local accumulation of pathogens is a common cause of negative feedback in nature and is important in grassland (92) and forest systems (7, 105). Although more work is needed on the dynamics of soil pathogens in unmanaged systems, studies of soil-borne diseases in agricultural systems show that a great taxonomic diversity of microbial pathogens have sufficient host specificity to generate negative feedbacks.

Despite the specificity of these interactions, plant root systems are exposed to numerous microbes simultaneously, and the competitive dynamics among these microorganisms likely influences net feedbacks on plants. Epidemiological models predict that multiple infections can result in greater virulence when virulence is correlated with increased efficiency of exploitation of the host (2, 126). To the extent that this is true with root pathogens, we would expect competition among multiple infecting pathogens to increase negative feedbacks provided that different plant species or genotypes are differentially affected by the pathogen. However, selective dynamics can lead to the evolution of either increased or decreased virulence, depending on the interplay of the selective pressures acting on the pathogens (1, 3, 21).

Dormancy:
a reversible state of extremely low metabolic activity that is characterized by the cessation of phenotypic development

Although costs associated with overcoming individual host defenses can drive differentiation of microbial communities on hosts (see Microbial Specialization on Plant Roots, above), these costs and other costs of virulence (110) can also alter the dynamics of pathogens within hosts. Once the host or a root of that host has been compromised by a virulent pathogen, other microbes that do not bear the costs of virulence may have a competitive advantage in exploiting that resource. For example, virulent *Agrobacterium tumefaciens* strains harbor costly virulence plasmids and are burdened by large costs of expressing the machinery required to infect plant tissues (110). These costs create a strong selective pressure favoring the evolution of avirulent, freeloading strains that have lost the ability to infect plants but maintain the ability to catabolize the opine nutrients exuded by infected plants (110, 111). One such avirulent freeloader is *Agrobacterium radiobacter* strain K84, which agriculturalists have been using for decades as a potent biocontrol agent of crown gall disease. Its freeloading makes K84 a resource competitor of *A. tumefaciens*, but in addition to this, K84 interferes with the pathogen via the production of a targeted toxin, allowing it to effectively suppress the disease on its host (67, 116).

The suppression of virulent pathogens through interactions with other microbes commonly occurs in agricultural systems with repeated monocropping. This phenomenon, called suppressive soil, has been observed with numerous soil-borne plant diseases such as take-all wheat (120), potato scab (80), and tobacco black root rot (72). Although trade-offs between virulence and saprophytic growth in facultative pathogens (68, 110) make virulent pathogens vulnerable to suppression, the antagonistic activities of members of the rhizosphere microbiome can be mediated by several phenotypes including the production of antibiotics, siderophores, and surfactants by microbes in the disease-suppressive community (68). Many bacterial taxa tend to be associated with disease-suppressive soils, indicating that this phenomenon depends on consortia of microbes (88, 120).

Establishment of Plant-Associated Microbial Diversity

The plant-soil microbial dynamics that we have described are dependent on and generate spatial structure in the microbial community. At the continental scale, the success of invasive species often depends on release from microbial antagonists (25, 117). Within communities, positive plant-soil microbial feedback reinforces spatial separation of microbial communities (93), and negative feedback results in plant replacement, which necessitates recolonization of locally novel roots. At the smallest scale within the root system of an individual plant, the interplay of plant defense, plant allocation, and microbial competition determines the direction of feedbacks. At each of these scales, several facets of microbial ecology and evolution determine the ability of prior or new microbial variants to establish themselves. We discuss the importance of dormancy and storage effects, dispersal, horizontal gene transfer, and mutation in this establishment process.

Dormancy. All soil microbes, ranging from oomycetes, nematodes, AM fungi, to bacteria, have the ability to enter a dormant state under stressful or unsuitable conditions (61, 129). Dormancy allows microbes to persist during unfavorable conditions, increasing local-scale microbial diversity. Surveys estimate that over 80% of the bacterial cells in the soil are dormant (77). Moreover, the community of physiologically active bacteria within the soil are distinct from those that are dormant (77). Similar distinctions are likely in other groups such as AM fungi (113). As a result, estimates of microbial composition using standard DNA extractions from soil may not provide measures that reflect the active players in the plant-microbe interaction, potentially obscuring field attempts to identify the agents of microbial feedbacks.

For plant-associated microbes, shifts into and out of dormancy may be determined by the availability of suitable plants. Dormancy can be triggered by resource deprivation, change in nutrient composition of the soil (increased carbon or phosphorus), or other environmental conditions (e.g., pH, water content), all factors that can be affected by plants. Interactions with other members of the microbial community also stimulate microbial dormancy, as competitors may deplete resources or inhibit growth through antibiotic production (35, 78). The ecology and evolution of microbial dormancy are also influenced by predation in the dormant state (61), which can be significant for groups with large, edible dormant structures, such as spores of AM fungi.

Dispersal. If plant-associated microbes are not already present when the plant begins to grow, then dispersal can introduce new microbes. Wind, water, animals, and insects are major dispersers of soil microorganisms (71). Over smaller scales, soil microbes can facilitate the spread of one another (53, 142). In addition, microbes have evolved a number of strategies to sense changes in the environment and move accordingly. In the rhizosphere, plant-exuded resources such as carbohydrates, amino acids, phenolics, and inorganic ions are accessible to the surrounding microflora, and bacteria will chemotax toward these root-associated exudates (33, 79). For example rhizobia chemotax toward legume-excreted flavonoids prior to the development of symbiotic nodules (60).

Horizontal gene transfer. Horizontal gene transfer is rampant in the microbial world, both within and between species, occurring at such high frequencies that the definition of a species can be blurred (102). A single conjugation event between two species of bacteria can alter the total genetic content by over 10% (18, 29). Many bacterial virulence determinants are associated with mobile genetic elements (46, 121), as are genes for symbiosis and the ability to overcome or even utilize plant exudates and defenses. In fact, facultative symbionts have the greatest concentration of mobile elements in their genome, suggesting that horizontal gene transfer can be particularly important to this group (99). Conjugation, transduction, or transformation can convert free-living microbes to plant pathogens or symbionts and generate novel combinations of specificity factors and virulence or mutualist functions (27, 40, 148). Genes for nitrogen fixation by the plant symbionts *Sinorhizobium meliloti* and *Rhizobium etli* are on mobile plasmids (97, 132), as are the genes encoding the effectors that determine host specificity of the pathogen *Pseudomonas syringae* (43, 54). Genetic transfers can be important in soil fungi as well. There is strong evidence that the pathogenicity genes of *Nectria haematococca*, the fungal causative agent of pea footrot disease, were horizontally acquired (81). Conjugation of plant-associated plasmids can be induced by proximity to hosts, which increases the likelihood that recombinants resulting from horizontal gene transfer will be important at shorter time frames (40).

Mutation. Mutation is another important force generating microbial variants that can interact differently with plant hosts. Mutations affecting virulence-associated genes have significant consequences for the evolution of virulence in a wide variety of pathogens (87, 127). One clear target of selection are mutants that can evade host defenses (109, 150), though other possible targets include mutations that influence the fitness of the pathogen in the rhizosphere, such as those conferring the ability to catabolize plant-produced resources (75). Similar effects are likely in mutualistic plant-microbe interactions. For example, sequence variation in the *nod* gene of *R. etli* determines the host range of this mutualist (123). Although mutations occur at low rates, microbial population sizes on plant roots are potentially large. Consequently, mutation combined with gene exchange may shape the evolution of host-pathogen interactions (82).

Relative importance of modes of microbe (re)introduction. At the scale of a single root, plant defense response and preferential allocation can change at rapid timescales—in as short as hours (16, 59, 66). At these small spatial and temporal scales, new variants are likely to be reintroduced by local dispersal and reactivation of dormant cells. Over the lifetime of an individual plant, these local processes are likely to be supplemented by evolutionary creation of new variants (109). Within agrobacteria, for example, mutation might create freeloading variants that could suppress virulent types (111). Over large spatial and temporal scales, evolution of local resident microbial populations can overcome the novel defenses of introduced plant species. This may have contributed to the increased negative feedbacks accumulated over hundreds of years following the invasion of *Cerastium alpinum* in New Zealand (34).

SIGNIFICANCE AND FUTURE DIRECTIONS

Microbial Drivers of Plant-Soil Feedback

In this review we have sketched major forces that drive microbial feedbacks on plant growth with the goal of working toward a predictive theory of plant-soil feedback. While doing so, we have identified several areas where further work is required to better understand the microbial drivers of feedbacks. For example, to what extent do plant secondary chemicals influence the relative growth rates and host-specific differentiation of antagonistic or beneficial symbiont species? And to what extent do processes and tensions central to within-host evolution of virulence or mutualism generate differential effects and net pairwise feedbacks between two plant species? To what extent are microbial phenotypes that drive feedbacks introduced through dispersal across space or time, or through mutation or recombination within the resident microbial community?

This review has focused primarily on microbes with direct effects on plant growth through mutualisms or pathogens. However, differentiation of saprophytic components of the soil microbial community may also generate feedbacks on plant growth. Saprophyte communities respond to changes in litter quality associated with plant secondary chemicals. Further, differences in rates of host tissue decomposition could accelerate nutrient cycling, thereby potentially altering plant-plant interactions if the plant species differed in their dependence on and uptake of these nutrients (4, 14, 90, 119). Further work is required to assess the importance of host-specific differentiation of the saprophytic community relative to that of pathogens and mutualists on net pairwise feedbacks.

Microbial Mediation of Feedbacks in the Plant Ecology Context

Much progress has been made through phenomenological investigations of plant-soil feedbacks (14). Given this progress, plant ecologists may ask why they should care about the details of the microbial population and community dynamics that generate these feedbacks. However, there are several important conceptual issues on the impact of microbes on plant ecology, which will not be addressed without understanding the details of microbial dynamics.

Given the evidence of the central role of microbial feedbacks in plant community structure, other aspects of microbial life history may have cascading impacts on plant ecology. Microbes vary in their tolerance to types of environmental stress, such as salt, drought, and temperature. For example, oomycetes depend on moisture to complete their life cycle. Because oomycetes are important host-specific pathogens that generate negative feedbacks in natural systems (92, 104), the strength of negative feedback may increase along a moisture gradient. Such constraints on microbial ecology could mediate the increase in conspecific negative density dependence observed in areas with greater plant productivity, which may contribute to continental patterns of tree diversity (58).

Moreover, the nuance of microbial dynamics on plant roots is critical to the underlying assumptions of the feedback framework. For example, the accumulation of competitively dominant saprophytes and antagonists can suppress root pathogens after repeated monocropping in agricultural systems (76). Such suppressive soils are unlikely to develop in communities of annual or short-lived perennial plants, as a particular host will likely die and be replaced by a second species prior to the buildup of suppressors of host-specific pathogens. It is possible, however, that suppression of host-specific pathogens is an important process that limits negative feedbacks on long-lived perennial plants. Conceptually, this may generate nonlinear temporal dynamics of negative feedbacks, in which strong negative feedbacks would be experienced at intermediate ages of plants, perhaps contributing to their replacement. However, in older plants the strength of negative feedback may be reduced.

SUMMARY POINTS

1. Negative soil microbial feedback on plant growth can contribute to the maintenance of plant species diversity, whereas positive soil community feedback can contribute to plant community conversion.
2. Host-specific differentiation of microbial communities can be driven by trade-offs in microbial responses to specific plant defenses and exudates.
3. The accumulation of microbial mutualists likely generates positive plant-soil feedback, whereas the accumulation of host-specific pathogens likely generates negative feedback.
4. Change in mutualist competition can generate negative feedback when the cost of mutualism dominates within host dynamics.
5. Preferential allocation can offset the cost of mutualism, potentially generating positive feedback through changes in mutualist composition.
6. Competition and trophic interactions within the microbial community can influence the fitness of host-specific pathogens and thereby potentially suppress negative feedbacks.
7. Feedbacks are driven by, and generate, spatial structure in microbial composition.
8. New microbial variants can arise within a plant-associated microbial community via evolution of the resident populations or dispersal across space or time.

DISCLOSURE STATEMENT

The authors are not aware of any affiliations, memberships, funding, or financial holdings that might be perceived as affecting the objectivity of this review.

ACKNOWLEDGMENTS

We acknowledge support from the National Science Foundation (DEB-0608155, DEB-1050237, and DEB-0919434) and the National Institutes of Health (R01 GM092660).

LITERATURE CITED

1. Alizon S, Hurford A, Mideo N, van Baalen M. 2009. Virulence evolution and the trade-off hypothesis: history, current state of affairs and the future. *J. Evol. Biol.* 22:245–59
2. Alizon S, van Baalen M. 2008. Multiple infections, immune dynamics, and the evolution of virulence. *Am. Nat.* 172:E150–68

3. Antolin MF. 2008. Unpacking beta: within-host dynamics and the evolutionary ecology of pathogen transmission. *Annu. Rev. Ecol. Evol. Syst.* 39:415–37
4. Ashton IW, Miller AE, Bowman WD, Suding KN. 2008. Nitrogen preferences and plant-soil feedbacks as influenced by neighbors in the alpine tundra. *Oecologia* 156:625–36
5. Badri DV, Quintana N, El Kassis EG, Kim HK, Choi YH, et al. 2009. An ABC transporter mutation alters root exudation of phytochemicals that provoke an overhaul of natural soil microbiota. *Plant Physiol.* 151:2006–17

> 6. Reviews role of plant root exudates in structuring microbial communities.

6. **Bais HP, Weir TL, Perry LG, Gilroy S, Vivanco JM. 2006. The role of root exudates in rhizosphere interactions with plants and other organisms.** *Annu. Rev. Plant Biol.* **57:233–66**
7. Bell T, Freckleton RP, Lewis OT. 2006. Plant pathogens drive density-dependent seedling mortality in a tropical tree. *Ecol. Lett.* 9:569–74
8. Bennett AE, Bever JD. 2009. Trade-offs between arbuscular mycorrhizal fungal competitive ability and host growth promotion in *Plantago lanceolata*. *Oecologia* 160:807–16
9. Bergelson J, Kreitman M, Stahl EA, Tian DC. 2001. Evolutionary dynamics of plant R-genes. *Science* 292:2281–85
10. Bever JD. 1994. Feedback between plants and their soil communities in an old field community. *Ecology* 75:1965–77
11. Bever JD. 1999. Dynamics within mutualism and the maintenance of diversity: inference from a model of interguild frequency dependence. *Ecol. Lett.* 2:52–62

> 12. Demonstrates negative feedback through changes in composition of microbial mutualists.

12. **Bever JD. 2002. Negative feedback within a mutualism: Host-specific growth of mycorrhizal fungi reduces plant benefit.** *Proc. Biol. Sci.* **269:2595–601**
13. Bever JD. 2003. Soil community feedback and the coexistence of competitors: conceptual frameworks and empirical tests. *New Phytol.* 157:465–73
14. Bever JD, Dickie IA, Facelli E, Facelli JM, Klironomos J, et al. 2010. Rooting theories of plant community ecology in microbial interactions. *Trends Ecol. Evol.* 25:468–78
15. Bever JD, Morton JB, Antonovics J, Schultz PA. 1996. Host-dependent sporulation and species diversity of arbuscular mycorrhizal fungi in a mown grassland. *J. Ecol.* 84:71–82
16. Bever JD, Richardson SC, Lawrence BM, Holmes J, Watson M. 2009. Preferential allocation to beneficial symbiont with spatial structure maintains mycorrhizal mutualism. *Ecol. Lett.* 12:13–21

> 17. Presents first model of plant-soil community feedback.

17. **Bever JD, Westover KM, Antonovics J. 1997. Incorporating the soil community into plant population dynamics: the utility of the feedback approach.** *J. Ecol.* **85:561–73**
18. Blanca-Ordóñez H, Oliva-García JJ, Pérez-Mendoza D, Soto MJ, Olivares J, et al. 2010. pSymA-dependent mobilization of the *Sinorhizobium meliloti* pSymB megaplasmid. *J. Bacteriol.* 192:6309–12
19. Broders KD, Lipps PE, Paul PA, Dorrance AE. 2007. Characterization of *Pythium* spp. associated with corn and soybean seed and seedling disease in Ohio. *Plant Dis.* 91:727–35
20. Broeckling CD, Broz AK, Bergelson J, Manter DK, Vivanco JM. 2008. Root exudates regulate soil fungal community composition and diversity. *Appl. Environ. Microbiol.* 74:738–44
21. Brown SP, Hochberg ME, Grenfell BT. 2002. Does multiple infection select for raised virulence? *Trends Microbiol.* 10:401–5
22. Bruns TD, Bidartondo MI, Taylor DL. 2002. Host specificity in ectomycorrhizal communities: What do the exceptions tell us? *Integr. Comp. Biol.* 42:352–59
23. Bullock DG. 1992. Crop-rotation. *Crit. Rev. Plant Sci.* 11:309–26
24. Callaway RM, Cipollini D, Barto K, Thelen GC, Hallett SG, et al. 2008. Novel weapons: Invasive plant suppresses fungal mutualists in America but not in its native Europe. *Ecology* 89:1043–55
25. Callaway RM, Thelan GC, Rodriquez A, Holben WE. 2004. Soil biota and exotic plant invasion. *Nature* 427:731–33
26. Castelli JP, Casper BB. 2003. Intraspecific AM fungal variation contributes to plant-fungal feedback in a serpentine grassland. *Ecology* 84:323–36
27. Cervantes L, Bustos P, Girard L, Santamaría RI, Dávila G, et al. 2011. The conjugative plasmid of a bean-nodulating *Sinorhizobium fredii* strain is assembled from sequences of two *Rhizobium* plasmids and the chromosome of a *Sinorhizobium* strain. *BMC Microbiol.* 11:149
28. Chase JM, Leibold MA. 2003. *Ecological Niches: Linking Classical and Contemporary Approaches*. Chicago: Univ. Chicago Press. 221 pp.

29. Chen LS, Chen YC, Wood DW, Nester EW. 2002. A new type IV secretion system promotes conjugal transfer in *Agrobacterium tumefaciens*. *J. Bacteriol.* 184:4838–45
30. Chesson P. 2000. Mechanisms of maintenance of species diversity. *Annu. Rev. Ecol. Syst.* 31:343–66
31. Chisholm ST, Coaker G, Day B, Staskawicz BJ. 2006. Host-microbe interactions: shaping the evolution of the plant immune response. *Cell* 124:803–14
32. Comita LS, Muller-Landau HC, Aguilar S, Hubbell SP. 2010. Asymmetric density dependence shapes species abundances in a tropical tree community. *Science* 329:330–32
33. Currier WW, Strobel GA. 1976. Chemotaxis of *Rhizobium* spp. to plant root exudates. *Plant Physiol.* 57:820–23
34. Diez JM, Dickie I, Edwards G, Hulme PE, Sullivan JJ, Duncan RP. 2010. Negative soil feedbacks accumulate over time for non-native plant species. *Ecol. Lett.* 13:803–9
35. Dörr T, Vulić M, Lewis K. 2010. Ciprofloxacin causes persister formation by inducing the TisB toxin in *Escherichia coli*. *PLoS Biol.* 8:e1000317
36. Dybzinski R, Tilman D. 2007. Resource use patterns predict long-term outcomes of plant competition for nutrients and light. *Am. Nat.* 170:305–18
37. Engelkes T, Morrien E, Verhoeven KJF, Bezemer TM, Biere A, et al. 2008. Successful range-expanding plants experience less above-ground and below-ground enemy impact. *Nature* 456:946–48
38. Eppinga MB, Rietkerk M, Dekker SC, De Ruiter PC, Van der Putten WH. 2006. Accumulation of local pathogens: a new hypothesis to explain exotic plant invasions. *Oikos* 114:168–76
39. Escobar MA, Dandekar AM. 2003. *Agrobacterium tumefaciens* as an agent of disease. *Trends Plant Sci.* 8:380–86
40. Fuqua WC, Winans SC. 1994. A LuxR-LuxI type regulatory system activates *Agrobacterium* Ti plasmid conjugal transfer in the presence of a plant tumor metabolite. *J. Bacteriol.* 176:2796–806
41. Grant MW, Archer SA. 1983. Calculation of selection coefficients against unnecessary genes for virulence from field data. *Phytopathology* 73:547–51
42. Gubry-Rangin C, Garcia M, Bena G. 2010. Partner choice in *Medicago truncatula-Sinorhizobium* symbiosis. *Proc. Biol. Sci.* 277:1947–51
43. Guttman DS, Greenberg JT. 2001. Functional analysis of the type III effectors AvrRpt2 and AvrRpm1 of *Pseudomonas syringae* with the use of a single-copy genomic integration system. *Mol. Plant-Microbe Interact.* 14:145–55
44. Guyon P, Chilton MD, Petit A, Tempe J. 1980. Agropine in "null-type" crown gall tumors: evidence for generality of the opine concept. *Proc. Natl. Acad. Sci. USA* 77:2693–97
45. Guyon P, Petit A, Tempe J, Dessaux Y. 1993. Transformed plants producing opines specifically promote growth of opine-degrading agrobacteria. *Mol. Plant-Microbe Interact.* 6:92–98
46. Hacker J, Kaper JB. 2000. Pathogenicity islands and the evolution of microbes. *Annu. Rev. Microbiol.* 54:641–79
47. Hartnett DC, Hetrick BAD, Wilson GWT, Gibson DJ. 1993. Mycorrhizal influence on intra- and interspecific neighbour interactions among co-occurring prairie grasses. *J. Ecol.* 81:787–95
48. Heath KD. 2010. Intergenomic epistasis and coevolutionary constraint in plants and rhizobia. *Evolution* 64:1446–58
49. Heath KD, Tiffin P. 2009. Stabilizing mechanisms in a legume-*Rhizobium* mutualism. *Evolution* 63:652–62
50. Heller G, Lundén K, Finlay RD, Asiegbu FO, Elsfstrand M. 2012. Expression analysis of Clavata1-like and Nodulin21-like genes from *Pinus sylvestris* during ectomycorrhiza formation. *Mycorrhiza* 22:271–77
51. Hillerislambers J, Clark JS, Beckage B. 2002. Density-dependent mortality and the latitudinal gradient in species diversity. *Nature* 417:732–35
52. Inderjit, van der Putten WH. 2010. Impacts of soil microbial communities on exotic plant invasions. *Trends Ecol. Evol.* 25:512–19
53. Ingham CJ, Kalisman O, Finkelshtein A, Ben-Jacob E. 2011. Mutually facilitated dispersal between the nonmotile fungus *Aspergillus fumigatus* and the swarming bacterium *Paenibacillus vortex*. *Proc. Natl. Acad. Sci. USA* 108:19731–36

54. Jackson RW, Athanassopoulos E, Tsiamis G, Mansfield JW, Sesma A, et al. 1999. Identification of a pathogenicity island, which contains genes for virulence and avirulence, on a large native plasmid in the bean pathogen *Pseudomonas syringae* pathovar phaseolicola. *Proc. Natl. Acad. Sci. USA* 96:10875–80
55. Janos DP. 1980. Mycorrhizae influence tropical succession. *Biotropica* 12:56–64
56. Janzen DH. 1970. Herbivores and the number of tree species in tropical forests. *Am. Nat.* 104:592–95
57. Jenner CE, Wang XW, Ponz F, Walsh JA. 2002. A fitness cost for Turnip mosaic virus to overcome host resistance. *Virus Res.* 86:1–6
58. Johnson DJ, Beaulieu WT, Bever JD, Clay K. 2012. Conspecific negative density dependence and forest diversity. *Science* 336:904–7
59. **Jones JDG, Dangl JL. 2006. The plant immune system. *Nature* 444:323–29**

59. Synthetic review of the plant immune system.

60. Jones KM, Kobayashi H, Davies BW, Taga ME, Walker GC. 2007. How rhizobial symbionts invade plants: the *Sinorhizobium-Medicago* model. *Nat. Rev. Microbiol.* 5:619–33
61. Jones SE, Lennon JT. 2010. Dormancy contributes to the maintenance of microbial diversity. *Proc. Natl. Acad. Sci. USA* 107:5881–86
62. Kardol P, Cornips NJ, van Kempen MML, Bakx-Schotman JMT, van der Putten WH. 2007. Microbe-mediated plant-soil feedback causes historical contingency effects in plant community assembly. *Ecol. Monogr.* 77:147–62
63. Kassen R, Rainey PB. 2004. The ecology and genetics of microbial diversity. *Annu. Rev. Microbiol.* 58:207–31
64. Kennedy PG, Hortal S, Bergemann SE, Bruns TD. 2007. Competitive interactions among three ectomycorrhizal fungi and their relation to host plant performance. *J. Ecol.* 95:1338–45
65. Kiers ET, Duhamel M, Beesetty Y, Mensah JA, Franken O, et al. 2011. Reciprocal rewards stabilize cooperation in the mycorrhizal symbiosis. *Science* 333:880–82
66. **Kiers ET, Rousseau RA, West SA, Denison RF. 2003. Host sanctions and the legume-*Rhizobium* mutualism. *Nature* 425:78–81**

66. Provides evidence of host sanctions of ineffective nodules.

67. Kim JG, Park YK, Kim SU, Choi D, Nahm BH, et al. 2006. Bases of biocontrol: Sequence predicts synthesis and mode of action of agrocin 84, the Trojan horse antibiotic that controls crown gall. *Proc. Natl. Acad. Sci. USA* 103:8846–51
68. **Kinkel LL, Bakker MG, Schlatter DC. 2011. A coevolutionary framework for managing disease-suppressive soils. *Annu. Rev. Phytopathol.* 49:47–67**

68. Reviews evidence for suppressive soils.

69. **Klironomos JN. 2002. Feedback with soil biota contributes to plant rarity and invasiveness in communities. *Nature* 417:67–70**

69. Finds correlation between strength of feedback and relative abundance.

70. Kulmatiski A, Beard KH, Stevens JR, Cobbold SM. 2008. Plant-soil feedbacks: a meta-analytical review. *Ecol. Lett.* 11:980–92
71. Kupriyanov AA, Kunenkova NN, van Bruggen AHC, Semenov AM. 2009. Translocation of bacteria from animal excrements to soil and associated habitats. *Eurasian Soil Sci.* 42:1263–69
72. Kyselkova M, Kopecky J, Frapolli M, Defago G, Sagova-Mareckova M, et al. 2009. Comparison of rhizobacterial community composition in soil suppressive or conducive to tobacco black root rot disease. *ISME J.* 3:1127–38
73. Laine AL, Burdon JJ, Dodds PN, Thrall PH. 2011. Spatial variation in disease resistance: from molecules to metapopulations. *J. Ecol.* 99:96–112
74. Lankau RA. 2011. Intraspecific variation in allelochemistry determines an invasive species' impact on soil microbial communities. *Oecologia* 165:453–63
75. Lapointe G, Nautiyal CS, Chilton WS, Farrand SK, Dion P. 1992. Spontaneous mutation conferring the ability to catabolize mannopine in *Agrobacterium tumefaciens*. *J. Bacteriol.* 174:2631–39
76. Larkin RP, Hopkins DL, Martin FN. 1993. Effect of successive watermelon plantings on *Fusarium oxysporum* and other micro-organisms in soils suppressive and conducive to *Fusarium* wilt of watermelon. *Phytopathology* 83:1097–105
77. **Lennon JT, Jones SE. 2011. Microbial seed banks: the ecological and evolutionary implications of dormancy. *Nat. Rev. Microbiol.* 9:119–30**

77. Demonstrates that a surprisingly high proportion of soil bacteria are dormant.

78. Lewis K. 2007. Persister cells, dormancy and infectious disease. *Nat. Rev. Microbiol.* 5:48–56
79. Lim WC, Lockwood JL. 1988. Chemotaxis of some phytopathogenic bacteria to fungal propagules in vitro and in soil. *Can. J. Microbiol.* 34:196–99

80. Liu DQ, Anderson NA, Kinkel LL. 1995. Biological control of potato scab in the field with antagonistic *Streptomyces scabies*. *Phytopathology* 85:827–31
81. Liu XG, Inlow M, VanEtten HD. 2003. Expression profiles of pea pathogenicity (PEP) genes in vivo and in vitro, characterization of the flanking regions of the PEP cluster and evidence that the PEP cluster region resulted from horizontal gene transfer in the fungal pathogen *Nectria haematococca*. *Curr. Genet.* 44:95–103
82. Ma WB, Dong FFT, Stavrinides J, Guttman DS. 2006. Type III effector diversification via both pathoadaptation and horizontal transfer in response to a coevolutionary arms race. *PLoS Genet.* 2:2131–42
83. Macarthur R, Levins R. 1967. The limiting similarity, convergence, and divergence of coexisting species. *Am. Nat.* 101:377–85
84. Mangan SA, Herre EA, Bever JD. 2010. Specificity between Neotropical tree seedlings and their fungal mutualists leads to plant-soil feedback. *Ecology* 91:2594–603
85. Mangan SA, Schnitzer SA, Herre EA, Mack KML, Valencia MC, et al. 2010. Negative plant-soil feedback predicts tree-species relative abundance in a tropical forest. *Nature* 466:752–55
86. Mangla S, Inderjit, Callaway RM. 2008. Exotic invasive plant accumulates native soil pathogens which inhibit native plants. *J. Ecol.* 96:58–67
87. McCann HC, Guttman DS. 2008. Evolution of the type III secretion system and its effectors in plant-microbe interactions. *New Phytol.* 177:33–47
88. Mendes R, Kruijt M, de Bruijn I, Dekkers E, van der Voort M, et al. 2011. Deciphering the rhizosphere microbiome for disease-suppressive bacteria. *Science* 332:1097–100
89. Middleton E, Bever JD. 2012. Inoculation with a native soil community advances succession in a grassland restoration. *Restor. Ecol.* 20:218–26
90. Miki T, Ushio M, Fukui S, Kondoh M. 2010. Functional diversity of microbial decomposers facilitates plant coexistence in a plant-microbe-soil feedback model. *Proc. Natl. Acad. Sci. USA* 107:14251–56
91. Miller TE, Burns JH, Munguia P, Walters EL, Kneitel JM, et al. 2005. A critical review of twenty years' use of the resource-ratio theory. *Am. Nat.* 165:439–48
92. Mills KE, Bever JD. 1998. Maintenance of diversity within plant communities: soil pathogens as agents of negative feedback. *Ecology* 79:1595–601
93. Molofsky J, Bever JD. 2002. A novel theory to explain species diversity in landscapes: positive frequency dependence and habitat suitability. *Proc. Biol. Sci.* 269:2389–93
94. Molofsky J, Bever JD, Antonovics J, Newman TJ. 2002. Negative frequency dependence and the importance of spatial scale. *Ecology* 83:21–27
95. Monod J. 1949. The growth of bacterial cultures. *Annu. Rev. Microbiol.* 3:371–94
96. Moora M, Zobel M. 1996. Effect of arbuscular mycorrhiza on inter- and intraspecific competition of two grassland species. *Oecologia* 108:79–84
97. Moriguchi K, Maeda Y, Satou M, Hardayani NSN, Kataoka M, et al. 2001. The complete nucleotide sequence of a plant root-inducing (Ri) plasmid indicates its chimeric structure and evolutionary relationship between tumor-inducing (Ti) and symbiotic (Sym) plasmids in *Rhizobiaceae*. *J. Mol. Biol.* 307:771–84
98. Nelson B, Helms T, Christianson T, Kural I. 1996. Characterization and pathogenicity of *Rhizoctonia* from soybean. *Plant Dis.* 80:74–80
99. Newton ILG, Bordenstein SR. 2011. Correlations between bacterial ecology and mobile DNA. *Curr. Microbiol.* 62:198–208
100. Nuñez MA, Horton TR, Simberloff D. 2009. Lack of belowground mutualisms hinders *Pinaceae* invasions. *Ecology* 90:2352–59
101. Nurnberger T, Brunner F, Kemmerling B, Piater L. 2004. Innate immunity in plants and animals: striking similarities and obvious differences. *Immunol. Rev.* 198:249–66
102. Ochman H, Lerat E, Daubin V. 2005. Examining bacterial species under the specter of gene transfer and exchange. *Proc. Natl. Acad. Sci. USA* 102:6595–99
103. Oono R, Anderson CG, Denison RF. 2011. Failure to fix nitrogen by non-reproductive symbiotic rhizobia triggers host sanctions that reduce fitness of their reproductive clonemates. *Proc. Biol. Sci.* 278:2698–703
104. Packer A, Clay K. 2000. Soil pathogens and spatial patterns of seedling mortality in a temperate tree. *Nature* 404:278–81

105. Packer A, Clay K. 2003. Soil pathogens and *Prunus serotina* seedling and sapling growth near conspecific trees. *Ecology* 84:108–19
106. Parker IM, Gilbert GS. 2004. The evolutionary ecology of novel plant-pathogen interactions. *Annu. Rev. Ecol. Evol. Syst.* 35:675–700
107. Perret X, Staehelin C, Broughton WJ. 2000. Molecular basis of symbiotic promiscuity. *Microbiol. Mol. Biol. Rev.* 64:180–201
108. Petermann JS, Fergus AJF, Turnbull LA, Schmid B. 2008. Janzen-Connell effects are widespread and strong enough to maintain diversity in grasslands. *Ecology* 89:2399–406
109. Pitman AR, Jackson RW, Mansfield JW, Kaitell V, Thwaites R, Arnold DL. 2005. Exposure to host resistance mechanisms drives evolution of bacterial virulence in plants. *Curr. Biol.* 15:2230–35
110. Platt TG, Bever JD, Fuqua C. 2012. A cooperative virulence plasmid imposes a high fitness cost under conditions which induce pathogenesis. *Proc. Biol. Sci.* 279:1691–99
111. Platt TG, Fuqua C, Bever JD. 2012. Resource and competitive dynamics shape the benefits of public goods cooperation in a plant pathogen. *Evolution* 66:1953–65
112. Porter PM, Chen SYY, Reese CD, Klossner LD. 2001. Population response of soybean cyst nematode to long term corn-soybean cropping sequences in Minnesota. *Agron. J.* 93:619–26
113. Pringle A, Bever JD. 2002. Divergent phenologies may facilitate the coexistence of arbuscular mycorrhizal fungi in a North Carolina grassland. *Am. J. Bot.* 89:1439–46
114. Pringle A, Bever JD, Gardes M, Parrent JL, Rillig MC, Klironomos JN. 2009. Mycorrhizal symbioses and plant invasions. *Annu. Rev. Ecol. Evol. Syst.* 40:699–715
115. Raaijmakers JM, Paulitz TC, Steinberg C, Alabouvette C, Moënne-Loccoz Y. 2009. The rhizosphere: a playground and battlefield for soilborne pathogens and beneficial microorganisms. *Plant Soil* 321:341–61
116. Reader JS, Ordoukhanian PT, Kim JG, de Crécy-Lagard V, Hwang I, et al. 2005. Major biocontrol of plant tumors targets tRNA synthetase. *Science* 309:1533
117. Reinhart KO, Packer A, Van der Putten WH, Clay K. 2003. Plant-soil biota interactions and spatial distribution of black cherry in its native and invasive ranges. *Ecol. Lett.* 6:1046–50
118. Reynolds HL, Haubensak KA. 2009. Soil fertility, heterogeneity, and microbes: towards an integrated understanding of grassland structure and dynamics. *Appl. Veg. Sci.* 12:33–44
119. Reynolds HL, Packer A, Bever JD, Clay K. 2003. Grassroots ecology: plant-microbe-soil interactions as drivers of plant community structure and dynamics. *Ecology* 84:2281–91
120. Sanguin H, Sarniguet A, Gazengel K, Moënne-Loccoz Y, Grundmann GL. 2009. Rhizosphere bacterial communities associated with disease suppressiveness stages of take-all decline in wheat monoculture. *New Phytol.* 184:694–707
121. Schmidt H, Hensel M. 2004. Pathogenicity islands in bacterial pathogenesis. *Clin. Microbiol. Rev.* 17:14–56
122. Schultz PA, Miller RM, Jastrow JD, Rivetta CV, Bever JD. 2001. Evidence of a mycorrhizal mechanism for the adaptation of *Andropogon gerardii* to high and low-nutrient prairies. *Am. J. Bot.* 88:1650–56
123. Schultze M, Quiclet-Sire B, Kondorosi E, Virelizer H, Glushka JN, et al. 1992. *Rhizobium meliloti* produces a family of sulfated lipooligosaccharides exhibiting different degrees of plant host specificity. *Proc. Natl. Acad. Sci. USA* 89:192–96
124. Sikorski J, Nevo E. 2005. Adaptation and incipient sympatric speciation of *Bacillus simplex* under microclimatic contrast at "Evolution Canyons" I and II, Israel. *Proc. Natl. Acad. Sci. USA* 102:15924–29
125. Simberloff D, Von Holle B. 1999. Positive interactions of nonindigenous species: invasional meltdown? *Biol. Invasions* 1:21–32
126. Smith J. 2011. Superinfection drives virulence evolution in experimental populations of bacteria and plasmids. *Evolution* 65:831–41
127. Sokurenko EV, Hasty DL, Dykhuizen DE. 1999. Pathoadaptive mutations: gene loss and variation in bacterial pathogens. *Trends Microbiol.* 7:191–95
128. **Stinson KA, Campbell SA, Powell JR, Wolfe BE, Callaway RM, et al. 2006. Invasive plant suppresses the growth of native tree seedlings by disrupting belowground mutualisms.** ***PLoS Biol.*** **4:727–31**

   128. Shows that an invasive plant inhibits natives by interfering with AM fungi.

129. Sussman AS, Douthit HA. 1973. Dormancy in microbial spores. *Annu. Rev. Plant Physiol. Plant Mol. Biol.* 24:311–52

130. Thrall PH, Burdon JJ. 2003. Evolution of virulence in a plant host-pathogen metapopulation. *Science* 299:1735–37

131. Tilman D. 1982. *Resource Competition and Community Structure*. Princeton, NJ: Princeton Univ. Press

132. Truchet G, Rosenberg C, Vasse J, Julliot JS, Camut S, Denarie J. 1984. Transfer of *Rhizobium meliloti* pSym genes into *Agrobacterium tumefaciens*: host-specific nodulation by atypical infection. *J. Bacteriol.* 157:134–42

133. Umbanhowar J, McCann K. 2005. Simple rules for the coexistence and competitive dominance of plants mediated by mycorrhizal fungi. *Ecol. Lett.* 8:247–52

134. van der Heijden MGA. 2006. Symbiotic bacteria as a determinant of plant community structure and plant productivity in dune grassland. *FEMS Microbiol. Ecol.* 56:178–87

135. van der Heijden MGA, Klironomos JN, Ursic M, Moutoglis P, Streitwolf-Engel R, et al. 1998. Mycorrhizal fungal diversity determines plant biodiversity, ecosystem variability and productivity. *Nature* 396:69–72

136. Van der Putten WH, Van Dijk C, Peters BAM. 1993. Plant-specific soil-borne diseases contribute to succession in foredune vegetation. *Nature* 362:53–56

137. Vanderplank JE. 1968. *Disease Resistance in Plants*. New York: Academic

138. Vitousek PM, Walker LR, Whiteaker LD, Mueller-Dombois D, Matson PA. 1987. Biological invasion by *Myrica faya* alters ecosystem development in Hawaii. *Science* 238:802–4

139. Vogelsang KM, Bever JD. 2009. Mycorrhizal densities decline in association with nonnative plants and contribute to plant invasion. *Ecology* 90:399–407

140. Vogelsang KM, Reynolds HL, Bever JD. 2006. Mycorrhizal fungal identity and richness determine the diversity and productivity of a tallgrass prairie system. *New Phytol.* 172:554–62

141. Walker TS, Bais HP, Grotewold E, Vivanco JM. 2003. Root exudation and rhizosphere biology. *Plant Physiol.* 132:44–51

142. Warmink JA, Nazir R, Corten B, van Elsas JD. 2011. Hitchhikers on the fungal highway: the helper effect for bacterial migration via fungal hyphae. *Soil Biol. Biochem.* 43:760–65

143. Westover KM, Bever JD. 2001. Mechanisms of plant species coexistence: roles of rhizosphere bacteria and root fungal pathogens. *Ecology* 82:3285–94

144. Wills C, Harms KE, Condit R, King D, Thompson J, et al. 2006. Nonrandom processes maintain diversity in tropical forests. *Science* 311:527–31

145. Wilson JB, Spijkerman E, Huisman J. 2007. Is there really insufficient support for Tilman's R* concept? A comment on Miller et al. *Am. Nat.* 169:700–6

146. Winans SC. 1992. Two-way chemical signaling in *Agrobacterium*-plant interactions. *Microbiol. Rev.* 56:12–31

147. Yang CH, Crowley DE. 2000. Rhizosphere microbial community structure in relation to root location and plant iron nutritional status. *Appl. Environ. Microbiol.* 66:345–51

148. Yang S, Wu Z, Gao W, Li J. 1993. Tn5-Mob transposon mediated transfer of salt tolerance and symbiotic characteristics between rhizobia genera. *Chin. J. Biotechnol.* 9:137–41

149. Zhang Q, Yang RY, Tang JJ, Yang HS, Hu SJ, Chen X. 2010. Positive feedback between mycorrhizal fungi and plants influences plant invasion success and resistance to invasion. *PLoS One* 5(8): e12380

150. Zhou HB, Morgan RL, Guttman DS, Ma WB. 2009. Allelic variants of the *Pseudomonas syringae* type III effector HopZ1 are differentially recognized by plant resistance systems. *Mol. Plant-Microbe Interact.* 22:176–89

---

130. Provides convincing evidence of costs of overcoming host defense and their effect on pathogen dynamics.

Bacterial Chemotaxis: The Early Years of Molecular Studies

Gerald L. Hazelbauer

Department of Biochemistry, University of Missouri, Columbia, Missouri 65211;
email: hazelbauerg@missouri.edu

Keywords

bacterial behavior, bacterial motility, signaling, transmembrane receptors, history of science

Abstract

This review focuses on the early years of molecular studies of bacterial chemotaxis and motility, beginning in the 1960s with Julius Adler's pioneering work. It describes key observations that established the field and made bacterial chemotaxis a paradigm for the molecular understanding of biological signaling. Consideration of those early years includes aspects of science seldom described in journals: the accidental findings, personal interactions, and scientific culture that often drive scientific progress.

Contents

OVERVIEW	286
STUDYING BEHAVIOR WITH BIOCHEMISTRY AND GENETICS	286
ELUCIDATING MOLECULAR MECHANISMS	287
THE MAJOR ISSUES	288
Identifying Attractants and Repellents	288
Generally Nonchemotactic Mutants and the Che Proteins	290
Flagellar Structure	290
Flagellar Function	290
Receptors	291
Methionine and Methylation	293
Methyl-Accepting Chemotaxis Proteins	293
Phosphorylation	294
THE CULTURE OF THE CHEMOTAXIS COMMUNITY	294
Annual Gatherings	296
Sensory Transduction in Microorganisms	297
Bacterial Locomotion and Sensory Transduction	298
FINAL THOUGHTS	299

OVERVIEW

Molecular studies of bacterial chemotaxis and motility began in the period 1965–1969 when Julius Adler and colleagues published pioneering studies with *Escherichia coli* (1–3, 7, 13–15). Over the following two decades, the foundation was laid for current understanding. This review focuses on those early years, when first one and then a few laboratories were involved. It includes aspects of science seldom described in journals: accidental findings, personal interactions, and scientific culture. For these I utilized material from colleagues and from my experiences in chemotaxis, which span the 44 years since I joined Adler's laboratory as a graduate student in January 1968. I thank my colleagues for their contributions and ask their understanding for my editorial decisions as I struggled to stay within the page limit.

Important companions to this review are the recent prefatory chapter by Julius Adler in the *Annual Review of Biochemistry* (6), and John S. (Sandy) Parkinson's article about early years in the Adler laboratory (75). The sidebar provides an overview of *Escherichia coli* chemotaxis.

STUDYING BEHAVIOR WITH BIOCHEMISTRY AND GENETICS

In 1960, new University of Wisconsin Assistant Professor Julius Adler began studying the behavior of *E. coli*. He was inspired by his interest in sensory phenomena, which he traces to observing butterflies as a boy (5), and by the nineteenth-century literature on bacterial taxis he discovered as a young scientist (6). Julius applied biochemical strategies and genetic approaches learned during training with Henry A. Lardy, Arthur Kornberg, and Dale Kaiser to molecular mechanisms of bacterial behavior (6). By the early 1960s, utilizing bacteria, particularly *E. coli*, to identify molecular mechanisms had proven impressively successful for macromolecular synthesis and control of gene expression. Several prominent senior scientists who had been part of these successes were turning their interest to neuroscience. However, for a new assistant professor to begin his independent research career by applying the notion "[a]nything found to be true of *E. coli* must also

A CHEMOTAXIS PRIMER

E. coli traces a random walk of straight runs, generated by counterclockwise rotation of flagellar rotary motors, and tumbling episodes, generated by clockwise rotation, that reorient the cell. The chemosensory system biases random walks by reducing probabilities of clockwise rotation for swims in favorable directions. The system comprises six cytoplasmic proteins and five transmembrane chemoreceptors plus four periplasmic ligand-binding proteins. Chemoreceptors Tsr and Tar mediate responses by binding serine and aspartate, respectively. Tar, Tap, and Trg detect maltose, dipeptides, and galactose or ribose through interaction with ligand-occupied forms of periplasmic binding proteins.

Receptors plus the autophosphorylating histidine kinase CheA and coupling protein CheW form signaling complexes, activating kinase ∼1000-fold and placing it under receptor control. Phosphoryl groups are transferred from the kinase to the response regulator CheY. CheY-P binds the flagellar rotary motor, inducing clockwise rotation and thus tumbles. Inherent instability and phosphatase CheZ make CheY-P short-lived. Binding of attractant to receptors inhibits kinase activity, reducing CheY-P and tumble frequency. The system adapts to persistent simulation via receptor methylation by methyltransferase CheR (positive stimuli) and demethylation by methylesterase CheB (negative stimuli). Kinase activities altered by changes in receptor occupancy are restored to prestimulus levels by adjusting receptor methylation. For instance, attractant occupancy enhances methylation and inhibits demethylation, generating increased methylation and thereby restoring kinase activity to prestimulus levels. Ligand-induced changes in signaling complex conformation and kinase activity are rapid, whereas methylation changes are slower. A resulting time disparity of ∼4 s creates a molecular memory that provides sensing of temporal gradients. *E. coli* chemotaxis exhibits high sensitivity and wide dynamic range, properties not explained by the simple model outlined above. These properties are thought to reflect interactions in extended arrays of signaling complexes, which vary in size and position but are largest at cell poles (43, 47, 48, 60).

---

be true of elephants" (64) to something as ephemeral as bacterial behavior was more than a little daring. Yet, Julius' earliest chemotaxis research had a receptive audience. For instance, his first chemotaxis publication was the written version of an invited talk at the 1965 Cold Spring Harbor Symposium (1).

Admiring comments from members of that audience directed students and junior scientists to bacterial chemotaxis. For instance, Howard Berg reports that he started working on a microscope for tracking swimming bacteria in September 1968, after Max Delbruck told him he would work on bacteria to study behavior if only he knew how to "tame" them. For me, in fall 1966 Julius' friend and postdoctoral mentor, Dale Kaiser, remarked to Harriet Ephrussi-Taylor, then a faculty member at Case Western Reserve University, that if he were a beginning graduate student he would join Julius Adler. As a beginning graduate student at Case Western Reserve, I was hoping to work with Dr. Ephrussi-Taylor and was in her office during that conversation. Within a year she became too sick to take students, so after a visit to Madison during Thanksgiving 1967, I began with Julius in January.

ELUCIDATING MOLECULAR MECHANISMS

In his 1969 *Science* paper, Julius (3) demonstrated that chemotactic responses of *E. coli* to galactose or its structural analogs did not require metabolism or transport. Thus, response was not the mechanistic result of benefit provided by the compound, a concept about bacterial behavior common in the literature, but instead was generated by recognizing the compound itself. This implied a specific system for recognition and response. It initiated contemporary studies of bacterial chemotaxis.

Wilhelm Pfeffer: a prominent nineteenth- and early-twentieth-century German botanist and microbiologist who studied microbial chemotaxis and many aspects of botany

The 1969 paper did more. It demonstrated the utility of investigating a sensory phenomenon using a model organism in which mutant derivatives could be generated, genetic manipulations were facile, and a substantial community of scientists could provide genetic and biochemical tools. Julius utilized nonmetabolizable analogs of bacterial attractants and mutant bacteria unable to metabolize or even concentrate an attractant to demonstrate that the chemotactic response was independent of benefit. This seminal study profited from the ready availability of *E. coli* mutants. Subsequent progress was greatly enhanced by generation of many chemotaxis mutants and Julius' practice of providing mutants freely, whether described in publications or not, to interested researchers. This practice became a feature of the field. Particularly notable is Sandy Parkinson, who began mutational analysis of *E. coli* chemotaxis as a postdoc with Julius in 1970. As an independent investigator, Sandy became the primary source of mutant strains, generously making available his vast strain collection and his knowledge and thus enhancing our collective progress.

Choosing *E. coli* as the model organism had an unanticipated benefit. We now know that the number of taxis components varies with species. Five core components are common to almost all systems, but many systems have one or more additional components (95). Also, a species can contain two or more sets of core components. Additional components often provide alternative or partially redundant activities; additional full sets often control responses other than movement (44, 94). Furthermore, the number of chemoreceptors can vary drastically, from a few to more than 50 (95). Serendipitously, *E. coli* contains a single set of the five core components plus only one additional component (a phosphatase) and five receptors. This relative simplicity greatly facilitated the identification of components and their roles, particularly prior to facile gene cloning and sequencing. As is often the case, chance enhanced scientific progress.

THE MAJOR ISSUES

In January 1968 when I joined Julius' laboratory, his group had established the experimental foundations for studying *E. coli* motility and chemotaxis and described the first chemotaxis mutants (1, 2, 7, 9, 13, 15). Julius was preparing what would become the 1969 *Science* paper. Yet we knew almost nothing about machinery or mechanisms. Each member of the lab (**Figure 1**) investigated a major issue: (*a*) attractants and repellents (Marge Dahl, Julius' long-serving and very effective research assistant, plus graduate students Bob Mesibov and me); (*b*) the common machinery (student John Armstrong); (*c*) structure of the motility apparatus (student Melvin Depamphilis) and its mode of function (occasionally by short-term lab members); (*d*) receptors (me and Bob Mesibov); (*e*) the role of methionine (John Armstrong and subsequently others); and (*f*) signaling, the mechanism(s) coupling receptors to flagella (much discussed but not then actively investigated). Here's what happened.

Identifying Attractants and Repellents

Identification of attractants and repellents provided tools for investigating chemotaxis machinery and mechanisms. Over multiple years, Marge Dahl, with some help from others, used the "Pfeffer Assay" to survey innumerable compounds. The assay was essentially Wilhelm Pfeffer's from the 1880s (76). A 1-µL glass capillary tube containing a chemical solution was placed in a bacterial suspension, and cell accumulation in response to the resulting diffusion gradient was determined. Pfeffer assessed accumulation under the microscope. Marge plated capillary contents (4). The data for *E. coli* attractants (8, 63) and repellents (93) remain the most extensive profile of chemotactic sensitivities for any bacterium. In related studies exploring parallels between the behavior of *E. coli* and higher organisms, several neuroactive compounds that Julius obtained from colleagues were

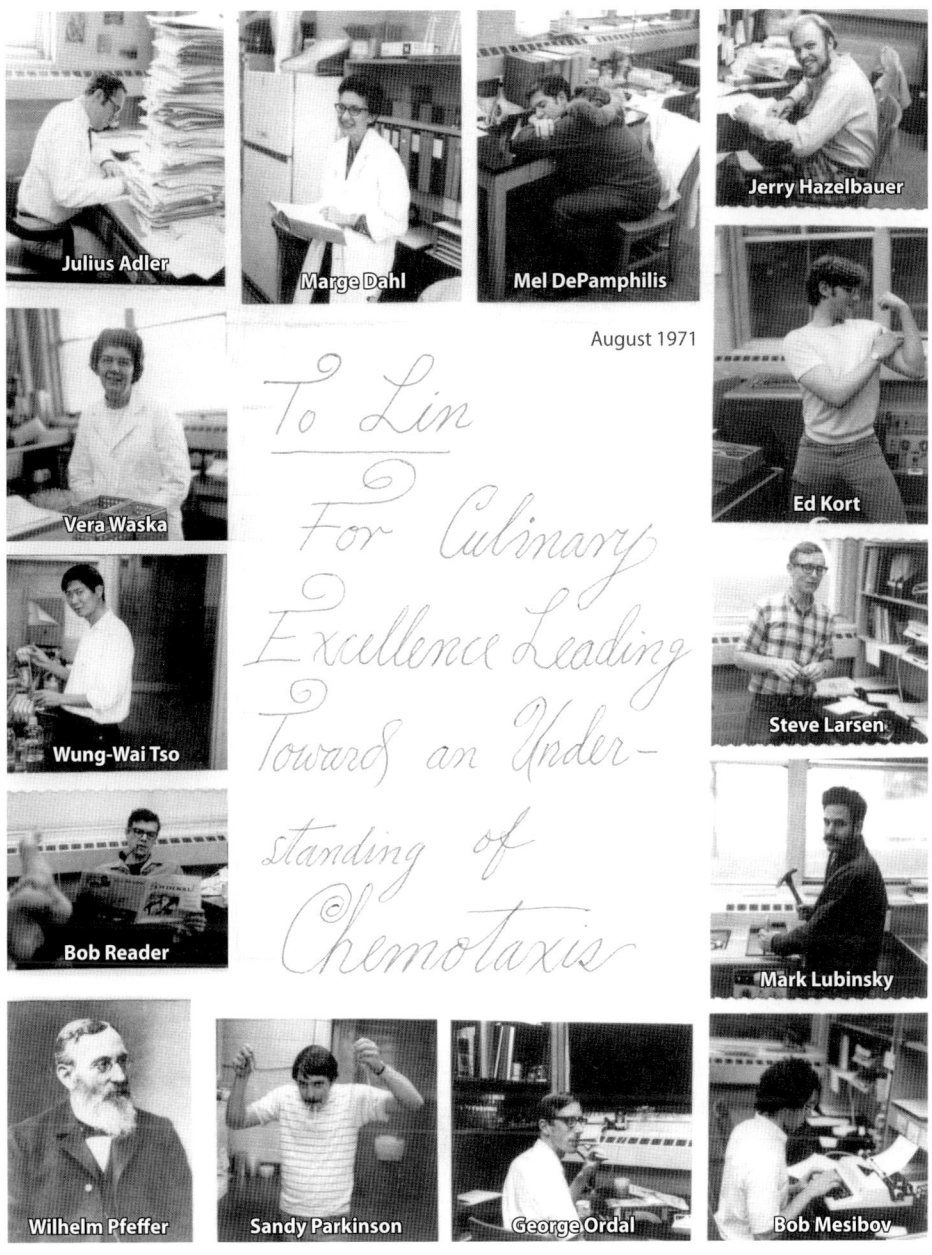

Figure 1

The Adler laboratory, summer 1971. Photos of lab members (plus Wilhelm Pfeffer) arrayed around a certificate of appreciation to Linda Randall, who as Jerry Hazelbauer's spouse had made gourmet desserts for weekly Adler group meetings for the previous year and was being thanked as the two departed for postdocs at the Pasteur Institute in Paris.

surveyed, usually by microscopic observation. These compounds included mescaline and lysergic acid diethylamide (LSD) (obtained from Timothy Leary when still legal). I recall LSD caused *E. coli* to tumble (see also Reference 75).

Generally Nonchemotactic Mutants and the Che Proteins

John Armstrong, Julius' first chemotaxis graduate student, isolated mutants unable to respond to any chemoeffector yet motile and without additional defects. He used semisolid agar plates containing tryptone, a sufficiently low agar concentration such that *E. coli* could swim through water channels in the matrix. As cells multiplied, they consumed amino acids in the tryptone according to their metabolic preferences, serine and then aspartate, creating sequential gradients to which they responded, forming respective rings of cells following those gradients. Using the cells that remained at the point of inoculation to inoculate a new plate and repeating this process multiple times enriched for motility and chemotaxis mutants (15). This yielded mutants defective in each nonreceptor component of the *E. coli* taxis system (14, 15, 33, 71–74). Subsequently, Melvin Simon's laboratory utilized then-new techniques of gene cloning to identify the gene products, the Che (chemotaxis) proteins, which were the common components of the chemotaxis system (80, 81, 83, 85).

Flagellar Structure

Adler's research group investigated motility as well as chemotaxis (7, 9, 13, 79). In the 1960s, little was known about flagellar structure besides what extended from the cell surface. In fall 1964, graduate student Mel DePamphilis began investigations. After many false starts, he isolated intact flagella (see A Breakthrough with No One to Tell). Using a relatively dated electron microscope, he recorded images (34–36) that persisted for 18 years as the best views of the core structure of what we later learned was the flagellar rotary motor.

Flagellar Function

How did the chemosensory system direct motility? In April 1970, Howard Berg got his tracking microscope working. By 1972, he found that cells in buffer traced random walks and that gradients

---

A BREAKTHROUGH WITH NO ONE TO TELL

In 1966 Mel DePamphilis, an almost third-year graduate student, had little indication his project would be successful. Mel recounts the breakthrough:

11 PM one evening I experienced the most exciting moment of my scientific career. I was at Biochemistry's Siemens Elmiskop I electron microscope to check my latest attempt to isolate intact flagella. The microscope required ~40 minutes to load a sample, create a vacuum inside and align the beam. Then I had to work quickly since the sample deteriorated rapidly under the intense beam. Moreover, the camera had only 10 sheets of film and exposures were done manually: opening the shutter, counting three seconds and closing. That evening, I saw the usual spaghetti-like flagellar preparation. Tracing individual strands to search for unusual structures at an end, I froze. Several had them! These were too remarkable and too reproducible to be anything but the long-sought basal body! I was so excited I could not take photos but kept thinking, "My God, I'm the first person to see this." I had to tell someone. After running all over the building, I phoned Julius, sure he would be excited. "Hello," answered a sleepy voice. "Julius, it worked!"; "What worked?"; "The method for isolating intact flagella; it worked!" "That's nice. You can tell me all about it in the morning."

of attractant biased those walks (19). In the same year, Macnab & Koshland (59) published a study using strobe images of swimming bacteria to conclude that gradients were detected by temporal sensing; i.e., there was a "memory" that stored information about the recent past and compared it to the present. The two studies provided the foundation for understanding the strategy by which the chemosensory system guides cells to favorable environments.

Yet we did not understand how flagella generated swimming or how the sensory system biased the random walk. In 1973, three labs provided the answers. First Berg & Anderson (18) concluded from published observations and conceptual arguments that flagella rotate. In 1972, Mike Silverman in Mel Simon's lab saw the consequences of rotation (86). Mike made antiserum to the flagellar hook protein, which makes the universal joint connecting the flagellar filament to the motor. When he added antiserum to a nonmotile mutant strain devoid of filaments but with elongated hooks, he saw cells apparently tethered to the microscope slide rotating first one direction and then the other, and suspended cells tethered to each other and counterrotating. My understanding is that Mike went excitedly to Mel only to hear it could be an artifact. Additional experiments showed that flagella rotation was real (82). Description of the results to Adler and Berg resulted in complementary experiments and three back-to-back *Nature* papers documenting flagellar rotation and identifying the control of rotational direction as the way the sensory system directed motility (16, 58, 82). Thus, the flagellar motor was identified as the first rotary motor in biology. In parallel, its source of energy was identified as protonmotive force (57).

Receptors

Existence of specific chemoreceptors was a fundamental prediction from Julius' 1969 *Science* paper. Potential receptors were identified by a combination of our list of attractants and repellents; the results of "jamming" experiments, which were determinations of response to a gradient of one compound in the presence of a uniformly high concentration of a second; and the information the laboratory had collected about inducible responses, ones greatly enhanced by growth in a specific condition. Because the 1969 paper focused on responses to galactose, the putative galactose

SEEING ISN'T NECESSARILY BELIEVING

Mike Silverman and Mel Simon were convinced they had demonstrated that bacterial flagella rotate (82). Mel reports:

> It fell to me to 'take the story on the road'. I went to a Structural Biology meeting at Lake Tahoe, showed films of rotating bacteria and concluded the flagellum rotated. A very distinguished structural biologist stood up and proclaimed, 'I don't believe it. There is no rotary motion in biology.' I said to the projectionist, 'Run the film again.' He did and our distinguished colleague said, 'It is precessing not rotating.' I had the film run a third time and pointed out that based on the motion, lack of foreshortening, etc., it could not be precession. I showed evidence for rotation of individual flagella and for rotation driven from the basal element. Our questioner ended the discussion by proclaiming, 'I still don't believe it.' While there were such naysayers (29), the chemotaxis community was receptive. After I spoke at Wisconsin, Adler's group used our approach to demonstrate that attractants and repellents controlled the probability clockwise rotation (58) and Howard Berg characterized rotation and provided a conceptual basis for thinking about it (16).

Twenty-three years later ATP synthase became the second documented biological rotary motor (67), and subsequently the subject of a Nobel Prize (28).

receptor was a prime candidate. The sidebars "Isolating" Receptor Mutants from the Strain Collection and Identification of a Protein Component with a Little Help from our Friends describe the unconventional path to its identification.

Identification of the galactose-binding protein as the galactose receptor (more accurately the galactose recognition component) was the first identification of a bacterial sensory protein and one of the first sensory proteins identified for any receptor system, after rhodopsin, as well as the

"ISOLATING" RECEPTOR MUTANTS FROM THE STRAIN COLLECTION

Upon joining Adler's laboratory, my project was to isolate the first chemoreceptor mutant. My interview the previous November had ended with Julius saying something like, "If we're correct, and *E. coli* is attracted to galactose by recognizing its molecular shape, there must be a specific galactose receptor, likely a protein, and thus a gene product. Given a gene, you should be able to isolate specific galactose taxis mutants and use them to identify the receptor. With that, you would earn a Ph.D." I succeeded, but not in the way anticipated. By late spring, neither my initial approach of enrichment using semisolid agar plates nor multiple variations succeeded. Apparently I looked discouraged because Marge Dahl inquired. I explained I had failed to find "galactose-blind mutants," laboratory jargon for galactose receptor mutants, and that such mutants would grow on galactose, not respond to galactose, but respond to other attractants. Marge replied: "We have one!" It was one among many galactose metabolism/transport mutants obtained from colleagues and tested for chemotaxis. Only one, from Esther Lederberg, was galactose taxis negative and derivatives that grew on galactose remained taxis negative. Marge recalled Julius saying save it because it could be useful. It was. I obtained a galactose taxis-positive revertant by selecting for formation of a galactose chemotactic ring; showing that mutant, but not revertant, was defective in galactose transport at submillimolar concentrations; and modifying the enrichment procedure to isolate additional galactose-blind mutants by compensating for the transport defect. That first galactose-blind mutant and a serine-blind mutant isolated by Marge, and discovered in our stain collection by Bob Mesibov soon after I found the galactose-blind mutant, were described in 1969 (49).

IDENTIFICATION OF A PROTEIN COMPONENT WITH A LITTLE HELP FROM OUR FRIENDS

As I began the search for the protein altered in galactose-blind mutants (see "Isolating" Receptor Mutants from the Strain Collection), a daunting task before genes could be cloned, Julius began receiving telephone calls from the distinguished Danish scientist and Harvard professor Herman Kalckar. Kalckar could be difficult to understand because of his strong Danish accent and a tendency to start in the middle of an explanation (53). After each phone call, Julius would emerge from his office unsure of what Kalckar said. Finally, Julius deduced Kalckar thought a galactose transport protein his lab studied was involved in galactose taxis but not why or what he wanted from us. After multiple calls, I was put on the phone with Kalckar's colleague, Winfried Boos. My memory is that Winfried expressed significant doubt about bacterial "behavior" but was extremely helpful in describing the galactose-binding protein and transport defects of its mutants, defects that mimicked those of galactose-blind mutants! Winfried offered to send a mutant and transport-positive parent. Most importantly, he mentioned an osmotic shock procedure that released this periplasmic protein. Off the telephone, I immediately started bacterial cultures and soon was testing osmotic shock fluid from a galactose-blind mutant and taxis-positive parent for radiolabeled galactose binding. When the scintillation counter display showed binding by parent but not mutant extract, I knew we had the galactose receptor. Work remained before publication (45), but the utility of mutants in identifying sensory proteins had been validated, thanks to significant help from our colleagues.

first chemical receptor protein. Importantly, binding proteins could be purified and characterized biochemically. Testing osmotic shock fluids for other predicted receptors revealed previously unknown ribose-binding and maltose-binding activities and indicated these newly discovered periplasmic binding proteins were the respective sugar receptors (45). I was disappointed I didn't find binding activities for anticipated serine or aspartate receptors, but later work demonstrated these should not have been released by osmotic shock (31). Periplasmic binding proteins were water soluble and so could not by themselves signal across the membrane. Thus, we postulated "transducers," transmembrane signaling proteins (45). Transducers were discovered by solving the methionine problem.

MCP: methyl-accepting chemotaxis protein

Taxis toward serine and some repellents (Tsr): a transmembrane chemoreceptor

Taxis toward aspartate and some repellents (Tar): a transmembrane chemoreceptor

Taxis toward ribose and galactose (Trg): a transmembrane chemoreceptor

Methionine and Methylation

Marge and Julius had shown methionine was required for chemotaxis (7). John Armstrong, Julius' first chemotaxis graduate student, pursued the issue in his thesis work (10). He continued these studies after graduating in 1968 and implicated S-adenosylmethionine (11, 12). However, its specific role, presumably as a methyl donor, remained unidentified. Seven years and multiple tries after John's departure, graduate student Ed Kort made the pivotal discovery that tritiated-methyl methionine radiolabeled a 60-kDa *E. coli* membrane protein. Ed combined forces with graduate students Michael Goy and Steve Larsen to demonstrate that the 60-kDa protein they named methyl-accepting chemotaxis protein (MCP) was involved in chemotaxis (55). Parallel studies led by postdoc Marty Springer showed methionine was required for tumbling, thus correlating the protein modification with chemotactic behavior (89). Soon thereafter, Ed, Steve, and postdoc Bob Reader moved on, leaving Michael and Marty to pursue methylation. They did so in a very productive partnership, establishing that protein methylation mediated the central process of sensory adaptation (41, 42, 87).

Methyl-Accepting Chemotaxis Proteins

Initial studies of MCP involved the original version of sodium dodecyl sulfate polyacrylamide gel electrophoresis (SDS-PAGE) in individual tube gels, cutting gels into slices and determining radioactivity by scintillation counting. When Michael and Marty turned to the newer technique of thin slab gels and detecting radioactivity by fluorography, the higher resolution immediately revealed that the 60-kDa region contained multiple methyl-labeled bands. In parallel, Mike Silverman in Mel's laboratory used then-new techniques of gene cloning to identify motility and chemotaxis gene products, including an MCP (80, 83–85). Complementary studies from the Adler and Simon laboratories, published as adjacent articles in the *Proceedings of the National Academy of Sciences USA*, identified the multiple bands as belonging to two parallel pathways of signal processing, each mediated by an MCP: Tsr for serine and certain repellents and Tar for aspartate and other repellents (84, 88).

Because the two MCPs handled responses to multiple ligands and a mutant missing both MCPs was generally nonchemotactic, it was thought there was a receptor resembling the galactose-binding protein for each ligand and that MCPs were downstream (88). However, a third MCP, Trg, was identified by my and Julius's labs as the transmembrane component for taxis to galactose and ribose (46, 54), and my lab showed that the three MCPs provided parallel signaling pathways (46). In addition, Tsr and Tar were found by Dan Koshland's lab to bind, respectively, serine and aspartate (31). Thus, the ligands for MCPs could be amino acids or ligand-occupied binding proteins. They were the transmembrane receptors of chemotaxis.

MULTIPLE FORMS OF MCPS

MCPs appear as multiple bands on SDS gels because the proteins are multiply methylated and those modifications increase mobility in SDS-PAGE (27, 30, 32, 37). The laboratories of Dahlquist, Hazelbauer, Koshland, and Simon deduced this essentially simultaneously and all came to the 1980 Gordon Conference on Sensory Transduction in Microorganisms to report. What transpired illustrates a characteristic combination of competition and cooperation. Within 24 hours we realized we had all come to the same conclusion, with different, albeit overlapping, approaches. This was reassuring because methylation added only 14 Da to a 60-kDa protein and separation of methylated forms should have been impossible using SDS-PAGE. We got together the first afternoon, summarized our respective results, designed a sequence of talks to inform the community, and cleared rearrangement of the published program with session chairs. I was to summarize the overall conclusion and describe our results. Alan Boyd from Mel Simon's lab would report theirs and Rick Dahlquist theirs. Dan Koshland indicated it was not necessary for him to talk about their results; I could summarize them. I asked what to say. Dan replied, in his usual witty style, "Tell them we did all that the rest did, only better." In my talk, I said exactly that. Laughter was substantial. This was one of a few times I got the better of Dan in a public intellectual fencing match, an activity at which he was a master.

Phosphorylation

How do receptors communicate with the flagellar motor? In the early 1980s my lab showed that receptors were not clustered around flagellar basal ends (38), indicating signaling could not involve direct physical contact but rather must occur over a distance. In higher organisms such signaling is often accomplished by ion fluxes across membranes. Julius was particularly interested in this possibility. He investigated extensively, first in his own laboratory (77, 90) and then in collaboration with faculty colleague Ching Kung (62). Their collaborative work led to the discovery of mechanosensitive ion channels in bacterial membranes (61, 78). Julius' lab investigated other candidates, including cyclic nucleotides (20, 21) and calcium ions (91, 92). Working on *Bacillus subtilis*, George Ordal found indications of calcium involvement (70). Although findings were never published, several laboratories investigated the possibility that protein phosphorylation was involved, but without success.

The breakthrough came from Mel Simon's laboratory, where the *che* genes had been cloned and placed in vectors producing the respective proteins. As I understand it, new postdoc Fred Hess, familiar with protein purification, proceeded to purify the Che proteins and screen for activities. Adding γ-^{32}P-ATP, he observed autophosphorylation of CheA and correlated it to chemotaxis (52). In rapid succession, publications from the Simon lab (22–24, 50, 51, 68) and one from Jeff Stock's lab (96) defined chemotaxis-linked phosphorylation, demonstrating that it was indeed the signaling mechanism between receptors and the flagellar motor. Beginning with the initial publications (52, 96), it was clear that this phosphotransfer and the proteins involved were members of a family of recently identified two-component regulatory systems (65, 66). These proved to be ubiquitous bacterial signaling elements coupling recognition of the environment and cellular response (25, 26). This meant that information about chemotaxis signaling was relevant to mechanisms of control of gene expression.

THE CULTURE OF THE CHEMOTAXIS COMMUNITY

Intertwined with increased understanding of bacterial chemotaxis was the development of an interactive community. This community continues to flourish, almost 40 years after it began. It is notable for cooperation and congenial relationships, regular meetings at which most gather, and encouragement of young scientists and those new to the field. I believe the community and its attitudes have contributed much to our collective progress. Our field is not without competition, but cooperation and even friendship persist.

This scientific culture owes much to its initial prominent members. Julius Adler was (and still is) thoughtful, thorough, deliberate, intensely concerned with achieving the highest standards of data and interpretation, interested in considering all explanations and possibilities even those "knowledgeable" scientists might consider unlikely, prone to share, and genuinely encouraging to others. These attributes had substantial positive influences. For a number of years, his was the only chemotaxis laboratory. In the early 1970s, three additional research groups, directed by Howard Berg, Daniel E. Koshland, Jr., and Melvin Simon, respectively, began to publish about bacterial chemotaxis and motility (**Figure 2**). Each is or was (Dan passed away in 2007) a well-respected scientist. At the time he began to study bacterial chemotaxis, Dan, already a member of the National Academy of Sciences, was a major figure in biochemistry and enzymology particularly because of the induced-fit concept of enzyme action. Julius was elected to the Academy in 1978, Berg and Simon in 1984 and 1985, respectively. These four were quite different, but each focused on important issues and aimed to provide convincing conclusions. Thus, chemotaxis and motility publications from their laboratories were relatively few in number but of notable quality and impact. Those of us trained in those laboratories have aimed to emulate this pattern. Publishing few but substantive papers is prevalent in the field to this day.

The first additions were Howard Berg and Dan Koshland. Howard, on his way from Harvard to a faculty position at the University of Colorado, spent the summer of 1970 in the Adler laboratory learning how we handled *E. coli* from Marge Dahl. He was a welcome guest. It was the first time for us that someone else was studying chemotaxis. Moreover, Howard's physics background meant a distinctly different perspective. His ability to explain his insights was invaluable and has continued to be over multiple decades.

In the late 1960s, Dan Koshland became interested in bacterial chemotaxis. As I understand from those who were in his laboratory, Dan was considering several possibilities for molecular approaches to neuroscience. Bacterial chemotaxis was championed by postdoc Frederick (Rick) Dahlquist and research assistant Peter Lovely. Soon thereafter, postdoc Robert (Bob) Macnab, with a chemistry and chemical engineering background, joined the Koshland laboratory. The result was full entry into our field and provision of a distinctly biochemical orientation to chemotaxis. Barry Taylor, a Koshland postdoc a few years later, reports that Dan was not comfortable with genetic analysis but encouraged his people to learn all they could from faculty colleague Bruce Ames. Dan used his prominence and considerable promotional skills to argue that bacterial chemotaxis was a prime system for addressing fundamental questions in neuroscience. We all benefited. He published an entire book, *Bacterial Chemotaxis as a Model Behavioral System* (56), based on a Distinguished Lectureship of the Society of General Physiologists. Activities like these brought substantial visibility to the field and enhanced our respective chances for funding. Dan also brought a style honed by many years in very competitive areas of biochemistry. This style was quite different from Julius' and led to tensions when the styles clashed. Both were professional, so tensions might not have been noticed by outsiders but were quite apparent to those from the respective research groups. Yet those tensions did not create significant extended problems. Communication and cooperation among alumni of the two laboratories have flourished.

The early years conferred not only high standards but also a pattern of encouraging young scientists or those new to the field. Adler, Berg, Koshland, and Simon all gave their students and postdocs not only credit for contributions but also opportunities to present their work. For instance, as a graduate student I spoke at two Gordon Conferences about my work, once in place

January 1977

September 1978

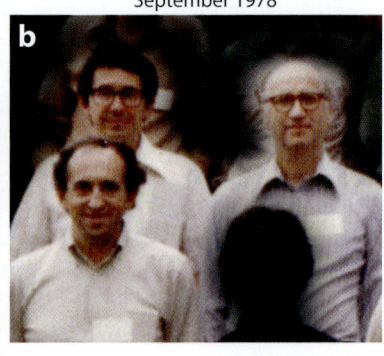

March 1983

Figure 2

Key investigators in the early days of chemotaxis research. (*a, left to right*) Howard Berg, Dan Koshland, and Julius Adler at the January 1977 Gordon Conference on Bacterial Cell Surfaces, Miramar Hotel, Santa Barbara, California. The image has been extracted from the participants' photograph and manipulated to show only those three. Photograph reprinted with permission from the Director of the Gordon Research Conferences. (*b, clockwise*) Julius Adler, Mel Simon, and Dan Koshland at the September 1978 conference Flagellar Motility in Hakone, Japan. The image has been extracted from the participants' photograph and manipulated to show only those three. (*c, clockwise*) George Ordal, Jerry Hazelbauer, Howard Berg, Bob Macnab, Sandy Parkinson, Barry Taylor, and Rick Dahlquist at dinner during the Table Ronde Roussel Uclaf Chemotaxis meeting, March 1983 in Paris, France. The image has been edited to show only researchers studying bacterial chemotaxis.

of Julius and the other because he split his speaking time with me. Similar situations occurred regularly in each group.

Annual Gatherings

A central factor in sustaining and expanding the bacterial chemotaxis community was the development of regular gatherings. By the mid-1970s, the field had expanded from Adler, Berg, Koshland,

and Simon to include laboratories of the initial generation of their chemotaxis offspring: Jerry Hazelbauer (Adler), Sandy Parkinson (Adler), George Ordal (Adler), Rick Dahlquist (Koshland), and Bob Macnab (Koshland). Bob's untimely death in 2003 was a great loss. Various combinations of these nine were regularly speakers in bacterial chemotaxis sessions at meetings such as Gordon Conferences on Bacterial Cell Surfaces or Molecular Pharmacology. This was valuable for us junior scientists, giving exposure to the wider scientific community and contact with the leaders of our growing field. I suspect our invited participation was often suggested by the senior four.

In the early 1980s, the bacterial chemotaxis community found a long-term home in a Gordon Conference. In 1991, our community created, and has since maintained, a companion meeting occurring in alternate years from that Gordon Conference. With one or the other meeting every January for over 20 years, it has become a habit to attend each January meeting (even if we are not invited to speak) and to bring as many members of our research group as space and finances allow.

Sensory transduction in microorganisms (STIM): a biennial Gordon Conference

Sensory Transduction in Microorganisms

The inaugural meeting of the Gordon Conference on Sensory Transduction in Microorganisms (STIM) was in California over New Year's week 1976 (**Figure 3**). It was organized by Bodo Diehn, from the University of Toledo in Ohio. Like Diehn, most speakers and participants studied behavior of eukaryotic microorganisms. Yet there was interest in bacterial chemotaxis and motility: The subjects occupied one day of the four-and-a-half-day conference. Adler, Berg, Koshland, and Parkinson gave major talks; Julius and Dan shared their speaking time with a member of their research groups, graduate student Michael Goy and postdoc Ruth Zukin, respectively. Rick Dahlquist, Bob Macnab, George Ordal, and I gave short presentations. Rick's graduate student Dan Chelsky attended the meeting, so 11 of 73 attendees (15%) worked on bacterial chemotaxis and motility but gave ~25% of the talks.

Unfortunately, the meeting had uncomfortable aspects. One was Diehn's frequently voiced opinion that everyone should use what he considered proper terminology in describing sensory

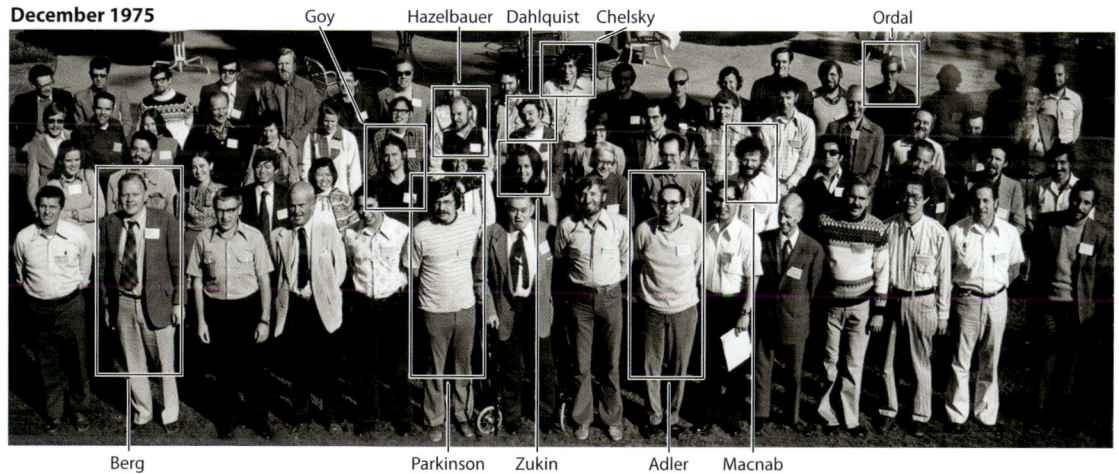

Figure 3

Participant photograph, first Gordon Research Conference on Sensory Transduction in Microorganisms, December 29, 1975 to January 2, 1976, at the Miramar Hotel in Santa Barbara, California. Participants from chemotaxis laboratories are boxed and labeled. Dan Koshland attended the meeting but did not arrive in time to be in the photograph. Photograph reprinted with permission from the Director of the Gordon Research Conferences.

> **Bacterial locomotion and sensory transduction (BLAST):** a biennial independent conference for researchers studying bacterial taxis and motility

behavior, the terminology of Fraenkel & Gunn (39, 40). He would interrupt speakers if he felt they erred. Perhaps in response, Julius utilized the terminology in his introduction to the session "Chemoaccumulation and Chemotransduction," but with the comic timing of Jack Benny. Most found this hilarious. Julius noted that Wilhelm Pfeffer's use of chemotaxis, a term that Diehn considered inappropriate for the tactic behavior of *E. coli*, had almost 60 years of precedence over Fraenkel and Gunn's term reverse klinokinesis. Those studying bacterial chemotaxis still use Pfeffer's term.

In addition, tensions reflected the times. The spectacular successes of molecular biology had resulted in many areas of biology being investigated not only by those who had been trained in an area but also by apparent amateurs. Tensions between traditionalists and molecular biologists occurred in many academic life sciences departments. It was present at this Gordon Conference, even though several working on bacterial chemotaxis, most prominently Julius and Howard, knew and respected the history of studying microbial behavior (6, 17) and several scientists from traditional areas welcomed our participation.

Perhaps as a result of tensions at the 1976 meeting only half of the eight bacterial chemotaxis principal investigators attended the second meeting in 1978 and attendees working on bacteria remained at 11. This changed in 1980, perhaps because Vice-Chair Winslow Briggs particularly valued our community. One-third of the talks were about bacterial motility or taxis and 26 of 110 attendees (24%) were in the field. Importantly, Bob Macnab, who had spoken at all three conferences, was elected vice-chair for 1982. From then on our field has been a central part of this meeting, to our great benefit. 1982 established a pattern of chairs alternating between a scientist studying bacteria and one studying eukaryotic microorganisms, with the vice-chair from the complementary area, and of a program equally distributed between the two areas.

Bacterial Locomotion and Sensory Transduction

In the 1980s, the study of bacterial taxis and motility expanded, as did molecular and mechanistic studies of taxis and motility in the eukaryotic microorganisms, particularly *Dictyostelium discoideum*, making it difficult to fit all desirable topics in one biennial Gordon Conference. The bacterial community began discussing the creation of their own Gordon Conference. The situation came to a head at the 1990 STIM Conference. In planning that meeting, Peter Devreotes and I, chair and vice-chair, respectively, realized coverage of newly developing areas would require fewer talks about topics usually covered. It seemed best to cover some every second meeting, so we identified one prokaryotic and one eukaryotic topic to cover in 1992 but not in 1990. The prokaryotic topic was bacterial flagella. I contacted the major investigators to explain. Bob Macnab was upset. At the meeting, Phil Matsumura mediated at an after-the-evening-session tasting of single malt scotch. Phil posited that having to skip a topic was a positive development because it indicated growth in our areas and thus the basis for creating a second meeting exclusively for bacterial taxis and motility. Bob strongly agreed. Impressively, Phil, in partnership with Sandy Parkinson, proceeded to do so. Thus, the Bacterial Locomotion and Sensory Transduction (BLAST) meeting was born, to be held on alternate years from the STIM Gordon Conference. The field owes this important development to the combination of Sandy's and Phil's visions and initiatives. Sandy recalls spending a morning concocting acronyms for the inaugural meeting. When he posed them to Bob Macnab, Bob's response was, "Let's have a BLAST." So we did.

BLAST is notable because it is a grassroots organization, with no wider affiliation. Phil incorporated BLAST as a nonprofit organization with Board of Directors Phil, Sandy, Mike Manson, and Joseph Falke. It has organized 11 successful meetings over 20 years, maintaining financial stability and greatly enhancing scientific progress. Much is owed to the organizational talents of

Phil and his effective assistant, Peggy O'Neill. Crucial features of BLAST are (*a*) no attendance cap and a sufficiently low cost such that many laboratories can bring multiple members; (*b*) a program for oral presentations constructed from submitted abstracts; and (*c*) a preference for talks by junior scientists, students, postdoctoral researchers, and those new to the field. These features have worked very well. Almost every lab in the field is represented at each meeting. In essence, we all come. The first meeting, 1991 in Austin, Texas, set the tone. We guessed 50 to 60 would attend. Amazingly, 128 registered for a meeting that was designed to pack the most science in the fewest number of days, thus reducing costs. From Thursday evening through Sunday noon, 54 20-minute talks were scheduled: one slot for each attending laboratory. Only a storm disrupting air travel from Europe saved us from complete exhaustion by keeping some European colleagues from attending and thus freeing slots that could be rearranged to create a talk-free Saturday night, when attendees could party or sleep. From that auspicious beginning, the meeting flourished. Some years later, an NSF program officer attending a BLAST meeting remarked that few scientific fields "take care of their young" and that our field reminded him of the special environment of the bacteriophage lambda field in the 1960s.

FINAL THOUGHTS

Studying bacterial chemotaxis and motility has been notably fruitful. *E. coli* chemotaxis became a paradigm for the understanding of molecular mechanisms in biological signaling. The field has expanded to encompass multiple species, each with informative variations on themes identified in *E. coli*. In this regard, George Ordal was a pioneer, beginning molecularly oriented studies of chemotaxis in a distinctly different bacterium, *B. subtilis*, in the 1970s (69). Signaling in chemotaxis is now understood as a prominent example of a much wider phenomenon, two-component signaling, which couples environmental detection to response across bacterial diversity (26). For systems biology, bacterial chemotaxis is providing a tractable example of a sophisticated signaling system for modeling cellular behavior with detailed biochemical parameters.

There is much in the history of our field that could not be included in an article of limited length. Furthermore, the field continues to flourish. Over my 44 years as a participant, each time a major advance generated the worry that everything interesting had been done, new observations revealed additional fundamental questions and wider significance. I am confident this pattern will continue.

SUMMARY POINTS

1. Molecular studies of bacterial chemotaxis and motility began in the 1960s with pioneering work from Julius Adler. His laboratory identified the major areas of investigation pursued in the following two decades, and made multiple seminal observations.
2. In the 1970s, three additional research groups, directed by Howard Berg, Daniel E. Koshland, Jr., and Melvin Simon, began to publish in the area. Their respective expertise enriched the research environment. The four senior scientists contributed not only scientific advances but also an emphasis on quality and impact.
3. As often in science, progress included contributions from accidental findings, personal interactions, and scientific culture.
4. The cooperative and interactive nature of the chemotaxis community contributed much to our collective progress.

5. An important feature of this community has been two alternating biennial meetings at which the field gathers annually.

6. *E. coli* chemotaxis became a paradigm for the understanding of molecular mechanisms in biological signaling.

DISCLOSURE STATEMENT

The author is not aware of any affiliations, memberships, funding, or financial holdings that might be perceived as affecting the objectivity of this review.

ACKNOWLEDGMENTS

I thank the numerous colleagues who provided commentaries, stories, photos, information, and/or comments on draft manuscripts. These include Julius Adler, Howard Berg, Mel Simon, Mel DePamphilis, Bob Mesibov, Sandy Parkinson, George Ordal, Winfried Boos, Ed Kort, Marty Springer, Barry Taylor, Winslow Briggs, Kenji Oosawa, and Igor Zhulin. Research in my laboratory is funded by grant GM29963 from the National Institute of General Medical Sciences.

LITERATURE CITED

1. Adler J. 1965. Chemotaxis in *Escherichia coli*. *Cold Spring Harb. Symp. Quant. Biol.* 30:289–92
2. Adler J. 1966. Chemotaxis in bacteria. *Science* 153:708–16
3. Adler J. 1969. Chemoreceptors in bacteria. *Science* 166:1588–97
4. Adler J. 1973. A method for measuring chemotaxis and use of the method to determine optimum conditions for chemotaxis by *Escherichia coli*. *J. Gen. Microbiol.* 74:77–91
5. Adler J. 2000. All because of a butterfly. In *Many Faces, Many Microbes: Personal Reflections in Microbiology*, ed. RM Atlas, pp. 253–57. Washington, DC: ASM Press
6. Adler J. 2011. My life with nature. *Annu. Rev. Biochem.* 80:42–70
7. Adler J, Dahl MM. 1967. A method for measuring the motility of bacteria and for comparing random and non-random motility. *J. Gen. Microbiol.* 46:161–73
8. Adler J, Hazelbauer GL, Dahl MM. 1973. Chemotaxis toward sugars in *Escherichia coli*. *J. Bacteriol.* 115:824–47
9. Adler J, Templeton B. 1967. The effect of environmental conditions on the motility of *Escherichia coli*. *J. Gen. Microbiol.* 46:175–84
10. Armstrong JB. 1968. *Chemotaxis in Escherichia coli*. PhD thesis. Univ. Wisconsin, Madison. 292 pp.
11. Armstrong JB. 1972. An *S*-adenosylmethionine requirement for chemotaxis in *Escherichia coli*. *Can. J. Microbiol.* 18:1695–701
12. Armstrong JB. 1972. Chemotaxis and methionine metabolism in *Escherichia coli*. *Can. J. Microbiol.* 18:591–96
13. Armstrong JB, Adler J. 1967. Genetics of motility in *Escherichia coli*: complementation of paralysed mutants. *Genetics* 56:363–73
14. Armstrong JB, Adler J. 1969. Complementation of nonchemotactic mutants of *Escherichia coli*. *Genetics* 61:61–66
15. Armstrong JB, Adler J, Dahl MM. 1967. Nonchemotactic mutants of *Escherichia coli*. *J. Bacteriol.* 93:390–98
16. Berg HC. 1974. Dynamic properties of bacterial flagellar motors. *Nature* 249:77–79
17. Berg HC. 1975. Chemotaxis in bacteria. *Annu. Rev. Biophys. Bioeng.* 4:119–36
18. Berg HC, Anderson RA. 1973. Bacteria swim by rotating their flagellar filaments. *Nature* 245:380–82
19. Berg HC, Brown DA. 1972. Chemotaxis in *Escherichia coli* analysed by three-dimensional tracking. *Nature* 239:500–4

20. Black RA, Hobson AC, Adler J. 1980. Involvement of cyclic GMP in intracellular signaling in the chemotactic response of *Escherichia coli*. *Proc. Natl. Acad. Sci. USA* 77:3879–83
21. Black RA, Hobson AC, Adler J. 1983. Adenylate cyclase is required for chemotaxis to phosphotransferase system sugars by *Escherichia coli*. *J. Bacteriol.* 153:1187–95
22. Borkovich KA, Alex LA, Simon MI. 1992. Attenuation of sensory receptor signaling by covalent modification. *Proc. Natl. Acad. Sci. USA* 89:6756–60
23. Borkovich KA, Kaplan N, Hess JF, Simon MI. 1989. Transmembrane signal transduction in bacterial chemotaxis involves ligand-dependent activation of phosphate group transfer. *Proc. Natl. Acad. Sci. USA* 86:1208–12
24. Borkovich KA, Simon MI. 1990. The dynamics of protein phosphorylation in bacterial chemotaxis. *Cell* 63:1339–48
25. Bourret RB, Hess JF, Borkovich KA, Pakula AA, Simon MI. 1989. Protein phosphorylation in chemotaxis and two-component regulatory systems of bacteria. *J. Biol. Chem.* 264:7085–88
26. Bourret RB, Silversmith RE. 2010. Two-component signal transduction. *Curr. Opin. Microbiol.* 13:113–15
27. Boyd A, Simon MI. 1980. Multiple electrophoretic forms of methyl-accepting chemotaxis proteins generated by stimulus-elicited methylation in *Escherichia coli*. *J. Bacteriol.* 143:809–15
28. Boyer PD. 1997. The ATP synthase—a splendid molecular machine. *Annu. Rev. Biochem.* 66:717–49
29. Calladine CR. 1974. Bacteria can swim without rotating flagellar filaments. *Nature* 249:385
30. Chelsky D, Dahlquist FW. 1980. Structural studies of methyl-accepting chemotaxis proteins of *Escherichia coli*: evidence for multiple methylation sites. *Proc. Natl. Acad. Sci. USA* 77:2434–38
31. Clarke S, Koshland DE Jr. 1979. Membrane receptors for aspartate and serine in bacterial chemotaxis. *J. Biol. Chem.* 254:9695–702
32. DeFranco AL, Koshland DE Jr. 1980. Multiple methylation in processing of sensory signals during bacterial chemotaxis. *Proc. Natl. Acad. Sci. USA* 77:2429–33
33. DeFranco AL, Parkinson JS, Koshland DE Jr. 1979. Functional homology of chemotaxis genes in *Escherichia coli* and *Salmonella typhimurium*. *J. Bacteriol.* 139:107–14
34. DePamphilis ML, Adler J. 1971. Attachment of flagellar basal bodies to the cell envelope: specific attachment to the outer, lipopolysaccharide membrane and the cytoplasmic membrane. *J. Bacteriol.* 105:396–407
35. DePamphilis ML, Adler J. 1971. Fine structure and isolation of the hook-basal body complex of flagella from *Escherichia coli* and *Bacillus subtilis*. *J. Bacteriol.* 105:384–95
36. DePamphilis ML, Adler J. 1971. Purification of intact flagella from *Escherichia coli* and *Bacillus subtilis*. *J. Bacteriol.* 105:376–83
37. Engström P, Hazelbauer GL. 1980. Multiple methylation of methyl-accepting chemotaxis proteins during adaptation of *E. coli* to chemical stimuli. *Cell* 20:165–71
38. Engström P, Hazelbauer GL. 1982. Methyl-accepting chemotaxis proteins are distributed in the membrane independently from basal ends of bacterial flagella. *Biochim. Biophys. Acta* 686:19–26
39. Fraenkel GS, Gunn DL. 1940. *The Orientation of Animals*. London: Oxford Univ. Press. 352 pp.
40. Fraenkel GS, Gunn DL. 1961. *The Orientation of Animals*. New York: Dover. 376 pp. 2nd ed.
41. Goy MF, Springer MS, Adler J. 1977. Sensory transduction in *Escherichia coli*: role of a protein methylation reaction in sensory adaptation. *Proc. Natl. Acad. Sci. USA* 74:4964–68
42. Goy MF, Springer MS, Adler J. 1978. Failure of sensory adaptation in bacterial mutants that are defective in a protein methylation reaction. *Cell* 15:1231–40
43. Greenfield D, McEvoy AL, Shroff H, Crooks GE, Wingreen NS, et al. 2009. Self-organization of the *Escherichia coli* chemotaxis network imaged with super-resolution light microscopy. *PLoS Biol.* 7:e1000137
44. Hamer R, Chen P-Y, Armitage J, Reinert G, Deane C. 2010. Deciphering chemotaxis pathways using cross species comparisons. *BMC Syst. Biol.* 4:3
45. Hazelbauer GL, Adler J. 1971. Role of the galactose binding protein in chemotaxis of *Escherichia coli* toward galactose. *Nat. New Biol.* 230:101–4
46. Hazelbauer GL, Engström P. 1980. Parallel pathways for transduction of chemotactic signals in *Escherichia coli*. *Nature* 283:98–100
47. Hazelbauer GL, Falke JJ, Parkinson JS. 2008. Bacterial chemoreceptors: high-performance signaling in networked arrays. *Trends Biochem. Sci.* 33:9–19

48. Hazelbauer GL, Lai W-C. 2010. Bacterial chemoreceptors: providing enhanced features to two-component signaling. *Curr. Opin. Microbiol.* 13:124–32
49. Hazelbauer GL, Mesibov RE, Adler J. 1969. *Escherichia coli* mutants defective in chemotaxis toward specific chemicals. *Proc. Natl. Acad. Sci. USA* 64:1300–7
50. Hess JF, Bourret RB, Simon MI. 1988. Histidine phosphorylation and phosphoryl group transfer in bacterial chemotaxis. *Nature* 336:139–43
51. Hess JF, Oosawa K, Kaplan N, Simon MI. 1988. Phosphorylation of three proteins in the signaling pathway of bacterial chemotaxis. *Cell* 53:79–87
52. Hess JF, Oosawa K, Matsumura P, Simon MI. 1987. Protein phosphorylation is involved in bacterial chemotaxis. *Proc. Natl. Acad. Sci. USA* 84:7609–13
53. Kennedy EP. 1996. Herman Moritz Kalckar. In *Biographical Memoirs*, Vol. 69, pp. 148–65. Washington, DC: Nat. Acad. Press
54. Kondoh H, Ball CB, Adler J. 1979. Identification of a methyl-accepting chemotaxis protein for the ribose and galactose chemoreceptors of *Escherichia coli*. *Proc. Natl. Acad. Sci. USA* 76:260–64
55. Kort EN, Goy MF, Larsen SH, Adler J. 1975. Methylation of a membrane protein involved in bacterial chemotaxis. *Proc. Natl. Acad. Sci. USA* 72:3939–43
56. Koshland DE Jr. 1980. *Bacterial Chemotaxis as a Model Behavior System*. New York: Raven Press. 193 pp.
57. Larsen SH, Adler J, Gargus JJ, Hogg RW. 1974. Chemomechanical coupling without ATP: the source of energy for motility and chemotaxis in bacteria. *Proc. Natl. Acad. Sci. USA* 71:1239–43
58. Larsen SH, Reader RW, Kort EN, Tso WW, Adler J. 1974. Change in direction of flagellar rotation is the basis of the chemotactic response in *Escherichia coli*. *Nature* 249:74–77
59. Macnab RM, Koshland DE Jr. 1972. The gradient-sensing mechanism in bacterial chemotaxis. *Proc. Natl. Acad. Sci. USA* 69:2509–12
60. Maddock JR, Shapiro L. 1993. Polar location of the chemoreceptor complex in the *Escherichia coli* cell. *Science* 259:1717–23
61. Martinac B, Adler J, Kung C. 1990. Mechanosensitive ion channels of *E. coli* activated by amphipaths. *Nature* 348:261–63
62. Martinac B, Buechner M, Delcour AH, Adler J, Kung C. 1987. Pressure-sensitive ion channel in *Escherichia coli*. *Proc. Natl. Acad. Sci. USA* 84:2297–301
63. Mesibov R, Adler J. 1972. Chemotaxis toward amino acids in *Escherichia coli*. *J. Bacteriol.* 112:315–26
64. Monod J, Jacob F. 1961. General conclusions: teleonomic mechanisms in cellular metabolism, growth, and differentiation. *Cold Spring Harb. Symp. Quant. Biol.* 26:389–401
65. Ninfa AJ, Magasanik B. 1986. Covalent modification of the glnG product, NRI, by the glnL product, NRII, regulates the transcription of the glnALG operon in *Escherichia coli*. *Proc. Natl. Acad. Sci. USA* 83:5909–13
66. Nixon BT, Ronson CW, Ausubel FM. 1986. Two-component regulatory systems responsive to environmental stimuli share strongly conserved domains with the nitrogen assimilation regulatory genes *ntrB* and *ntrC*. *Proc. Natl. Acad. Sci. USA* 83:7850–54
67. Noji H, Yasuda R, Yoshida M, Kinosita K. 1997. Direct observation of the rotation of F1-ATPase. *Nature* 386:299–302
68. Oosawa K, Hess JF, Simon MI. 1988. Mutants defective in bacterial chemotaxis show modified protein phosphorylation. *Cell* 53:89–96
69. Ordal G, Goldman D. 1975. Chemotaxis away from uncouplers of oxidative phosphorylation in *Bacillus subtilis*. *Science* 189:802–5
70. Ordal GW. 1977. Calcium ion regulates chemotactic behaviour in bacteria. *Nature* 270:66–67
71. Parkinson JS. 1975. Genetics of chemotactic behavior in bacteria. *Cell* 4:183–88
72. Parkinson JS. 1976. cheA, cheB, and cheC genes of *Escherichia coli* and their role in chemotaxis. *J. Bacteriol.* 126:758–70
73. Parkinson JS. 1977. Behavioral genetics in bacteria. *Annu. Rev. Genet.* 11:397–414
74. Parkinson JS. 1978. Complementation analysis and deletion mapping of *Escherichia coli* mutants defective in chemotaxis. *J. Bacteriol.* 135:45–53
75. Parkinson JS. 1987. Doing behavioral genetics with bacteria. *Genetics* 116:499–500

76. Pfeffer W. 1884. Locomotorische Richtungsbewegungen durch chemische Reise. *Untersuch. Bot. Inst. Tübingen* 1:363–482
77. Repaske DR, Adler J. 1981. Change in intracellular pH of *Escherichia coli* mediates the chemotactic response to certain attractants and repellents. *J. Bacteriol.* 145:1196–208
78. Saimi Y, Martinac B, Gustin MC, Culbertson MR, Adler J, Kung C. 1988. Ion channels in *Paramecium*, yeast and *Escherichia coli*. *Trends Biochem. Sci.* 13:304–9
79. Schade SZ, Adler J. 1967. Purification and chemistry of bacteriophage χ. *J. Virol.* 1:591–98
80. Silverman M, Matsumura P, Draper R, Edwards S, Simon MI. 1976. Expression of flagellar genes carried by bacteriophage lambda. *Nature* 261:248–50
81. Silverman M, Matsumura P, Hilmen M, Simon M. 1977. Characterization of lambda *Escherichia coli* hybrids carrying chemotaxis genes. *J. Bacteriol.* 130:877–87
82. Silverman M, Simon M. 1974. Flagellar rotation and the mechanism of bacterial motility. *Nature* 249:73–74
83. Silverman M, Simon M. 1976. Operon controlling motility and chemotaxis in *E. coli*. *Nature* 264:577–80
84. Silverman M, Simon M. 1977. Chemotaxis in *Escherichia coli*: methylation of *che* gene products. *Proc. Natl. Acad. Sci. USA* 74:3317–21
85. Silverman M, Simon M. 1977. Identification of polypeptides necessary for chemotaxis in *Escherichia coli*. *J. Bacteriol.* 130:1317–25
86. Silverman MR, Simon MI. 1972. Flagellar assembly mutants in *Escherichia coli*. *J. Bacteriol.* 112:986–93
87. Springer MS, Goy MF, Adler J. 1977. Sensory transduction in *Escherichia coli*: a requirement for methionine in sensory adaptation. *Proc. Natl. Acad. Sci. USA* 74:183–87
88. Springer MS, Goy MF, Adler J. 1977. Sensory transduction in *Escherichia coli*: two complementary pathways of information processing that involve methylated proteins. *Proc. Natl. Acad. Sci. USA* 74:3312–16
89. Springer MS, Kort EN, Larsen SH, Ordal GW, Reader RW, Adler J. 1975. Role of methionine in bacterial chemotaxis: requirement for tumbling and involvement in information processing. *Proc. Natl. Acad. Sci. USA* 72:4640–44
90. Szmelcman S, Adler J. 1976. Change in membrane potential during bacterial chemotaxis. *Proc. Natl. Acad. Sci. USA* 73:4387–91
91. Tisa LS, Adler J. 1992. Calcium ions are involved in *Escherichia coli* chemotaxis. *Proc. Natl. Acad. Sci. USA* 89:11804–8
92. Tisa LS, Sekelsky JJ, Adler J. 2000. Effects of organic antagonists of Ca^{2+}, Na^+, and K^+ on chemotaxis and motility of *Escherichia coli*. *J. Bacteriol.* 182:4856–61
93. Tso WW, Adler J. 1974. Negative chemotaxis in *Escherichia coli*. *J. Bacteriol.* 118:560–76
94. Wuichet K, Alexander RP, Zhulin IB. 2007. Comparative genomic and protein sequence analyses of a complex system controlling bacterial chemotaxis. *Methods Enzymol.* 422:3–31
95. Wuichet K, Zhulin IB. 2010. Origins and diversification of a complex signal transduction system in prokaryotes. *Sci. Signal.* 3:ra50
96. Wylie D, Stock A, Wong CY, Stock J. 1988. Sensory transduction in bacterial chemotaxis involves phosphotransfer between Che proteins. *Biochem. Biophys. Res. Commun.* 151:891–96

RNA Interference Pathways in Fungi: Mechanisms and Functions

Shwu-Shin Chang, Zhenyu Zhang, and Yi Liu

Department of Physiology, The University of Texas Southwestern Medical Center, Dallas, Texas 75390; email: Yi.Liu@UTsouthwestern.edu

Keywords

small noncoding RNA, *Schizosaccharomyces pombe*, *Neurospora*

Abstract

RNA interference (RNAi) is a conserved eukaryotic gene regulatory mechanism that uses small noncoding RNAs to mediate posttranscriptional/transcriptional gene silencing. The fission yeast *Schizosaccharomyces pombe* and the filamentous fungus *Neurospora crassa* have served as important model systems for RNAi research. Studies on these two organisms and other fungi have contributed significantly to our understanding of the mechanisms and functions of RNAi in eukaryotes. In addition, surprisingly diverse RNAi-mediated processes and small RNA biogenesis pathways have been discovered in fungi. In this review, we give an overview of different fungal RNAi pathways with a focus on their mechanisms and functions.

Contents

INTRODUCTION	306
QUELLING IN *NEUROSPORA*	307
The Discovery of Quelling	307
Molecular Mechanism of Quelling	307
qiRNA: A Type of sRNA Induced by DNA Damage	310
MEIOTIC SILENCING BY UNPAIRED DNA	310
RNAi-MEDIATED HETEROCHROMATIN FORMATION IN *SCHIZOSACCHAROMYCES POMBE*	312
RNAi IN BUDDING YEASTS	315
miRNA-LIKE sRNA IN *NEUROSPORA*	315
RNAi AS A HOST DEFENSE MECHANISM IN FUNGI	316
Transposon Control in *Neurospora*	316
An Antiviral Mechanism in *Cryphonectria parasitica* and *Aspergillus nidulans*	316
dsRNA/Viral Infection–Induced Transcriptional Response in *Neurospora* and *C. parasitica*	317
RNA SILENCING AND SMALL RNA STUDIES ON OTHER FILAMENTOUS FUNGI	317
Gene Silencing and Small RNAs in *Mucor circinelloides*	317
RNAi Studies on the Human Fungal Pathogen *Cryptococcus neoformans*	318
Small RNAs from *Magnaporthe oryzae*	318

INTRODUCTION

RNA interference (RNAi) in *Caenorhabditis elegans* was first described in 1998 as a phenomenon triggered by double-stranded RNA (dsRNA) that results in the silencing of the genes complementary to the dsRNA (31, 72). This process was initially thought to be a host defense mechanism against invading viruses and transposons (80); however, studies over the past decade have demonstrated that RNAi pathways use small noncoding RNAs (sRNAs) to regulate diverse cellular, developmental, and physiological processes. RNAi pathways are conserved in eukaryotes, and among the identified sRNAs, there are three major classes: small interfering RNA (siRNA), microRNA (miRNA), and PIWI-interacting RNA (piRNA) (10, 34, 54, 73, 100). Each class of sRNAs is produced differently and has diverse functions. In the siRNA and miRNA pathways, the RNase-III-like Dicer enzymes cleave dsRNA/hairpin RNA to generate 21- to 25-nt mature siRNA or miRNA duplexes. The duplex is incorporated into an effector complex, the RNA-induced silencing complex (RISC), containing an Argonaute protein with slicer activity. The passenger strand of the sRNA duplex is removed and the remaining guide RNA directs the RISC to complementary mRNA sequences, resulting in silencing by either mRNA cleavage or translational repression.

The fungi kingdom is a large and diverse group of eukaryotic organisms estimated to contain more than three million species, and has diverse and vital roles in the ecosystem (81). The RNAi-related phenomena were first described in plants and fungi (76, 85). Quelling, the RNAi-related posttranscriptional gene-silencing phenomenon in *Neurospora* was reported in 1992 (85). Since then, the identification of quelling components and the analysis of the pathway have contributed significantly to our understanding of RNAi-mediated gene silencing at the posttranscriptional level. RNAi was first found to be required for heterochromatin formation in the fission yeast

---

dsRNA: double-stranded RNA

RNA interference (RNAi): a conserved eukaryotic gene regulatory mechanism that uses small RNA to mediate gene silencing

sRNA: small noncoding RNA

microRNAs (miRNAs): a type of small RNAs that are generated from precursor RNAs with stem-loop structures. microRNAs can silence gene expression at posttranscriptional and translational levels

Schizosaccharomyces pombe (105), and this system remains the best-understood pathway of RNAi-mediated transcriptional gene silencing. In addition, RNAi components and their functions have been characterized in many fungal species (7). Furthermore, many classes of sRNAs have been found in different fungal species, revealing diverse sRNA biogenesis pathways. In this review, we focus on the mechanisms and functions of different fungal RNAi pathways.

QUELLING IN *NEUROSPORA*

The Discovery of Quelling

Soon after the discovery of cosuppression in plants (76), Romano & Macino (85) discovered a similar transgene-induced gene-silencing phenomenon in *Neurospora crassa* called quelling. Quelling was discovered by transforming *Neurospora* with DNA containing *albino-1* or *albino-3* genes, which are involved in carotenoid biosynthesis. After transformation, white/pale-yellow transformants with reduced *al* mRNA levels were obtained, indicating the silencing of endogenous *al* genes by the transgenes (9, 20, 32, 85). High copy numbers of the transgenes correlated with the quelling efficiency, suggesting that the presence of repetitive transgenes causes silencing. In addition, the quelled transformants spontaneously and progressively reverted to wild type or intermediate phenotypes owing to reductions in copy number of the transgene.

The phenotype of *al* quelled strains is dominant, suggesting that diffusible and *trans*-acting silencing molecules are involved in quelling (20). In addition, a sense RNA derived from the transgene was specifically detected in quelled strains, implying that aberrant transcription of transgenes is involved in silencing (20). Furthermore, quelling did not affect the levels of mRNA precursors, indicating that quelling is a posttranscriptional gene silencing (PTGS) mechanism (20). Together, these results led to the hypothesis that the production of aberrant RNA (aRNA) in the presence of multicopy transgenes causes PTGS.

> **Quelling:** an RNAi-related gene-silencing mechanism in *Neurospora* that is induced by repetitive transgenic sequences. Quelling occurs in the vegetative growth stage
>
> **Heterochromatin:** chromosome regions that are tightly packed throughout the cell cycle and are mostly genetically inactive
>
> **PTGS:** posttranscriptional gene silencing
>
> **QDE-2:** Quelling-Deficient 2

Molecular Mechanism of Quelling

Cogoni & Macino (21) isolated *quelling-defective* (*qde*) mutants, which were mapped to three complementation groups, *qde-1*, *qde-2*, and *qde-3*. The cloning and studies of these *qde* genes contributed significantly to our understanding of an RNAi pathway that is well conserved across eukaryotes (13, 14, 21, 25). QDE-1, which encodes a cellular RNA-dependent RNA polymerase (RdRP), was the first eukaryotic RNAi component identified (22); it was discovered soon after the silencing effect of dsRNA was demonstrated in *C. elegans* (31). The requirement of an RdRP in quelling shows that dsRNA is an essential intermediate of PTGS in vivo.

QDE-2 encodes an Argonaute protein that is homologous to RDE-1 (RNAi-Deficient-1) in *C. elegans*, which is required for dsRNA-mediated silencing (11). The identification of QDE-2 provided the first evidence that RNAi in *C. elegans* and quelling share a common genetic mechanism.

QDE-3 belongs to the RecQ helicase family and is homologous to the mammalian Bloom's and Werner's syndrome proteins (23). RecQ helicases are involved in homologous recombination, DNA replication, and DNA repair. As expected, QDE-3 contributes to the DNA repair process in *Neurospora* (23, 51). The requirement of QDE-3 in quelling raises the possibility that repetitive transgenes may form aberrant DNA structures that are recognized by QDE-3 to promote aRNA and sRNA production. Interestingly, OsRecQ1, a RecQ helicase homolog in rice, is required for inverted-repeat-induced RNA silencing (16). rRecQ-1, a homolog of QDE-3 in rats, is associated with the piRNA-binding complex (58).

Expression of an inverted-repeat-containing transgene can bypass QDE-1 and QDE-3 to trigger gene silencing (14, 61), indicating that QDE-1 and QDE-3 function in producing aRNA

qiRNA: A Type of sRNA Induced by DNA Damage

> **Meiotic silencing by unpaired DNA (MSUD):** an RNAi-related gene silencing mechanism that occurs in the presence of unpaired homologous DNA sequences during meiosis

Quelling triggered by repetitive transgenes occurs in the vegetative *Neurospora* culture under normal growth conditions. Intriguingly, we recently uncovered a class of QDE-2-interacting sRNA, designated qiRNA, that is specifically induced after *Neurospora* is treated with a DNA-damaging agent (61). qiRNAs originate mostly from the rDNA locus, which is the only highly repetitive sequence in the wild-type *Neurospora* genome. These sRNAs are ~21–23 nt in length and generally have a 5′ uridine. qiRNA biogenesis requires QDE-1, QDE-3, and the Dicers, indicating that qiRNAs are specific sRNA species made by the RNAi machinery upon DNA damage (61). qiRNAs originate from aRNA precursors. In fact, aRNAs ranging in size from 500 bp to 2 kb from the intergenic spacer regions accumulate in the *dcl* double mutant. The production of aRNA is abolished in the *qde-3* mutant, inferring that the RecQ helicase QDE-3 is required for the biogenesis of DNA-damage-induced aRNA. In addition, QDE-1 is essential for aRNA production, but the conventional RNA polymerases (Pol I, Pol II, and Pol III) are not, suggesting that QDE-1 acts as a DdRP to produce the DNA-damage-induced aRNAs.

DNA damage results in cell arrest, decrease of DNA replication, and protein synthesis. In mutants that are defective in qiRNA synthesis, DNA-damage-induced reduction of protein synthesis is blunted (61). QDE-3 plays a role in DNA damage repair (23, 51). Furthermore, *qde-1* and *dcl* mutants also exhibit increased sensitivity to DNA-damaging agents, suggesting that qiRNAs contribute to DNA damage checkpoints by inhibiting protein synthesis (61). In *Arabidopsis*, RNAi components are enriched in the nucleolus, and rDNA-derived siRNAs facilitate heterochromatin formation (83). Furthermore, the *Drosophila* RNAi-deficient mutants have disorganized nucleoli and rDNA loci (82). These studies suggest that rDNA-derived sRNAs have a shared role in eukaryotes to maintain genome integrity.

Both quelling and qiRNA pathways appear to require the same components, suggesting that they are mechanistically similar (60, 61). The repetitive nature of transgenes and the rDNA locus is very likely to be the common trigger for quelling and qiRNA production. The production of qiRNA is induced by DNA damage, implying that replication stress or double-stranded breaks trigger formation of aberrant DNA structures. Although quelling occurs under normal growth conditions, it is likely that the transgene integration loci are fragile DNA sites with elevated levels of replication stress (28, 35). If so, quelling should also be the result of DNA damage at the transgene integration loci trigged by the presence of repetitive DNA sequences.

MEIOTIC SILENCING BY UNPAIRED DNA

Meiotic silencing by unpaired DNA (MSUD) was originally discovered during studies of the *ascospore maturation 1* gene (*asm-1*) in *Neurospora* (3, 93, 94). Unlike quelling, which takes place in the vegetative stage, MSUD occurs during meiosis. *Neurospora* is a haploid ascomycete and becomes transiently diploid during meiosis following the fusion of two haploid nuclei from opposite mating types. MSUD occurs in prophase I of meiosis when unpaired DNA sequences are present and leads to the silencing of all homologous genes in the diploid ascus cell (94). It has been proposed that *trans*-sensing and meiotic silencing are two related but distinct steps in MSUD (53, 84). The *trans*-sensing step should occur in the nucleus, where the presence of unpaired DNA during meiosis is sensed and likely triggers the production of unpaired DNA-specific aRNA, whereas the meiotic silencing step converts the aRNA to dsRNA and sRNA, resulting in the sRNA-mediated gene silencing. Although, the mechanism of *trans*-sensing is still largely unclear, DNA methylation appears to influence *trans*-sensing without affecting silencing (84). MSUD also operates in the homothallic fungus *Gibberella zeae* (96).

A breakthrough in understanding the mechanism of MSUD was achieved by forward genetic screens for mutants that suppress meiotic silencing (94). The first two genes identified to be required for MSUD are *sad-1* (*suppressor of ascus dominance 1*), which encodes an RdRP and is a paralog of QDE-1 (93, 94), and *sad-2*, which does not contain any conserved domain (5, 95). SAD-1 and SAD-2 colocalize at the perinuclear region and interact physically. A functional *sad-2* gene is required for the proper localization of SAD-1, indicating that SAD-2 might function to recruit SAD-1 to the perinuclear region. The fact that SAD-1 encodes a putative cellular RdRP indicates that MSUD is an RNAi-related phenomenon. In addition, *sms-2* (*suppressor of meiotic silencing 2*), an Argonaute protein homolog in *Neurospora*, is also required for MSUD (59). Although an *sms-2* mutant did not show obvious defects during vegetative growth, the homozygous crosses of *sms-2* mutants are completely barren. In contrast, QDE-2 is not required for MSUD, indicating that there are two parallel RNAi pathways functioning separately in the vegetative and meiotic stages (59).

MSUD was further established as an RNAi-related pathway from studies on the role of *Neurospora* Dicer proteins in meiotic silencing (1). Unlike the single mutants of *sad-1* and *sad-2*, the *dcl-1* single mutants did not function as dominant suppressors of meiotic silencing (1, 94, 95). However, when DCL-1 is inactivated at a later stage of the sexual cycle, meiotic silencing is suppressed, indicating a critical role for DCL-1 in MSUD and therefore a requirement for dsRNA and sRNA in meiotic silencing. In addition, DCL-1 colocalizes with other MSUD components, including SAD-1, SAD-2, and SMS-2, at the nuclear periphery (1, 95). Interestingly, DCL-2, the major Dicer in the quelling pathway, is not required for MSUD (14). QIP, the exonuclease required for RISC activation in the quelling pathway, is also involved in MSUD (42, 60, 109). Like DCL-1, QIP localizes at the perinuclear region with other meiotic silencing components (109).

Recently, a putative RNA/DNA helicase, SAD-3, was demonstrated to be required for meiotic silencing (42). SAD-3 is homologous to *S. pombe* Hrr1, which is a component of the RNA-directed RNA polymerase complex required for the RNAi-mediated heterochromatin formation in fission yeast (75). This link raises the possibility that MSUD and the RNAi-induced heterochromatin assembly may be mechanistically related. SAD-3 is also found in the perinuclear region and associates with SAD-1, SAD-2, and SMS-2 (42). The fact that all the known components involved in MSUD are localized at the perinuclear region and associate with each other suggests that they form a large meiotic silencing complex in this region. It is likely that this complex functions by converting the aRNA produced in the nucleus to dsRNA and then sRNA to trigger gene silencing.

A simple model of the MSUD pathway is shown in **Figure 1**. During meiosis, the pairing of the homologous chromosomes leads to the detection of the unpaired DNA region and production of aRNA transcripts. In the perinuclear region, aRNA transcripts are converted to dsRNA by SAD-1, with the help of SAD-2 and SAD-3. DCL-1 processes dsRNA into sRNAs, which are then loaded onto an SMS-2/QIP-based RISC to execute posttranscriptional silencing of homologous genes. Despite the identification of many components involved in the meiotic silencing step of the MSUD pathway, little is known about the *trans*-sensing mechanism. First, what is the mechanism that scans and senses the unpaired DNA in homologous chromosomes during meiosis? DNA methylation is known to be involved at this step (84), suggesting that chromatin structures are important. Second, what is the DdRP that produces the aRNA and how is the specificity achieved? In quelling, QDE-1 serves as both DdRP and RdRP (60). It will be interesting to test whether QDE-1/SAD-1 has a similar role in MSUD. In addition, some genes appear to be immune to MSUD, such as rDNA clusters and the mating-type genes *mat* A and *mat* a, which have very different DNA sequences (92, 94). Thus, there must be a mechanism that protects certain genes from silencing during meiosis in *Neurospora*. Finally, since the discovery of meiotic silencing in *Neurospora*, unpaired homologous

DNA-triggered silencing during meiosis in *C. elegans*, *D. melanogaster*, and mammals has been described (53). Although the silencing mechanisms in *D. melanogaster* and mammals appear to be independent of RNAi, it is still not clear whether MSUD in *Neurospora* is mechanistically similar to the silencing phenomena in animals.

> **RNA-induced transcriptional silencing (RITS):** an RNAi-related mechanism that triggers or maintains the formation of heterochromatin, resulting in the silencing of gene expression at the level of transcription

RNAi-MEDIATED HETEROCHROMATIN FORMATION IN *SCHIZOSACCHAROMYCES POMBE*

Heterochromatin was initially considered to be transcriptionally inert. Heterochromatin in the fission yeast *S. pombe* is distributed in three different loci: telomeres, centromeres, and mating-type loci. The centromeres of fission yeast are similar to those of humans in their structure and epigenetic modifications. The central core region sequence is unique for every centromere, and it is at this region where the kinetochore binds. It is flanked by two types of repeated DNA sequences, the innermost (*imr*) and outermost (*otr*) repeats. The outermost pericentromeric repeats are further composed of the *dg* and *dh* sequences; these sequences are coated with histone H3 that is methylated at lysine 9 (H3K9). These specialized chromatin modifications provide docking sites for chromodomain proteins that maintain the transcriptionally silent status of the heterochromatin (29).

A direct link between heterochromatin formation and an RNAi pathway in the fission yeast was first established in 2002, when Volpe et al. (105) showed that RNAi components are required for heterochromatin formation in the centromeric regions. In addition, transcription is detected in the heterochromatin region and the heterochromatic RNAs accumulate in mutants of the RNAi pathway, including *RNA-dependent RNA polymerase* (*rdp1*), *dicer* (*dcr1*), and *argonaute* (*ago1*). The transcripts arising from the heterochromatin regions are mapped to both strands of DNA, indicating a dsRNA precursor. Reduced H3K9 methylation and increased H3K4 methylation are also observed in these mutants. Furthermore, sRNAs derived from the centromeric region are detected, and Rdp1 is enriched at the centromere (105). Together, these results demonstrated the requirement of the *S. pombe* RNAi pathway in heterochromatin formation.

In fission yeast, heterochromatin at the centromeric regions is formed through a self-enforcing loop orchestrated by the RNAi components and chromatin-modifying components (7, 37, 65, 73, 103, 104) (**Figure 2**). This self-enforcing loop is brought about by the transcription of the pericentromeric nascent transcripts by RNA Pol II during the S phase of the cell cycle. Rdp1 converts the nascent transcripts to dsRNA, which Dcr1 recognizes and processes into siRNA. The siRNA is then loaded onto Ago1 to form the RNA-induced transcriptional silencing (RITS) complex (101, 102).

RNA Pol II appears to be the DdRP responsible for the production of centromeric siRNA in *S. pombe*. The role of Pol II in this process was supported by the identification of two point mutations in the RNA Pol II subunits (Rpb2 and Rpb7) (52, 114). These mutations resulted in reduced levels of heterochromatic marks and siRNA production. Genome-wide transcriptional analyses suggest that these defects are not due to an indirect effect caused by changes in global gene expression. Furthermore, neither mutation altered the association of Pol II with the centromeric DNA repeats, suggesting that the defects in siRNA production in these mutants are not due to impaired recruitment of Pol II to centromeric DNA.

The identification of the RITS complex provided a mechanistic link between siRNA and heterochromatin assembly (101). The RITS complex consists of Ago1, Chp1, and Tas3. Chp1 is a chromodomain-containing protein that binds the histone H3 lysine 9 methylation (H3K9me), which recruits the RITS to the pericentromeric region. The active Ago1 bound to a single-stranded

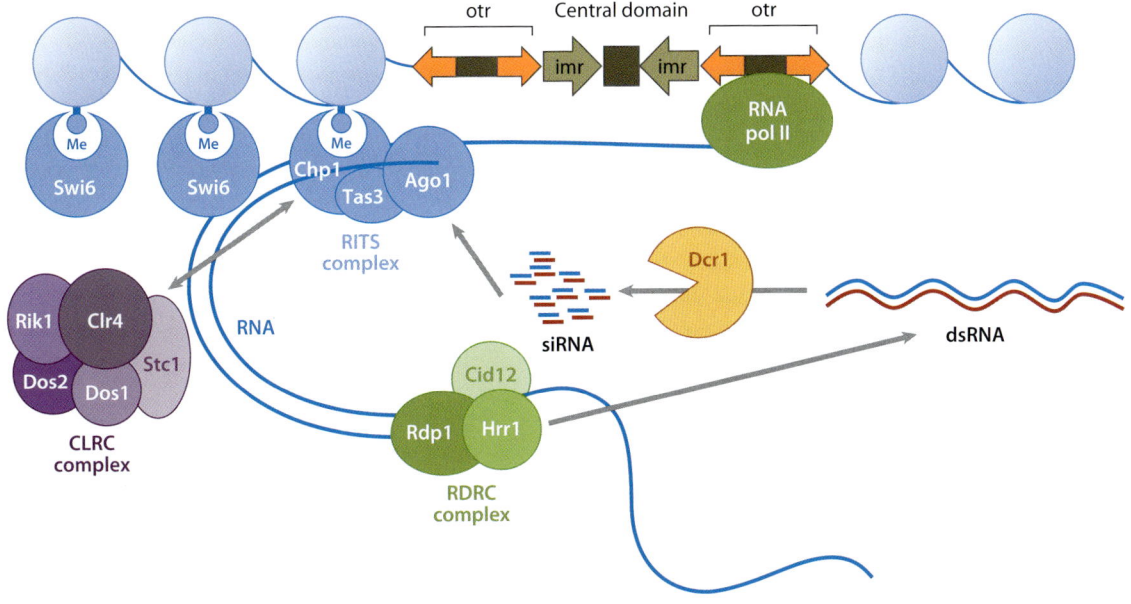

Figure 2

Model for RNAi-mediated heterochromatin formation in *Schizosaccharomyces pombe*. The siRNA-guided RITS complex targets the centromeric nascent transcript synthesized by RNA Pol II during the S phase of the cell cycle. RITS associates with the RDRC complex, where the RdRP activity of the RDRC complex generates more dsRNA. Dcr1 processes the dsRNA into siRNA, feeding into the loop. The CLRC complex associates with the RITS complex, which comes in close proximity to chromatin. Clr4 methylates H3K9 and allows Swi6 (HP1 homolog) to dock at the modified chromatin, thereby forming heterochromatin. Abbreviations: RNAi, RNA interference; siRNA, small interfering RNA; RITS, RNA-induced transcriptional silencing; Pol, polymerase; RDRC, RNA-dependent RNA polymerase complex; RdRP, RNA-dependent RNA polymerase; dsRNA, double-stranded RNA; Dcr1, Dicer 1; CLRC, Clr4-Rik1-Cul4 complex; otr, outer most repeat; imr, inner most repeat; Ago1, Argonaute 1; Tas3, tyrosine auxotrophy suppressor 3; Clr4, calcitonin-like receptor 4; Chp1, chromodomain-containing protein 1; Swi6, SWItch/sucrose nonfermentable 6; Rik1, RS2-interacting KH domain protein 1; Dos1/2, delocalization of Swi6 1/2; Stc1, siRNA to chromatin; Rdp1, RNA-dependent RNA polymerase 1; Hrr1, helicase required for RNAi-mediated heterochromatin assembly; Cid12, caffeine-induced death resistant 12.

guide siRNA also reinforces this interaction through homology-dependent association with the centromeric nascent RNA transcripts.

The production of the nascent transcripts provides a template for Rdp1 (75). Rdp1 is part of the RNA-dependent RNA polymerase complex (RDRC) that also contains Hrr1 and Cid12 (75). Both Rdp1 and Hrr1 associate with heterochromatin and centromeric (*cen*) RNA transcripts (*cen* RNA). Rdp1 generates dsRNA in a primer-independent manner from single-stranded *cen* RNA, and it has been proposed that the RITS complex acts as a priming complex for RDRC to localize at the centromeric DNA. Dicer cleaves the dsRNA into siRNA and greatly enhances the activity of RDRC in siRNA production in vitro (24). Although Dcr1 is present mostly in the cytoplasm (43), a study showed that Dicer is also present in the nucleus, suggesting that Dcr1-mediated siRNA production occurs in *cis* with transcription (30). Dcr1 may shuttle between the nucleus and cytoplasm, and the retention of Dcr1 in the nucleus is essential for the production of centromeric siRNAs.

The recruitment of RDRC to the centromeric region is dependent on Clr4 (75). Clr4, a histone methyltransferase, is recruited to the pericentromeric region, where it methylates H3K9. The methylated chromatin further recruits Clr4 and the chromodomain proteins Swi6, Chp1, and Chp2 to the region (86, 111). siRNA production is greatly reduced in the *clr4* mutants. Clr4 associates with Rik1, Dos1, Dos2, and Cul4 in a complex called Clr4-Rik1-Cul4 complex (CLRC); all the components in this complex are required for heterochromatin assembly (44, 45, 48, 66). The association of Swi6 to the heterochromatin allows the recruitment of cohesion, whereas the binding of Clr4 and Chp1 stabilizes CLRC and RITS at the heterochromatin. Together, this allows the reinforcement of silencing signals and spreading of heterochromatin (**Figure 2**).

Two models have been proposed to explain how RITS is recruited to specific chromosome regions. In the first model, RITS is guided by siRNA to the centromeric DNA through direct base-pairing, which would require unwinding of the DNA double helix. In the second model, the siRNA is targeted to the nascent transcript produced from the pericentromeric region. RNA immunoprecipitation demonstrated that RITS and RDRC are associated with the noncoding *cen* RNA and that this association requires Clr4 methyltransferase and Dcr1 (75). These results suggest that recruitment of RITS to the centromeric regions is not direct but rather is mediated through the siRNA-targeted binding of AGO1 to the nascent transcripts (7, 8, 107). This RNA platform model was tested in vivo by tethering Tas3 (a component of the RITS complex) to the RNA transcript of the normally active $ura4^+$. In this system, recruitment of RITS to the RNA led to transcriptional silencing of $ura4^+$ (4, 7). This study demonstrates that the association of RITS with the nascent transcript is sufficient to trigger H3K9me and transcriptional gene silencing.

The formation of heterochromatin in the centrometric regions is cell cycle dependent (16, 56). The DNA replication of the heterochromatin occurs in early S phase. This is also precisely the time during the cell cycle when pericentromeric *dg/dh* transcripts are detected. In addition, Pol II localization to the centromeric regions is detected in early S phase when the level of H3K9 methylation is at its lowest level. In S phase, RNAi components accumulate at the pericentromeric region and H3K9 methylation marks are detected. By the end of replication, the epigenetic marks of heterochromatin are faithfully passed on to the sister chromatids. The H3K9me levels further increase as the cell cycle progresses and peak during the G2 phase of the cell cycle. Thus, the RNAi components in fission yeast aid in the formation of heterochromatin and assist the cell in completing DNA synthesis in heterochromatic regions (110). Cotranscriptional RNAi releases the RNA Pol II from the centromeric regions, which allows the completion of DNA replication and subsequent spreading of histone modifications.

Recent advances in our understanding of the RNAi-mediated heterochromatin formation have demonstrated how RNAi and chromatin-modifying components are recruited specifically to the centromeric loci, and have revealed the time frame in which these processes occur. However, several recent studies have raised questions regarding the initial trigger that results in heterochromatin formation at the centromeric regions. Halic & Moazed (39) proposed that the production of primal RNAs (priRNAs) are initiators of this process. These priRNAs are thought to be the products of degraded transcripts, are generated independently of Dcr1 and RDRC, and are loaded onto Ago1. The primed Ago1-priRNAs complexes target the centromeric loci and further recruit Clr4, which methylates H3K9 and initiates the formation of RNAi-mediated heterochromatin. Results of another study led to the alternative hypothesis that the CLRC complex is recruited to the centromeric regions independently of RNAi components (91). Consistent with this hypothesis, overexpression of Clr4 in a *clr4 ago1* double mutant induced de novo H3K9me2 at the centromeric regions, suggesting RNAi may be dispensable for initiating heterochromatin formation but critical in maintaining the heterochromatic status. Forcibly tethering the Clr4 methyltransferase to a euchromatic region is also sufficient to establish heterochromatin formation in the absence of RNAi components (50). Although a tremendous amount of progress has been made, further studies are needed to fully understand the mechanism of RNAi-mediated heterochromatin formation in the fission yeast.

RNAi IN BUDDING YEASTS

Although the RNA-silencing pathway is highly conserved throughout the fungal kingdom, some budding yeasts such as *Saccharomyces cerevisiae* lack homologs of Argonaute, Dicer, and RdRP. However, sequence homologs of Argonaute but not other RNAi components are present in some of the budding yeasts, such as *Saccharomyces castellii*, *Kluyveromyces polysporus*, and *Candida albicans* (27). These yeasts use noncanonical Dicer proteins to produce siRNAs, which map mostly to transposable elements and subtelomeric repeats. Remarkably, the expression of the *S. castellii argonaute* and *dicer* genes in *S. cerevisiae* activates an RNAi pathway that inhibits transposon transposition. These results suggest that the RNAi pathway is an ancient host defense mechanism that has been lost in some fungal species.

milRNA: microRNA-like small RNA

miRNA-LIKE sRNA IN *NEUROSPORA*

miRNAs are endogenous ~21-nt RNA duplexes processed by Dicers from ssRNA precursors with a hairpin structure. miRNAs in animals, plants, and algae have been discovered and regulate physiological and developmental processes by targeting mRNA for cleavage or translational repression (2, 6, 38, 57, 63, 64, 69, 74). miRNA was thought to be absent in fungi until our recent identification of miRNA-like sRNAs (milRNAs) in *Neurospora* (62). At least 25 *Neurospora* milRNA-producing loci have been identified. Several lines of evidence suggest that milRNAs in *Neurospora* regulate gene expression similar to animal and plant miRNAs. First, the mRNA of predicated target genes of *milR-1* are upregulated in the *dcl* and *qde-2* mutants. Second, the expression of a reporter gene containing an *milR-1* target site is increased in the *qde-2* strain. Third, QDE-2 specifically associates with the predicted *milR-1* target mRNAs. Interestingly, milRNAs do not appear to play critical roles in *Neurospora* growth or developmental processes because *qde-2* or milRNA knock-out strains grow normally. Future studies are needed to establish the physiological importance of milRNAs in *Neurospora*. Although currently no miRNA-like sRNA has been experimentally confirmed in other fungal species, a study of the wheat pathogen *Mycosphaerella graminicola* found 65 potential milRNA-producing loci that produce sRNAs (36).

Unlike the animal and plant miRNAs, the *Neurospora* milRNAs are produced by at least four different biogenesis pathways that utilize different combinations of components, including Dicers, QDE-2, QIP, and an RNase III domain-containing protein, MRPL3 (62). Of the four miRNAs that were investigated in detail, only the biogenesis of *milR-3* is mechanistically similar to that of miRNAs in plants; it requires only Dicer for the generation of pre-milRNA and milRNA (49). The production of *milR-4* is only partially dependent on Dicers, indicating the involvement of other nucleases in milRNA biogenesis. *milR-1* is the most abundant sRNA-producing locus in the *Neurospora* genome, and production of milR-1 requires Dicers, QDE-2, and catalytically active QIP. The primary (pri)-*milR-1* is first processed by Dicer to produce pre-*milR-1*. Afterward, QDE-2 binds to pre-*milR-1* and recruits the exonuclease QIP to process the pre-*milR-1* into mature *milR-1*. In this pathway, the Argonaute QDE-2 functions as a scaffold for *milR-1* maturation.

Although biogenesis of *milR-1*, *milR-3*, and *milR-4* are completely or partially dependent on Dicers, the biogenesis of *milR-2* is completely independent of Dicer. Instead, the production of *milR-2* requires catalytically active QDE-2. Our results suggest that QDE-2 binds to the premilRNA and cleaves the passenger strand using its slicer activity. The *milR-2* biogenesis pathway is the first known example of a Dicer-independent—but Argonaute-dependent—mechanism for sRNA production. Soon after our study of *milR-2* biogenesis, *miR-451* in both mouse and zebrafish was shown to be produced by a similar mechanism (15, 19), suggesting that this is a conserved sRNA biogenesis pathway. The surprisingly diverse milRNA pathways in *Neurospora* shed important light on the biogenesis and evolution of eukaryotic miRNAs.

RNAi AS A HOST DEFENSE MECHANISM IN FUNGI

Transposon Control in *Neurospora*

Quelling results in the silencing of repetitive transgenes in *Neurospora*, suggesting that quelling is a mechanism that silences transposons. In most *Neurospora* strains, there are no functional transposons because of repeat-induced point mutation, a process that can mutate repetitive sequences during meiosis (33). However, a functional long interspersed element (LINE)-like transposon was previously identified in an African strain (55), and Nolan et al. (77) introduced the Tad transposon into laboratory *Neurospora* strains and showed that QDE-2 and Dicer, but not QDE-1 and QDE-3, are required for the suppression of transposon replication. These results suggest that transposition may generate inverted repeats that lead to the production of dsRNA, resulting in the silencing of transposon.

An Antiviral Mechanism in *Cryphonectria parasitica* and *Aspergillus nidulans*

Virus infection often leads to the production of siRNAs derived from viral dsRNA, suggesting a role for RNAi as part of the innate immunity mechanism against viral replication (26). Plants, arthropods, nematodes, and some other animals rely on the RNAi pathway to protect them from viral infection. To counter the effects of RNAi, viruses often express viral suppressors of RNA silencing that can inhibit RNAi. In fungi, an RNAi-based antiviral mechanism was first observed in the ascomycete filamentous fungus *Cryphonectria parasitica* (90, 97, 112, 113). *C. parasitica* is a rich resource for the study of basic aspects of virus-fungus interactions because it can support the replication of five virus families and is a well-established experimental system for the study of mycoviruses.

Like *Neurospora*, *C. parasitica* has two *dicer*-like genes, *dcl1* and *dcl2*. Hypovirus or mycoreovirus infection of the *dcl2* mutant results in a severe growth phenotype and elevated viral RNA levels (90), indicating that DCL2 is the major Dicer in *C. parasitica*. Consistently, 21- to 22-nt virus-derived sRNAs were abolished in the *dcl2* mutant (113). Although there are four Argonaute-like genes in *C. parasitica*, only *agl2* is required for the antiviral defense response (97). Together, these observations demonstrate that RNAi acts as a defense mechanism against viruses.

The mycovirus *Cryphonectria* hypovirus 1 (CHV1) expresses p29, a papain-like protease similar to the plant potyvirus-encoded suppressor of RNA silencing HC-Pro (17, 89, 98, 99). Deletion of p29 results in the reduction of viral RNA levels. p29 suppresses hairpin RNA-induced, virus-induced, and agroinfiltration-induced RNA silencing, indicating that it is a suppressor of RNA silencing. When *C. parasitica* is infected by a mutant hypovirus without p29, the mRNAs of *dcl2* and *agl2* accumulate to high levels, suggesting that p29 represses RNAi by inhibiting the expression of RNAi genes.

In addition to their roles in silencing viral replication by degrading viral RNAs, RNAi components also promote viral RNA recombination in *C. parasitica*, which leads to the production of hypovirus defective-interfering (DI) RNAs (97, 112). DI RNAs are a result of deletion events during recombination; DI RNAs suppress parental RNA accumulation, causing symptoms to attenuate. Importantly, DI RNAs do not accumulate in *agl1*, *agl2*, or *dcl2* mutants, indicating that these RNAi components are also involved in hypoviral RNA recombination and DI RNA production. The RNAi components may contribute to viral recombination by providing 5' and 3' fragments of viral RNA to the viral RdRP, but studies are needed to validate such a mechanism.

Studies of the RNAi pathway in *Aspergillus nidulans* also provided insights into the fungus-virus interaction. *A. nidulans* has one Argonaute gene and one Dicer gene; both are required for dsRNA-triggered gene silencing in the organism (41). Although there are two RdRP genes in the genome, neither is required for dsRNA-induced gene silencing, suggesting the absence of an sRNA amplification step by RdRPs (41). Infection with the *Aspergillus* virus 1816 suppresses dsRNA-induced RNA silencing (40), indicating the existence of an RNA-silencing suppressor encoded by this virus. In addition, the virus 341–derived siRNA was detected at a high level in the Argonaute mutant, indicating that this virus is targeted by the RNAi machinery (40).

dsRNA/Viral Infection–Induced Transcriptional Response in *Neurospora* and *C. parasitica*

Innate immunity, triggered after an organism detects dsRNA or foreign particles, is an important part of the eukaryotic host defense mechanism. The mechanism activates a signaling cascade, which in turn activates the transcription of effector genes that defend against the foreign invader. In *Neurospora*, a similar transcriptional response is observed upon the induction of dsRNA expression, which induces the expression of key RNAi components, QDE-2 and DCL2, and other putative antiviral genes are upregulated (18). The transcriptional response is retained in the *dcl1 dcl2* double mutant, indicating that the response is induced by dsRNA, not siRNA. Furthermore, the known RNAi components in *Neurospora* are not required for this response. Interestingly, dsRNA significantly induces the expression of QDE2 both transcriptionally and posttranscriptionally. In the *dcl* double mutant, despite the induction of *qde-2* transcription, the QDE-2 protein levels stay constant after dsRNA expression. This suggests that siRNAs are important for stabilizing the QDE2 protein. The induction of QDE-2 by dsRNA is required for efficient RNAi because mutants that lack this response have lower RNAi efficiency.

Genome-wide analyses identified approximately 60 genes that are activated by dsRNA. In addition to RNAi components, many of these genes are homologs of antiviral and interferon-stimulated genes, suggesting that the dsRNA-induced transcriptional response is part of an ancient host defense response. Because *Neurospora* does not encode homologs of the known mammalian dsRNA sensors, the transcriptional program triggered by dsRNA should represent a novel signaling cascade.

Interestingly, in *C. parasitica*, the expression of *dcl2* and *agl2* transcripts is strongly induced upon viral infection and expression of hairpin RNA (112, 113), indicating that this antiviral transcriptional activation response is conserved in filamentous fungi. In addition, the deletion of p29, also results in upregulation of both *dcl2* and *agl2*. Moreover, by using a GFP reporter fused with the promoter of *agl2*, Sun et al. (97) demonstrated that the induction of *agl2* by viral infection is regulated at the transcriptional level. Interestingly, the viral infection–induced *dcl-2* expression is blocked in the *agl2* mutant, suggesting that AGL2 plays a regulatory role in this gene activation pathway. These results suggest that the transcriptional regulation of RNAi components is important for the host antiviral response.

RNA SILENCING AND SMALL RNA STUDIES ON OTHER FILAMENTOUS FUNGI

Gene Silencing and Small RNAs in *Mucor circinelloides*

Mucor circinelloides, which causes invasive maxillofacial zygomycosis, is an important model system for studying RNAi in the zygomycete clade. Transformation with hairpin RNA-producing constructs and self-replicative plasmids shows that this organism can silence gene expression. Silencing results in the accumulation of two size classes of siRNAs (21 and 25 nt). *M. circinelloides* has two redundant *dicer* (*dcl1* and *dcl2*) genes and two *rdrp* genes (*rdrp1* and *rdrp2*); DCL-2 plays the major role in the RNA-silencing pathway and is induced by both sense and hairpin transgenes. Secondary sRNA were detected, suggesting an sRNA amplification step in the RNA-silencing pathway.

Detailed examination of the sRNA profile in this fungus revealed four classes of endogenous small RNA (esRNA). Interestingly, the esRNAs detected in *Mucor* are derived from exons and can regulate mRNA accumulation. The largest class of these exon-derived siRNAs (ex-siRNA) is dependent on DCL-2 and RDRP1, suggesting that mRNAs from these loci are converted to dsRNA by RdRP1 and processed by DCL-2. The second class of ex-siRNAs requires DCL-2 and RdRP2 for biogenesis. The third class requires both RdRP1 and RdRP2, and the two Dicers have redundant roles in their biogenesis. The fourth class requires DCL1, but not DCL2, and the two RdRPs are required. These results indicate that different RNAi components are used to produce different ex-siRNAs.

RNAi Studies on the Human Fungal Pathogen *Cryptococcus neoformans*

Cryptococcus neoformans, which causes fatal meningoencephalitis, has Argonaute, Dicer, and RdRP genes (47, 67). The RNAi pathway in *C. neoformans* acts in a process called sex-induced silencing (SIS) (106). SIS, which is similar to quelling, requires multiple copies of transgenes. In addition, the efficiency of SIS increases dramatically during the sexual cycle, which is correlated with increased expression of known RNAi components. SIS and SIS-associated sRNA production is abolished in RNAi mutants, indicating the role of RNAi in this silencing process during the sexual cycle. Examination of the sRNA profile revealed that some of the sRNAs, which were not derived from the transgenes, mapped to repetitive transposable elements and presumptive centromeric regions, suggesting that the RNAi pathway is a genome defense mechanism. Supporting this notion, in the *rdp1* mutant, retrotransposons are highly expressed with higher transposition rates.

Small RNAs from *Magnaporthe oryzae*

Magnaporthe oryzae is a model organism for the study of pathogen-host interactions because it is a primary plant pathogen of rice and wheat. A recent sRNA analysis revealed that a large variety of sRNAs map to loci, including tRNA loci, rRNA loci, coding regions, and intergenic loci (79). In the mycelia, sRNAs mostly come from intergenic regions and repetitive elements, such as long terminal repeat retrotransposons. However, sRNAs in the appressoria are enriched for tRNA-derived RNA fragments. The differential distribution of sRNAs in different tissues suggests these sRNAs have roles in the regulation of growth and development of this pathogen.

SUMMARY POINTS

1. *Neurospora crassa* has multiple RNAi-related silencing pathways, including quelling, qiRNA, and MSUD, which are triggered by multiple copies of transgenes, DNA damage, and unpaired DNA, respectively.
2. *Schizosaccharomyces pombe* requires an RNAi pathway for heterochromatin formation.
3. There are multiple sRNA pathways in fungi. milRNAs with distinct biogenesis pathways that require combinations of different components have been found in *Neurospora*.
4. RNAi is an important fungal host defense mechanism against transposon and viral invasion.

DISCLOSURE STATEMENT

The authors are not aware of any affiliations, memberships, funding, or financial holdings that might be perceived as affecting the objectivity of this review.

ACKNOWLEDGMENTS

We thank the members of the Liu laboratory for discussion and comments. This work was supported by grants from the National Institutes of Health and the Welch Foundation to Yi Liu (I-1560).

LITERATURE CITED

1. Alexander WG, Raju NB, Xiao H, Hammond TM, Perdue TD, et al. 2008. DCL-1 colocalizes with other components of the MSUD machinery and is required for silencing. *Fungal Genet. Biol.* 45:719–27
2. Ambros V, Bartel B, Bartel DP, Burge CB, Carrington JC, et al. 2003. A uniform system for microRNA annotation. *RNA* 9:277–79
3. Aramayo R, Metzenberg RL. 1996. Meiotic transvection in fungi. *Cell* 86:103–13
4. **Bühler M, Verdel A, Moazed D. 2006. Tethering RITS to a nascent transcript initiates RNAi- and heterochromatin-dependent gene silencing. *Cell* 125:873–86**
5. Bardiya N, Alexander WG, Perdue TD, Barry EG, Metzenberg RL, et al. 2008. Characterization of interactions between and among components of the meiotic silencing by unpaired DNA machinery in *Neurospora crassa* using bimolecular fluorescence complementation. *Genetics* 178:593–96
6. Bartel DP. 2004. MicroRNAs: genomics, biogenesis, mechanism, and function. *Cell* 116:281–97
7. Buhler M, Moazed D. 2007. Transcription and RNAi in heterochromatic gene silencing. *Nat. Struct. Mol. Biol.* 14:1041–48
8. Cam HP, Chen ES, Grewal SIS. 2009. Transcriptional scaffolds for heterochromatin assembly. *Cell* 136:610–14
9. Carattoli A, Cogoni C, Morelli G, Macino G. 1994. Molecular characterization of upstream regulatory sequences controlling the photoinduced expression of the *albino-3* gene of *Neurospora crassa*. *Mol. Microbiol.* 13:787–95
10. Carthew RW, Sontheimer EJ. 2009. Origins and mechanisms of miRNAs and siRNAs. *Cell* 136:642–55
11. **Catalanotto C, Azzalin G, Macino G, Cogoni C. 2000. Gene silencing in worms and fungi. *Nature* 404:245**
12. Catalanotto C, Azzalin G, Macino G, Cogoni C. 2002. Involvement of small RNAs and role of the *qde* genes in the gene silencing pathway in *Neurospora*. *Genes Dev.* 16:790–95
13. Catalanotto C, Nolan T, Cogoni C. 2006. Homology effects in *Neurospora crassa*. *FEMS Microbiol. Lett.* 254:182–89
14. Catalanotto C, Pallotta M, ReFalo P, Sachs MS, Vayssie L, et al. 2004. Redundancy of the two Dicer genes in transgene-induced posttranscriptional gene silencing in *Neurospora crassa*. *Mol. Cell. Biol.* 24:2536–45
15. Cheloufi S, Dos Santos CO, Chong MM, Hannon GJ. 2010. A Dicer-independent miRNA biogenesis pathway that requires Ago catalysis. *Nature* 465:584–89
16. **Chen ES, Zhang K, Nicolas E, Cam HP, Zofall M, Grewal SIS. 2008. Cell cycle control of centromeric repeat transcription and heterochromatin assembly. *Nature* 451:734–37**
17. Choi GH, Pawlyk DM, Nuss DL. 1991. The autocatalytic protease p29 encoded by a hypovirulence-associated virus of the chestnut blight fungus resembles the potyvirus-encoded protease HC-Pro. *Virology* 183:747–52
18. Choudhary S, Lee H-C, Maiti M, He Q, Cheng P, et al. 2007. A double-stranded-RNA response program important for RNA interference efficiency. *Mol. Cell. Biol.* 27:3995–4005
19. Cifuentes D, Xue H, Taylor DW, Patnode H, Mishima Y, et al. 2010. A novel miRNA processing pathway independent of Dicer requires Argonaute2 catalytic activity. *Science* 328:1694–98

4. Demonstrates that the targeting of RITS complex via RNA to a chromosomal locus is sufficient to induce transcriptional gene silencing.

11. Identifies the Argonaute protein QDE-2 as a factor required for quelling, indicating that dsRNA-induced gene silencing in animals and quelling in *Neurospora* share a common mechanism.

16. Discovered that siRNA production and heterochromatin formation are regulated by the cell cycle and suggested a role for RNAi in the cell cycle.

20. Cogoni C, Irelan JT, Schumacher M, Schmidhauser TJ, Selker EU, Macino G. 1996. Transgene silencing of the *al-1* gene in vegetative cells of *Neurospora* is mediated by a cytoplasmic effector and does not depend on DNA-DNA interactions or DNA methylation. *EMBO J.* 15:3153–63
21. Cogoni C, Macino G. 1997. Isolation of quelling-defective (*qde*) mutants impaired in posttranscriptional transgene-induced gene silencing in *Neurospora crassa*. *Proc. Natl. Acad. Sci. USA* 94:10233–38
22. **Cogoni C, Macino G. 1999. Gene silencing in *Neurospora crassa* requires a protein homologous to RNA-dependent RNA polymerase. *Nature* 399:166–69**
23. Cogoni C, Macino G. 1999. Posttranscriptional gene silencing in *Neurospora* by a RecQ DNA helicase. *Science* 286:2342–44
24. Colmenares SU, Buker SM, Buhler M, Dlakić M, Moazed D. 2007. Coupling of double-stranded RNA synthesis and siRNA generation in fission yeast RNAi. *Mol. Cell* 27:449–61
25. Dang Y, Yang Q, Xue Z, Liu Y. 2011. RNA interference in fungi: pathways, functions and applications. *Eukaryot. Cell* 10:1148–55
26. Ding SW, Lu R. 2011. Virus-derived siRNAs and piRNAs in immunity and pathogenesis. *Curr. Opin. Virol.* 1:533–44
27. Drinnenberg IA, Weinberg DE, Xie KT, Mower JP, Wolfe KH, et al. 2009. RNAi in budding yeast. *Science* 326:544–50
28. Durkin SG, Glover TW. 2007. Chromosome fragile sites. *Annu. Rev. Genet.* 41:169–92
29. Ebert A, Lein S, Schotta G, Reuter G. 2006. Histone modification and the control of heterochromatic gene silencing in *Drosophila*. *Chromosome Res.* 14:377–92
30. Emmerth S, Schober H, Gaidatzis D, Roloff T, Jacobeit K, Bühler M. 2010. Nuclear retention of fission yeast Dicer is a prerequisite for RNAi-mediated heterochromatin assembly. *Dev. Cell* 18:102–13
31. Fire A, Xu S, Montgomery MK, Kostas SA, Driver SE, Mello CC. 1998. Potent and specific genetic interference by double-stranded RNA in *Caenorhabditis elegans*. *Nature* 391:806–11
32. Fulci V, Macino G. 2007. Quelling: post-transcriptional gene silencing guided by small RNAs in *Neurospora crassa*. *Curr. Opin. Microbiol.* 10:199–203
33. Galagan JE, Selker EU. 2004. RIP: the evolutionary cost of genome defense. *Trends Genet.* 20:417–23
34. Ghildiyal M, Xu J, Seitz H, Weng Z, Zamore PD. 2010. Sorting of *Drosophila* small silencing RNAs partitions microRNA* strands into the RNA interference pathway. *RNA* 16:43–56
35. Glover TW. 2006. Common fragile sites. *Cancer Lett.* 232:4–12
36. Goodwin SB, Ben M'Barek S, Dhillon B, Wittenberg AHJ, Crane CF, et al. 2011. Finished genome of the fungal wheat pathogen *Mycosphaerella graminicola* reveals dispensome structure, chromosome plasticity, and stealth pathogenesis. *PLoS Genet.* 7:e1002070
37. Grewal SI. 2010. RNAi-dependent formation of heterochromatin and its diverse functions. *Curr. Opin. Genet. Dev.* 20:134–41
38. Grimson A, Srivastava M, Fahey B, Woodcroft BJ, Chiang HR, et al. 2008. Early origins and evolution of microRNAs and Piwi-interacting RNAs in animals. *Nature* 455:1193–97
39. Halic M, Moazed D. 2010. Dicer-independent primal RNAs trigger RNAi and heterochromatin formation. *Cell* 140:504–16
40. Hammond TM, Andrewski MD, Roossinck MJ, Keller NP. 2008. *Aspergillus* mycoviruses are targets and suppressors of RNA silencing. *Eukaryot. Cell* 7:350–57
41. Hammond TM, Keller NP. 2005. RNA silencing in *Aspergillus nidulans* is independent of RNA-dependent RNA polymerases. *Genetics* 169:607–17
42. Hammond TM, Xiao H, Boone EC, Perdue TD, Pukkila PJ, Shiu PK. 2011. SAD-3, a putative helicase required for meiotic silencing by unpaired DNA, interacts with other components of the silencing machinery. *Genes Genomes Genet.* 1:369–76
43. Hayashi A, Da-Qiao D, Tsutsumi C, Chikashige Y, Masuda H, et al. 2009. Localization of gene products using a chromosomally tagged GFP-fusion library in the fission yeast *Schizosaccharomyces pombe*. *Genes Cells* 14:217–25
44. Hong E-JE, Villén J, Moazed D. 2005. A cullin E3 ubiquitin ligase complex associates with Rik1 and the Clr4 histone H3-K9 methyltransferase and is required for RNAi-mediated heterochromatin formation. *RNA Biol.* 2:106–11

---

22. Identified the first eukaryotic RNAi component and suggested that dsRNA is an endogenous trigger for posttranscriptional gene silencing.

45. Horn PJ, Bastie J-N, Peterson CL. 2005. A Rik1-associated, cullin-dependent E3 ubiquitin ligase is essential for heterochromatin formation. *Genes Dev.* 19:1705–14
46. Hsieh J, Fire A. 2000. Recognition and silencing of repeated DNA. *Annu. Rev. Genet.* 34:187–204
47. Idnurm A, Bahn Y-S, Nielsen K, Lin X, Fraser JA, Heitman J. 2005. Deciphering the model pathogenic fungus *Cryptococcus neoformans*. *Nat. Rev. Micro* 3:753–64
48. Jia S, Noma K-I, Grewal SIS. 2004. RNAi-independent heterochromatin nucleation by the stress-activated ATF/CREB family proteins. *Science* 304:1971–76
49. Jones-Rhoades MW, Bartel DP, Bartel B. 2006. MicroRNAs and their regulatory roles in plants. *Annu. Rev. Plant Biol.* 57:19–53
50. Kagansky A, Folco HD, Almeida R, Pidoux AL, Boukaba A, et al. 2009. Synthetic heterochromatin bypasses RNAi and centromeric repeats to establish functional centromeres. *Science* 324:1716–19
51. Kato A, Akamatsu Y, Sakuraba Y, Inoue H. 2004. The *Neurospora crassa mus-19* gene is identical to the *qde-3* gene, which encodes a RecQ homologue and is involved in recombination repair and postreplication repair. *Curr. Genet.* 45:37–44
52. Kato H, Goto DB, Martienssen RA, Urano T, Furukawa K, Murakami Y. 2005. RNA polymerase II is required for RNAi-dependent heterochromatin assembly. *Science* 309:467–69
53. Kelly WG, Aramayo R. 2007. Meiotic silencing and the epigenetics of sex. *Chromosome Res.* 15:633–51
54. Kim VN, Han J, Siomi MC. 2009. Biogenesis of small RNAs in animals. *Nat. Rev. Mol. Cell Biol.* 10:126–39
55. Kinsey JA. 1990. Tad, a LINE-like transposable element of *Neurospora*, can transpose between nuclei in heterokaryons. *Genetics* 126:317–23
56. Kloc A, Zaratiegui M, Nora E, Martienssen R. 2008. RNA interference guides histone modification during the S phase of chromosomal replication. *Curr. Biol.* 18:490–95
57. Lagos-Quintana M, Rauhut R, Lendeckel W, Tuschl T. 2001. Identification of novel genes coding for small expressed RNAs. *Science* 294:853–58
58. Lau NC, Seto AG, Kim J, Kuramochi-Miyagawa S, Nakano T, et al. 2006. Characterization of the piRNA complex from rat testes. *Science* 313:363–67
59. Lee DW, Pratt RJ, McLaughlin M, Aramayo R. 2003. An Argonaute-like protein is required for meiotic silencing. *Genetics* 164:821–28
60. Lee HC, Aalto AP, Yang Q, Chang SS, Huang G, et al. 2010. The DNA/RNA-dependent RNA polymerase QDE-1 generates aberrant RNA and dsRNA for RNAi in a process requiring replication protein A and a DNA helicase. *PLoS Biol.* 8:e1000496
61. **Lee HC, Chang SS, Choudhary S, Aalto AP, Maiti M, et al. 2009. qiRNA is a new type of small interfering RNA induced by DNA damage. *Nature* 459:274–77**
62. **Lee HC, Li L, Gu W, Xue Z, Crosthwaite SK, et al. 2010. Diverse pathways generate microRNA-like RNAs and Dicer-independent small interfering RNAs in fungi. *Mol. Cell* 38:803–14**
63. Lee RC, Ambros V. 2001. An extensive class of small RNAs in *Caenorhabditis elegans*. *Science* 294:862–64
64. Lee RC, Feinbaum RL, Ambros V. 1993. The *C. elegans* heterochronic gene *lin-4* encodes small RNAs with antisense complementarity to *lin-14*. *Cell* 75:843–54
65. Lejeune E, Bayne EH, Allshire RC. 2010. On the connection between RNAi and heterochromatin at centromeres. *Cold Spring Harb. Symp. Quant. Biol.* 75:275–83
66. Li F, Goto DB, Zaratiegui M, Tang X, Martienssen R, Cande WZ. 2005. Two novel proteins, Dos1 and Dos2, interact with Rik1 to regulate heterochromatic RNA interference and histone modification. *Curr. Biol.* 15:1448–57
67. Lin X, Heitman J. 2006. The biology of the *Cryptococcus neoformans* species complex. *Annu. Rev. Microbiol.* 60:69–105
68. Liu Y, Ye X, Jiang F, Liang C, Chen D, et al. 2009. C3PO, an endoribonuclease that promotes RNAi by facilitating RISC activation. *Science* 325:750–53
69. Llave C, Kasschau KD, Rector MA, Carrington JC. 2002. Endogenous and silencing-associated small RNAs in plants. *Plant Cell* 14:1605–19
70. Maiti M, Lee HC, Liu Y. 2007. QIP, a putative exonuclease, interacts with the *Neurospora* Argonaute protein and facilitates conversion of duplex siRNA into single strands. *Genes Dev.* 21:590–600

61. Suggests that DNA damage is an important trigger for sRNA production and proposes that the RNA-dependent RNA polymerase also functions to produce aberrant RNAs.

62. Discovered the miRNA-like RNAs in *Neurospora* and uncovered surprisingly diverse biogenesis mechanisms for RNA production.

71. Matzke M, Kanno T, Daxinger L, Huettel B, Matzke AJ. 2009. RNA-mediated chromatin-based silencing in plants. *Curr. Opin. Cell Biol.* 21:367–76
72. Mello CC, Conte D. 2004. Revealing the world of RNA interference. *Nature* 431:338–42
73. Moazed D. 2009. Small RNAs in transcriptional gene silencing and genome defence. *Nature* 457:413–20
74. Molnar A, Schwach F, Studholme DJ, Thuenemann EC, Baulcombe DC. 2007. miRNAs control gene expression in the single-cell alga *Chlamydomonas reinhardtii*. *Nature* 447:1126–29

> 75. Discovered the critical link between the siRNA production process and heterochromatin assembly machinery.

75. **Motamedi MR, Verdel A, Colmenares SU, Gerber SA, Gygi SP, Moazed D. 2004. Two RNAi complexes, RITS and RDRC, physically interact and localize to noncoding centromeric RNAs. *Cell* 119:789–802**
76. Napoli C, Lemieux C, Jorgensen R. 1990. Introduction of a chimeric chalcone synthase gene into petunia results in reversible co-suppression of homologous genes in *trans*. *Plant Cell* 2:279–85
77. Nolan T, Braccini L, Azzalin G, De Toni A, Macino G, Cogoni C. 2005. The post-transcriptional gene silencing machinery functions independently of DNA methylation to repress a LINE1-like retrotransposon in *Neurospora crassa*. *Nucleic Acids Res.* 33:1564–73
78. Nolan T, Cecere G, Mancone C, Alonzi T, Tripodi M, et al. 2008. The RNA-dependent RNA polymerase essential for post-transcriptional gene silencing in *Neurospora crassa* interacts with replication protein A. *Nucleic Acids Res.* 36:532–38
79. Nunes C, Gowda M, Sailsbery J, Xue M, Chen F, et al. 2011. Diverse and tissue-enriched small RNAs in the plant pathogenic fungus, *Magnaporthe oryzae*. *BMC Genomics* 12:288
80. Obbard DJ, Gordon KHJ, Buck AH, Jiggins FM. 2009. The evolution of RNAi as a defence against viruses and transposable elements. *Philos. Trans. R. Soc. B* 364:99–115
81. O'Brien HE, Parrent JL, Jackson JA, Moncalvo JM, Vilgalys R. 2005. Fungal community analysis by large-scale sequencing of environmental samples. *Appl. Environ. Microbiol.* 71:5544–50
82. Peng JC, Karpen GH. 2007. H3K9 methylation and RNA interference regulate nucleolar organization and repeated DNA stability. *Nat. Cell Biol.* 9:25–35
83. Pontes O, Li CF, Costa Nunes P, Haag J, Ream T, et al. 2006. The *Arabidopsis* chromatin-modifying nuclear siRNA pathway involves a nucleolar RNA processing center. *Cell* 126:79–92
84. Pratt RJ, Lee DW, Aramayo R. 2004. DNA methylation affects meiotic trans-sensing, not meiotic silencing, in *Neurospora*. *Genetics* 168:1925–35

> 85. Describe the discovery of quelling, one of the first known RNAi-related phenomena, in *Neurospora*.

85. **Romano N, Macino G. 1992. Quelling: transient inactivation of gene expression in *Neurospora crassa* by transformation with homologous sequences. *Mol. Microbiol.* 6:3343–53**
86. Sadaie M, Iida T, Urano T, Nakayama J-I. 2004. A chromodomain protein, Chp1, is required for the establishment of heterochromatin in fission yeast. *EMBO J.* 23:3825–35
87. Salgado PS, Koivunen MR, Makeyev EV, Bamford DH, Stuart DI, Grimes JM. 2006. The structure of an RNAi polymerase links RNA silencing and transcription. *PLoS Biol.* 4:e434
88. Schramke V, Sheedy DM, Denli AM, Bonila C, Ekwall K, et al. 2005. RNA-interference-directed chromatin modification coupled to RNA polymerase II transcription. *Nature* 435:1275–79
89. Segers GC, van Wezel R, Zhang X, Hong Y, Nuss DL. 2006. Hypovirus papain-like protease p29 suppresses RNA silencing in the natural fungal host and in a heterologous plant system. *Eukaryot. Cell* 5:896–904
90. Segers GC, Zhang X, Deng F, Sun Q, Nuss DL. 2007. Evidence that RNA silencing functions as an antiviral defense mechanism in fungi. *Proc. Natl. Acad. Sci. USA* 104:12902–6
91. Shanker S, Job G, George OL, Creamer KM, Shaban A, Partridge JF. 2010. Continuous requirement for the Clr4 complex but not RNAi for centromeric heterochromatin assembly in fission yeast harboring a disrupted RITS complex. *PLoS Genet.* 6:e1001174
92. Shiu PK, Glass NL. 2000. Cell and nuclear recognition mechanisms mediated by mating type in filamentous ascomycetes. *Curr. Opin. Microbiol.* 3:183–88
93. Shiu PK, Metzenberg RL. 2002. Meiotic silencing by unpaired DNA: properties, regulation and suppression. *Genetics* 161:1483–95

> 94. Demonstrates that meiotic silencing in *Neurospora* is triggered by unpaired DNA and is an RNAi-related phenomenon.

94. **Shiu PK, Raju NB, Zickler D, Metzenberg RL. 2001. Meiotic silencing by unpaired DNA. *Cell* 107:905–16**

95. Shiu PK, Zickler D, Raju NB, Ruprich-Robert G, Metzenberg RL. 2006. SAD-2 is required for meiotic silencing by unpaired DNA and perinuclear localization of SAD-1 RNA-directed RNA polymerase. *Proc. Natl. Acad. Sci. USA* 103:2243–48
96. Son H, Min K, Lee J, Raju NB, Lee YW. 2011. Meiotic silencing in the homothallic fungus *Gibberella zeae*. *Fungal Biol.* 115:1290–302
97. Sun Q, Choi GH, Nuss DL. 2009. A single Argonaute gene is required for induction of RNA silencing antiviral defense and promotes viral RNA recombination. *Proc. Natl. Acad. Sci. USA* 106:17927–32
98. Suzuki N, Chen B, Nuss DL. 1999. Mapping of a hypovirus p29 protease symptom determinant domain with sequence similarity to potyvirus HC-Pro protease. *J. Virol.* 73:9478–84
99. Suzuki N, Maruyama K, Moriyama M, Nuss DL. 2003. Hypovirus papain-like protease p29 functions in trans to enhance viral double-stranded RNA accumulation and vertical transmission. *J. Virol.* 77:11697–707
100. Thomson T, Lin H. 2009. The biogenesis and function of PIWI proteins and piRNAs: progress and prospect. *Annu. Rev. Cell Dev. Biol.* 25:355–76
101. **Verdel A, Jia S, Gerber S, Sugiyama T, Gygi S, et al. 2004. RNAi-mediated targeting of heterochromatin by the RITS complex. *Science* 303:672–76**
102. Verdel A, Moazed D. 2005. RNAi-directed assembly of heterochromatin in fission yeast. *FEBS Lett.* 579:5872–78
103. Volpe T, Martienssen RA. 2011. RNA interference and heterochromatin assembly. *Cold Spring Harb. Perspect. Biol.* 3:a003731
104. Volpe T, Schramke V, Hamilton GL, White SA, Teng G, et al. 2003. RNA interference is required for normal centromere function in fission yeast. *Chromosome Res.* 11:137–46
105. **Volpe TA, Kidner C, Hall IM, Teng G, Grewal SIS, Martienssen RA. 2002. Regulation of heterochromatic silencing and histone H3 lysine-9 methylation by RNAi. *Science* 297:1833–37**
106. Wang X, Hsueh Y-P, Li W, Floyd A, Skalsky R, Heitman J. 2010. Sex-induced silencing defends the genome of *Cryptococcus neoformans* via RNAi. *Genes Dev.* 24:2566–82
107. White SA, Allshire RC. 2008. RNAi-mediated chromatin silencing in fission yeast. *Curr. Top. Microbiol. Immunol.* 320:157–83
108. Wierzbicki AT, Haag JR, Pikaard CS. 2008. Noncoding transcription by RNA polymerase Pol IVb/Pol V mediates transcriptional silencing of overlapping and adjacent genes. *Cell* 135:635–48
109. Xiao H, Alexander WG, Hammond TM, Boone EC, Perdue TD, et al. 2010. QIP, a protein that converts duplex siRNA into single strands, is required for meiotic silencing by unpaired DNA. *Genetics* 186:119–26
110. Zaratiegui M, Castel SE, Irvine DV, Kloc A, Ren J, et al. 2011. RNAi promotes heterochromatic silencing through replication-coupled release of RNA Pol II. *Nature* 479:135–38
111. Zhang K, Mosch K, Fischle W, Grewal SIS. 2008. Roles of the Clr4 methyltransferase complex in nucleation, spreading and maintenance of heterochromatin. *Nat. Struct. Mol. Biol.* 15:381–88
112. Zhang X, Nuss DL. 2008. A host Dicer is required for defective viral RNA production and recombinant virus vector RNA instability for a positive sense RNA virus. *Proc. Natl. Acad. Sci. USA* 105:16749–54
113. Zhang X, Segers GC, Sun Q, Deng F, Nuss DL. 2008. Characterization of hypovirus-derived small RNAs generated in the chestnut blight fungus by an inducible DCL-2-dependent pathway. *J. Virol.* 82:2613–19
114. Djupedal I, Portoso M, Spåhr H, Bonilla C, Gustafsson CM, et al. 2005. RNA Pol II subunit Rpb7 promotes centromeric transcription and RNAi-directed chromatin silencing. *Genes Dev.* 19(19):2301–6

101. Identifies the RITS complex, which is the link between the RNAi machinery and heterochromatin assembly factors in *S. pombe*.

105. The first study to demonstrate that RNAi components are required for heterochromatin formation in *S. pombe*.

Evolution of Two-Component Signal Transduction Systems

Emily J. Capra[1] and Michael T. Laub[1,2]

[1]Department of Biology, [2]Howard Hughes Medical Institute, Massachusetts Institute of Technology, Cambridge, Massachusetts 02139; email: laub@mit.edu

Keywords

duplication and divergence, cross-talk, histidine kinase, response regulator, lateral gene transfer

Abstract

To exist in a wide range of environmental niches, bacteria must sense and respond to a variety of external signals. A primary means by which this occurs is through two-component signal transduction pathways, typically composed of a sensor histidine kinase that receives the input stimuli and then phosphorylates a response regulator that effects an appropriate change in cellular physiology. Histidine kinases and response regulators have an intrinsic modularity that separates signal input, phosphotransfer, and output response; this modularity has allowed bacteria to dramatically expand and diversify their signaling capabilities. Recent work has begun to reveal the molecular basis by which two-component proteins evolve. How and why do orthologous signaling proteins diverge? How do cells gain new pathways and recognize new signals? What changes are needed to insulate a new pathway from existing pathways? What constraints are there on gene duplication and lateral gene transfer? Here, we review progress made in answering these questions, highlighting how the integration of genome sequence data with experimental studies is providing major new insights.

Contents

OVERVIEW... 326
THE TWO-COMPONENT SIGNAL TRANSDUCTION PARADIGM.......... 327
EVOLUTION OF GENOME CONTENT AND GENE NUMBER.............. 328
MECHANISMS FOR EVOLVING CHANGES IN TWO-COMPONENT
    SIGNALING GENE CONTENT.. 331
GENE FUSIONS, REARRANGEMENTS, AND DUPLICATIONS.............. 332
EVOLUTION OF SIGNALING PROTEIN STRUCTURE
    AND FUNCTION.. 333
    Histidine Kinase Sensory Domain Evolution.................................. 333
    Divergence and Evolution of Pathway Outputs................................ 334
    Evolution of Phosphotransfer Specificity and the Insulation of Pathways........... 337
DIMERIZATION SPECIFICITY... 340
FINAL PERSPECTIVE.. 342

OVERVIEW

Two-component signal transduction systems are a predominant means by which bacteria sense and respond to their environments. These systems generally comprise a receptor histidine kinase that senses a specific signal and translates that input into a desired output through the phosphorylation of its cognate response regulator. The success of two-component signaling systems as a strategy for coupling changes in the environment to changes in cellular physiology is underscored by their prevalence throughout the bacterial kingdom. These signaling proteins have been found in the genomes of nearly all sequenced bacteria, with the majority of species encoding dozens, and sometimes hundreds, of two-component proteins. They have been uncovered in countless genetic screens and respond to an enormous range of signals and stressors (for reviews, see References 42 and 77).

Although tremendous progress has been made in understanding the structure and function of some individual systems, two additional aspects of these pathways have garnered significant interest. First, how does a single cell coordinate so many highly related signaling pathways? The kinases and regulators encoded by a given organism are often similar at the sequence and structural levels, yet cells are able to match specific inputs to the desired output. How is unwanted cross-talk avoided? Do cells leverage the similarity of these proteins to integrate signals or diversify responses? The second related area of interest centers on understanding the evolution of these systems. How do two-component pathways evolve and how are new pathways introduced? How are new pathways insulated from one another? And finally, how do new pathways lead to new functions and new signaling capabilities?

Here, we review recent progress in tackling these questions, focusing in particular on the evolution of two-component signaling proteins. Gene duplication and lateral (horizontal) gene transfer provide the raw materials for producing new pathways and, in either case, the introduction of new signaling proteins requires a flurry of changes if the new proteins are to be maintained over the course of evolution. The new pathway must gain a new function to provide a selective advantage and to warrant maintenance in the genome. Domain shuffling likely plays a critical role and recent work has begun to reveal how, at a mechanistic level, this process occurs. New pathways must also avoid cross-talk with other pathways, and vice versa, leading to changes in the

Cross-talk:
detrimental communication between two different signaling pathways

Gene duplication:
a process that produces two copies of a gene or set of genes

Lateral (horizontal) gene transfer:
a process in which genetic material is transferred from one organism to another, but not through vertical inheritance

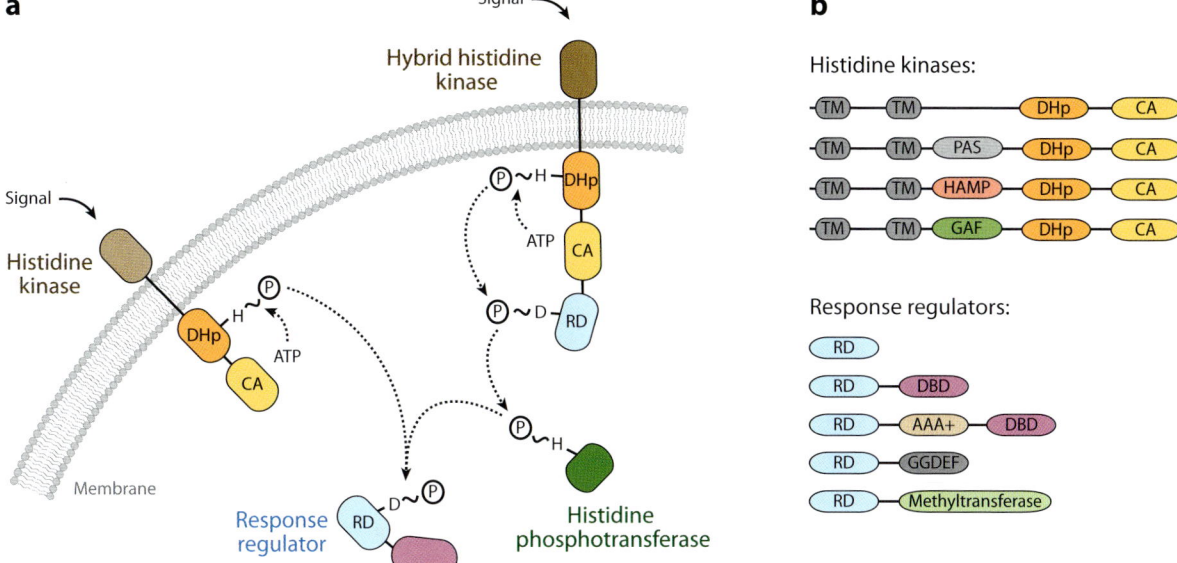

Figure 1

Overview of two-component signal transduction. (*a*) In the prototypical two-component pathway (*left*), the CA domain of a histidine kinase binds ATP and autophosphorylates a conserved histidine in the DHp domain. The phosphoryl group is then transferred to an aspartate in the RD of the cognate response regulator, activating its output domain to effect cellular changes, often through changes in transcription. In a phosphorelay system (*right*), a hybrid histidine kinase autophosphorylates and transfers its phosphoryl group intramolecularly to a RD. A histidine phosphotransferase then shuttles the phosphoryl group to a soluble response regulator that effects a pathway output. (*b*) Common domain organizations of histidine kinases and response regulators. For histidine kinases, the DHp and CA domains are shown with common intracellular domains PAS, HAMP, and GAF. Note that some kinases have multiple copies of such domains. Two TM domains are shown on the kinases, but kinases can harbor from 1 to 13 TM domains. A wide range of sensory domains (not shown) are often found in the periplasmic portions of membrane-bound histidine kinases. For response regulators, the conserved receiver domain is shown alone or with the common output domains: a DBD, a AAA+ and DBD, a GGDEF domain involved in cyclic-di-GMP synthesis, or a CheB-like methyltransferase domain. Abbreviations: CA, catalytic and ATPase; DHp, dimerization and histidine phosphotransferase; RD, receiver domain; PAS, Per Arnt Sim; HAMP, histidine kinase, adenyl cyclase, methyl-accepting proteins, and phosphatase; GAF, cGMP-specific phosphodiesterase, adenylyl cyclase, and FhlA; TM, transmembrane; DBD, DNA-binding domain.

specificity determinants of these pathways at multiple levels, including receptor dimerization and kinase-substrate partnering.

Although much of what is known about the evolution of two-component signaling is based on sequence and phylogenetic analyses, there are new efforts to integrate experimental approaches. Throughout the review we highlight ways in which computational, phylogenetic, and experimental studies have been combined to provide new insights and also highlight areas ripe for future investigation. We begin with a brief review of the structure and function of two-component signaling; for more comprehensive reviews of the mechanistic aspects of two-component signaling, see References 12, 24, 25, 77, and 86.

THE TWO-COMPONENT SIGNAL TRANSDUCTION PARADIGM

The eponymous two-component signaling pathway contains a sensor histidine kinase and a cognate response regulator (**Figure 1a**). Upon receipt of a stimulus, the histidine kinase typically catalyzes an autophosphorylation reaction on a conserved histidine residue. This phosphoryl group is then

Autophosphorylation: a reaction in which a protein kinase covalently attaches the γ-phosphoryl group from ATP to itself

Phosphatase: a histidine kinase that also stimulates the dephosphorylation of a cognate response regulator

Phosphotransfer: the transfer of a phosphoryl group from an autophosphorylated histidine kinase to a conserved aspartate on a response regulator

DHp: dimerization and histidine phosphotransfer

CA: catalytic and ATP binding

PAS: Per Arnt Sim

HAMP: histidine kinases, adenyl cyclases, methyl-accepting proteins, and phosphatases

GAF: cGMP-specific phosphodiesterases, adenylyl cyclases, and FhlA

Phosphorelay: a variant of the canonical two-component signaling pathway in which a phosphoryl group is transferred successively from a histidine kinase to a response regulator, then to a histidine phosphotransferase, and finally to a second, terminal response regulator containing an output domain

transferred to a conserved aspartate on a cognate response regulator. Phosphorylation of the regulator usually drives a conformational change that activates its output response, often leading to changes in gene expression. These systems thus represent versatile, powerful ways to couple changes in external or environmental conditions to corresponding changes in cellular physiology and gene expression. In most cases, histidine kinases are bifunctional such that, when not stimulated to autophosphorylate, they act as phosphatases for their cognate response regulators; thus, it is ultimately the ratio of kinase to phosphatase activity that is responsible for modulating the output response (36, 94). In some cases, input signals may promote the phosphatase state rather than stimulate autophosphorylation (65).

All histidine kinases contain two highly conserved domains, the dimerization and histidine phosphotransfer (DHp) domain, which harbors the conserved histidine that is the site of both the autophosphorylation and phosphotransfer reactions, and the catalytic and ATP binding (CA) domain. Histidine kinases also usually contain at least one (and often several) additional domains N-terminal to the DHp domain (**Figure 1b**). For the vast majority of kinases this includes 1 to 13 transmembrane domains (20) with signal recognition occurring primarily in the periplasmic or extracellular portion of the protein. Although some common domains have been noted, signal recognition domains tend to be more variable than the other domains. Most kinases also have at least one domain between the transmembrane and DHp domains, with PAS, HAMP, and GAF domains by far the most common (23). These domains can either relay signals from the periplasmic sensory domains to the DHp and CA domains or, in some cases, directly recognize cytoplasmic signals (52, 57).

Response regulators share a common, well-conserved receiver domain that catalyzes phosphotransfer from its cognate histidine kinase. Phosphorylation then promotes a conformational change on one face of the receiver domain, which in turn effects an output (24). In single-domain response regulators, the conformational change in the receiver domain allows the protein to directly produce an output response. Most response regulators, however, contain a DNA-binding output domain (21) (**Figure 1b**). For these regulators, phosphorylation induces homodimerization of the receiver domain, stimulating DNA binding and leading to transcriptional changes. Other common output domains include diguanylate cyclases and methyltransferases.

A common variant of the two-component paradigm is the so-called phosphorelay (8) (**Figure 1a**). These extended pathways typically initiate with a hybrid kinase, which is a histidine kinase with a receiver domain fused to its C terminus. After autophosphorylation and an intramolecular phosphotransfer to the receiver domain, the phosphoryl group is shuttled to a histidine phosphotransferase, and from there to a terminal response regulator that effects an output. Nearly 25% of all histidine kinases are hybrids (15), suggesting that phosphorelays are common.

EVOLUTION OF GENOME CONTENT AND GENE NUMBER

Two-component signaling proteins are among the most prevalent bacterial genes, and histidine kinases and response regulators constitute the two largest paralogous gene families in bacteria (20). Both kinases and regulators are easily identified by sequence homology, in contrast to many eukaryotic signaling systems in which protein kinases are easily identified but their substrates are not. Many histidine kinases are encoded in the same operon as their cognate regulators, allowing for cognate pairs to be identified through sequence analysis. Census taking is thus straightforward and easily applied to fully sequenced bacterial genomes (**Figure 2a**). Such analyses have revealed that the total number of two-component genes per genome typically grows as a square of the genome size (20) (**Figure 2b**). In addition, the number of two-component genes appears to correlate strongly with ecological and environmental niches (1, 20, 23, 38). Bacteria that live

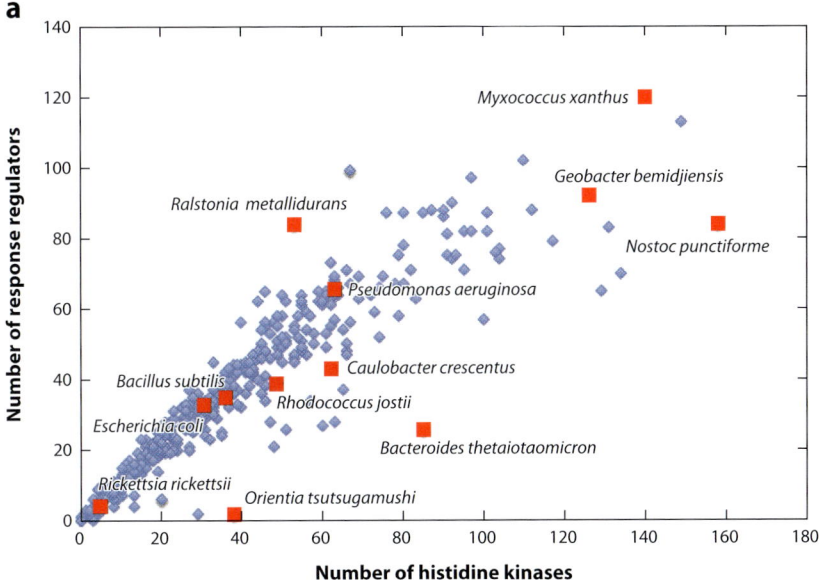

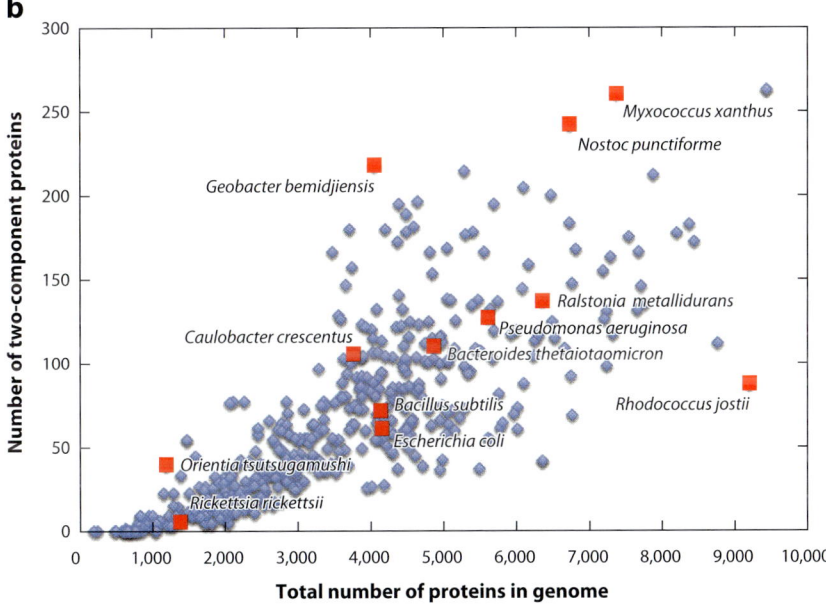

Figure 2

Diversity of two-component signaling gene content in bacterial genomes. (*a*) Plot showing the number of histidine kinases and response regulators in a range of organisms. Generally, most genomes contain equal numbers of kinases and regulators, as pathways typically comprise one kinase and one cognate regulator. When the ratio is not 1:1, there are usually more kinases than regulators, suggesting that response regulators may sometimes integrate signals from multiple kinases. (*b*) Plot showing the number of two-component proteins as a function of genome size for the same organisms as in panel *a*. Each plot is based on 504 bacterial genomes (22). A handful of well-studied and notable species are marked with red squares.

primarily in constant environments typically encode relatively few two-component signaling genes, even taking into account their smaller genome sizes and characteristic reductive genome evolution. In the extreme, many obligate intracellular parasites and endosymbionts harbor only a few pathways or sometimes none at all, as with *Mycoplasma* and *Amoebophilus*. By contrast, bacteria that inhabit rapidly changing or diverse environments typically encode large numbers of these signaling proteins. Extreme cases include *Myxococcus xanthus*, with 136 histidine kinases and 127 response regulators and *Nostoc punctiforme*, with 160 kinases and 98 regulators (81) (**Figure 2**). In some species, nearly 3% of the genome encodes for histidine kinases alone (20). These patterns of gene content strongly suggest that organisms expand their set of two-component signaling genes to adapt to fluctuations in their environment.

Although most abundant in the genomes of gram-negative bacteria and cyanobacteria, two-component signaling genes are found in all three domains of life (38, 71). However, they are considerably less abundant in archaea and eukaryotes. The majority of systems found in eukaryotes involve hybrid kinases and phosphorelays; whether there is selective pressure against canonical systems is unknown. Many of the archaeal and eukaryotic systems likely originated through multiple, independent lateral transfers from bacteria (37, 38); plants likely gained two-component pathways through the integration of chloroplast genes into the nuclear genome (48). In plants, the two-component genes obtained through lateral transfer likely expanded through duplication and diversification and now play integral roles in diverse developmental pathways (69).

Whereas two-component genes are found in yeasts, filamentous fungi, slime molds, and plants, they are conspicuously absent from higher eukaryotes and metazoans. The absence of two-component signaling proteins from humans, combined with their well-documented role in bacterial pathogenesis (28, 49), has made these proteins attractive new targets for antibiotic development (30). Evolutionarily, their absence begs the question of why they were supplanted as the primary means of signaling by pathways employing serine, threonine, and tyrosine phosphorylation. Although a definitive answer is lacking, we speculate that the intrinsic lability of phosphoryl groups on aspartates may have contributed. In eukaryotes, a need for longer, more stable outputs may have been desirable, and perhaps necessary, for transmitting signals from the cell membrane to the nucleus without signal loss en route in the form of phosphoryl group hydrolysis. Consistent with this idea, many of the two-component pathways in eukaryotes do not regulate transcription and instead target other cytoplasmic proteins. For example, in *Saccharomyces cerevisiae*, the Sln1-Ypd1-Ssk1 phosphorelay modulates the activity of a MAP kinase pathway that is also located in the cell membrane (61). Nevertheless, some eukaryotic response regulators directly affect transcription, particularly in plants. However, in these cases a histidine phosphotransferase typically shuttles phosphoryl groups from a cytoplasmic hybrid histidine kinase to a response regulator in the nucleus that is constitutively associated with the DNA (31, 34). Signal transmission may be successful in these cases because a histidyl-phosphate moiety is considerably more stable than an aspartyl-phosphate moiety.

Where did two-component signaling pathways, in any organism, evolve from in the first place? Given their ancient origin, an unequivocal answer to this question may not be attainable. However, one clue is that histidine kinases share distant homology in their ATP-binding domains with Hsp90, the mismatch repair protein MutL, and type II topoisomerases (17, 18). These proteins, members of the so-called GHKL superfamily, are thought to bind ATP in similar ways and share significant structural similarities; in some cases this domain is used to drive ATP hydrolysis, and in the case of histidine kinases the γ-phosphoryl group is transferred to a histidine in the DHp domain. It is thus plausible that histidine kinases emerged from one of these ATPases. In contrast to histidine kinases, there are no such weak homologies for response regulators and their origin remains a mystery.

There are likely two sources of histidine phosphotransferases. Some, particularly those that are monomeric (80, 92), may have evolved de novo from a range of other proteins, as there are few structural and sequence requirements to function as a phosphotransferase beyond a phosphorylatable histidine within an α-helical bundle. Others are dimeric and may have evolved through the degeneration of histidine kinases. For example, *Bacillus subtilis* Spo0B has two domains with significant similarity to those in histidine kinases (83, 95). The domain that contains the crucial histidine is structurally similar to the DHp domain of histidine kinases, and the other is topologically and structurally similar to a CA domain but lacks key residues usually involved in ATP binding. A similar scenario of recruitment and degeneration of a histidine kinase may hold for the phosphotransferase ChpT in *Caulobacter crescentus* (7). In general, however, evolutionary analysis of histidine phosphotransferases has been limited by the difficulty of identifying these proteins from sequence alone, in contrast to histidine kinases and response regulators.

Lineage-specific expansion: the growth of a paralogous gene family through duplication and subsequent vertical inheritance of the duplicated genes

Vertical inheritance: a process in which genetic material is transmitted from an organism to its offspring or progeny

MECHANISMS FOR EVOLVING CHANGES IN TWO-COMPONENT SIGNALING GENE CONTENT

Given the prevalence of two-component signaling pathways in bacterial genomes, it is natural to ask how new proteins and pathways arise. The possibilities fall into two broad categories: gene duplication and divergence, sometimes also referred to as lineage-specific expansion, and lateral gene transfer. To assess the contributions made by these two mechanisms, one study systematically examined the origins of histidine kinases from 207 genomes, using BLAST to identify the closest homologs of each kinase (1). For those most closely related to a kinase within the same genome, gene duplication, or lineage-specific expansion, was inferred as the source. If the closest homolog was from a closely related species, and if a gene tree built from all homologs matched a species tree, the kinase was classified as ancient and vertically transmitted. If, however, the closest homolog for a given kinase was from a distantly related species, lateral gene transfer was invoked. This interpretation assumes that multiple gene losses are less parsimonious and hence less likely to have occurred. However, gene loss occurs at very high rates in bacteria. In addition, inferences of lateral gene transfer can be confounded by the inaccuracy of sequence-based distances and heterotachy, the notion that substitution rates in different lineages often vary significantly (41).

Nevertheless, lateral transfer of two-component pathways undoubtedly has occurred and these systematic studies provide a general sense of the frequency, both across all species and within individual genomes (1). Overall, lineage-specific expansion, or gene duplication, appears to explain the origin of the vast majority of kinases. However, the relative balance of duplication and lateral transfer varies substantially from species to species. For example, in *Streptomyces coelicolor*, essentially all of its 140 histidine kinases appear to be ancient or derived from lineage-specific expansions. By contrast, in *Pseudomonas syringae* and *Ralstonia solanacearum*, many of the recently derived kinases probably came from lateral transfer events.

The lateral transfer of genes in bacteria can occur several ways, including through phage and plasmids, by direct conjugation, or by competence and the direct uptake of extracellular DNA. There are examples of two-component signaling genes encoded on plasmids, such as the VanR-VanS system found in enterococci that senses and responds to vancomycin (2, 90). In *R. solanacearum*, many of the laterally derived histidine kinases are encoded on a megaplasmid that may have moved laterally (70). There are also cases of two-component signaling proteins encoded on pathogenicity islands, such as the SpiR-SsrB system in *Salmonella*, which frequently move through conjugation (16). However, for many chromosomally encoded two-component genes derived by lateral transfer, the mechanism of transfer remains difficult to infer.

Both gene duplication and lateral gene transfer events are likely to have occurred more frequently than suggested by phylogenetic analyses. However, in most cases the newly introduced genes were likely eliminated from the genome and thus are no longer present in extant species. Bacteria typically have high rates of gene loss through mutation and deletion. Indeed, histidine kinases and response regulators are among the most common pseudogenes present in bacterial genomes (45); these pseudogenes likely arose through relatively recent duplications or lateral transfers, and were then inactivated, but have not yet been removed from the genome. To be fixed in a population, duplicated or laterally transferred genes must provide a substantial selective advantage within a relatively short period, as gene loss and pseudogeneization occur rapidly in bacteria (33, 40).

The function of a particular two-component system can also influence its evolutionary history. For example, a recent analysis of six species of *Xanthomonas* compared the complement of signaling genes present in each genome and found extensive gene loss (63). Notably, those pathways involved in *Xanthomonas* pathogenesis were never lost or duplicated, whereas other, presumably less critical pathways experienced more flux. Similarly, in *C. crescentus*, where two-component signaling proteins play important roles in cell cycle progression and development, pathways essential for viability are highly conserved in other *Alphaproteobacteria*, whereas those that are nonessential in *C. crescentus* are less well conserved (74). In most species there is probably a core set of two-component proteins that is maintained and relatively fixed and an additional set that can be lost, or modified, more easily.

This notion of fixed core signaling genes and malleable auxiliary factors has been well characterized in the context of bacterial chemotaxis, which centers on a two-component pathway, CheA-CheY. In *Escherichia coli*, in which chemotaxis has been best studied, signal recognition requires a methyl-accepting chemoreceptor protein (MCP) and an adaptor protein, CheW. Virtually all chemotactic bacteria encode orthologs of these core components: MCP, CheW, CheA, and CheY (91). In contrast, many of the auxiliary components, including the methyltransferase CheR and the methylesterase CheB, that influence signal adaptation are not universally conserved and are often missing or replaced by other types of regulators (91).

GENE FUSIONS, REARRANGEMENTS, AND DUPLICATIONS

Many two-component genes are encoded in operons as cognate kinase-regulator pairs, allowing for the duplication or lateral transfer of an intact signaling pathway. It is rare to see operon shuffling and the mixing and matching of genes encoded in operons. Hence, for a given kinase-regulator pair, the orthologs are also usually found together in an operon and in the same relative order (38, 87). Fusions of kinases and regulators to create hybrid kinases also seem to be rare, but there are some examples. For instance, analysis of six species of *Xanthomonas* found that the individual domains of a hybrid histidine kinase in one species were most similar to, and likely derived from, an operonic kinase-regulator pair encoded as separate open reading frames in a closely related species (63). Such fusions probably occur through the mutation of stop codons in operons where the histidine kinase is encoded upstream of the response regulator, although hybrid kinases may also form through the fusion of previously separated genes (63, 87, 97). As might be expected, fusion events that create hybrid kinases are rare for response regulators that contain DNA-binding output domains (15, 97). There are, however, examples of such hybrid kinases (75), but the mechanism by which these systems regulate transcription remains unclear.

Although *E. coli* encodes 55 of its 62 two-component genes in operons, many organisms encode a substantial fraction of their two-component genes as orphans. Frequently, only one gene from an operon is duplicated (or both are duplicated and one is lost), resulting in the production of orphan two-component signaling genes. An orphan kinase, however, may retain the ability to phosphorylate the regulator in the operon from which it was derived. Such duplication events, coupled with a change in kinase input domain, may be a primary mechanism for generating cross-regulated systems in which multiple, independent signals can trigger the same response. A classic example is in *B. subtilis*, in which each of the five orphan kinases KinA/B/C/D/E, which probably evolved through duplication, can phosphorylate Spo0F and initiate the sporulation phosphorelay (76). Similarly, duplication of only the response regulator from a given kinase-regulator pair can lead to a scenario in which a single sensor kinase can drive multiple outputs. For example, in cyanobacteria NblS-RpaB forms an essential two-component system. During divergence of the cyanobacteria in the clade including *Synechococcus* species, a duplication of RpaB produced a second response regulator, SrrA. This regulator retained the ability to be phosphorylated by NblS, but appears to affect transcription in a manner different from the way RpaB does (46).

Cross-regulation: communication between distinct signaling pathways that provides a physiological benefit to the organism

EVOLUTION OF SIGNALING PROTEIN STRUCTURE AND FUNCTION

Gene duplication and lateral gene transfer ultimately provide the raw material for generating new two-component signaling pathways. But what happens immediately after new signaling genes are introduced? Owing to large population sizes and selective pressure to minimize genome size (50), new signaling proteins presumably must quickly gain new functions to be retained. There are undoubtedly many mutations that must occur to produce a pathway that can respond to a new input or effect a new output. These mutations presumably include single amino acid substitutions, although rapid changes in function may rely heavily on larger-scale rearrangements such as domain shuffling. Below we summarize our current understanding of how cells generate new signaling functions from duplicated genes, focusing on (*a*) changes in kinase sensory domains and pathway inputs, (*b*) changes in response regulators and pathway outputs, and (*c*) changes required to insulate new pathways from existing pathways.

Histidine Kinase Sensory Domain Evolution

After the duplication of a histidine kinase, whether alone or with a cognate response regulator, the duplicate kinases must differentiate themselves and find new roles within the signaling network of a cell. One mechanism to accomplish this is through changes in the sensory domains of one or both kinases (14, 39). For most orthologous kinases, the sensory domains are less well-conserved than their catalytic domains. The ability to sense a new signal often arises via domain shuffling, which may occur coincident with, or shortly after, a duplication. Over 70% of recently duplicated histidine kinases show an input domain structure different from that of their closest paralog (1). Domain shuffling can occur between histidine kinases and other proteins. Sequence analyses indicate that the sensory domains of some histidine kinases are closely related to domains found on serine/threonine kinases (100), chemotaxis proteins, and diguanylate cyclases (98).

The domain shuffling observed in histidine kinases suggests that these proteins are intrinsically modular and, consequently, that the rational design of new kinases may be possible. Indeed, several groups have successfully fused the conserved phosphotransfer and catalytic domains from a histidine kinase to the sensory domain of another kinase, or even the sensory domain of completely unrelated proteins. The first such example, named Taz, is a chimeric protein that fused the sensory domain of the aspartate chemoreceptor Tar with the DHp and CA domains of the model histidine kinase EnvZ, producing an aspartate-responsive kinase (82). In addition to demonstrating the fundamental modularity of histidine kinases, Taz has been used to dissect the functions and activities of EnvZ in vivo (19, 36, 99). Other functional chemoreceptor-EnvZ constructs have also been made (4, 68).

How does domain shuffling, either during evolution or during rational construction of chimeric proteins, produce successful signal-responsive proteins? Is there a particular way in which sensory domains must be fused to the catalytic domains to function? These questions were recently examined in the context of a chimeric protein that fused a light-sensing PAS domain, taken from the *B. subtilis* protein YtvA (which is not a kinase), with the DHp and CA domains of the histidine kinase FixL from *Bradyrhizobium japonicum*. Successful fusions of the PAS domain to FixL led to light-responsive changes in FixL signaling and FixL-FixJ-dependent gene expression (51). Intriguingly, successful fusions had linkers, which form coiled coils, separating the PAS and DHp domains that differed in length by exactly seven amino acids. Inspection of other histidine kinases containing PAS domains further revealed that the linkers are of variable lengths but often differ by multiples of seven. Together, these results suggest that maintaining the heptad periodicity of the coiled-coil linker may be critical to the construction of functional chimeras, either during evolution or for rational engineering purposes. Further work demonstrated that, by following similar rules, multiple PAS domains could be engineered into the same kinase, allowing it to integrate multiple signals (53). Naturally occurring histidine kinases also often have multiple input domains, suggesting that partial gene duplications, in which only a single input domain is duplicated, may be a common mechanism for generating input diversity. In sum, these efforts to engineer novel proteins are not only producing valuable tools but also providing important new insights into how domain shuffling occurs and how it contributes to the origin of new two-component signaling pathways in nature.

An additional mechanism for acquiring new input signals is through accumulated substitutions in a sensory domain rather than its complete replacement. A prime example comes from the NarX and NarQ sensor kinases in *E. coli*. A gene duplication event led to the emergence of these two related kinases, although which is more ancestral is unclear. Nevertheless, studies of signal recognition have demonstrated that NarQ responds to both nitrate and nitrite, whereas NarX responds preferentially to nitrate (64). Although the periplasmic domains of NarQ and NarX are significantly diverged, they do share substantial similarity, particularly in a region critical to ligand binding (13). Notably, a single point mutation in this region of NarX that substitutes a lysine with an isoleucine, as found at the equivalent position in NarQ, reduced the ability of NarX to discriminate between nitrate and nitrite, rendering a more NarQ-like response pattern (88). This study highlights how the accumulation of single point mutations is a plausible means of rapidly generating new and different inputs to two-component signaling pathways.

Divergence and Evolution of Pathway Outputs

Within a two-component signaling pathway, the response regulator is the ultimate arbiter of physiological change. How does the output of a response regulator evolve, and how are new output responses generated by regulators after they emerge through duplication or following lateral gene transfer? As the majority of response regulators direct changes in gene expression, the evolution of pathway outputs can be easily studied by following changes in target genes.

One of the best-studied examples is the PhoQ-PhoP system found in the *Enterobacteriaceae*. In response to low extracellular concentrations of Mg^{2+}, the histidine kinase PhoQ drives phosphorylation of PhoP, which then regulates gene expression. The direct regulon of PhoP has been mapped in both *Salmonella enterica* serovar Typhimurium and *Yersinia pestis* (60), which probably shared a common ancestor ~200 mya. Strikingly, only three genes were directly regulated by PhoP in both species: the autoregulated *phoQ* and *phoP* genes and *slyB*, which encodes a lipoprotein

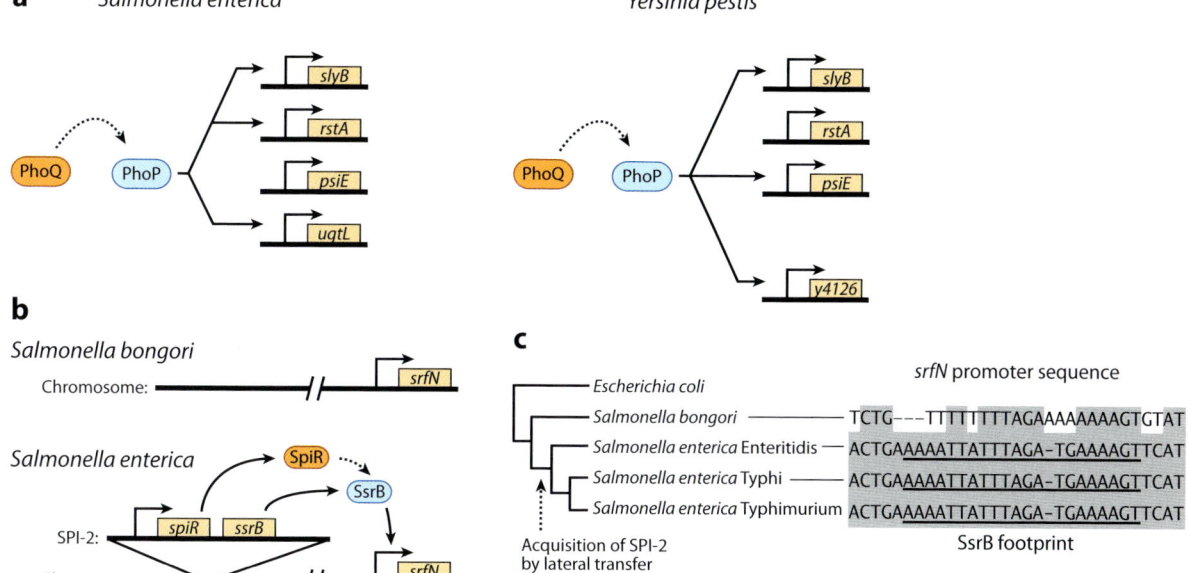

Figure 3

Evolution of transcriptional circuits controlled by two-component pathways. (*a*) Examples of genes directly regulated by the two-component pathway PhoQ-PhoP in *Salmonella enterica* and *Yersinia pestis*. *slyB* is conserved and directly regulated by PhoP in both species. *rstA* and *psiE* are conserved but directly regulated by PhoP in only one of the two species. *ugtL* and *y4126* are directly regulated and are unique to *S. enterica* and *Y. pestis*, respectively. (*b*) Schematic of *Salmonella bongori* and *S. enterica* chromosomes, each harboring a *srfN* ortholog. The horizontally acquired SpiR-SsrB system, encoded on *Salmonella* pathogenicity island 2 (SPI-2) in *S. enterica* but not *S. bongori*, evolved to transcriptionally activate *srfN*. (*c*) De novo evolution of a response regulator–binding site. SPI-2 encodes the two-component pathway SpiR-SsrB, which was acquired after the divergence of *S. enterica* from *S. bongori*. The gene *srfN*, ancestral to the *Salmonella* lineage, accumulated promoter mutations that enabled activation by SsrB, a transcriptional link that contributes to *Salmonella* virulence. The relevant portion of the *srfN* promoter is shown with conserved positions shaded gray and the region bound by SsrB in *S. enterica* underlined.

thought to be a critical regulator of PhoQ activity (**Figure 3a**). There were also some genes, such as *pbgP* and *ugd*, that were directly regulated in one species but indirectly regulated in the other; the overall regulatory logic for these genes was thus conserved, but the precise mechanism has changed. Despite these examples, the vast majority of genes directly regulated by PhoP in each organism were not conserved. Instead, transcriptional rewiring appears to have been prevalent since the divergence of *Salmonella* and *Yersinia*, leading to the gain and loss of PhoP-regulated genes in each species (**Figure 3a**). It is tempting to speculate that these changes have tailored the response of each species to magnesium limitation.

Notably, the change in PhoP regulons between *Salmonella* and *Yersinia* may not always result from a simple gain or loss of PhoP-binding sites. In some cases, regulon differences may reflect (*a*) changes in the orientation of, and distance between, a PhoP-binding site and the transcriptional start site and (*b*) concomitant changes in how PhoP recruits RNA polymerase. For instance, the PhoP-binding site in the promoter of *mgtC* in *Yersinia* is located in a position and orientation that enables gene activation by *Yersinia* PhoP, but not by *Salmonella* PhoP, even though *Salmonella* PhoP can bind the *Yersinia mgtC* promoter (59). The ability to change the targets of a response regulator without necessarily changing DNA-binding sites is also seen in *Desulfovibrio*, in which two recently duplicated response regulators share DNA-binding motifs but regulate nonoverlapping target genes (66). Point mutations in OmpR have also been identified that allow it to activate the *kdpABC* operon, usually activated by KdpE, not by changing DNA binding but by changing the ability to interact with RNA polymerase while bound to the promoter (55). Thus with a single point mutation, and without any changes needed in the promoters of target genes, a duplicated response regulator can regulate a new set of target genes. Collectively, these studies demonstrate that two-component pathway outputs can evolve through changes in the DNA-binding sites of response regulators or through changes in how response regulators interact with RNA polymerase. They also highlight the critical need to couple computational analyses of binding sites with experimental studies to reveal the functional and evolutionary consequences of binding site conservation or loss.

Changes in response regulator outputs may also frequently occur after duplication or lateral gene transfer events. For gene duplication, changes in the output response of one or both regulators is likely a critical step in the establishment of new functions and, consequently, the maintenance of the duplicated proteins. For instance, in *E. coli*, a duplication event likely gave rise to the paralogous systems NarX-NarL and NarQ-NarP, which respond to nitrate and nitrite in anaerobic conditions (64). While the regulators NarP and NarL share significant similarity and even recognize highly similar consensus binding sites, divergent evolution has enabled each regulator to recognize different promoter architectures and activate different genes (62). The duplication of the Nar two-component system has thus led to an increase in complexity of the transcriptional control of genes necessary for growth in anaerobic conditions.

The evolution of response regulator outputs following lateral gene transfer has also been explored recently. A particularly illuminating example comes from studies of *Salmonella* pathogenicity island 2 (SPI-2), which encodes a two-component signaling system called SpiR-SsrB (**Figure 3b,c**). In addition to regulating the expression of other SPI-2-encoded genes, the response regulator SsrB directly regulates the expression of genes outside SPI-2 (89), indicating that SsrB-binding sites probably evolved de novo within the promoters of these genes. This hypothesis was tested by examining the evolution of a *Salmonella* gene, *srfN* (56). This gene is ancestral to the *Salmonella* lineage and present in both *S. enterica* and *S. bongori*. By contrast, SPI-2 and SsrB are found in *S. enterica* but not *S. bongori* (**Figure 3b**). A comparison of the *cis*-regulatory regions of *srfN* indicated that the binding site for SsrB was not present in *S. bongori*, meaning it likely arose in the lineage leading to *S. enterica* (**Figure 3c**). Importantly, this recruitment of an ancestral gene into the regulon of a horizontally acquired response regulator provided *S. enterica* with an adaptive advantage as a pathogen. When the promoter of *S. enterica srfN* was replaced with that found in *S. bongori*, cells were rendered significantly less virulent compared to the wild type.

Conversely, the genes encoded on SPI-2 have evolved to be regulated by ancestral two-component pathways. A case in point is the expression of *ssrB* and *spiR*, which are themselves regulated by OmpR and PhoP, two response regulators found throughout the *Gammaproteobacteria* (6, 44). By controlling *spiR* and *ssrB*, these ancestral regulators likely help ensure that virulence genes are maximally expressed when *Salmonella* enters host cells. For instance, the PhoQ-PhoP system is activated by the low-magnesium conditions that *Salmonella* experiences inside host macrophages; the consequent activation of *ssrB* and *spiR* would then drive the expression of virulence genes.

Although we have discussed only a few cases, it is clear that response regulator outputs can, and do, change rapidly. The observed changes to transcriptional circuitry suggest that bacteria are resilient to, and are capable of, transcriptional rewiring (58). This notion was tested systematically by artificially rewiring transcriptional connections; promoters for 26 different sigma and transcription factors (including some response regulators) were combined with the open-reading frames of 23 of these transcriptional regulators and introduced into *E. coli* cells on a high-copy plasmid (35). Strikingly, over 95% of these constructs were tolerated, with little to no growth defect under standard laboratory conditions. One implication of this study is that after a new DNA-binding response regulator is introduced by gene duplication or lateral transfer, there is time to scan different regulatory possibilities. A new combination that yields even a slight benefit could then be selected and rapidly fixed in a population. Finally, the evolvability of response regulators and their outputs may also benefit from the fact that most prokaryotic transcription factors regulate only a few genes, either directly or indirectly (47), decreasing the number of binding sites that would need to coevolve with the DNA-binding domain of a response regulator, thereby increasing the likelihood that they can change (67).

Evolution of Phosphotransfer Specificity and the Insulation of Pathways

The flow of information through two-component signaling pathways depends critically on the transfer of phosphoryl groups from a histidine kinase to its cognate response regulator. Despite early suggestions of rampant cross-talk, there is little evidence for such promiscuity in vivo, with most kinases having one response regulator substrate or, on occasion, two or three (43, 74, 93). This in vivo preference is mirrored in vitro, with histidine kinases harboring a strong kinetic preference for phosphotransfer to their in vivo partner. For example, a systematic, global study of phosphotransfer from *E. coli* EnvZ to each of the 32 response regulators in *E. coli* demonstrated that OmpR was the preferred substrate. EnvZ transferred to other substrates only after extended incubation times (74). These in vitro studies demonstrate that the specificity of two-component signaling pathways is based primarily on molecular recognition rather than reliance on scaffolds or other cellular strategies. This observation further suggests that the information necessary for promoting the correct, or desired, interaction and preventing incorrect interactions is encoded at the sequence level (10, 73).

A consequence of relying on molecular recognition for specificity is that, during the course of evolution, any mutation in a residue contributing to a kinase-regulator interaction may disrupt signaling and place cells at a strong fitness disadvantage. Survival would then depend on reversion of the mutation or a compensatory mutation in the partner protein. Consistently, computational analyses of large sets of cognate kinase-regulator pairs have revealed extensive amino acid coevolution (9, 10, 73, 85). Conspicuously, the most significant coevolving pairs of residues map to the molecular interface formed during phosphotransfer (11), suggesting they mediate the specificity of this protein-protein interaction (**Figure 4a,b**). Using *E. coli* EnvZ as a model kinase, Skerker et al. (73) showed that a subset of these residues was sufficient, when mutated, to reprogram substrate specificity both in vitro and in vivo. For example, mutating as few as three residues in EnvZ to match those found at equivalent positions in RstB led EnvZ to preferentially phosphorylate RstA instead of OmpR (**Figure 4a,b**, also see **Figure 5b**). Similarly, response regulators, including OmpR from *E. coli* and CheY from *Rhodobacter sphaeroides*, have been rationally rewired to interact with noncognate kinases by mutating the coevolving, specificity-determining residues (5, 10). Directed evolution has also been used to rewire two-component specificity. For example, mutants of the *E. coli* kinase CpxA that phosphorylate and dephosphorylate OmpR were selected; many of the mutated residues were the same as those identified in the studies of kinase-regulator coevolution (72).

Although specificity-determining residues do coevolve, these correlated changes appear to be rare events as the specificity residues of many kinase-regulator systems are nearly invariant over relatively long timescales. So when and why do specificity residues change and coevolve? One strong possibility is that substitutions occur following gene duplication, helping insulate the duplicate kinase-regulator pairs from each other (**Figure 5a**). That is, a series of mutations presumably must occur to prevent cross-talk between two duplicated pathways while maintaining the interaction within each pair. Such an accumulation of changes in specificity residues, however,

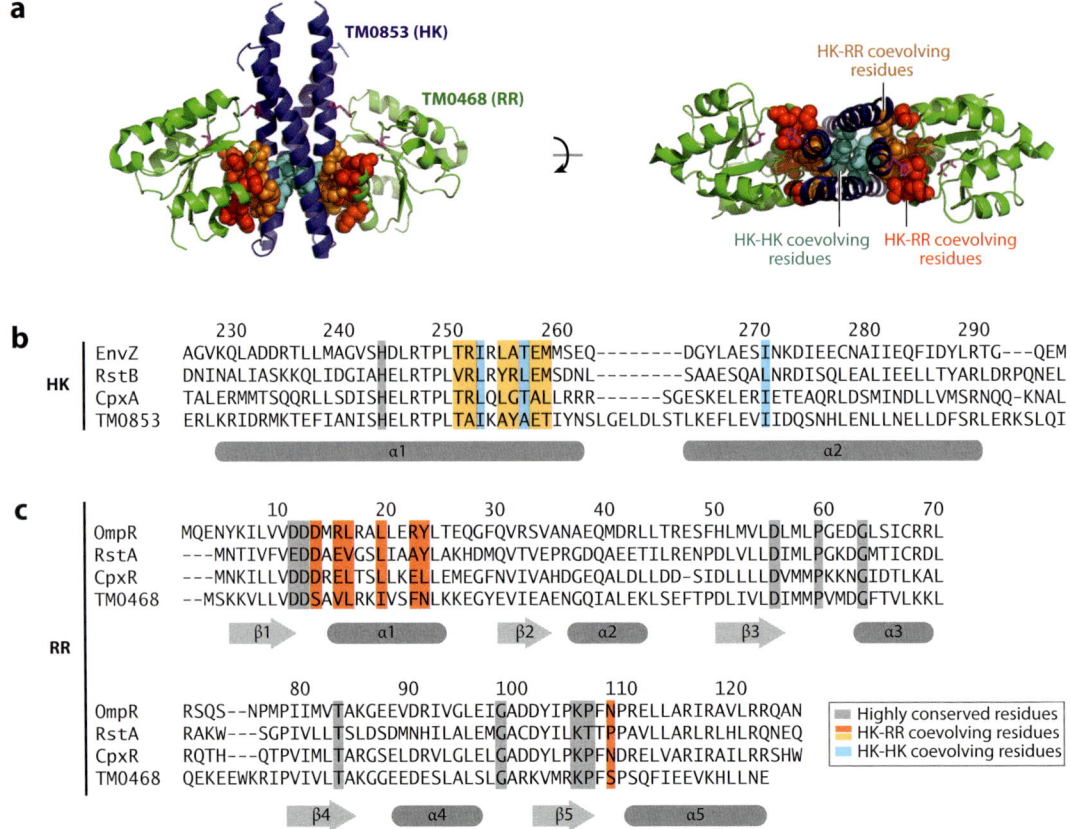

Figure 4

Amino acid coevolution in two-component signaling proteins. (*a*) Residues that coevolve in cognate pairs of histidine kinases (HKs) and response regulators (RRs) are shown with space-filling on the crystal structure of the *Thermotoga maritima* kinase TM0853 bound to its cognate regulator TM0468. The histidine and the aspartate that are involved in phosphotransfer are shown as sticks. Coevolving residues on the histidine kinase and response regulator are shown with space-filling and colored red and orange, respectively. Residues in histidine kinases that coevolve strongly with other kinase residues are also shown with space-filling and colored cyan. (*b,c*) Coevolving residues from panel *a* are shown on (*b*) a sequence alignment of TM0853 with three *Escherichia coli* kinases, EnvZ, RstB, and CpxA, and (*c*) an alignment of TM0468 with three *E. coli* regulators, OmpR, RstA, and CpxR. Secondary structure elements are indicated beneath the primary sequence.

is inherently risky business for a bacterium. Due to large population sizes, even slightly deleterious mutations are likely to be quickly removed from the population. Hence, for a new kinase-regulator pair to be maintained in the genome, the mutational intermediates between its initial state and its final, insulated state must be neutral or nearly neutral. In other words, cognate kinase-regulator pairs must retain their ability to interact as the specificity residues coevolve and find a region of sequence space in which they are insulated from other two-component proteins within the cell. Similarly, after a lateral gene transfer event involving two-component signaling genes, the newly introduced kinase-regulator pair may need to accumulate substitutions in phosphotransfer specificity residues to avoid cross-talk with existing systems, thereby maintaining the fidelity of information flow within the cell.

What mutational trajectories do two-component proteins follow during evolution? In particular, how do duplicated proteins move through the sequence space defined by the specificity-determining residues of histidine kinases and response regulators? Answering these questions through sequence analysis alone is problematic because transient intermediates may not be captured in extant sequences and the behavior of ancestral or intermediate states is difficult to infer from sequence alone. To circumvent these issues, one recent study experimentally examined all possible specificity intermediates between two *E. coli* histidine kinases, EnvZ and RstB (10). These kinases likely arose through duplication and divergence, and as noted above, the specificity of each can be converted to that of the other by just three mutations (10, 73) (**Figure 5b**). Similarly, their cognate response regulators, OmpR and RstA, can be rewired, with respect to partner specificity, through a small number of mutations. It was thus feasible to build all possible mutational intermediates and characterize their phosphotransfer specificity in vitro.

The results of this systematic study demonstrated that a cognate kinase-regulator pair can, in fact, move in sequence space from the region occupied by EnvZ-OmpR to that occupied by RstB-RstA while (*a*) introducing only one mutation at a time, (*b*) maintaining the interaction between the kinase and the regulator, and (*c*) avoiding the introduction of cross-talk to other closely related pathways such as CpxA-CpxR (10) (**Figure 5b**). Notably though, only a small fraction of all possible mutational trajectories satisfies these criteria, indicating that the evolution of signaling proteins postduplication may be fundamentally constrained. These studies further suggest that kinase-regulator pairs occupying relatively isolated regions of specificity sequence space may be easier to duplicate and retain. In addition, for cases in which a severely limited number of trajectories are accessible, there may be cases of convergent evolution in which independent duplication events have led to similar mutational trajectories.

The notion of insulation, or orthogonality, in sequence space can be extended from individual, recently duplicated pairs of signaling proteins to the entire complement of two-component signaling proteins in a given organism. For example, all 29 histidine kinases in *E. coli* ultimately arose through some combination of gene duplication and lateral gene transfer. The net result is a system of signaling pathways that are, with a few exceptions, insulated from one another in sequence space and with respect to phosphotransfer, as observed by global phosphotransfer profiling (**Figure 5c**). This system-wide insulation suggests that negative selection and the avoidance of cross-talk are powerful forces influencing the evolution of two-component signaling proteins. Negative selection has been suggested to influence other paralogous signaling protein families, such as SH3-domain-containing proteins found in eukaryotes (96). Two notable exceptions to the orthogonality of phosphotransfer specificity in *E. coli* are the kinases NarQ and NarX, which share significant similarity in terms of phosphotransfer specificity residues and, consistently, both phosphorylate the response regulators NarL and NarP, although with different kinetic preferences (54). Other exceptions to the orthogonality of specificity residues include hybrid histidine kinases, which phosphotransfer intramolecularly to an attached receiver domain. This spatial arrangement may enforce the specificity of phosphotransfer in hybrid kinases and, consequently, their specificity-determining residues may not be under the same pressure to avoid cross-talk with other two-component pathways. Consistent with this notion, hybrid kinases lacking their receiver domains can often phosphorylate noncognate response regulators as well as or better than their own receiver domains (7, 84).

Much remains to be understood about how kinase-regulator interactions evolve. Has the distribution of signaling proteins in specificity sequence space been optimized? How dense is sequence space and how does this affect mutational trajectories, both in the absence of duplication and postduplication? Is there a fundamental limit to the number of pathways an organism can have while avoiding cross-talk? Does gene loss lead to relaxed selection on the specificity residues of the remaining signaling proteins? Investigations into many of these questions will benefit from a systematic mutational analysis of kinase and regulator specificity residues. Most of what is known about kinase-regulator interactions comes from limited alanine-scanning or targeted mutagenesis (10, 79); the application of new, systematic mutagenesis techniques (32) promises to shed significant new light on the specificity and evolution of phosphotransfer in two-component signaling.

DIMERIZATION SPECIFICITY

The generation of new pathways by duplication and divergence also requires changes to the residues that mediate homodimerization of histidine kinases and response regulators. To establish new and insulated pathways, substitutions are needed that eliminate heterodimerization of the diverging paralogous proteins while maintaining homodimerization.

Most, if not all, histidine kinases form homodimers in order to autophosphorylate. There is almost no evidence of physiologically relevant heterodimerization, with one exception in *Pseudomonas aeruginosa* (29), indicating that histidine kinases harbor a set of amino acids that enforce homodimerization. Many of these residues are likely to reside in the DHp domain, although upstream domains, such as PAS and HAMP domains, could also contribute to dimerization specificity and stability. To better pinpoint the residues mediating specificity, one recent study looked for coevolving residues in a set of more than 15,000 histidine kinase sequences (3). As with the kinase-regulator interaction, this approach revealed a small set of strongly coevolving residues that mapped primarily to the DHp domain and mostly within the lower half of the four-helix bundle. These dimerization specificity residues map to the same general region as the phosphotransfer specificity residues, but the dimerization specificity residues are buried in the four-helix bundle and the kinase-regulator specificity residues are in solvent-exposed positions (**Figure 4**). As with kinase-regulator interactions, homodimerization specificity could be changed through directed mutagenesis of these homodimerization specificity residues (3).

Figure 5

Insulation of two-component pathways following gene duplication. (*a*) Schematic of major steps in the insulation of two pathways following a duplication event. The duplication of an ancestral pathway initially produces two identical pathways that cross-talk at the level of phosphotransfer. Through the accumulation of mutations in specificity-determining residues, the two pathways can become insulated. A similar process must occur, but is not shown, at the levels of kinase and regulator homodimerization. (*b*) Phosphotransfer specificity of EnvZ, RstB, and various RstB mutants. Each kinase was autophosphorylated and tested for transfer to each of three response regulators, RstA, OmpR, and CpxR. Data are from Reference 10. The wild-type RstB is shown at far left. The phosphotransfer specificity can be converted to that of EnvZ, shown at far right, by mutating three of its six specificity-determining residues to match those found in EnvZ to match those found in EnvZ (the other three sites are already identical between EnvZ and RstB; see **Figure 4***b*). This triple mutant of RstB as well as each single and double mutant intermediate is labeled on the basis of the identity of the three specificity residues, with blue text indicating identity with the wild-type RstB and red text indicating identity with the wild-type EnvZ. Notably, some intermediates do not phosphorylate any of the regulators, whereas some phosphorylate all three. (*c*) Schematic summarizing the distribution of histidine kinases in the sequence space defined by their specificity-determining residues. Each sphere represents the set of response regulators that a given kinase phosphorylates. With the exception of NarQ and NarX, these spheres are presented as nonoverlapping to reflect the minimal cross-talk between pathways. The relative positions of spheres are based on the ability of individual kinases to phosphorylate the cognate response regulators of other kinases after extended times in vitro (74, 93). Positions are approximate and the diagram is intended only to convey a general sense of how kinases are distributed in sequence space. Spheres are colored according to the subfamily of each kinase's response regulator. Spheres with dashed outlines indicate kinases for which no data exist to infer relative positions. Hybrid histidine kinases are excluded.

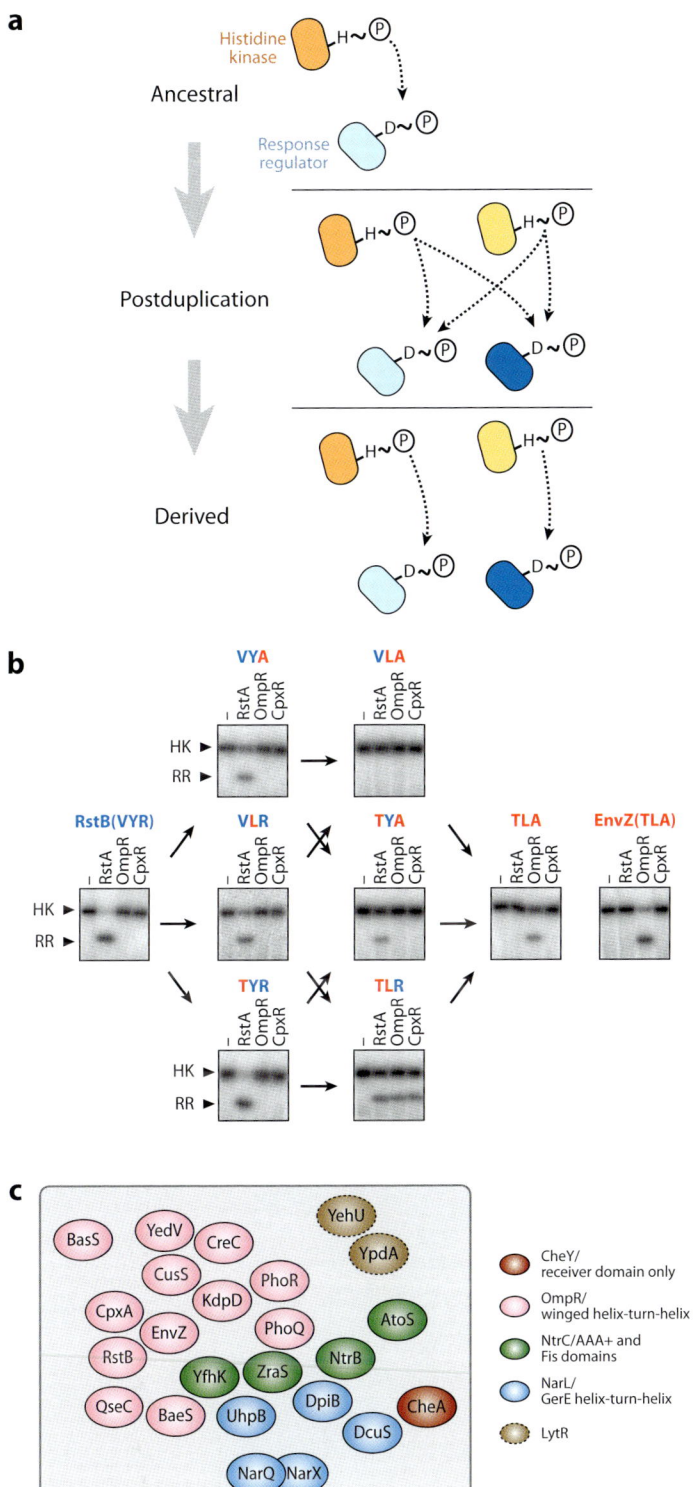

Nearly 50% of response regulators, including all members of the OmpR family (26), form homodimers upon phosphorylation. Homodimerization is often crucial for producing an output response, as many response regulators have DNA-binding domains and recognize tandem or inverted repeat elements within target promoters. A systematic study of the 17 OmpR family response regulators from *E. coli* demonstrated that essentially all of them specifically homodimerize (27). Although intermolecular interactions on the dimer interface involve highly conserved residues within the receiver domain, some interfacial residues do vary, perhaps providing a mechanism for ensuring homodimerization and excluding heterodimerization (78). As with kinase dimerization and kinase-regulator interaction, amino acid coevolution studies have identified a subset of interfacial residues that may help enforce homodimerization and prevent heterodimerization (85). These residues are likely to change following gene duplication as a means of insulating paralogous response regulators from one another, thereby enabling distinct outputs to result from the phosphorylation of each regulator.

FINAL PERSPECTIVE

Bacteria can survive and thrive in a bewildering array of environments and in the face of enormous competition. They have, consequently, evolved elegant mechanisms for sensing and responding to their environments, often involving two-component signal transduction pathways. The prevalence and diversity of these signaling proteins in the bacterial kingdom underscore the roles they have played in the adaptation of bacteria to a range of ecological niches. Studies of how two-component signaling pathways evolve are only in their infancy but promise to be an active area of exploration in the coming years. The diversity and depth of genome sequences, coupled with the ability to experimentally manipulate bacteria, should help further reveal, at a detailed molecular level, how these pathways are created, lost, rearranged, and integrated into complex regulatory circuits.

SUMMARY POINTS

1. Two-component signal transduction proteins are prevalent throughout the bacterial kingdom and are found in some archaea, plants, and lower eukaryotes; they were lost in metazoans.

2. Most bacteria encode dozens, and sometimes hundreds, of these signaling proteins. The number of proteins encoded in individual genomes typically scales with genome size and diversity of the environment in which organisms live.

3. Bacteria expand their repertoire of two-component signaling proteins through a combination of gene duplication and lateral gene transfer.

4. Domain shuffling is rampant among two-component signaling proteins, particularly after gene duplication events, enabling the rapid gain of new sensory and regulatory functions.

5. Studies of how domain shuffling occurs at the molecular level are enabling the rational design of new sensor kinases.

6. The transcriptional outputs of two-component signaling pathways show enormous plasticity, with gain and loss of *cis*-regulatory elements often driving rapid diversification of a response regulator's transcriptional program.

7. Cognate histidine kinases and response regulators coevolve to maintain their interaction and to avoid cross-talk with other pathways. Studies of amino acid coevolution in large sets of cognate kinase-regulator pairs have revealed the key specificity-determining residues.

8. Specificity residues may change to accommodate the emergence of new pathways that arise through duplication or lateral gene transfer; the reconstruction of ancestral and intermediate states of specificity residues in model kinase-regulator pairs is shedding new light on the mutational trajectories that occurred and the constraints that influenced them.

DISCLOSURE STATEMENT

The authors are not aware of any affiliations, memberships, funding, or financial holdings that might be perceived as affecting the objectivity of this review.

ACKNOWLEDGMENTS

Support was provided by the National Institutes of Health and the National Science Foundation. M.T.L. is an Early Career Scientist at the Howard Hughes Medical Institute.

LITERATURE CITED

1. Alm E, Huang K, Arkin A. 2006. The evolution of two-component systems in bacteria reveals different strategies for niche adaptation. *PLoS Comput. Biol.* 2:e143
2. Arthur M, Molinas C, Courvalin P. 1992. The VanS-VanR two-component regulatory system controls synthesis of depsipeptide peptidoglycan precursors in *Enterococcus faecium* BM4147. *J. Bacteriol.* 174:2582–91
3. Ashenberg O, Rozen-Gagnon K, Laub MT, Keating AE. 2011. Determinants of homodimerization specificity in histidine kinases. *J. Mol. Biol.* 413:222–35
4. Baumgartner JW, Kim C, Brissette RE, Inouye M, Park C, Hazelbauer GL. 1994. Transmembrane signalling by a hybrid protein: communication from the domain of chemoreceptor Trg that recognizes sugar-binding proteins to the kinase/phosphatase domain of osmosensor EnvZ. *J. Bacteriol.* 176:1157–63
5. Bell CH, Porter SL, Strawson A, Stuart DI, Armitage JP. 2010. Using structural information to change the phosphotransfer specificity of a two-component chemotaxis signalling complex. *PLoS Biol.* 8:e1000306
6. Bijlsma JJ, Groisman EA. 2005. The PhoP/PhoQ system controls the intramacrophage type three secretion system of *Salmonella enterica*. *Mol. Microbiol.* 57:85–96
7. Biondi EG, Reisinger SJ, Skerker JM, Arif M, Perchuk BS, et al. 2006. Regulation of the bacterial cell cycle by an integrated genetic circuit. *Nature* 444:899–904
8. Burbulys D, Trach KA, Hoch JA. 1991. Initiation of sporulation in *B. subtilis* is controlled by a multicomponent phosphorelay. *Cell* 64:545–52
9. Burger L, van Nimwegen E. 2008. Accurate prediction of protein-protein interactions from sequence alignments using a Bayesian method. *Mol. Syst. Biol.* 4:165
10. **Capra EJ, Perchuk BS, Lubin EA, Ashenberg O, Skerker JM, Laub MT. 2010. Systematic dissection and trajectory-scanning mutagenesis of the molecular interface that ensures specificity of two-component signaling pathways. *PLoS Genet.* 6:e1001220**
11. **Casino P, Rubio V, Marina A. 2009. Structural insight into partner specificity and phosphoryl transfer in two-component signal transduction. *Cell* 139:325–36**
12. Casino P, Rubio V, Marina A. 2010. The mechanism of signal transduction by two-component systems. *Curr. Opin. Struct. Biol.* 20:763–71

1. Computational inference of lateral transfer and duplication rates of histidine kinase genes across 207 genomes.

10. Experimental analysis of mutational trajectories separating extant two-component signaling pathways.

11. The first crystal structure of a histidine kinase in complex with its cognate response regulator.

13. Cheung J, Hendrickson WA. 2009. Structural analysis of ligand stimulation of the histidine kinase NarX. *Structure* 17:190–201
14. Cheung J, Hendrickson WA. 2010. Sensor domains of two-component regulatory systems. *Curr. Opin. Microbiol.* 13:116–23
15. Cock PJ, Whitworth DE. 2007. Evolution of prokaryotic two-component system signaling pathways: gene fusions and fissions. *Mol. Biol. Evol.* 24:2355–57
16. Deiwick J, Nikolaus T, Erdogan S, Hensel M. 1999. Environmental regulation of *Salmonella* pathogenicity island 2 gene expression. *Mol. Microbiol.* 31:1759–73
17. Dutta R, Inouye M. 2000. GHKL, an emergent ATPase/kinase superfamily. *Trends Biochem. Sci.* 25:24–28
18. Dutta R, Qin L, Inouye M. 1999. Histidine kinases: diversity of domain organization. *Mol. Microbiol.* 34:633–40
19. Dutta R, Yoshida T, Inouye M. 2000. The critical role of the conserved Thr247 residue in the functioning of the osmosensor EnvZ, a histidine kinase/phosphatase, in *Escherichia coli*. *J. Biol. Chem.* 275:38645–53
20. Galperin MY. 2005. A census of membrane-bound and intracellular signal transduction proteins in bacteria: bacterial IQ, extroverts and introverts. *BMC Microbiol.* 5:35
21. Galperin MY. 2006. Structural classification of bacterial response regulators: diversity of output domains and domain combinations. *J. Bacteriol.* 188:4169–82
22. Galperin MY, Higdon R, Kolker E. 2010. Interplay of heritage and habitat in the distribution of bacterial signal transduction systems. *Mol. Biosyst.* 6:721–28
23. Galperin MY, Nikolskaya AN, Koonin EV. 2001. Novel domains of the prokaryotic two-component signal transduction systems. *FEMS Microbiol. Lett.* 203:11–21
24. Gao R, Mack TR, Stock AM. 2007. Bacterial response regulators: versatile regulatory strategies from common domains. *Trends Biochem. Sci.* 32:225–34
25. Gao R, Stock AM. 2009. Biological insights from structures of two-component proteins. *Annu. Rev. Microbiol.* 63:133–54
26. Gao R, Stock AM. 2010. Molecular strategies for phosphorylation-mediated regulation of response regulator activity. *Curr. Opin. Microbiol.* 13:160–67
27. Gao R, Tao Y, Stock AM. 2008. System-level mapping of *Escherichia coli* response regulator dimerization with FRET hybrids. *Mol. Microbiol.* 69:1358–72
28. Gooderham WJ, Hancock RE. 2009. Regulation of virulence and antibiotic resistance by two-component regulatory systems in *Pseudomonas aeruginosa*. *FEMS Microbiol. Rev.* 33:279–94
29. Goodman AL, Merighi M, Hyodo M, Ventre I, Filloux A, Lory S. 2009. Direct interaction between sensor kinase proteins mediates acute and chronic disease phenotypes in a bacterial pathogen. *Genes Dev.* 23:249–59
30. Gotoh Y, Eguchi Y, Watanabe T, Okamoto S, Doi A, Utsumi R. 2010. Two-component signal transduction as potential drug targets in pathogenic bacteria. *Curr. Opin. Microbiol.* 13:232–39
31. Grefen C, Harter K. 2004. Plant two-component systems: principles, functions, complexity and cross talk. *Planta* 219:733–42
32. Hietpas RT, Jensen JD, Bolon DN. 2011. Experimental illumination of a fitness landscape. *Proc. Natl. Acad. Sci. USA* 108:7896–901
33. Hooper SD, Berg OG. 2003. On the nature of gene innovation: duplication patterns in microbial genomes. *Mol. Biol. Evol.* 20:945–54
34. Imamura A, Yoshino Y, Mizuno T. 2001. Cellular localization of the signaling components of *Arabidopsis* His-to-Asp phosphorelay. *Biosci. Biotechnol. Biochem.* 65:2113–17
35. Isalan M, Lemerle C, Michalodimitrakis K, Horn C, Beltrao P, et al. 2008. Evolvability and hierarchy in rewired bacterial gene networks. *Nature* 452:840–45
36. Jin T, Inouye M. 1993. Ligand binding to the receptor domain regulates the ratio of kinase to phosphatase activities of the signaling domain of the hybrid *Escherichia coli* transmembrane receptor, Taz1. *J. Mol. Biol.* 232:484–92
37. Kim D, Forst S. 2001. Genomic analysis of the histidine kinase family in bacteria and archaea. *Microbiology* 147:1197–212

38. Koretke KK, Lupas AN, Warren PV, Rosenberg M, Brown JR. 2000. Evolution of two-component signal transduction. *Mol. Biol. Evol.* 17:1956–70
39. Krell T, Lacal J, Busch A, Silva-Jimenez H, Guazzaroni ME, Ramos JL. 2010. Bacterial sensor kinases: diversity in the recognition of environmental signals. *Annu. Rev. Microbiol.* 64:539–59
40. Kuo CH, Ochman H. 2010. The extinction dynamics of bacterial pseudogenes. *PLoS Genet.* 6:e1001050
41. Kurland CG, Canback B, Berg OG. 2003. Horizontal gene transfer: a critical view. *Proc. Natl. Acad. Sci. USA* 100:9658–62
42. Laub MT. 2011. The role of two-component signal transduction systems in bacterial stress responses. In *Bacterial Stress Responses*, ed. G Storz, R Hengge, pp. 45–58. Washington, DC: ASM Press
43. Laub MT, Goulian M. 2007. Specificity in two-component signal transduction pathways. *Annu. Rev. Genet.* 41:121–45
44. Lee AK, Detweiler CS, Falkow S. 2000. OmpR regulates the two-component system SsrA-SsrB in *Salmonella* pathogenicity island 2. *J. Bacteriol.* 182:771–81
45. Liu Y, Harrison PM, Kunin V, Gerstein M. 2004. Comprehensive analysis of pseudogenes in prokaryotes: widespread gene decay and failure of putative horizontally transferred genes. *Genome Biol.* 5:R64
46. Lopez-Redondo ML, Moronta F, Salinas P, Espinosa J, Cantos R, et al. 2010. Environmental control of phosphorylation pathways in a branched two-component system. *Mol. Microbiol.* 78:475–89
47. Madan Babu M, Teichmann SA. 2003. Evolution of transcription factors and the gene regulatory network in *Escherichia coli*. *Nucleic Acids Res.* 31:1234–44
48. Martin W, Rujan T, Richly E, Hansen A, Cornelsen S, et al. 2002. Evolutionary analysis of *Arabidopsis*, cyanobacterial, and chloroplast genomes reveals plastid phylogeny and thousands of cyanobacterial genes in the nucleus. *Proc. Natl. Acad. Sci. USA* 99:12246–51
49. Miller SI, Kukral AM, Mekalanos JJ. 1989. A two-component regulatory system (phoP phoQ) controls *Salmonella typhimurium* virulence. *Proc. Natl. Acad. Sci. USA* 86:5054–58
50. Mira A, Ochman H, Moran NA. 2001. Deletional bias and the evolution of bacterial genomes. *Trends Genet.* 17:589–96
51. **Moglich A, Ayers RA, Moffat K. 2009. Design and signaling mechanism of light-regulated histidine kinases. *J. Mol. Biol.* 385:1433–44**
52. Moglich A, Ayers RA, Moffat K. 2009. Structure and signaling mechanism of Per-ARNT-Sim domains. *Structure* 17:1282–94
53. Moglich A, Ayers RA, Moffat K. 2010. Addition at the molecular level: signal integration in designed Per-ARNT-Sim receptor proteins. *J. Mol. Biol.* 400:477–86
54. **Noriega CE, Lin HY, Chen LL, Williams SB, Stewart V. 2010. Asymmetric cross-regulation between the nitrate-responsive NarX-NarL and NarQ-NarP two-component regulatory systems from *Escherichia coli* K-12. *Mol. Microbiol.* 75:394–412**
55. Ohashi K, Yamashino T, Mizuno T. 2005. Molecular basis for promoter selectivity of the transcriptional activator OmpR of *Escherichia coli*: isolation of mutants that can activate the non-cognate kdpABC promoter. *J. Biochem.* 137:51–59
56. **Osborne SE, Walthers D, Tomljenovic AM, Mulder DT, Silphaduang U, et al. 2009. Pathogenic adaptation of intracellular bacteria by rewiring a cis-regulatory input function. *Proc. Natl. Acad. Sci. USA* 106:3982–87**
57. Parkinson JS. 2010. Signaling mechanisms of HAMP domains in chemoreceptors and sensor kinases. *Annu. Rev. Microbiol.* 64:101–22
58. Perez JC, Groisman EA. 2009. Evolution of transcriptional regulatory circuits in bacteria. *Cell* 138:233–44
59. Perez JC, Groisman EA. 2009. Transcription factor function and promoter architecture govern the evolution of bacterial regulons. *Proc. Natl. Acad. Sci. USA* 106:4319–24
60. **Perez JC, Shin D, Zwir I, Latifi T, Hadley TJ, Groisman EA. 2009. Evolution of a bacterial regulon controlling virulence and Mg^{2+} homeostasis. *PLoS Genet.* 5:e1000428**
61. Posas F, Wurgler-Murphy SM, Maeda T, Witten EA, Thai TC, Saito H. 1996. Yeast HOG1 MAP kinase cascade is regulated by a multistep phosphorelay mechanism in the SLN1-YPD1-SSK1 "two-component" osmosensor. *Cell* 86:865–75

---

51. Construction of a light-sensing histidine kinase through rational domain shuffling.

54. Demonstration of cross-regulation between two signaling pathways at the level of phosphotransfer.

56. Evidence that promoter mutations led to regulation of an ancestral gene by a horizontally acquired two-component system.

60. Mapped the evolution of a response regulator's regulon from *Salmonella* to *Yersinia*.

62. Price MN, Dehal PS, Arkin AP. 2008. Horizontal gene transfer and the evolution of transcriptional regulation in *Escherichia coli*. *Genome Biol.* 9:R4
63. Qian W, Han ZJ, He C. 2008. Two-component signal transduction systems of *Xanthomonas* spp.: a lesson from genomics. *Mol. Plant Microbe Interact.* 21:151–61
64. Rabin RS, Stewart V. 1993. Dual response regulators (NarL and NarP) interact with dual sensors (NarX and NarQ) to control nitrate- and nitrite-regulated gene expression in *Escherichia coli* K-12. *J. Bacteriol.* 175:3259–68
65. Raivio TL, Silhavy TJ. 1997. Transduction of envelope stress in *Escherichia coli* by the Cpx two-component system. *J. Bacteriol.* 179:7724–33
66. Rajeev L, Luning EG, Dehal PS, Price MN, Arkin AP, Mukhopadhyay A. 2011. Systematic mapping of two component response regulators to gene targets in a model sulfate reducing bacterium. *Genome Biol.* 12:R99
67. Rajewsky N, Socci ND, Zapotocky M, Siggia ED. 2002. The evolution of DNA regulatory regions for proteo-gamma bacteria by interspecies comparisons. *Genome Res.* 12:298–308
68. Rampersaud A, Utsumi R, Delgado J, Forst SA, Inouye M. 1991. Ca^{2+}-enhanced phosphorylation of a chimeric protein kinase involved with bacterial signal transduction. *J. Biol. Chem.* 266:7633–37
69. Ren B, Liang Y, Deng Y, Chen Q, Zhang J, et al. 2009. Genome-wide comparative analysis of type-A *Arabidopsis* response regulator genes by overexpression studies reveals their diverse roles and regulatory mechanisms in cytokinin signaling. *Cell Res.* 19:1178–90
70. Salanoubat M, Genin S, Artiguenave F, Gouzy J, Mangenot S, et al. 2002. Genome sequence of the plant pathogen Ralstonia solanacearum. *Nature* 415:497–502
71. Schaller GE, Shiu SH, Armitage JP. 2011. Two-component systems and their co-option for eukaryotic signal transduction. *Curr. Biol.* 21:R320–30
72. Siryaporn A, Perchuk BS, Laub MT, Goulian M. 2010. Evolving a robust signal transduction pathway from weak cross-talk. *Mol. Syst. Biol.* 6:452

73. Identified specificity-determining residues in two-component proteins by analyzing amino acid coevolution and rationally rewiring histidine kinases.

73. **Skerker JM, Perchuk BS, Siryaporn A, Lubin EA, Ashenberg O, et al. 2008. Rewiring the specificity of two-component signal transduction systems. *Cell* 133:1043–54**
74. Skerker JM, Prasol M, Perchuk BS, Biondi EG, Laub MT. 2005. Two-component signal transduction pathways regulating growth and cell cycle progression in a bacterium: a system-level analysis. *PLoS Biol.* 3:e334
75. Sonnenburg ED, Sonnenburg JL, Manchester JK, Hansen EE, Chiang HC, Gordon JI. 2006. A hybrid two-component system protein of a prominent human gut symbiont couples glycan sensing in vivo to carbohydrate metabolism. *Proc. Natl. Acad. Sci. USA* 103:8834–39
76. Stephenson K, Hoch JA. 2002. Evolution of signalling in the sporulation phosphorelay. *Mol. Microbiol.* 46:297–304
77. Stock AM, Robinson VL, Goudreau PN. 2000. Two-component signal transduction. *Annu. Rev. Biochem.* 69:183–215
78. Toro-Roman A, Wu T, Stock AM. 2005. A common dimerization interface in bacterial response regulators KdpE and TorR. *Protein Sci.* 14:3077–88
79. Tzeng YL, Hoch JA. 1997. Molecular recognition in signal transduction: the interaction surfaces of the Spo0F response regulator with its cognate phosphorelay proteins revealed by alanine scanning mutagenesis. *J. Mol. Biol.* 272:200–12
80. Ulrich DL, Kojetin D, Bassler BL, Cavanagh J, Loria JP. 2005. Solution structure and dynamics of LuxU from *Vibrio harveyi*, a phosphotransferase protein involved in bacterial quorum sensing. *J. Mol. Biol.* 347:297–307
81. Ulrich LE, Zhulin IB. 2010. The MiST2 database: a comprehensive genomics resource on microbial signal transduction. *Nucleic Acids Res.* 38:D401–7
82. Utsumi R, Brissette RE, Rampersaud A, Forst SA, Oosawa K, Inouye M. 1989. Activation of bacterial porin gene expression by a chimeric signal transducer in response to aspartate. *Science* 245:1246–49
83. Varughese KI, Madhusudan, Zhou XZ, Whiteley JM, Hoch JA. 1998. Formation of a novel four-helix bundle and molecular recognition sites by dimerization of a response regulator phosphotransferase. *Mol. Cell* 2:485–93

84. Wegener-Feldbrügge S, Søgaard-Andersen L. 2009. The atypical hybrid histidine protein kinase RodK in *Myxococcus xanthus*: Spatial proximity supersedes kinetic preference in phosphotransfer reactions. *J. Bacteriol.* 191:1765–76
85. Weigt M, White RA, Szurmant H, Hoch JA, Hwa T. 2009. Identification of direct residue contacts in protein-protein interaction by message passing. *Proc. Natl. Acad. Sci. USA* 106:67–72
86. West AH, Stock AM. 2001. Histidine kinases and response regulator proteins in two-component signaling systems. *Trends Biochem. Sci.* 26:369–76
87. Whitworth DE, Cock PJ. 2009. Evolution of prokaryotic two-component systems: insights from comparative genomics. *Amino Acids* 37:459–66
88. Williams SB, Stewart V. 1997. Discrimination between structurally related ligands nitrate and nitrite controls autokinase activity of the NarX transmembrane signal transducer of *Escherichia coli* K-12. *Mol. Microbiol.* 26:911–25
89. Worley MJ, Ching KH, Heffron F. 2000. *Salmonella* SsrB activates a global regulon of horizontally acquired genes. *Mol. Microbiol.* 36:749–61
90. Wright GD, Holman TR, Walsh CT. 1993. Purification and characterization of VanR and the cytosolic domain of VanS: a two-component regulatory system required for vancomycin resistance in *Enterococcus faecium* BM4147. *Biochemistry* 32:5057–63
91. **Wuichet K, Zhulin IB. 2010. Origins and diversification of a complex signal transduction system in prokaryotes.** *Sci. Signal.* **3:ra50**
92. Xu Q, Carlton D, Miller MD, Elsliger MA, Krishna SS, et al. 2009. Crystal structure of histidine phosphotransfer protein ShpA, an essential regulator of stalk biogenesis in *Caulobacter crescentus*. *J. Mol. Biol.* 390:686–98
93. Yamamoto K, Hirao K, Oshima T, Aiba H, Utsumi R, Ishihama A. 2005. Functional characterization in vitro of all two-component signal transduction systems from *Escherichia coli*. *J. Biol. Chem.* 280:1448–56
94. Yang Y, Inouye M. 1993. Requirement of both kinase and phosphatase activities of an *Escherichia coli* receptor (Taz1) for ligand-dependent signal transduction. *J. Mol. Biol.* 231:335–42
95. Zapf J, Sen U, Madhusudan, Hoch JA, Varughese KI. 2000. A transient interaction between two phosphorelay proteins trapped in a crystal lattice reveals the mechanism of molecular recognition and phosphotransfer in signal transduction. *Structure* 8:851–62
96. Zarrinpar A, Park SH, Lim WA. 2003. Optimization of specificity in a cellular protein interaction network by negative selection. *Nature* 426:676–80
97. Zhang W, Shi L. 2005. Distribution and evolution of multiple-step phosphorelay in prokaryotes: lateral domain recruitment involved in the formation of hybrid-type histidine kinases. *Microbiology* 151:2159–73
98. Zhang Z, Hendrickson WA. 2010. Structural characterization of the predominant family of histidine kinase sensor domains. *J. Mol. Biol.* 400:335–53
99. Zhu Y, Inouye M. 2003. Analysis of the role of the EnvZ linker region in signal transduction using a chimeric Tar/EnvZ receptor protein, Tez1. *J. Biol. Chem.* 278:22812–19
100. Zhulin IB, Nikolskaya AN, Galperin MY. 2003. Common extracellular sensory domains in transmembrane receptors for diverse signal transduction pathways in bacteria and archaea. *J. Bacteriol.* 185:285–94

91. Comprehensive survey of the evolution of chemotaxis systems across bacteria.

The Unique Paradigm of Spirochete Motility and Chemotaxis

Nyles W. Charon,[1] Andrew Cockburn,[1] Chunhao Li,[3] Jun Liu,[4] Kelly A. Miller,[1] Michael R. Miller,[2] Md. A. Motaleb,[5] and Charles W. Wolgemuth[6]

[1]Department of Microbiology, Immunology, and Cell Biology, [2]Department of Biochemistry, West Virginia University, Health Sciences Center, Morgantown, West Virginia 26506-9177; email: ncharon@hsc.wvu.edu, acockbur@mail.wvu.edu, kamiller@hsc.wvu.edu, mmiller@hsc.wvu.edu

[3]Department of Oral Biology, The State University of New York at Buffalo, New York 14214-3092; email: cli9@buffalo.edu

[4]The University of Texas–Houston Medical School, Department of Pathology and Laboratory Medicine, Houston, Texas 77030; email: jun.liu.1@uth.tmc.edu

[5]Department of Microbiology and Immunology, Brody School of Medicine, East Carolina University, Greenville, North Carolina 27834; email: motalebm@ecu.edu

[6]Department of Cell Biology and Center for Cell Analysis and Modeling, University of Connecticut Health Center, Farmington, Connecticut 06030-3505; email: cwolgemuth@uchc.edu

Keywords

Borrelia, Lyme disease, motor, flagella

Abstract

Spirochete motility is enigmatic: It differs from the motility of most other bacteria in that the entire bacterium is involved in translocation in the absence of external appendages. Using the Lyme disease spirochete *Borrelia burgdorferi* (*Bb*) as a model system, we explore the current research on spirochete motility and chemotaxis. *Bb* has periplasmic flagella (PFs) subterminally attached to each end of the protoplasmic cell cylinder, and surrounding the cell is an outer membrane. These internal helix-shaped PFs allow the spirochete to swim by generating backward-moving waves by rotation. Exciting advances using cryoelectron tomography are presented with respect to in situ analysis of cell, PF, and motor structure. In addition, advances in the dynamics of motility, chemotaxis, gene regulation, and the role of motility and chemotaxis in the life cycle of *Bb* are summarized. The results indicate that the motility paradigms of flagellated bacteria do not apply to these unique bacteria.

Contents

INTRODUCTION	350
B. burgdorferi Life Cycle, Lyme Disease, and Genomics	350
Structure of a Spirochete	351
MORPHOLOGY	351
B. burgdorferi as a Model Spirochete	351
Ultrastructure of *B. burgdorferi*	352
Periplasmic Flagella In Situ	352
Periplasmic Flagella Structure and Composition	353
Cytoskeletal Function of the Periplasmic Flagellar Filaments	353
Why is *B. burgdorferi* a Flat Wave?	354
Hook and Motor Structure	354
DYNAMICS OF MOTILITY	356
Swimming In Vitro	356
Swimming In Vivo	358
CHEMOTAXIS	359
TRANSCRIPTIONAL AND TRANSLATIONAL REGULATION OF MOTILITY AND CHEMOTAXIS GENES	361
VIRULENCE AS RELATED TO CHEMOTAXIS AND MOTILITY	363
CONCLUSION	364

INTRODUCTION

A century ago Clifford Dobell (23) said "the movements of the Spirochaets are still surrounded in mystery." A half century later Claes Weibull (101) quoted Dobell and went on to say "it could be asked whether the situation has changed very much since those days." After yet another half century, although progress has been made, many intriguing questions about spirochete motility remain unanswered and some obscurity about their elegant motions still persists. Genomic analysis indicates that spirochetes are a monophyletic clade (72), so we expect that spirochete motility has some similarity across taxa. Several reviews have been published on spirochete motility and the reader is referred to these articles (15, 34, 51, 102). However, there has been some fresh and robust progress in several areas that focus on particular aspects of spirochete motility, especially on the spirochete *Borrelia burgdorferi* (*Bb*). This progress has been spurred by the increase in research interest in *Bb* because of the importance of the disease it causes and recent breakthroughs in genetic manipulation. Furthermore, owing to its small diameter, *Bb* is optimal for analysis utilizing the groundbreaking methodology of cryoelectron tomography (cryo-ET).

B. burgdorferi Life Cycle, Lyme Disease, and Genomics

Bb: *Borrelia burgdorferi*

Cryo-ET: cryoelectron tomography

Bb is the causative agent of the zoonosis called Lyme disease (81, 86, 89). Small mammals such as mice and voles, and specific species of birds, serve as reservoirs of infection. Humans are accidental hosts. Transmission occurs via the bite of hard shell ticks of certain *Ixodes* species. In humans, Lyme disease has many manifestations, including a spreading rash (erythema migrans), acute and chronic arthritis, a skin disease (acrodermatitis), neurologic problems, and heart block. In the United States, Lyme disease is the most prevalent arthropod-borne human infection. Ten different species in *Bb*

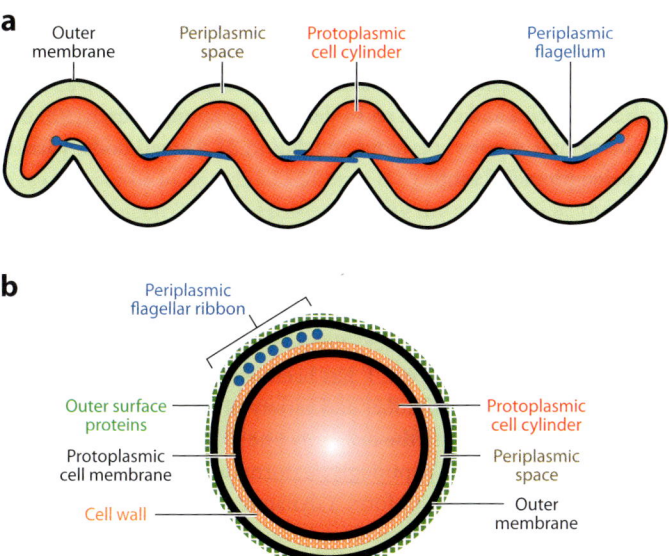

Figure 1
(*a*) Longitudinal diagram of a typical spirochete. Note that the periplasmic flagella overlap in the cell center. (*b*) Cross-section diagram of *Borrelia burgdorferi* illustrating the component parts. Note that seven periplasmic flagella form a tightly packed ribbon that causes the outer membrane to bulge.

have been identified and are called the senso lato species complex. The species that is the most studied is *Bb* senso stricto (hereafter referred to as *Bb*); it is the most prevalent species associated with disease in North America. *Borrelia afzelii* and *Borrelia garinii* are the species most commonly associated with Lyme disease in Eurasia. The genomes of *Bb* senso lato consist of one small, linear chromosome (approximately 950 kb) and a variable number (7–21) of circular and linear plasmids. The plasmids compose approximately one-third of the spirochete's total genetic material. Because of its small genome, *B. burgdorferi* senso lato nutritional requirements is complex.

Structure of a Spirochete

What is a spirochete? In general, spirochetes have a unique and distinct structure; spirochetes are one of the few phyla of bacteria that can be recognized on the basis of morphology (72) (**Figure 1*a***). Outermost is a lipid bilayer outer membrane (OM), and internal to the OM is the protoplasmic cell cylinder (PC). At each end of the PC are subterminally attached periplasmic flagella (PFs) that reside in the periplasmic space (PS). The PFs rotate in a manner similar to the flagella of externally flagellated bacteria (15, 16, 33). The final shape of the cell, depending on the species, is either a helix or a flat wave. The size of the spirochete, the number of PFs attached at each end, and whether the PFs overlap at the middle of the cell vary from species to species. The overall morphology of spirochetes, with their flagella located internally, raises the long-standing questions: How do these bacteria swim and how do they carry out chemotaxis?

MORPHOLOGY

B. burgdorferi as a Model Spirochete

Bb fits the morphological description of a spirochete (**Figure 1*b***). Both light-microscopy and high-voltage transmission electron microscopy (TEM) indicate that *Bb* has a flat-wave morphology (32, 33) (**Supplemental Figure 1**, **Supplemental Movie 1**; follow the **Supplemental Material link**

OM: outer membrane

PC: protoplasmic cell cylinder

PF: periplasmic flagellum

PS: periplasmic space

TEM: transmission electron microscopy

produces PFs that are considerably longer than those of the wild type. Cryo-ET analysis revealed that the flagellar ribbons in *csrA* mutants wrap around the PC more tightly than do ribbons in wild-type cells with a smaller helix pitch and diameter. In addition, the two ribbons interdigitate with one another. These results indicate that increasing the length of the PFs leads to cells with an altered flat-wave morphology (see below). Taken together, the results with the many filament-deficient mutants and the *csrA* deletion mutant strongly support the concept that the PFs influence cell shape by having a skeletal function.

Why is *B. burgdorferi* a Flat Wave?

How do the PFs exert their skeletal function? Bacterial cells and flagella are elastic materials. When forces are applied these structures deform, and when the forces are removed these structures revert back to their original shape. Because the PC of *Bb* is a straight rod when the PFs are not present, and purified PFs are helical, evidently the interaction between the two components leads to the flat-wave morphology. The helix pitch of the PFs in situ (2.83 μm) is markedly different from those that are purified away from the cell body (1.43 μm and 2.0 μm) (16, 24, 32). For the PFs to be contained within the PS, these structures must deform. The PC produces the force that bends the PFs. Concomitantly, the PFs exert an equal but opposite force back onto the PC, which causes the cell to bend into its characteristic shape. Mathematical modeling, which takes into consideration the elastic properties of the PFs and PC, has shown that the balance between these two forces is a cell with a flat-wave shape—it is not intuitive! However, this force balance conspires to produce the correct wavelength and amplitude only if the PFs are in the helical conformation that is observed less frequently in purified PFs (helix diameter of 0.8 μm and a pitch of 2.0 μm) (24). How does the *csrA* deletion mutant that has longer PFs cause the cell to have a shorter wavelength and cell amplitude (93)? One possible explanation is that the increase in length of the PFs, and perhaps the interdigitation characteristic of the PF ribbons in the mutant, causes the PFs to flip into the helical state that is observed more frequently in purified PFs. This change in preferred conformation of the PFs is predicted to produce a flat-wave shape with a smaller wavelength and amplitude.

Hook and Motor Structure

The flagellar hook region in *Bb* and certain other spirochetes has some unique attributes. In other bacteria, the hook serves as a universal joint connecting the flagellar filament to the motor (85). In *Bb*, the hook is a 61-nm-long hollow tube consisting of approximately 133 FlgE units (84). As mentioned above, inactivation of *flgE* results in cells that lack flagellar filaments, are nonmotile, and are rod shaped (84). In most bacteria, the FlgE subunits are held together by protein-protein interactions that are readily dissociated by denaturing agents. Surprisingly, FlgE in *Bb* and other spirochetes forms a stable, high-molecular-weight complex (52, 84). Several lines of evidence indicate that FlgE subunits in *Bb* are covalently cross-linked (84) (K. Miller, M. Miller & N. Charon, unpublished data). The flagellar hook and filament are likely to be under greater stress in spirochetes than in other bacteria, as the PFs deform the PC while they rotate (see below, Dynamics of Motility). Perhaps the hook proteins are cross-linked in spirochetes for added structural strength to perform this function.

The incredible methodology of cryo-ET is being exploited to analyze the intact *Bb* motor (44, 54), as well as other spirochetal motors (53, 69). The overall structures of spirochetal motors are similar to each other, consisting of the MS ring, the C ring, the rod, the export apparatus, and the stator (**Figure 3**) (**Supplemental Movie 2**). *Bb* exhibits a relatively large C ring, ~57 nm in diameter, compared to the 45-nm C ring of *E. coli* and *Salmonella enterica* serovar Typhimurium

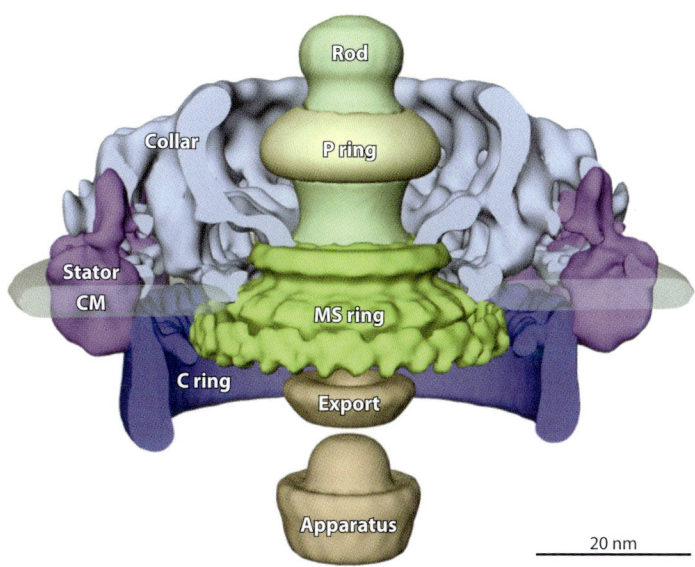

Figure 3

A three-dimensional reconstruction of the *Borrelia burgdorferi* flagellar motor. The major components (the rod, the stator, the P ring, and the MS ring) are labeled. The C ring is composed of FliG, FliM, and FliN. The "collar" is a spirochete-specific feature. The export apparatus is divided into two separate densities, although the boundary between the MS ring and the export apparatus is not well defined. Abbreviation: CM, protoplasmic cell membrane.

(95). This enlarged C ring shares similar features with the *S. enterica* structure: a cylindrical structure with a slightly bulky bottom and a Y-shaped extension at the top (95). Thus, the C ring is likely composed of multiple copies of FliG, FliM, and FliN proteins arranged in a manner similar to that of *E. coli* and *S. enterica*.

The stator, encoded by the *motA* and *motB* genes, is the motor force generator embedded in the protoplasmic cell membrane. The visualization of the stators in many species of bacteria is difficult (19), and this difficulty may in part be the result of the stator being dynamic and that its functional units freely interchange between the motors circulating in the protoplasmic cell membrane (46). In contrast, the stators of *Bb* and other spirochetes are clearly visualized (44, 53, 54, 69). Cryo-ET analysis of *motA* and *motB* mutants of *Bb* provides the first structural evidence that these genes encode the 16 stators radially arranged around the rotor (**Figure 3**) (**Supplemental Movie 2**) (X. Zhao, T. Boquoi, M.A. Motaleb & J. Liu, unpublished data). Conceivably, the stators in *Bb* and other spirochetes are relatively stable, or unit interchange occurs without disrupting the overall arrangement of the stator. Remarkably, the cytoplasmic domain of the stator is adjacent to the C-terminal domain of the C-ring rotor protein FliG. This stator-rotor interaction evidently induces an unexpected conformational change in FliG, and this change is likely to be a fundamental mechanism for flagellar rotation (X. Zhao, T. Boquoi, M.A. Motaleb & J. Liu, unpublished data).

The *Bb* flagellar motor is a remarkable and complicated nanomachine consisting of at least 20 different proteins (**Figure 3**) (**Supplemental Movie 2**). To understand flagellar motor assembly and function, the structural and functional roles of all flagellar proteins must be determined. Fortunately, many mutants in motor genes have been constructed by either transposon mutagenesis or targeted mutagenesis (**Supplemental Figure 2**, **Supplemental Table 1**). *fliE* mutants are defective in rod assembly, and *fliM* mutants are deficient in the middle and the bottom of the C ring (X. Zhao, T. Boquoi, A. Manne, M.A. Motaleb & J. Liu, unpublished data). *fliI* and *fliH* mutants are defective in the assembly of the bottom part of the export apparatus (X. Zhao, T. Lin & S. Norris, unpublished data). FliL proteins localize between the stator and rotor, and these proteins are evidently involved in the proper orientation of the PFs within the PS (65). *flgI* mutants fail to form a hollow, torus-shaped structure around the rod. However, in contrast to other bacteria, these *flgI* mutants are still fully motile; evidently, the P ring is not required for flagellar rotation (54). Instead, a spirochete-specific "collar" may function as a bushing that facilitates the rotation of the flagellum without disrupting the surrounding peptidoglycan layer. The identities of the specific genes encoding the "collar" have yet to be identified. In sum, the structure of the *Bb* flagellar motor is being characterized in some detail, and specific motor proteins and function are being correlated with specific genes in situ for the first time.

CCW: counterclockwise

DYNAMICS OF MOTILITY

Swimming In Vitro

Given the structure of *Bb*, the obvious first question with respect to their motility is, What do swimming cells look like? Also, given that the PFs are located in the PS, how does PF rotation drive motility? The swimming behavior of *Bb* is more complicated than that of other bacteria as a result of their complex geometry. *Bb* has four motility modes based on the direction of flagellar rotation: two translational modes (with either end leading) and two nontranslational modes (**Figure 4**) (**Supplemental Movie 3**). In an isotropic environment, a given cell runs at least 90% of the time (M.A. Motaleb & N. Charon, unpublished data). During a run, waves are propagated from the front to the back at a frequency of 5 to 10 Hz at room temperature. In contrast to eukaryotic flagella, propagating waves are full size at the anterior end instead of starting as small bends and increasing in size. These waves propagate at a speed of 34 $\mu m\ s^{-1}$ relative to the cell (33). The cells swim at a mean speed of 4.25 $\mu m\ s^{-1}$ in a pure liquid (33). The ratio of the swim speed to wave propagation is 0.12; i.e., in the time it takes a wave to travel the length of the cell, the cell advances 12% of its length through the medium. Thus, as *Bb* swims, waves are clearly evident as they swim in a given direction.

How are the waves generated, and what is the basis for the nontranslational forms? Everything points to the PFs. We know they are essential for motility, they rotate, they are helix shaped, and they have a skeletal function. Although it has not been directly proven, the following model is proposed (**Supplemental Movie 4**) (15, 32): During a run, the PFs of the anterior ribbon are predicted to rotate counterclockwise (CCW), and those of the posterior ribbon rotate in the CW direction (as a frame of reference, the PFs are viewed from their distal end to where they insert into the PC). Thus, the PFs of the two ribbons rotate asymmetrically relative to one another during a run. This rotation causes backward-moving waves to be propagated down the length of the cell. Reversals occur when the PFs of both ribbons change the direction of rotation. In addition, recent mathematical modeling suggests that during translation the rotating PFs are not in direct contact with the PC, but rather a thin layer of fluid separates these filaments from the PC (107). Forces between the PFs and the PC are therefore mediated by viscous forces in the fluid present in the PS. This hypothesis is based on the known elasticity of the PFs, PC, and the dynamics of swimming cells. The modeling predicts that the thin layer of fluid in the PS is essential for the smooth backward-propagating waves noted on translating cells. Otherwise, if the PFs interacted directly with the PC, the PFs would be predicted to easily tangle up with one another and the propagating wave would not be as regular as observed. Several questions arise with the overall model of translating *Bb*. If the PFs are rotating CCW along the axis of the cell as viewed from behind the cell, is the cell rolling about the body axis CW to balance that torque as predicted

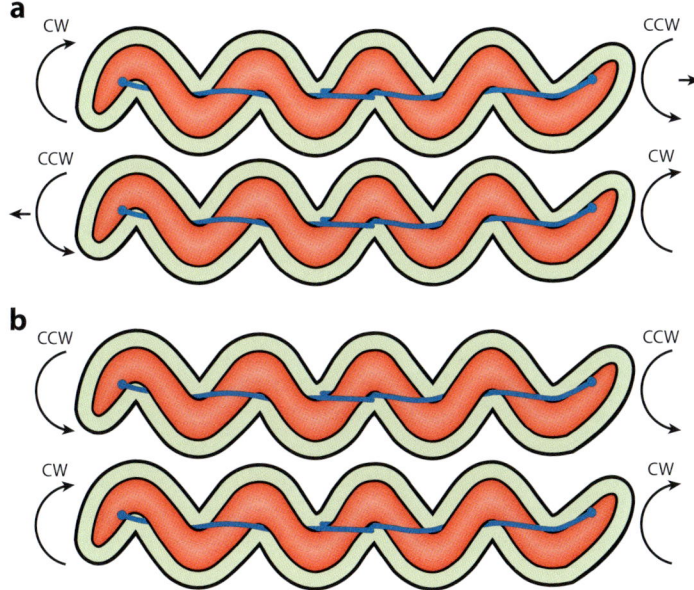

Figure 4

Swimming cells of *Borrelia burgdorferi* as a function of the direction in which the periplasmic flagella (PFs) rotate. Straight arrows at cell ends indicate direction of swimming. Curved arrows indicate the direction in which the PFs rotate. For simplification, only one PF is shown attached at each end of protoplasmic cell cylinder. Panel *a* shows translational forms, and panel *b* shows nontranslational forms. Reprinted and modified with permission from Reference 15. Abbreviations: CW, clockwise; CCW, counterclockwise.

(15, 32)? What is happening to the OM? Cryo-ET indicates that it is held very close to the PC and even bulges in the vicinity of the PF ribbon. How is the tight fitting achieved, especially in light of evidence that the OM is a lipid bilayer? Are there bonds that form between the OM and PC, and are those bonds broken as the PFs rotate?

In the nontranslational modes, the prediction is that the PFs from both ribbons rotate in the same direction: both rotate either CW or CCW. This stopped interval is referred to as a flex and is thought to be analogous to the tumble seen in externally flagellated bacteria. The cells often become distorted during this interval and may bend in the middle. It is difficult to distinguish cells with PFs at both ends rotating CCW from cells with PFs at both ends rotating CW. It is not clear exactly what is occurring in the PS that causes the cell to form a distorted morphology. Do the PFs from the opposite end wind around one another to cause the distorted shape? In other spirochete species, the distortion associated with flexing occurs only if the PFs overlap in the center of the cell (33), so this is a possibility.

The description of swimming given above comes largely from motility assays in liquid environments. Many of the natural environments that *Bb* encounters do not behave like a liquid. For example, spirochetes are deposited into the dermis of the mammal via the saliva of the tick. The dermis of the mammal is composed largely of a cross-linked collagen network. This gel-like environment responds to external forces from the bacterium with a combination of elastic and viscous forces; i.e., it is a viscoelastic material. Liquid media, on the other hand, only responds with viscous forces. One of the first investigations into the effect of viscoelasticity on the motility of *Bb* examined swimming in solutions of methylcellulose or hyaluronic acid (41). Swimming speed increases

GFP: green fluorescent protein

substantially as the concentration of methylcellulose or hyaluronic acid is increased. The increase in swimming speed with the viscoelasticity of the environment suggested that *Bb* motility may be optimized for migration through host tissue. In addition, *Bb*, like several other spirochete species, can translate in highly viscoelastic medium that markedly slows or stops the motility of other flagellated bacteria. Liquid media and methylcellulose solutions, though, are poor facsimiles for many of the natural environments that *Bb* encounters. For example, collagen in the dermis forms a gel-like network (i.e., it is more elastic than viscous), whereas low-concentration methylcellulose solutions (less than a few percent) are viscoelastic fluids (i.e., they are more viscous than elastic) (12). The natural environments are differentiated further from liquid media and methylcellulose solutions because they contain cells and various extracellular matrix components, such as collagen, fibronectin, and decorin, to which *Bb* binds.

Bb in gelatin exhibited four motility states, which are determined by transient adhesion between the bacterium and the matrix (36). In addition to adhering to these substrates, spirochetes also migrate through the matrix, even though the pores in the gelatin matrices are significantly smaller than the diameter of the bacteria. As in previous reports, the undulation and migration speed of the bacteria depend strongly on the physical properties of the environment; however, the bacteria always swim slower in gelatin than in liquid media. Therefore, the unique motility mechanism of *Bb* enables it to penetrate dense tissues in its hosts, but the speed of the bacteria may not be enhanced in these natural environments.

Swimming In Vivo

We are at an early stage in sorting out the role of *Bb* motility in vivo. One of the major breakthroughs is the ability to track green fluorescent protein (GFP)-tagged virulent *Bb* in vivo using intravital microscopy and fluorescence spinning disc confocal microscopy. Soon after injection of *Bb* into mice, spirochetes are seen interacting with capillaries and veins (61). The movement of the organisms is stop and go, with the stops characterized by adherence to the endothelium. Approximately 90% of interactions with the endothelium are less than 1 s, and about 10% are 1–20 s. These interactions are mediated in part by the spirochete surface protein BBK32 and host fibronectin and glycosaminoglycans (70). In contrast, approximately 1% of the cells are stationary, with one end attached and partially embedded in the endothelium, often at the endothelial cell junctures, and the other end gyrating. These stationary cells resemble the often seen tethered spirochetes whereby the cells adhere to stationary ligands via an OM surface-associated component yet remain motile (18). The attached spirochetes eventually escape the endothelial cell lining and penetrate into connective tissue. In the liver, the spirochetes first interact with Kupffer cells, which then trap, immobilize, and engulf the spirochetes, and then present antigens to the invariant natural killer T cells (47). The role of motility and chemotaxis in both the adherence and penetration through blood vessels, and in the interaction of immune cells in the liver, is likely to be important. Future experiments with specific motility and chemotaxis mutants should be definitive.

Some remarkable imaging studies have examined GFP-tagged *Bb* in the tick host both before and after feeding on mice (26). In unfed nymphs, *Bb* cells are nonmotile and distributed throughout the midgut. By 72 h after feeding on mice, the density of *Bb* is high; however, the spirochetes remain nonmotile but are viable. Remarkably, the presence of a diffusible factor(s) in the tick midgut is likely to be responsible for their nonmotility. At 72 h, only a small number of the spirochetes traverse the midgut basement membrane, enter the hemocoel, and colonize the salivary glands for transmission to mammals. These invasive spirochetes are motile. Why is *Bb* motility specifically inhibited in the tick midgut? Perhaps the tick developed a means of defense to keep the infection localized by inhibiting *Bb* motility. Future experiments to characterize the inhibitory factors should be enlightening.

CHEMOTAXIS

Chemotaxis (movement toward or away from a chemical stimulus) in spirochetes is unique. Bacteria undergo a biased random walk during chemotaxis, and this walk is achieved by modulating the direction or speed of flagellar rotation (76, 88, 99). A two-component regulatory system mediates the biased walk up a gradient of attractant or away from a repellent. In the paradigmatic chemotaxis model of *E. coli* and *S. enterica*, variation in motor direction is determined as follows: The response regulator CheY is phosphorylated by the histidine kinase CheA to form CheY-P. CheA is part of the polarly located methyl-accepting chemotaxis protein (MCP) receptor signal complexes. The probability that CheA phosphorylates CheY is a function of the occupancy level of an attractant or repellent that binds directly or indirectly to the MCPs in the PS. CheY-P diffuses through the cytoplasm and interacts with the flagellar switch protein FliM, causing the motor rotational biases to shift from the default rotation of CCW to CW. If all the motors rotate CCW, the cell swims. If one of the motors rotates CW, the cell tumbles (97). The motors are within 1 μm of the MCP complexes. Thus, CheY-P can diffuse and bind to the motors within 0.1 s (98). The CheY-P formed has phosphatase activity associated with it. However, dephosphorylation of CheY-P, which restores the default CCW behavior such that the cell can immediately respond to changes in the environment, is enhanced by the action of the CheY-P-specific phosphatase CheZ.

Chemotaxis in *Bb* is different from this and other well-studied bacterial chemotaxis models. Two conundrums are evident. First, these spirochetes have two sets of flagellar motors: one at each cell end. As noted above, these sets of motors are located at a considerable distance from one another (between 10 and 20 μm). In addition, recent results using fluorescent antibodies and cryo-ET indicate that the MCP complexes are subpolarly located at each cell end (105) (**Figure 2**). If CheY-P is formed at one end of the cell at these complexes, it could readily diffuse to the adjacent motors. However, given a cell length of 10 μm, it would take at least 10 s for CheY-P to diffuse to the motors at the other end of the cell (67). Because the speed of these bacteria is at least 4 μm s^{-1} (33), simple diffusion of CheY-P to the opposite end of the cell to coordinate the rotation of the PFs is unlikely. Furthermore, in *Bb*, instead of runs and tumbling, there are four swim configurations based on the direction of flagellar rotation (**Figure 4**), all of which contribute to a higher level of complexity in swim behavior than that of most other bacteria. Thus, the first conundrum is, How does *Bb* coordinate the rotation of the PFs within the ribbons at both cell ends to effect chemotaxis?

Second, for the bacteria to swim in a given direction, the motors at each end must rotate in the opposite direction of those at the other end; i.e., if the PFs in one ribbon rotate CCW, the others rotate CW. This asymmetrical rotation is markedly different from the swimming behavior of *E. coli* and *S. enterica* whereby all the motors rotate in the same direction during translation. Furthermore, as discussed below, in the absence of a functional CheY response regulator, *Bb* constantly runs. Thus, the second conundrum is, What is the basis for asymmetrical rotation of the PFs during translational swimming in the absence of CheY?

Although significant progress has been made in understanding *Bb* chemotaxis, a palpable molecular model putting it all together is still in the developmental stage. To address the above questions, a capillary tube chemotaxis assay was developed using flow cytometry to enumerate cells, and several chemically defined chemoattractants were identified (4, 63). These attractants include N-acetylglucosamine, glucosamine, glucosamine dimers, and glutamate. The assay allowed

MCP: methyl-accepting chemotaxis protein

researchers to test whether specific genes were involved in chemotaxis. *Bb* has multiple copies of chemotaxis genes, including six *mcp* (two lack membrane-spanning regions), two *cheA*, three *cheY*, two *cheB*, two *cheR*, and three *cheW* (29). It has only one CheY-P phosphatase gene commonly found in bacteria, *cheX*, which has an activity similar to that of *cheZ* found in *E. coli* (64, 74). The results of extensive gene targeting analysis indicated that the chemotaxis pathway involves MCPs, CheW3, CheA2, CheY3, and CheX (**Supplemental Figure 2, Supplemental Table 1**). Biochemical analysis supports the conclusion that CheA2 readily phosphorylates CheY3 (67) and CheX dephosphorylates CheY3-P (64, 74). Interestingly, the half-life of the CheY3-P is 10 min (67), which is considerably longer than the half-lives of most CheY-P of other bacteria such as *E. coli*, which is a few seconds (37). Most important, because CheY plays such a pivotal role in the chemotaxis of all known species of bacteria, CheY3 is concluded to be the key chemotaxis response regulator in *Bb*.

An analysis of the swim behaviors of the chemotaxis mutants may yield a clue to the basis of asymmetrical rotation of the flagellar ribbons during translational motility. Both *cheA2* and *cheY3* mutants constantly run in one direction (48, 67) (**Supplemental Table 1, Supplemental Movie 3**). These results are similar to those of other bacteria such as *E. coli* and *S. enterica*; *cheA* and *cheY* mutants in these bacteria also constantly run. However, in *Bb* the motors rotate asymmetrically at one end relative to the other during a run, whereas in *E. coli* all the motors are rotating CCW. Evidently, in the default state, i.e., when no CheY3-P are present, the motors at one cell end are different from those at the other, as they rotate in opposite directions. Perhaps there is a protein that interacts with motors at one cell end such that in the default state the PFs rotate in the opposite direction relative to those at the other end. There is precedent for protein localization of this type, as some bacteria-specific proteins localize at the old cell end and not at the new cell end (6). A candidate protein is FliG1 (50). *Bb* and many other spirochetes have two *fliG* genes encoded in their genomes, *fliG1* and *fliG2* (29, 50). FliG is involved in motor rotation in other bacteria, is essential for flagellar assembly and motility, and plays a major role in determining the direction of motor rotation (55). *fliG2* likely plays this role in *Bb*, as mutants in this gene are nonmotile and lack PFs; FliG2 has all the essential sequence and structural domains common to other bacterial FliGs (50). However, FliG1 lacks some of the essential residues common to FliG of other bacteria. *fliG1* mutants are still motile but are unable to swim in highly viscous media containing methylcellulose; only one of the cell ends is able to gyrate. Interestingly, GFP-tagged FliG1 localizes at only one cell end (50). It will be interesting to determine whether FliG1 is partly responsible for the asymmetrical rotation of the PFs.

The phenotypes of other specific mutants complicate this scenario. The *cheX* mutant constantly flexes (64), which is analogous to the constantly tumbling *cheZ* mutants of *E. coli* (10). A *cheX* mutant is predicted to have a higher CheY3-P concentration than the wild type does. One expectation of the protein localization model is that cells with a high CheY3-P concentration should also run, not flex. For example, those motors that run CW in the default state should reverse and rotate CCW under a high CheY3-P concentration, and those motors that run CCW in the default state should run CW. Perhaps there is another CheY-P phosphatase that has not been identified in *Bb*, and as result, the *cheX* mutant could have an intermediate level of CheY3-P that results in flexing. In *B. subtilis*, FliY and CheC have CheY-P phosphatase activity, but *Bb* lacks homologs to these genes (68).

Bb also has a gene encoding CheD. CheD in other bacteria augments the phosphatase activity of CheC, binds to and deamidates glutamine residues on MCPs, and regulates the activity of CheA kinase. *cheD* mutants in other bacteria have a decreased activity of CheA kinase, leading to lower levels of CheY-P (78). CheD may have a similar function in *Bb*, as a *cheD* mutant has a nonchemotactic, constantly running phenotype, similar to the *cheY3* and *cheA2* mutants that are also expected to decrease CheY3-P concentration (M.A. Motaleb & N. Charon, unpublished data). However, because CheD in other bacteria has functions in addition to activating the CheA kinase, its role in *Bb* chemotaxis is presently unclear.

The second messenger, 3′,5′-cyclic-diguanosine monophosphate (c-di-GMP), is emerging as a major factor influencing motility and possibly chemotaxis (see Transcriptional and Translational Regulation of Motility and Chemotaxis Genes, below). Certain mutants involved in c-di-GMP synthesis, degradation, and effector binding have aberrant swim behavior, such that some mutants (e.g., *pdeA*) constantly run, whereas others (e.g., *pdeB*) have a high flex rate (90, 91). c-di-GMP and its effector protein bind to the flagellar motor protein in other bacteria and influence the direction and speed of flagellar rotation (9, 21, 27, 73). At this time it is difficult to determine how c-di-GMP influences *Bb* swimming behavior at a molecular level.

c-di-GMP: 3′,-5′-cyclic-diguanosine monophosphate

The results accumulated indicate the following conclusions: First, CheY3 is the major response regulator, and the pathway leading to its phosphorylation and dephosphorylation involves MCPs, CheW3, CheA2, CheX, and possibly CheD. Second, the asymmetrical rotation of the PFs in the default state can best be explained by differences in the motors at each end of the cell, but support for this hypothesis awaits experimental evidence. Third, it is too early to understand the basis of flagellar coordination mediated by CheY3 that results in chemotaxis, although many hypotheses are conceivable. As previously mentioned, diffusion of CheY3-P from one cell end to the other is too slow to coordinate the rotation of the PFs at the distal end. As an alternative, perhaps there is a cytoskeletal structure whereby CheY-P moves rapidly from one end to the other. There is precedence for this possibility: In *Myxococcus xanthus*, the protein AglZ mediates gliding motility by moving from one cell end to the other by way of the MreB cytoskeleton (58). Because completion of this internal cell migration of AglZ is on the order of several minutes, a CheY3 transport system in *Bb* would have to be considerably faster.

Another model possibility for coordinating the PFs at each end relates to a mechanosensing mechanism. Perhaps there is an interaction between the PFs at one end of the cell and those at the other such that rotation of the PFs at one end influences the rotation of the PFs at the other end. This possibility is conceivable, as the PFs in *Bb* overlap in the center of the cell. However, this hypothesis does not apply to all spirochetes, as *Leptospira interrogans* has PFs that are short and do not overlap (102), yet these spirochetes are chemotactic (110) (N. Takahashi & N. Charon, unpublished data). CheY3-P could act at another, unknown site such that the membrane potential is altered when the cell is undergoing chemotaxis, and this change in potential might allow coordination of the PFs (35). Alternatively, perhaps CheY3-P together with c-di-GMP coordinates PF rotation.

Finally, conceivably there is no internal signal that coordinates the motors at both cell ends: Flagellar coordination and chemotaxis are achieved by the attractant binding to either one or both of the MCP clusters at the cell ends. The change in CheY3-P concentration generated by this binding specifically affects the direction of rotation of the motors that are adjacent to those MCPs. For example, if the attractant binds the MCPs at one cell end, it causes the motors only at that end of the cell to change their direction of rotation, and the cell flexes. In contrast, if attractant molecules simultaneously bind to the MCPs at both cell ends, the motors at both ends change directions and the cell runs. In closing, now that specific compounds that serve as attractants are known, and that CheY3 is the functional response regulator, the basis for coordinating PF rotation for chemotaxis can finally be determined.

TRANSCRIPTIONAL AND TRANSLATIONAL REGULATION OF MOTILITY AND CHEMOTAXIS GENES

The genetic map indicating the genes that are involved in motility and chemotaxis is presented in **Supplemental Figure 2**. Most motility genes are present in single copies (except for *fliG*, the motor control protein). As noted above, there are multiple copies of chemotaxis genes. The analysis of mutants described above indicates that the *Bb* cluster *cheA2-cheW3-cheX-cheY3* is involved in chemotaxis under standard laboratory conditions. This cluster is closely related to chemotaxis gene systems in the other spirochetes and is probably inherited from a common ancestor (15, 48, 103). The other cluster consisting of *cheW2-bb0566-cheA1-cheB2-bb0569-cheY2* may be a recent gene transfer from the *Proteobacteria* (48, 103). The function of this second cluster is not known, but it may be involved in chemotaxis under different environmental conditions.

The regulation of the motility and chemotaxis genes of *Bb* is unique. In other bacteria, there is cascade control of gene regulation of motility gene expression (20). For example, at least 50 genes are involved in the motility and chemotaxis of *E. coli* and *S. enterica*, and these genes fall into three classes (class I, II, and III), which are under the tight regulation of a transcriptional hierarchy. Within this regulatory cascade, the class I master regulator (FlhDC), in conjunction with the housekeeping sigma factor (σ^{70}), directs RNA polymerase to initiate the transcription of the class II genes. Class II genes encode the structural proteins involved in the motor-hook complexes, and two regulatory elements: the flagellum-specific sigma factor FliA (σ^{28}) and the antagonist of FliA (anti-σ^{28}), FlgM. Prior to the completion of the hook assembly, FlgM binds to FliA to prevent premature synthesis of the class III genes encoding the flagellin and the chemotaxis proteins. Upon completion of the hook structure, FlgM is excreted by the flagellum-export apparatus, thereby allowing FliA to initiate the transcription of the class III genes. The last step allows for completion of flagellar assembly and chemotaxis gene expression. In contrast, in silico analysis indicates that no FliA, FlgM, or σ^{28} promoter consensus sequences are present in the genome of *Bb* (15, 29). All the motility and chemotaxis genes identified thus far fall under the regulation of σ^{70} (**Supplemental Figure 2**). These and other results indicate that *Bb* does not employ a transcription cascade to regulate its motility genes. Instead, these genes are regulated primarily by a posttranscriptional mechanism. Two key studies that lead to this conclusion stand out. First, in a *flaB* filament mutant, the amount of FlaA synthesized is only 13% of the wild-type level despite equivalent levels of *flaA* transcript (66). Second, in a *flgE* mutant, FlaA and FlaB accumulation are decreased by more than 80%, whereas the levels of their respective encoding mRNA are equivalent to those of the wild type (84). Furthermore, although FlaA is slowly degraded in the *flgE* mutant, there is no turnover of FlaB. The conclusion is that translational control and not protein turnover is responsible for the lack of accumulation of FlaB and possibly that of FlaA in the *flgE* mutant.

Recent experiments indicate that the *Bb* carbon storage regulator A (CsrA) is a major regulator of translational control of FlaB (93). The Csr system is present in many bacterial species (3, 80), and its importance in carbon metabolism, virulence, biofilm formation, and motility is well established (2, 56, 80). In *E. coli*, the Csr system comprises an RNA binding protein, CsrA; two noncoding RNAs, CsrB and CsrC; and a regulatory protein, CsrD. CsrA functions through binding to the consensus sequence RUACARGGAUGU, which is present within the leader region of its targeted transcripts and subsequently regulates gene expression posttranscriptionally (25, 60). In *E. coli*, CsrA positively regulates flagellar synthesis by serving as an activator of *flhDC* expression (100). However, in *B. subtilis*, CsrA has an opposite effect—it negatively regulates flagellin synthesis at a posttranscriptional level by binding to mRNA and inhibiting translation (106). In a similar manner, *Bb* CsrA is a negative regulator of FlaB. CsrA binds to two sites present in the leader region of the *flaB* transcript, with one of them overlapping the Shine-Dalgarno sequence. Binding of CsrA to these regions leads to translation inhibition of FlaB, presumably via blocking ribosomes from accessing the Shine-Dalgarno sequence of *flaB* mRNA. Thus, the amount of FlaB in a cell may be controlled at the translational level by CsrA (93).

The two-component regulatory system called HK2/Rrp2 is involved in motility gene expression in *Bb* (42, 71, 104). Acetyl-phosphate, an intermediate product from acetate metabolism, autophosphorylates Rrp2, the response regulator of the HK2/Rrp2, and in doing so activates RpoN, a σ^{54} transcription factor. RpoN in turn upregulates RpoS (the Rrp2-RpoN-RpoS pathway) (104). CsrA serves as a repressor of phosphoacetyltransferase (Pta), one of the key proteins in acetate metabolism, and indirectly modulates the level of acetyl-phosphate in *Bb* as well as the subsequent activation of the Rrp2-RpoN-RpoS pathway (92). Microarray analysis of *rrp2*, *rpoN*, and *rpoS* mutants has revealed a potential role for the Rrp2-RpoN-RpoS network in the regulation of chemotaxis gene expression (14, 28, 71). For instance, the transcription of eight chemotaxis genes is regulated by the Rrp2-RpoN-RpoS network (71). Interestingly, Rrp2-RpoN-RpoS activity is maximally induced under conditions that mimic the mammalian host environment (108). Regulation of chemotaxis proteins via the Rrp2 pathway may allow the spirochete to modulate its chemotaxis genes expression while being transmitted between different hosts to aid in colonization as well as dissemination.

Recent studies in *Bb* indicate a link between the signaling molecule second-messenger c-di-GMP and cell motility (42, 75, 90, 91). In *Bb*, the c-di-GMP metabolism pathway consists of Rrp1, a sole diguanylate cyclase (42, 83); two phosphodiesterases, PdeA and PdeB (90, 91); and PlzA (30, 75), a c-di-GMP binding protein. A mutation in any of the genes of the c-di-GMP pathway alters both cell motility and chemotaxis, but many different phenotypes are generated. For example, disruption of Rrp1 causes cells to constantly run and have an attenuated chemotactic response (42). Mutations in the phosphodiesterase genes result in two different phenotypes: A *pdeA* mutant runs and pauses but fails to reverse, and a *pdeB* mutant has increased flexing frequency (90, 91). A *plzA* mutant has a defect in motility in agar but has the wild-type swimming behavior (75). Remarkably, the *plzApdeB* double mutant constantly flexes (90). Taken together, c-di-GMP clearly affects motor function in *Bb*, but as previously stated, how this occurs is unclear.

In summary, our understating of the regulation of chemotaxis and motility gene expression in *Bb* is still in its infancy. Translational control of motility gene expression appears to be crucial, with CsrA and c-di-GMP emerging as major regulators of chemotaxis and motility. Many questions remain unanswered. What is the interplay between the identified regulatory elements in modulating gene expression? How do environmental cues play a role in differential gene expression within the two hosts, and what are those cues?

VIRULENCE AS RELATED TO CHEMOTAXIS AND MOTILITY

The role of motility and chemotaxis during the infection and disease processes has been examined in several species of pathogenic bacteria. In many species, mutant analysis indicates that motility and chemotaxis are essential for infection and invasion (13, 40, 59, 109). Among the spirochetes, motility-deficient mutants of *Brachyspira hyodysenteriae* are attenuated in a mouse model of swine dysentery (82). In *Treponema denticola*, motility- and chemotaxis-deficient mutants are less invasive than their parent in a human gingival keratinocyte monolayer cell penetration model (57). Recent results suggest that motility is essential for *L. interrogans* to cause disease in the hamster model for leptospirosis (45; E. Wunder & A. Ko, unpublished data).

Motility and chemotaxis are likely to be important factors for *Bb* to survive in nature. Genomic analysis suggests that over 50 genes (5–6% of the genome) are potentially involved in motility and chemotaxis (15, 29). In addition, approximately 10% of the total cellular protein is FlaB (66). These results imply *Bb* dedicates a significant proportion of its energy to motility and chemotaxis, and they further support the concept that these processes are necessary for the survival of the spirochete. In addition, during mammalian infection, and when growing *Bb* under conditions that mimic in vivo conditions, motility and chemotaxis proteins are upregulated and some are among the most potent immunogens (5, 79, 96). Evidently the synthesis of these proteins is not turned off soon after infection as in some species of bacteria (1).

Motility and chemotaxis are likely to be important for *Bb* to participate in several steps in completing the host-vector cycle. First, after transmission from the tick to a vertebrate host, *Bb* disseminates through skin and migrates to an appropriate target tissue such as the joint. These sites allow for persistence and evasion of the expanding adaptive immune response. Second, after residing in the vertebrate host for weeks to years, *Bb* possibly detects the presence of feeding ticks, and then migrates to those sites to enter the blood meal of those ticks. Finally, during the blood meal of an infected tick, the spirochetes need to migrate from the tick gut to the salivary glands to restart the cycle.

Although using a genetic approach to determine the contribution of motility and chemotaxis to the life cycle of *Bb* is difficult (15, 65), significant progress is being made. One key study focused on the presumed motor protein *fliG1* gene from a virulent strain of *Bb* (11, 50). Mutations in this gene result in motility-deficient cells and are unable to establish an infection in mice. These studies are the first to show that motility is involved in the virulence of *Bb*. In another ongoing study, targeted *flaB* mutants from a virulent strain of *Bb* are nonmotile as expected, but they are also noninfectious in mice (S. Sultan, M.A. Motaleb, P. Stewart, P. Rosa & N. Charon, unpublished data). These preliminary results, coupled with those with the *fliG1* mutant, suggest that motility plays a critical role in the disease process.

Recent results suggest that chemotaxis may also be involved in *Bb* virulence. *Bb* is attracted to tick salivary extract (87). In a recent study, a nonchemotactic *cheA2* mutant failed to infect either immunocompetent or immunodeficient mice and was quickly eliminated from the initial inoculation sites. Furthermore, tick-mouse infection studies revealed that although the mutant was able to survive in ticks, it failed to establish a new infection in mice via tick bites (94). The altered phenotypes were restored when the mutant was complemented. In addition, both a nonchemotactic *cheY3* mutant (M.A. Motaleb, unpublished data) and a c-di-GMP *pdeA* mutant that constantly runs were unable to infect mice (91). Collectively, these data demonstrate that *Bb* needs chemotaxis to establish mammalian infection and to accomplish its natural enzootic cycle. We expect that the analysis of other motility and chemotaxis mutants will lead to a better understanding of these processes with respect to tick transmission and mammalian infection and disease.

CONCLUSION

In this review we summarize the developments made in understanding spirochete motility and chemotaxis with *Bb* chosen as the model. Significant progress has been made since the times of Dobell and Weibull, but many questions still remain. Spirochetes cause dreadful diseases prevalent throughout the world including syphilis, leptospirosis, relapsing fever, and Lyme disease. The tragedy is that so little is known about these terrifying bacteria: We cannot even grow the syphilis spirochete in the laboratory, and only a handful of laboratories in the world are doing basic studies on *Treponema pallidum*. We can only hope that in another half century (or sooner!) sufficient progress in our understanding of the biology of these enigmatic pathogens will lead to new means of disease prevention and treatment such that these diseases are eliminated.

SUMMARY POINTS

1. Cryo-ET allows for exquisitely detailed analysis of the *Bb* flagellar motor.
2. When *Bb* swims in one direction in the absence of the response regulator CheY3-P, it rotates the PFs of the polar ribbons asymmetrically.

3. *Bb* is so long that chemotaxis models as described in other bacteria do not directly apply.
4. *Bb* lacks the cascade control of motility and chemotaxis gene expression and relies on translational control.
5. Initial gene targeting experiments indicate that motility and chemotaxis play important roles in the life cycle of *Bb* in both tick and mammalian hosts.

FUTURE ISSUES

1. What are the functions of each motor protein in generating flagellar rotation?
2. What is the molecular basis for asymmetrical flagellar rotation in the absence of CheY3-P?
3. What is the molecular mechanism of chemotaxis in *Bb* and how is the direction of flagellar rotation controlled?
4. What are the details of motility and chemotaxis gene expression in *Bb*, and how do CsrA and c-di-GMP exert their activities?
5. What are the precise points of involvement of motility and chemotaxis in the life cycle of *Bb*?

DISCLOSURE STATEMENT

The authors are not aware of any affiliations, memberships, funding, or financial holdings that might be perceived as affecting the objectivity of this review.

ACKNOWLEDGMENTS

The authors dedicate this review to Professor Stuart F. Goldstein in honor of his forthcoming retirement from the University of Minnesota. His contributions have been crucial in understanding spirochete motility and in moving the field forward. We appreciate the comments and suggestions by J. Coburn, S. Goldstein, M. James, R. Silversmith, V. Sourjik, and M. Wooten. The research in this review was supported by Public Health Service grants AI078958 to C.L., GM0072004 to C.W., AR060834 to M.A.M., AI29743 and AI093917 to N.W.C. and M.R.M., and AI087946 and a Welch Foundation grant (AU-1714) to J.L. K.A.M. is supported by an American Heart Association graduate fellowship and WVNano.

LITERATURE CITED

1. Akerley BJ, Cotter PA, Miller JF. 1995. Ectopic expression of the flagellar regulon alters development of the *Bordetella*-host interaction. *Cell* 80:611–20
2. Babitzke P, Baker CS, Romeo T. 2009. Regulation of translation initiation by RNA binding proteins. *Annu. Rev. Microbiol.* 63:27–44
3. Babitzke P, Romeo T. 2007. CsrB sRNA family: sequestration of RNA-binding regulatory proteins. *Curr. Opin. Microbiol.* 10:156–63

4. Bakker RG, Li C, Miller MR, Cunningham C, Charon NW. 2007. Identification of specific chemoattractants and genetic complementation of a *Borrelia burgdorferi* chemotaxis mutant: a flow cytometry-based capillary tube chemotaxis assay. *Appl. Environ. Microbiol.* 73:1180–88
5. Barbour AG, Jasinskas A, Kayala MA, Davies DH, Steere AC, et al. 2008. A genome-wide proteome array reveals a limited set of immunogens in natural infections of humans and white-footed mice with *Borrelia burgdorferi*. *Infect. Immun.* 76:3374–89
6. Bardy SL, Maddock JR. 2007. Polar explorations: recent insights into the polarity of bacterial proteins. *Curr. Opin. Microbiol.* 10:617–23
7. Beck G, Benach JL, Habicht GS. 1990. Isolation, preliminary chemical characterization, and biological activity of *Borrelia burgdorferi* peptidoglycan. *Biochem. Biophys. Res. Commun.* 167:89–95
8. Ben-Menachem G, Kubler-Kielb J, Coxon B, Yergey A, Schneerson R. 2003. A newly discovered cholesteryl galactoside from *Borrelia burgdorferi*. *Proc. Natl. Acad. Sci. USA* 100:7913–18
9. Boehm A, Kaiser M, Li H, Spangler C, Kasper CA, et al. 2010. Second messenger-mediated adjustment of bacterial swimming velocity. *Cell* 141:107–16
10. Boesch KC, Silversmith RE, Bourret RB. 2000. Isolation and characterization of nonchemotactic CheZ mutants of *Escherichia coli*. *J. Bacteriol.* 182:3544–52
11. Botkin DJ, Abbott AN, Stewart PE, Rosa PA, Kawabata H, et al. 2006. Identification of potential virulence determinants by Himar1 transposition of infectious *Borrelia burgdorferi* B31. *Infect. Immun.* 74:6690–99
12. Brau RR, Ferrer JM, Lee H, Castro E, Tam BK, et al. 2007. Passive and active microrheology with optical tweezers. *J. Opt. A* 9:S103–12
13. Butler SM, Camilli A. 2005. Going against the grain: chemotaxis and infection in *Vibrio cholerae*. *Nat. Rev. Microbiol.* 3:611–20
14. Caimano MJ, Iyer R, Eggers CH, Gonzalez C, Morton EA, et al. 2007. Analysis of the RpoS regulon in *Borrelia burgdorferi* in response to mammalian host signals provides insight into RpoS function during the enzootic cycle. *Mol. Microbiol.* 65:1193–217

15. **Charon NW, Goldstein SF. 2002. Genetics of motility and chemotaxis of a fascinating group of bacteria: the spirochetes. *Annu. Rev. Genet.* 36:47–73**

> 15. Provides a general review of motility and chemotaxis of spirochetes.

16. Charon NW, Goldstein SF, Block SM, Curci K, Ruby JD, et al. 1992. Morphology and dynamics of protruding spirochete periplasmic flagella. *J. Bacteriol.* 174:832–40
17. Charon NW, Goldstein SF, Marko M, Hsieh C, Gebhardt LL, et al. 2009. The flat-ribbon configuration of the periplasmic flagella of *Borrelia burgdorferi* and its relationship to motility and morphology. *J. Bacteriol.* 191:600–7
18. Charon NW, Lawrence CW, O'Brien S. 1981. Movement of antibody-coated latex beads attached to the spirochete *Leptospira interrogans*. *Proc. Natl. Acad. Sci. USA* 78:7166–70
19. Chen S, Beeby M, Murphy GE, Leadbetter JR, Hendrixson DR, et al. 2011. Structural diversity of bacterial flagellar motors. *EMBO J.* 30:2972–81

20. **Chevance FF, Hughes KT. 2008. Coordinating assembly of a bacterial macromolecular machine. *Nat. Rev. Microbiol.* 6:455–65**

> 20. Reviews the control of assembly of the flagellar motor.

21. Christen M, Christen B, Allan MG, Folcher M, Jeno P, et al. 2007. DgrA is a member of a new family of cyclic diguanosine monophosphate receptors and controls flagellar motor function in *Caulobacter crescentus*. *Proc. Natl. Acad. Sci. USA* 104:4112–17
22. Cox CD. 1972. Shape of *Treponema pallidum*. *J. Bacteriol.* 109:943–44
23. Dobell C. 1912. Researches on the spirochaets and related organisms. *Arch. Protistenk.* 26:117–239
24. Dombrowski C, Kan W, Motaleb MA, Charon NW, Goldstein RE, Wolgemuth CW. 2009. The elastic basis for the shape of *Borrelia burgdorferi*. *Biophys. J.* 96:4409–17
25. Dubey AK, Baker CS, Romeo T, Babitzke P. 2005. RNA sequence and secondary structure participate in high-affinity CsrA-RNA interaction. *RNA* 11:1579–87
26. Dunham-Ems SM, Caimano MJ, Pal U, Wolgemuth CW, Eggers CH, et al. 2009. Live imaging reveals a biphasic mode of dissemination of *Borrelia burgdorferi* within ticks. *J. Clin. Invest.* 119:3652–65
27. Fang X, Gomelsky M. 2010. A post-translational, c-di-GMP-dependent mechanism regulating flagellar motility. *Mol. Microbiol.* 76:1295–305

28. Fisher MA, Grimm D, Henion AK, Elias AF, Stewart PE, et al. 2005. *Borrelia burgdorferi* sigma^{54} is required for mammalian infection and vector transmission but not for tick colonization. *Proc. Natl. Acad. Sci. USA* 102:5162–67
29. Fraser CM, Casjens S, Huang WM, Sutton GG, Clayton R, et al. 1997. Genomic sequence of a Lyme disease spirochaete, *Borrelia burgdorferi*. *Nature* 390:580–86
30. Freedman JC, Rogers EA, Kostick JL, Zhang H, Iyer R, et al. 2010. Identification and molecular characterization of a cyclic-di-GMP effector protein, PlzA (BB0733): additional evidence for the existence of a functional cyclic-di-GMP regulatory network in the Lyme disease spirochete, *Borrelia burgdorferi*. *FEMS Immunol. Med. Microbiol.* 58:285–94
31. Ge Y, Li C, Corum L, Slaughter CA, Charon NW. 1998. Structure and expression of the FlaA periplasmic flagellar protein of *Borrelia burgdorferi*. *J. Bacteriol.* 180:2418–25
32. Goldstein SF, Buttle KF, Charon NW. 1996. Structural analysis of *Leptospiraceae* and *Borrelia burgdorferi* by high-voltage electron microscopy. *J. Bacteriol.* 178:6539–45
33. Goldstein SF, Charon NW, Kreiling JA. 1994. *Borrelia burgdorferi* swims with a planar waveform similar to that of eukaryotic flagella. *Proc. Natl. Acad. Sci. USA* 91:3433–37
34. Goldstein SF, Li C, Liu J, Miller MR, Motaleb MA, et al. 2010. The chic motility and chemotaxis of *Borrelia burgdorferi*. In *Borrelia: Molecular Biology, Host Interaction, and Pathogenesis*, ed. D Scott Samuels, JD Radolf, pp. 167–87. Norfolk, UK: Calister Acad.
35. Goulbourne EA Jr, Greenberg EP. 1981. Chemotaxis of *Spirochaeta aurantia*: involvement of membrane potential in chemosensory signal transduction. *J. Bacteriol.* 148:837–44
36. Harman MW, Dunham-Ems SM, Caimano MJ, Belperron AA, Bockenstedt LK, et al. 2012. The heterogeneous motility of the Lyme disease spirochete in gelatin mimics dissemination through tissue. *Proc. Natl. Acad. Sci. USA.* 109:3059–64
37. Hess JF, Oosawa K, Kaplan N, Simon MI. 1988. Phosphorylation of three proteins in the signaling pathway of bacterial chemotaxis. *Cell* 53:79–87
38. Hovind-Hougen K. 1984. Ultrastructure of spirochetes isolated from *Ixodes ricinus* and *Ixodes dammini*. *Yale J. Biol. Med.* 57:543–48
39. Izard J, Renken C, Hsieh C, Desrosiers DC, Dunham-Ems S, et al. 2009. Cryo-electron tomography elucidates the molecular architecture of *Treponema pallidum*, the syphilis spirochete. *J. Bacteriol.* 191:7566–80
40. Josenhans C, Suerbaum S. 2002. The role of motility as a virulence factor in bacteria. *Int. J. Med. Microbiol.* 291:605–14
41. Kimsey RB, Spielman A. 1990. Motility of Lyme disease spirochetes in fluids as viscous as the extracellular matrix. *J. Infect. Dis.* 162:1205–8
42. Kostick JL, Szkotnicki LT, Rogers EA, Bocci P, Raffaelli N, Marconi RT. 2011. The diguanylate cyclase, Rrp1, regulates critical steps in the enzootic cycle of the Lyme disease spirochetes. *Mol. Microbiol.* 81:219–31
43. Kudryashev M, Cyrklaff M, Baumeister W, Simon MM, Wallich R, Frischknecht F. 2009. Comparative cryo-electron tomography of pathogenic Lyme disease spirochetes. *Mol. Microbiol.* 71:1415–34
44. Kudryashev M, Cyrklaff M, Wallich R, Baumeister W, Frischknecht F. 2010. Distinct in situ structures of the *Borrelia* flagellar motor. *J. Struct. Biol.* 169:54–61
45. Lambert A, Picardeau M, Haake DA, Sermswan RW, Srikram A, et al. 2012. FlaA proteins in *Leptospira interrogans* are essential for motility and virulence but are not required for formation of the flagellum sheath. *Infect. Immun.* 80:2019–25
46. Leake MC, Chandler JH, Wadhams GH, Bai F, Berry RM, Armitage JP. 2006. Stoichiometry and turnover in single, functioning membrane protein complexes. *Nature* 443:355–58
47. Lee WY, Moriarty TJ, Wong CH, Zhou H, Strieter RM, et al. 2010. An intravascular immune response to *Borrelia burgdorferi* involves Kupffer cells and iNKT cells. *Nat. Immunol.* 11:295–302
48. Li C, Bakker RG, Motaleb MA, Sartakova ML, Cabello FC, Charon NW. 2002. Asymmetrical flagellar rotation in *Borrelia burgdorferi* nonchemotactic mutants. *Proc. Natl. Acad. Sci. USA* 99:6169–74
49. Li C, Wolgemuth CW, Marko M, Morgan DG, Charon NW. 2008. Genetic analysis of spirochete flagellin proteins and their involvement in motility, filament assembly, and flagellar morphology. *J. Bacteriol.* 190:5607–15

50. Li C, Xu H, Zhang K, Liang FT. 2010. Inactivation of a putative flagellar motor switch protein FliG1 prevents *Borrelia burgdorferi* from swimming in highly viscous media and blocks its infectivity. *Mol. Microbiol.* 75:1563–76
51. Limberger RJ. 2004. The periplasmic flagellum of spirochetes. *J. Mol. Microbiol. Biotechnol.* 7:30–40
52. Limberger RJ, Slivienski LL, Samsonoff WA. 1994. Genetic and biochemical analysis of the flagellar hook of *Treponema phagedenis*. *J. Bacteriol.* 176:3631–37
53. Liu J, Howell JK, Bradley SD, Zheng Y, Zhou ZH, Norris SJ. 2010. Cellular architecture of *Treponema pallidum*: novel flagellum, periplasmic cone, and cell envelope as revealed by cryo electron tomography. *J. Mol. Biol.* 403:546–61
54. **Liu J, Lin T, Botkin DJ, McCrum E, Winkler H, Norris SJ. 2009. Intact flagellar motor of Borrelia burgdorferi revealed by cryo-electron tomography: evidence for stator ring curvature and rotor/C-ring assembly flexion. *J. Bacteriol.* 191:5026–36**

> 54. Detailed examination of the *Bb* motor in situ at a resolution of 3.5 nm using cryo-ET.

55. Lloyd SA, Tang H, Wang X, Billings S, Blair DF. 1996. Torque generation in the flagellar motor of *Escherichia coli*: evidence of a direct role for FliG but not for FliM or FliN. *J. Bacteriol.* 178:223–31
56. Lucchetti-Miganeh C, Burrowes E, Baysse C, Ermel G. 2008. The post-transcriptional regulator CsrA plays a central role in the adaptation of bacterial pathogens to different stages of infection in animal hosts. *Microbiology* 154:16–29
57. Lux R, Miller JN, Park NH, Shi W. 2001. Motility and chemotaxis in tissue penetration of oral epithelial cell layers by *Treponema denticola*. *Infect. Immun.* 69:6276–83
58. Mauriello EM, Mignot T, Yang Z, Zusman DR. 2010. Gliding motility revisited: How do the myxobacteria move without flagella? *Microbiol. Mol. Biol. Rev.* 74:229–49
59. McGee DJ, Langford ML, Watson EL, Carter JE, Chen YT, Ottemann KM. 2005. Colonization and inflammation deficiencies in Mongolian gerbils infected by *Helicobacter pylori* chemotaxis mutants. *Infect. Immun.* 73:1820–27
60. Mercante J, Edwards AN, Dubey AK, Babitzke P, Romeo T. 2009. Molecular geometry of CsrA (RsmA) binding to RNA and its implications for regulated expression. *J. Mol. Biol.* 392:511–28
61. Moriarty TJ, Norman MU, Colarusso P, Bankhead T, Kubes P, Chaconas G. 2008. Real-time high resolution 3D imaging of the Lyme disease spirochete adhering to and escaping from the vasculature of a living host. *PLoS Pathog.* 4:e1000090
62. Motaleb MA, Corum L, Bono JL, Elias AF, Rosa P, et al. 2000. *Borrelia burgdorferi* periplasmic flagella have both skeletal and motility functions. *Proc. Natl. Acad. Sci. USA* 97:10899–904
63. Motaleb MA, Miller MR, Bakker RG, Li C, Charon NW. 2007. Isolation and characterization of chemotaxis mutants of the Lyme disease spirochete *Borrelia burgdorferi* using allelic exchange mutagenesis, flow cytometry, and cell tracking. *Methods Enzymol.* 422:419–37
64. Motaleb MA, Miller MR, Li C, Bakker RG, Goldstein SF, et al. 2005. CheX is a phosphorylated CheY phosphatase essential for *Borrelia burgdorferi* chemotaxis. *J. Bacteriol.* 187:7963–69
65. Motaleb MA, Pitzer JE, Sultan SZ, Liu J. 2011. A novel gene inactivation system reveals altered periplasmic flagellar orientation in a *Borrelia burgdorferi fliL* mutant. *J. Bacteriol.* 193:3324–31
66. Motaleb MA, Sal MS, Charon NW. 2004. The decrease in FlaA observed in a *flaB* mutant of *Borrelia burgdorferi* occurs posttranscriptionally. *J. Bacteriol.* 186:3703–11
67. Motaleb MA, Sultan SZ, Miller MR, Li C, Charon NW. 2011. CheY3 of *Borrelia burgdorferi* is the key response regulator essential for chemotaxis and forms a long-lived phosphorylated intermediate. *J. Bacteriol.* 193:3332–41
68. Muff TJ, Ordal GW. 2008. The diverse CheC-type phosphatases: chemotaxis and beyond. *Mol. Microbiol.* 70:1054–61
69. **Murphy GE, Leadbetter JR, Jensen GJ. 2006. In situ structure of the complete Treponema primitia flagellar motor. *Nature* 442:1062–64**

> 69. First examination of the flagellar motor in situ using cryo-ET.

70. Norman MU, Moriarty TJ, Dresser AR, Millen B, Kubes P, Chaconas G. 2008. Molecular mechanisms involved in vascular interactions of the Lyme disease pathogen in a living host. *PLoS Pathog.* 4:e1000169
71. Ouyang Z, Blevins JS, Norgard MV. 2008. Transcriptional interplay among the regulators Rrp2, RpoN and RpoS in *Borrelia burgdorferi*. *Microbiology* 154:2641–58

72. Paster BJ. 2011. Phylum XV. Spirochaetes Garrity and Holt 2001. In *Bergey's Manual of Systematic Bacteriology*, ed. NR Krieg, W Ludwig, WB Whitman, BP Hedlund, BJ Paster, et al., 4:471–566. New York: Springer
73. Paul K, Nieto V, Carlquist WC, Blair DF, Harshey RM. 2010. The c-di-GMP binding protein YcgR controls flagellar motor direction and speed to affect chemotaxis by a "backstop brake" mechanism. *Mol. Cell* 38:128–39
74. Pazy Y, Motaleb MA, Guarnieri MT, Charon NW, Zhao R, Silversmith RE. 2010. Identical phosphatase mechanisms achieved through distinct modes of binding phosphoprotein substrate. *Proc. Natl. Acad. Sci. USA* 107:1924–29
75. Pitzer JE, Sultan SZ, Hayakawa Y, Hobbs G, Miller MR, Motaleb MA. 2011. Analysis of the *Borrelia burgdorferi* cyclic-di-GMP-binding protein PlzA reveals a role in motility and virulence. *Infect. Immun.* 79:1815–25
76. **Porter SL, Wadhams GH, Armitage JP. 2011. Signal processing in complex chemotaxis pathways. *Nat. Rev. Microbiol.* 9:153–65**
77. Raddi G, Morado DR, Yan J, Haake DA, Yang XF, Liu J. 2012. Three-dimensional structures of pathogenic and saprophytic *Leptospira* species revealed by cryo-electron tomography. *J. Bacteriol.* 194:1299–306
78. Rao CV, Glekas GD, Ordal GW. 2008. The three adaptation systems of *Bacillus subtilis* chemotaxis. *Trends Microbiol.* 16:480–87
79. Revel AT, Talaat AM, Norgard MV. 2002. DNA microarray analysis of differential gene expression in *Borrelia burgdorferi*, the Lyme disease spirochete. *Proc. Natl. Acad. Sci. USA* 99:1562–67
80. Romeo T. 1998. Global regulation by the small RNA-binding protein CsrA and the non-coding RNA molecule CsrB. *Mol. Microbiol.* 29:1321–30
81. **Rosa PA, Tilly K, Stewart PE. 2005. The burgeoning molecular genetics of the Lyme disease spirochaete. *Nat. Rev. Microbiol.* 3:129–43**
82. Rosey EL, Kennedy MJ, Yancey RJ Jr. 1996. Dual *flaA1 flaB1* mutant of *Serpulina hyodysenteriae* expressing periplasmic flagella is severely attenuated in a murine model of swine dysentery. *Infect. Immun.* 64:4154–62
83. Ryjenkov DA, Tarutina M, Moskvin OV, Gomelsky M. 2005. Cyclic diguanylate is a ubiquitous signaling molecule in bacteria: insights into biochemistry of the GGDEF protein domain. *J. Bacteriol.* 187:1792–98
84. Sal MS, Li C, Motalab MA, Shibata S, Aizawa S, Charon NW. 2008. *Borrelia burgdorferi* uniquely regulates its motility genes and has an intricate flagellar hook-basal body structure. *J. Bacteriol.* 190:1912–21
85. Samatey FA, Matsunami H, Imada K, Nagashima S, Shaikh TR, et al. 2004. Structure of the bacterial flagellar hook and implication for the molecular universal joint mechanism. *Nature* 431:1062–68
86. **Samuels DS, Radolf JD, eds. 2010. *Borrelia: Molecular Biology, Host Interaction and Pathogenesis*. Norfolk, UK: Calister Acad. 548 pp.**
87. Shih CM, Chao LL, Yu CP. 2002. Chemotactic migration of the Lyme disease spirochete (*Borrelia burgdorferi*) to salivary gland extracts of vector ticks. *Am. J. Trop. Med. Hyg.* 66:616–21
88. Sourjik V, Armitage JP. 2010. Spatial organization in bacterial chemotaxis. *EMBO J.* 29:2724–33
89. Stanek G, Wormser GP, Gray J, Strle F. 2011. Lyme borreliosis. *Lancet* 379:461–73
90. Sultan SZ, Pitzer JE, Boquoi T, Hobbs G, Miller MR, Motaleb MA. 2011. Analysis of the HD-GYP domain cyclic dimeric GMP phosphodiesterase reveals a role in motility and the enzootic life cycle of *Borrelia burgdorferi*. *Infect. Immun.* 79:3273–83
91. Sultan SZ, Pitzer JE, Miller MR, Motaleb MA. 2010. Analysis of a *Borrelia burgdorferi* phosphodiesterase demonstrates a role for cyclic-di-guanosine monophosphate in motility and virulence. *Mol. Microbiol.* 77:128–42
92. Sze CW, Li C. 2011. Inactivation of *bb0184*, which encodes carbon storage regulator A, represses the infectivity of *Borrelia burgdorferi*. *Infect. Immun.* 79:1270–79
93. Sze CW, Morado DR, Liu J, Charon NW, Xu H, Li C. 2011. Carbon storage regulator A (CsrA$_{Bb}$) is a repressor of *Borrelia burgdorferi* flagellin protein FlaB. *Mol. Microbiol.* 82:851–64
94. Sze CW, Zhang K, Kariu T, Pal U, Li C. 2012. *Borrelia burgdorferi* needs chemotaxis to establish infection in mammals and to accomplish its enzootic cycle. *Infect. Immun.* 80:2485–92

76. Reviews chemotaxis in different species of bacteria.

81. Discusses how recent breakthroughs in the molecular genetics of *Bb* are leading to a better understanding of the pathogenesis of Lyme disease.

86. Gives a comprehensive analysis of the biology of *Bb* and Lyme disease.

95. Thomas DR, Francis NR, Xu C, DeRosier DJ. 2006. The three-dimensional structure of the flagellar rotor from a clockwise-locked mutant of *Salmonella enterica* serovar Typhimurium. *J. Bacteriol.* 188:7039–48
96. Tokarz R, Anderton JM, Katona LI, Benach JL. 2004. Combined effects of blood and temperature shift on *Borrelia burgdorferi* gene expression as determined by whole genome DNA array. *Infect. Immun.* 72:5419–32
97. Turner L, Ryu WS, Berg HC. 2000. Real-time imaging of fluorescent flagellar filaments. *J. Bacteriol.* 182:2793–801
98. Vaknin A, Berg HC. 2004. Single-cell FRET imaging of phosphatase activity in the *Escherichia coli* chemotaxis system. *Proc. Natl. Acad. Sci. USA* 101:17072–77
99. Wadhams GH, Armitage JP. 2004. Making sense of it all: bacterial chemotaxis. *Nat. Rev. Mol. Cell Biol.* 5:1024–37
100. Wei BL, Brun-Zinkernagel AM, Simecka JW, Pruss BM, Babitzke P, Romeo T. 2001. Positive regulation of motility and *flhDC* expression by the RNA-binding protein CsrA of *Escherichia coli*. *Mol. Microbiol.* 40:245–56
101. Weibull C. 1960. Movement. In *The Bacteria, A Treatise on Structure and Function*, ed. IC Gunsalas, RY Stanier, 1:153–234. New York/ London: Academic
102. Wolgemuth CW, Charon NW, Goldstein SF, Goldstein RE. 2006. The flagellar cytoskeleton of the spirochetes. *J. Mol. Microbiol. Biotechnol.* 11:221–27
103. **Wuichet K, Zhulin IB. 2010. Origins and diversification of a complex signal transduction system in prokaryotes. *Sci. Signal.* 3:ra50**
104. Xu H, Caimano MJ, Lin T, He M, Radolf JD, et al. 2010. Role of acetyl-phosphate in activation of the Rrp2-RpoN-RpoS pathway in *Borrelia burgdorferi*. *PLoS. Pathog.* 6:e1001104
105. **Xu H, Raddi G, Liu J, Charon NW, Li C. 2011. Chemoreceptors and flagellar motors are subterminally located in close proximity at the two cell poles in spirochetes. *J. Bacteriol.* 193:2652–56**
106. Yakhnin H, Pandit P, Petty TJ, Baker CS, Romeo T, Babitzke P. 2007. CsrA of *Bacillus subtilis* regulates translation initiation of the gene encoding the flagellin protein (*hag*) by blocking ribosome binding. *Mol. Microbiol.* 64:1605–20
107. Yang J, Huber G, Wolgemuth CW. 2012. Forces and torques on rotating spirochete flagella. *Phys. Rev. Lett.* 107:268101
108. Yang X, Goldberg MS, Popova TG, Schoeler GB, Wikel SK, et al. 2000. Interdependence of environmental factors influencing reciprocal patterns of gene expression in virulent *Borrelia burgdorferi*. *Mol. Microbiol.* 37:1470–79
109. Young GM, Badger JL, Miller VL. 2000. Motility is required to initiate host cell invasion by *Yersinia enterocolitica*. *Infect. Immun.* 68:4323–26
110. Yuri K, Takamoto Y, Okada M, Hiramune T, Kikuchi N, Yanagawa R. 1993. Chemotaxis of leptospires to hemoglobin in relation to virulence. *Infect. Immun.* 61:2270–72

---

103. Provides an extensive genomic analysis of the evolution of chemotaxis genes in bacteria.

105. Uses light microscopy and cryo-EM to analyze one end of a *Bb* cell.

RELATED RESOURCES

Brisson D, Drecktrah D, Eggers CH, Samuels DS. 2012. Genetics of *Borrelia burgdorferi*. *Annu. Rev. Genet.* 46:In press

Hazelbauer GL. 2012. Bacterial chemotaxis: the early years of molecular studies. *Annu. Rev. Microbiol.* 66:285–303

Paul K, Gonzalez-Bonet G, Bilwes AM, Crane BR, Blair D. 2011. Architecture of the flagellar rotor. *EMBO J.* 30:2962–71

Samuels DS. 2011. Gene regulation in *Borrelia burgdorferi*. *Annu. Rev. Microbiol.* 65:479–99

Tilly K, Rosa PA, Stewart PE. 2008. Biology of infection with *Borrelia burgdorferi*. *Infect. Dis. Clin. North Am.* 22:217–34

Vaginal Microbiome: Rethinking Health and Disease

Bing Ma,[1] Larry J. Forney,[2] and Jacques Ravel[1]

[1]Institute for Genome Sciences, University of Maryland School of Medicine, Baltimore, Maryland 21201; email: bma@som.umaryland.edu, jravel@som.umaryland.edu

[2]Department of Biological Sciences and the Initiative for Bioinformatics and Evolutionary Studies, University of Idaho, Moscow, Idaho 83844; email: lforney@uidaho.edu

Keywords

vaginal microbiota, vaginal ecosystem, bacterial vaginosis

Abstract

Vaginal microbiota form a mutually beneficial relationship with their host and have a major impact on health and disease. In recent years our understanding of vaginal bacterial community composition and structure has significantly broadened as a result of investigators using cultivation-independent methods based on the analysis of 16S ribosomal RNA (rRNA) gene sequences. In asymptomatic, otherwise healthy women, several kinds of vaginal microbiota exist, the majority often dominated by species of *Lactobacillus*, while others are composed of a diverse array of anaerobic microorganisms. Bacterial vaginosis is the most common vaginal condition and is vaguely characterized as the disruption of the equilibrium of the normal vaginal microbiota. A better understanding of normal and healthy vaginal ecosystems that is based on their true function and not simply on their composition would help better define health and further improve disease diagnostics as well as the development of more personalized regimens to promote health and treat diseases.

Contents

INTRODUCTION	372
COMPOSITION AND STRUCTURE OF THE VAGINAL MICROBIOTA	373
Culture-Dependent and Culture-Independent Approaches to Survey Microbial Community Composition and Structure	373
Lactobacillus-Dominated Vaginal Microbiota	374
Lactobacillus Species' Antimicrobial Substances Production	376
Other Types of Vaginal Microbiota	377
BACTERIAL VAGINOSIS	377
The Bacterial Vaginosis Enigma	377
Complex Etiology of Bacterial Vaginosis	379
RETHINKING NORMAL AND HEALTHY	379
Temporal Dynamics of Vaginal Communities	380
Toward a System-Level Understanding of the Vaginal Ecosystem	381
CONCLUDING REMARKS	383

INTRODUCTION

The microbiota normally associated with the human body have an important influence on human development, physiology, immunity, and nutrition (18, 23, 65, 66, 70, 111). The vast majority of these indigenous microbiota exist in a mutualistic relationship with their human host, although few are opportunistic pathogens that can cause both chronic infections and life-threatening diseases. These microbial communities are believed to constitute the first line of defense against infection by competitively excluding invasive nonindigenous organisms that cause diseases. Despite their importance, surprisingly little is known about how these communities differ between individuals in composition and function and, more importantly, how their constituent members interact with each other and the host to form a dynamic ecosystem that responds to environmental disturbances. Major efforts are now under way to better understand the true role of these communities in health and diseases (84).

The human vagina and the bacterial communities that reside therein are an example of this finely balanced mutualistic association. In this relationship, the host provides benefit to the microbial communities in the form of the nutrients needed to support bacterial growth. This is of obvious importance because bacteria are continually shed from the body in vaginal secretions, and bacterial growth must occur to replenish their numbers. Some of the required nutrients are derived from sloughed cells, while others are from glandular secretions. The indigenous bacterial communities, on the other hand, play a protective role in preventing colonization of the host by potentially pathogenic organisms, including those responsible for symptomatic bacterial vaginosis, yeast infections, sexually transmitted infections (STIs), and urinary tract infections (42, 47, 96, 98, 113, 118). Lactobacilli have long been thought to be the keystone species of vaginal communities in reproductive-age women. These microorganisms benefit the host by producing lactic acid as a fermentation product that lowers the vaginal pH to ∼3.5–4.5 (12). Although a wide range of other species are members of vaginal bacterial communities, their ecological roles and influences on the overall community dynamics and function are largely undetermined. The vaginal ecosystem is thought to have been shaped by coevolutionary processes between the human host and specific

Microbiota: microbial community composition and structure

microbial partners, although the selective forces (traits) behind this mutualistic association are still not clear.

The development of culture-independent approaches has greatly facilitated comprehensive surveys of the composition of vaginal microbial communities. These studies have shown that several distinct kinds of vaginal communities with markedly different species composition occur and that the frequency of these types of microbiota varies in different ethnic groups (86, 122–124). It is hypothesized that differences in species composition may correlate with how vaginal communities respond to disturbances (52, 104, 115, 123). Conceptually this is important because vaginal communities continually experience various kinds of chronic and acute disturbances caused by human behaviors, such as the use of antibiotics, hormonal contraceptives and other methods of birth control, sexual activity, vaginal lubricants, douching, and so forth, in addition to many other intrinsic factors such as the innate and adaptive immune systems of hosts (64, 88, 110). Further, a disturbed state itself may constitute the clinical syndrome known as bacterial vaginosis (BV), which as a disruption of ecological equilibria is believed to increase the risk of invasion by infectious agents. Although knowledge accumulated over the past few decades has provided some insights into the vaginal ecosystem, there remains a need to define and better understand factors that affect the composition and dynamics of vaginal microbiota, including the role of human genetics and physiology in both health and diseases. This knowledge will facilitate the development of new strategies for disease diagnosis and personalized treatments to promote health and improve the quality of women's lives. This cannot be accomplished without addressing the fundamental issue of what constitutes a normal and healthy vaginal microbiota and understanding its function in health and diseases.

BV: bacterial vaginosis

Phylotype: community members represented by a set of phylogenetically related 16S rRNA gene sequences

COMPOSITION AND STRUCTURE OF THE VAGINAL MICROBIOTA

Comprehensive surveys of vaginal microbial communities using culture-independent approaches have revealed that *Lactobacillus* species are the dominant vaginal bacterial species in a majority of women. However, an appreciable proportion of asymptomatic, otherwise healthy individuals have vaginal microbiota lacking significant numbers of *Lactobacillus* spp. and harboring a diverse array of facultative and strictly anaerobic microorganisms.

Culture-Dependent and Culture-Independent Approaches to Survey Microbial Community Composition and Structure

Most of our knowledge of the composition, metabolic function, and ecology of indigenous microbial communities associated with humans has come from studies that depended on cultivating microbial populations. Hence, our current understanding of microbe-host interactions is limited and skewed because the overwhelming majority of microbial species (>99%) resist cultivation in the laboratory (8). Our limited ability to culture may result from strict, yet unknown, growth requirements, such as the optimal combination of nutrients, growth temperatures, and dissolved-oxygen levels, or potentially the need to cocultivate with key microbial partners (3, 27). Our knowledge of microbial diversity has expanded enormously through the use of culture-independent approaches based on the analysis of 16S rRNA gene sequences (50, 107). These strategies circumvent the need to cultivate organisms by directly extracting genetic materials from environmental or biological samples. This is followed by amplification of the 16S rRNA genes using primers that anneal to highly conserved regions of the gene, followed by sequencing and classification of the phylotypes present. This constitutes an efficient way to comprehensively characterize microbial diversity. The development of next-generation sequencing technologies, including the use of massively parallel

Community class: clusters of community state types' profiles that have similar temporal patterns of bacterial community dynamics (applies to time series)

Community state types: clusters of community states that have similar phylotype composition and relative abundance

Community states: phylotype composition and relative abundance of a single sample or a sample in a time series

DNA sequencing of short, hypervariable regions of the 16S rRNA gene, now affords us the opportunity to obtain detailed surveys of microbial communities, including the identification of taxa present in low abundance that compose the rare biospheres (27, 102). Other conserved genes such as *cpn60*, *rpoC*, *uvrB*, and *recA* have also been used for these purposes (92, 114).

Culture-independent methods have demonstrated that, when surveyed cross-sectionally, several kinds of vaginal communities (community state types) exist in normal and otherwise healthy women, each with a markedly different bacterial species composition. These communities either are dominated by one of four common *Lactobacillus* species (*L. crispatus*, *L. iners*, *L. gasseri*, and *L. jensenii*) or do not contain significant numbers of lactobacilli, but instead have a diverse array of strict and facultative anaerobes (86).

Lactobacillus-Dominated Vaginal Microbiota

Members of the genus *Lactobacillus* are commonly identified as the hallmark of a normal or healthy vagina (25, 42, 69, 98). Since they were first identified by cultivation in vaginal secretion in the late nineteenth century by Donderlein (24, 90, 106), *Lactobacillus* spp. have been thought to play a major role in protecting the vaginal environment from nonindigenous and potentially harmful microorganisms. This is accomplished through the production of lactic acid, resulting in a low and protective pH (3.5–4.5) (1, 11, 12, 54, 87, 91). Interestingly, lactic acid is more effective than acidity alone as a microbicide against HIV or against pathogens such as *Neisseria gonorrhoeae* (38, 60). Exposure to gram-negative bacteria, in the presence of lactic acid, is believed to have stimulatory effects on the host innate immune defense system (120). A recent study using in vitro colonization of vaginal epithelial cell monolayers with common bacteria such as *L. crispatus*, *Prevotella bivia*, and *Atopobium vaginae* demonstrated that these key vaginal bacteria appear to regulate the epithelial innate immunity in a species-specific manner (32).

L. crispatus was previously thought to be one of the most common species of lactobacilli in the vagina (5). However, the application of the culture-independent method has identified *L. iners*, an organism that is difficult to cultivate and does not grow on traditional culture media, as the most prevalent vaginal bacterial species (30, 122). In these studies, vaginal microbiota of 42% (17) and 66% (123) of the reproductive-age women sampled were dominated by *L. iners*. A recent large-scale cross-sectional study of 396 healthy asymptomatic women revealed that *L. iners* was detected in 83.5% of the subjects and dominated 34.1% of the communities analyzed (86), and that *L. crispatus*, *L. gasseri*, and *L. jensenii* were present in 64.5, 42.9, and 48.2% of the subjects and dominated in 26.2, 6.3, and 5.3% of the samples, respectively (86). This large study showed that vaginal bacterial communities that had similar species composition and abundance could be classified into five groups, which are referred to as community state types (**Figure 1**). The four community state types dominated by *Lactobacillus* spp. represented 73% of the samples, which supports the prevailing view that *Lactobacillus* spp. are important members of vaginal microbiota. The remaining 27% represented communities that lacked significant numbers of *Lactobacillus* spp. but instead were composed of a diverse array of facultative or strictly anaerobic bacteria. Interestingly, the distribution of *Lactobacillus* spp.–dominated community state types varies significantly among individuals with different ethnic backgrounds (86, 123, 124). White and Asian women are more likely than Hispanic and Black women to have vaginal communities dominated by lactobacilli (86). When a *Lactobacillus* species is present, vaginal communities of Hispanic and Black women are more often dominated by *L. iners* (86). The study also noted a higher average pH in Black and Hispanic women, 4.7 and 5.0 respectively, compared to 4.4 and 4.2 for Asian and White women. This observation supports the hypothesis that host factors may play an important role in determining vaginal microbial community composition and structure.

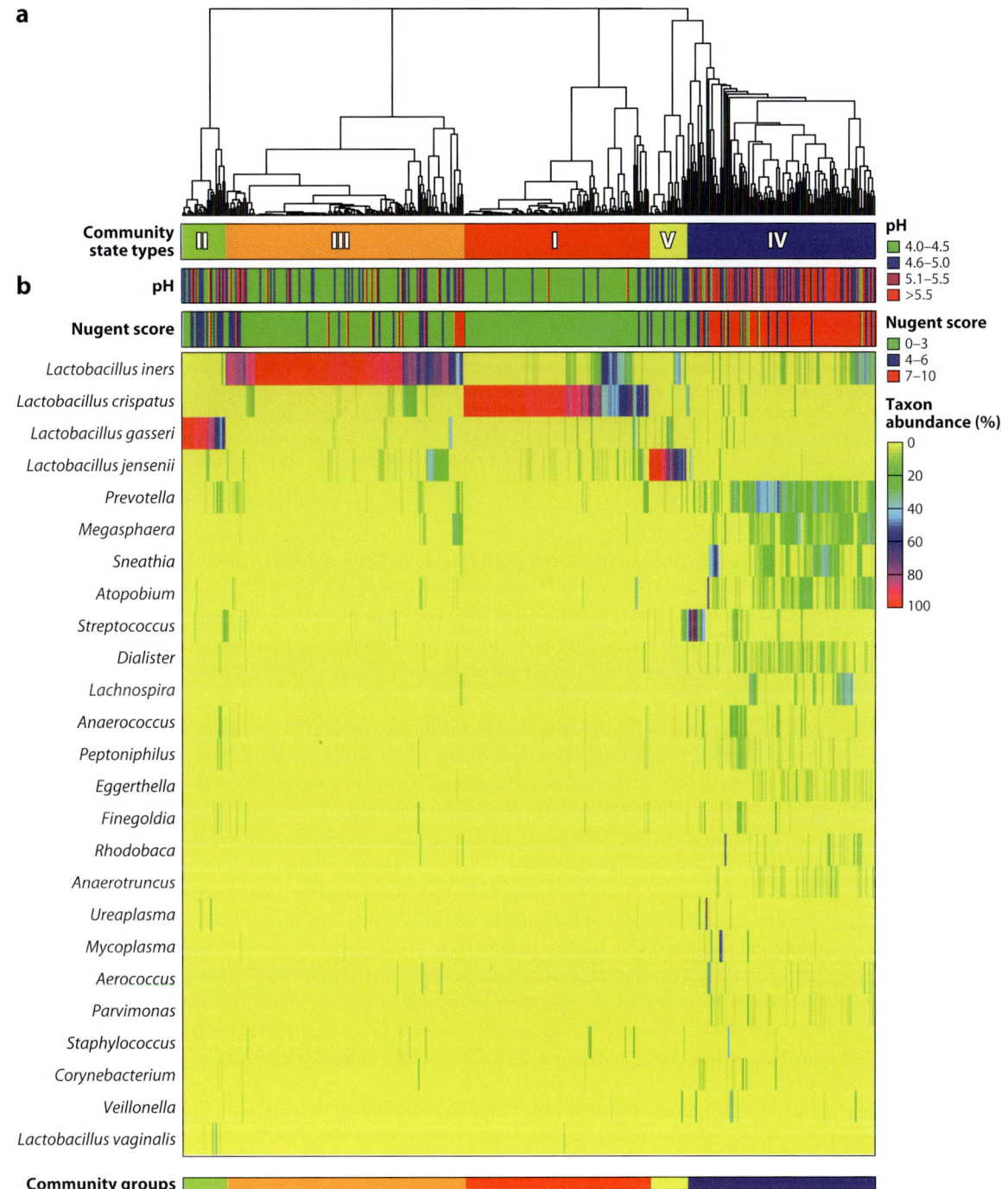

Figure 1

Heatmap of percentage abundance of microbial taxa found in the vaginal microbial communities of 394 reproductive-age women. (*a*) Complete linkage clustering of samples based on species composition and abundance in communities defining five community state types (CST I–V). (*b*) Nugent scores and pH measurements for each of the 394 samples. Adapted from Reference 86.

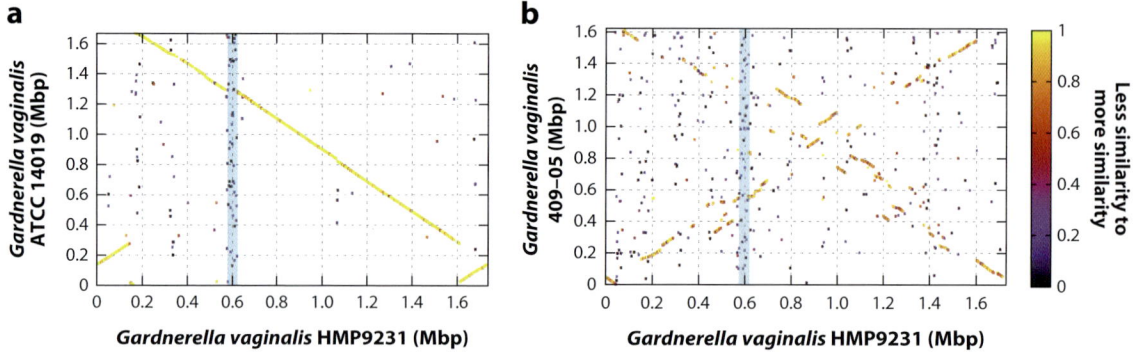

Figure 2
Whole-genome comparative analysis of *Gardnerella vaginalis* using BLAST score ratio analysis. A protein match between two genomes is represented by a point plotted using the genomic coordinate of both matched proteins as *x* and *y* coordinates. The level of protein sequence similarity is represented by the color of the points (see scale at right). (*a*) High degree of protein similarity and synteny are observed between *G. vaginalis* strains HMP9231 and ATCC 14019. (*b*) Lack of synteny and low degree of protein similarity are observed between *G. vaginalis* strains HMP9231 and 409-05. The vertical blue bar highlights a set of syntenic genes that is unique to the *G. vaginalis* HMP9231 genome and is not present in the other two genomes.

These findings highlight potential differences in the protective capabilities of vaginal *Lactobacillus* species. Statistically significant differences have been observed in the ability of different *Lactobacillus* species to lower pH between different community state types. *L. crispatus*–dominated communities are able to acidify the vaginal milieu to pH 4.0; communities dominated by other species achieved pH ranging from 4.4 to 5.0 (86). Although *Lactobacillus* spp. drive these processes, other community members can contribute by either producing or utilizing lactic acid. Moreover, it is anticipated that strains of the same species will also demonstrate genomic differences that will result in specific physiological and biochemical traits. Previous comparative genomic analyses have identified a high level of genetic diversity and varied metabolic potential of closely related bacterial species or strains of the same species. For example, *Escherichia coli* strains can vary by as much as 25% in their gene content (79), and strains of the same serovars of *Salmonella enterica* can vary by more than 20% (61). Much of the variation occurs in the form of large genomic islands or lineage-specific regions that may be involved in adaptation to the host microenvironment (10, 82). No comparative genomic studies of vaginal *Lactobacillus* spp. have been reported. However, in **Figure 2** we show strains of *Gardnerella vaginalis*, which is commonly found in the vaginal microbiota and associated with bacterial vaginosis (73, 116), can differ by 31% in gene content and gene order (synteny). Advanced knowledge of genetic variation among *Lactobacillus* species (or strains) may provide further insight into their functional potential, which may have significant implications for health and diseases. Because of species-level or strain-level genomic heterogeneity, the analysis of 16S rRNA gene sequences, though taxonomically informative, is not able to identify functional differences without considerable speculation, and attempts to infer the function of any bacterial community knowing only "who is there" should be made with caution.

Lactobacillus Species' Antimicrobial Substances Production

Vaginal *Lactobacillus* species produce antimicrobial compounds in addition to lactic acid, including target-specific bacteriocins (2, 6) and broad-spectrum hydrogen peroxide (28, 43). Bacteriocins are proteinaceous, bactericidal substances synthesized by bacteria that have a narrow spectrum of activity (53). Their antimicrobial activity is usually based on permeabilization of the target cell membrane (81). In the vagina, bacteriocins could play a major role in fending off the growth of nonindigenous or pathogenic organisms (26). Many *Lactobacillus* spp. produce hydrogen peroxide in vitro under aerobic conditions, which could inhibit colonization of potential pathogenic bacteria in vivo (43, 48, 112). However, the vagina is virtually an anaerobic environment wherein dissolved oxygen levels are low. Therefore, it is unlikely that significant amounts of hydrogen peroxide are produced and accumulate to a toxic level. Further, a recent study showed that physiological concentrations of hydrogen peroxide have no detectable effect on 17 BV-associated bacteria (BVAB) under anaerobic growth conditions and the presence of vaginal fluid can actually block its antimicrobial activity (80). In addition, it appears that high concentrations of hydrogen peroxide are even more toxic to vaginal *Lactobacillus* than to BVAB (80). Interestingly, some *Lactobacillus* species, such as *L. iners*, fail to produce hydrogen peroxide. This feature has been used to differentiate beneficial versus nonbeneficial vaginal *Lactobacillus* isolates (5), and it has been suggested that hydrogen peroxide–producing vaginal *Lactobacillus* spp. are more likely to be protective against acquisition of BV (43). Given the information summarized above, it is more likely that in vitro hydrogen peroxide production is not a significant factor in preventing the emergence of disease-causing organisms; however, it could be a surrogate marker for other yet unknown biochemical or physiological traits. Overall, this work suggests that lactic acid, not hydrogen peroxide, is more likely to contribute to the protective role of vaginal microbiota.

Other Types of Vaginal Microbiota

Recent studies have found that 20–30% of asymptomatic, otherwise healthy women harbor vaginal communities that lack appreciable numbers of *Lactobacillus* but include a diverse array of facultative or strictly anaerobic bacteria that are associated with a somewhat higher pH (5.3–5.5) (86, 122–124). This proportion of communities can reach 40% among Black and Hispanic women (86). These microbiota include members of the genera *Atopobium*, *Corynebacterium*, *Anaerococcus*, *Peptoniphilus*, *Prevotella*, *Gardnerella*, *Sneathia*, *Eggerthella*, *Mobiluncus*, and *Finegoldia*, among others (52, 86, 115, 122–124). These findings challenge the common wisdom that the occurrence of high numbers of lactobacilli and a vaginal pH of <4.5 are synonymous with "normal" and "healthy." Previous studies have hypothesized non-*Lactobacillus*-dominant vaginal microbiota may be nonetheless able to maintain functional vaginal ecosystems by preserving lactic acid production and possibly other important functions (36, 86, 122). Many underappreciated microorganisms, such as members from *Atopobium*, *Streptococcus*, *Staphylococcus*, *Megasphaera*, and *Leptotrichia*, are capable of homolactic or heterolactic acid fermentations (89, 122). The highly diversified microbial community may have accommodated functional redundancy, allowing for the function of the ecosystem to persist in the face of perturbations (117). In the absence of symptomology, these types of vaginal bacterial communities might be considered normal and healthy, even though the composition of these communities closely resembles those associated with symptomatic BV.

BACTERIAL VAGINOSIS

BV is a highly prevalent vaginal disorder in reproductive-age women, but its diagnostics and treatment are disappointingly ineffective. BV is often vaguely characterized as the disruption of the equilibrium of the normal vaginal ecosystem.

The Bacterial Vaginosis Enigma

BV is the most frequently cited cause of vaginal discharge and malodor, and it is the most common vaginal condition of reproductive-age women, resulting in millions of health care visits annually

Nugent-score BV:
BV that is diagnosed based on Nugent Gram stain test; symptomology is not taken into account

in the United States alone (99). In a cross-sectional study of reproductive-age women in 2001, the National Health and Nutrition Examination Survey found that the prevalence of BV in the United States was 29.2% (59). BV is an independent risk factor for the acquisition of STIs (19, 69, 83, 118), the acquisition and transmission of HIV (20–22, 69, 103, 105), the development of pelvic inflammatory disease (76), as well as for reproductive tract and obstetric sequelae (37, 39, 49, 71, 72). Numerous investigations have identified factors that increase a woman's risk for BV. Menstrual blood, a new sexual partner, vaginal douching, smoking, and lack of condom use are among the strongest risk factors for BV (9, 15, 44, 45, 51, 57, 75, 95, 119). In general, these suspected factors often manifest themselves as relatively minor risks in clinical studies, and many women without the above risk factors have BV. In most women, the symptoms of BV resolve on their own without intervention (58). When necessary, the treatment of BV typically includes antibiotics such as metronidazole (oral tablets or topical vaginal gel) or clindamycin vaginal cream (121). However, recurrence of BV after treatment is common: Fifteen to 30% of women have symptomatic BV 30 to 90 days following antibiotic therapy; 70% of patients experience a recurrence within nine months (13, 62, 101). Strategies for managing recurrent BV are not standardized (100), and because the etiology of BV remains unknown, the causes of relapse remain unclear (13, 97).

The confusion about BV stems in part from the subjective diagnostic criteria used. In clinical settings, BV is commonly diagnosed on the basis of the clinical criteria described by Amsel et al. (4), wherein three of the following four symptoms must be evident: (*a*) a homogenous, white, noninflammatory discharge that smoothly coats the vaginal walls; (*b*) the presence of clue cells (squamous epithelial cells covered with adherent bacteria) on microscopic examination; (*c*) a vaginal fluid pH over 4.5; and (*d*) a fishy odor of vaginal discharge before or after addition of 10% KOH (potassium hydroxide). The reliability of the Amsel criteria has been subject to debate, particularly in reference to pregnancy, given the increased vaginal discharge that is often experienced by pregnant women, and to the variation of pH depending on how and where samples are taken (41). However, not all symptoms are observed in every case (56), and because the diagnosis is subjective, controversy persists about the definition of BV. The sensitivity and specificity of the Amsel criteria are 70% and 94%, respectively (93), when compared to another diagnostic assay, the Nugent Gram stain score, which is used in research and laboratories. In these settings, BV is traditionally diagnosed by scoring a Gram-stained vaginal smear using the criteria defined by Nugent et al. (77). The Nugent score reflects the relative abundance of large gram-positive rods (lactobacilli), gram-negative rods and cocci, gram-variable rods and cocci (e.g., *G. vaginalis*, *Prevotella*, *Porphyromonas*, and *Peptostreptococcus*), and curved gram-negative rods (*Mobiluncus*). This technique permits assessment of relative numbers of bacterial morphotypes and other cellular elements, allowing for a rough evaluation of bacterial load, as well as the presence of polymorphonuclear leukocytes, candidal spores, fungal hyphae, and sperm. It is based on a linear scale ranging from 0 to 10. A score of 0–3 is normal, 4–6 is intermediate, and 7–10 is considered BV. Although the Nugent criteria are commonly used to assess BV, the scoring of specimens can be subjective. Nonetheless, with a sensitivity of 89% and specificity of 83% (93) compared to Amsel criteria, the Nugent Gram stain test remains the preferred diagnostic tool (41, 59), and it can be performed on self-collected vaginal smears (74), thus facilitating longitudinal field-based studies (14, 94). Interestingly, as much as 50% of all women with BV (as defined by Nugent score) are asymptomatic (4), which led to the use of the term Nugent-score BV (85). It is unclear whether these women are truly without symptoms or whether the symptoms were poorly recognized or underreported. The meaning and implications of asymptomatic BV are not known. Even so, and because of growing concern for the complications linked to BV, there is a practice of treating asymptomatic disease under certain circumstances, such as prior to a hysterectomy procedure or in women at high risk for preterm birth (121).

Complex Etiology of Bacterial Vaginosis

Despite decades of research, attempts to find a single causative agent for BV have failed. Consequently, Koch's postulates are not fulfilled, in which the etiologic agent is both necessary and sufficient to cause disease and should not be found in subjects without disease (29, 35). Indeed, there is growing evidence that BV is characterized, and perhaps caused, by disruption of the vaginal ecosystem, which is reflected in alterations to the composition and structure of vaginal microbial communities, such that the numbers of lactic-acid-producing bacteria are decreased and the diversity and numbers of strictly anaerobic bacteria are increased, including species of *Gardnerella*, *Atopobium*, *Mobiluncus*, and *Prevotella*, as well as other taxa of the order *Clostridiales* (34). The vaginal microbial community composition associated with BV is somewhat similar to the community state type described above that is found in asymptomatic healthy women that lack a significant number of *Lactobacillus* species. Culture-independent methods have identified potentially BV-associated bacteria (BVAB) that could not be identified by traditional culture-based methods (31, 34). BVAB are distantly related to known species of the phyla *Actinobacteria* and *Firmicutes*. However, the significance of these findings remains unclear, as it is not known whether these microorganisms are pathogens that cause BV or whether they simply are opportunistic organisms that take advantage of the temporarily higher pH environment and thus increase in numerical dominance. Overall, these molecular studies have shown that the diversity, composition, and relative abundances of microbial species in the vagina vary dramatically in both normal, healthy women and women presenting with BV. These diverse organisms accumulate to form different communities, or profiles, which support the hypothesis that BV is not a single entity, but a syndrome linked to various community types that cause somewhat similar physiological symptoms. This suggests that a yet unknown common community function may account for BV and the differing responses to antibiotic therapies.

However, because these studies often rely on a single sample collected from women presenting to their physician with symptomatic BV, it is not possible to elucidate the causes of BV (microbiological, biochemical, molecular, or behavioral) without access to samples collected prior to the diagnosis and during the events leading to BV. Prospective longitudinal studies, in which samples are collected frequently along with detailed behavioral metadata, are necessary to understand the causes of BV. Such information is expected to suggest methods to identify women who are at risk of acquiring symptomatic BV, to identify new targets for intervention and prevention strategies, and to enable development of more accurate diagnostic criteria.

RETHINKING NORMAL AND HEALTHY

The paradigm that healthy women are always colonized with high numbers of *Lactobacillus* species (46, 48) has previously been challenged as discussed above. Although numerous studies have shown that women with abundant *Lactobacillus* species do not have BV, the corollary that women whose vaginal communities have few or no *Lactobacillus* species have BV is faulty logic. Unfortunately, the commonly used diagnostic criteria (both Amsel and Nugent), wherein the degree of healthiness is in part assessed by scoring the abundance of *Lactobacillus* morphotypes, tends to overdiagnose BV. This could account at least partly for the reported high incidence (as high as 42%; 56) of so-called asymptomatic BV in reproductive-age women, as defined by a positive Nugent score and no reported vaginal symptoms. It could also explain a portion of BV treatment failures and

Stability: the capability of an ecosystem to resist change in the face of disturbance

apparent recurrences of BV (33). To better understand symptomatic BV, its causes, and the factors predisposing or triggering the condition, it is essential to apply molecular analyses of vaginal communities to a statistically significant number of women sampled longitudinally and prospectively in order to define the diversity and dynamics of the vaginal ecosystem in the general populace. These studies would help us better understand what constitutes normal and healthy vaginal microbiota and the fluctuations that commonly occur in normal and healthy communities. As suggested by Marrazzo et al. (67), only by using such approaches will we be able to change and refine the definition, etiology, and epidemiology of BV.

The fine line that separates a normal and healthy vaginal microbiota from one that is abnormal and unhealthy is further complicated by potential confusion between health and the predisposition to diseases such as STIs, and by the lack of a complete understanding of the functional intricacies of the host and its vaginal microbiota. It is difficult to envision the evolutionary processes that led to a vaginal microbiota with the sole function of protecting the host from STIs, mainly because only a small proportion of women have been or are exposed to STI pathogens. Interestingly, humans are among the very few mammals with a vaginal microbiota often dominated by *Lactobacillus* spp., and with such a low pH. Hence, it appears that other yet unknown functions must have driven the composition of the human vaginal microbiota. For example, one potential function could relate to immune stimulation or microbial protection of the newborn in the first days of life. Without more knowledge of these functions, one might consider separating the concept of health from the concept of resistance to STIs. That said, it is essential to understand the factors that increase the risks of acquiring STIs and the community types that might be more susceptible to infection. It is conceivable that from time to time a dynamic system such as the vaginal microbiota might enter states (defined by their composition or their function) that would increase the risk of infection. The frequency and duration of these states might represent better predictors of risk of infection than the abundance of a single community member or a given microbial community profile.

With knowledge of the factors driving these dynamics and a better understanding of the function of the vaginal microbiota, novel prevention strategies could be developed to lower the risks. These strategies might include driving the maintenance of more protective and highly functional vaginal microbiota, or possibly using more personalized probiotics or prebiotic mixtures. In addition, these risks are at play only when a woman has the potential to be exposed to the pathogens. If exposure is not likely (perhaps through the practice of celibacy or monogamy), it might not be appropriate to define the healthiness of a woman's vaginal microbiota by factors associated with their predisposition to infections. Hence, a new thinking would involve dissociating the concept of normal and healthy vaginal microbiota from that of predisposition to STIs. A healthy vaginal microbiota could then be defined as a microbial community with a functional output that is adequately beneficial to the host and not solely defined by its composition, and this function could be provided by several kinds of vaginal communities. In this context, different types of vaginal microbiota could be considered healthy in the absence of symptoms, with or without lactobacilli, while having differing degrees of predisposition to infections by sexually transmitted pathogens.

Temporal Dynamics of Vaginal Communities

To date most studies of vaginal microbiology have employed cross-sectional designs in which samples are obtained from individuals at a single time point or with long intervals between sampling times (weeks or months). Although these studies have provided important information on the species composition of vaginal communities, they yield little insight into the normal temporal dynamics of these bacterial communities within individuals and do not provide an estimate of community stability. Daily fluctuations in the composition of the vaginal microbiota have been previously documented by microscopy (16, 44, 55, 95). Even more recently, a longitudinal study conducted by our group described the temporal dynamics of vaginal community composition in 32 healthy reproductive-age women sampled twice weekly over a 16-week period (36). The study showed that some communities changed markedly over a short period, whereas others were relatively stable, including communities lacking a significant number of *Lactobacillus* spp. (**Figure 3**). In an effort to model the dependence of vaginal bacterial community stability on the time in the menstrual cycle and other time-varying factors, menses was identified as having the most negative effect on stability, along with the type of communities and sexual activity to a lesser extent. On the other hand, time periods of the menstrual cycle corresponding to a high level of estrogen or estrogen and progesterone were associated with higher stability. This study highlights the great potential of prospective longitudinal studies to elucidate the cause and etiology of multifactorial diseases such as BV compared with studies that often rely on a single sample collected from women presenting with symptomatic BV. Longitudinal study designs, in which samples are collected frequently along with detailed behavioral metadata, would afford access to samples collected prior to the diagnosis and during the events leading to BV. The knowledge gained from such studies is expected to elucidate factors that govern this dynamic ecosystem, to forecast symptomatic BV susceptibility, and to enable the development of innovative diagnostic, intervention, and prevention strategies.

Microbiome: microbial community gene content

Toward a System-Level Understanding of the Vaginal Ecosystem

Although comprehensive molecular community surveys have provided a great deal of information about the composition of the vaginal microbiota, its role and the intrinsic dynamics that drive its interaction with its host are still unknown. In order to have translational impact on women's health, it is essential to develop an understanding of healthy and disease states that includes knowledge of the functional characteristics of the vaginal microbiota and the types of vaginal microbiota that can provide the needed functions. For example, the notion of an enterotype in the gut microbiome is defined not by the presence of a core set of organisms, but by a core set of available conserved genes that are involved in critical metabolic pathways (7, 86, 108, 109). This notion is informative for subject stratification, but it is still limited to an understanding of the functional potential of a microbial community, and not of its true function and benefit to the host.

State-of-the-art -omics technologies combined with a statistical modeling framework offer an opportunity to develop a systems-level understanding of the vaginal ecosystem by measuring biological components of a system to derive functional modules that reflect specific phenotypic traits. This could be accomplished by using a multilevel approach based on (*a*) metagenomics to catalog the relative abundance of all microbial and, to some extent, human genes and their polymorphisms, and the functional potentials and their degree of redundancy; and (*b*) metatranscriptomics and metaproteomics to assess levels of differential expression of microbial and host genes in healthy and disease states or in response to various perturbations. This multilevel approach would also provide insight into the functional interaction between the vaginal microbiota and the host by using metabolomics to characterize the products of ecosystem-level physiological processes and metabolic output. Predictive statistical models used in a systems biology framework could be used to integrate these various datasets and to quantitatively assess critical biological processes, environmental conditions, or behaviors associated with healthy states, as well as disease initiation, progression, and symptomatology (40, 63, 68, 78). The goal of these efforts would be to develop a systems-level understanding of the molecular events that promote health or lead to disease, and to spur the development of novel diagnostic screens and enable more holistic prevention and treatment regimens.

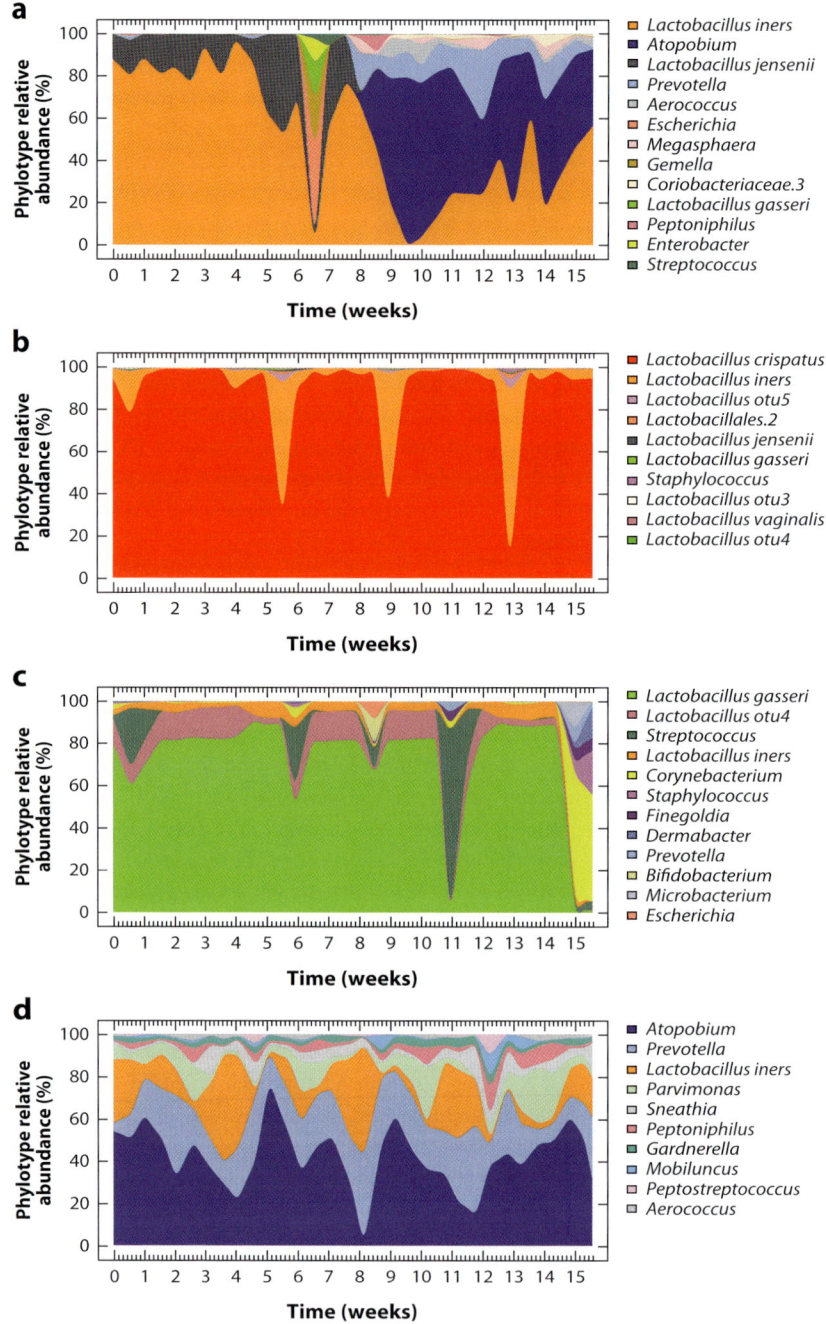

Figure 3

Temporal dynamics of vaginal bacterial communities in women sampled twice weekly over 16 weeks. Interpolated bar plot of the relative abundance of phylotypes from four subjects (*a–d*) with different community dynamics profiles. Color key for each phylotype is shown at the top of each graph.

CONCLUDING REMARKS

The dynamics of vaginal bacterial communities during the menstrual cycle, and the dramatic changes associated with transitions between the physiological stages of a woman's life span, from the first week of life to puberty, reproductive years, and menopause, are a reflection of the interplay of the mutualistic relationship between the vaginal microbiota and its human host. The ever-changing yet finely tuned vaginal ecosystem is a result of adaptive coevolutionary processes that integrate many different aspects such as sexual hormone levels, features of host physiology, and composition and functional output of the vaginal microbiota. Study of the systems-level temporal dynamics of the vaginal ecosystem and its functional output will contribute to our understanding of what truly constitutes normal and healthy. The application of the modern -omics technologies to the study of the vaginal ecosystem is expected to translate to better diagnostics and improved personalized treatments.

SUMMARY POINTS

1. The development of culture-independent community surveys has greatly advanced our understanding of the composition and structure of vaginal microbiota.

2. *Lactobacillus* species dominate vaginal microbiota in most normal and healthy women. However, an appreciable proportion of asymptomatic, otherwise healthy individuals have vaginal microbiota that lack significant numbers of *Lactobacillus* spp. and harbor a diverse array of facultative and strictly anaerobic microorganisms, challenging the conventional wisdom that the presence of lactobacilli equates to normal and healthy vaginal microbiota.

3. Neither clinical criteria (using either the Amsel or the Nugent scoring systems) nor community composition and structure can fully explain symptomatic BV, which appears to be a multifactorial clinical syndrome with complex and still unknown etiologies.

4. The concept of normal and healthy vaginal microbiota is difficult to define without a complete understanding of its true function(s) and its effect on host physiology. One might envision separating the concept of normal and healthy from predisposition to diseases such as STIs.

FUTURE ISSUES

1. Study of the vaginal ecosystem using prospective and frequent sampling study design allows the analysis of samples collected before, during, and after a disease event.

2. There is a need to further characterize the true function of the vaginal microbiota in the context of vaginal health to better understand disease.

3. The role of host genotype in community assembly, composition, and dynamics requires further examination.

4. A systems-level model of the vaginal ecosystem should be developed in order to characterize the functional interaction between the vaginal microbiota and the host.

5. It is expected that future effort should be made to translate current knowledge to the development of personalized preventive or curative regimens (based on vaginal community types), including probiotics and prebiotics.

DISCLOSURE STATEMENT

The authors are not aware of any affiliations, memberships, funding, or financial holdings that might be perceived as affecting the objectivity of this review.

ACKNOWLEDGMENTS

This work was supported by the National Institute of Allergies and Infectious Diseases, National Institutes of Health (grant numbers U19 AI084044, UO1 AI070921, and UH2 AI083264).

LITERATURE CITED

1. Alakomi HL, Skytta E, Saarela M, Mattila-Sandholm T, Latva-Kala K, Helander IM. 2000. Lactic acid permeabilizes gram-negative bacteria by disrupting the outer membrane. *Appl. Environ. Microbiol.* 66:2001–5
2. Alpay-Karaoglu S, Aydin F, Kilic SS, Kilic AO. 2002. Antimicrobial activity and characteristics of bacteriocins produced by vaginal lactobacilli. *Turk. J. Med. Sci.* 33:7–12
3. Amann RI, Ludwig W, Schleifer KH. 1995. Phylogenetic identification and in situ detection of individual microbial cells without cultivation. *Microbiol. Rev.* 59:143–69
4. Amsel R, Totten PA, Spiegel CA, Chen KC, Eschenbach D, Holmes KK. 1983. Nonspecific vaginitis: diagnostic criteria and microbial and epidemiologic associations. *Am. J. Med.* 74:14–22
5. Antonio MA, Hawes SE, Hillier SL. 1999. The identification of vaginal *Lactobacillus* species and the demographic and microbiologic characteristics of women colonized by these species. *J. Infect. Dis.* 180:1950–56
6. Aroutcheva A, Gariti D, Simon M, Shott S, Faro J, et al. 2001. Defense factors of vaginal lactobacilli. *Am. J. Obstet. Gynecol.* 185:375–79
7. Arumugam M, Raes J, Pelletier E, Le Paslier D, Yamada T, et al. 2011. Enterotypes of the human gut microbiome. *Nature* 473:174–80
8. Bakken LR. 1985. Separation and purification of bacteria from soil. *Appl. Environ. Microbiol.* 49:1482–87
9. Beigi RH, Wiesenfeld HC, Hillier SL, Straw T, Krohn MA. 2005. Factors associated with absence of H_2O_2-producing *Lactobacillus* among women with bacterial vaginosis. *J. Infect. Dis.* 191:924–29
10. Beres SB, Sylva GL, Barbian KD, Lei B, Hoff JS, et al. 2002. Genome sequence of a serotype M3 strain of group A *Streptococcus*: phage-encoded toxins, the high-virulence phenotype, and clone emergence. *Proc. Natl. Acad. Sci. USA* 99:10078–83
11. Boskey ER, Cone RA, Whaley KJ, Moench TR. 2001. Origins of vaginal acidity: high D/L lactate ratio is consistent with bacteria being the primary source. *Hum. Reprod.* 16:1809–13
12. Boskey ER, Telsch KM, Whaley KJ, Moench TR, Cone RA. 1999. Acid production by vaginal flora in vitro is consistent with the rate and extent of vaginal acidification. *Infect. Immun.* 67:5170–75
13. Bradshaw CS, Morton AN, Hocking J, Garland SM, Morris MB, et al. 2006. High recurrence rates of bacterial vaginosis over the course of 12 months after oral metronidazole therapy and factors associated with recurrence. *J. Infect. Dis.* 193:1478–86
14. Brotman RM, Ghanem KG, Klebanoff MA, Taha TE, Scharfstein DO, Zenilman JM. 2008. The effect of vaginal douching cessation on bacterial vaginosis: a pilot study. *Am. J. Obstet. Gynecol.* 198:628.e1–7
15. Brotman RM, Klebanoff MA, Nansel TR, Andrews WW, Schwebke JR, et al. 2008. A longitudinal study of vaginal douching and bacterial vaginosis—a marginal structural modeling analysis. *Am. J. Epidemiol.* 168:188–96
16. **Brotman RM, Ravel J, Cone RA, Zenilman JM. 2010. Rapid fluctuation of the vaginal microbiota measured by Gram stain analysis. Sex. Transm. Infect. 86:297–302**
17. Burton JP, Cadieux PA, Reid G. 2003. Improved understanding of the bacterial vaginal microbiota of women before and after probiotic instillation. *Appl. Environ. Microbiol.* 69:97–101
18. Cash HL, Whitham CV, Behrendt CL, Hooper LV. 2006. Symbiotic bacteria direct expression of an intestinal bactericidal lectin. *Science* 313:1126–30

16. Observes rapid fluctuation over time of vaginal microbiota using Nugent scores, and suggests the importance of prospective longitudinal studies with frequent sampling.

19. Cherpes TL, Meyn LA, Krohn MA, Lurie JG, Hillier SL. 2003. Association between acquisition of herpes simplex virus type 2 in women and bacterial vaginosis. *Clin. Infect. Dis.* 37:319–25
20. Cohen CR, Duerr A, Pruithithada N, Rugpao S, Hillier S, et al. 1995. Bacterial vaginosis and HIV seroprevalence among female commercial sex workers in Chiang Mai, Thailand. *AIDS* 9:1093–97
21. Coleman JS, Hitti J, Bukusi EA, Mwachari C, Muliro A, et al. 2007. Infectious correlates of HIV-1 shedding in the female upper and lower genital tracts. *AIDS* 21:755–59
22. Cu-Uvin S, Hogan JW, Caliendo AM, Harwell J, Mayer KH, Carpenter CC. 2001. Association between bacterial vaginosis and expression of human immunodeficiency virus type 1 RNA in the female genital tract. *Clin. Infect. Dis.* 33:894–96
23. Dethlefsen L, McFall-Ngai M, Relman DA. 2007. An ecological and evolutionary perspective on human-microbe mutualism and disease. *Nature* 449:811–18
24. Döderlein A. 1892. Das Scheidensekret und Seine Bedeutung für DasPuerperalfieber. *Zbl. Bakteriol.* 11:699
25. Donders GG, Bosmans E, Dekeersmaecker A, Vereecken A, Van Bulck B, Spitz B. 2000. Pathogenesis of abnormal vaginal bacterial flora. *Am. J. Obstet. Gynecol.* 182:872–78
26. Dover SE, Aroutcheva AA, Faro S, Chikindas ML. 2008. Natural antimicrobials and their role in vaginal health: a short review. *Int. J. Probiotics Prebiotics* 3:219–30
27. Eckburg PB, Bik EM, Bernstein CN, Purdom E, Dethlefsen L, et al. 2005. Diversity of the human intestinal microbial flora. *Science* 308:1635–38
28. Eschenbach DA, Davick PR, Williams BL, Klebanoff SJ, Young-Smith K, et al. 1989. Prevalence of hydrogen peroxide-producing *Lactobacillus* species in normal women and women with bacterial vaginosis. *J. Clin. Microbiol.* 27:251–56
29. Evans M, Dyson P. 1993. Pulsed-field gel electrophoresis of *Streptomyces lividans* DNA. *Trends Genet.* 9:72
30. Falsen E, Pascual C, Sjoden B, Ohlen M, Collins MD. 1999. Phenotypic and phylogenetic characterization of a novel *Lactobacillus* species from human sources: description of *Lactobacillus iners* sp. nov. *Int. J. Syst. Bacteriol.* 49(Pt. 1):217–21
31. Ferris MJ, Masztal A, Aldridge KE, Fortenberry JD, Fidel PL Jr, Martin DH. 2004. Association of *Atopobium vaginae*, a recently described metronidazole resistant anaerobe, with bacterial vaginosis. *BMC Infect. Dis.* 4:5
32. Fichorova RN, Yamamoto HS, Delaney ML, Onderdonk AB, Doncel GF. 2011. Novel vaginal microflora colonization model providing new insight into microbicide mechanism of action. *mBio* 2:e00168–11
33. Forney LJ, Foster JA, Ledger W. 2006. The vaginal flora of healthy women is not always dominated by *Lactobacillus* species. *J. Infect. Dis.* 194:1468–69
34. **Fredricks DN, Fiedler TL, Marrazzo JM. 2005. Molecular identification of bacteria associated with bacterial vaginosis. *N. Engl. J. Med.* 353:1899–911**
35. Fredricks DN, Relman DA. 1996. Sequence-based identification of microbial pathogens: a reconsideration of Koch's postulates. *Clin. Microbiol. Rev.* 9:18–33
36. **Gajer P, Brotman RM, Bai G, Sakamoto J, Schutte U, et al. 2012. Temporal dynamics of the human vaginal microbiota. *Sci. Transl. Med.* 4:132ra–52**
37. Goldenberg RL, Andrews WW, Yuan AC, MacKay HT, St. Louis ME. 1997. Sexually transmitted diseases and adverse outcomes of pregnancy. *Clin. Perinatol.* 24:23–41
38. Graver MA, Wade JJ. 2011. The role of acidification in the inhibition of *Neisseria gonorrhoeae* by vaginal lactobacilli during anaerobic growth. *Ann. Clin. Microbiol. Antimicrob.* 10:8
39. Gravett MG, Nelson HP, DeRouen T, Critchlow C, Eschenbach DA, Holmes KK. 1986. Independent associations of bacterial vaginosis and *Chlamydia trachomatis* infection with adverse pregnancy outcome. *JAMA* 256:1899–903
40. Greenblum S, Turnbaugh PJ, Borenstein E. 2012. Metagenomic systems biology of the human gut microbiome reveals topological shifts associated with obesity and inflammatory bowel disease. *Proc. Natl. Acad. Sci. USA* 109:594–99
41. Guise JM, Mahon SM, Aickin M, Helfand M, Peipert JF, Westhoff C. 2001. Screening for bacterial vaginosis in pregnancy. *Am. J. Prev. Med.* 20:62–72

---

34. Identifies three novel uncultivated phylotypes that appear to be associated with BV.

36. A longitudinal study of 32 women sampled twice-weekly that describes the temporal dynamics of vaginal community composition.

42. Gupta K, Stapleton AE, Hooton TM, Roberts PL, Fennell CL, Stamm WE. 1998. Inverse association of H_2O_2-producing lactobacilli and vaginal *Escherichia coli* colonization in women with recurrent urinary tract infections. *J. Infect. Dis.* 178:446–50
43. Hawes SE, Hillier SL, Benedetti J, Stevens CE, Koutsky LA, et al. 1996. Hydrogen peroxide-producing lactobacilli and acquisition of vaginal infections. *J. Infect. Dis.* 174:1058–63
44. Hay PE, Ugwumadu A, Chowns J. 1997. Sex, thrush and bacterial vaginosis. *Int. J. STD AIDS* 8:603–8
45. Hellberg D, Nilsson S, Mardh PA. 2000. Bacterial vaginosis and smoking. *Int. J. STD AIDS* 11:603–6
46. Hillier SL. 2007. Normal genital flora. In *Sexually Transmitted Diseases*, ed. KK Holmes, PF Sparling, WE Stamm, P Piot, JN Wasserheit, et al., 18:289–308. New York: McGraw-Hill Medical
47. Hillier SL, Krohn MA, Klebanoff SJ, Eschenbach DA. 1992. The relationship of hydrogen peroxide-producing lactobacilli to bacterial vaginosis and genital microflora in pregnant women. *Obstet. Gynecol.* 79:369–73
48. Hillier SL, Krohn MA, Rabe LK, Klebanoff SJ, Eschenbach DA. 1993. The normal vaginal flora, H_2O_2-producing lactobacilli, and bacterial vaginosis in pregnant women. *Clin. Infect. Dis.* 16(Suppl. 4):S273–81
49. Hillier SL, Nugent RP, Eschenbach DA, Krohn MA, Gibbs RS, et al. 1995. Association between bacterial vaginosis and preterm delivery of a low-birth-weight infant. *N. Engl. J. Med.* 333:1737–42
50. Hugenholtz P, Goebel BM, Pace NR. 1998. Impact of culture-independent studies on the emerging phylogenetic view of bacterial diversity. *J. Bacteriol.* 180:4765–74
51. Hutchinson KB, Kip KE, Ness RB. 2007. Condom use and its association with bacterial vaginosis and bacterial vaginosis-associated vaginal microflora. *Epidemiology* 18:702–8
52. Hyman RW, Fukushima M, Diamond L, Kumm J, Giudice LC, Davis RW. 2005. Microbes on the human vaginal epithelium. *Proc. Natl. Acad. Sci. USA* 102:7952–57
53. Jack RW, Tagg JR, Ray B. 1995. Bacteriocins of gram-positive bacteria. *Microbiol. Rev.* 59:171–200
54. Kashket ER. 1987. Bioenergetics of lactic acid bacteria: cytoplasmic pH and osmotolerance. *FEMS Microbiol.* 46:233–44
55. Keane FE, Ison CA, Taylor-Robinson D. 1997. A longitudinal study of the vaginal flora over a menstrual cycle. *Int. J. STD AIDS* 8:489–94
56. Klebanoff MA, Schwebke JR, Zhang J, Nansel TR, Yu KF, Andrews WW. 2004. Vulvovaginal symptoms in women with bacterial vaginosis. *Obstet. Gynecol.* 104:267–72
57. Koumans EH, Markowitz LE, Berman SM, St Louis ME. 1999. A public health approach to adverse outcomes of pregnancy associated with bacterial vaginosis. *Int. J. Gynaecol. Obstet.* 67(Suppl. 1):S29–33
58. Koumans EH, Markowitz LE, Hogan V, Group CBW. 2002. Indications for therapy and treatment recommendations for bacterial vaginosis in nonpregnant and pregnant women: a synthesis of data. *Clin. Infect. Dis.* 35:S152–72
59. Koumans EH, Sternberg M, Bruce C, McQuillan G, Kendrick J, et al. 2007. The prevalence of bacterial vaginosis in the United States, 2001–2004; associations with symptoms, sexual behaviors, and reproductive health. *Sex. Transm. Dis.* 34:864–69
60. Lai SK, Hida K, Shukair S, Wang YY, Figueiredo A, et al. 2009. Human immunodeficiency virus type 1 is trapped by acidic but not by neutralized human cervicovaginal mucus. *J. Virol.* 83:11196–200
61. Lan R, Reeves PR. 2000. Intraspecies variation in bacterial genomes: the need for a species genome concept. *Trends Microbiol.* 8:396–401
62. Larsson PG. 1992. Treatment of bacterial vaginosis. *Int. J. STD AIDS* 3:239–47
63. Lewis NE, Hixson KK, Conrad TM, Lerman JA, Charusanti P, et al. 2010. Omic data from evolved *E. coli* are consistent with computed optimal growth from genome-scale models. *Mol. Syst. Biol.* 6:390
64. Ley RE, Hamady M, Lozupone C, Turnbaugh PJ, Ramey RR, et al. 2008. Evolution of mammals and their gut microbes. *Science* 320:1647–51
65. Ley RE, Peterson DA, Gordon JI. 2006. Ecological and evolutionary forces shaping microbial diversity in the human intestine. *Cell* 124:837–48
66. Ley RE, Turnbaugh PJ, Klein S, Gordon JI. 2006. Microbial ecology: human gut microbes associated with obesity. *Nature* 444:1022–23
67. Marrazzo J. 2006. The vaginal flora of healthy women is not always dominated by *Lactobacillus* species—reply to Forney et al. *J. Infect. Dis.* 194:1469–70

68. Martin FP, Dumas ME, Wang Y, Legido-Quigley C, Yap IK, et al. 2007. A top-down systems biology view of microbiome-mammalian metabolic interactions in a mouse model. *Mol. Syst. Biol.* 3:112
69. Martin HL, Richardson BA, Nyange PM, Lavreys L, Hillier SL, et al. 1999. Vaginal lactobacilli, microbial flora, and risk of human immunodeficiency virus type 1 and sexually transmitted disease acquisition. *J. Infect. Dis.* 180:1863–68
70. Mazmanian SK, Liu CH, Tzianabos AO, Kasper DL. 2005. An immunomodulatory molecule of symbiotic bacteria directs maturation of the host immune system. *Cell* 122:107–18
71. McDonald HM, O'Loughlin JA, Jolley P, Vigneswaran R, McDonald PJ. 1992. Prenatal microbiological risk factors associated with preterm birth. *Br. J. Obstet. Gynaecol.* 99:190–96
72. Meis PJ, Goldenberg RL, Mercer B, Moawad A, Das A, et al. 1995. The Preterm Prediction Study: significance of vaginal infections. *Am. J. Obstet. Gynecol.* 173:1231–35
73. Menard JP, Fenollar F, Henry M, Bretelle F, Raoult D. 2008. Molecular quantification of *Gardnerella vaginalis* and *Atopobium vaginae* loads to predict bacterial vaginosis. *Clin. Infect. Dis.* 47:33–43
74. Morgan DJ, Aboud CJ, McCaffrey IM, Bhide SA, Lamont RF, Taylor-Robinson D. 1996. Comparison of Gram-stained smears prepared from blind vaginal swabs with those obtained at speculum examination for the assessment of vaginal flora. *Br. J. Obstet. Gynaecol.* 103:1105–8
75. Ness RB, Hillier S, Richter HE, Soper DE, Stamm C, et al. 2003. Can known risk factors explain racial differences in the occurrence of bacterial vaginosis? *J. Natl. Med. Assoc.* 95:201–12
76. Ness RB, Kip KE, Hillier SL, Soper DE, Stamm CA, et al. 2005. A cluster analysis of bacterial vaginosis-associated microflora and pelvic inflammatory disease. *Am. J. Epidemiol.* 162:585–90
77. Nugent RP, Krohn MA, Hillier SL. 1991. Reliability of diagnosing bacterial vaginosis is improved by a standardized method of Gram stain interpretation. *J. Clin. Microbiol.* 29:297–301
78. Oberhardt MA, Palsson BO, Papin JA. 2009. Applications of genome-scale metabolic reconstructions. *Mol. Syst. Biol.* 5:320
79. Ochman H, Jones IB. 2000. Evolutionary dynamics of full genome content in *Escherichia coli*. *EMBO J.* 19:6637–43
80. **O'Hanlon DE, Moench TR, Cone RA. 2011. In vaginal fluid, bacteria associated with bacterial vaginosis can be suppressed with lactic acid but not hydrogen peroxide. *BMC Infect. Dis.* 11:200**
81. Oscariz JC, Pisabarro AG. 2001. Classification and mode of action of membrane-active bacteriocins produced by gram-positive bacteria. *Int. Microbiol.* 4:13–19
82. Paulsen IT, Banerjei L, Myers GS, Nelson KE, Seshadri R, et al. 2003. Role of mobile DNA in the evolution of vancomycin-resistant *Enterococcus faecalis*. *Science* 299:2071–74
83. Peters SE, Beck-Sague CM, Farshy CE, Gibson I, Kubota KA, et al. 2000. Behaviors associated with *Neisseria gonorrhoeae* and *Chlamydia trachomatis*: cervical infection among young women attending adolescent clinics. *Clin. Pediatr.* 39:173–77
84. Peterson J, Garges S, Giovanni M, McInnes P, Wang L, et al. 2009. The NIH Human Microbiome Project. *Genome Res.* 19:2317–23
85. Rauch M, Lynch S. 2012. The potential for probiotic manipulation of the gastrointestinal microbiome. *Curr. Opin. Biotechnol.* 23:192–201
86. **Ravel J, Gajer P, Abdo Z, Schneider GM, Koenig SS, et al. 2011. Vaginal microbiome of reproductive-age women. *Proc. Natl. Acad. Sci. USA* 108(Suppl. 1):4680–87**
87. Redondo-Lopez V, Cook RL, Sobel JD. 1990. Emerging role of lactobacilli in the control and maintenance of the vaginal bacterial microflora. *Rev. Infect. Dis.* 12:856–72
88. Relman DA. 2008. 'Til death do us part': coming to terms with symbiotic relationships. Forward. *Nat. Rev. Microbiol.* 6:721–24
89. Rodriguez Jovita M, Collins MD, Sjoden B, Falsen E. 1999. Characterization of a novel *Atopobium* isolate from the human vagina: description of *Atopobium vaginae* sp. nov. *Int. J. Syst. Bacteriol.* 49(Pt. 4):1573–76
90. Rogosa M, Sharpe ME. 1960. Species differentiation of human vaginal lactobacilli. *J. Gen. Microbiol.* 23:197–201
91. Russell JB, Diez-Gonzalez F. 1998. The effects of fermentation acids on bacterial growth. *Adv. Microb. Physiol.* 39:205–34

---

80. Provides experimental evidence obtained under anaerobic conditions suggesting that lactic acid, not hydrogen peroxide, likely plays a protective role by suppressing BV-associated bacteria.

86. Classifies vaginal bacteria composition profiles into five community state types, and establishes differences in community state types frequencies in four ethnic groups.

92. Schellenberg J, Links MG, Hill JE, Dumonceaux TJ, Peters GA, et al. 2009. Pyrosequencing of the chaperonin-60 universal target as a tool for determining microbial community composition. *Appl. Environ. Microbiol.* 75:2889–98
93. Schwebke JR, Hillier SL, Sobel JD, McGregor JA, Sweet RL. 1996. Validity of the vaginal Gram stain for the diagnosis of bacterial vaginosis. *Obstet. Gynecol.* 88:573–76
94. Schwebke JR, Morgan SC, Weiss HL. 1997. The use of sequential self-obtained vaginal smears for detecting changes in the vaginal flora. *Sex. Transm. Dis.* 24:236–39
95. Schwebke JR, Richey CM, Weiss HL. 1999. Correlation of behaviors with microbiological changes in vaginal flora. *J. Infect. Dis.* 180:1632–36
96. Sewankambo N, Gray RH, Wawer MJ, Paxton L, McNaim D, et al. 1997. HIV-1 infection associated with abnormal vaginal flora morphology and bacterial vaginosis. *Lancet* 350:546–50
97. Sobel JD. 1997. Vaginitis. *N. Engl. J. Med.* 337:1896–903
98. Sobel JD. 1999. Is there a protective role for vaginal flora? *Curr. Infect. Dis. Rep.* 1:379–83
99. Sobel JD. 2005. What's new in bacterial vaginosis and trichomoniasis? *Infect. Dis. Clin. N. Am.* 19:387–406
100. Sobel JD, Ferris D, Schwebke J, Nyirjesy P, Wiesenfeld HC, et al. 2006. Suppressive antibacterial therapy with 0.75% metronidazole vaginal gel to prevent recurrent bacterial vaginosis. *Am. J. Obstet. Gynecol.* 194:1283–89
101. Sobel JD, Schmitt C, Meriwether C. 1993. Long-term follow-up of patients with bacterial vaginosis treated with oral metronidazole and topical clindamycin. *J. Infect. Dis.* 167:783–84
102. Sogin ML, Morrison HG, Huber JA, Mark Welch D, Huse SM, et al. 2006. Microbial diversity in the deep sea and the underexplored "rare biosphere". *Proc. Natl. Acad. Sci. USA* 103:12115–20
103. Spear GT, St. John E, Zariffard MR. 2007. Bacterial vaginosis and human immunodeficiency virus infection. *AIDS Res. Ther.* 4:25
104. Sundquist A, Bigdeli S, Jalili R, Druzin ML, Waller S, et al. 2007. Bacterial flora-typing with targeted, chip-based pyrosequencing. *BMC Microbiol.* 7:108
105. Taha TE, Hoover DR, Dallabetta GA, Kumwenda NI, Mtimavalye LA, et al. 1998. Bacterial vaginosis and disturbances of vaginal flora: association with increased acquisition of HIV. *AIDS* 12:1699–706
106. Thomas S. 1928. Doderlein's bacillus: *Lactobacillus acidophilus*. *J. Infect. Dis.* 43:219–27
107. Torsvik V, Ovreas L. 2002. Microbial diversity and function in soil: from genes to ecosystems. *Curr. Opin. Microbiol.* 5:240–45
108. Turnbaugh PJ, Gordon JI. 2009. The core gut microbiome, energy balance and obesity. *J. Physiol.* 587:4153–58
109. Turnbaugh PJ, Hamady M, Yatsunenko T, Cantarel BL, Duncan A, et al. 2009. A core gut microbiome in obese and lean twins. *Nature* 457:480–84
110. Turnbaugh PJ, Ley RE, Hamady M, Fraser-Liggett CM, Knight R, Gordon JI. 2007. The Human Microbiome Project. *Nature* 449:804–10
111. Turnbaugh PJ, Ley RE, Mahowald MA, Magrini V, Mardis ER, Gordon JI. 2006. An obesity-associated gut microbiome with increased capacity for energy harvest. *Nature* 444:1027–31
112. Vallor AC, Antonio MA, Hawes SE, Hillier SL. 2001. Factors associated with acquisition of, or persistent colonization by, vaginal lactobacilli: role of hydrogen peroxide production. *J. Infect. Dis.* 184:1431–36
113. van De Wijgert JH, Mason PR, Gwanzura L, Mbizvo MT, Chirenje ZM, et al. 2000. Intravaginal practices, vaginal flora disturbances, and acquisition of sexually transmitted diseases in Zimbabwean women. *J. Infect. Dis.* 181:587–94
114. van der Lelie D, Lesaulnier C, McCorkle S, Geets J, Taghavi S, Dunn J. 2006. Use of single-point genome signature tags as a universal tagging method for microbial genome surveys. *Appl. Environ. Microbiol.* 72:2092–101
115. Verhelst R, Verstraelen H, Claeys G, Verschraegen G, Delanghe J, et al. 2004. Cloning of 16S rRNA genes amplified from normal and disturbed vaginal microflora suggests a strong association between *Atopobium vaginae*, *Gardnerella vaginalis* and bacterial vaginosis. *BMC Microbiol.* 4:16
116. Verstraelen H, Verhelst R, Claeys G, Temmerman M, Vaneechoutte M. 2004. Culture-independent analysis of vaginal microflora: the unrecognized association of *Atopobium vaginae* with bacterial vaginosis. *Am. J. Obstet. Gynecol.* 191:1130–32

117. Wardle DA. 2000. Stability of ecosystem properties in response to above-ground functional group richness and composition. *Oikos* 89:11–23
118. Wiesenfeld HC, Hillier SL, Krohn MA, Landers DV, Sweet RL. 2003. Bacterial vaginosis is a strong predictor of *Neisseria gonorrhoeae* and *Chlamydia trachomatis* infection. *Clin. Infect. Dis.* 36:663–68
119. Wilson JD, Lee RA, Balen AH, Rutherford AJ. 2007. Bacterial vaginal flora in relation to changing oestrogen levels. *Int. J. STD AIDS* 18:308–11
120. Witkin SS, Alvi S, Bongiovanni AM, Linhares IM, Ledger WJ. 2011. Lactic acid stimulates interleukin-23 production by peripheral blood mononuclear cells exposed to bacterial lipopolysaccharide. *FEMS Immunol. Med. Microbiol.* 61:153–58
121. Workowski KA, Berman SM. 2006. Sexually transmitted diseases treatment guidelines, 2006. *MMWR Recomm. Rep.* 55:1–94
122. Zhou X, Bent SJ, Schneider MG, Davis CC, Islam MR, Forney LJ. 2004. Characterization of vaginal microbial communities in adult healthy women using cultivation-independent methods. *Microbiology* 150:2565–73
123. Zhou X, Brown CJ, Abdo Z, Davis CC, Hansmann MA, et al. 2007. Differences in the composition of vaginal microbial communities found in healthy Caucasian and black women. *ISME J.* 1:121–33
124. Zhou X, Hansmann MA, Davis CC, Suzuki H, Brown CJ, et al. 2010. The vaginal bacterial communities of Japanese women resemble those of women in other racial groups. *FEMS Immunol. Med. Microbiol.* 58:169–81

123. Demonstrates differences in the vaginal microbiota structure in Caucasian and Black women.

124. Shows variation in composition and structure of vaginal bacterial community of women in different ethnic groups, and suggests that host factors might contribute to shaping the vaginal microbiota.

Electromicrobiology

Derek R. Lovley

Department of Microbiology, University of Massachusetts, Amherst, Massachusetts 01003; email: dlovley@microbio.umass.edu

Keywords

Geobacter, *Shewanella*, microbial nanowires, microbial fuel cells, microbial electrosynthesis, bioelectronics

Abstract

Electromicrobiology deals with the interactions between microorganisms and electronic devices and with the novel electrical properties of microorganisms. A diversity of microorganisms can donate electrons to, or accept electrons from, electrodes without the addition of artificial electron shuttles. However, the mechanisms for microbe-electrode electron exchange have been seriously studied in only a few microorganisms. *Shewanella oneidensis* interacts with electrodes primarily via flavins that function as soluble electron shuttles. *Geobacter sulfurreducens* makes direct electrical contacts with electrodes via outer-surface, *c*-type cytochromes. *G. sulfurreducens* is also capable of long-range electron transport along pili, known as microbial nanowires, that have metallic-like conductivity similar to that previously described in synthetic conducting polymers. Pili networks confer conductivity to *G. sulfurreducens* biofilms, which function as a conducting polymer, with supercapacitor and transistor functionalities. Conductive microorganisms and/or their nanowires have a number of potential practical applications, but additional basic research will be necessary for rational optimization.

Contents

INTRODUCTION	392
WHY MICROORGANISMS MIGHT INTERACT ELECTRONICALLY WITH ELECTRODES	393
MECHANISMS FOR ELECTRON TRANSFER TO ELECTRODES	394
Electron Transfer via Soluble Electron-Shuttling Molecules	394
Short-Range Direct Electron Transfer via Redox-Active Proteins	396
Long-Range Electron Transport via Conductive Pili	397
FEEDING ELECTRONS TO MICROBES	400
BIOELECTRONICS	401
FUTURE DIRECTIONS	402

INTRODUCTION

Electromicrobiology is the study of microbial electron exchange with external electronic devices and functionalities of microorganisms that have the potential to contribute to the emerging field of bioelectronics. A wide diversity of microorganisms have the ability to exchange electrons with electrodes, which contribute to a broad range of practical applications (4, 30, 58–60, 62, 63, 101, 102, 108). Furthermore, some microorganisms have surprising electronic characteristics. For example, biofilms of *Geobacter* species have conductivities that rival those of conductive polymers (74), and can function as supercapacitors (71) or transistors (74). The pili of these organisms are capable of long-range (>1 cm) electron transport via a metallic-like conductivity, not previously observed in a biological material (74).

Many of the recent advances in electromicrobiology have arisen from the study of microbial fuel cells, devices initially designed for harvesting electricity from organic matter (58, 59, 101). Difficulties in scaling up microbial fuel cells for extracting energy on an industrial scale have greatly limited their short-term practical use for current production to niche applications, such as harvesting organic matter from aquatic sediments to power electronic monitoring devices (65). Some of the most attractive practical applications of the microbial fuel cell concept are those in which there is no need to harvest current. These include monitoring rates of microbial metabolism in subsurface environments (137) and providing electrodes as an electron acceptor to stimulate the degradation of organic contaminants in sediments (63, 142). Another promising application of the ability of microorganisms to transfer electrons to electrodes is the potential for balancing electron flow within microbial cells by removing excess electrons in order to promote the synthesis of desired products (29).

In a similar manner, new potential applications for electron flow in the reverse direction, i.e., from electrodes to cells, are rapidly emerging (64, 102, 127). Feeding electrons to microorganisms living on electrode surfaces has significant potential to contribute to bioremediation of a diversity of contaminants, including radioactive and toxic metals (38, 124), chlorinated compounds (1, 12, 119, 122, 128), and nitrate (37, 96). Microorganisms have the potential to catalyze the production of hydrogen and methane with electrons derived from electrodes (14, 35, 64, 133, 134). Electrons derived from electrodes can potentially serve as the reductant for effecting microbial reduction of organic compounds to more desirable organic commodities, or for altering fermentation pathways in desired directions (26, 44, 95, 118). Furthermore, with the newly developed process of microbial electrosynthesis (87, 93), it is possible to electrically power the microbial reduction of carbon

---

Electromicrobiology: the study of microbial electron exchange with electronic devices or the investigation of the electronic properties of microorganisms

Bioelectronics: technology incorporating electronics in biological applications or developing electronic devices from biological components or electronic components that mimic biological materials

Microbial fuel cell: device for harvesting electricity from organic matter in which microorganisms are the catalyst for oxidizing the organic matter

dioxide to liquid transportation fuels and other useful organic commodities (63, 64). When driven with electricity generated from solar technologies, microbial electrosynthesis is functionally an artificial form of photosynthesis with the potential to be much more efficient and environmentally sustainable than biomass-based strategies for fuel and chemical production.

Findings from the study of microbe-electrode exchange have also led to new insights into the functioning of anaerobic ecosystems (63). For over 40 years it has been considered that microorganisms in methanogenic environments exchange electrons primarily via interspecies transfer of hydrogen, with the electron-donating microorganism disposing of electrons by reducing protons to hydrogen and an electron-accepting methanogen oxidizing hydrogen with the reduction of methane (81, 117). However, it is feasible for different species of microorganisms to forge direct electrical connections, similar to those that they establish with electrodes (123), and direct interspecies electron transfer can be the primary mechanism for electron exchange in microbial aggregates converting wastes to methane (86).

These developments demonstrate that a better understanding of the mechanisms by which microorganisms exchange electrons with electrodes could benefit the development of various new technologies as well as provide a better understanding of anaerobic microbial ecology. As more is learned, additional applications will probably emerge. This review summarizes current knowledge on microbe-electrode interactions and the novel electronic materials that some microorganisms can produce.

Microbial electrosynthesis: strategy for directly converting carbon dioxide to transportation fuels or other organic commodities in which microorganisms accept electrons from an electrode to reduce the carbon dioxide to the desired product, which is excreted from the cell

Direct interspecies electron transfer: microbial syntrophy in which microorganisms establish electrical connections for the transfer of electrons from one syntrophic partner to another

WHY MICROORGANISMS MIGHT INTERACT ELECTRONICALLY WITH ELECTRODES

A key to understanding the mechanisms of microbe-electrode interactions may be to elucidate how this capability evolved. For example, short-term adaptive evolution studies have provided insight into the mechanisms for electron transfer to electrodes and Fe(III) oxides (132, 141). The mere fact that microorganisms are able to exchange electrons with electrodes to produce electric current and can consume current to power their respiration is fascinating, especially when it is considered that electrodes, per se, are not a part of the natural environment. It has been suggested that microbe-electrode exchange is a fortuitous result of the fact that some microorganisms have developed over billions of years of evolution effective strategies for extracellular electron exchange with insoluble minerals and related natural extracellular electron acceptors or donors (59, 64).

However, there is a significant difference between insoluble minerals and electrodes. Electrodes provide a surface with long-term electron-accepting or electron-donating capacity, whereas the ability of individual insoluble minerals to accept or donate electrons is eventually depleted. Hence, the relationship between cells and electrodes is different from that between cells and minerals. This is readily apparent with *Geobacter* species. When actively reducing Fe(III) oxides, *Geobacter* species express flagella and are motile, presumably because they need to continually search for new sources of Fe(III) (15, 63). In contrast, when *Geobacter* species are oxidizing organic compounds with electron transfer to electrodes, they are not planktonic. The cells firmly attach to the electrode surface and form thick (>80 μm) metabolically active biofilms (32, 88, 104). The physiological status of these sessile cells packed in biofilms and provided with a constant electron acceptor is expected to be much different from that of planktonic cells actively hunting for minerals on which to dump electrons. Electron transfer to electrodes can be much faster than electron transfer to Fe(III) oxides, proceeding at rates comparable to the reduction of soluble, chelated Fe(III) (4). Is there a better natural analog for electrodes than small individually dispersed minerals?

One option may be graphite deposits. These deposits, which have significant conductivity, can span distances up to 1 km and appear able to transmit electrons between anaerobic and oxic zones in the subsurface, producing a geobattery (3). In this initial geobattery concept, electron transfer to the graphite in the anaerobic zone was discussed as abiotic, with reduced chemical species such as ferrous iron donating electrons to the graphite (3). In a similar manner, it was initially proposed that electron transfer to electrodes was abiotic in benthic microbial fuel cells designed to harvest electricity from anaerobic marine sediments (105). However, subsequent studies (5, 41, 125) revealed a specific enrichment of *Geobacteraceae* microorganisms on the surface of the electron-accepting electrodes (the anodes), which could be attributed to the ability of these microorganisms to oxidize organic compounds with direct electron transfer to electrodes. Graphite is a preferred electrode material. The ability of the electron-accepting end of geobattery graphite deposits to serve as a consistent, long-term electron sink would provide an environment highly analogous to the graphite anodes of microbial fuel cells. Therefore, it seems likely that microorganisms that are highly effective in current production in microbial fuel cells may have first perfected this capability as catalysts promoting electron flow in geobatteries. Additional study of these, and possibly other natural analogs for electrodes, could enhance our understanding of microbe-electrode interactions.

> **Anode:** electrode that accepts electrons
>
> **Electron shuttle:** compound that facilitates electron transfer between microorganisms and electrodes
>
> **Long-range electron transport:** transport of electrons over multiple cell lengths

MECHANISMS FOR ELECTRON TRANSFER TO ELECTRODES

As previously reviewed in detail (59), many microorganisms can exchange electrons with electrodes when artificial electron shuttles are provided. Shuttles such as methylviologen, neutral red, or thionine can accept electrons from redox-active moieties within cells and transfer the electrons to electrodes. However, the practical benefit of this type of electrical interaction has yet to be proved; electron shuttles are often unstable and toxic, are uneconomical in large-scale processes, and cannot be employed in open environments.

Microorganisms transfer electrons to an electrode without the addition of an artificial electron shuttle in three ways (**Figure 1**): (*a*) electron transfer via microbially produced soluble redox-active molecules, (*b*) short-range direct electron transfer between redox-active molecules on the outer cell surface and the electrode; and (*c*) long-range electron transport through conductive biofilms (62). As detailed below, *Geobacter sulfurreducens*, the microorganism that produces the highest currents in pure culture (92, 141), appears to accomplish this with a combination of long-range electron transport through thick, conductive biofilms and short-range electron transfer between the conductive biofilm and the electrode that is mediated by an extracellular *c*-type cytochrome.

Electron Transfer via Soluble Electron-Shuttling Molecules

A diversity of both gram-negative and gram-positive microorganisms have the ability to produce electron shuttles to promote electron transfer to electrodes (101). The concept of self-produced electron shuttles facilitating electron transfer to electrodes follows previous studies that demonstrated that some microorganisms produce shuttles that promote electron transfer between cells and insoluble Fe(III) oxides (54, 90, 91, 94). For example, *Geothrix fermentans*, which can reduce Fe(III) oxide enclosed in porous alginate beads via a shuttle (90), also appeared to release an electron shuttle to promote electron transfer to electrodes (7).

Shewanella species have a similar ability to reduce Fe(III) with which they are not in direct contact (54, 91); this is attributed to the release of flavin in *S. oneidensis* cultures (135). The finding that cells of *S. oneidensis* were primarily planktonic in microbial fuel cells suggested that an electron shuttle was also involved in electron transfer to electrodes (50). The role of flavins in promoting electron transfer to electrodes with *S. oneidensis* has been well established by electrochemical studies (2, 78). *S. oneidensis* can reduce flavins at the outer cell surface with the *c*-type cytochrome

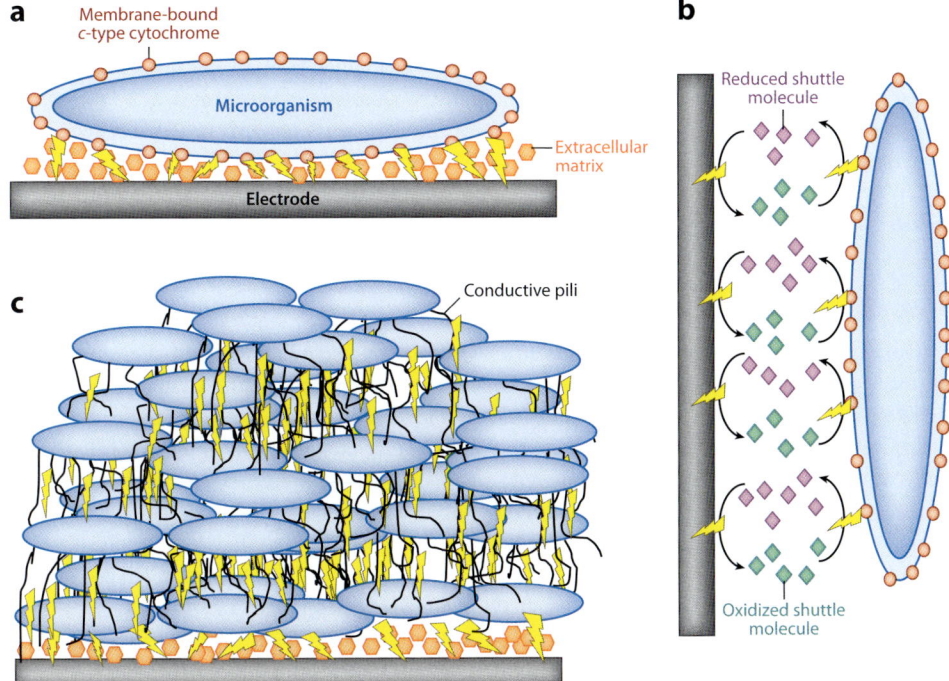

Figure 1

Potential mechanisms for microorganisms to transfer electrons to electrodes. (*a*) Short-range electron transfer by microorganisms in close association with the electrode surface through redox-active proteins, such as *c*-type cytochromes associated with the outer cell surface or in the extracellular matrix. (*b*) Electron transfer via reduction of soluble electron shuttles released by the cell. Oxidized shuttle molecules are reduced at the outer cell surface, and the reduced shuttle molecules donate electrons to the electrode. (*c*) Long-range electron transport through a conductive biofilm via electrically conductive pili, accompanied by short-range electron transfer from the biofilm to the electron mediated by extracellular cytochromes as in panel *a*.

MtrC (18), which is part of a multiprotein complex that transports electrons from the periplasm to the outer surface of the cell (16, 39).

Electrons can hop directly from MtrC to an electrode (2, 55). However, direct electron transfer in intact cells was possible only when anodes were artificially poised at positive potentials significantly higher than those typical of microbial fuel cells, and the rates of electron transfer were much faster in the presence of flavin. These results suggest that electron transfer via a flavin is the preferred route of electron transfer in *S. oneidensis* microbial fuel cells. Furthermore, elegant studies in which direct contact between *S. oneidensis* and electrodes could be prevented with a nonconducting mask with nanohole openings demonstrated that current was produced when the possibility for contact was eliminated as well as when cells established contact (48). The conclusion from these studies was that electron transfer via an electron shuttle was the predominant means of electron transfer even when cells were in contact with the electrode. In a similar manner, it is possible for MtrC to transfer electrons directly to Fe(III) oxides, but the rates of electron transfer are too low to account for observed rates of Fe(III) oxide reduction (109). MtrC serves as flavin reductase (18), and only in the presence of flavin can MtrC transfer electrons to Fe(III) oxide at physiologically relevant rates (109).

The maximum current densities produced by microorganisms that rely on electron shuttling to transfer electrons to electrodes are much lower than those for microorganisms capable of long-range electron transport through thick conductive biofilms because the slow diffusive flux of the shuttle is a major limitation (129). Although the shuttling mechanism may be somewhat effective in closed laboratory systems, in open environments this approach suffers from losses of the shuttle from the immediate microbe-electrode interface. For these and other reasons (59), it is not surprising that *Shewanella* species have never been found to be important constituents of anodes harvesting electricity from complex organic matter in open environments (49, 63).

Short-Range Direct Electron Transfer via Redox-Active Proteins

Evidence for direct electron transfer to electrodes has been presented for several microorganisms (6, 13, 40, 77, 139, 140). The mechanisms for direct electron transfer to electrodes have been studied most extensively in *G. sulfurreducens*. *G. sulfurreducens* is closely related to the *Geobacter* species that typically predominate on electrodes harvesting current from organic matter, especially when oxygen intrusions are eliminated so that organic substrates are efficiently converted to current, and when the electrode potential is not artificially poised with electronics (49, 66). Early investigations suggested that, just as *Geobacter* species do not use shuttles to reduce Fe(III) oxide (89), shuttles are not involved in electron transfer to electrodes (6). This was subsequently demonstrated more definitively by electrochemical studies (9, 10, 79, 80, 106).

G. sulfurreducens has a wide diversity of *c*-type cytochromes (84), many of which are exposed on the outer surface of the cell (20, 45, 53, 83, 99). The outer-surface *c*-type cytochromes that have been purified can reduce known extracellular electron acceptors in vitro (45, 67, 99). Gene deletion studies suggest that these same *c*-type cytochromes transfer electrons to a diversity of extracellular electron acceptors in vivo (51, 52, 83, 112, 136). Numerous studies of current-producing *G. sulfurreducens* biofilms have demonstrated that *c*-type cytochromes are in electrochemical communication with the anode (9, 10, 11, 27, 34, 47, 56, 57, 79, 80, 85, 106, 116, 121). In some instances the cytochromes are positioned close enough to the electrode surface for direct electron transfer from the cytochromes to the electrode (9) and hence function as the electrochemical gate between cells in contact with the electrode and the electrode surface (21).

Comparison of gene expression in current-producing cells versus expression in cells growing on alternative electron acceptors, as well as gene deletion studies, identified several candidate outer-surface *c*-type cytochromes that might help *G. sulfurreducens* make electrical contacts with electrodes (42, 88). OmcS was implicated in thin biofilms generating low levels of current (42), but OmcZ appears to be the most important cytochrome in biofilms producing high levels of current (88). OmcZ is a hydrophobic protein with a molecular mass of 30 kDa (45). It has eight hemes, which cover a wide range of redox potentials (-420 mV to -60 mV). The midpoint potential is -220 mV. Deletion of *omcZ* greatly inhibited current production (88), as did deletion of another gene that significantly reduced the abundance of OmcZ on the outer cell surface (107). Cyclic voltammetry demonstrated an increased resistance to electron transfer to electrodes in the OmcZ-deficient strain. Deleting genes for other outer-surface *c*-type cytochromes did not yield a similar response (106).

Immunogold labeling of current-producing biofilms demonstrated that significant quantities of OmcZ accumulated at the biofilm/anode interface, indicating that it was ideally positioned to facilitate electron transfer to electrodes (45). This accumulation of OmcZ was not observed in biofilms grown on the electrode material, but with fumarate serving as the electron acceptor (45).

Thus, multiple lines of evidence suggest that OmcZ is the key cytochrome for electron transfer between *G. sulfurreducens* biofilms and anodes. Further investigation of the properties that uniquely suit OmcZ for this purpose is required. Also, although it is possible for *G. sulfurreducens* to overcome deletion of other genes that are highly expressed in current-producing biofilms, such as the gene for the outer-membrane-bound *c*-type cytochrome OmcB (88), this does not mean that electron flow through these components is not important in wild-type cells, because cells may adapt with increased expression of other cell components. A better understanding of the role of other outer-surface components, such as putative multi-copper proteins (43, 82), is also required.

Of the other microorganisms that appear to make direct electrical contact with electrodes, some of the most surprising are the gram-positive species of the genus *Therminocola* (77, 139, 140). The abundant *c*-type cytochromes in *T. potens* were involved in extracellular electron transfer and might be the electrical contacts with electrodes (140). Only cells in direct contact with the electrode appeared to contribute to current production, suggesting that a mechanism for long-range electron transport was absent (140).

Long-Range Electron Transport via Conductive Pili

The current production capability of a monolayer of cells in direct contact with an electrode surface is limited by the space available for microorganisms to directly access the electrode surface. Higher-current densities (current produced per surface area of electrode) are possible from electrically conductive biofilms, which permit multiple layers of cells to contribute to current production (104). As previously reviewed (74), the biofilms of most microorganisms appear to act as insulators rather than conductors and the concept of a conductive biofilm is still rather new and controversial (70).

Conductive pili and biofilms of *Geobacter sulfurreducens*. The possibility of a conductive biofilm was first proposed in studies on current-producing biofilms of *G. sulfurreducens* (104) and subsequently confirmed by direct measurements (74). Conductive biofilms have been invoked or inferred in other studies (56, 75, 97, 129). However, failure to measure conductivity, as well as highly speculative and unsubstantiated models for conductivity, has led to significant debate about the mechanisms for long-range electron transport through biofilms (70, 73), which can be resolved only by additional direct measurements of conductivity and rigorous experimentation.

The high conductivity of current-producing biofilms of *G. sulfurreducens* (74) allows cells at distances of multiple cell lengths from the anode to contribute to current production (31, 104). The available evidence suggests that the conductivity of the biofilms can be attributed to a dense network of pili with metallic-like conductivity (74).

Initial interest in the type pili of *Geobacter* species came from the observation that *Geobacter metallireducens* expressed pili when growing on insoluble electron acceptors, such as Fe(III) or Mn(IV) oxides, but not when grown with soluble Fe(III) citrate (15), even though Fe(III) citrate is also an extracellular electron acceptor (17). Increased pili production was associated with higher expression of the gene for PilA, the structural protein for type IV pili (15), which are ubiquitous in gram-negative bacteria (19). *G. sulfurreducens* pili have a gross morphology (3–5 nm in width and up to 10–20 μm in length) that is similar to that of other type IV pili, but so far only the pili of *G. sulfurreducens* have been shown to be conductive.

Deletion of the gene for PilA inhibited the capacity for Fe(III) oxide reduction, but not the reduction of Fe(III) citrate (103). Addition of anthraquinone-2,6-disulfonate as a soluble electron shuttle alleviated the inhibition of Fe(III) oxide reduction in the *pilA* mutant. These results suggested that the PilA pili were required specifically for Fe(III) oxide reduction, but not for electron transfer to the outer surface of the cell.

Conductivity across the diameter of individual, chemically fixed pili was observed with conducting atomic force microscopy (103). Additional cellular material was often associated with the

Microbial nanowires: pili capable of long-range electron transport

pili and acted as insulators for current flow between the conducting tip and the graphite. This observation led to the suggestion that the pili themselves were conductive, rather than the alternative that the pili served as a scaffold for electron-hopping between pilin-associated proteins, and that conduction along the length of the pili permitted *G. sulfurreducens* to greatly extend the potential distance for extracellular electron transfer (103). Thus, the pili were termed microbial nanowires.

Circumstantial evidence that the pili could carry out long-range electron transport came from studies with *G. sulfurreducens* growing on graphite electrodes serving as an electron acceptor. Viability staining indicated that cells at distance from the electrode were viable, and a direct correlation between the extent of current production and biofilm biomass suggested that the cells not in direct contact with the electrode were contributing as much to current production as cells at the electrode surface (104). The simplest explanation for these observations was that long-range electron transport through the biofilm was possible. The finding that a *pilA* mutant did not form thick biofilms on electrodes suggested that a network of microbial nanowires was responsible for the conduction through the biofilm (104). Subsequent observations demonstrated that *pilA* was one of the most highly upregulated genes in current-producing biofilms, providing further circumstantial evidence for the role of pili in current production (88).

Direct measurements of the conductivity of live *G. sulfurreducens* biofilms growing on two gold electrodes that converged across a nonconductive gap demonstrated that the biofilms were conductive, with conductivities rivaling those of synthetic organic conducting polymers (74). Evaluation of different strains of *G. sulfurreducens* revealed significant differences in biofilm conductivity (74); strains that produced more conductive biofilms generated higher current densities in microbial fuel cells (72). There was a strong correlation between conductivity levels and expression of PilA. For example, a strain of *G. sulfurreducens* that was selected specifically for its capacity for high current production and expressed more pili (141) also formed biofilms with the highest conductivity (74).

Surprisingly, the conductivity of the *G. sulfurreducens* biofilms exhibited properties consistent with metallic-like conductivity (74). For example, conductivity initially increased exponentially with a decrease in temperature, a hallmark characteristic of metallic-like conductivity that was previously observed in conducting organic polymers. A similar metallic-like conductivity was observed when pili preparations were spotted on the two-electrode system, forming a network that bridged the nonconducting gap between the electrodes. X-ray diffraction analysis of pili suggested π-π stacking, similar to that previously documented in the organic metal polyaniline (74). Thus, a working hypothesis is that aromatic amino acids are aligned along the outer surface of the pili to provide the apparent π-π stacking. Another similarity to polyaniline was that the addition of protons to the pili preparations greatly increased their conductivity (74).

The apparent metallic-like conductivity along the pili of *G. sulfurreducens* is in marked contrast to previously described biological electron transfer via electron hopping or tunneling. It is well known that electrons associated with a discrete molecule, such as a cytochrome, can move to another molecule if the two molecules are sufficiently close (<20 Å) (23, 115). However, in metallic-like conductivity the electrons are delocalized. The possibility of delocalized electron transfer in biomolecules has previously been dismissed "due to their lack of periodicity, random fluctuations, and limited conductance values from experiments" (115).

The metallic-like conductive properties of the pili of *G. sulfurreducens* rule out the possibility that electrons are conducted along the length of pili via electron hopping between discrete electron-carrier molecules associated with the pili, such as cytochromes. Furthermore, denaturing cytochromes in pili preparations had no impact on conductivity (74). However, the multiheme *c*-type cytochrome OmcS (100) is specifically associated with the pili of *G. sulfurreducens* (53). Initial observations with immunogold labeling suggested that the OmcS molecules were spaced too far apart for electron hopping between OmcS molecules to account for electron transport along the pili (53), and this finding has subsequently been confirmed by atomic force microscopy (N. Malvankar, unpublished data). OmcS is required for Fe(III) oxide reduction (83). Therefore, it has been proposed that there are barriers to direct electron transfer from pili to Fe(III) oxides and the hypothesized role of OmcS is to facilitate electron transfer from the pili to Fe(III) oxides (63).

In a similar manner, the absolute need for OmcZ, as well as pili, for the production of the highest-current densities in *G. sulfurreducens* biofilms (88, 106), coupled with the localization of OmcZ at the biofilm/anode interface (45), suggests a two-phase electron transport process to electrodes, in which long-range electron transport through the biofilm is along the pili network and OmcZ facilitates the electron transfer from the biofilm to the electrode (63). It is conceivable that with a change in environmental conditions and/or electrode materials that electrochemical gates other than OmcZ may become important. Monitoring electron transfer between specific cytochromes and electrodes as well as cytochrome-to-cytochrome electron transfer may become possible as new tools for simultaneously monitoring the redox status of cytochromes and electron transfer to electrodes become available (57).

The present model for long-range electron transport in *Geobacter* biofilms suggests that one avenue to increase current production might be to increase biofilm conductivity. Comparison of direct measurements of biofilm conductivity and the amount of current produced in microbial fuel cells with different strains of *G. sulfurreducens* demonstrated that there was a direct correlation between biofilm conductivity and current production (72). Furthermore, strains with higher biofilm conductivities had lower resistance to electron transfer across the biofilm-anode interface, presumably because electrons were delivered to the interface at a lower potential when resistance to transport through the biofilm was lower.

However, with the best current-producing isolate, strain KN400, the relationship between biofilm conductivity and current production deviated from the strong linear relationship observed with other strains (72). This result suggests that as the current-production capacity of microorganisms is increased, factors other than the maximum potential respiration rate and capacity for long-range electron transport of the organisms begin to limit current production. One possibility is that the protons that must be released from cells during extracellular electron transfer (68) accumulate to levels within the biofilm that inhibit microbial activity (32, 76, 130).

It is not yet known whether the pili of other *Geobacter* species are electronically conductive, but it has been demonstrated that the PilA pili of *G. metallireducens* are required for optimal Fe(III) oxide reduction and current production (131). The metallic-like conductivity of methane-producing aggregates from a wastewater treatment plant suggested that *Geobacter* species, and possibly other organisms in this mixed natural community, were capable of producing conductive filaments (86).

Putative conductive filaments in other microorganisms. Electrically conductive pili greatly benefit *Geobacter* species in their ability to electronically interact with their extracellular environment, and it would be surprising if other microorganisms had not adopted a similar strategy. In fact, preliminary evidence, based on scanning tunneling microscopy, has suggested that a wide diversity of microorganisms produce conductive filaments (36). However, there was significant uncertainty about the filament structure and the mechanisms for conductivity.

The studies focused primarily on filaments of *S. oneidensis*. The diameter of the filaments (50 to >150 nm) was much too broad for the filaments to be type IV pili. Furthermore, direct examination of a role of pili in extracellular electron transfer suggested they are not important in extracellular electron transfer in *S. oneidensis*, as strains that could not produce pili filaments continued to produce electrical current better than wild-type strains (8).

It was suggested that cytochromes associated with *S. oneidensis* filaments conferred conductivity because conductive filaments could not be detected in a mutant strain that did not produce two outer-surface *c*-type cytochromes or a mutant strain deficient in a type II secretion system required for cytochrome export (36). Cytochrome-based conductivity was also inferred from studies using conducting tip atomic force microscopy (24). However, no direct evidence for the association of cytochromes with the filaments has ever been reported and it is generally regarded that the cytochromes in question are associated with the outer surface of the cell body rather than filaments (113). Furthermore, it seems unlikely that cytochromes could be packed tightly enough along pili to confer conductivity via cytochrome-to-cytochrome electron hopping.

As noted above, there is substantial evidence that much of the extracellular electron transfer in *S. oneidensis* is likely to proceed via soluble electron shuttles. The fact that *S. oneidensis* cannot form thick biofilms on electrodes under strict anaerobic conditions (50, 74) further suggests that it is not capable of long-range electron transfer via conductive pili. Therefore, even though conductance could be measured along the length of a filament of *S. oneidensis* (25), the available evidence suggests that it is unlikely that long-range electron transport along conductive filaments is a significant process in *S. oneidensis*.

Scanning tunneling microscopy also suggested that a strain of *Synechocystis*, a phototrophic cyanobacterium, produced conductive filaments and that the thermophilic fermentative bacterium *Pelotomaculum thermopropionicum* produced conductive filaments that established connections with the methanogen *Methanothermobacter thermautotrophicus* (36). A physiological role for the filaments of *Synechocystis* has yet to be determined. Subsequent studies with the *P. thermopropionicum–M. thermautotrophicus* coculture identified the filament spanning between the two organisms as a flagellum (114), suggesting that the role of the filament is to establish contact between the two microorganisms, not to mediate electron transfer (81). As previously discussed in detail (61), long-range electron transport via microbial nanowires should not be invoked without evidence for conduction along the length of the proposed nanowires and a demonstration that the filaments are required for the reduction of the proposed electron acceptor.

FEEDING ELECTRONS TO MICROBES

As limitations in producing electrical current with microbial fuel cells have become apparent, there has been a significant shift in focus toward the development of practical applications in which electrons flow from electrodes to microorganisms (64, 65, 102). Feeding electrons to microbes typically involves an input of energy that can help alleviate many of the limitations that arise when trying to extract energy with microbial fuel cells.

As previously reviewed (127), electrons can be supplied indirectly to microorganisms via artificial electron shuttles, but this approach has the same limitations for practical applications that were discussed above for current production. It is also possible to electrochemically reduce protons to hydrogen gas, but the low solubility and explosive nature of hydrogen gas limit the usefulness of this approach for most applications (127). Furthermore, efficient production of hydrogen gas typically requires expensive metallic catalysts or substantial inputs of energy to overcome sluggishness in proton reduction at electrode surfaces. Therefore, direct electron transfer from electrodes to microorganisms is expected to be the best choice for most applications as long as sufficiently high rates of electron transfer can be established (127).

The possibility that direct electron transfer from electrodes to microbes could drive microbial respiration was first noted in *Geobacter* species that have the capacity to reduce fumarate (37), nitrate (37), uranium (38), and chlorinated compounds (119, 122) with an electrode as the sole electron donor. A series of control studies demonstrated that hydrogen gas was not an intermediate in electron transfer between the electrode and the cells, and all the available evidence suggested that the *Geobacter* species were accepting electrons directly from the electrode (37). Gene expression patterns in biofilms of *G. sulfurreducens* reducing fumarate with an electrode as the electron donor were significantly different from those in *G. sulfurreducens* biofilms producing current (120). Deletion of genes, such as *omcZ* and *pilA*, that are essential for current production, had no impact on current consumption, whereas deletion of a cytochrome gene that is essential for current consumption had no impact on electron transfer to electrodes (120). These results suggest that the route for electron transfer from electrodes into *Geobacter* species is different from that for electron transfer in the opposite direction, a conclusion that is also supported by electrochemical studies (22). In contrast, electrons provided to *S. oneidensis* for fumarate reduction appear to enter via the same Mtr pathway that is responsible for electron flow to the outer cell surface (110).

Cathode: electrode that donates electrons

Mechanisms for energy conservation in cells receiving electrons from electrodes are poorly understood (64, 108). A potential source of energy conservation is the proton gradient across the inner membrane that should be generated when protons are consumed to reduce electron acceptors in the cytoplasm (64). Biofilms of current-consuming *Geobacter* species are much thinner than current-producing biofilms, suggesting that energy conservation is poorer for current consumption than for current generation.

Only a few pure cultures other than *Geobacter* and *Shewanella* species have been shown to carry out anaerobic respiration on cathodes. It was suggested that the methanogen *Methanobacterium palustre* (14) was capable of accepting electrons directly from electrodes, but there was the possibility of significant hydrogen production under the conditions employed (64, 134). *Anaeromyxobacter dehalogenans* could reduce fumarate and reductively dehalogenated 2-chlorophenol to phenol with an electrode serving as the sole electron donor (119). Proof-of-concept studies for microbial electrosynthesis, the process in which microorganisms use electrons derived from electrodes for the reduction of carbon dioxide to organic products, demonstrated that a number of acetogenic microorganisms accepted electrons for the reduction of carbon dioxide to acetate at potentials too high for hydrogen to serve as the intermediate for electron transfer (87, 93). Deleting the gene for the hydrogen-uptake hydrogenase in one of these acetogens, *Clostridium ljungdahlii*, had no impact on current consumption, further suggesting that electrons are transferred from the electrode directly to the cells (T. Ueki & K.P. Nevin, unpublished data).

BIOELECTRONICS

Electrically active microorganisms have the potential to make significant contributions to the emerging field of bioelectronics (138). For example, the ability of microorganisms to sense a wide diversity of chemicals and environmental conditions, coupled with the possibility of translating a response into an electrical signal, suggests many possibilities for the development of biological sensors and biocomputing (126). Furthermore, the discovery of unexpected electronic properties of conductive biofilms and the prospect of long-range electron transport through networks of conductive pili have opened the possibility that these materials may serve as models for the development of new synthetic electronic materials or may even be used directly in novel "living" electronic devices.

Electronics grown or constructed from living materials have the potential benefits that they can be produced from inexpensive feedstocks with little waste generation and avoid the use of toxic compounds. If the living microorganisms and their components are part of the electronic application, they can have the capacity for self-repair and replication. Charge can be transmitted and stored underwater. Furthermore, the demonstrated ability of some microorganisms and/or their extracellular components to make electrical contacts with electrodes suggests that they may be ideal tools for establishing electrical connections between abiological and biological components in medical devices and sensors.

In addition to their high conductivity, biofilms of *G. sulfurreducens* can function as supercapacitors (71). The abundant *c*-type cytochromes in the biofilms (28, 57, 61, 111) provide a capacitance comparable to that of synthetic supercapacitors with low self-discharge rates (71). The likelihood of manipulating both conductivity and capacitance with genetic engineering (71, 74), as well as improving the cohesiveness and other beneficial properties of the biofilms (C. Leang, unpublished data), demonstrates the potential for further developing these materials for practical applications. *G. sulfurreducens* biofilms can function as transistors, offering the possibility of developing field-effect transistors and other logic devices based on microbial nanowires (74, 98).

The metallic-like conductivity of pili offers the possibility of mass-producing novel wires for electronics. As the mechanisms for conductivity are elucidated, it should be feasible to modify the properties of the wires for specific applications. If the concept that cytochromes associated with pili facilitate electrical connections between pili and external electron acceptors/donors is correct, then it can be envisioned that it may be possible to genetically modify the structure of the electrical connection or to introduce new connectors to provide new functionalities.

FUTURE DIRECTIONS

There are many promising future research avenues in electromicrobiology. Our understanding of how microorganisms donate electrons to electrodes is still rather superficial and even less is known about electron transfer from electrodes to cells. Furthermore, previous study on this topic has been limited to a few microbes. It is remarkable that the organisms that have been studied in detail, *T. potens*, *S. oneidensis*, and *G. sulfurreducens*, have significantly different approaches for transferring electrons to electrodes. The intensive study of the electrophysiology of *Shewanella* and *Geobacter* species has been possible only because of substantial earlier investments that facilitated systems-scale investigations (33, 69). Similar in-depth investigations of other organisms are warranted. Elucidation of the mechanisms for metallic-like conductivity along pili and further investigation into the diversity of microorganisms that possess conductive filaments, the function of those filaments, and their mechanisms for conduction are needed.

Electromicrobiology has the potential to alleviate pressing societal needs. Although the justification for many of the early studies in electromicrobiology was further optimization of microbial fuel cells for energy harvesting, many more promising concepts for applications for microbe-electrode interactions have recently emerged and undoubtedly more will be envisioned. For example, as noted in the Introduction, the ability to favorably alter microbial fermentation with a supply of electrons from electrodes has been demonstrated, but only with the addition of soluble mediators. Developing a system for direct electron transfer might make this technology practical at a large scale. The rationale development of any of these technologies will depend on continued study of basic mechanisms of electromicrobiology.

SUMMARY POINTS

1. Electromicrobiology is a rapidly emerging field of microbiology.
2. Some microorganisms have the ability to either donate electrons to, or accept electrons from, electrodes.
3. Some microorganisms exchange electrons with electrodes via soluble molecules that facilitate electron transfer between the cells and the electrodes.

4. Other microorganisms can directly exchange electrons with electrodes via outer-surface, redox-active proteins, such as c-type cytochromes.

5. *Geobacter* species can produce thick, conductive biofilms with supercapacitor and transistor properties and conductivities that rival those of synthetic conductive polymers.

6. The pili of *G. sulfurreducens* can function as electrical wires, transporting electrons with metallic-like conductivity, a property not previously observed in biological materials.

7. Microbe-electrode exchange offers a number of potential practical applications in bioenergy, sensing, and bioremediation and serves as a model for important natural phenomena, such as interspecies electron transfer in anaerobic environments.

8. The electronic materials that microorganisms can produce have the potential to be incorporated into novel electronic devices or serve as models for the production of new synthetic electronic materials.

DISCLOSURE STATEMENT

The author is not aware of any affiliations, memberships, funding, or financial holdings that might be perceived as affecting the objectivity of this review.

ACKNOWLEDGMENTS

Electromicrobiology research in my laboratory is supported by the Advanced Research Projects Agency–Energy (ARPA-E), U.S. Department of Energy, under awards no. DE-AR0000087 and DE-AR0000159; the Office of Naval Research grants no. N00014-10-1-0084 and N00014-12-1-0229; and the Office of Science (BER), U.S. Department of Energy, under awards no. DE-SC0004114 and DE-SC000448 and cooperative agreement no. DE-FC02-02ER63446.

LITERATURE CITED

1. Aulenta F, Canosa A, Reale P, Rossetti S, Panero S, Majone M. 2009. Microbial reductive dechlorination of trichloroethene to ethene with electrodes serving as electron donors without the external addition of redox mediators. *Biotechnol. Bioeng.* 101:85–91
2. Baron DB, LaBelle E, Coursolle D, Gralnick JA, Bond DR. 2009. Electrochemical measurements of electron transfer kinetics by *Shewanella oneidensis* MR-1. *J. Biol. Chem.* 284:28865–73
3. Bigalke J, Grabner EW. 1997. The geobattery model: a contribution to large scale electrochemistry. *Electrochim. Acta* 42:3443–52
4. Bond DR. 2010. Electrodes as electron acceptors and the bacteria who love them. In *Geomicrobiology: Molecular and Environmental Perspectives*, ed. L Barton, M Mandl, A Loy, pp. 385–99. New York: Springer
5. **Bond DR, Holmes DE, Tender LM, Lovley DR. 2002. Electrode-reducing microorganisms that harvest energy from marine sediments. *Science* 295:483–85**
6. Bond DR, Lovley DR. 2003. Electricity production by *Geobacter sulfurreducens* attached to electrodes. *Appl. Environ. Microbiol.* 69:1548–55
7. Bond DR, Lovley DR. 2005. Evidence for involvement of an electron shuttle in electricity generation by *Geothrix fermentans*. *Appl. Environ. Microbiol.* 71:2186–89
8. Bouhenni RA, Vora GJ, Biffinger JC, Shirodkar S, Brockman K, et al. 2010. The role of *Shewanella oneidensis* MR-1 outer surface structures in extracellular electron transfer. *Electroanalysis* 22:856–64

5. First demonstration that microorganisms could oxidize organic compounds to carbon dioxide with direct electron transfer to electrodes.

8. Illustrates a genetic approach to directly test electron transfer concepts.

9. Busalmen JP, Esteve-Nunez A, Berna A, Feliu JM. 2008. *C*-type cytochromes wire electricity-producing bacteria to electrodes. *Agnew. Chem. Int. Ed.* 47:4874–77

10. Busalmen JP, Esteve-Nunez A, Berna A, Feliu JM. 2010. ATR-SEIRAs characterization of surface redox processes in *G. sulfurreducens*. *Bioelectrochemistry* 78:25–29

10. Good example of sophisticated approaches that are becoming available for studying microbe-electrode interactions.

11. Busalmen JP, Esteve-Nunez A, Feliu JM. 2008. Whole cell electrochemistry of electricity-producing microorganisms evidence an adaptation for optimal exocellular electron transport. *Envrion. Sci. Technol.* 42:2445–50

12. Butler C, Clauwaert P, Green SJ, Verstraete W, Nerenberg R. 2010. Bioelectrochemical perchlorate reduction in a microbial fuel cell. *Environ. Sci. Technol.* 44:4685–91

13. Chaudhuri SK, Lovley DR. 2003. Electricity generation by direct oxidation of glucose in mediatorless microbial fuel cells. *Nat. Biotechnol.* 21:1229–32

14. Cheng S, Xing D, Call DF, Logan BE. 2009. Direct biological conversion of electrical current into methane by electromethanogenesis. *Environ. Sci. Technol.* 43:3953–58

15. Childers SE, Ciufo S, Lovley DR. 2002. *Geobacter metallireducens* accesses insoluble Fe(III) oxide by chemotaxis. *Nature* 416:767–69

16. Clarke TA, Edwards MJ, Gates AJ, Hall A, White GF, et al. 2011. Structure of a cell surface decaheme electron conduit. *Proc. Natl. Acad. Sci. USA* 108:9384–9

16. Provides important insight into how Shewanella oneidensis transports electrons to the outer cell surface (see also Reference 39).

17. Coppi MV, O'Neil RA, Leang C, Kaufmann F, Methé BA, et al. 2007. Involvement of *Geobacter sulfurreducens* SfrAB in acetate metabolism rather than intracellular Fe(III) reduction. *Microbiology* 153:3572–85

18. Coursolle D, Baron DB, Bond DR, Gralnick JA. 2010. The Mtr respiratory pathway is essential for reducing flavins and electrodes in *Shewanella oneidensis*. *J. Bacteriol.* 192:467–74

19. Craig L, Piquie ME, Tainer JA. 2004. Type IV pilus structure and bacterial pathogenicity. *Nat. Rev. Microbiol.* 2:363–78

20. Ding YHR, Hixson KK, Giometti CS, Stanley A, Esteve-Nunez A, et al. 2006. The proteome of dissimilatory metal-reducing microorganism *Geobacter sulfurreducens* under various growth conditions. *Biochim. Biophys. Acta* 1764:1198–206

21. Dumas C, Basseguy R, Bergel A. 2008. Electrochemical activity of *Geobacter sulfurreducens* biofilms on stainless steel anodes. *Electrochim. Acta* 53:5235–41

22. Dumas C, Basseguy R, Bergel A. 2008. Microbial electrocatalysis with *Geobacter sulfurreducens* biofilm on stainless steel cathodes. *Electrochm. Acta* 53:2494–500

23. Edwards PP, Gray HB, Lodge MTJ, Williams RJP. 2008. Electron transfer and electronic conduction through an intervening medium. *Angew. Chem. Int. Ed.* 47:6758–65

24. El-Naggar MY, Gorby YA, Xia W, Nealson KH. 2008. The molecular density states in bacterial nanowires. *Biophys. J.* 95:L10–12

25. El-Naggar MY, Wanger G, Leung KM, Yuzvinsky TD, Southam G, et al. 2010. Electrical transport along bacterial nanowires from *Shewanella oneidensis*. *Proc. Natl. Acad. Sci. USA* 107:18127–31

26. Emde R, Schink B. 1990. Enhanced propionate formation by *Propionibacterium freudenreichii* subsp. *freudenreichii* in a three-electrode amperometric culture system. *Appl. Environ. Microbiol.* 56:2771–76

27. Esteve-Nunez A, Busalmen JP, Berna A, Gutierrez-Garran C, Feliu JM. 2011. Opportunities behind the unusual ability of *Geobacter sulfurreducens* for exocellular respiration and electricity production. *Energy Environ. Sci.* 4:2066–69

28. Esteve-Nunez A, Sosnik J, Visconti P, Lovley DR. 2008. Fluorescent properties of *c*-type cytochromes reveal their potential role as an extracytoplasmic electron sink in *Geobacter sulfurreducens*. *Environ. Microbiol.* 10:497–505

29. Flynn JM, Ross DE, Hunt KA, Bond DR, Gralnick JA. 2010. Enabling unbalanced fermentations by using engineered electrode-interfaced bacteria. *mBio* 1:e00190–10

29. Demonstrates a promising biotechnological application for microbe-electrode electron exchange.

30. Franks AE, Nevin KP. 2010. Microbial fuel cells, a current review. *Energies* 3:899–919

31. Franks AE, Nevin KP, Glaven RH, Lovley DR. 2010. Microtoming coupled to microarray analysis to evaluate the spatial metabolic status of *Geobacter sulfurreducens* biofilms. *ISME J.* 4:509–19

32. Franks AE, Nevin KP, Jia H, Izallalen M, Woodard TL, Lovley DR. 2009. Novel strategy for three-dimensional real-time imaging of microbial fuel cell communities: monitoring the inhibitory effects of proton accumulation within the anode biofilm. *Energy Environ. Sci.* 2:113–19

33. Fredrickson JK, Romine MF, Beliaev AS, Auchtung JM, Driscoll ME, et al. 2008. Towards environmental systems biology of *Shewanella*. *Nat. Rev. Microbiol.* 6:592–603
34. Fricke K, Harnisch F, Schroder U. 2008. On the use of cyclic voltammetry for the study of anodic electron transfer in microbial fuel cells. *Energy Environ. Sci.* 1:144–47
35. Geelhoed JS, Hamelers HVM, Stams AJM. 2010. Electricity-mediated biological hydrogen production. *Curr. Opin. Microbiol.* 13:307–15
36. Gorby YA, Yanina S, McLean JS, Rosso KM, Moyles D, et al. 2006. Electrically conductive bacterial nanowires produced by *Shewanella oneidensis* strain MR-1 and other microorganisms. *Proc. Natl. Acad. Sci. USA* 103:11358–63
37. Gregory KB, Bond DR, Lovley DR. 2004. Graphite electrodes as electron donors for anaerobic respiration. *Environ. Microbiol.* 6:596–604
38. Gregory KB, Lovley DR. 2005. Remediation and recovery of uranium from contaminated subsurface environments with electrodes. *Environ. Sci. Technol.* 39:8943–47
39. Hartshorne RS, Reardon CL, Ross D, Nuester J, Clarke TA, et al. 2009. Characterization of an electron conduit between bacteria and the extracellular environment. *Proc. Natl. Acad. Sci. USA* 106:22169–74
40. Holmes DE, Bond DR, Lovley DR. 2004. Electron transfer by *Desulfobulbus propionicus* to Fe(III) and graphite electrodes. *Appl. Environ. Microbiol.* 70:1234–37
41. Holmes DE, Bond DR, O'Neil RA, Reimers CE, Tender LR, Lovley DR. 2004. Microbial communities associated with electrodes harvesting electricity from a variety of aquatic sediments. *Microb. Ecol.* 48:178–90
42. Holmes DE, Chaudhuri SK, Nevin KP, Mehta T, Methe BA, et al. 2006. Microarray and genetic analysis of electron transfer to electrodes in *Geobacter sulfurreducens*. *Environ. Microbiol.* 8:1805–15
43. Holmes DE, Mester T, O'Neil RA, Larrahondo MJ, Adams LA, et al. 2008. Genes for two multicopper proteins required for Fe(III) oxide reduction in *Geobacter sulfurreducens* have different expression patterns both in the subsurface and on energy-harvesting electrodes. *Microbiology* 145:1422–35
44. Hongo M, Iwahara M. 1979. Application of electro-energizing method to L-glutamic acid fermentation. *Agric. Biol. Chem.* 10:2075–81
45. Inoue K, Leang C, Franks AE, Woodard TL, Nevin KP, Lovley DR. 2011. Specific localization of the *c*-type cytochrome OmcZ at the anode surface in current-producing biofilms of *Geobacter sulfurreducens*. *Environ. Microbiol. Rep.* 3:211–17
46. Inoue K, Qian X, Morgado L, Kim BC, Mester T, et al. 2010. Purification and characterization of OmcZ, an outer-surface, octaheme *c*-type cytochrome essential for optimal current production by *Geobacter sulfurreducens*. *Appl. Environ. Microbiol.* 76:3999–4007
47. Jain A, Gazzola G, Panzera A, Zanoni M, Marsili E. 2011. Visible spectroelectrochemical characterization of *Geobacter sulfurreducens* biofilms on optically transparent indium tin oxide electrode. *Electrochim. Acta* 47:12530–32
48. **Jiang X, Hu J, Fitzgerald LA, Biffinger JC, Xie P, et al. 2010. Probing electron transfer mechanisms in *Shewanella oneidensis* MR-1 using a nanoelectrode platform and single-cell imaging. *Proc. Natl. Acad. Sci. USA* 107:16806–10**
49. Kiely PD, Regan JM, Logan BE. 2011. The electric picnic: synergistic requirements for exoelectrogenic microbial communities. *Curr. Opin. Biotechnol.* 22:378–85
50. Lanthier M, Gregory KB, Lovley DR. 2008. Growth with high planktonic biomass in *Shewanella oneidensis* fuel cells. *FEMS Microbiol. Lett.* 278:29–35
51. Leang C, Adams LA, Chin K-J, Nevin KP, Methé BA, et al. 2005. Adaption to disruption of electron transfer pathway for Fe(III) reduction in *Geobacter sulfurreducens*. *J. Bacteriol.* 187:5918–26
52. Leang C, Coppi MV, Lovley DR. 2003. OmcB, a *c*-type polyheme cytochrome, involved in Fe(III) reduction in *Geobacter sulfurreducens*. *J. Bacteriol.* 185:2096–103
53. Leang C, Qian X, Mester T, Lovley DR. 2010. Alignment of the *c*-type cytochrome OmcS along pili of *Geobacter sulfurreducens*. *Appl. Environ. Microbiol.* 76:4080–84
54. Lies DP, Hernandez ME, Kappler A, Mielke RE, Gralnick JA, Newman DK. 2005. *Shewanella oneidensis* MR-1 uses overlapping pathways for iron reduction at a distance and by direct contact under conditions relevant for biofilms. *Appl. Environ. Microbiol.* 71:4414–26

48. Innovative investigation that helped resolve the mechanisms for electron transfer to electrodes in *S. oneidensis*.

55. Liu H, Newton GJ, Nakamura R, Hashimoto K, Nakanishi. 2010. Electrochemical characterization of a single electricity-producing bacterial cell of *Shewanella* using optical tweezers. *Angew. Chem. Int. Ed.* 49:6596–99
56. Liu Y, Kim H, Franklin R, Bond DR. 2010. Gold line array electrodes increase substrate affinity and current density of electricity-producing *G. sulfurreducens* biofilms. *Energy Environ. Sci.* 3:1782–88
57. Liu Y, Kim H, Franklin RR, Bond DR. 2011. Linking spectral and electrochemical analysis to monitor *c*-type cytochrome redox status in living *Geobacter sulfurreducens* biofilms. *ChemPhysChem* 12:2235–41
58. Logan BE. 2009. Exoelectrogenic bacteria that power microbial fuel cells. *Nat. Rev. Microbiol.* 7:375–81
59. Lovley DR. 2006. Bug juice: harvesting electricity with microorganisms. *Nat. Rev. Microbiol.* 4:497–508
60. Lovley DR. 2006. Microbial fuel cells: novel microbial physiologies and engineering approaches. *Curr. Opin. Biotechnol.* 17:327–32
61. Lovley DR. 2008. Extracellular electron transfer: wires, capacitors, iron lungs, and more. *Geobiology* 6:225–31
62. Lovley DR. 2008. The microbe electric: conversion of organic matter to electricity. *Curr. Opin. Biotechnol.* 19:564–71
63. Lovley DR. 2011. Live wires: direct extracellular electron exchange for bioenergy and the bioremediation of energy-related contamination. *Energy Environ. Sci.* 4:4896–906
64. Lovley DR. 2011. Powering microbes with electricity: direct electron transfer from electrodes to microbes. *Environ. Microbiol. Rep.* 3:27–35
65. Lovley DR, Nevin KP. 2011. A shift in the current: new applications and concepts for microbe-electrode electron exchange. *Curr. Opin. Biotechnol.* 22:441–48
66. Lovley DR, Ueki T, Zhang T, Malvankar NS, Shrestha PM, et al. 2012. *Geobacter*: the microbe electric's physiology, ecology, and practical applications. *Adv. Microb. Physiol.* 59:1–100
67. Magnuson TS, Isoyama N, Hodges-Myerson AL, Davidson G, Maroney MJ, et al. 2001. Isolation, characterization and gene sequence analysis of a membrane-associated 89 kDa Fe(III) reducing cytochrome *c* from *Geobacter sulfurreducens*. *Biochem. J.* 359:147–52
68. Mahadevan R, Bond DR, Butler JE, Esteve-Nunez A, Coppi MV, et al. 2006. Characterization of metabolism in the Fe(III)-reducing organism *Geobacter sulfurreducens* by constraint-based modeling. *Appl. Environ. Microbiol.* 72:1558–68
69. Mahadevan R, Palsson BO, Lovley DR. 2011. In situ to in silico and back: elucidating the physiology and ecology of *Geobacter* spp. using genome-scale modelling. *Nat. Rev. Microbiol.* 9:39–50
70. Malvankar NS, Lovley DR. 2012. Microbial nanowires: a new paradigm for biological electron transfer and bioelectronics. *ChemSusChem* 5:1039–46
71. Malvankar NS, Mester T, Tuominen MT, Lovley DR. 2012. Supercapacitors based on *c*-type cytochromes using conductive nanostructured networks of living bacteria. *ChemPhysChem* 13:463–68
72. Malvankar NS, Tuominen MT, Lovley DR. 2012. Biofilm conductivity as a decisive variable for the high-current-density *Geobacter sulfurreducens* microbial fuel cells. *Energy Environ. Sci.* 5:5790–97
73. Malvankar NS, Tuominen MT, Lovley DR. 2012. Comment on "On electrical conductivity of microbial nanowires and biofilms" by S.M. Strycharz-Glaven, R.M. Snider, A. Guiseppi-Elie and L. M. Tender, *Energy Environ. Sci.*, 2011, 4, 4366. *Energy Environ. Sci.* 5:6247–49
74. **Malvankar NS, Vargas M, Nevin KP, Franks AE, Leang C, et al. 2011. Tunable metallic-like conductivity in nanostructured biofilms comprised of microbial nanowires. *Nat. Nanotechnol.* 6:573–79**
75. Marcus AK, Torres CI, Rittmann BE. 2007. Conduction-based modeling of the biofilm anode of a microbial fuel cell. *Biotechnol. Bioeng.* 98:1171–82
76. Marcus AK, Torres CI, Rittmann BE. 2011. Analysis of microbial electrochemical cell using the proton condition in a biofilm (PCBIOFILM) model. *Bioresour. Technol.* 102:253–62
77. Marshall C, May H. 2009. Electrochemical evidence of direct electrode reduction by a thermophilic gram-positive bacterium *Thermincola ferriacetica*. *Energy Environ. Sci.* 2:699–705
78. Marsili E, Baron DB, Shikhare I, Coursolle D, Gralnick JA, Bond DR. 2008. *Shewanella* secretes flavins that mediate extracellular electron transfer. *Proc. Natl. Acad. Sci. USA* 105:3968–73

---

74. Demonstrates metallic-like conductivity for transport along pili and through biofilms offers an unprecedented mechanism for long-range electron transport in biology.

79. Marsili E, Rollefson JB, Baron DB, Hozalski RM, Bond DR. 2008. Microbial biofilm voltammetry: direct electrochemical characterization of catalytic electrode-attached biofilms. *Appl. Environ. Microbiol.* 74:7329–37
80. Marsili E, Sun J, Bond DR. 2010. Voltammetry and growth physiology of *Geobacter sulfurreducens* biofilms as a function of growth stage and imposed potential. *Electroanalysis* 22:865–74
81. McInerney MJ, Sieber JR, Gunsalus RP. 2009. Syntrophy in anaerobic global carbon cycles. *Curr. Opin. Biotechnol.* 20:623–32
82. Mehta T, Childers SE, Glaven R, Lovley DR, Mester T. 2006. A putative multicopper protein secreted by an atypical type II secretion system involved in the reduction of insoluble electron acceptors in *Geobacter sulfurreducens*. *Microbiology* 152:2257–64
83. Mehta T, Coppi MV, Childers SE, Lovley DR. 2005. Outer membrane c-type cytochromes required for Fe(III) and Mn(IV) oxide reduction in *Geobacter sulfurreducens*. *Appl. Environ. Microbiol.* 71:8634–41
84. Methé BA, Nelson KE, Eisen JA, Paulsen IT, Nelson W, et al. 2003. The genome of *Geobacter sulfurreducens*: insights into metal reduction in subsurface environments. *Science* 302:1967–69
85. Millo D, Harnisch F, Patil SA, Ly HK, Schröder U, Hildebrandt P. 2011. In situ spectroelectrochemical investigation of electrocatalytic microbial biofilms by surface-enhanced resonance Raman spectroscopy. *Angew. Chem. Int. Ed.* 50:2625–27
86. Morita M, Malvankar NS, Franks AE, Summers ZM, Giloteaux L, et al. 2011. Potential for direct interspecies electron transfer in methanogenic wastewater digester aggregates. *mBio* 2:e00159–11
87. Nevin KP, Hensley SA, Franks AE, Summers ZM, Ou J, et al. 2011. Electrosynthesis of organic compounds from carbon dioxide is catalyzed by a diversity of acetogenic microorganisms. *Appl. Environ. Microbiol.* 77:2882–86
88. Nevin KP, Kim BC, Glaven RH, Johnson JP, Woodard TL, et al. 2009. Anode biofilm transcriptomics reveals outer surface components essential for high density current production in *Geobacter sulfurreducens* fuel cells. *PLoS One* 4:e5628
89. Nevin KP, Lovley DR. 2000. Lack of production of electron-shuttling compounds or solubilization of Fe(III) during reduction of insoluble Fe(III) oxide by *Geobacter metallireducens*. *Appl. Environ. Microbiol.* 66:2248–51
90. Nevin KP, Lovley DR. 2002. Mechanisms for accessing insoluble Fe(III) oxide during dissimilatory Fe(III) reduction by *Geothrix fermentans*. *Appl. Environ. Microbiol.* 68:2294–99
91. Nevin KP, Lovley DR. 2002. Mechanisms for Fe(III) oxide reduction in sedimentary environments. *Geomicrobiol. J.* 19:141–59
92. Nevin KP, Richter H, Covalla SF, Johnson JP, Woodard TL, et al. 2008. Power output and columbic efficiencies from biofilms of *Geobacter sulfurreducens* comparable to mixed community microbial fuel cells. *Environ. Microbiol.* 10:2505–14
93. **Nevin KP, Woodard TL, Franks AE, Summers ZM, Lovley DR. 2010. Microbial electrosynthesis: feeding microbes electricity to convert carbon dioxide and water to multicarbon extracellular organic compounds. *mBio* 1:e00103–10**
94. Newman DK, Kolter R. 2000. A role for excreted quinones in extracellular electron transfer. *Nature* 405:93–97
95. Park DH, Laivenieks M, Guettler MV, Jain MK, Zeikus JG. 1999. Microbial utilization of electrically reduced neutral red as the sole electron donor for growth and metabolite production. *Appl. Environ. Microbiol.* 65:2912–17
96. Park HI, Kim DK, Choi Y-J, Pak D. 2005. Nitrate reduction using an electrode as direct electron donor in a biofilm-electrode reactor. *Process Biochem.* 40:3383–88
97. Picioreanu C, Head IM, Katuri KP, van Loosdrecht MCM, Scott K. 2007. A computational model for biofilm-based microbial fuel cells. *Water Res.* 41:2921–40
98. Qian F, Li Y. 2011. A natural source of nanowires. *Nat. Nanotechnol.* 6:538–9
99. Qian X, Reguera G, Mester T, Lovley DR. 2007. Evidence that OmcB and OmpB of *Geobacter sulfurreducens* are outer membrane surface proteins. *FEMS Microbiol. Lett.* 277:21–27
100. Qian XL, Mester T, Morgado L, Arakawa T, Sharma ML, et al. 2011. Biochemical characterization of purified OmcS, a c-type cytochrome required for insoluble Fe(III) reduction in *Geobacter sulfurreducens*. *Biochim. Biophys. Acta* 1807:404–12

93. Proof of concept for an artificial form of photosynthesis with many potential advantages over biomass-based strategies for the production of transportation fuels and other organic commodities.

101. Rabaey K, Rodriguez J, Blackall LL, Keller J, Gross P, et al. 2007. Microbial ecology meets electrochemistry: electricity-driven and driving communities. *ISME J.* 1:9–18
102. Rabaey K, Rozendal RA. 2010. Microbial electrosynthesis—revisiting the electrical route for microbial production. *Nat. Rev. Microbiol.* 8:706–16
103. Reguera G, McCarthy KD, Mehta T, Nicoll JS, Tuominen MT, Lovley DR. 2005. Extracellular electron transfer via microbial nanowires. *Nature* 435:1098–101
104. Reguera G, Nevin KP, Nicoll JS, Covalla SF, Woodard TL, Lovley DR. 2006. Biofilm and nanowire production leads to increased current in *Geobacter sulfurreducens* fuel cells. *Appl. Environ. Microbiol.* 72:7345–48
105. Reimers CE, Tender LM, Fertig S, Wang W. 2001. Harvesting energy from the marine sediment-water interface. *Environ. Sci. Technol.* 35:192–95
106. Richter H, Nevin KP, Jia H, Lowy DA, Lovley DR, Tender LM. 2009. Cyclic voltammetry of biofilms of wild type and mutant *Geobacter sulfurreducens* on fuel cell anodes indicates possible roles of OmcB, OmcZ, type IV pili, and protons in extracellular electron transfer. *Energ. Environ. Sci.* 2:506–16
107. Rollefson JB, Stephen CS, Tien M, Bond DR. 2011. Identification of an extracellular polysaccharide network essential for cytochrome anchoring and biofilm formation in *Geobacter sulfurreducens*. *J. Bacteriol.* 193:1023–33
108. Rosenbaum M, Aulenta F, Villano M, Angenent LT. 2011. Cathodes as electron donors for microbial metabolism: Which extracellular electron transfer mechanisms are involved? *Biores. Technol.* 102:324–33
109. Ross DE, Brantley SL, Tien M. 2009. Kinetic characterization of OmcA and MtrC, terminal reductases involved in respiratory electron transfer for dissimilatory iron reduction in *Shewanella oneidensis* MR-1. *Appl. Environ. Microbiol.* 75:5218–26
110. Ross DE, Flynn JM, Baron DB, Gralnick JA, Bond DR. 2011. Towards electrosynthesis in *Shewanella*: energetics of reversing the Mtr pathway for reductive metabolism. *PLoS One* 6:e16649
111. Schrott GD, Bonnani PS, Robuschi L, Esteve-Núñez A, Busalmen JP. 2011. Electrochemical insight into the mechanism of electron transport in biofilms of *Geobacter sulfurreducens*. *Electrochim. Acta* 56:10791–95
112. Shelobolina ES, Coppi MV, Korenevsky AA, DiDonato LN, Sullivan SA, et al. 2007. Importance of *c*-type cytochromes for U(VI) reduction by *Geobacter sulfurreducens*. *BMC Microbiol.* 7:16
113. Shi L, Squier TC, Zachara JM, Fredrickson JK. 2007. Respiration of metal (hydr)oxides by *Shewanella* and *Geobacter*: a key role for multihaem *c*-type cytochromes. *Mol. Microbiol.* 65:12–20
114. Shimoyama T, Kato S, Ishii S, Watanabe K. 2009. Flagellum mediates symbiosis. *Science* 323:1574
115. Shinwari MW, Deen MJ, Starikov EB, Cuniberti G. 2010. Electrical conductance in biological molecules. *Adv. Funct. Mater.* 20:1865–83
116. Srikanth S, Marsili E, Flickinger MC, Bond DR. 2008. Electrochemical characterization of *Geobacter sulfurreducens* cells immobilized on graphite paper anodes. *Biotechnol. Bioeng.* 99:1065–73
117. Stams AJ, Plugge CM. 2009. Electron transfer in syntrophic communities of anaerobic bacteria and archaea. *Nat. Rev. Microbiol.* 7:568–77
118. Steinbusch KJJ, Hamelers HVM, Schaap JD, Kampman C, Buisman CJN. 2010. Bioelectrochemical ethanol production through mediated acetate reduction by mixed cultures. *Environ. Sci. Technol.* 44:513–17
119. Strycharz SM, Gannon SM, Boles AR, Nevin KP, Franks AE, Lovley DR. 2010. *Anaeromyxobacter dehalogenans* interacts with a poised graphite electrode for reductive dechlorination of 2-chlorophenol. *Environ. Microbiol. Rep.* 2:289–294
120. Strycharz SM, Glaven RH, Coppi MV, Gannon SM, Perpetua LA, et al. 2011. Gene expression and deletion analysis of mechanisms for electron transfer from electrodes to *Geobacter sulfurreducens*. *Bioelectrochemistry* 80:142–50
121. Strycharz SM, Malanoski AP, Snider RM, Yi H, Lovley DR, Tender LM. 2011. Application of cyclic voltammetry to investigate enhanced catalytic current generation by biofilm-modified anodes of *Geobacter sulfurreducens* strain DL1 versus variant strain KN400. *Energy Environ. Sci.* 4:896–913
122. Strycharz SM, Woodward TL, Johnson JP, Nevin KP, Sanford RA, et al. 2008. Graphite electrode as a sole electron donor for reductive dechlorination of tetrachlorethene by *Geobacter lovleyi*. *Appl. Environ. Microbiol.* 74:5943–47

123. Summers ZM, Fogarty HE, Leang C, Franks AE, Malvankar NS, Lovley DR. 2010. Direct exchange of electrons within aggregates of an evolved syntrophic coculture of anaerobic bacteria. *Science* 330:1413–15
124. Tandukar M, Huber SJ, Onodera T, Pavlostathis SG. 2009. Biological chromium(VI) reduction in the cathode of a microbial fuel cell. *Environ. Sci. Technol.* 43:8159–65
125. Tender LM, Reimers CE, Stecher HA, Holmes DE, Bond DR, et al. 2002. Harnessing microbially generated power on the seafloor. *Nat. Biotechnol.* 20:821–25
126. **TerAvest MA, Li Z, Angenent LT. 2011. Bacteria-based biocomputing with cellular computing circuits to sense, decide, signal, and act. *Energy Environ. Sci.* 4:4907–16**
127. Thrash JC, Coates JD. 2008. Review: direct and indirect electrical stimulation of microbial metabolism. *Environ. Sci. Technol.* 42:3921–31
128. Thrash JC, Van Trump IV, Weber KA, Miller E, Achenbach LA, Coates JD. 2007. Electrochemical stimulation of microbial perchlorate reduction. *Environ. Sci. Technol.* 41:1740–46
129. **Torres CI, Marcus AK, Lee H-S, Parameswaran P, Krajmalnik-Brown R, Rittmann BE. 2010. A kinetic perspective on extracellular electron transfer by anode-respiring bacteria. *FEMS Microbiol. Rev.* 34:3–17**
130. **Torres CI, Marcus AK, Rittmann BE. 2008. Proton transport inside the biofilm limits electrical current generation by anode-respiring bacteria. *Biotechnol. Bioeng.* 100:872–81**
131. Tremblay P-L, Aklujkar M, Leang C, Lovley DR. 2011. A genetic system for *Geobacter metallireducens*: role of flagella and pili in extracellular electron transfer. *Environ. Microbiol. Rep.* 4:82–88
132. Tremblay P-L, Summers ZM, Glaven RH, Nevin KP, Zengler K, et al. 2011. A *c*-type cytochrome and a transcriptional regulator responsible for enhanced extracellular electron transfer in *Geobacter sulfurreducens* uncovered by adaptive evolution. *Environ. Microbiol.* 13:13–23
133. Van Eerten-Jansen MCAA, Heijne AT, Buisman CJN, Hamelers HVM. 2012. Microbial electrolysis cells for production of methane from CO_2: long-term performance and perspectives. *Int. J. Energy Res.* 36:809–19
134. Villano M, Aulenta F, Ciucci C, Ferri T, Giuliano A, Majone M. 2010. Bioelectrochemical reduction of CO_2 to CH_4 via direct and indirect extracellular electron transfer by a hydrogenophilic methanogenic culture. *Biores. Technol.* 101:3085–90
135. von Canstein H, Ogawa J, Shimizu S, Lloyd JR. 2008. Secretion of flavins by *Shewanella* species and their role in extracellular electron transfer. *Appl. Environ. Microbiol.* 74:615–23
136. Voordeckers JW, Izallalen M, Kim B-C, Lovley DR. 2010. Role of *Geobacter sulfurreducens* outer surface *c*-type cytochromes in the reduction of soil humic acid and the humics analog anthraquinone-2,6-disulfonate. *Appl. Environ. Microbiol.* 76:2371–75
137. Williams KN, Nevin KP, Franks AE, Englert A, Long PE, Lovley DR. 2010. Electrode-based approach for monitoring in situ microbial activity during subsurface bioremediation. *Environ. Sci. Technol.* 44:47–54
138. Willner I, Katz E, eds. 2005. *Bioelectronics: From Theory to Applications*. Weinheim: Wiley-VCH
139. Wrighton KC, Agbo P, Warnecke F, Weber KA, Brodie EL, et al. 2008. A novel ecological role of the *Firmicutes* identified in thermophilic microbial fuel cells. *ISME J.* 2:1146–56
140. Wrighton KC, Thrash JC, Melnyk RA, Bigi JP, Byrne-Bailey KG, et al. 2011. Evidence for direct electron transfer by a gram-positive bacterium isolated from a microbial fuel cell. *Appl. Environ. Microbiol.* 77:7633–39
141. Yi H, Nevin KP, Kim B-C, Franks AE, Klimes A, et al. 2009. Selection of a variant of *Geobacter sulfurreducens* with enhanced capacity for current production in microbial fuel cells. *Biosens. Bioelectron.* 24:3498–503
142. Zhang T, Gannon SM, Nevin KP, Franks AE, Lovley DR. 2010. Stimulating the anaerobic degradation of aromatic hydrocarbons in contaminated sediments by providing an electrode as the electron acceptor. *Environ. Microbiol.* 12:1011–20

126. Forward-looking discussion of exciting possibilities in electromicrobiology.

129. Excellent discussion of the role of biofilm conductivity in generating high-current densities in microbial fuel cells.

130. First study emphasizing the importance of proton accumulation as a factor limiting current production in anode biofilms.

RELATED RESOURCES

Geobacter Project. http://www.geobacter.org/
www.electrofuels.org

Origin and Diversification of Eukaryotes

Laura A. Katz

Department of Biological Sciences, Smith College, Northampton, Massachusetts 01063; email: lkatz@smith.edu

Program in Organismic and Evolutionary Biology, University of Massachusetts, Amherst, Massachusetts 01003

Keywords

eukaryotic diversity, protists, tree of life, nucleus, cytoskeleton, mitochondria

Abstract

The bulk of the diversity of eukaryotic life is microbial. Although the larger eukaryotes—namely plants, animals, and fungi—dominate our visual landscapes, microbial lineages compose the greater part of both genetic diversity and biomass, and contain many evolutionary innovations. Our understanding of the origin and diversification of eukaryotes has improved substantially with analyses of molecular data from diverse lineages. These data have provided insight into the nature of the genome of the last eukaryotic common ancestor (LECA). Yet, the origin of key eukaryotic features, namely the nucleus and cytoskeleton, remains poorly understood. In contrast, the past decades have seen considerable refinement in hypotheses on the major branching events in the evolution of eukaryotic diversity. New insights have also emerged, including evidence for the acquisition of mitochondria at the time of the origin of eukaryotes and data supporting the dynamic nature of genomes in LECA.

Contents

INTRODUCTION	412
PART I: ORIGIN OF EUKARYOTES AND FEATURES OF THE LAST EUKARYOTIC COMMON ANCESTOR	414
Origin of Eukaryotic Genomes	414
Origin of the Nucleus	416
Origin of the Cytoskeleton	417
Origin of Mitochondria	417
PART II: EVOLUTION OF PHOTOSYNTHESIS WITH EUKARYOTES	418
PART III: RELATIONSHIPS AMONG MAJOR LINEAGES	418
Root of the Eukaryotic Tree of Life	419
Major Eukaryotic Clades	420
CONCLUSION	422

INTRODUCTION

Eukaryote: a cell with a nucleus

Last eukaryotic common ancestor (LECA): lineage that gave rise to extant eukaryotes

Cytoskeleton: complex structure in eukaryotes that provides for shape and motility

We live on a microbial planet. Microbes have dominated Earth's history and continue to represent the majority of both biodiversity and biomass on our planet. Two of the three domains of life, the Bacteria and Archaea, are virtually exclusively microbial, and microbial forms dominate among the third domain, Eukaryota, which is the focus of this review. Yet despite their importance, much remains to be learned about microbial life in terms of discovering new forms, understanding major innovations, and incorporating the biology of microorganisms into theories and models across disciplines within biology.

Eukaryotes are named for one of their defining features—the presence of a nucleus (*eu*, "true," and *karyo*, "kernel" or "seed"). A defining feature is one that is found in every eukaryote and that was present in the last eukaryotic common ancestor (LECA). A second defining feature is the presence of a cytoskeleton, which is a complex set of structures underlain by a tremendous diversity of proteins (e.g., actins, tubulins, dyneins). The cytoskeleton gives eukaryotes their diverse morphologies (**Figure 1**), variable motility, and ability to engulf other organisms.

Early attempts to reconstruct the tree of life focused on macroscopic organisms, first dividing living things between Plantae and Animalia, and then adding Protista as a grab bag of

---

Figure 1

Representative eukaryotic lineages, with quotes around taxon names that are either controversial or as yet lack robust support, following suggestions in References 88 and 126: (*a–c*) 'Plantae.' (*a*) *Eremosphaera viridis*, a green alga. (*b*) *Cyanidium* sp., a red alga. (*c*) *Cyanophora* sp., a glaucophyte. (*d*) *Chroomonas* sp., a cryptomonad. (*e*) *Emiliania huxleyi*, a haptophyte. (*f–m*) 'SAR' (Stramenopila, Alveolata, and Rhizaria). (*f*) *Akashiwo sanguinea*, a dinoflagellate. (*g*) *Trithigmostoma cucullulus*, a ciliate. (*h*) *Colpodella perforans*, an apicomplexan. (*i*) *Thalassionema* sp., a colonial diatom. (*j–m*) 'Rhizaria.' (*j*) *Chlorarachnion reptans*, a core cercozoan. (*k*) *Acantharea* sp., formerly known as a radiolarian. (*l*) *Ammonia beccarii*, a calcareous foraminiferan. (*m*) *Corallomyxa tenera*, a reticulate rhizarian amoeba. (*n–p*) 'Excavata.' (*n*) *Jakoba* sp., a jakobid with two flagella. (*o*) *Chilomastix cuspidata*, a flagellate in Fornicata. (*p*) *Euglena sanguinea*, an autotrophic Euglenozoa. (*q–s*) 'Amoebozoa.' (*q*) *Trichosphaerium* sp., a naked stage (lacking surface spicules) of an unusual amoeba with alternation of generations, one naked and one with spicules. (*r*) *Stemonitis axifera*, a dictyostelid. (*s*) *Arcella hemisphaerica*, a testate amoeba in Tubulinea. (*t–w*) Opisthokonta. (*t*) *Homo sapiens*, animal. (*u*) *Campyloacantha* sp., a choanoflagellate. (*v*) *Amanita flavoconia*, a basidiomycete fungus. (*w*) *Chytriomyces* sp., a chytrid. All images are provided by micro*scope (http://starcentral.mbl.edu/microscope/portal.php) except panel *t*, which is provided by the author. Redrawn from Reference 116, *BioScience* 59(6), Copyright 2009, American Institute of Biological Sciences.

organisms that did not clearly fit in either category (reviewed in Reference 103). Beginning in the mid-twentieth century, biodiversity was seen as belonging to five kingdoms: macroscopic plants, animals, and fungi, and microscopic monera (bacteria) and protists (80, 121, 122). With the advent of better microscopes and, more recently, the explosion of molecular studies, the tree of life has been divided into three major domains—Bacteria, Archaea, and Eukaryota (124, 125)—with a still-disputed number of major clades within each.

This review discusses current ideas on the origin and diversification of eukaryotes through evaluation of evidence, review of recent hypotheses, and indication of open questions. To this end, I focus on three topics: the origin of eukaryotes based on insights from analyses of features present in LECA, the acquisition of photosynthesis among eukaryotes, and the relationships among extant eukaryotes.

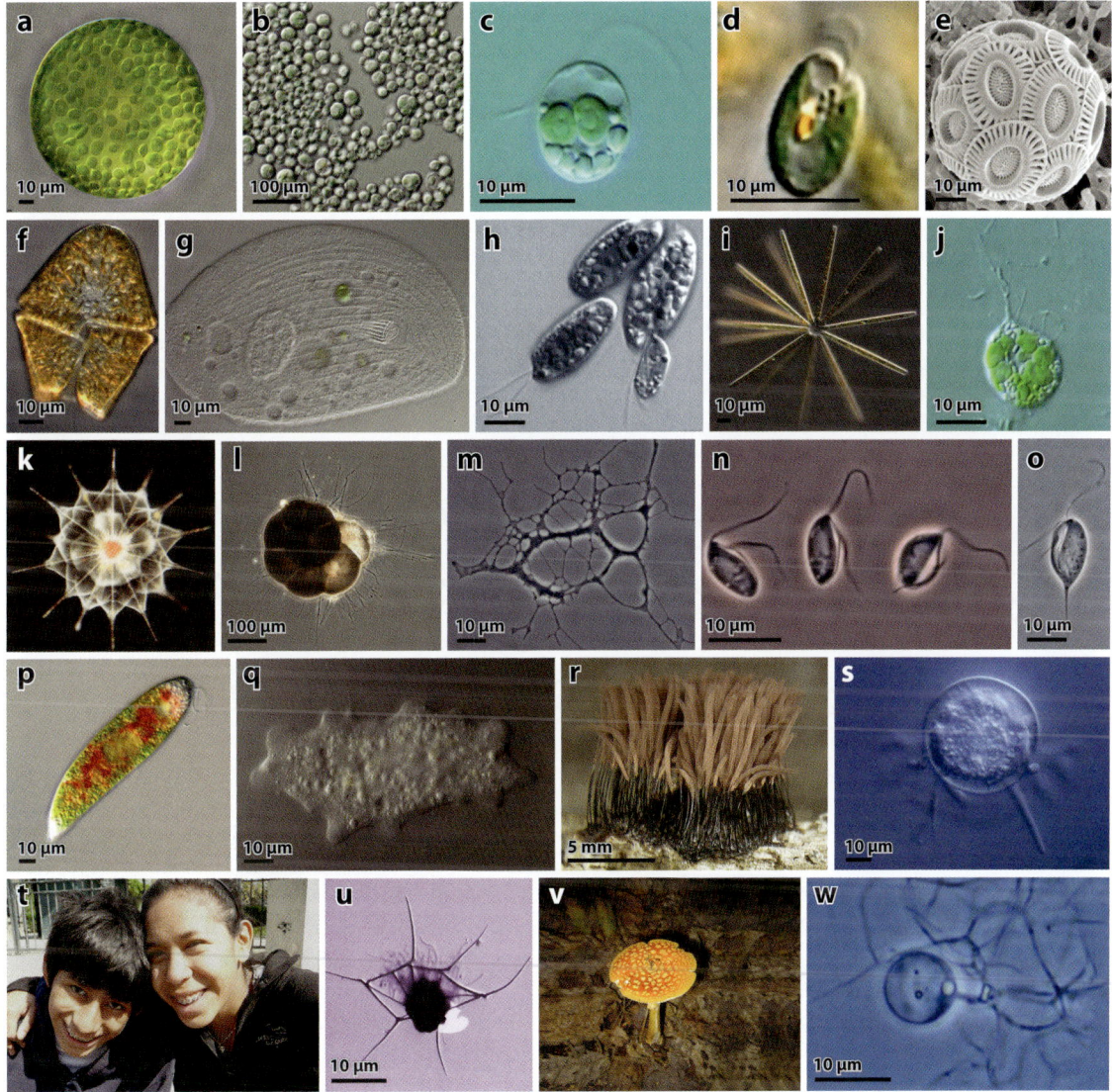

PART I: ORIGIN OF EUKARYOTES AND FEATURES OF THE LAST EUKARYOTIC COMMON ANCESTOR

LGT: lateral gene transfer, also called horizontal gene transfer

Chimerism: the presence of genes of varying ancestries within eukaryotic genomes

Homologs: shared characteristics present in last common ancestor of a group of organisms

Beyond the ubiquitous nucleus and cytoskeleton, we can infer that LECA was complex in terms of its morphology and genome. Insights into LECA emerge from a long history of study of diverse eukaryotic organisms coupled with more recent inferences from molecular data. These data reveal that LECA had complex morphology, with a nucleus, mitochondria, and a cytoskeleton plus associated features (e.g., flagella). As argued in detail below, LECA also had a genome that was both chimeric, with respect to bacteria and archaea, and dynamic, with epigenetic phenomena playing key roles during life cycles.

Origin of Eukaryotic Genomes

Evidence of the evolutionary history of eukaryotic genomes provides a backdrop for interpretation of all other eukaryotic features. For example, the genomes of extant eukaryotes are chimeric, containing genes with ancestries among both the bacteria and archaea (38, 44, 45, 47, 117). Interpreting the history of lineages that contributed to LECA's genome is complicated given the extensive lateral gene transfer (LGT) that occurred before and after the origin of eukaryotes. In a recent genome-scale study, eukaryotic genes were related most frequently to either *Euryarchaeota* or *Alphaproteobacteria*, but there were many other sister relationships that reflect the complex history of LGT across the tree of life (117).

Models of the origins of eukaryotes account for this chimerism by hypothesizing a fusion or similar event between an archaeon and a bacterium. In the simplest forms, these models refer to just the fusion of unspecified bacterial and archaeal lineages (127) or, in the case of the hypothesis on the ring of life, a fusion between a proteobacterium and an archaeal eocyte (96). In their more refined forms, such models aim to explain multiple features of eukaryotes beyond the origin of the chimeric genome, including the acquisition of mitochondria (**Figure 2**); under such versions, the players are generally an archaeon and a proteobacterium (83).

LGTs that occurred after the origin of eukaryotes have also contributed to the chimeric nature of eukaryotic genomes. In contrast to previous beliefs that LGT was only a property of bacterial and archaeal life, recent analyses of individual genes and complete genomes indicate that eukaryotes have also been impacted by LGT because eukaryotic genomes contain genes transferred from bacteria, archaea, and other eukaryotes (e.g., 4, 5, 63, 66, 69). The transfer of genes is likely enhanced by the ability of eukaryotes to engulf other organisms, and this feature inspired the application of the phrase "you are what you eat" as a descriptor of the mechanism underlying the chimeric nature of eukaryotic genomes (43) (**Figure 2**).

Beyond chimerism, we can infer that LECA had a complex genome including spliceosomal introns and diverse epigenetic mechanisms (70, 116), which suggest an important role for RNAs in shaping eukaryotic genome structure. The presence of a spliceosome (a complex structure made of both RNA and proteins) in LECA is supported by the broad distribution of spliceosomal introns across the eukaryotic tree of life (reviewed in Reference 100). Homologs to many spliceosomal components are not apparent in bacteria or archaea, and it is not clear how these complex structures evolved (97, 119). Similarly, the machinery for RNAi and other epigenetic phenomena was also likely present in LECA, as components such as Dicer and Argonaute are widespread among eukaryotes (35, 104).

The genome of LECA was likely also dynamic. Features such as cyclic polyploidization, extrachromosomal DNA, and life-cycle-dependent chromosomal rearrangements are widespread among extant eukaryotes (91, 92, 128). For example, extrachromosomal ribosomal DNAs are

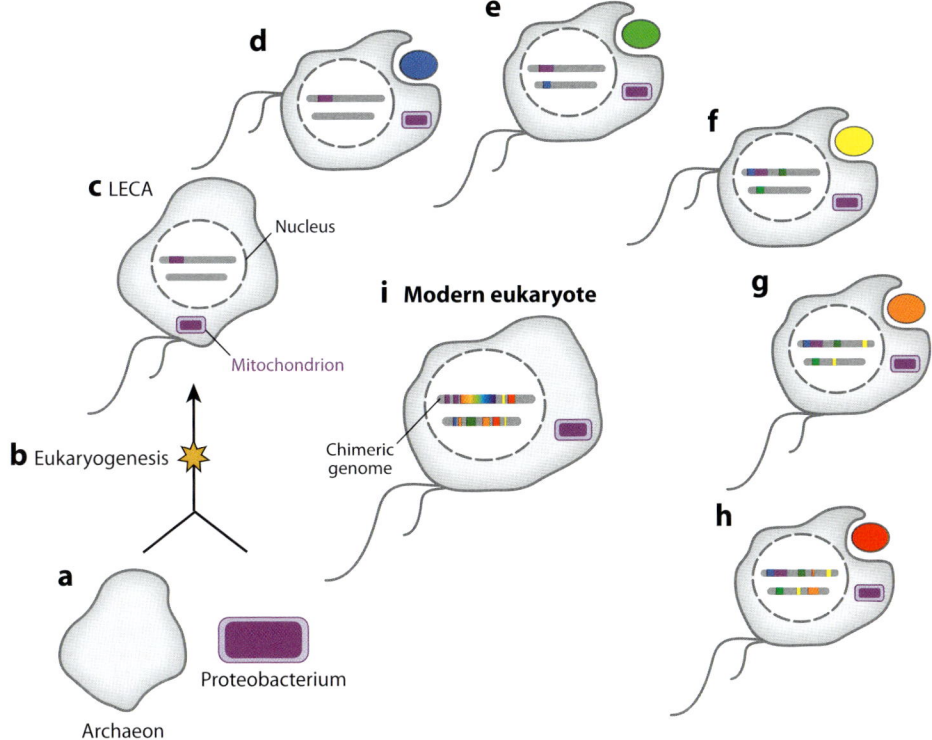

Figure 2

The chimeric nature of the genome in extant eukaryotes (center image, *i*) is consistent with a fusion of an archaeon and a bacterium at the time of the origin of eukaryotes (83) coupled with subsequent aberrant lateral transfers of genes from food items (43). (*a*) An archaeon and a proteobacterium that are potential symbiotic partners in the origin of eukaryotes. (*b*) Eukaryogenesis, the origin of the nucleus, cytoskeleton, and mitochondria through as yet unknown mechanisms and events. (*c*) Last eukaryotic common ancestor (LECA) with nucleus, mitochondria, and chimeric genome (i.e., purple portion of chromosome). (*d–h*) Repeated engulfment of food and incorporation of genes into the host nucleus. (*i*) Modern eukaryote whose chimeric genome is the product of panels *a–h*. Redrawn from Reference 116, *BioScience* 59(6), Copyright 2009, American Institute of Biological Sciences.

found in diverse lineages such as *Euglena* ('Excavata'), *Dictyostelium* ('Amoebozoa'), ciliates ('SAR,' Stramenopila, Alveolata, and Rhizaria), and *Xenopus* (Opisthokonta), among others (128). The presence of extrachromosomal copies of other genes also appears widespread, at least among plants and animals (36), and has been hypothesized to exist among foraminiferans ('SAR'; 54, 92). Life-cycle-dependent chromosomal rearrangements occur in ciliates ('SAR'), flax ('Plantae'), and some animal lineages including copepods, nematodes, and hagfish (92, 128). A special case of genome dynamics involves antigenic variation in parasites seeking to escape host immune systems [e.g., trypanosomes ('Excavata'); 115] and in the adaptive immune responses of the host genomes [e.g., V(D)J systems in vertebrates; 60]; here, a combination of DNA rearrangements and epigenetic mechanisms govern these dynamic genome features.

The existence of dynamic features across diverse eukaryotes suggests that all eukaryotes are able to distinguish the portion of their genome that will be inherited (i.e., a germline genome) from the remainder of the genome that is more malleable (i.e., a somatic genome; **Table 1**). This

'SAR': hypothesized major clade of eukaryotes containing Stramenopila, Alveolata, and Rhizaria

Table 1 Evidence for germline/soma distinctions among diverse eukaryotes

Category	Feature	Exemplar taxa	References
Chromosomal rearrangements	Extrachromosomal rDNA	Many including *Euglena*, *Entamoeba*, *Dictyostelium*, and *Xenopus*	128
	Other extrachromosomal DNAs	Various plants and animals	36
	Antigenic variation	*Trypanosoma*	115
	Adaptive immune response [V(D)J]	Vertebrates	60
Distinct germline and somatic nuclei	Sequestered germline	Triploblast animals, ciliates, and some foraminiferans	49, 50, 65, 87
	Processing of somatic chromosomes	Ciliates, nematodes, copepods, and hagfish	15, 86, 92, 128
Changes in DNA content	Zerfall (see text)	Various foraminiferans	14, 54, 90
	Cyclical polyploidization	Various lineages including Foraminifera, Phaeodaria, some Lobosea, Oxymonadida, and Apicomplexa	92

distinction is most obvious in eukaryotes such as triploblast animals, which generally sequester their germline genomes early in development (49, 50), and in ciliates as well as some foraminiferans, which have distinct germline and somatic nuclei within a single cell (65). Beyond these sequestered genomes, a long list of broadly distributed lineages with features such as cyclical polyploidy, extrachromosomal gene copies, and developmentally regulated genome rearrangements demonstrates the considerable flexibility among eukaryotic genomes (91, 92, 128) (**Table 1**). For example, during the life cycle of foraminiferans, portions of the genome are eliminated prior to nuclear division; this process, termed zerfall, may be indicative of the removal of amplified chromosomes or portions of chromosomes prior to the separation of germline material (14, 54, 91). We hypothesize that eukaryotes use epigenetic markers to distinguish the germline genome from the somatic genome that is marked by the dynamic processes described above (91, 92, 128).

Origin of the Nucleus

The origin of the nucleus, the feature that gives eukaryotes their name, remains a mystery. The nucleus is a complex structure with an outer membrane that is generally contiguous with the endoplasmic reticulum and has a system of multiprotein pores that enable transport from cytoplasm to nucleoplasm. As with the origin of eukaryotes themselves, hypotheses on the origin of the nucleus can be divided into those that focus on endosymbiosis and those that focus on autogenous origins (reviewed in Reference 81). Few data support either hypothesis, and phylogenomic studies have not provided much help as most genes encoding the nuclear proteome lack clear homologs in bacteria and archaea.

Several recent hypotheses have emerged supporting the autogenous origin of the nucleus. Cavalier-Smith (30) presents an extensive discussion on the evolution of the nucleus that builds from details on both the cell and the molecular biology of eukaryotic cells. Another hypothesis focuses on the potential benefit in separating nucleoplasm and cytoplasm functions following expansion of the number of group II introns in the genome (82). Under this hypothesis, the nuclear compartment evolved as a way of separating the processing of RNAs (e.g., removal of expanding

Endosymbiosis: symbiosis in which one organism lives within another

numbers of introns) from translation; these two processes can occur nearly simultaneously in bacteria and archaea.

Origin of the Cytoskeleton

As with the nucleus, few data support existing hypotheses on the origin of the eukaryotic cytoskeleton and its many diverse proteins. Perhaps even more important for cell function than the nucleus, the cytoskeleton provides eukaryotic cells with their diverse structures, forms of motility, and the ability to engulf other cells. Margulis and colleagues (e.g., 77–79) argued that the eukaryotic cytoskeleton is specifically descended from structures in spirochetes. In this scenario, early eukaryotic cells moved first in a loose association with these highly motile bacteria, and this association later transformed into an endosymbiosis, with the spirochete proteins evolving into the eukaryotic cytoskeleton (79). This hypothesis is not supported by available data, as there is no strong footprint of spirochete ancestry among eukaryotic cytoskeletal genes (7). Instead, eukaryotic cytoskeletal proteins either lack bacterial/archaeal homologs or have distant homologs as is the case for actin/MreB and tubulin/FtsZ (123).

Amitochondriate: describes organisms lacking mitochondria

Origin of Mitochondria

In contrast with knowledge on the origin of the nucleus and cytoskeleton, both the timing of and the source for the acquisition of mitochondria are now well understood. Mitochondria are derived from an alphaproteobacterium, as evidenced by similarities in their morphology and genomes (48, 55). Phylogenetic analyses of mitochondrial genes place eukaryotic mitochondria as a single clade nested among extant *Alphaproteobacteria* (52, 98). The bacterium that gave rise to mitochondria may have been either a parasite, related to the extant alphaproteobacterial lineage *Rickettsia*, or a partner in early symbiosis (**Figure 2**).

The phylogenetic distribution of mitochondria plus mitochondrial-derived organelles indicates that mitochondria were acquired prior to the divergence of extant eukaryotes (105). Numerous amitochondriate lineages, including the parasitic genera *Trichomonas* and *Giardia*, some free-living ciliates (e.g., *Trimyema* and *Metopus*), and several fungal genera (e.g., *Neocallimastix*, *Encephalitozoon*), are restricted largely to anaerobic or microaerophilic environments (46, 58, 61, 85, 105). All these lineages nest within larger clades of mitochondria-containing lineages. For example, *Trichomonas* and *Giardia* are placed within the 'Excavata,' which include numerous mitochondria-containing lineages such as Euglenozoa, Jakobida, and Heterolobosea. Moreover, mitochondrial-derived organelles have been found in many lineages that had previously been considered amitochondriate; these organelles are alternatively called hydrogenosomes [e.g., in *Neocallimastix*, *Nyctotherus* (ciliate), and *Trichomonas*] or mitosomes (e.g., in *Encephalitozoon*, *Giardia*, and *Cryptosporidium*) (61, 85, 105). Finally, genes of alphaproteobacterial origin have been found in the nuclear genomes of amitochondriate eukaryotes, again consistent with secondary loss of this organelle (61, 105). However, interpretation of the last observation is complicated by the tremendous exchange of genes among lineages across the tree of life (117).

Following the acquisition of mitochondria, there have been complex patterns of gene retention, gene transfer to nucleus, and redirection of non-alphaproteobacterial proteins to the mitochondrial proteome (71, 113). Gene number within mitochondrial genomes is small (i.e., 7–90 genes) (16) relative to the complexity of mitochondrial proteome, and mitochondrial genomes are often reduced to a handful of proteins, most of which are involved in cellular respiration (55). Phylogenetic reconstructions indicate that there have been many parallel losses of genes from mitochondria since LECA (55). Today, the proteome of mitochondria is derived from the handful of mitochondrially encoded genes, a relatively small number of genes of alphaproteobacterial origin that are now encoded in the nucleus, and numerous other nuclear genes from varying sources whose products have been redirected to the mitochondria (113). This complexity underlying the mitochondrial proteome may have provided the energetic leap required for the evolution of complex and ultimately macrobial eukaryotic lineages (74).

> **Monophyly:** a group of organisms that includes an ancestor and all descendants

PART II: EVOLUTION OF PHOTOSYNTHESIS WITH EUKARYOTES

Plastids, the generic name for chloroplasts, are present in a diverse array of lineages across the eukaryotic tree of life (**Figure 3**) and were likely acquired after diversification of major lineages (93). Plastids are derived from cyanobacteria, as evidenced by both structural similarities and sequence analysis of plastid genomes (24). There is still, however, debate on the number and timing of the acquisitions of this organelle and on the contributions of genes from other photosynthetic lineages in shaping photosynthesis among eukaryotes (12, 51, 75). The current popular view is that there was a single primary acquisition of chloroplasts in the last common ancestor of a clade alternatively named 'Plantae' or 'Archaeplastida,' which includes green algae, red algae, and glaucophytes (2, 28, 39). Evidence for a single primary acquisition of plastids includes the phylogeny of plastid genes, which tend to form a monophyletic group within extant cyanobacteria (37, 42), and a shared Tic-Toc transport system for moving proteins across plastid membranes (64).

Contradictory evidence does exist, including the multiple origins of the key photosynthetic enzyme RuBisCo and the complex origins of the varying plastid pigments (12, 40, 75, 114). Further, there appears to have been a major bottleneck among cyanobacteria after the acquisition of plastids in eukaryotes, making it impossible to distinguish between single and multiple origins by looking at the history of plastid genes, as close relatives of potential donor lineages may have gone extinct (37). An alternative model for the evolution of photosynthesis among eukaryotes, termed the shopping bag model, suggests that photosynthesis among eukaryotes relies on the products of genes acquired from multiple sources by LGT over evolutionary time (75).

The remaining lineages of photosynthetic eukaryotes (e.g., diatoms, brown algae, euglenids, cryptomonads, haptophytes, and dinoflagellates) acquired plastids by engulfing either a red or green alga in a process called secondary endosymbiosis (6, 39). That secondary endosymbioses have occurred is indisputable, but the number of these events is more contentious. Two lineages, cryptomonads and chlorarachniophytes (core Cercozoa), have retained a remnant nucleus (nucleomorph) from a red and green algal endosymbiont, respectively (6, 25). Sequencing of these remnant nuclei reveals that these highly reduced genomes contain few genes involved in plastid function (8, 53, 84).

The history of plastid acquisition among lineages that lack a nucleomorph remains debated. One convenient hypothesis, which has now been rejected by many independent analyses, is that there was a single acquisition of a red algal symbiont at the base of the 'Chromalveolate' clade, which was originally described to include alveolates, stramenopiles, cryptomonads, and haptophytes (23). However, numerous analyses of these host genomes fail to support the monophyly of this group (13, 56, 62, 89). Multigene analyses indicate stramenopiles and alveolates fall in a clade with the Rhizaria (i.e., SAR), and there is no compelling evidence of an ancestral red algal plastid within the Rhizaria (18, 19, 57). Finally, as discussed below, the position of haptophytes and cryptomonads remains uncertain, as the relationships of these lineages are unstable in many analyses (88, 89).

PART III: RELATIONSHIPS AMONG MAJOR LINEAGES

Because of their incredible morphological diversity, eukaryotic microorganisms (protists) have been the subject of intense study since the time of the earliest microscopes. These earliest studies focused primarily on describing taxa rather than estimating higher-level relationships (1, 103). My interpretation of this rich history is that many of the shallower (i.e., more recent) clades defined by morphology and/or ultrastructural features have remained robust to more recent molecular analyses (88, 89, 94), though exceptions certainly exist.

With the advent of phylogenetic analyses based on DNA sequence data, the field of eukaryotic systematics has gone through considerable turmoil, though hypotheses do seem to be coalescing in recent years. Molecular analyses of eukaryotic diversity were launched with analyses of the ubiquitous small-subunit ribosomal RNA (ssu-rRNA, and later ssu-rDNA) sequences, which initially suggested the idea that eukaryotic diversity consisted of a base of microbial lineages topped by a crown of plants, animals, and fungi and their microbial relatives (109, 118). As additional genes were sequenced and revealed conflicting topologies, there was a brief period in which arguments were made for why one gene was better than another, and then the field launched into combined analyses of multiple genes (11). In most of these analyses, an individual sequence is chosen to represent each taxon (i.e., paralogs removed) and these sequences are concatenated to yield many characters per taxon. Such multigene studies have led to a plethora of hypotheses about eukaryotic diversity (17–19, 28, 29, 33, 34, 59, 76, 101, 111).

Paralogs: duplicated copies of a gene

ssu-rDNA: small subunit ribosomal DNA

One somewhat discouraging aspect of eukaryotic systematics is the heterogeneity in philosophy for naming higher taxa (i.e., more inclusive clades), which I describe in overly simplistic terms below to highlight the differences in approaches. There is a tendency to subscribe to what could be called the "Chupacabra" approach, whereby clades are named on the basis of only very limited data and upon first sighting. (The Chupacabra is a mythical creature that has been reported in the Americas, parallel to sightings of Big Foot in the Pacific Northwest of the United States). A second discouraging approach is one whereby researchers set out to find data to support a hypothesis, often of the Chupacabra type. Here, a researcher might sift among thousands of observations (e.g., expressed sequence tags, genome sequences) to find a half dozen or so that support a hypothesis and then use these data to conclude that a hypothesis is correct regardless of the insights from the remaining observations. Fortunately, the approaches described above are not adopted by the majority of the field, as most focus on analyses of all available data and draw conclusions based on the preponderance of evidence in manuscripts that discuss both caveats and alternative hypotheses.

Because the field of eukaryotic systematics is in flux, I focus on a subset of hypotheses below, including those that are best supported by current data. Numerous reviews exist for readers wanting to know more about specific lineages and hypotheses (1, 2, 10, 18, 27–30, 68, 73, 88, 99, 108, 112).

Root of the Eukaryotic Tree of Life

The root of the eukaryotic tree of life remains unknown, largely because numerous characters conserved among eukaryotes lack homologs in bacteria and archaea. As a result, it is difficult to find appropriate outgroup sequences for most molecular studies. Added to this complexity is the impact of LGT on the history of individual genes within bacteria, archaea, and those eukaryotes that lack sequestered germline genomes (4, 69). Hypotheses on the location of the root either have focused on characters argued to be primitive (e.g., lack of organelle) or have emerged from analyses of molecular data. For example, the "Archezoa" hypothesis (21, 31) argued that the root of eukaryotes lay among putatively primitive amitochondriate lineages such as *Trichomonas* and *Giardia*, which are now known to be nested in clades of mitochondria-containing lineages. Similarly, a root between 'Amoebozoa' + Opisthokonta (the so-called unikonts, as many members with flagella have only a single flagellum) and all remaining eukaryotes was proposed on the basis of a gene fusion event (26, 111). Alternatively, several studies suggest that the root of the eukaryotic tree of life lies between Opisthokonta and all remaining eukaryotes (9, 41, 110), which is what we found based on gene tree parsimony analyses of ~20 genes (67). Understanding of the location of the root of the eukaryotic tree of life will likely change with the addition of data, from both genes and yet unsampled taxa, and the development of new analytical tools.

Major Eukaryotic Clades

Estimates on the nature of major clades in the eukaryotic tree of life have stabilized in recent years (**Figure 3**). Early molecular studies led to a plethora of hypotheses, often based on very few data, that have seen varying fates with collection of additional data. These major clades have been named supergroups (2, 108), creating a novel taxonomic hierarchy that lacks rigorous definition. In more recent years, these major clades have been subjected to more evaluation, moving beyond resampling the same data (i.e., ssu-rDNA and a few genes) to more gene- and taxon-rich approaches (18, 19, 59, 89, 126). Comprehensive taxon sampling is key to characterizing the eukaryotic tree of life, and we can anticipate changes as additional lineages are sampled. On the basis of the current understanding of molecular and morphological characters, several major eukaryotic lineages have emerged, albeit with varying levels of support. Details on membership and support for these groups have been reviewed elsewhere (1, 88, 108), so I provide only highlights here, with quotes around taxon names that either are controversial or lack robust support.

The best supported of the major clades is the Opisthokonta, which unites animals, fungi, and their microbial relatives (32, 112, 120). The name Opisthokonta reflects the posterior ("opistho") position of the flagellum ("kont") in lineages that have maintained flagella (32). The monophyly of this group is supported by numerous molecular characters (59, 88, 89, 112). The strong support for this clade is also consistent with hypotheses that place the root of the eukaryotic tree of life between opisthokonts and all remaining eukaryotes (9, 41, 67, 110).

The 'Amoebozoa' were first proposed from early molecular analyses (22), and this clade has generally remained robust in light of additional sampling of genes and taxa (59, 72, 89, 95). This diverse clade contains the classic lobose amoeba (e.g., *Amoeba proteus*), the beautiful testate (shelled) amoebae in the Arcellinida, the slime molds (e.g., *Physarum* and *Dictyostelium*), and the causative agent of dysentery (*Entamoeba histolytica*).

The 'Excavata' were hypothesized on the basis of a morphological feature, an excavate grove, in the last common ancestor of this clade (26, 106, 107). Whereas early molecular trees failed to support this clade (88), more recent analyses do recover 'Excavata,' albeit with low support at deep nodes (59, 89). Many members of this clade, such as the human parasites *Giardia lamblia* and *Trichomonas vaginalis*, have elevated rates of evolution across their genomes, which likely contributes to the instability of the 'Excavata.'

The placement of photosynthetic lineages remains more problematic, likely due to the combination of multiple secondary (and tertiary and quaternary) endosymbiotic transfer events and endosymbiotic gene transfer from plastid to host (6, 13, 62, 102). As discussed above, the 'Plantae' (or 'Archaeplastida') (2) unite three lineages—green algae, red algae, and glaucophytes—that are believed to be descended from the eukaryote that first evolved plastids through symbiosis with a cyanobacterium (20, 39). Evidence in support of the hypothesis of a single primary endosymbiosis at this time includes the shared machinery for transport across plastid membranes. However, phylogenies based on genes in the nucleus are not consistent in recovering the monophyly of these lineages, as red algae often fall outside of this clade (89).

Perhaps most unstable in recent years has been our understanding of relationships among members of what has recently been called the 'SAR' clade: Stramenopila, Alveolata, and Rhizaria (18, 19, 57). Each of these three clades appears to represent diverse monophyletic assemblages,

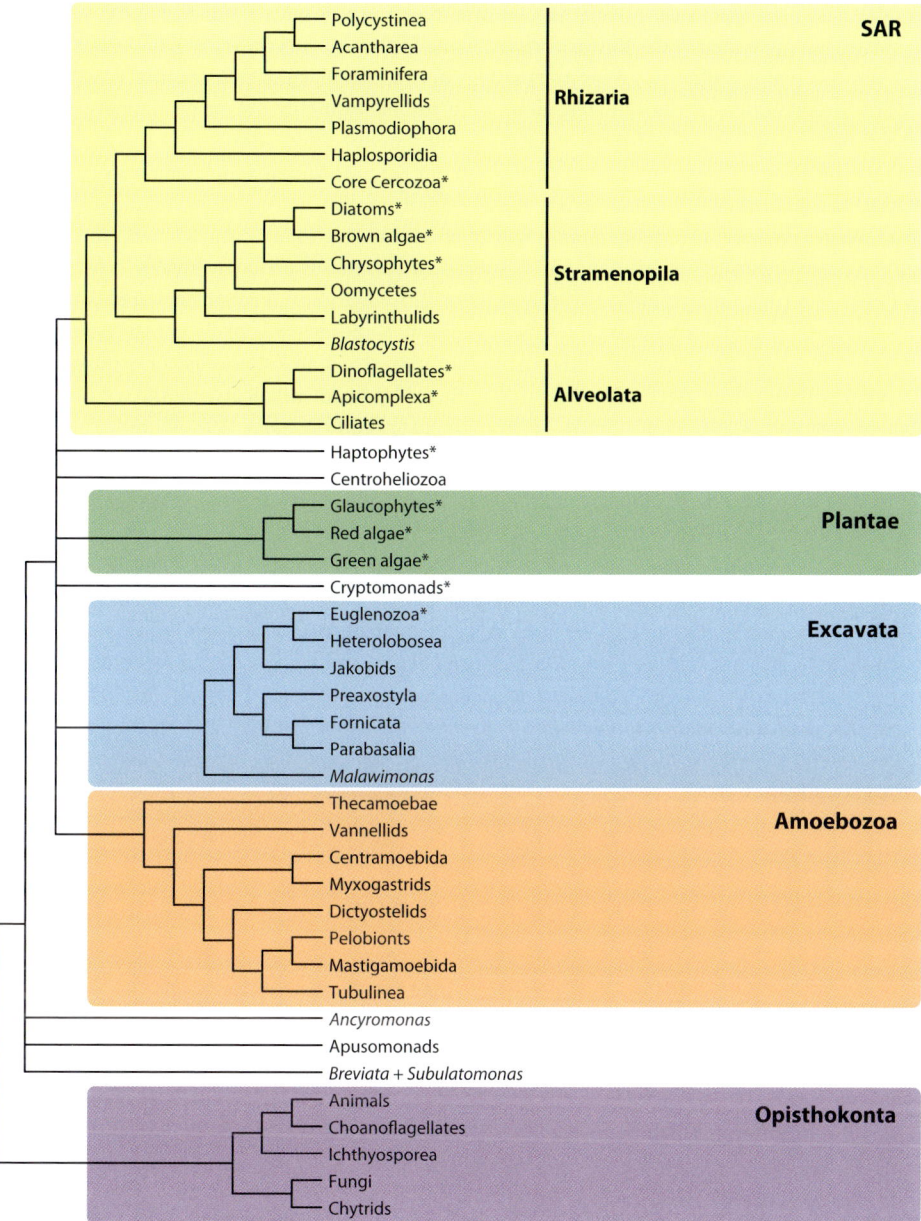

Figure 3
A hypothesis of phylogenetic relationships among representatives of major lineages of eukaryotes. Lineages with members that have plastids are marked with an asterisk. The figure synthesizes information from literature discussed in the text.

with greater support for the first two than for Rhizaria. Yet, relationships among these lineages and their putative relatives have been more controversial. The monophyly of both stramenopiles and alveolates are supported by morphological and molecular data, and both represent diverse assemblages of photosynthetic and nonphotosynthetic lineages. Synapomorphies for these clades include hair-like structures on one flagellum in the stramenopiles (3) and alveolar sacs in the alveolates (94). In contrast, the Rhizaria were originally proposed on the basis of molecular analyses (57) and this major clade lacks clear morphological or molecular synapomorphies.

> **Synapomorphy:** a shared, derived characteristic marking a monophyletic group

The stramenopiles and alveolates, along with the haptophytes and cryptomonads, had been argued to be part of the 'Chromalveolata' on the basis of a hypothesis that the last common ancestor of this group engulfed a red algal symbiont (23). Although a few genes can be found to support this hypothesis, the preponderance of evidence fails to support the monophyly of these host lineages. Instead, the stramenopiles and alveolates appear to fall in a clade with the Rhizaria, and the placement of cryptomonads and haptophytes remains as yet unclear (89).

Given that our understanding of the structure of eukaryotic diversity is dependent on the available taxon sampling to date, data from additional lineages will likely transform our views on deep relationships. Additional taxon sampling may also stabilize the list of orphan lineages (lineages without clear sister taxa and without clear placement in eukaryotic tree of life), which includes cryptomonads, haptophytes, centroheliozoans, and breviates (89). Moreover, there are likely additional, as yet undiscovered lineages to be added to the eukaryotic tree of life.

CONCLUSION

The greatest diversity of eukaryotes on Earth exists among microbial lineages, and analyses of these lineages have yielded many insights into basic principles of biology. Innovations among eukaryotes include the acquisition of organelles through endosymbioses (i.e., mitochondria and plastids), dynamic genomes marked by chromosomal rearrangements and cyclical polyploidization, and myriad diverse morphologies underlain by a complex cytoskeleton. With the rise of studies on eukaryotic diversity coupled with powerful tools in molecular biology and microscopy, we are poised to collect additional data that will illuminate details on the origin and diversification of eukaryotic life on Earth.

SUMMARY POINTS

1. The bulk of eukaryotic diversity is microbial, with lineages marked by a dazzling array of morphological and genomic innovations.
2. Despite recent advances, many aspects of the origin of eukaryotes (e.g., origin of nucleus and cytoskeleton) remain unknown.
3. The last common ancestor of extant eukaryotes had a complex and dynamic genome that may have been able to distinguish between DNA to be passed on to future generations (i.e., germline) and more flexible 'somatic' DNA.
4. The shape of the eukaryotic tree of life has come into clearer focus in recent years, although many open questions remain.

DISCLOSURE STATEMENT

The author is not aware of any affiliations, memberships, funding, or financial holdings that might be perceived as affecting the objectivity of this review.

ACKNOWLEDGMENTS

I am grateful to numerous students and colleagues, including Jessica Grant, Daniel Lahr, Bill Martin, and Laura Wegener Parfrey, for discussions of the concepts in this manuscript. This work was supported by grants from the National Science Foundation (OCE-0648713, ATOL-043115, DEB-0919152) and the National Institutes of Health (1R15GM081865-01).

LITERATURE CITED

1. Adl SM, Leander BS, Simpson AG, Archibald JM, Anderson OR, et al. 2007. Diversity, nomenclature, and taxonomy of protists. *Syst. Biol.* 56:684–89
2. Adl SM, Simpson AGB, Farmer MA, Andersen RA, Anderson OR, et al. 2005. The new higher level classification of eukaryotes with emphasis on the taxonomy of protists. *J. Eukaryot. Microbiol.* 52:399–451
3. Andersen RA. 2004. Biology and systematics of heterokont and haptophyte algae. *Am. J. Bot.* 91:1508–22
4. Andersson JO. 2005. Lateral gene transfer in eukaryotes. *Cell. Mol. Life Sci.* 62:1182–97
5. Andersson JO, Hirt RP, Foster PG, Roger AJ. 2006. Evolution of four gene families with patchy phylogenetic distributions: influx of genes into protist genomes. *BMC Evol. Biol.* 6:18
6. Archibald JM. 2009. The puzzle of plastid evolution. *Curr. Biol.* 19:R81–88
7. Archibald JM. 2011. Origin of eukaryotic cells: 40 years on. *Symbiosis* 54:69–86
8. Archibald JM, Lane CE. 2009. Going, going, not quite gone: nucleomorphs as a case study in nuclear genome reduction. *J. Hered.* 100:582–90
9. Arisue N, Hasegawa M, Hashimoto T. 2005. Root of the Eukaryota tree as inferred from combined maximum likelihood analyses of multiple molecular sequence data. *Mol. Biol. Evol.* 22:409–20
10. Baldauf SL. 2003. The deep roots of eukaryotes. *Science* 300:1703–6
11. Baldauf SL, Roger AJ, Wenk-Siefert I, Doolittle WF. 2000. A kingdom-level phylogeny of eukaryotes based on combined protein data. *Science* 290:972–77
12. Bodyl A, Mackiewicz P, Stiller JW. 2009. Early steps in plastid evolution: current ideas and controversies. *BioEssays* 31:1219–32
13. Bodyl A, Stiller JW, Mackiewicz P. 2009. Chromalveolate plastids: direct descent or multiple endosymbioses? *Trends Ecol. Evol.* 24:119–21
14. Bowser SS, Habura A, Pawlowski J. 2006. Molecular evolution of Foraminifera. In *Genome Evolution in Eukaryotic Microbes*, ed. LA Katz, D Bhattacharya, pp. 78–93. Oxford: Oxford Univ. Press
15. Bron JE, Frisch D, Goetze E, Johnson SC, Lee CE, Wyngaard GA. 2011. Observing copepods through a genomic lens. *Front. Zool.* 8:22
16. Burger G, Gray MW, Lang BF. 2003. Mitochondrial genomes: anything goes. *Trends Genet.* 19:709–16
17. Burki F, Inagaki Y, Brate J, Archibald JM, Keeling PJ, et al. 2009. Large-scale phylogenomic analyses reveal that two enigmatic protist lineages, Telonemia and Centroheliozoa, are related to photosynthetic chromalveolates. *Genome Biol. Evol.* 1:231–38
18. Burki F, Shalchian-Tabrizi K, Minge M, Skjaeveland A, Nikolaev SI, et al. 2007. Phylogenomics reshuffles the eukaryotic supergroups. *PLoS One* 2:e790
19. Burki F, Shalchian-Tabrizi K, Pawlowski J. 2008. Phylogenomics reveals a new 'megagroup' including most photosynthetic eukaryotes. *Biol. Lett.* 4:366–69
20. Cavalier-Smith T. 1981. Eukaryote kingdoms: seven or nine? *Biosystems* 14:461–81
21. Cavalier-Smith T. 1995. Cell-cycles, diplokaryosis and the archezoan origin of sex. *Arch. Protistenkd.* 145:189–207
22. Cavalier-Smith T. 1997. Amoeboflagellates and mitochondrial cristae in eukaryote evolution: megasystematics of the new protozoan subkingdoms Eozoa and Neozoa. *Arch. Protistenkd.* 147:237–58
23. Cavalier-Smith T. 1999. Principles of protein and lipid targeting in secondary symbiogenesis: euglenoid, dinoflagellate, and sporozoan plastid origins and the eukaryote family tree. *J. Eukaryot. Microbiol.* 46:347–66
24. Cavalier-Smith T. 2000. Membrane heredity and early chloroplast evolution. *Trends Plant Sci.* 5:173–82
25. Cavalier-Smith T. 2002. Nucleomorphs: enslaved algal nuclei. *Curr. Opin. Microbiol.* 5:612–19
26. Cavalier-Smith T. 2002. The phagotrophic origin of eukaryotes and phylogenetic classification of protozoa. *Int. J. Syst. Evol. Microbiol.* 52:297–354
27. Cavalier-Smith T. 2004. Only six kingdoms of life. *Proc. R. Soc. Lond. Ser. B* 271:1251–62

28. Cavalier-Smith T. 2009. Megaphylogeny, cell body plans, adaptive zones: causes and timing of eukaryote basal radiations. *J. Eukaryot. Microbiol.* 56:26–33
29. Cavalier-Smith T. 2010. Kingdoms Protozoa and Chromista and the eozoan root of the eukaryotic tree. *Biol. Lett.* 6:342–45
30. Cavalier-Smith T. 2010. Origin of the cell nucleus, mitosis and sex: roles of intracellular coevolution. *Biol. Direct* 5:7
31. Cavalier-Smith T, Chao EE. 1996. Molecular phylogeny of the free-living archezoan *Trepomonas agilis* and the nature of the first eukaryote. *J. Mol. Evol.* 43:551–62
32. Cavalier-Smith T, Chao EEY. 1995. The opalozoan *Apusomonas* is related to the common ancestor of animals, fungi, and choanoflagellates. *Proc. R. Soc. Lond. Ser. B* 261:1–6
33. Cavalier-Smith T, Chao EEY. 2003. Phylogeny and classification of phylum Cercozoa (Protozoa). *Protist* 154:341–58
34. Cavalier-Smith T, Chao EEY, Oates B. 2004. Molecular phylogeny of 'Amoebozoa' and the evolutionary significance of the unikont *Phalansterium*. *Eur. J. Protistol.* 40:21–48
35. Cerutti H, Casas-Mollano JA. 2006. On the origin and functions of RNA-mediated silencing: from protists to man. *Curr. Genet.* 50:81–99
36. Cohen S, Segal D. 2009. Extrachromosomal circular DNA in eukaryotes: possible involvement in the plasticity of tandem repeats. *Cytogenet. Genome Res.* 124:327–38
37. Criscuolo A, Gribaldo S. 2011. Large-scale phylogenomic analyses indicate a deep origin of primary plastids within cyanobacteria. *Mol. Biol. Evol.* 28:3019–32
38. Dagan T, Martin W. 2006. The tree of one percent. *Genome Biol.* 7:118
39. Delwiche CF. 1999. Tracing the tread of plastid diversity through the tapestry of life. *Am. Nat.* 154:S164–77
40. Delwiche CF, Palmer JD. 1996. Rampant horizontal transfer and duplication of rubisco genes in eubacteria and plastids. *Mol. Biol. Evol.* 13:873–82
41. Derelle R, Lang BF. 2011. Rooting the eukaryotic tree with mitochondrial and bacterial proteins. *Mol. Biol. Evol.* 29:1277–89
42. Deusch O, Landan G, Roettger M, Gruenheit N, Kowallik KV, et al. 2008. Genes of cyanobacterial origin in plant nuclear genomes point to a heterocyst-forming plastid ancestor. *Mol. Biol. Evol.* 25:748–61
43. Doolittle WF. 1998. You are what you eat: A gene transfer ratchet could account for bacterial genes in eukaryotic nuclear genomes. *Trends Genet.* 14:307–11
44. Doolittle WF. 1999. Phylogenetic classification and the universal tree. *Science* 284:2124–28
45. Doolittle WF, Brown JR. 1999. Gene descent, duplication, and horizontal transfer in the evolution of glutamyl- and glutaminyl-tRNA synthetases. *J. Mol. Evol.* 49:485–95
46. Embley TM. 2006. Multiple secondary origins of the anaerobic lifestyle in eukaryotes. *Philos. Trans. R. Soc. Lond. B* 361:1055–67
47. Embley TM, Martin W. 2006. Eukaryotic evolution, changes and challenges. *Nature* 440:623–30
48. Embley TM, van der Giezen M, Horner DS, Dyal PL, Bell S, Foster PG. 2003. Hydrogenosomes, mitochondria and early eukaryotic evolution. *IUMBM Life* 55:387–95
49. Extavour CG, Akam M. 2003. Mechanisms of germ cell specification across the metazoans: epigenesis and preformation. *Development* 130:5869–84
50. Extavour CGM, Wilkins AS. 2008. Evolution of the metazoan germline: a unifying hypothesis. *Biol. Reprod.* 78:282
51. Falkowski PG, Katz ME, Knoll AH, Quigg A, Raven JA, et al. 2004. The evolution of modern eukaryotic phytoplankton. *Science* 305:354–60
52. Fitzpatrick DA, Creevey CJ, McInerney JO. 2006. Genome phylogenies indicate a meaningful α-proteobacterial phylogeny and support a grouping of the mitochondria with the *Rickettsiales*. *Mol. Biol. Evol.* 23:74–85
53. Gilson PR, McFadden GI. 2002. Jam packed genomes: a preliminary, comparative analysis of nucleomorphs. *Genetica* 115:13–28
54. Goldstein ST. 1997. Gametogenesis and the antiquity of reproductive pattern in the Foraminiferida. *J. Foraminifer. Res.* 27:319–28

55. Gray MW, Lang BF, Burger G. 2004. Mitochondria of protists. *Annu. Rev. Genet.* 38:477–524
56. Grzebyk D, Schofield O, Vetriani C, Falkowski PG. 2003. The Mesozoic radiation of eukaryotic algae: the portable plastid hypothesis. *J. Phycol.* 39:259–67
57. Hackett JD, Yoon HS, Li S, Reyes-Prieto A, Rümmele SE, Bhattacharya D. 2007. Phylogenomic analysis supports the monophyly of cryptophytes and haptophytes and the association of 'Rhizaria' with Chromalveolates. *Mol. Biol. Evol.* 24:1702–13
58. Hackstein JHP, Tjaden J, Huynen M. 2006. Mitochondria, hydrogenosomes and mitosomes: products of evolutionary tinkering! *Curr. Genet.* 50:225–45
59. Hampl V, Hug L, Leigh JW, Dacks JB, Lang BF, et al. 2009. Phylogenomic analyses support the monophyly of 'Excavata' and resolve relationships among eukaryotic "supergroups". *Proc. Natl. Acad. Sci. USA* 106:3859–64
60. Hirano M, Das S, Guo P, Cooper MD. 2011. The evolution of adaptive immunity in vertebrates. *Adv. Immunol.* 109:125–57
61. Hjort K, Goldberg AV, Tsaousis AD, Hirt RP, Embley TM. 2010. Diversity and reductive evolution of mitochondria among microbial eukaryotes. *Philos. Trans. R. Soc. Lond. B* 365:713–27
62. Howe CJ, Barbrook AC, Nisbet RE, Lockhart PJ, Larkum AW. 2008. The origin of plastids. *Philos. Trans. R. Soc. Lond. B* 363:2675–85
63. Hug LA, Stechmann A, Roger AJ. 2010. Phylogenetic distributions and histories of proteins involved in anaerobic pyruvate metabolism in eukaryotes. *Mol. Biol. Evol.* 27:311–24
64. Kalanon M, McFadden GI. 2008. The chloroplast protein translocation complexes of *Chlamydomonas reinhardtii*: a bioinformatic comparison of Toc and Tic components in plants, green algae and red algae. *Genetics* 179:95–112
65. Katz LA. 2001. Evolution of nuclear dualism in ciliates: a reanalysis in light of recent molecular data. *Int. J. Syst. Evol. Microbiol.* 51:1587–92
66. Katz LA. 2002. Lateral gene transfers and the evolution of eukaryotes: theories and data. *Int. J. Syst. Evol. Microbiol.* 52:1893–900
67. Katz LA, Grant JR, Parfrey LW, Burleigh JG. 2012. Turning the crown upside down: gene tree parsimony roots the eukaryotic tree of life. *Syst. Biol.* doi: 10.1093/sysbio/sys026
68. Keeling PJ, Burger G, Durnford DG, Lang BF, Lee RW, et al. 2005. The tree of eukaryotes. *Trends Ecol. Evol.* 20:670–66
69. Keeling PJ, Palmer JD. 2008. Horizontal gene transfer in eukaryotic evolution. *Nat. Rev. Genet.* 9:605–18
70. Koonin EV. 2010. The origin and early evolution of eukaryotes in the light of phylogenomics. *Genome Biol.* 11:209
71. Kurland CG, Andersson SGE. 2000. Origin and evolution of the mitochondrial proteome. *Microbiol. Mol. Biol. Rev.* 64:786–820
72. Lahr DJG, Grant J, Nguyen T, Lin JH, Katz LA. 2011. Comprehensive phylogenetic reconstruction of 'Amoebozoa' based on concatenated analyses of SSU-rDNA and actin genes. *PLoS One* 6:e22780
73. Lane CE, Archibald JM. 2008. The eukaryotic tree of life: endosymbiosis takes its TOL. *Trends Ecol. Evol.* 23:268–75
74. Lane N, Martin W. 2010. The energetics of genome complexity. *Nature* 467:929–34
75. Larkum AW, Lockhart PJ, Howe CJ. 2007. Shopping for plastids. *Trends Plant Sci.* 12:189–95
76. Lecroq B, Gooday AJ, Cedhagen T, Sabbatini A, Pawlowski J. 2009. Molecular analyses reveal high levels of eukaryotic richness associated with enigmatic deep-sea protists (Komokiacea). *Mar. Biodiv.* 39:45–55
77. Margulis L. 1993. Serial endosymbiosis theory. In *Symbiosis in Cell Evolution: Microbial Communities in the Archean and Proterozoic Eons*, pp. 1–18. New York: Freeman. 2nd ed.
78. Margulis L. 1996. Archaeal-eubacterial mergers in the origin of Eukarya: phylogenetic classification of life. *Proc. Natl. Acad. Sci. USA* 93:1071–76
79. Margulis L, Dolan MF, Guerrero R. 2000. The chimeric eukaryote: origin of the nucleus from the karyomastigont in amitochondriate protists. *Proc. Natl. Acad. Sci. USA* 97:6954–59
80. Margulis L, Schwartz K. 1988. *Five Kingdoms: An Illustrated Guide to the Phyla of Life on Earth*. New York: Freeman. 2nd ed.
81. Martin W. 2005. Archaebacteria (Archaea) and the origin of the eukaryotic nucleus. *Curr. Opin. Microbiol.* 8:630–37

82. Martin W, Koonin EV. 2006. Introns and the origin of nucleus-cytosol compartmentalization. *Nature* 440:41–45
83. Martin W, Müller M. 1998. The hydrogen hypothesis for the first eukaryote. *Nature* 392:37–41
84. Moore CE, Archibald JM. 2009. Nucleomorph genomes. *Annu. Rev. Genet.* 43:251–64
85. Müller M, Mentel M, van Hellemond J, Henze K, Wöhle C, et al. 2012. Biochemistry and evolution of anaerobic energy metabolism in eukaryotes. *Microbiol. Mol. Biol Rev.* In press
86. Nakai Y, Kubota S, Kohno S. 1991. Chromatin diminution and chromosome elimination in four Japanese hagfish species. *Cytogenet. Cell Genet.* 56:196–98
87. Orias E. 1991. On the evolution of the karyorelict ciliate life cycle: heterophasic ciliates and the origin of ciliate binary fission. *Biosystems* 25:67–73
88. Parfrey LW, Barbero E, Lasser E, Dunthorn M, Bhattacharya D, et al. 2006. Evaluating support for the current classification of eukaryotic diversity. *PLoS Genet.* 2:e220
89. Parfrey LW, Grant J, Tekle YI, Lasek-Nesselquist E, Morrison HG, et al. 2010. Broadly sampled multigene analyses yield a well-resolved eukaryotic tree of life. *Syst. Biol.* 59:518–33
90. Parfrey LW, Grant JR, Katz LA. 2012. Ribosomal DNA is differentially amplified across life cycle stages in the foraminifer *Allogromia laticollaris* strain CSH. *J. Foraminifer. Res.* In press
91. Parfrey LW, Katz LA. 2010. Dynamic genomes of eukaryotes and the maintenance of genomic integrity. *Microbe* 5:156–64
92. Parfrey LW, Lahr DJG, Katz LA. 2008. The dynamic nature of eukaryotic genomes. *Mol. Biol. Evol.* 25:787–94
93. Parfrey LW, Lahr DJG, Knoll AH, Katz LA. 2011. Estimating the timing of early eukaryotic diversification with multigene molecular clocks *Proc. Natl. Acad. Sci. USA* 108:13624–29
94. Patterson DJ. 1999. The diversity of eukaryotes. *Am. Nat.* 154:S96–124
95. Pawlowski J. 2008. The twilight of Sarcodina: a molecular perspective on the polyphyletic origin of amoeboid protists. *Protistology* 5:281–302
96. Rivera MC, Lake JA. 2004. The ring of life provides evidence for a genome fusion origin of eukaryotes. *Nature* 431:152–55
97. Rodriguez-Trelles F, Tarrio R, Ayala FJ. 2006. Origins and evolution of spliceosomal introns. *Annu. Rev. Genet.* 40:47–76
98. Roger AJ. 1999. Reconstructing early events in eukaryotic evolution. *Am. Nat.* 154:S146–63
99. Roger AJ, Hug LA. 2006. The origin and diversification of eukaryotes: problems with molecular phylogenetics and molecular clock estimation. *Philos. Trans. R. Soc. Lond. B* 361:1039–54
100. Roy SW, Irimia M. 2009. Splicing in the eukaryotic ancestor: form, function and dysfunction. *Trends Ecol. Evol.* 24:447–55
101. Ruiz-Trillo I, Roger AJ, Burger G, Gray MW, Lang BF. 2008. A phylogenomic investigation into the origin of Metazoa. *Mol. Biol. Evol.* 25:664–72
102. Sanchez-Puerta MV, Bachvaroff TR, Delwiche CF. 2007. Sorting wheat from chaff in multi-gene analyses of chlorophyll c-containing plastids. *Mol. Phylogenet. Evol.* 44:885–97
103. Sapp J. 2009. *The New Foundations of Evolution: On the Tree of Life*. New York: Oxford Univ. Press. 425 pp.
104. Shabalina SA, Koonin EV. 2008. Origins and evolution of eukaryotic RNA interference. *Trends Ecol. Evol.* 23:578–87
105. Shiflett AM, Johnson PJ. 2010. Mitochondrion-related organelles in eukaryotic protists. *Annu. Rev. Microbiol.* 64:409–29
106. Simpson AGB. 2003. Cytoskeletal organization, phylogenetic affinities and systematics in the contentious taxon 'Excavata' (Eukaryota). *Int. J. Syst. Evol. Microbiol.* 53:1759–77
107. Simpson AGB, Patterson DJ. 1999. The ultrastructure of *Carpediemonas membranifera* (Eukaryota) with reference to the "Excavate hypothesis". *Eur. J. Protistol.* 35:353–70
108. Simpson AGB, Roger AJ. 2004. The real 'kingdoms' of eukaryotes. *Curr. Biol.* 14:R693–96
109. Sogin ML, Gunderson JH, Eldwood HJ, Alonso RA, Peattie DA. 1989. Phylogenetic meaning of the kingdom concept: an unusual ribosomal RNA from the *Giardia lamblia*. *Science* 243:75–67
110. Stechmann A, Cavalier-Smith T. 2002. Rooting the eukaryote tree by using a derived gene fusion. *Science* 297:89–91

111. Stechmann A, Cavalier-Smith T. 2003. Phylogenetic analysis of eukaryotes using heat-shock protein Hsp90. *J. Mol. Evol.* 57:408–19
112. Steenkamp ET, Wright J, Baldauf SL. 2006. The protistan origins of animals and fungi. *Mol. Biol. Evol.* 23:93–106
113. Szklarczyk R, Huynen MA. 2010. Mosaic origin of the mitochondrial proteome. *Proteomics* 10:4012–24
114. Tabita FR, Hanson TE, Li H, Satagopan S, Singh J, Chan S. 2007. Function, structure, and evolution of the RubisCO-like proteins and their RubisCO homologs. *Microbiol. Mol. Biol. Rev.* 71:576–99
115. Taylor JE, Rudenko G. 2006. Switching trypanosome coats: What's in the wardrobe? *Trends Genet.* 22:614–20
116. Tekle YI, Parfrey LW, Katz LA. 2009. Molecular data are transforming hypotheses on the origin and diversification of eukaryotes. *Bioscience* 59:471–81
117. Thiergart T, Landan G, Schenk M, Dagan T, Martin WF. 2012. An evolutionary network of genes present in the eukaryote common ancestor polls genomes on eukaryotic and mitochondrial origin. *Genome Biol. Evol.* 4:466–85
118. Van de Peer Y, De Wachter R. 1997. Evolutionary relationships among the eukaryotic crown taxa taking into account site-to-site variation in 18S rRNA. *J. Mol. Evol.* 45:619–30
119. Vesteg M, Krajcovic J. 2011. The falsifiability of the models for the origin of eukaryotes. *Curr. Genet.* 57:367–90
120. Wainright PO, Hinkle G, Sogin ML, Stickel SK. 1993. Monophyletic origins of the Metazoa: an evolutionary link with fungi. *Science* 260:340–42
121. Whittaker RH. 1969. New concepts of kingdoms of organisms. *Science* 163:150–60
122. Whittaker RH, Margulis L. 1978. Protist classification and the kingdoms of organisms. *Biosystems* 10:3–18
123. Wickstead B, Gull K. 2011. The evolution of the cytoskeleton. *J. Cell Biol.* 194:513–25
124. Woese CR, Fox GE. 1977. Phylogenetic structure of the prokaryotic domain: the primary kingdoms. *Proc. Natl. Acad. Sci. USA* 74:5088–90
125. Woese CR, Kandler O, Wheelis ML. 1990. Towards a natural system of organisms: proposal for the domains Archaea, Bacteria, and Eucarya. *Proc. Natl. Acad. Sci. USA* 87:4576–79
126. Yoon HS, Grant J, Tekle YI, Wu M, Chaon BC, et al. 2008. Broadly sampled multigene trees of eukaryotes. *BMC Evol. Biol.* 8:14
127. Zillig W, Klenk HP, Palm P, Leffers H, Puhler G, et al. 1989. Did eukaryotes originate by a fusion event. *Endocytobiosis Cell Res.* 6:1–25
128. Zufall RA, Robinson T, Katz LA. 2005. Evolution of developmentally regulated genome rearrangements in eukaryotes. *J. Exp. Zool. Part B* 304B:448–55

Genomic Insights into Syntrophy: The Paradigm for Anaerobic Metabolic Cooperation

Jessica R. Sieber,[1] Michael J. McInerney,[1] and Robert P. Gunsalus[2]

[1]Department of Botany and Microbiology, University of Oklahoma, Norman, Oklahoma 73019; email: jrsieber@illinois.edu, mcinerney@ou.edu

[2]Department of Microbiology, Immunology, and Molecular Genetics, University of California, Los Angeles, California 90095; email: robg@microbio.ucla.edu

Keywords

methanogenesis, hydrogen, formate, mutualism, biodegradation, reverse electron transfer, consortium

Abstract

Syntrophy is a tightly coupled mutualistic interaction between hydrogen-/formate-producing and hydrogen-/formate-using microorganisms that occurs throughout the microbial world. Syntrophy is essential for global carbon cycling, waste decomposition, and biofuel production. Reverse electron transfer, e.g., the input of energy to drive critical redox reactions, is a defining feature of syntrophy. Genomic analyses indicate multiple systems for reverse electron transfer, including ion-translocating ferredoxin:NAD$^+$ oxidoreductase and hydrogenases, two types of electron transfer flavoprotein:quinone oxidoreductases, and other quinone reactive complexes. Confurcating hydrogenases that couple the favorable production of hydrogen from reduced ferredoxin with the unfavorable production of hydrogen from NADH are present in almost all syntrophic metabolizers, implicating their critical role in syntrophy. Transcriptomic analysis shows upregulation of many genes without assigned functions in the syntrophic lifestyle. High-throughput technologies provide insight into the mechanisms used to establish and maintain syntrophic consortia and conserve energy from reactions that operate close to thermodynamic equilibrium.

Contents

INTRODUCTION	430
DEFINING SYNTROPHY	432
Phylogenetic and Metabolic Diversity of Syntrophic Metabolism	435
Interspecies Electron Transfer	437
REDOX REACTIONS CRITICAL TO SYNTROPHIC METABOLISM	438
Electron Confurcation	439
Membrane-Associated Reverse Electron Transfer	439
EXAMPLES OF SYNTROPHIC METABOLISM	441
Syntrophic Ethanol, Amino Acid, and Glyoxylic Acid Metabolism	441
Syntrophic Lactate Metabolism	441
Syntrophic Butyrate Metabolism	442
Syntrophic Propionate Metabolism	444
SYNTROPHIC LIFESTYLE	444
CONCLUDING REMARKS	445

INTRODUCTION

Syntrophy is a tightly coupled mutualistic interaction where the pool size of intermediates that are exchanged between the partners must be kept very low for efficient cooperation among the partners to occur. The first example of syntrophy was the interaction between phototrophic green sulfur bacteria and chemolithotrophic, sulfur-reducing bacteria that exchange sulfur compounds between the partners (6). Subsequently, tightly coupled, syntrophic associations between fermentative bacteria and methanogenic archaea were discovered that involved the exchange of hydrogen or formate between the partners (10, 56). The fermentative syntrophic metabolizer produces hydrogen and formate from its growth substrate (e.g., propionate, butyrate, and benzoate) (60, 87). The methanogenic partner consumes these products, keeping them at concentrations low enough so the overall degradative reaction is thermodynamically favorable. For example, the syntrophic degradation of butyrate to acetate and H_2 at pH 7, 1 atm H_2 (101 kPa), and 1 M of acetate and butyrate is thermodynamically unfavorable with a Gibbs free energy change of +48.6 kJ per mole of butyrate metabolized.

$$\text{Butyrate}^- + 2H_2O \rightarrow 2\,\text{Acetate}^- + H^+ + 2H_2 \qquad 1.$$

When hydrogen and formate concentrations are kept low by methanogens or other hydrogen/formate users, butyrate degradation is thermodynamically favorable with a Gibbs free energy change of −39.2 kJ per mole of butyrate (1 Pa H_2 partial pressure and concentrations of butyrate and acetate at 0.1 mM apiece).

Microbial syntrophy is an essential process in the global carbon cycle. The degradation of natural polymers such as polysaccharides, proteins, nucleic acids, and lipids to CO_2 and CH_4 involves a complex microbial community that interacts syntrophically (87) (**Figure 1**). In a three-stage process, fermentative bacteria first hydrolyze the polymeric substrates and ferment the hydrolysis products to acetate and longer-chain fatty acids, CO_2, formate, and hydrogen. Propionate, longer-chain fatty acids, alcohols, some amino acids, and aromatic compounds are then syntrophically metabolized to the methanogenic substrates hydrogen, formate, and acetate (56, 87). Last, two different groups of methanogens, hydrogenotrophic and acetotrophic methanogens, complete

Kilopascal (kPa): a derived unit of pressure where 1 kPa equals 10^3 Pa (Pascal units) and is approximately equal to 0.01 atm of the indicated gas

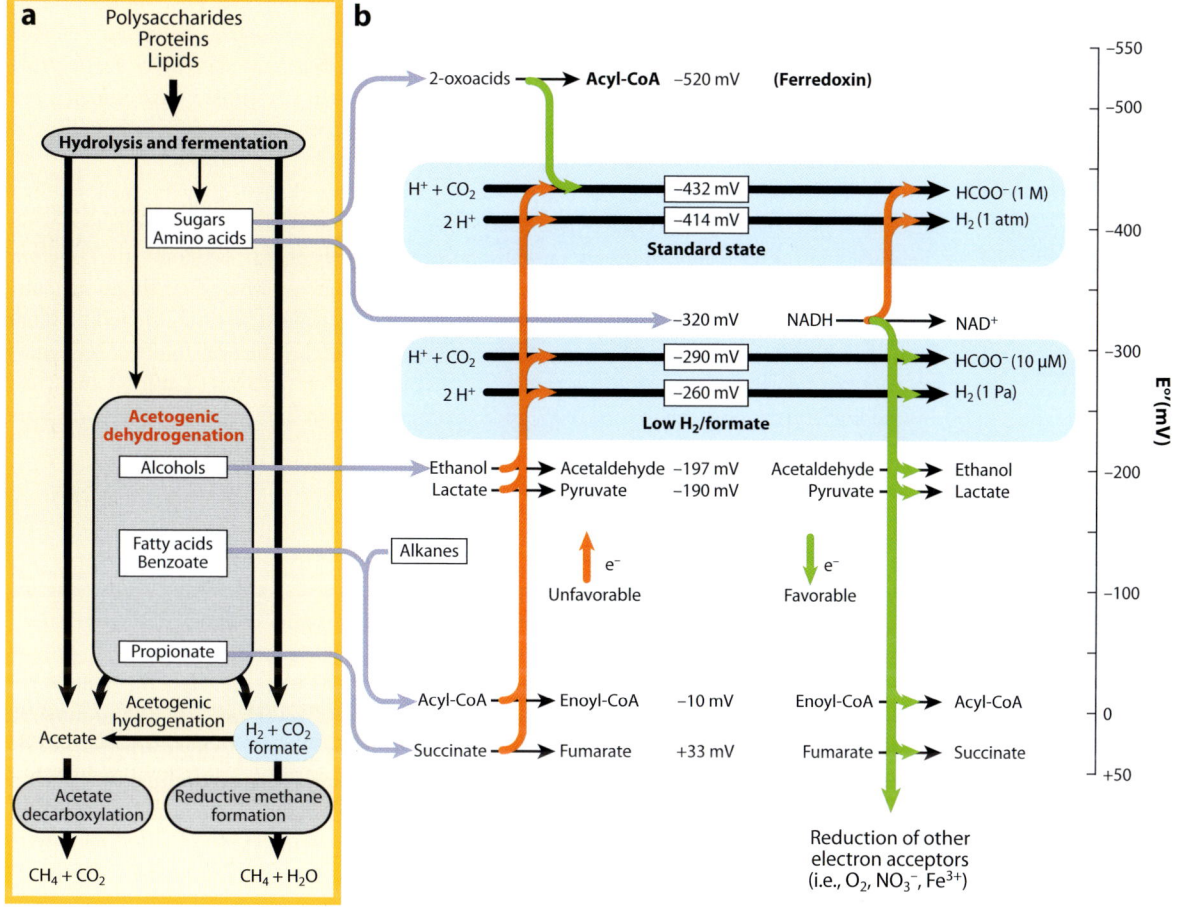

Figure 1

(*a*) Major metabolic processes involved in the conversion of organic matter to methane and (*b*) the critical redox reactions involved in syntrophic metabolism. Acetogenic dehydrogenation involves syntrophic consortia. Key intermediates are shown in boxes. At low hydrogen and formate concentrations, the redox potential for hydrogen and formate production is lowered as shown in the blue highlighted regions. Favorable electron flow is illustrated in green and unfavorable electron flow, which would require reverse electron transfer, is in orange. Figure modified from Reference 58. Reprinted with permission from the American Society for Microbiology (*Microbe*, November 2011, pp. 479–85).

the anaerobic recycling process by converting hydrogen, formate, and acetate made by other microorganisms to methane and CO_2. Syntrophic fatty acid and aromatic acid metabolism accounts for much of the carbon flux in methanogenic environments and is often rate limiting in these environments (55, 87, 88, 111). Many other compounds, including sugars, amino acids, alcohols, organic acids, aromatic compounds, and hydrocarbons, are also syntrophically metabolized (59, 60).

Syntrophic metabolism occurs in sulfate-reducing environments as well. The flux of methane from gas hydrates in ocean sediments is controlled by anaerobic methane-oxidizing archaea that frequently form syntrophic associations with sulfate-reducing *Deltaproteobacteria* (8). 16S ribosomal RNA sequences related to *Syntrophomonadaceae*, whose members are known to syntrophically metabolize fatty acids, were abundant in butyrate-degrading, sulfate-reducing aquifer sediments (102)

NAD+/NADH: oxidized and reduced forms of the electron carrier nicotinamide adenine dinucleotide

E′: the mid-potential value or standard redox potential of a redox reaction that is often expressed in milivolts

Redox: oxidation reduction

and fatty acid-degrading sewage sludge exposed to sulfate for long periods (99). Addition of the hydrogenase inhibitor, carbon monoxide, disrupted butyrate metabolism in sulfate-reducing aquifer sediments, implicating the need for hydrogen exchange and syntrophy for butyrate metabolism (102). The ubiquity of syntrophic metabolism in many anoxic environments emphasizes that metabolic cooperation among microbial species is often the rule rather than the exception and that the consortium is the catalytic unit of anaerobic metabolism.

Even under optimal syntrophic growth conditions, free energy changes are very small, <-20 kJ per mole (1, 36, 91), and the available free energy must be shared by the different organisms (87). Microbial syntrophy truly represents an extreme existence, an energy economy that operates close to equilibrium (57). How syntrophic consortia conserve energy when their thermodynamic driving force is very low and what molecular mechanisms are needed to establish and maintain the syntrophic lifestyle are important questions for future research. The availability of sequenced genomes of a number of microorganisms capable of syntrophic metabolism (**Tables 1** and **2**) has allowed high-throughput techniques, e.g., microarray and mutant library analyses, to unravel the intricacies involved in the syntrophic lifestyle.

DEFINING SYNTROPHY

Fermentative bacteria grow in pure culture on sugars, amino acids, and organic acids. The fermentation of these compounds generates compounds with carbonyl moieties (e.g., pyruvate or acetaldehyde) that serve as the electron acceptor for the reoxidation of the nicotinamide adenine dinucleotide (NADH) pool. The oxidation of NADH (E′ of -320 mV) coupled to the reduction of acetaldehyde (E′ of -197 mV), pyruvate (E′ of -190 mV), enoyl-CoA (E′ of -10 mV), or fumarate (E′ of $+33$ mV) is energetically favorable, allowing fermentative bacteria to form ethanol, lactate, butyrate, or propionate, respectively. When hydrogen-/formate-using microorganisms such as methanogens are active, hydrogen and formate concentrations are low, <10 Pa and <10 μM, respectively. At these low levels, the redox (oxidation reduction) potential for hydrogen and formate production changes from -410 and -420 mV to -260 and -290 mV, respectively (**Figure 1**). Thus, hydrogen and formate production from NADH becomes thermodynamically favorable. In the presence of hydrogen-/formate-using microorganisms, fermentative bacteria make little to no ethanol, lactate, propionate, or butyrate (55).

Some syntrophic metabolizers, such as *Syntrophomonas wolfei* (97), *Syntrophus aciditrophicus* (57), and *Pelotomaculum thermopropionicum* (42), appear to be hardwired for syntrophy. They are metabolic specialists that lack the ability to use alternate electron acceptors or to reoxidize their reduced cofactors by making products such as ethanol, lactate, propionate, or butyrate (**Table 1**). Thus, they must rely on the reduction of protons or CO_2, forming hydrogen or formate, respectively, to reoxidize their reduced cofactors. Many bacteria capable of syntrophic metabolism grow fermentatively in pure culture with a more oxidized derivative of their ecologically relevant substrate such as crotonate for fatty acid degraders or fumarate for propionate degraders (5, 106). Others, such as *Syntrophobacter* spp., *Desulfovibrio* spp., *Geobacter* spp., and *Pelobacter* spp., have alternative metabolic pathways and can grow fermentatively with certain substrates or by anaerobic respiration using electron acceptors such as sulfate, iron(III), or sulfur (**Table 1**). A few species (*Pelotomaculum schinkii*, *Syntrophomonas zehnderi*, *Syntrophorhabdus aromaticivorans*, and *Pelotomaculum isophthalicicum*) grow only syntrophically (17, 80, 81, 100). Future genome sequencing studies will reveal whether these latter organisms have undiscovered metabolic potential that would allow them to grow in pure culture or whether they are truly obligately syntrophic.

Table 1 Substrates used in pure culture and syntrophically by organisms capable of syntrophic metabolism whose genomes have been sequenced

Organism	Substrates used		GenBank ID
	Pure culture	Syntrophically	
Firmicutes			
Clostridium sporogenes ATCC 15579	Amino acids	Some amino acids	NZ_ABKW00000000
Syntrophomonas wolfei Göttingen	C4:1-C6:1[a]	C4-C10; iB	NC_007759
Syntrophothermus lipocalidus DSM 12680	C4:1	C4-C10; iB	NC_014220
Syntrophobotulus glycolicus DSM 8271	Glycolic acid	Glyoxylic acid	Not submitted
Pelotomaculum thermopropionicum SI	Pyruvate, fumarate	C3, lactate, alcohols	NC_009454
Proteobacteria			
Syntrophus aciditrophicus SB	C4:1, benzoate, cyclohex-1-ene-1-carboxylate	Benzoate, alicyclic compounds, fatty acids	NC_008346
Syntrophobacter fumaroxidans MPOB	$C3 + SO_4^-, SO_3^-$, or $S_2O_3^-$; fumarate, malate, or pyruvate	C3	NC_008554
Desulfatibacillum alkenivorans AK-01	Sulfate respiration	Hexadecane	NC_011768
Pelobacter carbinolicus DSM 2380	2,3-Butanediol, acetoin, ethylene glycol	C2 to C4 alcohols	NC_007498
Geobacter sulfurreducens PCA	Metal and sulfur respiration, organic acids	Acetate	NC_002939
Desulfovibrio alaskensis G20	Sulfate respiration	Ethanol, lactate	NC_007519
Desulfovibrio vulgaris Hildenborough	Sulfate respiration	Ethanol, lactate	NC_002937
Synergistetes			
Aminobacterium colombiense DSM 12261	Amino acids, peptides, organic acids	Ala, Glu, Leu, Ile, Val, Asp, Met	NC_014011
Aminomonas paucivorans GLU-3, DSM 12260	Arg, His, Gln, Thr, Gly	Arg, His, Glu	AEIV00000000
Thermoanaerovibrio acidaminovorans DSM 6589	Sugars, amino acids, organic acids	Sugars, amino acids, organic acids	NC_013522
WWE1			
"*Candidatus* Cloacamonas acidaminovorans"	Not known	Inferred use of amino acids	NS_000195

[a]The number of carbons in the compound is indicated; the number following the colon is the number of unsaturated bonds; i refers to iso. When a range is given, this means that the organism can use compounds within the indicated range of carbon numbers, but not all possibilities were tested.

Table 2 Enzymes potentially involved in reverse electron transfer or direct electron transfer in microorganisms capable of syntrophic metabolism whose genomes have been sequenced

Organism	FeS oxidoreductase	Fnr	Fix	Confurcating H$_2$ase	Other soluble H$_2$ase	Membrane-bound H$_2$ase	NADH-linked FDH	Other FDH	Membrane-bound FHD	Direct electron transfer
Aminobacterium colombiense (Amico_)[a]	—	0105-110[b]	—	1558-55	—	1275-62	—	—	—	—
Aminomonas paucivorans (Apau_)	—	2016-11	—	0245-48	0524-28	—	0238-41	—	—	—
"*Candidatus* C. acidaminovorans" (CLOAM)	—	1528-23	—	1184-80	—	1183-86	1684-86	—	—	—
Clostridium sporogenes (CLOSPO_)	—	00568-73	—	03146-48	—	—	—	—	—	—
Desulfatibacillum alkenivorans (Dalk_)	3215-17, 3318-20	4354-58	—	—	2272-69, 4967-62	2276-75	2597, 4343	—	1854-56	—
Desulfovibrio alaskensis (Dde_)	—	0581-0586	—	—	—	0725, 2281-80, 3754-56, 2134-35	0972-75	—	0716-18	—
Desulfovibrio vulgaris (DVU_)	—	2792-97	—	—	0429-34, 2286-93	1769-70, 1921-22, 2524-26, 1917-18	—	—	0587-88, 2481-84, 2809-12	—
Geobacter sulfurreducens (GSU)	2795-97	—	—	—	2722-18 2417-20	0123-21, 0782-85	—	—	0777-81	2503-04 (*omcT,S*), 0618 (*omcE*), 1496 (*pilA*)
Pelobacter carbinolicus (Pcar_)	—	0269-65	—	1602-05	—	2502	0833-35, 1846-43	—	—	—
Pelotomaculum thermopropionicum (PTH_)	1552-54	—	0016-19, 0597-600, 1765-68, 2430-32	2010-12, 1377-79	—	1701-04	2645-51	1711-14	—	—
Syntrophobacter fumaroxidans (Sfum_)	—	2694-99	3926-30	0846-44	2220-22, 2713-16	2952-53, 3535-37, 3954-56	2703-07	0030-31	0035-37, 1273-75, 3509-11	—
Syntrophobotulus glycolicus (Sgly_)	—	—	1133-36	2916-14	—	—	2687-85	—	—	—
Syntrophomonas wolfei	0696-98	—	2121-24	1019-17	2436	1925-27	0656-58, 0783-86, 1024-30, 1829-31	—	0796-00, 1823-26	—
Syntrophothermus lipocalidus (Slip_)	1416-18	—	0457-60, 0875-78	0123-25	0543	1534-32	1031-32	—	1434-32	—
Syntrophus aciditrophicus (SYN_)	02636-38	01658-64	—	01369-70	02219-22	—	00629-31, 02138-40	—	00602-05, 00632-35	—
Thermoanaerovibrio acidaminovorans (Taci_)	—	0342-47	—	0147-50	0473-70	—	—	—	—	—

[a]The abbreviation given in parentheses is the abbreviation that begins each gene locus tag for this organism.
[b]Gene locus tag number for gene system is given.
Abbreviations: FeS oxidoreductase and Fix, putative electron transfer flavoprotein:quinone oxidoreductase; Fnr, ion-translocating ferredoxin:NAD$^+$ oxidoreductase; H$_2$ase, hydrogenase; FDH, formate dehydrogenase.

Phylogenetic and Metabolic Diversity of Syntrophic Metabolism

Syntrophic metabolism is not restricted to a distinct phylogenetic group of microorganisms; it occurs throughout the microbial world (**Figure 2**). Syntrophic metabolism is found in three groups within the phylum *Firmicutes* (**Figure 2**). Members of the family *Syntrophomonadaceae* are metabolic specialists that syntrophically metabolize fatty acids in association with hydrogen-/formate-using microorganisms (**Table 1**) (98). The second group of syntrophic species within the *Firmicutes* is found in the *Desulfotomaculum* lineage (**Table 1**) (**Figure 2**). *Desulfotomaculum thermocisternum* and *Desulfotomaculum thermobenzoicum* syntrophically oxidize propionate (69), while *D. thermobenzoicum* subsp. *thermosyntrophicum* also uses benzoate (75). *Pelotomaculum* spp. syntrophically metabolize propionate, alcohols, and aromatic compounds, but they do not use sulfate as an electron acceptor as do *Desulfotomaculum* spp. (17, 33, 81). *Sporotomaculum syntrophicum* syntrophically metabolizes benzoate (82), *Syntrophobotulus glycolicus* syntrophically oxidizes glycolate (26), and *Gelria glutamica* syntrophically degrades amino acids and several sugars (76). The third group in the *Firmicutes* with syntrophic species is the family *Thermoanaerobacteraceae*, which contains the syntrophic acetate oxidizers *Thermoacetogenium phaeum* (29) and *Syntrophaceticus schinkii* (109). Stable-isotope probing indicates that organisms related to *Thermacetogenium* species and members of the *Thermoanaerobacteraceae* are involved in syntrophic acetate oxidation in rice paddy soils (51). *Thermacetogenium* spp. have also been implicated in syntrophic acetate oxidation in crude oil-degrading microcosms (54). Like *Thermoacetogenium phaeum*, *Clostridium ultunense* and an unnamed strain, AOR (acetate-oxidizing rod), grow homoacetogenically or syntrophically (47, 90). Other homoacetogens such as *Sporomusa acidovorans* and *Moorella mulderi* likely syntrophically metabolize methanol (3, 14). *Moorella* sp. strain AMP and *Desulfovibrio* sp. strain G11 syntrophically metabolize formate (21).

Other *Firmicutes* capable of syntrophic metabolism include *Acidaminobacter hydrogenoformans*, *Eubacterium acidaminophilum*, and *Clostridium sporogenes*, which syntrophically oxidize amino acids (88), and *Tepidanaerobacter syntrophicus*, which syntrophically degrades lactate and alcohols (94) (**Table 1**). "*Candidatus* Contubernalis alkalaceticum," which syntrophically oxidizes acetate, isobutyrate, several alcohols, serine, and fructose, belongs within the family *Syntrophomonadaceae* (112). An unnamed strain, LC 13D, syntrophically degrades primary amines (cadaverine and putrascine), amino acids, and butyrate (84). Stable-isotope probing and substrate enrichment experiments implicate phylotypes related to *Pelotomaculum/Cryptanaerobacter* and Epsilonproteobacteria in syntrophic benzene degradation under sulfate-reducing conditions (30, 41, 45) and new lineages in the *Firmicutes* and Deltaproteobacteria in syntrophic fatty acid metabolism (13, 28).

Deltaproteobacteria species are also capable of syntrophic metabolism. Some species of *Desulfovibrio* syntrophically metabolize lactate and ethanol, *Geobacter* species syntrophically oxidize acetate and ethanol, and *Pelobacter* species syntrophically metabolize alcohols (**Table 1**) (60). The psychrophile, *Algorimarina butyrica*, syntrophically degrades butyrate (40). *Syntrophobacter* species syntrophically metabolize propionate (60). All four *Syntrophobacter* species grow in pure culture by oxidizing propionate with sulfate or fumarate as an electron acceptor or by fermenting organic acids (**Table 1**). *Smithella propionica* grows syntrophically with propionate, butyrate, malate, and fumarate, and in pure culture with crotonate (52). *Syntrophus* species syntrophically metabolize benzoate, other aromatic compounds, alicyclic compounds and fatty acids (**Table 1**) (60). *S. aciditrophicus* also grows by benzoate respiration or the fermentation of benzoate or crotonate (23, 64, 65). *Syntrophorhabdus aromaticivorans* syntrophically degrades various aromatic compounds (80), and *Desulfoglaeba alkanexedens* (15) and *Desulfatibacillum alkenivorans* AK-01 syntrophically degrade alkanes (12) (**Table 1**).

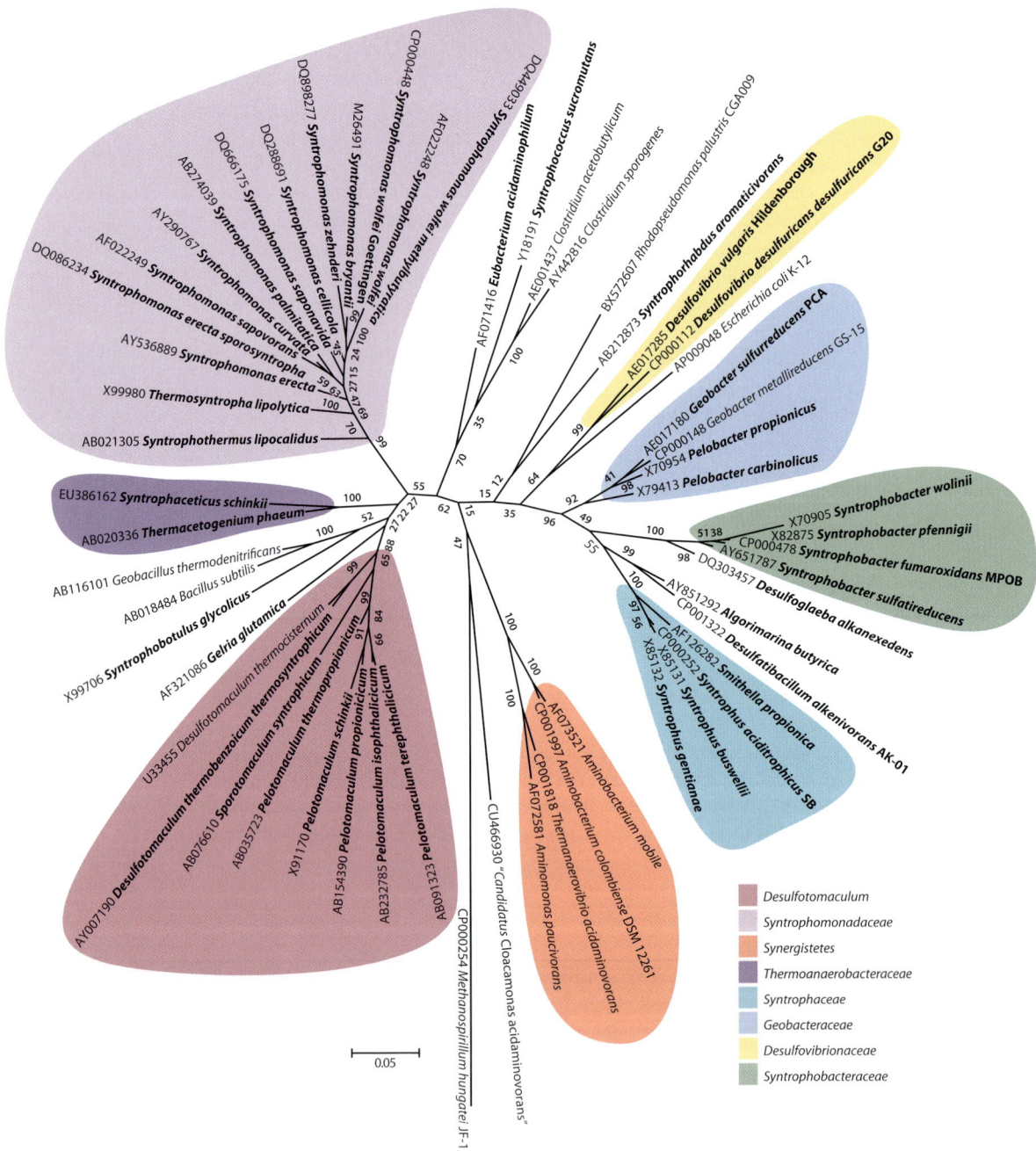

Figure 2

Phylogenetic distribution of microorganisms capable of syntrophic metabolism (shown in bold). Others organisms are added for comparative purposes. The maximum likelihood phylogenetic tree was constructed and bootstrapped 1,000 times using the RAxML v. 7.2.7 program (101) implemented on the CIPRES Web server (**http://www.phylo.org**). Full-length or near-full-length 16S RNA genes of known syntrophic organisms were downloaded and aligned from the Silva ribosomal RNA database (**http://www.arb-silva.de/**). Phylogenetic trees were visualized and edited using the MEGA 4 software package (105). Major groups containing microorganisms capable of syntrophic metabolism are highlighted in color.

"*Candidatus* Cloacamonas acidaminovorans" is a member of the candidate phylum WWE1, and metabolic reconstruction suggests that it is a syntrophic amino acid metabolizer because of the presence of genes for several hydrogenases; five different ferredoxin oxidoreductases, which are typically involved in amino acid fermentation; and the oxidative degradation of propionate (74) (**Table 1**). Another metagenomic study detected *Pelotomaculum*-related sequences in a terephthalate-degrading enrichment culture, which was presumably syntrophically degraded to hydrogen, CO_2, and acetate (53). Genomes related to members of the phylum *Thermotogae*, genus *Syntrophus*, and candidate phyla OP5 and WWE1 were also present, suggesting additional interactions during terephthalate degradation. The OP5 genome contains genes for anaerobic autotrophic butyrate production, and the *Thermotogae*, *Syntrophus*, and WWE1 genomes contain genes to oxidize butyrate to CO_2, hydrogen, and acetate.

Another phylum with syntrophic species is the *Synergistetes*, which comprise *Aminobacterium colombiense*, *Aminomonas paucivorans*, and *Thermoanaerovibrio acidaminovorans*, which syntrophically metabolize amino acids (**Table 1**) (**Figure 2**). *T. acidaminovorans* can also syntrophically metabolize sugars and organic acids (2, 78, 79) (**Table 1**). Microautoradiography–fluorescence in situ hybridization and stable-isotope probing of 16S ribosomal RNA implicated a bacterium belonging to *Synergistetes* group 4 as a dominant acetate-using microorganism in anaerobic digestor sludge (35).

Molecular phylogenetic analyses, ^{13}C lipid isotopic determinations, and microscopic mass spectrometric analyses have identified three clades of methanogen-related archaea that anaerobically oxidize methane: ANME-1, ANME-2, and ANME-3 (31, 71, 72). The close physical association between ANME-2 and ANME-3 and sulfate-reducing bacteria suggests a syntrophic relationship (8, 72). ANME-1 microorganisms are most often found independent of sulfate reducers, although they too can form loose associations with sulfate reducers (72). Although ANME microbes have not been isolated and the nature of metabolic interactions between ANME organisms and the sulfate reducer has not been elucidated, it does appear that archaeal ANME microorganisms are capable of syntrophic metabolism.

Ferredoxin: a low-molecular-weight protein that functions as a low-potential electron carrier

Interspecies hydrogen transfer: the mutualistic transfer of hydrogen between the syntrophic partner organisms

FDH: formate dehydrogenase

Interspecies Electron Transfer

The conclusion of many studies is that syntrophic metabolism can involve interspecies transfer of either hydrogen or formate. Syntrophic metabolism of propionate by *Pelotomaculum thermopropionicum* (34), glycolate by *Syntrophobotulus glycolicus* (26), glucose by *Syntrophococcus sucromutans* (44), and acetate by strain AOR (47) and *Thermoacetogenium phaeum* (29) involved hydrogen transfer, as the methanogen partner used only hydrogen. However, syntrophic amino acid metabolism by *Eubacterium acidaminophilum* involved formate transfer, as the sulfate-reducing partner used formate but not hydrogen (113). Very high levels of formate dehydrogenase (FDH) activity were detected in both *Syntrophobacter fumaroxidans* and its methanogenic partner during syntrophic propionate metabolism, arguing for the importance of formate exchange (19). Although flux analyses indicated that hydrogen diffused too slowly to account for the rate of syntrophic propionate and butyrate degradation (9, 19), other modeling results suggest that the exchange of hydrogen and formate are both equally likely (4).

Genomic analysis of microorganisms whose genomes have been sequenced supports the involvement of either compound in syntrophic metabolism (**Table 2**). Multiple FDH and hydrogenase genes are present in the genomes of syntrophic metabolizers except for *Aminobacterium colombiense*, *Clostridium sporogenes*, and *Thermanaerovibrio acidaminovorans*, which do not contain FDH genes (**Table 2**).

Molecules other than hydrogen or formate may be involved in interspecies electron transfer. An acetate-oxidizing coculture of *Geobacter sulfurreducens* and *Wolinella succinogenes* used cysteine as the

Interspecies electron carrier: a mutualistic transfer of hydrogen, formate, or another organic or inorganic molecule that carries electrons from one organism to another

Nanowire: an electron-conductive extracellular protein assembly able to transfer electrons between microbes

interspecies electron carrier (37). The cycling of sulfur compounds could be a common mechanism for syntrophic metabolism because sulfur cycling is believed to be required for the maintenance of several phototrophic consortia (73) and for the spatially structured association between the chemoautotrophic, sulfur-oxidizing *Acidithiobacillus* sp. and the heterotrophic *Acidiphilium* sp. (70).

Direct electron transfer between syntrophic partners by electron conductive pili or nanowires could also be involved in syntrophy (27, 83). *G. sulfurreducens* has two distinct *pilA* genes. When the shorter *pilA* was mutated, *G. sulfurreducens* did not reduce Fe(III) oxides or grow with iron(III) oxides as an electron acceptor, even though cells attached to the mineral surface (83). In addition to the pilus, mutational analysis showed that multiheme, outer membrane cytochromes, e.g., MtrC and OmcA in *Shewanella oneidensis* and OmcE and OmcS in *G. sulfurreducens*, were required for reduction of solid iron [Fe(III)] oxyhydroxides (61, 95). Electron conductive, nanowire-like structures connected the syntrophic propionate degrader *P. thermopropionicum* with its methanogenic partner (27, 34). However, a subsequent study showed that these structures were the flagella (96). To maximize the flux of hydrogen or formate, pili could function as an attachment mechanism to maintain the syntrophic partners in close proximity to each other. The metagenome of ANME-1 contains genes for an FeFe-hydrogenase, an FDH, and nine multiheme *c*-type cytochromes, some of which are predicted to be excreted (62). The secreted multiheme *c*-type cytochromes could be involved in interspecies electron transfer through interacting with conductive pili (46) or by associating with extracellular polymeric matrix.

Genetic evidence for the involvement of electrically conductive nanowires has been obtained for the syntrophic degradation of ethanol by the ethanol-using *G. metallireducens* and the fumarate-reducing *G. sulfurreducens* (103). A mutation in a gene encoding an RpoN-dependent enhancer-binding protein (*pilR*) enhanced aggregate formation and substrate use. The mutation of *pilR* enhanced the expression of *omcS*, which encodes one of the multiheme outer membrane cytochromes (OmcS) needed for solid Fe(III) mineral reduction. Deletion of the genes for *omcS* or *pilA* in *G. sulfurreducens* prevented growth of the coculture; however, deletion of *hyb*, the hydrogenase gene needed for hydrogen uptake by *G. sulfurreducens*, did not. Because *G. sulfurreducens* does not use formate and the Hyb mutant cannot use hydrogen, direct electron transfer appears to be the only mechanism available for syntrophic ethanol metabolism.

Natural microbial communities may be wired for direct electron transfer, as spatially separate (~12 mm) redox reactions were instantaneously coupled in sediments (68). The addition of electrically conductive iron minerals (hematite and magnetite), but not nonconductive ferrihydrite, accelerated acetate and ethanol degradation and methanogenesis, whereas the methanogenic inhibitor 2-bromoethanesulfonic acid inhibited acetate and ethanol degradation and methanogenesis by paddy soil enrichments, consistent with the involvement of direct interspecies electron transfer (38). The importance of direct interspecies electron transfer was also implicated for methanogenic ethanol and acetate degradation by wastewater aggregates (63).

Whether direct interspecies electron transfer is a general approach for syntrophic metabolism awaits further study. On the one hand, the genomes of all known syntrophic metabolizers have hydrogenase and/or FDH genes (**Table 2**), emphasizing the importance of these gene systems during syntrophy. On the other hand, genes for *pilA* and outer membrane cytochromes believed to be needed to transfer electrons to nanowires so far have been found only in species within genera *Geobacter* and *Shewanella* (95) (**Table 2**).

REDOX REACTIONS CRITICAL TO SYNTROPHIC METABOLISM

Some syntrophic redox reactions have very positive change in electric potentials ($\Delta E'$) even when hydrogen and formate concentrations are low (87) (**Figure 1**). For example, at 1 Pa hydrogen partial pressure, the $\Delta E'$ for hydrogen from electrons generated from the oxidation of acyl-CoA intermediates (E' of -10 mV) (86) or succinate (E' of $+33$ mV) is very positive, about 250 or 290 mV, respectively. Hydrogen partial pressures of 10^{-5} to 10^{-9} Pa would be needed to make these reactions thermodynamically favorable, e.g., negative $\Delta E'$ (87). Such low hydrogen partial pressures are not possible because hydrogenotrophic methanogenesis reaches thermodynamic equilibrium at 0.2 Pa of hydrogen. The only way that these redox reactions become favorable is by energy input through a process called reverse electron transfer. Genomic analysis indicates that several different kinds of gene systems are involved in syntrophic reverse electron transfer (**Table 2**).

Reverse electron transfer: the energetically unfavorable movement of electrons that requires the input of energy to drive a critical cellular oxidation/reduction reaction

Electron Confurcation

Hydrogen or formate production from electrons derived from NADH reoxidation is unfavorable at high hydrogen and formate concentrations (**Figure 1**). One way to overcome this energetic barrier is to couple hydrogen and formate production from NADH with an energetically favorable redox reaction, e.g., oxidation of reduced ferredoxin. The trimeric hydrogenase purified from *Thermotoga maritima* provides an example of such an approach because it uses the thermodynamically favorable production of hydrogen from reduced ferredoxin to drive the unfavorable production of hydrogen from NADH by a process called electron confurcation (67, 93) (**Figure 3**).

Electron confurcation: a coupled biochemical reaction using two dissimilar electron donors that generate a single product such as hydrogen gas

$$\text{NADH} + \text{Ferredoxin}_{\text{red}} + 3\text{H}^+ \rightarrow 2\text{H}_2 + \text{NAD}^+ + \text{Ferredoxin}_{\text{ox}} \qquad 2.$$

Genes homologous to those for the confurcating hydrogenase in *T. maritima* are present in the genomes of many organisms capable of syntrophic metabolism and anaerobes known to produce high molar ratios of hydrogen (>3 hydrogen per glucose) from glucose (93, 97) (**Figure 3**). Sieber et al. (97) also noted the association of the gene for the catalytic subunit of FDH with genes for NADH:quinone oxidoreductase subunits in genomes of several organisms capable of syntrophic metabolism, suggesting that formate production from NADH may also involve electron confurcation. Recently, an NADH:acceptor oxidoreductase was partially purified from cell extract of *Syntrophomonas wolfei* that contained subunits predicted to function as an NADH-linked FDH (encoded by Swol_0783, Swol_0785, and Swol_0786) and an NADH-linked hydrogenase (encoded by Swol_1017, Swol_1018, and Swol_1019) (66) (**Table 2**), providing experimental evidence for the involvement of confurcating hydrogenases and FDHs in syntrophic butyrate metabolism.

Confurcating: combining electrons from two dissimilar donors to generate a single product

Membrane-Associated Reverse Electron Transfer

Reverse electron transfer during syntrophic metabolism could also be driven by ion gradients (59) (**Table 2**). Many microorganisms capable of syntrophic metabolism contain menaquinones, which could function as the electron carrier between membrane complexes involved in oxidation of substrate-derived metabolites and hydrogenases, FDHs, or other membrane redox complexes (59). Alternatively, primary-ion-translocating membrane complexes such as ferredoxin:NAD$^+$ oxidoreductase could be involved. The ferredoxin:NAD$^+$ oxidoreductase may function as a reverse electron transfer complex, using the ion gradient to drive the unfavorable reduction of ferredoxin with NADH (7). The fact that ion gradients are needed for hydrogen production by *S. wolfei* and *Syntrophus buswellii* (108) and for glyoxylate metabolism by *Syntrophobotulus glycolicus* (24, 25) provides experimental support for reverse electron transport in syntrophic metabolism.

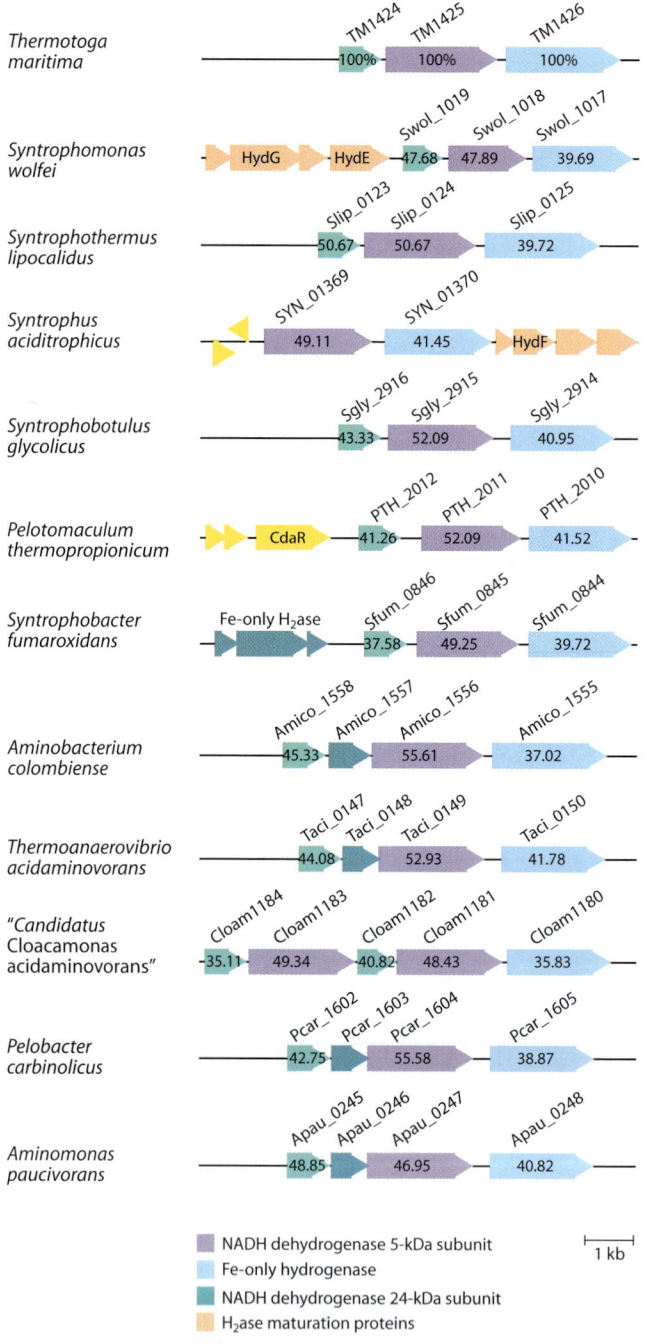

Figure 3

Confurcating [Fe]-hydrogenase gene clusters in bacteria capable of syntrophic metabolism. Numbers are percent identities at the amino acid level to the respective *Thermotoga maritima* gene product. Modified from Reference 97. Reprinted with permission from the Society for Applied Microbiology and Blackwell Publishing Ltd.

EXAMPLES OF SYNTROPHIC METABOLISM

Syntrophic Ethanol, Amino Acid, and Glyoxylic Acid Metabolism

Hydrogen or formate production from electrons derived from oxidation of ethanol to acetaldehyde requires energy input (**Figure 1**). The enzymology of syntrophic ethanol oxidation is poorly understood; the most likely scenario suggests ethanol is oxidized to acetaldehyde coupled to NADH formation and acetaldehyde is oxidized to acetate to form reduced ferredoxin (87). Comparative genomic analysis shows that the syntrophic ethanol oxidizer *Pelobacter carbinolicus*, but not *Pelobacter propionicus*, which is not a syntrophic metabolizer, contains genes for a membrane-bound, ion-translocating ferredoxin:NAD^+ oxidoreductase (11). However, *P. carbinolicus* contains genes for a confurcating hydrogenase that could directly couple the oxidation of NADH and reduced ferredoxin to produce hydrogen (Equation 2) (**Table 2**).

The genomes of syntrophic amino acid users *Aminobacterium colombiense*, *Aminomonas paucivorans*, *Clostridium sporogenes*, and *Thermanaerovibrio acidaminovorans*, and that of the presumed amino acid user "*Candidatus* Cloacamonas acidaminovorans," also contain genes for confurcating hydrogenase and the membrane-bound, ion-translocating ferredoxin:NAD^+ oxidoreductase (**Table 2**). Other potential reverse electron transfer systems include a membrane-bound hydrogenase in *A. colombiense* and "*Candidatus* C. acidaminovorans" and an NADH-linked FDH in *A. paucivorans* and "*Candidatus* C. acidaminovorans" (**Table 2**).

The genome of the syntrophic glyoxylic acid degrader *Syntrophobotulus glycolicus* contains genes for confurcating hydrogenase, NADH-linked FDH, and a membrane-bound, electron transfer flavoprotein:quinone oxidoreductase called Fix (**Table 2**). Fix is proposed to use the energy in the ion gradient to drive the unfavorable reduction of ferredoxin with electrons derived from fatty acid oxidation (22). Its role in the metabolism of glyoxylic acid is unclear, but Fix could supply reduced ferredoxin needed for biosynthesis. The presence of genes for confurcating hydrogenase in genomes of all sequenced microorganisms capable of syntrophic metabolism except *Desulfovibrio* spp. (**Table 2**) argues for the importance of confurcating hydrogenases in syntrophic metabolism.

Syntrophic Lactate Metabolism

The production of hydrogen or formate from electrons generated in the oxidation of lactate to pyruvate by lactate dehydrogenase requires reverse electron transfer (**Figure 1**). Genomic analysis shows that *Desulfovibrio vulgaris* Hildenborough and *D. alaskensis* G20 have genes for membrane-bound hydrogenases and FDHs and for a membrane-bound ferredoxin:NAD^+ oxidoreductase that could function in reverse electron transfer (**Table 2**). Neither genome contained genes for confurcating hydrogenases. *D. vulgaris* Hildenborough also contains genes for NADH-linked FDH. One hundred sixty-nine genes were upregulated when *D. vulgaris* Hildenborough was grown syntrophically with lactate compared to pure culture growth by sulfate respiration (107). Upregulated genes during syntrophic lactate metabolism included those for lactate degradation, ethanol production, and the formation of heterodisulfide reductase, a membrane-bound, ion-translocating hydrogenase (*coo*), a high-molecular-weight *c*-type cytochrome complex (*hmc*), a quinone-reducing complex (*qrc*), and two hydrogenases (*hyd* and *hyn*) (**Figure 4a**). Deletion of *coo* and *hmc* severely impaired syntrophic growth with lactate or pyruvate, implicating these two complexes in reverse electron transfer. *hmc* genes were downregulated when *D. vulgaris* Hildenborough transitioned from syntrophic growth to pure culture growth by sulfate respiration (77).

Mutational analysis in *Desulfovibrio alaskensis* G20 detected 20 mutants with impaired ability to grow syntrophically on lactate with *Methanospirillum hungatei* but not in pure culture by sulfate respiration (49, 50). Both *qrcB*, which encodes a quinone reductase complex, and *cycA*, which encodes the periplasmic tetraheme cytochrome c_3 (TpIc$_3$), were required by G20 to grow syntrophically with lactate. These two genes and *hydA*, which encodes an Fe-only hydrogenase, were required for hydrogen production from lactate by washed cells. The current model is that Qrc interacts with the menaquinone pool to use ion gradients to drive the reverse electron transfer of electrons derived from lactate oxidation to TpIc$_3$ and ultimately to periplasmic hydrogenases such as Hyd (**Figure 4a**).

Hyd: hydrogenase enzyme

Electron bifurcation: a biochemical reaction operating in the reverse direction to confurcation where two products are formed (e.g., NADH and reduced Fd are formed from butyryl-CoA)

Electron transfer flavoprotein (ETF): a protein involved in the oxidation of acyl-CoA intermediates

FeS complex: a membrane-bound iron-sulfur-containing oxidoreductase involved in oxidizing ETFs

Syntrophic Butyrate Metabolism

The oxidation of acyl-CoA to enoyl-CoA intermediates coupled to hydrogen or formate production is the critical redox reaction in syntrophic fatty acid, aromatic acid, and alkane metabolism (**Figure 1**). One possible mechanism for reverse electron transfer is electron bifurcation by butyryl-CoA dehydrogenase:electron transfer flavoprotein (Bcd/EtfAB) complex of *Clostridium kluyveri* (48). This enzyme couples the energetically favorable reduction of crotonyl-CoA to butyryl-CoA ($E^{o\prime} = -10$ mV) by NADH ($E^{o\prime} = -320$ mV) with the unfavorable reduction of ferredoxin ($E^{o\prime} = -410$ mV) by NADH. The reverse of this reaction (Equation 3) could drive the unfavorable reduction of NAD$^+$ by butyryl-CoA by using the favorable reduction of NAD$^+$ by reduced ferredoxin. *Syntrophomonas wolfei* has a gene cluster (Swol_0266-68) homologous to the *bcd/etfAB* complex from *C. kluyveri* (97). However, *S. wolfei* does not have a mechanism to generate reduced ferredoxin during syntrophic fatty acid metabolism (**Figure 4b**). Also, a recent study showed that the predominate Bcd of *S. wolfei* is not associated with EtfAB and probably does not function as an electron-bifurcating complex (66).

$$\text{Butyryl-CoA} + \text{Fd}_{red} + 2\text{NAD}^+ \rightarrow \text{Crotonyl-CoA} + \text{Fd}_{ox} + 2\text{NADH} + 2\text{H}^+ \qquad 3.$$

The genomes of the syntrophic fatty acid degraders *S. wolfei* and *Syntrophothermus lipocalidus*, the syntrophic alkane degrader *Desulfatibacillum alkenivorans*, and the syntrophic fatty and aromatic acid degrader *Syntrophus aciditrophicus* contain a gene for a membrane-bound iron-sulfur (FeS) oxidoreductase that is adjacent to the two genes for electron transfer flavoprotein (*etfAB*) (12, 57, 97) (**Table 2**) (**Figure 4b,c**). The C-terminal portion of the FeS oxidoreductase was detected in a partially purified acyl-CoA dehydrogenase activity from syntrophically grown *S. wolfei* cells (66). The FeS complex could funnel electrons from β-oxidation to membrane redox carriers (57, 97) (**Figure 4b**).

The genomes of *S. wolfei*, *S. lipocalidus*, *D. alkenivorans*, and *S. aciditrophicus* contain genes for NADH-linked FDHs and membrane-bound FDHs (**Table 2**) (**Figure 4b,c**). All the above genomes except the *D. alkenivorans* genome contain genes for confurcating hydrogenases, and all the above genomes except the *S. aciditrophicus* genome contain membrane-bound hydrogenases. The Firmicutes *S. wolfei* and *S. lipocalidus* have *fix* genes, and the Deltaproteobacteria *D. alkenivorans* and *S. aciditrophicus* contain genes for the membrane-bound, ion-translocating ferredoxin:NAD$^+$

Figure 4

Overview of the syntrophic metabolism. (*a*) Syntrophic lactate metabolism by *Desulfovibrio* species (50, 107). (*b*) Syntrophic butyrate metabolism by *Syntrophomonas wolfei* (97). (*c*) Syntrophic benzoate metabolism by *Syntrophus aciditrophicus* (57). (*d*) Syntrophic propionate metabolism by *Syntrophobacter fumaroxidans* (110). Abbreviations: CoA, coenzyme A; Coo, membrane-bound, ion-translocating hydrogenase; ETF, electron transfer flavoprotein; Fd, ferredoxin; FeS red, iron-sulfur oxidoreductase; FRD, fumarate reductase; Hdr, heterodisulfide reductase; Hmc, high-molecular-weight *c*-type cytochrome complex; Hyn, hydrogenase; LDH, lactate dehydrogenase; MK, menaquinone; ox, oxidized; QRC, quinone reductase complex; red, reduced; RNF (*Rhodobacter* nitrogen fixation), membrane-bound, ion-translocating ferredoxin:NAD$^+$ oxidoreductase; TpIC$_3$, type I, cytochrome c_3.

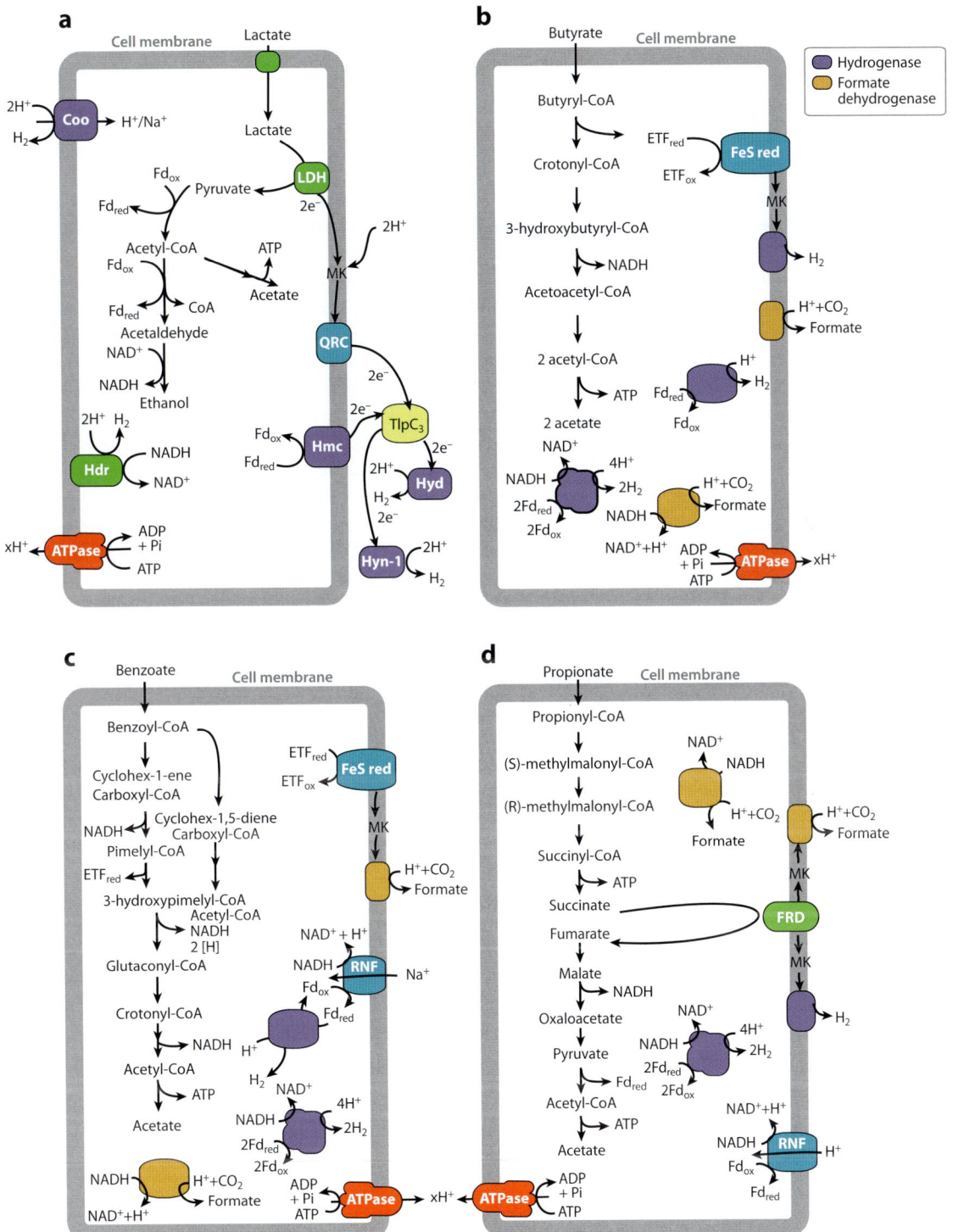

oxidoreductase (**Table 2**). The presence of an FeS oxidoreductase gene adjacent to *etfAB* in the genomes of *S. wolfei*, *S. lipocalidus*, *D. alkenivorans*, and *S. aciditrophicus* argues for its importance in syntrophic alkane and fatty acid and aromatic acid metabolism.

Syntrophic Propionate Metabolism

Many propionate-degrading organisms capable of syntrophy use the methylmanonyl-CoA pathway to oxidize propionate to acetate (32, 43) (**Figure 4d**). The key redox reactions are the oxidation of succinate to fumarate, malate to oxaloacetate, and pyruvate to acetyl-CoA and CO_2 (87). The conversion of pyruvate to acetyl-CoA can be coupled to hydrogen ($E^{o\prime} = -414$ mV) or formate production ($E^{o\prime} = -432$ mV) using a pyruvate:ferredoxin oxidoreductase. The oxidation of succinate to fumarate ($E^{o\prime} = +30$ mV) and malate to oxaloacetate ($E^{o\prime} = -176$ mV) would require an energy input via reverse electron transport (**Figure 1**).

The current model for reverse electron transport during the oxidation of succinate during syntrophic propionate metabolism involves coupling menaquinone reduction ($E^{o\prime}$ of -70 mV) with the oxidation of succinate to fumarate ($E^{o\prime}$ of $+33$ mV) by succinate dehydrogenase (**Figure 4d**). This system is similar to that found in *Bacillus subtilis* in which a membrane-bound succinate dehydrogenase complex reduces menaquinone (89). In organisms such as *Syntrophobacter fumaroxidans* and *Pelotomaculum thermopropionicum*, menaquinone oxidation could be linked to an externally oriented hydrogenase or FDH (**Figure 4c**). *S. fumaroxidans* has a membrane-bound succinate dehydrogenase encoded by Sfum_1998-2000, two cytoplasmic succinate dehydrogenases encoded by Sfum_0173-74 and Sfum_2103-04, and FDHs and hydrogenases that are membrane associated (16, 18, 20) (**Table 2**). The corresponding enzyme activities have been detected and are membrane associated (106). All the genes for the major subunits of the hydrogenases and FDHs in *S. fumaroxidans* were expressed under coculture and monoculture growth conditions, regardless of substrate (110). Significantly higher expression of the confurcating hydrogenase, a periplasmic FDH, and the hydrogen-formate lyase was observed during syntrophic growth versus monoculture growth. *P. thermopropionicum* contains a similar system with a membrane-associated succinate dehydrogenase encoded by PTH_1016-18 (42, 43). This succinate dehydrogenase was transcribed more highly during coculture growth on propionate (39). Many of the hydrogenases and FDHs in *P. thermopropionicum* were constitutively expressed or downregulated during syntrophic growth compared to axenic growth on fumarate, except for a soluble FDH, PTH_2645-2649, which was transcribed more highly under all syntrophic conditions except growth on lactate.

Multiple routes to accomplish for reverse electron transfer during syntrophic metabolism have been proposed. Common themes to syntrophic reverse electron transfer, despite the distinctly different phylogenetic lineages of these organisms, include the presence of confurcating-type hydrogenase genes in almost all microorganisms capable of syntrophic metabolism (**Figure 3**) (**Table 2**) and a gene for novel membrane-bound FeS oxidoreductase and adjacent to *etfAB* (57, 97), which may function to funnel electrons derived from oxidation of acyl-CoA intermediates to membrane carriers (**Figure 4b,c**) (**Table 2**).

SYNTROPHIC LIFESTYLE

Adaptation to a syntrophic lifestyle requires metabolic changes that might not be apparent when cells are growing in pure culture. One could therefore ask whether there are special biochemical pathways and associated genes vital for maintenance of syntrophic systems. Transcriptomic analyses found that many genes without functional assignment (e.g., hypothetical genes) were differentially regulated during syntrophic growth (39, 77, 107). A functionally unknown set of three genes, two of which encode iron-sulfur cluster-binding/ATPase domain proteins and the third of which encodes an MTH1175-like domain family protein, and a five-gene cluster encoding several lipoproteins and membrane-bound proteins were downregulated when *D. vulgaris* transitioned from syntrophic metabolism to sulfate respiration (77, 92).

Shifts in the expression of amino acid biosynthesis genes were detected in cocultures of termite gut spirochetes (85) and syntrophically grown *P. thermopropionicum* (39). Isoleucine/leucine/valine transport, tryptophan/phenylalanine/tyrosine biosynthesis, and methionine biosynthesis and transport genes were transcribed more highly in *Treponema primitia* when cocultured with *Treponema azotonutricium* (85). *P. thermopropionicum* requires yeast extract for pure culture growth but not for coculture growth. Transcriptional analysis of *P. thermopropionicum* revealed that amino acid transport and metabolism genes, particularly those involved in glutamine and glutamate synthesis, were more expressed under syntrophic conditions, suggesting that the syntrophic partners are transferring amino acids to each other (39). *P. thermopropionicum* also upregulated 16 hypothetical genes and altered the expression of many chemotaxis genes in response to the presence of its methanogenic partner.

It is unclear how syntrophic partners regulate their metabolisms to form a coordinated catalytic unit. One possibility is quorum signaling; however, cell-density-dependent gene systems were not found in the genomes of *S. wolfei* or *S. aciditrophicus* (57, 97). The genomes of *S. wolfei* and *S. aciditrophicus* contain many σ^{54}-interacting transcriptional regulators (57, 97). In *P. thermopropionicum*, putative σ^{54}-dependent promoter sequences were found immediately upstream of genes for the methylmalonyl-CoA degradation pathway (*mmc* gene cluster) (42). Many of the catabolic degradation genes in the genomes of syntrophic organisms are physically associated with genes for proteins that contain PAS domains or GAF domains, suggesting that these organisms respond to environmental conditions and/or global cellular situations. The addition of FliD (flagellar cap protein), but not FliC, of *P. thermopropionicum* to pure cultures of *Methanothermobacter thermautotrophicus* accelerated methanogenesis and altered the expression of about 50 genes (96). Thus, syntrophic partners sense the presence of each other by a mechanism similar to the Toll-like receptors of immune cells used to perceive flagella of pathogens (104).

CONCLUDING REMARKS

Syntrophic microbial processes are ubiquitous in natural and human-made habitats devoid of oxygen and/or other external electron acceptors needed to oxidize reduced organic compounds. The combined metabolic actions of syntrophic associations allow for the turnover of organic matter generated by photosynthesis back to simple carbon molecules (CO_2 and methane). Syntrophy generally involves two complementary microbial cell types whose combined metabolism performs reactions not possible by either alone. The biochemical and molecular understanding of these processes has been elusive over the years due to the very low growth rates and cell yields, and by the lack of specialized tools to study them. The availability of high-throughput microbial genome sequencing technology in recent years has provided a new approach to study microbial syntrophy. Genomic analysis has revealed the genetic blueprints of model syntrophic substrate-utilizing microorganisms, which in turn has provided insights into the metabolic pathways operative for carbon and electron flow to the end products acetate, hydrogen, and formate. The emerging biochemical paradigm includes the need for reverse electron transfer via membrane-bound as well as cytoplasmic confurcation-type enzymes that dispose of electrons as hydrogen and/or formate. Syntrophic strategies truly exist at the thermodynamic limits of life.

SUMMARY POINTS

1. Syntrophic metabolism occurs throughout the microbial world and is essential for anaerobic decomposition of many substrates.

2. Exchange of hydrogen and formate between syntrophic partners is common but direct electron transfer may be operative in some types of syntrophic metabolism.

3. Confurcating hydrogenases are present in many syntrophic metabolizers and most likely play a critical role in hydrogen production from NADH and reduced ferredoxin.

4. Multiple enzyme systems are involved in syntrophic reverse electron transfer including ion-translocating ferredoxin:NAD^+ oxidoreductase and hydrogenases, two types of electron transfer flavoprotein:quinone oxidoreductases, and other quinone reactive complexes.

5. The syntrophic lifestyle involves many genes without assigned functions and may not be used for pure culture growth.

6. Regulation of syntrophic metabolism may also involve regulators with PAS or GAF domains, σ^{54} factors, and Toll-like receptors.

7. Syntrophy operates close to thermodynamic equilibrium whereby both partners must share the limited energy released by their overall reactions.

FUTURE ISSUES

1. What is the extent of the phylogenetic diversity for syntrophic microorganisms?

2. Are the currently described pathways representative of the major routes for syntrophic metabolism that occur in nature?

3. How do syntrophic specialists fully metabolize aromatic compounds and how are these processes modulated?

4. Are there major and unrecognized genetic solutions to accomplish syntrophy?

5. How is growth of multispecies syntrophic consortia achieved at low thermodynamic driving forces?

6. What are the biochemical mechanisms needed to accomplish reverse electron transfer?

7. How widespread is syntrophy in opportunistic microbes in situations where obligate syntrophy does not occur?

DISCLOSURE STATEMENT

The authors are not aware of any affiliations, memberships, funding, or financial holdings that might be perceived as affecting the objectivity of this review.

ACKNOWLEDGMENTS

We thank N.Q. Wofford for artwork. This work was supported by grants from the National Science Foundation Award, NSF EF-0333294, and from the US Department of Energy, DE-FG02-96ER20214 (MM), DE-FG03-86ER13498 (RG), and DE-FC02-02ER6421 (RG).

LITERATURE CITED

1. Adams CJ, Redmond MC, Valentine DL. 2006. Pure-culture growth of fermentative bacteria, facilitated by H_2 removal: bioenergetics and H_2 production. *Appl. Environ. Microbiol.* 72:1079–85
2. Baena S, Fardeau ML, Woo TH, Ollivier B, Labat M, Patel BK. 1999. Phylogenetic relationships of three amino-acid-utilizing anaerobes, *Selenomonas acidaminovorans*, '*Selenomonas acidaminophila*' and *Eubacterium acidaminophilum*, as inferred from partial 16S rDNA nucleotide sequences and proposal of *Thermanaerovibrio acidaminovorans* gen. nov., comb. nov. and *Anaeromusa acidaminophila* gen. nov., comb. nov. *Int. J. Syst. Bacteriol.* 49(Pt. 3):969–74
3. Balk M, Weijma J, Friedrich MW, Stams AJ. 2003. Methanol utilization by a novel thermophilic homoacetogenic bacterium, *Moorella mulderi* sp. nov., isolated from a bioreactor. *Arch. Microbiol.* 179:315–20
4. Batstone DJ, Picioreanu C, van Loosdrecht MC. 2006. Multidimensional modelling to investigate interspecies hydrogen transfer in anaerobic biofilms. *Water Res.* 40:3099–108
5. Beaty PS, McInerney MJ. 1987. Growth of *Syntrophomonas wolfei* in pure culture on crotonate. *Arch. Microbiol.* 147:389–93
6. Biebl H, Pfennig N. 1978. Growth yields of green sulfur bacteria in mixed cultures with sulfur and sulfate reducing bacteria. *Arch. Microbiol.* 117:9–16
7. Biegel E, Müller V. 2010. Bacterial Na^+-translocating ferredoxin: NAD^+ oxidoreductase. *Proc. Natl. Acad. Sci. USA* 107:18138–42
8. Boetius A, Ravenschlag K, Schubert CJ, Rickert D, Widdel F, et al. 2000. A marine microbial consortium apparently mediating anaerobic oxidation of methane. *Nature* 407:623–26
9. Boone DR, Johnson RL, Liu Y. 1989. Diffusion of the interspecies electron carriers H_2 and formate in methanogenic ecosystems and its implications in the measurement of K_m for H_2 or formate uptake. *Appl. Environ. Microbiol.* 55:1735–41
10. Bryant MP, Wolin EA, Wolin MJ, Wolfe RS. 1967. *Methanobacillus omelianskii*, a symbiotic association of two species of bacteria. *Arch. Microbiol.* 59:20–31
11. Butler JE, Young ND, Lovley DR. 2009. Evolution from a respiratory ancestor to fill syntrophic and fermentative niches: comparative genomics of six *Geobacteraceae* species. *BMC Genomics* 10:103
12. Callaghan AV, Morris BE, Pereira IA, McInerney MJ, Austin RN, et al. 2011. The genome sequence of *Desulfatibacillum alkenivorans* AK-01: a blueprint for anaerobic alkane oxidation. *Environ. Microbiol.* 14:101–13
13. Chauhan A, Ogram A. 2006. Fatty acid-oxidizing consortia along a nutrient gradient in the Florida Everglades. *Appl. Environ. Microbiol.* 72:2400–6
14. Cord-Ruwisch R, Ollivier B. 1986. Interspecific hydrogen transfer during methanol degradation by *Sporomusa acidovorans* and hydrogenophilic anaerobes. *Arch. Microbiol.* 144:163–65
15. Davidova IA, Duncan KE, Choi OK, Suflita JM. 2006. *Desulfoglaeba alkanexedens* gen. nov., sp. nov., an *n*-alkane-degrading, sulfate-reducing bacterium. *Int. J. Syst. Evol. Microbiol.* 56:2737–42
16. de Bok FA, Hagedoorn PL, Silva PJ, Hagen WR, Schiltz E, et al. 2003. Two W-containing formate dehydrogenases (CO_2-reductases) involved in syntrophic propionate oxidation by *Syntrophobacter fumaroxidans*. *Eur. J. Biochem.* 270:2476–85
17. de Bok FA, Harmsen HJ, Plugge CM, de Vries MC, Akkermans AD, et al. 2005. The first true obligately syntrophic propionate-oxidizing bacterium, *Pelotomaculum schinkii* sp. nov., co-cultured with *Methanospirillum hungatei*, and emended description of the genus *Pelotomaculum*. *Int. J. Syst. Evol. Microbiol.* 55:1697–703
18. de Bok FA, Luijten ML, Stams AJ. 2002. Biochemical evidence for formate transfer in syntrophic propionate-oxidizing cocultures of *Syntrophobacter fumaroxidans* and *Methanospirillum hungatei*. *Appl. Environ. Microbiol.* 68:4247–52

6. First description of syntrophic bacterial associations based on sulfur cycling.

12. Genomic analysis of anaerobic alkane degrader led to the discovery that the bacterium is capable of syntrophy.

18. Combined enzymatic analyses and modeling to show the importance of formate transfer for syntrophic propionate metabolism.

19. de Bok FA, Plugge CM, Stams AJ. 2004. Interspecies electron transfer in methanogenic propionate degrading consortia. *Water Res.* 38:1368–75
20. de Bok FA, Roze EH, Stams AJ. 2002. Hydrogenases and formate dehydrogenases of *Syntrophobacter fumaroxidans*. *Antonie van Leeuwenhoek* 81:283–91
21. Dolfing J, Jiang B, Henstra AM, Stams AJ, Plugge CM. 2008. Syntrophic growth on formate: a new microbial niche in anoxic environments. *Appl. Environ. Microbiol.* 74:6126–31
22. Edgren T, Nordlund S. 2004. The *fixABCX* genes in *Rhodospirillum rubrum* encode a putative membrane complex participating in electron transfer to nitrogenase. *J. Bacteriol.* 186:2052–60
23. Elshahed MS, McInerney MJ. 2001. Benzoate fermentation by the anaerobic bacterium *Syntrophus aciditrophicus* in the absence of hydrogen-using microorganisms. *Appl. Environ. Microbiol.* 67:5520–25
24. Friedrich M, Schink B. 1993. Hydrogen formation from glycolate driven by reversed electron transport in membrane vesicles of a syntrophic glycolate-oxidizing bacterium. *Eur. J. Biochem.* 217:233–40
25. Friedrich M, Schink B. 1995. Electron transport phosphorylation driven by glyoxylate respiration with hydrogen as electron donor in membrane vesicles of a glyoxylate-fermenting bacterium. *Arch. Microbiol.* 163:268–75
26. Friedrich M, Springer N, Ludwig W, Schink B. 1996. Phylogenetic positions of *Desulfofustis glycolicus* gen. nov., sp. nov., and *Syntrophobotulus glycolicus* gen. nov., sp. nov., two new strict anaerobes growing with glycolic acid. *Int. J. Syst. Bacteriol.* 46:1065–69
27. Gorby YA, Yanina S, McLean JS, Rosso KM, Moyles D, et al. 2006. Electrically conductive bacterial nanowires produced by *Shewanella oneidensis* strain MR-1 and other microorganisms. *Proc. Natl. Acad. Sci. USA* 103:11358–63
28. Hatamoto M, Imachi H, Yashiro Y, Ohashi A, Harada H. 2007. Diversity of anaerobic microorganisms involved in long-chain fatty acid degradation in methanogenic sludges as revealed by RNA-based stable isotope probing. *Appl. Environ. Microbiol.* 73:4119–27
29. Hattori S, Kamagata Y, Hanada S, Shoun H. 2000. *Thermacetogenium phaeum* gen. nov., sp. nov., a strictly anaerobic, thermophilic, syntrophic acetate-oxidizing bacterium. *Int. J. Syst. Evol. Microbiol.* 50:1601–9
30. Herrmann S, Kleinsteuber S, Chatzinotas A, Kuppardt S, Lueders T, et al. 2010. Functional characterization of an anaerobic benzene-degrading enrichment culture by DNA stable isotope probing. *Environ. Microbiol.* 12:401–11
31. Hinrichs KU, Hayes JM, Sylva SP, Brewer PG, DeLong EF. 1999. Methane-consuming archaebacteria in marine sediments. *Nature* 398:802–5
32. Houwen FP, Plokker J, Stams AJM, Zehnder AJB. 1990. Enzymatic evidence for involvement of the methylmalonyl-CoA pathway in propionate oxidation by *Syntrophobacter wolinii*. *Arch. Microbiol.* 155:52–55
33. Imachi H, Sekiguchi Y, Kamagata Y, Hanada S, Ohashi A, Harada H. 2002. *Pelotomaculum thermopropionicum* gen. nov., sp. nov., an anaerobic, thermophilic, syntrophic propionate-oxidizing bacterium. *Int. J. Syst. Evol. Microbiol.* 52:1729–35
34. Ishii S, Kosaka T, Hori K, Hotta Y, Watanabe K. 2005. Coaggregation facilitates interspecies hydrogen transfer between *Pelotomaculum thermopropionicum* and *Methanothermobacter thermautotrophicus*. *Appl. Environ. Microbiol.* 71:7838–45
35. Ito T, Yoshioka H, Ariesyady HD, Okada H. 2011. Identification of a novel acetate-utilizing bacterium belonging to *Synergistes* group 4 in anaerobic digestor sludge. *ISME J.* 5:1844–56
36. Jackson BE, McInerney MJ. 2002. Anaerobic microbial metabolism can proceed close to thermodynamic limits. *Nature* 415:454–56
37. Kaden J, Galushko AS, Schink B. 2002. Cysteine-mediated electron transfer in syntrophic acetate oxidation by cocultures of *Geobacter sulfurreducens* and *Wolinella succinogenes*. *Arch. Microbiol.* 178:53–58
38. Kato S, Hashimoto K, Watanabe K. 2012. Methanogenesis facilitated by electric syntrophy via (semi)conductive iron-oxide minerals. *Environ. Microbiol.* 14:1646–54
39. Kato S, Kosaka T, Watanabe K. 2009. Substrate-dependent transcriptomic shifts in *Pelotomaculum thermopropionicum* grown in syntrophic co-culture with *Methanothermobacter thermautotrophicus*. *Microb. Biotechnol.* 2:575–84

40. Kendall MM, Liu Y, Boone DR. 2006. Butyrate- and propionate-degrading syntrophs from permanently cold marine sediments in Skan Bay, Alaska, and description of *Algorimarina butyrica* gen. nov., sp. nov. *FEMS Microbiol. Lett.* 262:107–14
41. Kleinsteuber S, Schleinitz KM, Breitfeld J, Harms H, Richnow H-H, Vogt C. 2008. Molecular characterization of bacterial communities mineralizing benzene under sulfate-reducing conditions. *FEMS Microbiol. Ecol.* 66:143–57
42. **Kosaka T, Kato S, Shimoyama T, Ishii S, Abe T, Watanabe K. 2008. The genome of *Pelotomaculum thermopropionicum* reveals niche-associated evolution in anaerobic microbiota. *Genome Res.* 18:442–48**

42. Suggests that microorganisms capable of syntrophic metabolism evolved as metabolite specialists by interacting with their partners.

43. Kosaka T, Uchiyama T, Ishii S, Enoki M, Imachi H, et al. 2006. Reconstruction and regulation of the central catabolic pathway in the thermophilic propionate-oxidizing syntroph *Pelotomaculum thermopropionicum*. *J. Bacteriol.* 188:202
44. Krumholz LR, Bryant MP. 1986. *Syntrophococcus sucromutans* sp. nov. gen. nov. uses carbohydrates as electron donors and formate, methoxymonobenzenoids or *Methanobrevibacter* as electron acceptor systems. *Arch. Microbiol.* 143:313–18
45. Laban NA, Selesi D, Jobelius C, Meckenstock RU. 2009. Anaerobic benzene degradation by gram-positive sulfate-reducing bacteria. *FEMS Microbiol. Ecol.* 68:300–11
46. Leang C, Qian X, Mester T, Lovley DR. 2010. Alignment of the c-type cytochrome OmcS along pili of *Geobacter sulfurreducens*. *Appl. Environ. Microbiol.* 76:4080–84
47. Lee MJ, Zinder SH. 1988. Isolation and characterization of a thermophilic bacterium which oxidizes acetate in syntrophic association with a methanogen and which grows acetogenically on H_2-CO_2. *Appl. Environ. Microbiol.* 54:124–29
48. **Li F, Hinderberger J, Seedorf H, Zhang J, Buckel W, Thauer RK. 2008. Coupled ferredoxin and crotonyl coenzyme A (CoA) reduction with NADH catalyzed by the butyryl-CoA dehydrogenase/Etf complex from *Clostridium kluyveri*. *J. Bacteriol.* 190:843–50**

48. First documentation of anaerobic, electron bifurcating enzyme reaction in which the favorable reduction of crotonyl-CoA by NADH drives the unfavorable reduction of ferredoxin by NADH.

49. Li X, Luo Q, Wofford NQ, Keller KL, McInerney MJ, et al. 2009. A molybdopterin oxidoreductase is involved in H_2 oxidation in *Desulfovibrio desulfuricans* G20. *J. Bacteriol.* 191:2675–82
50. Li X, McInerney MJ, Stahl DA, Krumholz LR. 2011. Metabolism of H_2 by *Desulfovibrio alaskensis* G20 during syntrophic growth on lactate. *Microbiology* 157:2912–21
51. Liu F, Conrad R. 2010. *Thermoanaerobacteriaceae* oxidize acetate in methanogenic rice field soil at 50°C. *Environ. Microbiol.* 12:2341–54
52. Liu Y, Balkwill DL, Aldrich HC, Drake GR, Boone DR. 1999. Characterization of the anaerobic propionate-degrading syntrophs *Smithella propionica* gen. nov., sp. nov. and *Syntrophobacter wolinii*. *Int. J. Syst. Bacteriol.* 49:545–56
53. Lykidis A, Chen CL, Tringe SG, McHardy AC, Copeland A, et al. 2011. Multiple syntrophic interactions in a terephthalate-degrading methanogenic consortium. *ISME J.* 5:122–30
54. Mayumi D, Mochimaru H, Yoshioka H, Sakata S, Maeda H, et al. 2010. Evidence for syntrophic acetate oxidation coupled to hydrogenotrophic methanogenesis in the high-temperature petroleum reservoir of Yabase oil field (Japan). *Environ. Microbiol.* 13:1995–2006
55. McInerney MJ, Bryant MP. 1981. Basic principles of anaerobic degradation and methane production. In *Biomass Conversion Processes for Energy and Fuels*, ed. SS Sofer, OR Zaborsky, pp. 277–96. New York: Plenum
56. McInerney MJ, Bryant MP, Pfennig N. 1979. Anaerobic bacterium that degrades fatty acids in syntrophic association with methanogens. *Arch. Microbiol.* 122:129–35
57. **McInerney MJ, Rohlin L, Mouttaki H, Kim U, Krupp RS, et al. 2007. The genome of *Syntrophus aciditrophicus*: life at the thermodynamic limit of microbial growth. *Proc. Natl. Acad. Sci. USA* 104:7600–5**

57. Shows unexpected ways to degrade benzoate and generate hydrogen and formate by reverse electron transfer.

58. McInerney MJ, Sieber JR, Gunsalus R. 2011. Microbial syntrophy: ecosystem-level biochemical cooperation. *Microbe* 6:479–85
59. McInerney MJ, Sieber JR, Gunsalus RP. 2009. Syntrophy in anaerobic global carbon cycles. *Curr. Opin. Biotechnol.* 20:623–32

60. McInerney MJ, Struchtemeyer CG, Sieber J, Mouttaki H, Stams AJM, et al. 2008. Physiology, ecology, phylogeny, and genomics of microorganisms capable of syntrophic metabolism. *Ann. NY Acad. Sci.* 1125:58–72
61. Mehta T, Coppi MV, Childers SE, Lovley DR. 2005. Outer membrane c-type cytochromes required for Fe(III) and Mn(IV) oxide reduction in *Geobacter sulfurreducens*. *Appl. Environ. Microbiol.* 71:8634–41
62. Meyerdierks A, Kube M, Kostadinov I, Teeling H, Glockner FO, et al. 2010. Metagenome and mRNA expression analyses of anaerobic methanotrophic archaea of the ANME-1 group. *Environ. Microbiol.* 12:422–39
63. Morita M, Malvankar NS, Franks AE, Summers ZM, Giloteaux L, et al. 2011. Potential for direct interspecies electron transfer in methanogenic wastewater digester aggregates. *mBio* 2:e00159–11
64. Mouttaki H, Nanny MA, McInerney MJ. 2007. Cyclohexane carboxylate and benzoate formation from crotonate in *Syntrophus aciditrophicus*. *Appl. Environ. Microbiol.* 73:930–38
65. Mouttaki H, Nanny MA, McInerney MJ. 2008. Use of benzoate as an electron acceptor by *Syntrophus aciditrophicus* grown in pure culture with crotonate. *Environ. Microbiol.* 10:3265–74
66. Müller N, Schleheck D, Schink B. 2009. Involvement of NADH:acceptor oxidoreductase and butyryl coenzyme A dehydrogenase in reversed electron transport during syntrophic butyrate oxidation by *Syntrophomonas wolfei*. *J. Bacteriol.* 191:6167–77
67. Müller N, Worm P, Schink B, Stams AJM, Plugge CM. 2010. Syntrophic butyrate and propionate oxidation processes: from genomes to reaction mechanisms. *Environ. Microbiol. Rep* 2:489–99
68. Nielsen LP, Risgaard-Petersen N, Fossing H, Christensen PB, Sayama M. 2010. Electric currents couple spatially separated biogeochemical processes in marine sediment. *Nature* 463:1071–74
69. Nilsen RK, Torsvik T, Lien T. 1996. *Desulfotomaculum thermocisternum* sp. nov., a sulfate reducer isolated from a hot North Sea oil reservoir. *Int. J. Syst. Bacteriol.* 46:397–402
70. Norlund KL, Southam G, Tyliszczak T, Hu Y, Karunakaran C, et al. 2009. Microbial architecture of environmental sulfur processes: a novel syntrophic sulfur-metabolizing consortia. *Environ. Sci. Technol.* 43:8781–86
71. Orphan VJ, House CH, Hinrichs KU, McKeegan KD, DeLong EF. 2001. Methane-consuming archaea revealed by directly coupled isotopic and phylogenetic analysis. *Science* 293:484–87
72. Orphan VJ, House CH, Hinrichs KU, McKeegan KD, DeLong EF. 2002. Multiple archaeal groups mediate methane oxidation in anoxic cold seep sediments. *Proc. Natl. Acad. Sci. USA* 99:7663–68
73. Overmann J, Schubert K. 2002. Phototrophic consortia: model systems for symbiotic interrelations between prokaryotes. *Arch. Microbiol.* 177:201–8
74. Pelletier E, Kreimeyer A, Bocs S, Rouy Z, Gyapay G, et al. 2008. "*Candidatus* Cloacamonas acidaminovorans": genome sequence reconstruction provides a first glimpse of a new bacterial division. *J. Bacteriol.* 190:2572–79
75. Plugge CM, Balk M, Stams AJ. 2002. *Desulfotomaculum thermobenzoicum* subsp. *thermosyntrophicum* subsp. nov., a thermophilic, syntrophic, propionate-oxidizing, spore-forming bacterium. *Int. J. Syst. Evol. Microbiol.* 52:391–99
76. Plugge CM, Balk M, Zoetendal EG, Stams AJ. 2002. *Gelria glutamica* gen. nov., sp. nov., a thermophilic, obligately syntrophic, glutamate-degrading anaerobe. *Int. J. Syst. Evol. Microbiol.* 52:401–7
77. Plugge CM, Scholten JCM, Culley DE, Nie L, Brockman FJ, Zhang W. 2011. Global transcriptomics analysis of the *Desulfovibrio vulgaris* change from syntrophic growth with *Methanosarcina barkeri* to sulfidogenic metabolism. *Microbiology* 156:2746–56
78. Plugge CM, Stams AJ. 2002. Enrichment of thermophilic syntrophic anaerobic glutamate-degrading consortia using a dialysis membrane reactor. *Microb. Ecol.* 43:378–87
79. Plugge CM, van Leeuwen JM, Hummelen T, Balk M, Stams AJ. 2001. Elucidation of the pathways of catabolic glutamate conversion in three thermophilic anaerobic bacteria. *Arch. Microbiol.* 176:29–36
80. Qiu YL, Hanada S, Ohashi A, Harada H, Kamagata Y, Sekiguchi Y. 2008. *Syntrophorhabdus aromaticivorans* gen. nov., sp. nov., the first cultured anaerobe capable of degrading phenol into acetate in obligate syntrophic associations with a hydrogenotrophic methanogen. *Appl. Environ. Microbiol.* 74:2051–58
81. Qiu YL, Sekiguchi Y, Hanada S, Imachi H, Tseng IC, et al. 2006. *Pelotomaculum terephthalicum* sp. nov. and *Pelotomaculum isophthalicum* sp. nov.: two anaerobic bacteria that degrade phthalate isomers in syntrophic association with hydrogenotrophic methanogens. *Arch. Microbiol.* 185:172–82

82. Qiu YL, Sekiguchi Y, Imachi H, Kamagata Y, Tseng IC, et al. 2003. *Sporotomaculum syntrophicum* sp. nov., a novel anaerobic, syntrophic benzoate-degrading bacterium isolated from methanogenic sludge treating wastewater from terephthalate manufacturing. *Arch. Microbiol.* 179:242–49
83. Reguera G, McCarthy KD, Mehta T, Nicoll JS, Tuominen MT, Lovley DR. 2005. Extracellular electron transfer via microbial nanowires. *Nature* 435:1098–101
84. Roeder J, Schink B. 2009. Syntrophic degradation of cadaverine by a defined methanogenic coculture. *Appl. Environ. Microbiol.* 75:4821–28
85. Rosenthal AZ, Matson EG, Eldar A, Leadbetter JR. 2011. RNA-seq reveals cooperative metabolic interactions between two termite-gut spirochete species in co-culture. *ISME J.* 5:1133–42
86. Sato K, Nishina Y, Setoyama C, Miura R, Shiga K. 1999. Unusually high standard redox potential of acrylyl-CoA/propionyl-CoA couple among enoyl-CoA/acyl-CoA couples: a reason for the distinct metabolic pathway of propionyl-CoA from longer acyl-CoAs. *J. Bacteriol.* 126:668–75
87. Schink B. 1997. Energetics of syntrophic cooperation in methanogenic degradation. *Microbiol. Mol. Biol. Rev.* 61:262–80
88. Schink B, Stams AJM. 2006. Syntrophism among prokaryotes. In *The Prokaryotes: An Evolving Electronic Resource for the Microbiological Community*, ed. M Dworkin, S Falkow, E Rosenberg, KH Schleifer, E Stackebrandt, pp. 309–35. New York: Springer-Verlag
89. Schirawski J, Unden G. 1998. Menaquinone-dependent succinate dehydrogenase of bacteria catalyzes reversed electron transport driven by the proton potential. *Eur. J. Biochem.* 257:210–15
90. Schnurer A, Schink B, Svensson BH. 1996. *Clostridium ultunense* sp. nov., a mesophilic bacterium oxidizing acetate in syntrophic association with a hydrogenotrophic methanogenic bacterium. *Int. J. Syst. Bacteriol.* 46:1145–52
91. Scholten JC, Conrad R. 2000. Energetics of syntrophic propionate oxidation in defined batch and chemostat cocultures. *Appl. Environ. Microbiol.* 66:2934–42
92. Scholten JC, Culley DE, Brockman FJ, Wu G, Zhang W. 2007. Evolution of the syntrophic interaction between *Desulfovibrio vulgaris* and *Methanosarcina barkeri*: involvement of an ancient horizontal gene transfer. *Biochem. Biophys. Res. Commun.* 352:48–54
93. **Schut GJ, Adams MW. 2009. The iron-hydrogenase of *Thermotoga maritima* utilizes ferredoxin and NADH synergistically: a new perspective on anaerobic hydrogen production. *J. Bacteriol.* 191:4451–57**
94. Sekiguchi Y, Imachi H, Susilorukmi A, Muramatsu M, Ohashi A, et al. 2006. *Tepidanaerobacter syntrophicus* gen. nov., sp. nov., an anaerobic, moderately thermophilic, syntrophic alcohol- and lactate-degrading bacterium isolated from thermophilic digested sludges. *Int. J. Syst. Evol. Microbiol.* 56:1621–29
95. Shi L, Richardson D, Wang Z, Kerisit S, Rosso K, et al. 2009. The roles of outer membrane cytochromes of *Shewanella* and *Geobacter* in extracellular electron transfer. *Environ. Microbiol. Rep.* 1:220–27
96. **Shimoyama T, Kato S, Ishii S, Watanabe K. 2009. Flagellum mediates symbiosis. *Science* 323:1574**
97. Sieber JR, Sims DR, Han C, Kim E, Lykidis A, et al. 2010. The genome of *Syntrophomonas wolfei*: new insights into syntrophic metabolism and biohydrogen production. *Environ. Microbiol.* 12:2289–301
98. Sobieraj M, Boone DR. 2006. Syntrophomonadaceae. In *The Prokaryotes: An Evolving Electronic Resource for the Microbiological Community*, ed. M Dworkin, S Falkow, E Rosenberg, KH Schleifer, E Stackebrandt, pp. 1041–46. New York: Springer-Verlag
99. Sousa DZ, Alves JI, Alves MM, Smidt H, Stams AJ. 2009. Effect of sulfate on methanogenic communities that degrade unsaturated and saturated long-chain fatty acids (LCFA). *Environ. Microbiol.* 11:68–80
100. Sousa DZ, Smidt H, Alves MM, Stams AJ. 2007. *Syntrophomonas zehnderi* sp. nov., an anaerobe that degrades long-chain fatty acids in co-culture with *Methanobacterium formicicum*. *Int. J. Syst. Evol. Microbiol.* 57:609–15
101. Stamatakis A, Hoover P, Rougemont J. 2008. A rapid bootstrap algorithm for the RAxML Web servers. *Syst. Biol.* 57:758–71
102. Struchtemeyer CG, Duncan KE, McInerney MJ. 2011. Evidence for syntrophic butyrate metabolism under sulfate-reducing conditions in a hydrocarbon-contaminated aquifer. *FEMS Microbiol. Ecol.* 76:289–300
103. Summers ZM, Fogarty HE, Leang C, Franks AE, Malvankar NS, Lovley DR. 2010. Direct exchange of electrons within aggregates of an evolved syntrophic coculture of anaerobic bacteria. *Science* 330:1413–15

93. Discovery of a bifurcating hydrogenase that couples the energetically unfavorable formation of hydrogen from NADH with the energetically favorable formation of hydrogen from reduced ferredoxin.

96. Evidence for contact-dependent communication between syntrophic partners.

104. Takeda K, Akira S. 2004. Microbial recognition by Toll-like receptors. *J. Dermatol. Sci.* 34:73–82
105. Tamura K, Dudley J, Nei M, Kumar S. 2007. MEGA4: Molecular Evolutionary Genetics Analysis (MEGA) Software Version 4.0. *Mol. Biol. Evol.* 24:1596–99
106. Van Kuijk BL, Schlosser E, Stams AJ. 1998. Investigation of the fumarate metabolism of the syntrophic propionate-oxidizing bacterium strain MPOB. *Arch. Microbiol.* 169:346–52
107. **Walker CB, He Z, Yang ZK, Ringbauer JA Jr, He Q, et al. 2009. The electron transfer system of syntrophically grown *Desulfovibrio vulgaris*. *J. Bacteriol.* 191:5793–801**

> 107. Genetic and transcriptomic analyses delineate the electron transfer, carbon metabolism, and energy-conserving systems that are unique to syntrophic metabolism.

108. Wallrabenstein CS, Schink B. 1994. Evidence of reversed electron transport in syntrophic butyrate or benzoate oxidation by *Syntrophomonas wolfei* and *Syntrophus buswellii*. *Arch. Microbiol.* 162:136–42
109. Westerholm M, Roos S, Schnurer A. 2010. *Syntrophaceticus schinkii* gen. nov., sp. nov., an anaerobic, syntrophic acetate-oxidizing bacterium isolated from a mesophilic anaerobic filter. *FEMS Microbiol. Lett.* 309:100–4
110. Worm P, Stams AJM, Cheng X, Plugge CM. 2011. Growth- and substrate-dependent transcription of formate dehydrogenase and hydrogenase coding genes in *Syntrophobacter fumaroxidans* and *Methanospirillum hungatei*. *Microbiology* 157:280–89
111. **Wust PK, Horn MA, Drake HL. 2009. Trophic links between fermenters and methanogens in a moderately acidic fen soil. *Environ. Microbiol.* 11:1395–409**

> 111. Redefines the classical three-stage trophic scheme for anaerobic biodegradation and introduces the concept of intermediary ecosystem metabolism.

112. Zhilina TN, Zavarzina DG, Kolganova TV, Tourova TP, Zavarzin GA. 2005. "*Candidatus* Contubernalis alkalaceticum," an obligately syntrophic alkaliphilic bacterium capable of anaerobic acetate oxidation in a coculture with *Desulfonatronum cooperativum*. *Microbiology* 74:695–703
113. Zindel U, Freudenberg W, Rieth M, Andreesen JR, Schnell J, Widdel F. 1988. *Eubacterium acidaminophilum* sp. nov., a versatile amino acid-degrading anaerobe producing or utilizing H_2 or formate. *Arch. Microbiol.* 150:254–66

Structure and Regulation of the Type VI Secretion System

Julie M. Silverman,[1] Yannick R. Brunet,[2] Eric Cascales,[2,*] and Joseph D. Mougous[1,*]

[1]Department of Microbiology, University of Washington, Seattle, Washington 98195; email: mougous@u.washington.edu

[2]Institut de Microbiologie de la Méditerranée, CNRS UMR7255, Aix-Marseille Université, Marseille, 13402 France; email: cascales@imm.cnrs.fr

Keywords

macromolecular, regulation, bacteriophage, interbacterial interactions, host-cell interactions

Abstract

The type VI secretion system (T6SS) is a complex and widespread gram-negative bacterial export pathway with the capacity to translocate protein effectors into a diversity of target cell types. Current structural models of the T6SS indicate that the apparatus is composed of at least two complexes, a dynamic bacteriophage-like structure and a cell-envelope-spanning membrane-associated assembly. How these complexes interact to promote effector secretion and cell targeting remains a major question in the field. As a contact-dependent pathway with specific cellular targets, the T6SS is subject to tight regulation. Thus, the identification of regulatory elements that control T6S expression continues to shape our understanding of the environmental circumstances relevant to its function. This review discusses recent progress toward characterizing T6S structure and regulation.

*Authors contributed equally to the manuscript

Contents

INTRODUCTION	454
ARCHITECTURE OF THE TYPE VI SECRETION SYSTEM	455
Core and Accessory Components	455
Bacteriophage-Like Subunits	456
Membrane-Associated Components	459
INPUTS THAT MODULATE TYPE VI SECRETION EXPRESSION AND ACTIVITY	460
Environmental Signals	460
Bacteria-Derived Signals	464
Quorum Sensing	465

INTRODUCTION

The type VI secretion system (T6SS) is a recently discovered mechanism in gram-negative bacteria that targets proteins to both prokaryotic and eukaryotic cells (59, 102). Like the type III and IV secretion (T3S and T4S) pathways, T6S translocates substrates directly into recipient cells in a contact-dependent manner (3, 27). Also similar to these pathways, the T6 pathway accomplishes this feat using a complex machine—13 core subunits are essential for basic secretory functions of the apparatus, and additional components may be involved in translocation (16). Although clear functional parallels exist between T6 and other secretory pathways, with few exceptions the proteins that constitute the T6S apparatus are novel and cannot easily be assigned roles in the machine. Thus, determining the composition of the T6S apparatus, how its components interact, and how accessory factors modulate the function of the system, have been active areas of research over the past several years.

The components of a T6SS are generally encoded by a group of tightly clustered genes. Such genetic clusters average over 20 kb and are prevalent in proteobacterial genomes (16, 103). In many instances, multiple T6S gene clusters can be identified in a single organism. On the basis of sequence divergence, phenotypic profiles, and coordinate differential regulation, such systems do not appear redundant (43, 69, 103). Rather, the plasticity of T6S clusters among high taxonomic ranks suggests that evolutionarily divergent systems are frequently acquired horizontally. Recent analyses indicate that T6S gene clusters separate into at least five distinct phylogenetic groups (13, 16).

Secretion systems: mechanistically distinct pathways for passaging proteins through membranes

T6SS: type VI secretion system

T6S: type VI secretion

Hcp: hemolysin coregulated protein

VgrG: valine-glycine repeat protein G

A T6SS can be recognized at a functional level by the robust transport of two proteins to the milieu, hemolysin coregulated protein (Hcp) and valine-glycine repeat protein G (VgrG). These proteins demonstrate codependency for export and together constitute part of or, perhaps, the entire extracellular portion of the T6S apparatus (46, 53, 89, 121). Structurally, Hcp and VgrG resemble bacteriophage tail tube and tailspike proteins, respectively (61, 68, 88). The specific evolutionary mechanism driving this relatedness has not been investigated in detail; however, the proteins do not share significant primary sequence homology with their phage counterparts. Whether this structural relationship translates into functional similarity is also not known. Nonetheless, a popular model depicting the T6SS as an inverted bacteriophage-like structure on the bacterial cell surface has emerged from these observations (30, 51, 61).

Functionally, T6SSs separate into four categories: (*a*) bacterial cell targeting, (*b*) eukaryotic cell targeting, (*c*) bacterial and eukaryotic cell targeting, and (*d*) other. The last includes systems implicated in processes such as conjugation, gene regulation, and cellular adhesion (4, 31, 37, 112). Importantly, it remains to be determined whether these effects are direct or indirect consequences of T6S. Because the genetic factors underlying the functional diversity of T6S are not yet defined, and few systematic studies of its function have been reported, it remains plausible that the system is considerably more promiscuous than currently appreciated.

Although many T6SSs have been studied and ascribed functions, the effector proteins involved have been identified in only a small subset of T6SSs (53, 79, 89, 98, 106, 111). Yet even from this small sampling it is evident that the T6S apparatus can accommodate a structurally and functionally diverse array of substrates. Moreover, these substrates can be configured either as specialized domains fused to apparatus components or as more canonical, independently exported proteins. Genetic data suggest that the latter grouping of substrates transit the apparatus in a one-step mechanism that avoids periplasmic intermediates (98).

In this review, we focus on two rapidly advancing topics in the field of T6S: the structure of the secretory apparatus and the conditions and signals that influence its expression and activation. We make no claims to have been exhaustive and we apologize in advance to our colleagues whose work we have omitted.

Effector proteins: proteins secreted by bacteria that influence target cell physiology

ARCHITECTURE OF THE TYPE VI SECRETION SYSTEM

The T6SS is composed of a minimal set of 13 different subunits, which are thought to form the core apparatus (13, 24). Although there is not yet a high-resolution structure of the whole or subcomplexes of the T6SS, such as those available for the T3SS and T4SS (109, 114), a comprehensive model of T6SS assembly has recently emerged. Biochemical and structural studies support a general model in which T6SS components form two subassemblies. Aside from these core components, T6SS gene clusters usually encode accessory subunits. The function of most of these proteins is not yet known; however, several proteins are required for proper assembly or function of the apparatus, whereas others have been shown to trigger its timely assembly (6, 105). In this section, we summarize recent findings pertaining to the topology, structure, protein-protein interaction, and function of individual T6S machine components. On the basis of these data, we propose a structure-function model of the apparatus.

Core and Accessory Components

Systematic mutagenesis studies performed in *Edwardsiella tarda* and *Vibrio cholerae* have shown that each of the 13 conserved T6S genes is required for function (120, 121). On the basis of bioinformatic approaches, these genes can be classified into three categories. The first category includes genes encoding membrane-associated proteins, either integral membrane (TssL, TssM) or lipoproteins (TssJ) (**Figure 1**, highlighted in green). The second category of genes encodes proteins with relatedness to tailed bacteriophage components (Hcp, VgrG, TssB, TssC, TssE) (**Figure 1**, highlighted in pink). For the most part, this homology is apparent at the structural level, not at the sequence level. The last category contains proteins for which no function can be inferred from in silico analyses (TssA, TssF, TssG, TssK). Although we do not yet have a high-resolution structure of the T6SS, the machine can be viewed as an assemblage of two distinct substructures—a bacteriophage-like structure and a membrane complex—that interact to form an inverted bacteriophage-like structure anchored to the cell envelope (15, 61, 94).

In addition to these core components, accessory genes are usually associated with T6SS gene clusters. Because the bacteria that carry T6S gene clusters can be found in different environments and the function of T6S is highly versatile, these accessory proteins might be linked to regulation

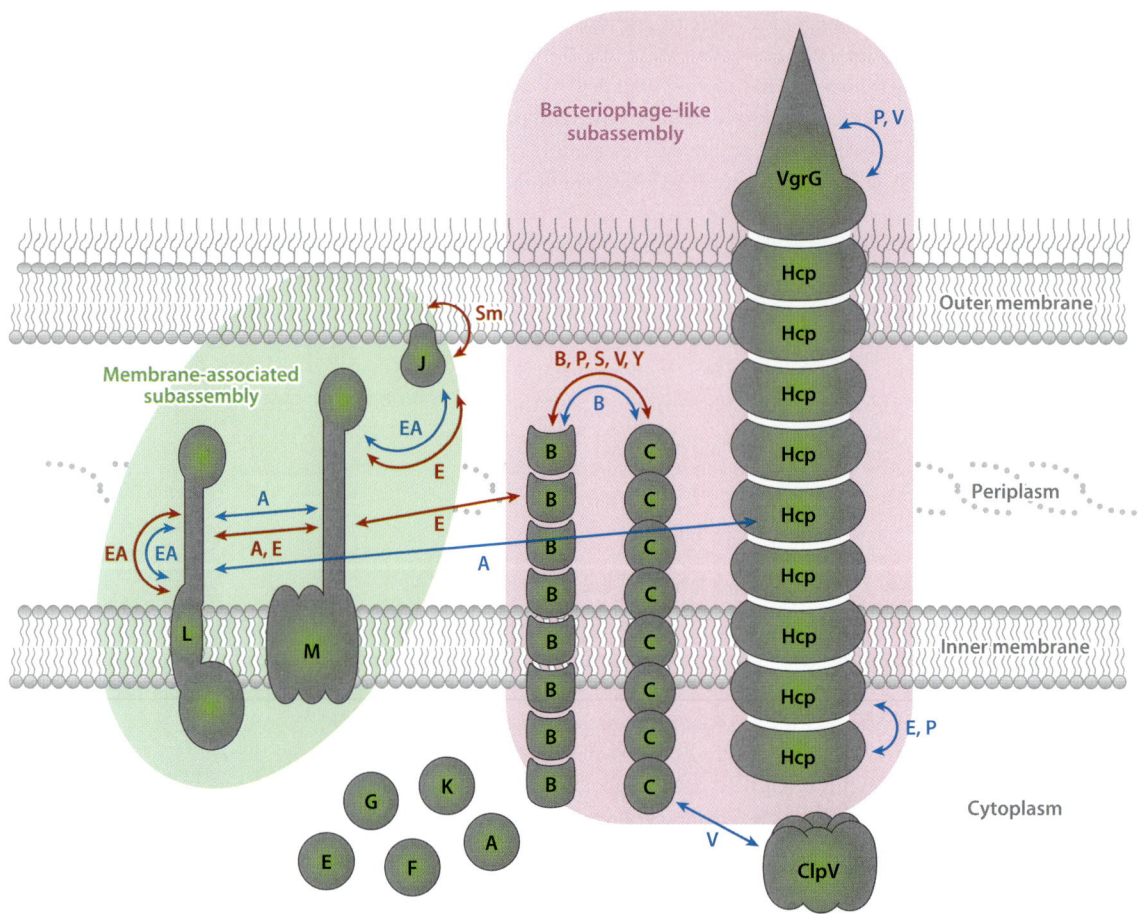

Figure 1

Protein interaction network between type VI secretion subunits. The localization and topologies of the core components of the T6SS are represented. Shapes containing letters indicate Tss designations of the subunits. Arrows indicate interactions detected among the subunits by biochemical/structural (*blue*) or two-hybrid approaches (*red*). The letter accompanying the arrow denotes the system where the interaction was detected. The membrane-associated subassembly and the bacteriophage-like subassembly are highlighted in green and pink, respectively. The question mark represents subunits for which the localization has not been investigated. Relevant studies are discussed in the text (8, 14, 19, 35, 39, 46, 60, 68, 74, 75, 81, 86, 89, 91, 121). Abbreviations: T6SS, type VI secretion system; A, *Agrobacterium tumefaciens*; B, *Burkholderia cenocepacia*; E, *Edwardsiella tarda*; EA, enteroaggregative *Escherichia coli*; Hcp, hemolysin coregulated protein; P, *Pseudomonas aeruginosa*; S, *Salmonella enterica*; Sm, *Serratia marcescens*; V, *Vibrio cholerae*; Y, *Yersinia pseudotuberculosis*.

or provide supplemental functions to the apparatus. For example, accessory components critical for transcriptional regulation or posttranslational activation of the system have been characterized (see below). Thus, even though these proteins are referred to as accessory, their functions can be essential for the proper production, assembly, or activity of the apparatus.

Bacteriophage-Like Subunits

Rapidly after the identification of the T6SS, it appeared that at least two subunits, Hcp and VgrG, share secondary structure similarity with bacteriophage components. These similarities were confirmed soon after by structural data, which demonstrated that the fold of Hcp is related to that of the major tail protein of phage λ, gpV (81, 88), and that VgrG assumes a fold and quaternary arrangement similar to the gp27/gp5 complex, the spike of bacteriophage T4 (62, 68).

During bacteriophage tail assembly, gpV oligomerizes into a filamentous tubular structure of specific length (1, 28, 63). X-ray crystal packing of Hcp homologs from *Pseudomonas aeruginosa* and *E. tarda* revealed that under certain conditions Hcp hexamers assemble in a head-to-head or head-to-tail fashion (60, 81, 86). Using disulfide bond engineering, Ballister et al. (9) reported that Hcp nanotubes of defined length can be constructed in vitro. Although the current data suggest that Hcp might form a tubular structure, it remains to be determined whether such a structure exists in vivo and whether its length is regulated.

In the bacteriophage T4, a sheath surrounds the tail filament. The sheath is a contractile structure composed of >100 gp18 protomers helically assembled (2) (**Figure 2a,b**). In 2009, a study reported that two T6S subunits, TssB and TssC, form tubular structures observable by electron microscopy. Image reconstruction of the particles showed that TssB and TssC assemble into structures that exhibit cogwheel-like cross sections resembling the bacteriophage sheath (14, 25). More recently, Basler et al. (10) demonstrated that TssB/TssC tubule structures exist in two conformations, likely corresponding to extended and contracted sheath-like structures. Time-lapse microscopy further showed that these tubules assemble dynamically in the cytoplasm and oscillate between the extended and contracted conformations through cycles of TssB/TssC assembly, disassembly, and recycling. Interestingly, the internal diameter of the TssB/TssC complex (∼10 nm) is sufficient to accommodate the Hcp tubule (external diameter of ∼9 nm) (14, 25). On the basis of these observations, it has been proposed that TssB/TssC subunits would form a structure enclosing a hypothetical Hcp tube. Similar to the bacteriophage T4 sheath, contraction of the TssB/TssC tubules might propel the Hcp tube toward the cell exterior prior to secretion (**Figure 2c,d**). This is in agreement with the observation that secretion by the T6S apparatus is a one-step mechanism, in which substrates are directly transported from the cytoplasm of the donor to the recipient cell without transient accumulation in the donor periplasm (98). Whether substrates are preloaded in the Hcp tube or transit into the tube upon target cell perforation remains to be answered. A specific interaction between the ClpV AAA$^+$-family ATPase and an N-terminal peptide of TssC has been defined, and ClpV disassembles TssB/TssC tubules in vitro (14). Given the findings that the purified tubules to which ClpV binds are in the contracted conformation and that TssB/TssC tubules are static in *clpV* muant cells, it has been proposed that ClpV is not responsible for TssB/TssC tubule assembly, but rather for recycling the TssB and TssC subunits upon contraction (10).

The VgrG protein forms a trimer that structurally resembles the (gp27/gp5)$_3$ complex of bacteriophage T4 (62, 68). The (gp27/gp5)$_3$ complex assembles at the tip of the tail, and the β-helical needle-like structure formed by the C-terminal domain of gp5 is used to perforate the outer membrane of the target bacteria (62). By analogy, it is supposed that VgrG is displayed at the tip of the Hcp tube and that the Hcp tube propels VgrG toward the target cell. This model is supported by several observations. First, Hcp is required for VgrG release in culture supernatants (46, 89, 121). Second, specialized VgrG proteins carry a C-terminal extension that possesses specific properties. The C-terminal domain is located downstream of the β-prism and therefore is the first domain to enter into the recipient target cell (**Figure 2**). Bioinformatic analyses predicted several activities such as fibronectin or peptidoglycan binding, general adhesion, and actin modification. Whereas most of these predictions remain to be experimentally tested, the *Vibrio cholerae* VgrG1 protein carries a domain that catalyzes actin cross-linking (73, 89), and a VgrG of *Aeromonas hydrophila* acts as an actin ADP-ribosylating enzyme (106).

On the basis of the phage model of the T6SS, one would predict a physical interaction between Hcp and VgrG. Although a direct interaction between the proteins has not been detected, several

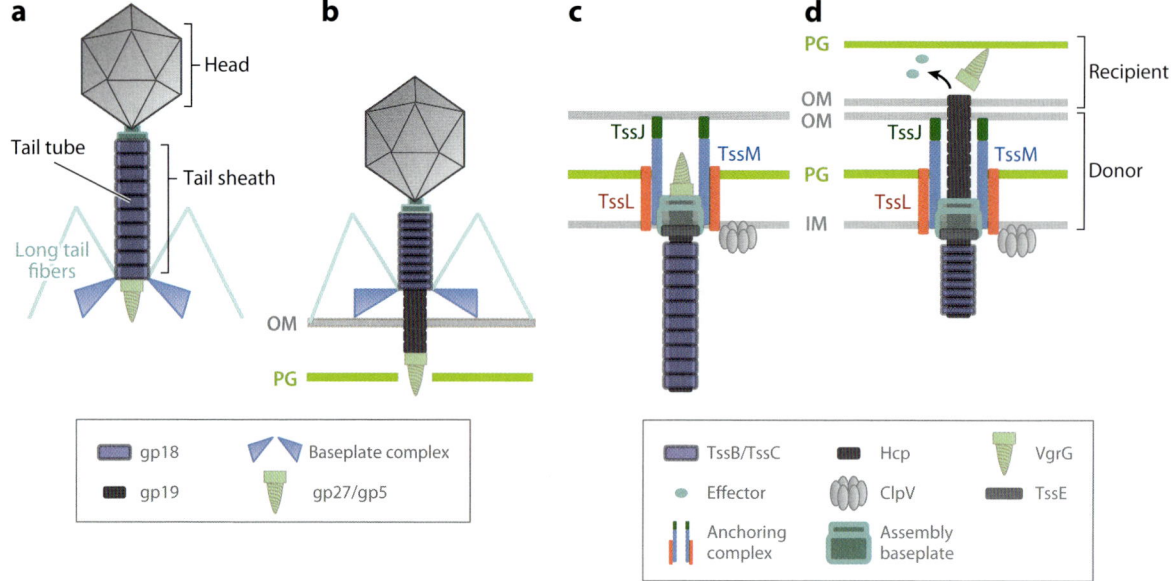

Figure 2

Schematic representation comparing proposed models of bacteriophage T4 and T6S. Homologous and analogous type VI secretion (T6S) are colored the same as their T4 phage counterparts. (*a*) The bacteriophage T4 tail tube is surrounded by the tail sheath and terminated by the cell-puncturing device (gp27/gp5). (*b*) Upon host cell binding, the bacteriophage T4 baseplate undergoes a conformational change that triggers tail sheath contraction and results in puncturing of the outer membrane (OM) and DNA delivery. Models of (*c*) inactivated and (*d*) activated states of T6S based on protein localization and interactions between T6S subunits. The three membrane-associated proteins TssL, TssM, and TssJ form a complex bound to the peptidoglycan (PG) layer via TssL. The T6S appendix formed by an Hcp tube and a VgrG trimer is thought to be anchored at the cell envelope by the membrane-associated complex. It has been hypothesized that an assembly baseplate can participate in T6S appendix assembly. (*c*) TssB and TssC may form a sheath-like structure enclosing the Hcp tube within the periplasmic space. (*d*) Activation of the T6SS results in effector delivery to a target cell through the Hcp tube. By analogy with bacteriophage T4, the sheath-like structure could propel, through contraction, the Hcp tube toward the cell exterior or directly to the target cell. Abbreviation: IM, inner membrane.

groups have reported that Hcp and VgrG exhibit codependency for secretion (53, 89, 90, 121). Given the relative positions of VgrG and Hcp in the proposed phage-like structure, this suggests that Hcp polymerization requires, or is triggered by, VgrG recruitment to the apparatus. Interestingly, such a mechanism is consistent with what has been described for the corresponding phage proteins during particle morphogenesis (65, 78, 117).

Recent microscopy images obtained by Basler et al. (10) showed that the base of the cytoplasmic T6SS sheath appears to be connected to the inner membrane by a large complex that may represent an assembly platform similar to the bacteriophage baseplate. The baseplate is responsible for proper assembly of the hub and tail tube, as well as for contraction of the tail sheath (97). Among the T6S core components, the cytoplasmic subunit, TssE, shares primary sequence homology with gp25, a structural component of the bacteriophage baseplate (13, 72). If the function of gp25 is conserved in the T6SS, TssE, in conjunction with a baseplate-like complex, might participate in assembly of the Hcp tube and the TssB/TssC sheath. In agreement with this hypothesis, the TssB/TssC tubules do not assemble in cells lacking *tssE* (10).

The current model argues that canonical protein substrates are transported through the Hcp tube. Whereas the internal diameter of the Hcp hexamer (∼4 nm) can accommodate a globular protein with a molecular mass less than approximately 50 kDa, the extremity of the VgrG β-prism is too narrow to allow substrate passage. One can speculate that the VgrG protein—as proposed for the bacteriophage T4—is released upon puncturing the target cell membrane, leaving the Hcp tube open and allowing delivery of substrates (**Figure 2**). Among the three characterized substrates of *P. aeruginosa*, Tse1 and Tse3 have peptidoglycan hydrolysis activities and are released in the periplasm of the bacterial target cell (98). By contrast, the biological target of Tse2, which is not yet identified, is located in the cytoplasm (53, 98).

EAEC: enteroaggregative *Escherichia coli*

Tag: type VI associated gene

PGB: peptidoglycan binding

TMS: transmembrane segment

Membrane-Associated Components

Recently, Aschtgen et al. (5) reported the isolation of a complex composed of four membrane proteins of the enteroaggregative *E. coli* (EAEC) Sci-1 T6SS: TssL, TssM, TssJ, and TagL. Further studies defined the localization and topology and delineated the interaction contacts of these different proteins. These and more recent biochemical and structural studies support a model in which these proteins form a complex that spans the cell envelope.

TagL is inserted into the inner membrane through three transmembrane segments (TMSs), and its large periplasmic domain carries a functional peptidoglycan-binding (PGB) motif. Mutagenesis studies have shown that TagL-mediated PGB is required for the activity of the T6SS; therefore, TagL may act as an anchor that tethers the apparatus to the cell wall (5). It is surprising that in the Sci-1 T6SS, the essential PGB activity is carried out by an accessory protein. Indeed, this is often not always the case, and the PGB domain is instead fused to TssL or other subunits (6).

TagL interacts directly with TssL, an inner membrane protein anchored through a single C-terminal TMS. The bulk of TssL localizes in the cytoplasm and adopts a unique fold resembling a hook (7, 35). TssL has strong homology with IcmH (or DotU), a component of the type IVB secretion system (T4bSS). In the T4bSS, IcmH interacts with IcmF to form a complex that stabilizes core components of the apparatus (83). In the T6SS, the IcmF homolog is TssM. TssM is anchored to the inner membrane through three TMSs. A cytoplasmic domain of ~30 kDa, which usually contains functional ATP-binding and hydrolysis Walker motifs, is located within the cytoplasmic loop flanked by TMS2 and TMS3. Mutagenesis studies have been performed to investigate the importance of the Walker motif. Interestingly, the outcome of these studies depends on the T6SS model. In *E. tarda*, the Walker motif is dispensable for T6SS assembly, as the secretion of Hcp and VgrG proteins is not affected by its inactivation (121). By contrast, in *Agrobacterium tumefaciens*, the Walker motif is critical for Hcp secretion (74). TssM undergoes a conformational change dependent on ATP binding and hydrolysis that allows for the recruitment of Hcp to the TssM/TssL complex (75). In several cases, Walker motifs cannot be identified in the cytoplasmic domain of TssM. It thus appears that TssM ATP binding and hydrolysis might be adapted to the specific needs of each T6SS.

The bulk of TssM (~80 kDa) is localized in the periplasm. It is composed of two subdomains: a large helical domain followed by a C-terminal β-domain (39). Yeast two-hybrid and coimmunoprecipitation experiments have shown that the periplasmic domain interacts with TssJ (39, 121). TssM associates with TssJ in a 1:1 stoichiometry and a K_d of 2–4 μM. Mapping studies showed that the TssM C-terminal β-domain is sufficient to interact with TssJ. TssJ is a lipoprotein that is anchored to the outer membrane through the acylation of its N-terminal cysteine residue (4). The crystal structures of the soluble domain of two TssJ proteins (from EAEC and *Serratia marcescens*) have been reported recently (39, 91). TssJ has a transthyretin fold (two parallel β-sheets) with additional elements, including a variable loop between β-strands 1 and 2 required for interaction with TssM. The conservation level of this loop is low, suggesting that it can act as a specificity determinant during assembly of the T6S apparatus (39). By virtue of interacting with TssL at the inner membrane and TssJ at the outer membrane, TssM spans the envelope. Rao et al. (91)

reported that the *S. marcescens* TssJ lipoprotein is engaged in homomeric interactions. Interestingly, the TssL protein also oligomerizes (35). It is therefore conceivable that the TssL-TssM-TssJ complex associates as a ring-like structure to shape a channel spanning the cell envelope, analogous to membrane-spanning structures that have been observed for the T3SS and T4SS (41, 114).

The observation that two distinct subassemblies exist in the T6SS raises the question of how these are connected. The periplasmic domain of the TssM protein interacts with TssB in a yeast two-hybrid assay, while Hcp coimmunoprecipitates with the periplasmic portion of the TssM-TssL complex (121, 75). These results are compatible with the current model in which the sheath structure surrounding the Hcp tube is embedded in a channel-like structure formed by the assembly of the periplasmic domains of TssM.

Ferric-uptake regulator (Fur): a conserved Fe(II)-dependent regulator of gene expression

INPUTS THAT MODULATE TYPE VI SECRETION EXPRESSION AND ACTIVITY

Even though our understanding of T6S function has grown considerably in recent years, its role in nature remains unclear. For instance, many bacteria of significant health concern possess one or more T6SSs; however, the pathogenic relevance of most of these systems is unknown. The majority of organisms with T6SSs are not pathogens and instead are found in marine environments, the rhizosphere, and soil, or they are associated with higher organisms as symbionts or commensals (13, 16). Defining the signals and conditions that control the expression and activation of T6S under the highly varied environments such bacteria occupy will be critical for revealing its role in these contexts. Below we provide examples of regulatory pathways and signals that modulate T6S expression. We discuss how environmental cues that influence these pathways might provide insights into the physiological function of the system. Our review of T6S regulation is not comprehensive; therefore, we refer readers interested in further details to more exhaustive recent reviews (12, 70).

Environmental Signals

Recent findings have demonstrated various classes of regulators sensitive to environmental cues that specifically modulate the activity of the T6SS. Further defining the signals that stimulate T6SSs is essential for understanding the physiological context in which these systems act.

Iron. The ferric-uptake regulator (Fur) protein is a key modulator of iron-dependent gene expression in bacteria (23). This regulator generally represses transcription through Fe(II)-dependent dimerization and subsequent DNA binding to a consensus sequence called the Fur box located within promoter regions. Transcriptional activation and iron-independent regulation by Fur also occur (22, 47). In addition to its important role in regulating iron acquisition and homeostasis, Fur regulates genes and processes that are not directly involved in iron metabolism. Some examples of these include toxins, adhesins, motility, and resistance to reactive oxygen species (21–23, 50, 93).

Expression of T6S in two opportunistic enteric pathogens, *E. tarda* and EAEC, is repressed directly at the transcriptional level by Fur (20, 26) (**Figure 3**). *E. tarda* is primarily a pathogen of fish; however, consumption of contaminated seafood can lead to gastroenteritis in humans (100). Infection models suggest that T6S plays an important role in the virulence of *E. tarda* and its close relative, *Edwardsiella ictaluri*, against fish (80, 96, 121). Deletion of genes encoding core components of the *E. tarda* Evp (*E. tarda* virulence proteins) T6SS, or insertional disruption of a gene encoding a putative effector of this system, *evpP*, attenuated the organism approximately 100-fold in blue gourami.

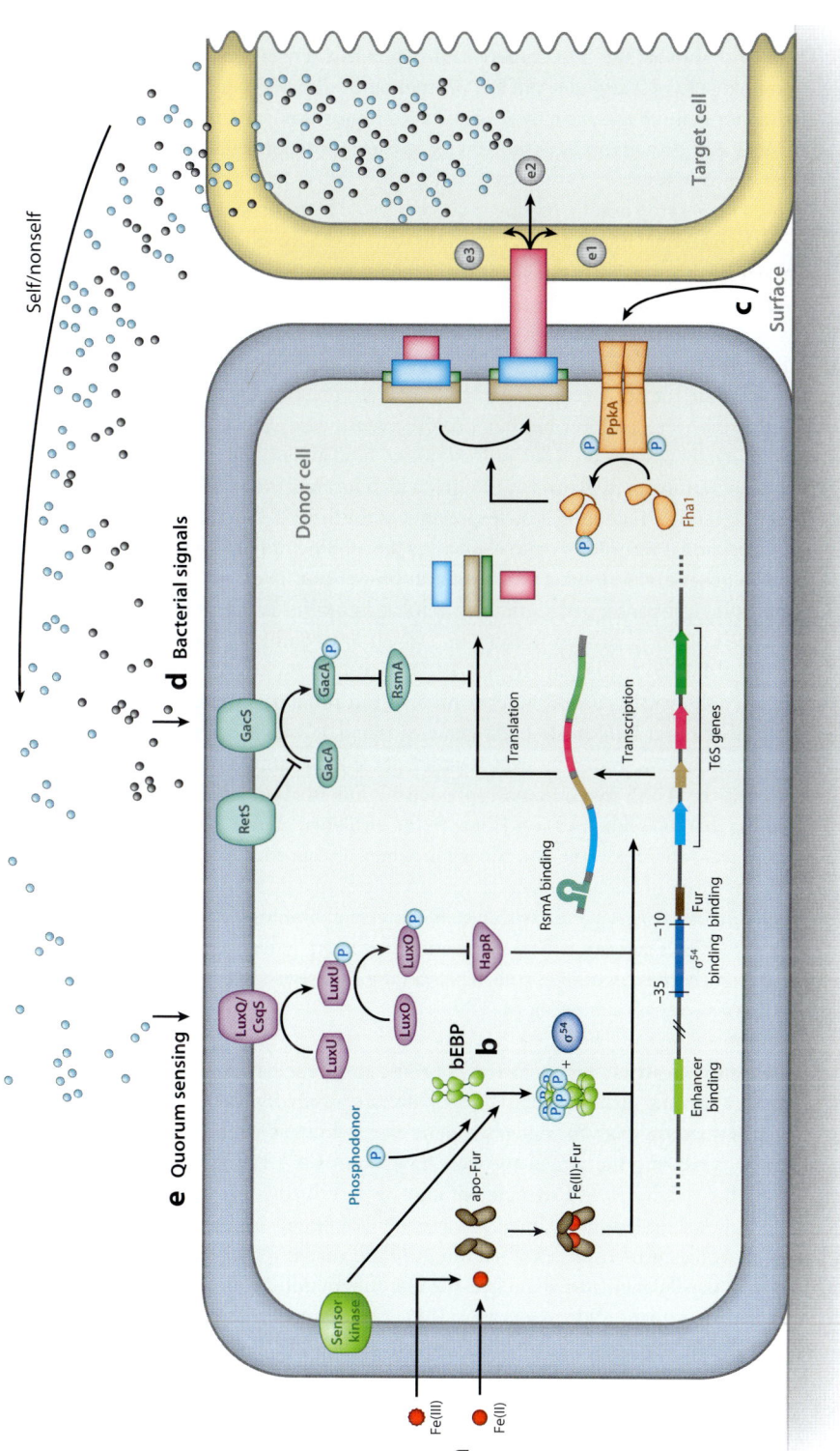

Figure 3

Schematic representation of the diverse regulatory systems that modulate T6S expression and activation in assorted bacteria. Only those regulatory pathways emphasized in this review are depicted. Pathways are labeled *a–e* corresponding to the order of their presentation in the text. (*a*) Fur represses T6S transcription in the presence of iron. (*b*) bEBPs function in conjunction with σ^{54} to activate T6S transcription. (*c*) The TPP posttranslationally activates T6S in response to surface association. Self- and nonself-derived bacterial signals modulate T6S (*d*) posttranscriptionally through the Gac/Rsm pathway or (*e*) transcriptionally via quorum sensing. At right is a target bacterium undergoing intoxication by Tse1–3 effectors (e1–3). Abbreviations: TPP, threonine phosphorylation pathway; bEBP, bacterial enhancer binding protein; T6S, type VI secretion.

σ^{54}: a sigma factor that promotes a stable, closed RNAP complex requiring activators to open; also known as RpoN

bEBP: bacterial enhancer-binding protein

Fur-dependent regulation of the Evp T6SS was recently demonstrated by Mok and colleagues (26). Their work showed that Fur confers iron-dependent repression of production and export of an Hcp homolog (EvpC) and that the Fur protein binds directly to a Fur box sequence upstream of *evpP*, the first gene in the *evp* cluster. Fur-based repression of the *evp* genes is consistent with the contribution of the system to pathogenesis, as Fur repression would likely be alleviated inside iron-depleted host tissues.

EAEC is an emerging enteric pathogen characterized by its propensity to self-adhere and form biofilms on the intestinal mucosa (84). Infections with EAEC result in diarrhea and can be acute or chronic in nature; it is currently the second most common cause of diarrhea in persons traveling to developing countries (48, 107). The Sci-1 T6SS of EAEC is required for biofilm formation and is regulated by iron availability through a pathway involving DNA adenine methyltransferase (Dam)-catalyzed methylation and Fur repression (20). Two Fur boxes and three Dam methylation sites are present upstream of the Sci-1 gene cluster. Interestingly, one of the Fur-binding sites overlaps with a Dam methylation site, and Fur binding prohibits methylase access to the site. In the absence of iron, Fur dissociates and allows RNA polymerase to bind and initiate transcription. Similarly, the loss of Fur also permits methylation at the site, which inhibits the reassociation of Fur. Thus, low cytoplasmic Fe(II) yields stable on-state expression of the Sci-1 T6SS. It is not currently understood how *sci-1* expression is returned to the off-state, as the binding of Fur to hemimethylated sequences that would be generated following DNA replication was not investigated (20).

The physiological consequences of *sci-1* regulation by Fur are not yet known and depend on the abundance and form of iron present in a given environment. The anaerobic environment of the intestinal lumen favors the ferrous [Fe(II)] form of the ion. In support of this, studies of *Salmonella* have shown that genes under control of Fur remain repressed prior to tissue invasion (55, 58). Furthermore, *Salmonella* pathogenicity island I, which encodes a T3SS required for *Salmonella* invasion, is activated by iron-bound Fur (36). Extending these findings to EAEC Sci-1 suggests that the T6SS may remain repressed by Fur in vivo. This is congruent with studies demonstrating that T6S does not contribute to the virulence of EAEC in animal infection models (4, 34). However, an important consideration when interpreting these data is that the animal models employed are unlikely to accurately recapitulate chronic EAEC infection. It is conceivable that ferrous iron becomes depleted within stable intestinal biofilm communities of EAEC, thereby leading to Sci-1 T6SS activation. Sci-1 activation mediated by Fur depression is likely to occur in an environmental context, where its role in promoting adhesion could be exploited as an adaptation to oxidative stress or iron starvation (4, 32).

σ^{54}-dependent activators. Sigma factor 54 (σ^{54})-dependent activator proteins, also termed bacterial enhancer-binding proteins (bEBPs), are a diverse group of proteins that mediate the translation of environmental signals to changes in gene expression in bacteria (92). bEBPs regulate gene expression by catalyzing the closed-to-open transition of σ^{54}-RNAP holoenzyme transcription complexes. This activity is ATP-dependent and, as described below, can be regulated by signal binding and phosphorylation. bEBPs are generally composed of three domains: an N-terminal regulatory domain, an internal AAA$^+$-family ATPase domain, and a C-terminal DNA-binding domain that typically mediates sequence-specific interactions with activator sequences 100 to 200 nucleotides upstream of the promoter (101). The N-terminal activation domain and the C-terminal DNA-binding domain are highly variable among bEBP homologs (104). This low conservation is partly responsible for the diversity of signals bEBP homologs detect and the distinct DNA sequences they bind.

A broad range of cellular processes, including nitrogen assimilation, motility, and virulence, are regulated by bEBPs (104). This class of proteins also serves a general role in the regulation of

T6S (**Figure 3**). Bioinformatic analyses identified bEBPs encoded within phylogenetically diverse T6S gene clusters (11, 16). These analyses have further revealed probable σ^{54}-binding sequences within the respective T6S promoter regions (11). A subset of T6S-associated bEBPs, including representatives from *V. cholerae*, *A. hydrophila*, *Pectobacterium atrosepticum*, and *Marinomonas* spp., were investigated using in vitro binding experiments and reconstituted transcriptional reporter assays. These studies confirmed the predicted role of the proteins in σ^{54}-dependent transcriptional activation of T6S promoters.

TPP: threonine phosphorylation pathway

The bEBP encoded within the virulence-associated secretion (*vas*) gene cluster, VasH, is a key regulator of *V. cholerae* T6S (66, 90). Reflecting its important regulatory role, T6S-dependent defense against *Dictyostelium discoideum* and killing of *E. coli* in this organism require VasH (76, 89). An analysis of *vasH* homologs in 26 *V. cholerae* strains identified an enrichment of nonsynonymous single nucleotide polymorphisms in the N-terminal regulatory domain (66). Amino acid changes within this domain could provide a mechanism for variable T6S expression response profiles within *V. cholerae* strains exposed to environmental signals. Interestingly, N-terminal domains of bEBPs can act as intramolecular activators or repressors (115). In *V. cholerae* V52, this domain appears to be a positive regulator, as *vas* T6S expression was decreased by an N-terminally truncated VasH allele (66). Notably, polymorphisms in *vasH* do not appear to underlie the significant differences in *vas* expression and activation observed between non-O1/non-O139 strains (e.g., V52) and pandemic disease strains (e.g., N16961, C6706, and A1552). Instead, this variability is explained likely by differences in the strength of repression through quorum-based signaling and by the novel regulator TsrA (type VI secretion system regulator A) (57, 122).

The activity of bEBPs can be regulated by specific phosphorylation events catalyzed by sensor histidine kinases and low-molecular-weight phosphodonors (77, 104). An example is the well-characterized bEBP NtrC (nitrogen regulatory protein C), which is activated through phosphorylation by NtrB (113). In addition, NtrC efficiently autophosphorylates in vitro in the presence of acetyl phosphate, and evidence suggests that acetyl phosphate is also relevant to its signaling properties in vivo (40). A sensor kinase that acts on T6S bEBPs has not been identified; however, these proteins autophosphorylate in vitro in the presence of acetyl phosphate (11). This finding suggests that upstream events such as environmental sampling by sensor kinases or changes in metabolism leading to the accumulation of low-molecular-weight phosphodonors could have a significant impact on σ^{54}-dependent T6S expression.

Surface association. Surface association can promote dramatic changes to bacterial cell physiology. An analysis of global gene expression changes in *Salmonella typhimurium* demonstrated that one-third of its genes are altered during surface-growth conditions (110). Such changes may be caused by signaling systems that directly detect surfaces or that respond to concomitant alterations in the local environment. Cells adhered to surfaces can develop into sessile communities, sometimes referred to as biofilms, in which long-term cell-cell contact is enhanced relative to planktonic cells (29). According to the current understanding of T6S-mediated effects, these conditions are favorable to its contact-dependent mechanism of effector delivery.

The HSI-I (Hcp secretion island-I)-encoded T6SS (H1-T6SS) of *P. aeruginosa* is posttranslationally activated by a threonine phosphorylation pathway (TPP) in response to surface growth of the organism (105) (**Figure 3**). Components of this pathway include PpkA, an inner-membrane-spanning serine/threonine kinase, and Fha1, a forkhead-associated domain-containing protein that is activated by PpkA via phosphorylation (54, 82). Activated Fha1 promotes H1-T6S-apparatus assembly and effector secretion. Unlike planktonically grown *P. aeruginosa*, strains placed on an agar surface for 4 h assemble an activated apparatus and contain elevated levels of phosphorylated Fha1 (105). Genetic analysis of the requirements for competitive fitness mediated by the H1-T6SS further established the role of the TPP in surface-dependent H1-T6SS activation and ruled out the involvement of a second phosphorylation-independent pathway. Surface activation of T6S by the TPP may be a general phenomenon, as components of the pathway are found in approximately 30% of identified T6S gene clusters (16).

Though the mechanism(s) is not yet clear, sessile growth may also serve as a cue for regulatory changes that elevate T6S expression. Recent findings demonstrated that cellular levels of TssC1, an H1-T6S component, are elevated in a biofilm compared with planktonically grown cells (118). Consistent with this observation, the protein Hcp is abundant in P. aeruginosa biofilms (99).

Bacteria-Derived Signals

Often functioning as a cell contact-dependent bacterial interaction pathway, it is not surprising that the sensing of other bacteria plays an important role in modulating T6S activity. Here we describe several examples of bacteria-derived signals that influence T6S expression.

The Gac/Rsm pathway. The Gac/Rsm signaling pathway couples extracellular bacteria-derived signals with marked changes in target mRNA translation (67). The pathway is initiated by the GacS/GacA two-component system, which upon stimulation leads to elevated expression of one or more small regulatory RNA (sRNA) molecules, variably termed *rsmB*, *rsmX*, *rsmY*, *rsmZ*, *csrB*, or *csrC*. These sRNA molecules interact with and sequester an mRNA-binding protein known as RsmA or CsrA. This protein generally acts as an inhibitor of translation by associating with sequences near or overlapping the ribosome-binding site. Thus, sRNA expression typically facilitates increased translation of specific mRNA targets. In some cases, mRNA binding by RsmA/CsrA can activate translation; however, the mechanism is less clear in these instances.

In the pseudomonads, the Gac/Rsm pathway is a key regulator of many important processes, including biocontrol and virulence factor production, cellular aggregation, and quorum sensing (44). Studies have revealed that P. aeruginosa, P. fluorescens, and P. syringae also use this pathway to regulate the expression of T6S (49, 81, 95) (**Figure 3**). This was first noted in P. aeruginosa, wherein microarray analyses of strains lacking *retS* or *ladS*, which encode a repressor and activator of GacA/S signaling, respectively, strongly implicated RsmA in the stability of HSI-I transcripts (45, 108). Later functional analyses demonstrated that assembly, activation, and effector secretion and targeting by the H1-T6SS are stimulated in the Δ*retS* background (53, 81, 82). The most definitive evidence for the involvement of the Gac/Rsm pathway in HSI-I regulation is found in recent work by Brencic & Lory (17), which demonstrates direct RsmA binding to the 5′ leader sequence of two HSI-I transcripts.

The incorporation of the H1-T6SS into the global regulon of the P. aeruginosa Gac/Rsm pathway has yielded valuable insights into the settings relevant to its function. In P. aeruginosa, the Gac/Rsm pathway directly or indirectly regulates the expression of approximately 500 genes (18, 45). Within this expansive set, researchers have noted reciprocal regulation of factors associated with planktonic and sessile growth—leading to the hypothesis that the Gac/Rsm pathway coordinates this physiological transition of the organism (116). In the Gac/Rsm regulon, H1-T6SS expression occurs coincident with factors having experimentally demonstrable roles in sessile community formation, such as two aggregation and adhesion-promoting exopolysaccharides, Pel and Psl. This early regulatory link implied that T6SS activity is relevant to closely interfacing bacteria; however, in what capacity remained unknown. This question has been answered—at least in part—by more recent studies demonstrating the role of the H1-T6SS in contact-dependent interbacterial interactions and the involvement of the system in biofilm-specific antibiotic resistance

(53, 98, 118). Notably, the Gac/Rsm pathway of *P. fluorescens*, which is closely related to that of *P. aeruginosa*, responds to signals generated by other pseudomonads and certain *Vibrio* spp. (33).

Quorum Sensing

Quorum sensing is a bacterial regulatory mechanism that modulates gene expression based on cell population density (42, 85). A quorum is sensed by the accumulation of diffusible signaling molecules, which are themselves typically under quorum control. Regulation of quorum-controlled genes is achieved either by direct or indirect effects that signal molecules impart on the DNA-binding properties of dedicated regulatory proteins. A prevailing model is that quorum sensing regulates social behavior of bacteria, both within and between species (87). One piece of evidence in support of this model is that secreted products are disproportionally abundant in the quorum regulon of many species (52). Given this, it is not surprising that many instances of quorum-sensing-regulated T6SSs have arisen in the literature (43, 56, 64, 69, 71, 119, 122).

In *V. cholerae*, two chemically distinct quorum-sensing systems, autoinducer-2 (AI-2) and cholera autoinducer-1 (CAI-1), collaborate to influence density-dependent gene expression (85) (**Figure 3**). Signal molecules from these pathways are detected by two sensor kinases, LuxQ and CqsS, respectively. The pathways converge on the phosphotransfer protein LuxU, which acts on LuxO, a DNA-binding response regulator protein. Phosphorylated LuxO activates the expression of sRNA molecules (*qrr1–4*) that in turn repress the production of HapR, a TetR-family global transcriptional regulator.

The majority of studies on *V. cholerae* T6S have been conducted in the serotype O37 strain V52. The V52 strain is a valuable model for studying T6S within the species, as Hcp and VgrG proteins are abundantly exported from the strain, its *vas* genes are highly expressed, and it exhibits strong Vas-dependent phenotypes against prokaryotic and eukaryotic cells (76, 89, 90). However, under similar in vitro conditions, most strains of *V. cholerae*, including O1 and O139 pandemic strains, display markedly lower *vas* expression and as a result do not exhibit Vas-dependent phenotypes (56, 57, 122).

Recent studies have suggested that differences in direct quorum-sensing-dependent expression of *vas* genes are partially responsible for T6S variability in *V. cholerae*. Ishikawa et al. (56) observed a striking correlation between HapR and Hcp expression among a panel of O1 isolates. Furthermore, a deletion of *luxO* was shown to strongly induce *vas* expression in two serotype O1 strains, A1552 and C6707 (56, 122). Consistent with current models of quorum-sensing circuitry in *V. cholerae*, activation of *vas* expression in Δ*luxO* required *hapR* and Δ*hfq* recapitulated the effects of the *luxO* deletion.

Despite robust *vas* expression in O1 strains lacking *luxO*, the secretion system can remain functionally quiescent in this background (122). Thus, levels of HapR do not fully reconcile T6S-related phenotypic differences observed between *V. cholerae* strains. Briefly, complete activation of T6S in pandemic *V. cholerae* strains appears to involve additional factors such as high osmolarity, low temperature, and relief of repression imposed on the system by the TsrA protein (57, 122).

The adaptive role of quorum control over T6S remains to be elucidated. In addition to its role in intraspecies sensing, quorum sensing might play a role in perceiving cells of other species (38). This seems particularly probable for the AI-2 pathway, as the dedicated signal synthase involved in AI-2 synthesis, LuxS, is widely conserved. Therefore, AI-2 signal levels could serve as a cue for *V. cholerae* to activate antibacterial defenses, such as its T6SS, in response to potential competitors. If the signal was self-derived, coregulated immunity proteins might ameliorate the detrimental consequences of self-targeting (98).

Not all T6SSs regulated by quorum sensing are induced at high cell densities. For example, in *P. aeruginosa* the H1-T6SS is repressed at high cell densities by a direct or indirect mechanism involving LasR, an acyl homoserine lactone-type quorum regulator (69). Interestingly, the two other T6SSs of *P. aeruginosa* are regulated reciprocally with the H1-T6SS by quorum sensing. The two T6SSs of *Vibrio parahaemolyticus* also display reciprocal regulation by quorum sensing (43). Differential regulation of T6S by quorum sensing, particularly those cases wherein this occurs within one bacterium, suggests that the system can act in a wide range of contexts and underscores its functional versatility.

SUMMARY POINTS

1. The T6SS is a multicomponent secretory machine that delivers effector proteins to both prokaryotic and eukaryotic cells in a contact-dependent manner.
2. The T6SS is composed of five bacteriophage-like proteins that likely form a cell-puncturing device and participate in apparatus assembly.
3. Interactions between four essential membrane-associated proteins suggest a cell-envelope-spanning complex is central to T6S function.
4. Expression and activation of T6S are tightly controlled by diverse regulatory systems.
5. Fur is an iron-responsive regulator that directly represses T6S expression in EAEC and *E. tarda*.
6. σ^{54}-dependent T6S expression is often controlled by bEBPs encoded within T6S gene clusters.
7. T6S accessory genes involved in signaling via threonine phosphorylation control post-translational activation of T6S in response to surface association.
8. Self and nonself signals influence expression of T6S through quorum sensing and, in the case of pseudomonads, the Gac/Rsm pathway.

FUTURE ISSUES

1. How similar is the mechanism of T6S effector secretion to bacteriophage DNA delivery to target cells?
2. What is the overall architecture of the T6SS? Can the entire apparatus be isolated in vitro or visualized in vivo?
3. How did two phylogenetically unrelated complexes, bacteriophage-like and bacterial membrane-associated, evolve to form the T6SS? How do the two subassemblies of the T6SS interact with each other?
4. What are the physiological relevant targets of T6SSs? Defining the regulatory networks that influence T6SS expression and the environmental context in which these systems are functional will advance our understanding of the role of T6SSs.
5. What are the upstream regulators and corresponding signals that activate dedicated T6S-associated regulators and regulatory systems such as bEBPs and the TPP?
6. What role does competitor sensing play in bacterial cell-targeting T6SSs? Is specificity determined by regulatory factors, effectors, or surface receptors?

DISCLOSURE STATEMENT

The authors are not aware of any affiliations, memberships, funding, or financial holdings that might be perceived as affecting the objectivity of this review.

ACKNOWLEDGMENTS

J.M.S. was supported in part by Public Health Service, National Research Service Award T32 GM07270, from the National Institute of General Medical Services. J.D.M. was funded by grants from the NIH (AI080609 and AI057141). J.D.M. holds an Investigator in the Pathogenesis of Infectious Disease Award from the Burroughs Wellcome Fund. Work in the E.C. laboratory is supported by the CNRS and funded by the Agence Nationale de la Recherche (ANR-10-JCJC-1303-03). Y.R.B. is supported by a doctoral fellowship from the French Ministry of Research. We thank members of the Mougous and Cascales laboratories and Pacome-Aude Lavieille for helpful discussions.

LITERATURE CITED

1. Abuladze NK, Gingery M, Tsai J, Eiserling FA. 1994. Tail length determination in bacteriophage T4. *Virology* 199:301–10
2. Aksyuk AA, Leiman PG, Kurochkina LP, Shneider MM, Kostyuchenko VA, et al. 2009. The tail sheath structure of bacteriophage T4: a molecular machine for infecting bacteria. *EMBO J.* 28:821–29
3. Alvarez-Martinez CE, Christie PJ. 2009. Biological diversity of prokaryotic type IV secretion systems. *Microbiol. Mol. Biol. Rev.* 73:775–808
4. Aschtgen MS, Bernard CS, De Bentzmann S, Lloubes R, Cascales E. 2008. SciN is an outer membrane lipoprotein required for type VI secretion in enteroaggregative *Escherichia coli*. *J. Bacteriol.* 190:7523–31
5. **Aschtgen MS, Gavioli M, Dessen A, Lloubes R, Cascales E. 2010. The SciZ protein anchors the enteroaggregative *Escherichia coli* type VI secretion system to the cell wall. *Mol. Microbiol.* 75:886–99**

    5. Provides evidence that four T6S proteins assemble a membrane-associated complex.

6. Aschtgen MS, Thomas MS, Cascales E. 2010. Anchoring the type VI secretion system to the peptidoglycan: TssL, TagL, TagP... what else? *Virulence* 1:535–40
7. Aschtgen MS, Zoued A, Lloubès R, Journet L, Cascales E. 2012. The C-tail anchored TssL subunit, an essential protein of the enteroaggregative *Escherichia coli* Sci-1 Type VI secretion system, is inserted by YidC. *Microbiol. Open* 1:71–82
8. Aubert D, MacDonald DK, Valvano MA. 2010. BcsKC is an essential protein for the type VI secretion system activity in *Burkholderia cenocepacia* that forms an outer membrane complex with BcsLB. *J. Biol. Chem.* 285:35988–98
9. Ballister ER, Lai AH, Zuckermann RN, Cheng Y, Mougous JD. 2008. In vitro self-assembly of tailorable nanotubes from a simple protein building block. *Proc. Natl. Acad. Sci. USA* 105:3733–38
10. **Basler M, Pilhofer M, Henderson GP, Jensen GJ, Mekalanos JJ. 2012. Type VI secretion requires a dynamic contractile phage tail-like structure. *Nature* 483:182–86**

    10. Describes a dynamic tubular structure associated with the T6S apparatus.

11. Bernard CS, Brunet YR, Gavioli M, Lloubes R, Cascales E. 2011. Regulation of type VI secretion gene clusters by σ^{54} and cognate enhancer binding proteins. *J. Bacteriol.* 193:2158–67
12. Bernard CS, Brunet YR, Gueguen E, Cascales E. 2010. Nooks and crannies in type VI secretion regulation. *J. Bacteriol.* 192:3850–60
13. Bingle LE, Bailey CM, Pallen MJ. 2008. Type VI secretion: a beginner's guide. *Curr. Opin. Microbiol.* 11:3–8
14. Bonemann G, Pietrosiuk A, Diemand A, Zentgraf H, Mogk A. 2009. Remodelling of VipA/VipB tubules by ClpV-mediated threading is crucial for type VI protein secretion. *EMBO J.* 28:315–25
15. Bonemann G, Pietrosiuk A, Mogk A. 2010. Tubules and donuts: a type VI secretion story. *Mol. Microbiol.* 76:815–21

16. Boyer F, Fichant G, Berthod J, Vandenbrouck Y, Attree I. 2009. Dissecting the bacterial type VI secretion system by a genome wide in silico analysis: What can be learned from available microbial genomic resources? *BMC Genomics* 10:104

17. Brencic A, Lory S. 2009. Determination of the regulon and identification of novel mRNA targets of *Pseudomonas aeruginosa* RsmA. *Mol. Microbiol.* 72:612–32

17. Shows that T6S in *P. aeruginosa* is directly regulated by the Gac/Rsm pathway.

18. Brencic A, McFarland KA, McManus HR, Castang S, Mogno I, et al. 2009. The GacS/GacA signal transduction system of *Pseudomonas aeruginosa* acts exclusively through its control over the transcription of the RsmY and RsmZ regulatory small RNAs. *Mol. Microbiol.* 73:434–45

19. Bröms JE, Lavander M, Sjöstedt A. 2009. A conserved α-helix essential for type VI secretion-like system on *Francisella tularensis*. *J. Bacteriol.* 8:2431–46

20. Brunet YR, Bernard CS, Gavioli M, Lloubes R, Cascales E. 2011. An epigenetic switch involving overlapping Fur and DNA methylation optimizes expression of a type VI secretion gene cluster. *PLoS Genet.* 7:e1002205

20. Demonstrates direct regulation of T6S by Fur and DNA methylation in EAEC.

21. Calderwood SB, Mekalanos JJ. 1987. Iron regulation of Shiga-like toxin expression in *Escherichia coli* is mediated by the fur locus. *J. Bacteriol.* 169:4759–64

22. Campoy S, Jara M, Busquets N, de Rozas AM, Badiola I, Barbe J. 2002. Intracellular cyclic AMP concentration is decreased in *Salmonella typhimurium fur* mutants. *Microbiology* 148:1039–48

23. Carpenter BM, Whitmire JM, Merrell DS. 2009. This is not your mother's repressor: the complex role of fur in pathogenesis. *Infect. Immun.* 77:2590–601

24. Cascales E. 2008. The type VI secretion toolkit. *EMBO Rep.* 9:735–41

25. Cascales E, Cambillau C. 2012. Structural biology of type VI secretion systems. *Philos. Trans. R. Soc. Lond. B Biol. Sci.* 367:1102–11

26. Chakraborty S, Sivaraman J, Leung KY, Mok YK. 2011. Two-component PhoB-PhoR regulatory system and ferric uptake regulator sense phosphate and iron to control virulence genes in type III and VI secretion systems of *Edwardsiella tarda*. *J. Biol. Chem.* 286:39417–30

27. Cornelis GR. 2006. The type III secretion injectisome. *Nat. Rev. Microbiol.* 4:811–25

28. Cornelis GR, Agrain C, Sorg I. 2006. Length control of extended protein structures in bacteria and bacteriophages. *Curr. Opin. Microbiol.* 9:201–6

29. Costerton JW, Stewart PS, Greenberg EP. 1999. Bacterial biofilms: a common cause of persistent infections. *Science* 284:1318–22

30. Cotter P. 2011. Microbiology: Molecular syringes scratch the surface. *Nature* 475:301–3

31. Das S, Chakrabortty A, Banerjee R, Chaudhuri K. 2002. Involvement of in vivo induced *icmF* gene of *Vibrio cholerae* in motility, adherence to epithelial cells, and conjugation frequency. *Biochem. Biophys. Res. Commun.* 295:922–28

32. de Pace F, Boldrin de Paiva J, Nakazato G, Lancellotti M, Sircili MP, et al. 2011. Characterization of IcmF of the type VI secretion system in an avian pathogenic *Escherichia coli* (APEC) strain. *Microbiology* 157:2954–62

33. Dubuis C, Haas D. 2007. Cross-species GacA-controlled induction of antibiosis in pseudomonads. *Appl. Environ. Microbiol.* 73:650–54

34. Dudley EG, Thomson NR, Parkhill J, Morin NP, Nataro JP. 2006. Proteomic and microarray characterization of the AggR regulon identifies a *pheU* pathogenicity island in enteroaggregative *Escherichia coli*. *Mol. Microbiol.* 61:1267–82

35. Durand E, Zoued A, Spinelli S, Watson PJ, Aschtgen MS, et al. 2012. Structural characterization and oligomerization of the TssL protein, a component shared by bacterial type VI and type IVb secretion systems. *J. Biol. Chem.* 287:14157–68

36. Ellermeier JR, Slauch JM. 2008. Fur regulates expression of the *Salmonella* pathogenicity island 1 type III secretion system through HilD. *J. Bacteriol.* 190:476–86

37. Enos-Berlage JL, Guvener ZT, Keenan CE, McCarter LL. 2005. Genetic determinants of biofilm development of opaque and translucent *Vibrio parahaemolyticus*. *Mol. Microbiol.* 55:1160–82

38. Federle MJ, Bassler BL. 2003. Interspecies communication in bacteria. *J. Clin. Invest.* 112:1291–99

39. Felisberto-Rodrigues C, Durand E, Aschtgen MS, Blangy S, Ortiz-Lombardia M, et al. 2011. Towards a structural comprehension of bacterial type VI secretion systems: characterization of the TssJ-TssM complex of an *Escherichia coli* pathovar. *PLoS Pathog.* 7:e1002386

40. Feng J, Atkinson MR, McCleary W, Stock JB, Wanner BL, Ninfa AJ. 1992. Role of phosphorylated metabolic intermediates in the regulation of glutamine synthetase synthesis in *Escherichia coli*. *J. Bacteriol.* 174:6061–70
41. Fronzes R, Christie PJ, Waksman G. 2009. The structural biology of type IV secretion systems. *Nat. Rev. Microbiol.* 7:703–14
42. Fuqua C, Parsek MR, Greenberg EP. 2001. Regulation of gene expression by cell-to-cell communication: acyl-homoserine lactone quorum sensing. *Annu. Rev. Genet.* 35:439–68
43. Gode-Potratz CJ, McCarter LL. 2011. Quorum sensing and silencing in *Vibrio parahaemolyticus*. *J. Bacteriol.* 193:4224–37
44. Gooderham WJ, Hancock RE. 2009. Regulation of virulence and antibiotic resistance by two-component regulatory systems in *Pseudomonas aeruginosa*. *FEMS Microbiol. Rev.* 33:279–94
45. Goodman AL, Kulasekara B, Rietsch A, Boyd D, Smith RS, Lory S. 2004. A signaling network reciprocally regulates genes associated with acute infection and chronic persistence in *Pseudomonas aeruginosa*. *Dev. Cell* 7:745–54
46. Hachani A, Lossi NS, Hamilton A, Jones C, Bleves S, et al. 2011. Type VI secretion system in *Pseudomonas aeruginosa*: secretion and multimerization of VgrG proteins. *J. Biol. Chem.* 286:12317–27
47. Hall HK, Foster JW. 1996. The role of fur in the acid tolerance response of *Salmonella typhimurium* is physiologically and genetically separable from its role in iron acquisition. *J. Bacteriol.* 178:5683–91
48. Harrington SM, Dudley EG, Nataro JP. 2006. Pathogenesis of enteroaggregative *Escherichia coli* infection. *FEMS Microbiol. Lett.* 254:12–18
49. Hassan KA, Johnson A, Shaffer BT, Ren Q, Kidarsa TA, et al. 2010. Inactivation of the GacA response regulator in *Pseudomonas fluorescens* Pf-5 has far-reaching transcriptomic consequences. *Environ. Microbiol.* 12:899–915
50. Hassett DJ, Sokol PA, Howell ML, Ma JF, Schweizer HT, et al. 1996. Ferric uptake regulator (Fur) mutants of *Pseudomonas aeruginosa* demonstrate defective siderophore-mediated iron uptake, altered aerobic growth, and decreased superoxide dismutase and catalase activities. *J. Bacteriol.* 178:3996–4003
51. Hayes CS, Aoki SK, Low DA. 2010. Bacterial contact-dependent delivery systems. *Annu. Rev. Genet.* 44:71–90
52. Hense BA, Kuttler C, Muller J, Rothballer M, Hartmann A, Kreft JU. 2007. Does efficiency sensing unify diffusion and quorum sensing? *Nat. Rev. Microbiol.* 5:230–39
53. Hood RD, Singh P, Hsu F, Guvener T, Carl MA, et al. 2010. A type VI secretion system of *Pseudomonas aeruginosa* targets a toxin to bacteria. *Cell Host Microbe* 7:25–37

    53. Reports the finding that the T6SS can target bacteria.
54. Hsu F, Schwarz S, Mougous JD. 2009. TagR promotes PpkA-catalysed type VI secretion activation in *Pseudomonas aeruginosa*. *Mol. Microbiol.* 72:1111–25
55. Ikeda JS, Janakiraman A, Kehres DG, Maguire ME, Slauch JM. 2005. Transcriptional regulation of sitABCD of *Salmonella enterica* serovar Typhimurium by MntR and Fur. *J. Bacteriol.* 187:912–22
56. Ishikawa T, Rompikuntal PK, Lindmark B, Milton DL, Wai SN. 2009. Quorum sensing regulation of the two *hcp* alleles in *Vibrio cholerae* O1 strains. *PLoS One* 4:e6734

    56. Demonstrates *V. cholerae* O1 T6S is negatively regulated by quorum sensing.
57. Ishikawa T, Sabharwal D, Broms J, Milton DL, Sjostedt A, et al. 2012. Pathoadaptive conditional regulation of the type VI secretion system in *Vibrio cholerae* O1 strains. *Infect. Immun.* 80:575–84
58. Janakiraman A, Slauch JM. 2000. The putative iron transport system SitABCD encoded on SPI1 is required for full virulence of *Salmonella typhimurium*. *Mol. Microbiol.* 35:1146–55
59. Jani AJ, Cotter PA. 2010. Type VI secretion: not just for pathogenesis anymore. *Cell Host Microbe* 8:2–6
60. Jobichen C, Chakraborty S, Li M, Zheng J, Joseph L, et al. 2010. Structural basis for the secretion of EvpC: a key type VI secretion system protein from *Edwardsiella tarda*. *PLoS One* 5:e12910
61. Kanamaru S. 2009. Structural similarity of tailed phages and pathogenic bacterial secretion systems. *Proc. Natl. Acad. Sci. USA* 106:4067–68
62. Kanamaru S, Leiman PG, Kostyuchenko VA, Chipman PR, Mesyanzhinov VV, et al. 2002. Structure of the cell-puncturing device of bacteriophage T4. *Nature* 415:553–57
63. Katsura I, Hendrix RW. 1984. Length determination in bacteriophage lambda tails. *Cell* 39:691–98
64. Khajanchi BK, Sha J, Kozlova EV, Erova TE, Suarez G, et al. 2009. N-acylhomoserine lactones involved in quorum sensing control the type VI secretion system, biofilm formation, protease production, and in vivo virulence in a clinical isolate of *Aeromonas hydrophila*. *Microbiology* 155:3518–31

65. King J. 1971. Bacteriophage T4 tail assembly: four steps in core formation. *J. Mol. Biol.* 58:693–709
66. Kitaoka M, Miyata ST, Brooks TM, Unterweger D, Pukatzki S. 2011. VasH is a transcriptional regulator of the type VI secretion system functional in endemic and pandemic *Vibrio cholerae*. *J. Bacteriol.* 193:6471–82
67. Lapouge K, Schubert M, Allain FH, Haas D. 2008. Gac/Rsm signal transduction pathway of γ-proteobacteria: from RNA recognition to regulation of social behaviour. *Mol. Microbiol.* 67:241–53

> 68. Together with Reference 88, this study illustrates the structural conservation between T6SS and phage-tail components.

68. **Leiman PG, Basler M, Ramagopal UA, Bonanno JB, Sauder JM, et al. 2009. Type VI secretion apparatus and phage tail-associated protein complexes share a common evolutionary origin. *Proc. Natl. Acad. Sci. USA* 106:4154–59**
69. Lesic B, Starkey M, He J, Hazan R, Rahme LG. 2009. Quorum sensing differentially regulates *Pseudomonas aeruginosa* type VI secretion locus I and homologous loci II and III, which are required for pathogenesis. *Microbiology* 155:2845–55
70. Leung KY, Siame BA, Snowball H, Mok YK. 2011. Type VI secretion regulation: crosstalk and intracellular communication. *Curr. Opin. Microbiol.* 14:9–15
71. Liu H, Coulthurst SJ, Pritchard L, Hedley PE, Ravensdale M, et al. 2008. Quorum sensing coordinates brute force and stealth modes of infection in the plant pathogen *Pectobacterium atrosepticum*. *PLoS Pathog.* 4:e1000093
72. Lossi NS, Dajani R, Freemont P, Filloux A. 2011. Structure-function analysis of HsiF, a *gp25*-like component of the type VI secretion system, in *Pseudomonas aeruginosa*. *Microbiology* 157:3292–305
73. Ma AT, McAuley S, Pukatzki S, Mekalanos JJ. 2009. Translocation of a *Vibrio cholerae* type VI secretion effector requires bacterial endocytosis by host cells. *Cell Host Microbe* 5:234–43
74. Ma LS, Lin JS, Lai EM. 2009. An IcmF family protein, ImpLM, is an integral inner membrane protein interacting with ImpKL, and its Walker A motif is required for type VI secretion system-mediated Hcp secretion in *Agrobacterium tumefaciens*. *J. Bacteriol.* 191:4316–29
75. Ma LS, Narberhaus F, Lai EM. 2012. IcmF family protein TssM exhibits ATPase activity and energizes type VI secretion. *J. Biol. Chem.* 287:15610–21
76. MacIntyre DL, Miyata ST, Kitaoka M, Pukatzki S. 2010. The *Vibrio cholerae* type VI secretion system displays antimicrobial properties. *Proc. Natl. Acad. Sci. USA* 107:19520–24
77. McCleary WR, Stock JB, Ninfa AJ. 1993. Is acetyl phosphate a global signal in *Escherichia coli*? *J. Bacteriol.* 175:2793–98
78. Meezan E, Wood WB. 1971. The sequence of gene product interaction in bacteriophage T4 tail core assembly. *J. Mol. Biol.* 58:685–92
79. Miyata ST, Kitaoka M, Brooks TM, McAuley SB, Pukatzki S. 2011. *Vibrio cholerae* requires the type VI secretion system virulence factor VasX to kill *Dictyostelium discoideum*. *Infect. Immun.* 79:2941–49
80. Moore MM, Fernandez DL, Thune RL. 2002. Cloning and characterization of Edwardsiella ictaluri proteins expressed and recognized by the channel catfish *Ictalurus punctatus* immune response during infection. *Dis. Aquat. Organ.* 52:93–107
81. Mougous JD, Cuff ME, Raunser S, Shen A, Zhou M, et al. 2006. A virulence locus of *Pseudomonas aeruginosa* encodes a protein secretion apparatus. *Science* 312:1526–30

> 82. Identifies T6S proteins that posttranslationally regulate the T6SS.

82. **Mougous JD, Gifford CA, Ramsdell TL, Mekalanos JJ. 2007. Threonine phosphorylation posttranslationally regulates protein secretion in *Pseudomonas aeruginosa*. *Nat. Cell Biol.* 9:797–803**
83. Nagai H, Kubori T. 2011. Type IVB secretion systems of *Legionella* and other gram-negative bacteria. *Front. Microbiol.* 2:136
84. Nataro JP, Kaper JB. 1998. Diarrheagenic *Escherichia coli*. *Clin. Microbiol. Rev.* 11:142–201
85. Ng WL, Bassler BL. 2009. Bacterial quorum-sensing network architectures. *Annu. Rev. Genet.* 43:197–222
86. Osipiuk J, Xu X, Cui H, Savchenko A, Edwards A, Joachimiak A. 2011. Crystal structure of secretory protein Hcp3 from *Pseudomonas aeruginosa*. *J. Struct. Funct. Genomics* 12:21–26
87. Parsek MR, Greenberg EP. 2005. Sociomicrobiology: the connections between quorum sensing and biofilms. *Trends Microbiol.* 13:27–33
88. Pell LG, Kanelis V, Donaldson LW, Howell PL, Davidson AR. 2009. The phage lambda major tail protein structure reveals a common evolution for long-tailed phages and the type VI bacterial secretion system. *Proc. Natl. Acad. Sci. USA* 106:4160–65

89. Pukatzki S, Ma AT, Revel AT, Sturtevant D, Mekalanos JJ. 2007. Type VI secretion system translocates a phage tail spike-like protein into target cells where it cross-links actin. *Proc. Natl. Acad. Sci. USA* 104:15508–13

    89. Describes the activity of a eukaryotic cell-targeting T6S effector.

90. Pukatzki S, Ma AT, Sturtevant D, Krastins B, Sarracino D, et al. 2006. Identification of a conserved bacterial protein secretion system in *Vibrio cholerae* using the *Dictyostelium* host model system. *Proc. Natl. Acad. Sci. USA* 103:1528–33
91. Rao VA, Shepherd SM, English G, Coulthurst SJ, Hunter WN. 2011. The structure of *Serratia marcescens* Lip, a membrane-bound component of the type VI secretion system. *Acta Crystallogr. D* 67:1065–72
92. Rappas M, Bose D, Zhang X. 2007. Bacterial enhancer-binding proteins: unlocking σ^{54}-dependent gene transcription. *Curr. Opin. Struct. Biol.* 17:110–16
93. Rashid RA, Tarr PI, Moseley SL. 2006. Expression of the *Escherichia coli* IrgA homolog adhesin is regulated by the ferric uptake regulation protein. *Microb. Pathog.* 41:207–17
94. Records AR. 2011. The type VI secretion system: a multipurpose delivery system with a phage-like machinery. *Mol. Plant Microbe Interact.* 24:751–57
95. Records AR, Gross DC. 2010. Sensor kinases RetS and LadS regulate *Pseudomonas syringae* type VI secretion and virulence factors. *J. Bacteriol.* 192:3584–96
96. Rogge ML, Thune RL. 2011. Regulation of the *Edwardsiella ictaluri* type III secretion system by pH and phosphate concentration through EsrA, EsrB, and EsrC. *Appl. Environ. Microbiol.* 77:4293–302
97. Rossmann MG, Mesyanzhinov VV, Arisaka F, Leiman PG. 2004. The bacteriophage T4 DNA injection machine. *Curr. Opin. Struct. Biol.* 14:171–80
98. **Russell AB, Hood RD, Bui NK, LeRoux M, Vollmer W, Mougous JD. 2011. Type VI secretion delivers bacteriolytic effectors to target cells. *Nature* 475:343–47**

    98. Describes the activity of cell wall–targeting antibacterial T6S effectors.

99. Sauer K, Camper AK, Ehrlich GD, Costerton JW, Davies DG. 2002. *Pseudomonas aeruginosa* displays multiple phenotypes during development as a biofilm. *J. Bacteriol.* 184:1140–54
100. Schlenker C, Surawicz CM. 2009. Emerging infections of the gastrointestinal tract. *Best Pract. Res. Clin. Gastroenterol.* 23:89–99
101. Schumacher J, Joly N, Rappas M, Zhang X, Buck M. 2006. Structures and organisation of AAA+ enhancer binding proteins in transcriptional activation. *J. Struct. Biol.* 156:190–99
102. Schwarz S, Hood RD, Mougous JD. 2010. What is type VI secretion doing in all those bugs? *Trends Microbiol.* 18:531–37
103. Schwarz S, West TE, Boyer F, Chiang WC, Carl MA, et al. 2010. *Burkholderia* type VI secretion systems have distinct roles in eukaryotic and bacterial cell interactions. *PLoS Pathog.* 6(8):e1001068
104. Shingler V. 1996. Signal sensing by σ^{54}-dependent regulators: derepression as a control mechanism. *Mol. Microbiol.* 19:409–16
105. Silverman JM, Austin LS, Hsu F, Hicks KG, Hood RD, Mougous JD. 2011. Separate inputs modulate phosphorylation-dependent and -independent type VI secretion activation. *Mol. Microbiol.* 82:1277–90
106. Suarez G, Sierra JC, Erova TE, Sha J, Horneman AJ, Chopra AK. 2010. A type VI secretion system effector protein VgrG1 from *Aeromonas hydrophila* that induces host cell toxicity by ADP-ribosylation of actin. *J. Bacteriol.* 192:155–68
107. Taylor DN, Bourgeois AL, Ericsson CD, Steffen R, Jiang ZD, et al. 2006. A randomized, double-blind, multicenter study of rifaximin compared with placebo and with ciprofloxacin in the treatment of travelers' diarrhea. *Am. J. Trop. Med. Hyg.* 74:1060–66
108. Ventre I, Goodman AL, Vallet-Gely I, Vasseur P, Soscia C, et al. 2006. Multiple sensors control reciprocal expression of *Pseudomonas aeruginosa* regulatory RNA and virulence genes. *Proc. Natl. Acad. Sci. USA* 103:171–76
109. Waksman G, Fronzes R. 2010. Molecular architecture of bacterial type IV secretion systems. *Trends Biochem. Sci.* 35:691–98
110. Wang Q, Frye JG, McClelland M, Harshey RM. 2004. Gene expression patterns during swarming in *Salmonella typhimurium*: genes specific to surface growth and putative new motility and pathogenicity genes. *Mol. Microbiol.* 52:169–87
111. Wang X, Wang Q, Xiao J, Liu Q, Wu H, et al. 2009. *Edwardsiella tarda* T6SS component *evpP* is regulated by *esrB* and iron, and plays essential roles in the invasion of fish. *Fish Shellfish Immunol.* 27:469–77

112. Weber B, Hasic M, Chen C, Wai SN, Milton DL. 2009. Type VI secretion modulates quorum sensing and stress response in *Vibrio anguillarum*. *Environ. Microbiol.* 11:3018–28
113. Weiss V, Magasanik B. 1988. Phosphorylation of nitrogen regulator I (NRI) of *Escherichia coli*. *Proc. Natl. Acad. Sci. USA* 85:8919–23
114. Worrall LJ, Lameignere E, Strynadka NC. 2011. Structural overview of the bacterial injectisome. *Curr. Opin. Microbiol.* 14:3–8
115. Xu H, Hoover TR. 2001. Transcriptional regulation at a distance in bacteria. *Curr. Opin. Microbiol.* 4:138–44
116. Yahr TL, Greenberg EP. 2004. The genetic basis for the commitment to chronic versus acute infection in *Pseudomonas aeruginosa*. *Mol. Cell* 16:497–98
117. Yap ML, Mio K, Ali S, Minton A, Kanamaru S, Arisaka F. 2010. Sequential assembly of the wedge of the baseplate of phage T4 in the presence and absence of *gp11* as monitored by analytical ultracentrifugation. *Macromol. Biosci.* 10:808–13
118. Zhang L, Hinz AJ, Nadeau JP, Mah TF. 2011. *Pseudomonas aeruginosa tssC1* links type VI secretion and biofilm-specific antibiotic resistance. *J. Bacteriol.* 193:5510–13
119. Zhang W, Xu S, Li J, Shen X, Wang Y, Yuan Z. 2011. Modulation of a thermoregulated type VI secretion system by AHL-dependent quorum sensing in *Yersinia pseudotuberculosis*. *Arch. Microbiol.* 193:351–63
120. Zheng J, Ho B, Mekalanos JJ. 2011. Genetic analysis of anti-amoebae and anti-bacterial activities of the type VI secretion system in *Vibrio cholerae*. *PLoS One* 6:e23876
121. Zheng J, Leung KY. 2007. Dissection of a type VI secretion system in *Edwardsiella tarda*. *Mol. Microbiol.* 66:1192–206
122. Zheng J, Shin OS, Cameron DE, Mekalanos JJ. 2010. Quorum sensing and a global regulator TsrA control expression of type VI secretion and virulence in *Vibrio cholerae*. *Proc. Natl. Acad. Sci. USA* 107:21128–33

Network News: The Replication of Kinetoplast DNA

Robert E. Jensen[1] and Paul T. Englund[2]

[1]Department of Cell Biology and [2]Department of Biological Chemistry, The Johns Hopkins University School of Medicine, Baltimore, Maryland 21205; email: robjensen@jhmi.edu, penglund@jhmi.edu

Keywords

trypanosomes, minicircle, maxicircle, kDNA replication, kDNA segregation

Abstract

One of the most fascinating and unusual features of trypanosomatids, parasites that cause disease in many tropical countries, is their mitochondrial DNA. This genome, known as kinetoplast DNA (kDNA), is organized as a single, massive DNA network formed of interlocked DNA rings. In this review, we discuss recent studies on kDNA structure and replication, emphasizing recent developments on replication enzymes, how the timing of kDNA synthesis is controlled during the cell cycle, and the machinery for segregating daughter networks after replication.

Contents

INTRODUCTION.. 474
    What Is kDNA?.. 475
    Why Is kDNA a Network?.. 477
KINETOPLAST DNA REPLICATION..................................... 478
    Highlights of kDNA Replication....................................... 478
    Two Different Mechanisms of kDNA Replication................ 478
    Ring Mechanism.. 479
    Polar Mechanism... 480
    kDNA Division... 480
    How Is the Minicircle Repertoire Preserved During Replication
        and Segregation?... 481
PROTEINS AND ENZYMES INVOLVED IN kDNA REPLICATION.......... 481
    Universal Minicircle Sequence Binding Protein................. 481
    p38 and p93.. 482
    Mitochondrial Topo II... 482
    Mitochondrial DNA Polymerases.................................... 482
    Mitochondrial PIF Helicases.. 483
    Mitochondrial Primases... 483
CONTROL OF KINETOPLAST DNA REPLICATION..................... 483
    Redox Regulation of UMSBP Binding.............................. 484
    HslVU Protease... 484
    mRNA Cycling... 484
THE ROLE OF THE TRIPARTITE ATTACHMENT COMPLEX
    IN KINETOPLAST DNA SEGREGATION............................ 484
    p166.. 485
    Mab22... 487
    AEP-1.. 487

INTRODUCTION

Trypanosomatids are protozoan parasites that cause several human and animal diseases. They are best known as agents of human African trypanosomiasis (sleeping sickness), Chagas disease, and leishmaniasis, all of which strike primarily in underdeveloped countries. These parasites are early-branching eukaryotes, and their ancient lineage reveals some unusual biological properties that are irresistible to biologists. One of their most striking features is their mitochondrial DNA, called kinetoplast DNA (kDNA). Members of our laboratory wrote two earlier reviews on kDNA in this series in 1984 and 1995 (59, 66). After providing background information, we emphasize some exciting developments since the last review. In some cases, we stress earlier work if it raises important unsolved questions. Many early biochemical studies utilized the insect parasite *Crithidia fasciculata* because large quantities could be grown inexpensively. Most recent studies focus on *Trypanosoma brucei* because of the available genetic tools, such as knockdown of gene function by RNA interference (RNAi). In addition, *T. brucei* goes through a complex life cycle in both the insect and mammalian hosts. Two of these stages, the procyclic form (which occurs in the insect

Kinetoplast DNA (kDNA): mitochondrial genome of the trypanosome; composed of a giant network of many interlocked maxicircles and minicircles

midgut) and the mammalian bloodstream form, which differ morphologically and metabolically, are easily cultured in the lab. For other reviews on kDNA, see References 33 and 67.

What Is kDNA?

A kDNA network consists of thousands of interlocked DNA rings of two types, minicircles and maxicircles. A *C. fasciculata* network has about 5,000 minicircles (each ~2.5 kb) and ~25 maxicircles (each ~39 kb) (16, 41). Isolated networks expand into an elliptical planar or cup-shaped structure, with the major axis ~15 μm and the minor axis ~10 μm. *T. brucei* networks are smaller, with ~1-kb minicircles and ~23-kb maxicircles. Studies with *Trypanosoma equiperdum*, a close relative of *T. brucei*, show not only that maxicircles are linked to minicircles, but that all the maxicircles

Minicircle: small (0.5 to 2.5 kbp) DNA ring that encodes most of the guide RNAs that control the specificity for editing maxicircle transcripts

Maxicircle: large (20 to 40 kbp) DNA ring that encodes rRNAs and some subunits of the mitochondrial bioenergetics machinery

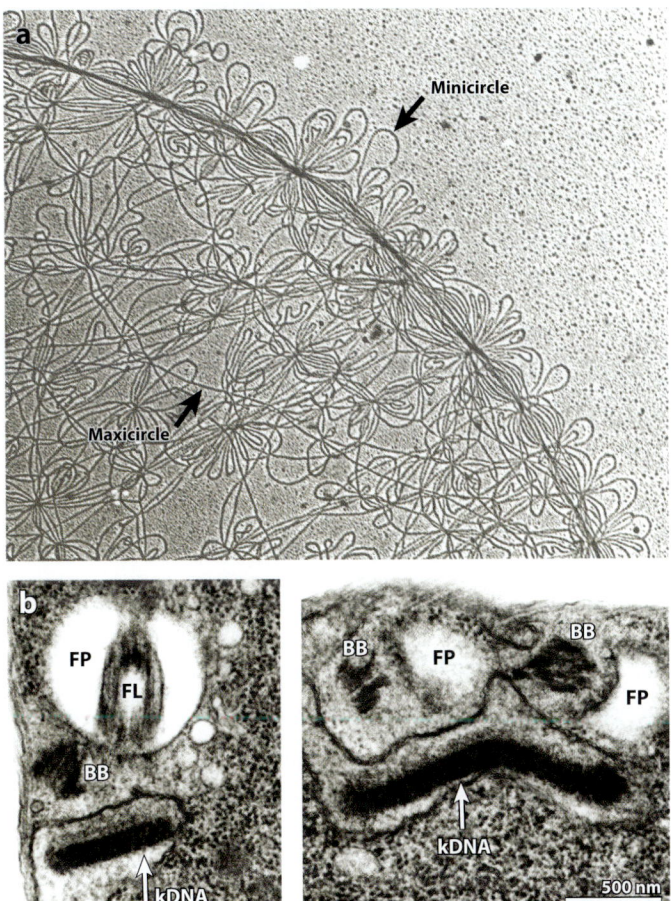

Figure 1

The kDNA network. (*a*) An EM of the edge of an isolated kDNA network from *Crithidia fasciculata*. Small DNA loops are minicircles, and long strands are parts of maxicircles. (*b*) Thin section EM of *Trypanosoma brucei* showing (*left*) the kinetoplast (*white arrow*) prior to replication and (*right*) a kDNA network that has nearly finished replication. The kDNA disk is composed primarily of aligned minicircles as shown in **Figure 2**. Abbreviations: BB, basal body; kDNA, kinetoplast DNA; EM, electron micrograph; FL, flagellum; FP, flagellar pocket.

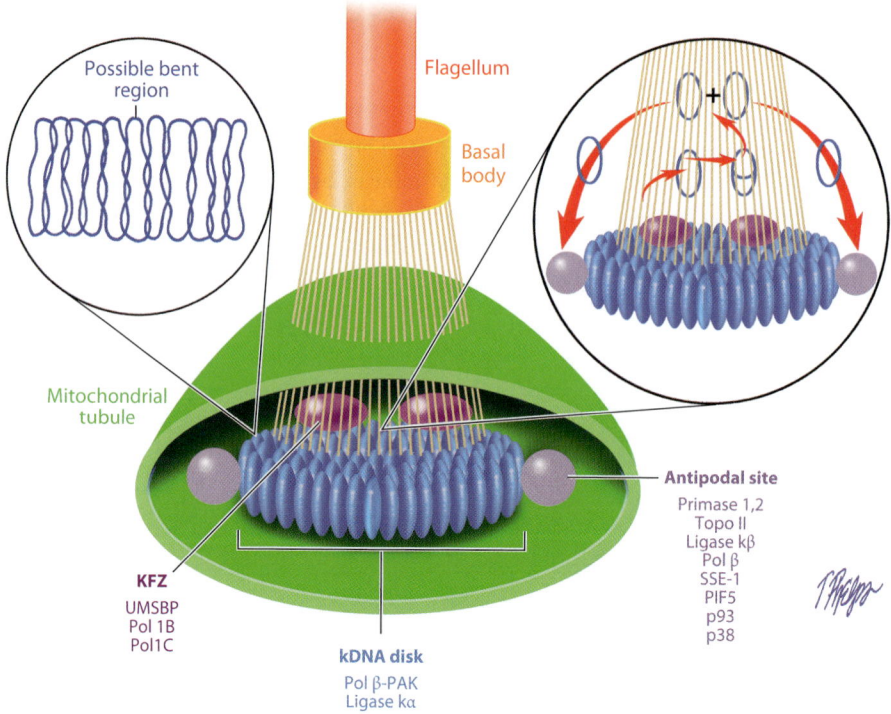

Figure 2

The current kDNA replication model. An oblique section of the end of the mitochondrial tubule (*green*) adjacent to the flagellar basal body shows the kDNA disk, which is organized with minicircles stacked parallel to its axis. Locations of critical replication proteins are also shown. (*Left inset*) An enlarged section of minicircles from the kinetoplast disk shows that they are stretched out, stand side by side, and are catenated to neighboring minicircles, as well as to maxicircles (not shown). kDNA-binding proteins that facilitate the compact organization of the network have been removed. (*Right inset*): For replication, covalently closed minicircles are released from the network into the KFZ, where they initiate replication as theta structures by UMSBP, Pol 1B, and other proteins. The sister minicircles are thought to migrate to the antipodal sites, where primer removal by SSE-1 and PIF5, gap filling by Pol β, and the sealing of most of the nicks by DNA ligase kβ occur. Minicircles are then attached to the network periphery by topo II. At the end of replication, minicircles containing at least one gap are repaired in the network by Pol β-PAK, DNA ligase kα, and other proteins. The filament system (tripartite attachment complex) linking the kDNA to the flagellar basal body is also shown. Abbreviations: kDNA, kinetoplast DNA; TAC, tripartite attachment complex; KFZ, kinetoflagellar zone; UMSBP, universal minicircle binding protein; PAK, proline, alanine and lysine domain; SSE, structure-specific endonuclease; Pol, polymerase; topo, topoisomerase.

Kinetoplast: disk-shaped DNA-protein structure within the mitochondrial matrix, containing a single kDNA network

interlock with each other. Thus, kDNA from this species, and probably others, forms "networks within networks" (65). **Figure 1a** shows an electron micrograph (EM) of part of a *C. fasciculata* network. Although EMs of isolated networks reveal that kDNA are planar, they condense in vivo into a disk-shaped structure in which the interlocked minicircles are stretched out and stand side by side, forming the kinetoplast disk (**Figure 1b** and **Figure 2**). The height of the disk is always approximately half the circumference of a minicircle (**Figure 2**).

Maxicircles, like mitochondrial DNAs from more conventional organisms, encode rRNAs and several proteins involved in energy transduction (including subunits of cytochrome oxidase,

NADH dehydrogenase, and the ATP synthase). Maxicircle transcripts in *T. brucei* are heavily edited, an amazing reaction in which uridylate residues are incorporated into or removed from specific internal sites within the transcript to create an open reading frame. Minicircles encode most of the small guide RNAs that are templates for editing specificity. Because the mitochondrial transcripts in *T. brucei* contain so many editing sites, they require many different guide RNAs to produce a set of functional mRNAs. Thus, the minicircle population is large and heterogeneous, with different DNA species encoding different guide RNAs. *C. fasciculata* minicircles are less complex, with one predominant minicircle class. Editing and other mitochondrial RNA-processing reactions are reviewed in Reference 3.

RNA editing: the insertion or deletion of uridylate residues to form functional messenger RNAs

Universal minicircle sequence (UMS): 12-nucleotide sequence conserved in minicircles that is part of the replication origin

Although the minicircles in a network are heterogeneous, they contain one or more conserved regions (roughly 100 to 200 base pairs). *T. brucei* minicircles contain one conserved region, whereas those from *C. fasciculata* contain two that are 180° apart. The conserved regions contain origins of replication, and when there are two origins on one minicircle, only one, which is randomly chosen, is used to initiate DNA synthesis (6, 23). Within the conserved region is the universal minicircle sequence (UMS) (6, 46), which is the initiation site for leading-strand synthesis. Another invariant site is the hexamer, the start site for the first Okazaki fragment (59, 67, 79). Another feature of minicircles is the presence of a bent DNA structure, caused by multiple A-tracts (each ~5 bp long) positioned in phase with the helical repeat (40). The most severely bent DNA, which has 26 A-tracts, is from *C. fasciculata*; *T. brucei* minicircles have a more modest bend with only 5 A-tracts. The reason for DNA bending is not known, but the altered structure may help minicircles organize in the kinetoplast (**Figure 2**).

Why Is kDNA a Network?

Because of the extensive editing of maxicircle transcripts, a large number of different minicircles are needed for the complete set of guide RNAs. If even one class of minicircles is lost, the parasite cannot completely edit the maxicircle transcripts and therefore dies. If minicircles were not in a network, random segregation would lead to the rapid loss of some classes (9). We discuss below how the network structure can ensure preservation of the minicircle repertoire during kDNA replication. Interestingly, some kinetoplastids do not contain networks (39) and appear

IS KINETOPLAST DNA A VALID DRUG TARGET?

Trypanosoma brucei bloodstream forms (BSFs) with partial or even complete loss of kDNA, termed dyskinetoplastidy (Dk) and akinetoplastidy (Ak), are found in nature and can be induced in the laboratory with DNA-binding drugs such as acriflavine or ethidium bromide (61). The viability of Dk and Ak strains seems to argue that the kinetoplast might be a poor target for antitrypanosome therapies (27). However, in the laboratory kDNA loss is difficult and often takes dozens of passages in the bloodstream of mice or rats. This difficulty is explained by the finding that a mutation in the nucleus-encoded γ-subunit of the F_1-ATPase, a very rare event, is needed to allow kDNA loss (61) (A. Schnaufer, personal communication). In nature most *T. brucei* strains contain a kinetoplast, and RNAi studies show that knockdown of kDNA replication and editing proteins is lethal to BSFs (61), suggesting that the kinetoplast is a valid drug target. Moreover, we have recently shown that minicircle replication is the most vulnerable target of ethidium bromide, which is still used to treat trypanosomiasis in African cattle (56). Because the kinetoplast has no counterpart in other eukaryotes, the breathtaking complexity of kDNA replication and segregation present myriad potential drug targets for future investigators.

Tripartite attachment complex (TAC): a system of filaments connecting the kinetoplast to the flagellar basal body required for positioning and segregating the kDNA network

to have invented some intriguing ways to segregate their kDNA. For example, the fish parasite *Trypanoplasma borreli* has long tandem arrays of minicircle-like sequences on 200-kb molecules (39). We speculate that others, such as *Bodo saltans*, form kDNA networks that are held together by proteins instead of by catenation. Other kinetoplastids, such as *Dimastigella mimosa*, have dispersed kDNAs and may replenish lost minicircles by genetic exchange during mating (19).

KINETOPLAST DNA REPLICATION

When kDNA networks were first discovered in the early 1970s, nothing was known about how such a complicated structure could be replicated. For example, how can a single minicircle replicate when it is interlocked with several neighbors? How are molecules that have replicated distinguished from those that have not? We have answers to these particular questions, but the studies have raised a lot more.

Highlights of kDNA Replication

During the mitochondrial S phase, the network doubles in size and then it splits in two, and the virtually identical, progeny networks segregate into daughter cells (16, 26). Positioning of the network in the mitochondrial matrix and segregation of the daughter networks are mediated by a cytoskeletal structure named the tripartite attachment complex (TAC) (48).

As shown in **Figure 2**, some replication proteins localize within the kinetoplast disk. Others are found at different sites surrounding the kinetoplast, such as the kinetoflagellar zone (KFZ), the region between the kDNA disk and the mitochondrial membrane near the flagellar basal body (2), and the antipodal sites, two protein assemblies 180° apart situated near the kinetoplast periphery (18, 42).

Minicircles replicate after vectorial release from the network into the KFZ (15). Replication of free minicircles, starting in the KFZ, is unidirectional via theta structures (58). Unlike most other organisms, RNA primers for Okazaki fragments are not immediately removed. Instead, replicated minicircles migrate from the KFZ to the antipodal sites. Here, most but not all primers are removed and the gaps between Okazaki fragments are repaired (57). A minicircle, containing at least one gap, is then reattached to the network periphery by topoisomerase (topo) II in the antipodal sites (42). Maxicircles also replicate unidirectionally via theta structures, but unlike minicircles, they replicate while still attached to the network (12).

To ensure that each minicircle and maxicircle replicates once, and only once each generation, the replication machinery must distinguish replicated from nonreplicated circles. The difference is that nonreplicated minicircles and maxicircles are covalently closed, and when replicated, they contain gaps (17, 35).

Two Different Mechanisms of kDNA Replication

To ensure that daughter cells receive virtually identical kDNA networks, it is critical that sister minicircles synthesized in the KFZ are attached at the antipodal sites to opposite sides of the replicating disk prior to kinetoplast division. Two fundamentally distinct mechanisms are used by different kinetoplastids (22). In the ring mechanism, used by *C. fasciculata*, *T. cruzi*, *Leishmania major*, *Leishmania tarentolae*, and *Phytomonas davidii*, the kDNA disk rotates with respect to the antipodal sites. *T. brucei* uses the polar mechanism, in which the antipodal sites remain fixed at the poles of the kDNA disk.

Ring Mechanism

The ring mechanism emerged over almost two decades, with key papers published by several laboratories (53, 69, 70). These studies used metabolic labeling of *C. fasciculata* with ³H-thymidine, isolating the networks, and then examined the ³H incorporated into DNA by EM autoradiography (**Figure 3a**). With a 3-min pulse, radioactivity is restricted to two sites on opposite sides of the network. This is consistent with the attachment of the minicircles by topo II at the antipodal sites

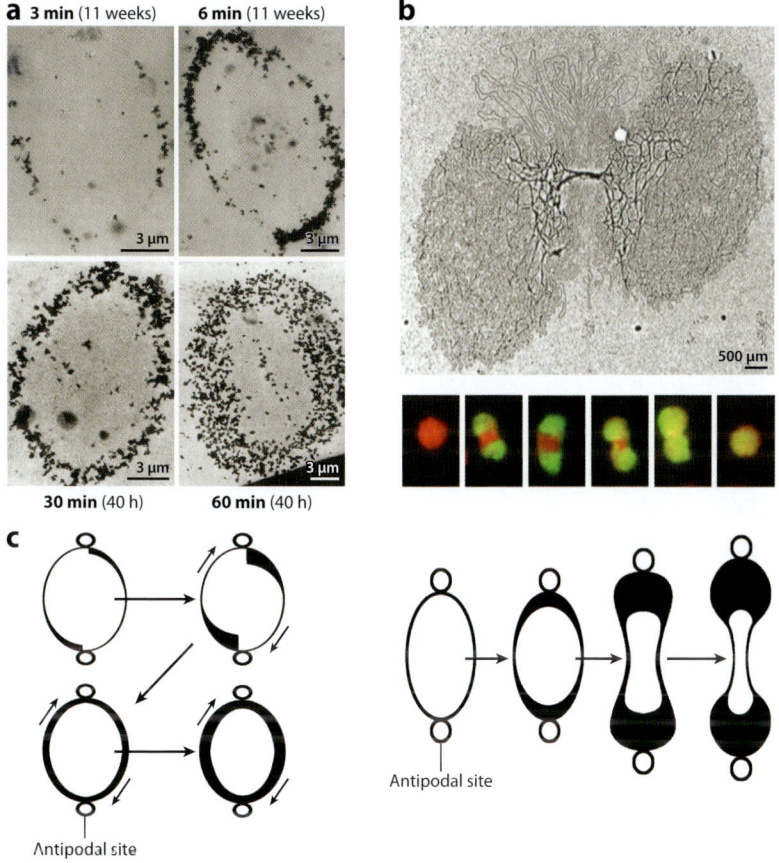

Figure 3

kDNA replication mechanisms of *Crithidia fasciculata* and *Trypanosoma brucei*. (*a*) EM autoradiograph of *C. fasciculata* networks from cells pulse-labeled with ³H-thymidine for 3, 6, 30, and 60 min (indicated in bold, with exposure times in parentheses). (*b*) Upper panel: EM of kDNA-isolated *T. brucei* that has finished replication but has not divided. Lower panels: DAPI-stained (*red*) *T. brucei* networks labeled in vitro with TdT (*green*) to show gapped minicircles as kDNA replication progresses. The first image is a kinetoplast prior to replication, the last image is a final product, but the minicircle gaps have not yet been completely repaired. (*c*) Diagram depicting the ring mechanism of *C. fasciculata*. Antipodal sites (*small circles*) flank the kinetoplast disk, and arrows indicate rotation of the kinetoplast. After one turn, the attachment continues in a spiral. (*d*) Diagram of the *T. brucei* polar mechanism. Newly synthesized minicircles shown in bold are attached at the antipodal sites, leading to a shrinking middle zone that contains an increasing concentration of maxicircles as replication progresses. Panel *a* reproduced with permission from Reference 53. Panel *b* reproduced with permission from 35. Abbreviations: kDNA, kinetoplast DNA; EM, electron micrograph; DAPI, 4′,6-diamidino-2-phenylindole; TdT, terminal deoxynucleotidyl transferase.

(42). As the exogenous ^{3}H-thymidine mixes with intracellular nonradioactive pools, the specific activity rises, leading to decreased exposure times. This rise occurring during the 6-min pulse leads to a gradient of silver grain density. Because sites 180° apart always have about the same silver grain density (specific radioactivity), they must have been synthesized at the same time. At later times (30 and 60 min), rings of labeling are seen. These results imply that there is progressive movement of the kDNA disk relative to the antipodal sites (**Figure 3c**). The antipodal sites could be moving, but more likely it is the kinetoplast that is rotating. How these incredible movements occur remains a complete mystery.

Polar Mechanism

Unlike the ring mechanism, which took years to elucidate, the major features of the polar mechanism were presented in a single paper (26). In *T. brucei*, when unit-size networks containing exclusively covalently closed minicircles (identified in EM by their twisting with ethidium bromide) initiated replication, a zone of gapped minicircles (not twisted by ethidium bromide) formed at each network pole. As replication continued, the polar zones enlarged, the central zone became smaller, and the network became elongated. The final stage, shown in the EM in **Figure 3b**, is dumbbell shaped, with all the maxicircles concentrated in the center. After segregation, the sister networks then remodel in a reaction that must be catalyzed by topo II, distributing the maxicircles throughout the two duplicated networks. Confirming evidence for this mechanism comes from terminal deoxynucleotidyl transferase studies, which are used to fluorescently label the newly replicated, gapped minicircles in isolated networks (**Figure 3b**). The gapped minicircles are always on the poles, and the central, unlabeled zone shrinks as replication proceeds (**Figure 3d**).

A recent publication presenting three-dimensional reconstructions from serial section EMs shows that during kinetoplast duplication in *T. brucei* the new basal body rotates around the old, dragging the kinetoplast with it (20) (**Figure 5**). However, in contrast to *C. fasciculata*, the antipodal sites in *T. brucei* are always situated at the poles of the disk during replication and do not change position relative to the kinetoplast. Therefore, it is unlikely that this 180° rotation in *T. brucei* is related to the rotating disk in *C. fasciculata*, which would need about seven to eight rotations to finish replication. A previous paper (38) suggested that the *T. brucei* kinetoplast undergoes oscillation during replication, but new studies suggest this may have been an artifact (75).

kDNA Division

The last step in kDNA replication is the division of the double-sized network. In *T. brucei*, at the start of S phase, maxicircles are uniformly distributed throughout a network of covalently closed minicircles. As the minicircles are released from the network, and as their gapped progeny are reattached to the poles of the network, the maxicircles are passively concentrated in the shrinking central zone of covalently closed minicircles (**Figure 3b**). When the last, unreplicated minicircle is released, the two progeny minicircle networks are effectively segregated. The daughter networks remain joined by a thin thread of 4′,6-diamidino-2-phenylindole (DAPI)-staining material, shown by fluorescence in situ hybridization to be maxicircles (20). Consequently, the last step in kinetoplast separation is to cleave the maxicircle thread, presumably by topo II, to unlink the sister networks.

C. fasciculata networks divide by a much different mechanism. EM autoradiography of isolated networks shows that after ^{3}H-thymidine labeling for 30 to 60 minutes, many networks have a fairly thick ring of silver grains around the periphery (52). However, the silver grains in some networks resembled horseshoes, suggesting that the double-sized, ring-shaped network was cut down the middle. Some of the horseshoe-shaped structures had an unlabeled, flat side, implying that the initial cut was a straight line. This reaction must require topo II.

Trypanosomes: protozoan, flagellated parasites in the family Trypanosomatidae and order Kinetoplastida that often cause disease in tropical regions

How Is the Minicircle Repertoire Preserved During Replication and Segregation?

Although the two antipodal sites positioned 180° apart on the kinetoplast are ideally poised to distribute daughter minicircles to opposite poles of the duplicating network, it is not clear how the bookkeeping is accomplished. In the first model, the sister minicircles are somehow targeted to opposite antipodal sites. This model could apply to either *T. brucei* or *C. fasciculata*. The long distance between the network center and the antipodal sites (**Figure 2**) suggests that there would also be a special mechanism for transport. The second model applies only to *C. fasciculata*, where a late theta structure would migrate to one antipodal site and finish replication there. Previous studies show that minicircles with the newly synthesized L-strand attach to the network immediately, whereas the H-strand, with multiple gaps, attaches after a delay (28). In this scenario, the kinetoplast would rotate halfway around for the sister minicircle to be attached on the opposite side of the network.

PROTEINS AND ENZYMES INVOLVED IN kDNA REPLICATION

The recent genetic and biochemical analysis of individual proteins and enzymes involved in kDNA replication, as well as determining their precise localization within the kinetoplast, has been critical for deciphering the replication mechanism. Because kDNA replication is so complex, we guess there could be as many as 150 proteins involved, but only ~30 have been fairly well characterized. One reason for this complexity is that trypanosomes often have multiple proteins, with similar activities but with different functions, whereas other eukaryotes can manage with a single enzyme. For example, yeast and human cells have one mitochondrial DNA polymerase, known as Pol γ; trypanosomes have at least seven. Similarly, the trypanosome has six mitochondrial proteins related to the yeast Pif1 helicase. Other cells have one such enzyme or, at most, two. We briefly highlight some key replication enzymes in the following sections.

Universal Minicircle Sequence Binding Protein

In minicircle DNA, the origin of replication contains the UMS, the binding site for the universal minicircle binding protein (UMSBP). UMSBP is located in two sites in the KFZ, where minicircle replication initiates (2). Surprisingly, UMSBP binds only to single-stranded DNA (74), but there is evidence that in intact minicircles the UMS either is partly denatured or has an unusual conformation (4). Another surprise is that UMSBP also binds to the hexamer, a sequence that is at the start of the first Okazaki fragment (1). However, whereas UMSBP binds to the template strand of the UMS, it binds to the complementary strand of the hexamer. It is possible that UMS is the only essential binding site in the origin and the hexamer has some other function in replication. For example, the gap flanking the first Okazaki fragment, which starts at the hexamer and ends at the UMS, is one of the last minicircle gaps repaired after replication. Perhaps UMSBP binds to the 5′ terminus of the Okazaki fragment region and protects it from premature repair.

UMSBP in *C. fasciculata* has 116 amino acid residues, with five CCHC-type zinc fingers. *T. brucei* has two *UMSBP* proteins, one with five zinc fingers and one with seven. Why *T. brucei* has two is not yet known. Knockdown of both UMSBPs by RNAi not only affects the initiation of minicircle replication, but also inhibits segregation of the daughter networks and blocks nuclear division (43). Thus, the role of UMSBP is likely much broader than originally thought.

p38 and p93

p38, an abundant *T. brucei* protein discovered in a proteomics assay, may be involved in the initiation of kDNA replication (34). RNAi knockdown of p38 causes kDNA loss and the accumulation of a new minicircle species, called fraction S. This fraction is composed of a family of highly underwound DNA. Fraction S also has regions of Z-DNA that compensate for the extreme unwinding. Remarkably, p38 binds to the replication origin, the template strands of both the UMS and the hexamer, raising the possibility that p38, rather than UMSBP, is the origin recognition protein. However, p38 localizes to the antipodal sites, instead of at the KFZ, where minicircle replication initiates. Localization is not always a good criterion for function because the two primases, also required for initiation, are also at the antipodal sites (24, 25). We speculate that p38 allows a helicase, together with a topoisomerase, to unwind the DNA helix. The unwinding possibly starts at the origin and proceeds along the minicircle strand. The gene for another replication protein, p93, was identified owing to the presence of sequences associated with S-phase expression (30). p93 localizes to antipodal sites, but only when kDNA is being replicated. RNAi resulted in the loss of nicked, gapped minicircle replication intermediates, suggesting a role for p93 in kDNA synthesis.

Mitochondrial Topo II

The trypanosome genome encodes three mitochondrial topoisomerases: a type IA enzyme (62), a type IB enzyme (5), and a single type II enzyme (72). We focus on topo II, which was first purified from *C. fasciculata* and shown to be localized primarily to the antipodal sites (42). When *T. brucei* topo II was depleted by RNAi, kDNA was lost and gapped minicircles accumulated, leading to the conclusion that a critical role for topo II was the reattachment of newly replicated minicircles to the network (76). Other studies showed that knockdown of topo II resulted in a smear of minicircles dimers known as fraction U, a family of multiply interlocked, dimeric minicircles that are probably derived from late replication intermediates (73). For unclear reasons, fraction U also forms after RNAi of Pol 1B and TbPIF1 (see below). Another striking phenotype observed following depletion of topo II was the appearance of holes in the network (32). Thus, another function of topo II must be to mend kDNA after releasing minicircles for replication. There are other likely functions for topo II, such as the cleavage of maxicircle "threads" and the release of minicircles from the network into the KFZ for replication, but they are not observed following RNAi because they may require much lower concentrations of the enzyme.

Mitochondrial DNA Polymerases

T. brucei has at least seven mitochondrial DNA polymerases. Four are related to bacterial DNA Pol I (29), two are related to DNA Pol β (because Pol βs are typically nuclear enzymes involved in base excision repair, these are the first examples of mitochondrial forms) (60), and one is related to Pol κ, a low-fidelity polymerase that functions to bypass DNA damage (such as 8-oxoguanine or abasic sites). Pol κ has been characterized from *T. cruzi* mitochondria (54) and the corresponding gene is present in *T. brucei*. In contrast to conventional eukaryotes, there is no Pol γ in *T. brucei* mitochondria.

RNAi knockdowns indicate that Pol IB, Pol IC, and Pol ID are required for viability and are likely replicative polymerases (13, 29). Similar to UMSBP, Pol 1B and Pol IC localize to two sites in the KFZ (29). Pol ID is distributed throughout the mitochondrial matrix during most of the cell cycle, except in S phase, where a substantial portion moves to the antipodal sites (J. Concepcion-Acevedo & M. Klingbeil, personal communication). Pol IA is also distributed throughout the mitochondrion, possibly acting as a repair enzyme (26). Because Pol IB synthesizes both the leading and lagging strands of minicircles, it probably functions as a homodimer (11). The roles of Pol IC and Pol ID are not known, but they are likely involved in maxicircle replication.

T. brucei has two Pol β–like enzymes that are thought to function in gap filling of minicircles at the end of replication (60). Pol β is in the antipodal sites and is believed, along with ligase kβ (14), to repair most of the gaps prior to their attachment to the network. Pol β-PAK has a 300-amino-acid N-terminal extension rich in prolines, alanines, and lysines (PAK domain), resides in the network, and with ligase k-α (14) likely repairs the final gaps of network-associated minicircles at the end of kDNA replication (60). Like Pol β in other eukaryotes, both recombinant proteins have 5′-deoxyribosephosphate lyase activity (60) in addition to polymerase activity, suggesting they may also play a role in base excision repair.

Mitochondrial PIF Helicases

The *T. brucei* genome encodes eight helicases related to the yeast Pif1 helicase (named for its discovery in the yeast petite integration frequency locus). Localization of green fluorescent protein fusions showed that six of them were in the mitochondrion (35). Of these six, three are essential (TbPIF1, TbPIF2, and TbPIF8) and TbPIF5 causes a growth phenotype when overexpressed. TbPIF1 is involved with minicircle replication, and *TbPIF1* RNAi causes the accumulation of fraction U discussed above (37). TbPIF2 regulates maxicircle replication and is discussed below. The TbPIF5 helicase functions in primer removal from Okazaki fragments (36). TbPIF8 is the smallest and most divergent of the PIFs and is unlikely to have helicase activity (75). Although TbPIF8 is essential, RNAi has little effect on kDNA replication intermediates, but it clearly affects kinetoplast structure. Thus, TbPIF8 might have an indirect role in kDNA synthesis, such as organizing parts of the replication machinery.

Mitochondrial Primases

T. brucei has two primases. Primase 1 is located at the antipodal sites, and RNAi causes maxicircle loss faster than minicircle loss. Thus, primase 1 could be a maxicircle primase (24). Primase 2 also localizes to the antipodal sites, and RNAi causes an increase in covalently closed free minicircles and a decrease in the free, gapped form, providing strong evidence that it is a minicircle primase (25).

CONTROL OF KINETOPLAST DNA REPLICATION

Trypanosomes are unusual (compared to other eukaryotes) in that timing of mitochondrial genome replication is regulated during the cell cycle, with kDNA synthesis initiating just prior to nuclear S phase and with kDNA network division occurring prior to nuclear division (77). Until recently, there have been no clues about what controls timing of kDNA replication. We discuss three intriguing control mechanisms below.

Redox Regulation of UMSBP Binding

Although redox pathways control biochemical processes such as transcription, redox regulation of UMSBP is the first example of redox control of DNA synthesis (49). UMSBP contains multiple cysteine-containing zinc finger domains, and the active form of the protein is fully reduced. Oxidation of -SH to S-S converts the monomeric, zinc-bound protein to a zinc-depleted form that does not bind the UMS. Studies with synchronized *C. fasciculata* cultures showed that the binding activity and the reduced, active form of UMSBP fluctuated in a cell cycle–dependent manner, with two peaks of activity in S phase, suggesting a new mechanism for the regulation of minicircle synthesis (63, 64). Trypanosomes use a unique redox system in which trypanothione replaces glutathione and tryparedoxin replaces thioredoxin (45). Using purified trypanothione, trypanothione reductase, tryparedoxin, and tryparedoxin peroxidase, investigators (64, 68) found that NADPH stimulated the reduction of UMSBP and increased its binding to the UMS, and that oxidation with cumene hydroperoxide inhibited binding.

HslVU Protease

Another potential regulatory mechanism for kDNA replication comes from the discovery of a proteasome-like, ATP-dependent protease in the *T. brucei* mitochondrion (31). This protease is named TbHslVU after its bacterial counterpart. TbHslVU plays a critical role in *T. brucei*, for which it is essential for cell growth, whereas the bacterial protease is needed only at elevated temperatures. In *T. brucei*, RNAi results in the accumulation of huge kinetoplasts (20-fold increase in minicircles and three- to fourfold increase in maxicircles) due to a block in kDNA segregation but not replication. TbHslVU has been proposed to regulate kDNA synthesis by degrading a positive regulator of DNA replication. RNAi knockdown of TbHslVU stabilizes the regulator and thus kDNA replication continues unrestrained.

One likely target of the HslVU protease is the TbPIF2 helicase. As mentioned above, TbPIF2 controls maxicircle replication. RNAi of this protein eliminates all maxicircles and overexpression causes a massive increase in their number (35). Knockdown of the HslVU protease causes an increase in TbPIF2 protein levels and consequently an accumulation of maxicircles (31). Other substrates, such as the regulator of minicircle replication, have not yet been found.

mRNA Cycling

mRNAs of proteins involved in DNA synthesis in both the nucleus (e.g., subunits of replication protein A and others) and the mitochondrion (e.g., topoisomerase II and others) vary in their level during the cell cycle. mRNAs are high just after the start of S phase and low during G2 and M phases (50). This effect is mediated by a *cis*-acting octameric sequence present in one or more copies within the 5′ or 3′ UTR of the mRNA (51). *Trans*-acting proteins bind the octamer and presumably regulate mRNA stability. Two of these proteins are phosphorylated with kinetics resembling that of mRNA cycling, suggesting that a protein kinase is the ultimate regulator (44).

THE ROLE OF THE TRIPARTITE ATTACHMENT COMPLEX IN KINETOPLAST DNA SEGREGATION

Historically, there have been several indications of a connection between the kinetoplast and the flagellum. In fact, the name kinetoplast implies that this structure was somehow associated with

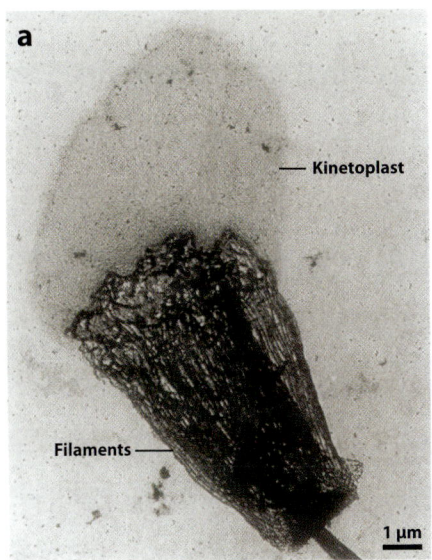 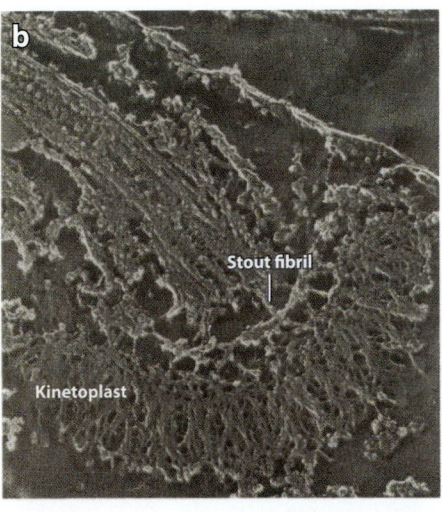

Figure 4

Electron micrographs showing connections between the kinetoplast and the flagellum. (*a*) A kinetoplast from *Crithidia luciliae* spread by the Miller technique and platinum stained (10) showing a filamentous connection to the flagellum. (*b*) Freeze-etch micrograph of a kinetoplast from a *Trypanosome cruzi* epimastigote (71). A stout fibril extends from the kinetoplast DNA to the basal body above. Magnification is $102,000 \times$. Panels *a* and *b* reproduced with permission from References 10 and 71, respectively.

cell movement. Several elegant EMs showed a striking connection between the kDNA and the end of the flagellum (**Figure 4**). Subsequent studies demonstrated a tight, biochemical association (molecular hard wiring) between the kDNA and the basal body–flagellar complex, providing a mechanism for correct positioning of kDNA in the mitochondrion and the separation of sister kinetoplasts during cytokinesis (55) (**Figure 5**). This connection is called TAC, which consists of series of filaments running from one side of the basal body, through the mitochondrial membranes, to the kDNA in the matrix (48). The three components of TAC (starting from outside the mitochondrion) are the exclusion-zone filaments, the differentiated mitochondrial membranes, and the unilateral filaments.

The exclusion-zone filaments traverse a region previously shown to be devoid of ribosomes. The differentiated mitochondrial membranes clearly differ from the surrounding membranes in that they are more electron dense and resist detergent extraction. The unilateral filaments run from the mitochondrial inner membrane to the proximal face of the kDNA disk. Different staining techniques showed that the unilateral filaments are composed of distinct inner and outer zones (21), and 3D reconstructions of electron micrographs confirm that the basal bodies drive the separation of daughter kDNA networks (20). Until recently, all studies of TAC and kDNA segregation had been morphological, but three candidate proteins have been identified, leading to molecular studies.

p166

The first protein component of TAC, p166, was discovered in an RNAi library screen for proteins whose knockdown causes loss of kDNA (78). For p166, some cells lost kDNA, whereas others

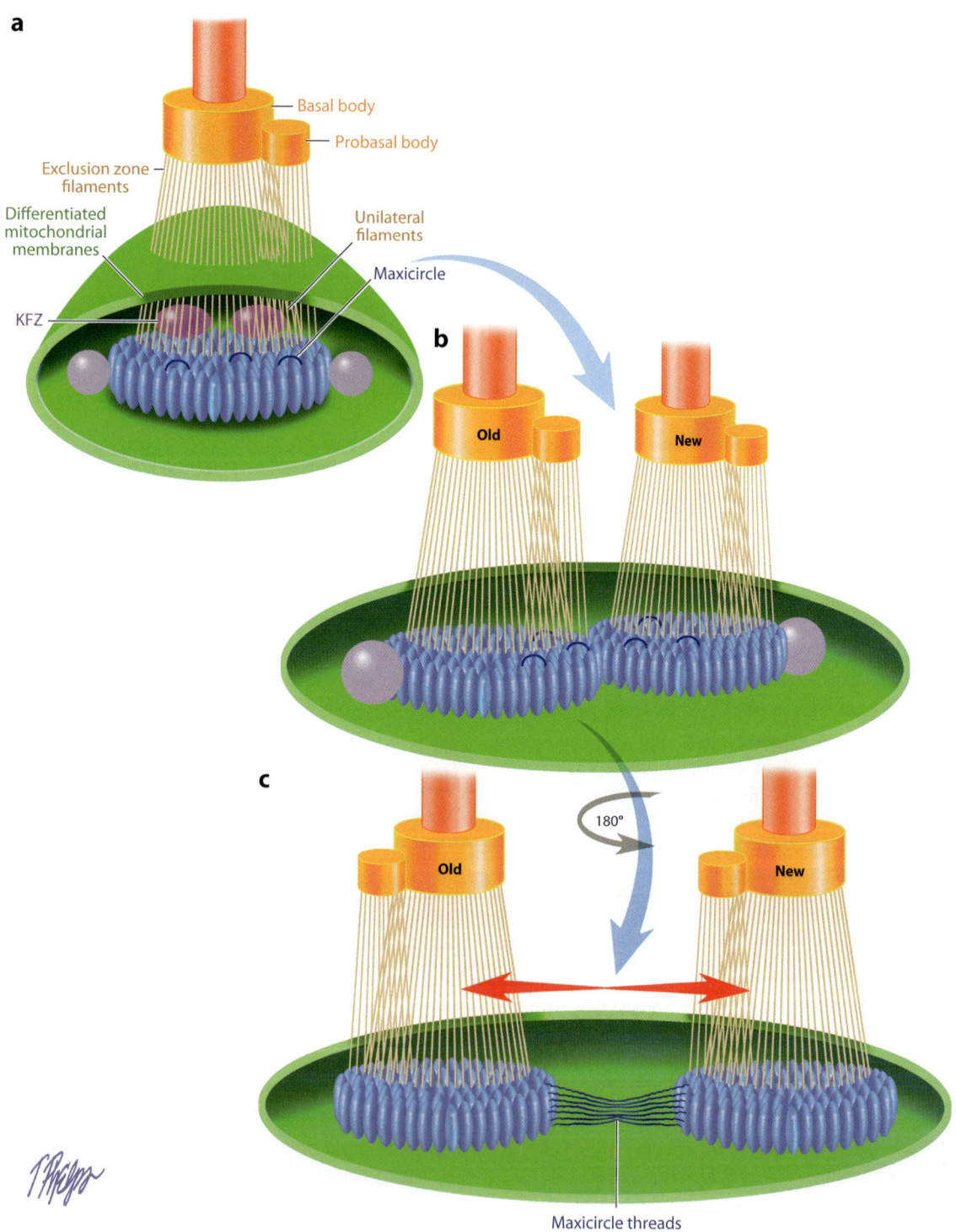

had giant networks, indicating that the cells were defective in kDNA segregation but not in replication. During RNAi, the kDNA networks continued to grow in size, but the entire network often remained in one daughter and attached to the old basal body. The other cell would not get kDNA. As replication continued in the absence of segregation, the networks grew up to 10 times their normal size, and the number of cells lacking kDNA increased.

p166 is a 166-kDa mitochondrial protein with a predicted transmembrane region at its carboxyl terminus, with 39 residues extending beyond that. Immunofluorescence showed that p166 localized between the kDNA disk and the flagellar basal body in intact cells, as well as to isolated kDNA-flagellar complexes. Immunogold electron microscopy revealed that p166 was very close to the kDNA, possibly in the region of the unilateral filaments. It is likely that one end of the protein is inserted into the mitochondrial inner membrane, with its ~150-kDa amino-terminal domain facing the matrix.

Mab22

A potential TAC component was discovered by using a collection of monoclonal antibodies generated to purified kDNA-TAC-flagellar complexes (7). The Mab22 antigen localizes to the exclusion zone, suggesting it is a component of these fibers. Surprisingly, Mab22 recognizes three proteins, one of which is the 69-kDa BILBO1, a flagellar pocket protein (8). So far, the exclusion-zone protein recognized by Mab22 has eluded identification, but this monoclonal antibody remains a promising tool for studying TAC.

AEP-1

The third TAC component was found during the investigation of mitochondrial RNA editing, with the unprecedented discovery of alternative editing (47). In particular, an alternatively edited mRNA for cytochrome oxidase subunit III (COXIII) produces a novel protein (214 amino acids) called AEP-1. AEP-1 has a unique amino-terminal region fused to the carboxyl-terminal portion of COXIII that contains five transmembrane segments. Antibodies to AEP-1 showed that the protein was distributed throughout the mitochondrion, but the signal was enhanced in the region of the kinetoplast. Immunofluorescence studies with isolated kDNA-TAC-flagellar complexes showed that the position of AEP-1 between the kDNA and the basal body was remarkably similar to that of p166. A truncated, dominant-negative version of AEP-1 expressed from the nucleus led to a growth defect and to an increase in cells with either no kDNA or large kinetoplasts. Thus, AEP-1 appears to play a critical role in kDNA segregation.

Figure 5

Segregation of kinetoplastid DNA (kDNA) by the tripartite attachment complex (TAC). (*a*) TAC connecting a kinetoplast (before replication) to the flagellar basal body is composed of unilateral filaments, differentiated mitochondrial membranes, and exclusion-zone filaments. (*b*) A kinetoplast, flanked by antipodal sites, undergoing replication. The flagellar system and TAC have duplicated, and a new basal body is shown at the beginning of a 180° rotation around the old basal body (20). (*c*) The new basal body has finished its rotation, dragging the two connected kDNA networks apart. Sister networks remain joined by a thin thread of maxicircles until cleaved by topoisomerase II. Abbreviation: KFZ, kinetoflagellar zone.

FUTURE ISSUES

1. For a more complete understanding of kDNA replication, many more components (e.g., replication enzymes, TAC proteins) must be uncovered. However, it is often much easier to discover a new protein than to determine its function.

2. How sister minicircles are attached on opposite sides of the network remains a mystery. Similarly, how maxicircles are equally partitioned during kDNA division is poorly understood (as are many other aspects of maxicircle replication).

3. Additional substrates of the HslVU protease (such as the minicircle regulator) remain unidentified. Is the protease itself under cell cycle control, or are HslVU substrates tagged for destruction at the appropriate times?

4. How is the complicated and compact structure of the kinetoplast and its associated machinery maintained? Does TAC play a role?

5. Some fantastic, large-scale movements (e.g., the rotating kinetoplast of *C. fasciculata*, the separation of sister networks by TAC) occur during kDNA replication, but little or nothing is known about the motors and mechanisms. Radically new approaches will be needed to address these issues.

DISCLOSURE STATEMENT

The authors are not aware of any affiliations, memberships, funding, or financial holdings that might be perceived as affecting the objectivity of this review.

ACKNOWLEDGMENTS

We thank Beiyu Liu, Dan Ray, Rahul Bakshi, and all members of the Jensen lab for valuable comments. We especially thank Jianyang Wang for her images used in **Figure 1** and **Figure 3**, and Tim Phelps (Johns Hopkins Art as Applied to Medicine) for the diagrams in **Figure 2** and **Figure 5**. This work was supported by NIH grant AI058613 to R.E.J. and P.T.E.

LITERATURE CITED

1. Abu-Elneel K, Kapeller I, Shlomai J. 1999. Universal minicircle sequence-binding protein, a sequence-specific DNA-binding protein that recognizes the two replication origins of the kinetoplast DNA minicircle. *J. Biol. Chem.* 274:13419–26
2. Abu-Elneel K, Robinson DR, Drew ME, Englund PT, Shlomai J. 2001. Intramitochondrial localization of universal minicircle sequence-binding protein, a trypanosomatid protein that binds kinetoplast minicircle replication origins. *J. Cell Biol.* 153:725–34
3. Aphasizhev R, Aphasizheva I. 2011. Mitochondrial RNA processing in trypanosomes. *Res. Microbiol.* 162:655–63
4. Avrahami D, Tzfati Y, Shlomai J. 1995. A single-stranded DNA binding protein binds the origin of replication of the duplex kinetoplast DNA. *Proc. Natl. Acad. Sci. USA* 92:10511–15
5. Bakshi RP, Shapiro TA. 2004. RNA interference of *Trypanosoma brucei* topoisomerase IB: Both subunits are essential. *Mol. Biochem. Parasitol.* 136:249–55
6. Birkenmeyer L, Sugisaki H, Ray DS. 1987. Structural characterization of site-specific discontinuities associated with replication origins of minicircle DNA from *Crithidia fasciculata*. *J. Biol. Chem.* 262:2384–92

7. Bonhivers M, Landrein N, Decossas M, Robinson DR. 2008. A monoclonal antibody marker for the exclusion-zone filaments of *Trypanosoma brucei*. *Parasit. Vectors* 1:21
8. Bonhivers M, Nowacki S, Landrein N, Robinson DR. 2008. Biogenesis of the trypanosome endo-exocytotic organelle is cytoskeleton mediated. *PLoS Biol.* 6:e105
9. Borst P. 1991. Why kinetoplast DNA networks? *Trends Genet.* 7:139–41
10. Borst P, Hoeijmakers JH. 1979. Kinetoplast DNA. *Plasmid* 2:20–40
11. Bruhn DF, Mozeleski B, Falkin L, Klingbeil MM. 2010. Mitochondrial DNA polymerase POLIB is essential for minicircle DNA replication in African trypanosomes. *Mol. Microbiol.* 75:1414–25
12. Carpenter LR, Englund PT. 1995. Kinetoplast maxicircle DNA replication in *Crithidia fasciculata* and *Trypanosoma brucei*. *Mol. Cell. Biol.* 15:6794–803
13. Chandler J, Vandoros AV, Mozeleski B, Klingbeil MM. 2008. Stem-loop silencing reveals that a third mitochondrial DNA polymerase, POLID, is required for kinetoplast DNA replication in trypanosomes. *Eukaryot. Cell* 7:2141–46
14. Downey N, Hines JC, Sinha KM, Ray DS. 2005. Mitochondrial DNA ligases of *Trypanosoma brucei*. *Eukaryot. Cell* 4:765–74
15. Drew ME, Englund PT. 2001. Intramitochondrial location and dynamics of *Crithidia fasciculata* kinetoplast minicircle replication intermediates. *J. Cell Biol.* 153:735–44
16. Englund PT. 1978. The replication of kinetoplast DNA networks in *Crithidia fasciculata*. *Cell* 14:157–68
17. Englund PT. 1979. Free minicircles of kinetoplast DNA in *Crithidia fasciculata*. *J. Biol. Chem.* 254:4895–900
18. Ferguson M, Torri AF, Ward DC, Englund PT. 1992. In situ hybridization to the *Crithidia fasciculata* kinetoplast reveals two antipodal sites involved in kinetoplast DNA replication. *Cell* 70:621–29
19. Gibson W, Garside L. 1990. Kinetoplast DNA minicircles are inherited from both parents in genetic hybrids of *Trypanosoma brucei*. *Mol. Biochem. Parasitol.* 42:45–53
20. Gluenz E, Povelones ML, Englund PT, Gull K. 2011. The kinetoplast duplication cycle in *Trypanosoma brucei* is orchestrated by cytoskeleton-mediated cell morphogenesis. *Mol. Cell. Biol.* 31:1012–21
21. Gluenz E, Shaw MK, Gull K. 2007. Structural asymmetry and discrete nucleic acid subdomains in the *Trypanosoma brucei* kinetoplast. *Mol. Microbiol.* 64:1529–39
22. Guilbride DL, Englund PT. 1998. The replication mechanism of kinetoplast DNA networks in several trypanosomatid species. *J. Cell Sci.* 111(Pt. 6):675–79
23. Hines JC, Ray DS. 2008. Structure of discontinuities in kinetoplast DNA-associated minicircles during S phase in *Crithidia fasciculata*. *Nucleic Acids Res.* 36:444–50
24. Hines JC, Ray DS. 2010. A mitochondrial DNA primase is essential for cell growth and kinetoplast DNA replication in *Trypanosoma brucei*. *Mol. Cell. Biol.* 30:1319–28
25. Hines JC, Ray DS. 2011. A second mitochondrial DNA primase is essential for cell growth and kinetoplast minicircle DNA replication in *Trypanosoma brucei*. *Eukaryot. Cell* 10:445–54
26. **Hoeijmakers JH, Weijers PJ. 1980. The segregation of kinetoplast DNA networks in *Trypanosoma brucei*. *Plasmid* 4:97–116**
27. Kaminsky R, Schmid C, Lun ZR. 1997. Susceptibility of dyskinetoplastic *Trypanosoma evansi* and *T. equiperdum* to isometamidium chloride. *Parasitol. Res.* 83:816–18
28. Kitchin PA, Klein VA, Englund PT. 1985. Intermediates in the replication of kinetoplast DNA minicircles. *J. Biol. Chem.* 260:3844–51
29. Klingbeil MM, Motyka SA, Englund PT. 2002. Multiple mitochondrial DNA polymerases in *Trypanosoma brucei*. *Mol. Cell* 10:175–86
30. Li Y, Sun Y, Hines JC, Ray DS. 2007. Identification of new kinetoplast DNA replication proteins in trypanosomatids based on predicted S-phase expression and mitochondrial targeting. *Eukaryot. Cell* 6:2303–10
31. **Li Z, Lindsay ME, Motyka SA, Englund PT, Wang CC. 2008. Identification of a bacterial-like HslVU protease in the mitochondria of *Trypanosoma brucei* and its role in mitochondrial DNA replication. *PLoS Pathog.* 4:e1000048**
32. Lindsay ME, Gluenz E, Gull K, Englund PT. 2008. A new function of *Trypanosoma brucei* mitochondrial topoisomerase II is to maintain kinetoplast DNA network topology. *Mol. Microbiol.* 70:1465–76

26. Works out the complete pathway of the polar mechanism by electron microscopy.

31. Identifies the mitochondrial HslVU protease and shows its role controlling kDNA replication.

33. Liu B, Liu Y, Motyka SA, Agbo EE, Englund PT. 2005. Fellowship of the rings: the replication of kinetoplast DNA. *Trends Parasitol.* 21:363–69
34. Liu B, Molina H, Kalume D, Pandey A, Griffith JD, Englund PT. 2006. Role of p38 in replication of *Trypanosoma brucei* kinetoplast DNA. *Mol. Cell. Biol.* 26:5382–93
35. **Liu B, Wang J, Yaffe N, Lindsay ME, Zhao Z, et al. 2009. Trypanosomes have six mitochondrial DNA helicases with one controlling kinetoplast maxicircle replication.** ***Mol. Cell* 35:490–501**
36. Liu B, Wang J, Yildirir G, Englund PT. 2009. TbPIF5 is a *Trypanosoma brucei* mitochondrial DNA helicase involved in processing of minicircle Okazaki fragments. *PLoS Pathog.* 5:e1000589
37. Liu B, Yildirir G, Wang J, Tolun G, Griffith JD, Englund PT. 2010. TbPIF1, a *Trypanosoma brucei* mitochondrial DNA helicase, is essential for kinetoplast minicircle replication. *J. Biol. Chem.* 285:7056–66
38. Liu Y, Englund PT. 2007. The rotational dynamics of kinetoplast DNA replication. *Mol. Microbiol.* 64:676–90
39. Lukes J, Guilbride DL, Votypka J, Zikova A, Benne R, Englund PT. 2002. Kinetoplast DNA network: evolution of an improbable structure. *Eukaryot. Cell* 1:495–502
40. Marini JC, Levene SD, Crothers DM, Englund PT. 1983. A bent helix in kinetoplast DNA. *Cold Spring Harb. Symp. Quant. Biol.* 47(Pt. 1):279–83
41. Marini JC, Miller KG, Englund PT. 1980. Decatenation of kinetoplast DNA by topoisomerases. *J. Biol. Chem.* 255:4976–79
42. **Melendy T, Sheline C, Ray DS. 1988. Localization of a type II DNA topoisomerase to two sites at the periphery of the kinetoplast DNA of *Crithidia fasciculata*. *Cell* 55:1083–88**
43. Milman N, Motyka SA, Englund PT, Robinson D, Shlomai J. 2007. Mitochondrial origin-binding protein UMSBP mediates DNA replication and segregation in trypanosomes. *Proc. Natl. Acad. Sci. USA* 104:19250–55
44. Mittra B, Ray DS. 2004. Presence of a poly(A) binding protein and two proteins with cell cycle-dependent phosphorylation in *Crithidia fasciculata* mRNA cycling sequence binding protein II. *Eukaryot. Cell* 3:1185–97
45. Müller S, Liebau E, Walter RD, Krauth-Siegel RL. 2003. Thiol-based redox metabolism of protozoan parasites. *Trends Parasitol.* 19:320–28
46. Ntambi JM, Englund PT. 1985. A gap at a unique location in newly replicated kinetoplast DNA minicircles from *Trypanosoma equiperdum*. *J. Biol. Chem.* 260:5574–79
47. Ochsenreiter T, Anderson S, Wood ZA, Hajduk SL. 2008. Alternative RNA editing produces a novel protein involved in mitochondrial DNA maintenance in trypanosomes. *Mol. Cell. Biol.* 28:5595–604
48. Ogbadoyi EO, Robinson DR, Gull K. 2003. A high-order *trans*-membrane structural linkage is responsible for mitochondrial genome positioning and segregation by flagellar basal bodies in trypanosomes. *Mol. Biol. Cell* 14:1769–79
49. **Onn I, Milman-Shtepel N, Shlomai J. 2004. Redox potential regulates binding of universal minicircle sequence binding protein at the kinetoplast DNA replication origin. *Eukaryot. Cell* 3:277–87**
50. Pasion SG, Brown GW, Brown LM, Ray DS. 1994. Periodic expression of nuclear and mitochondrial DNA replication genes during the trypanosomatid cell cycle. *J. Cell Sci.* 107:3515–20
51. Pasion SG, Hines JC, Ou X, Mahmood R, Ray DS. 1996. Sequences within the 5′ untranslated region regulate the levels of a kinetoplast DNA topoisomerase mRNA during the cell cycle. *Mol. Cell. Biol.* 16:6724–35
52. Pérez-Morga D, Englund PT. 1993. The structure of replicating kinetoplast DNA networks. *J. Cell Biol.* 123:1069–79
53. **Pérez-Morga DL, Englund PT. 1993. The attachment of minicircles to kinetoplast DNA networks during replication. *Cell* 74:703–11**
54. Rajão MA, Passos-Silva DG, DaRocha WD, Franco GR, Macedo AM, et al. 2009. DNA polymerase kappa from *Trypanosoma cruzi* localizes to the mitochondria, bypasses 8-oxoguanine lesions and performs DNA synthesis in a recombination intermediate. *Mol. Microbiol.* 71:185–97
55. **Robinson DR, Gull K. 1991. Basal body movements as a mechanism for mitochondrial genome segregation in the trypanosome cell cycle. *Nature* 352:731–33**

56. Roy Chowdhury A, Bakshi R, Wang J, Yildirir G, Liu B, et al. 2010. The killing of African trypanosomes by ethidium bromide. *PLoS Pathog.* 6:e1001226
57. Ryan KA, Englund PT. 1989. Replication of kinetoplast DNA in *Trypanosoma equiperdum*. Minicircle H strand fragments which map at specific locations. *J. Biol. Chem.* 264:823–30
58. Ryan KA, Englund PT. 1989. Synthesis and processing of kinetoplast DNA minicircles in *Trypanosoma equiperdum*. *Mol. Cell. Biol.* 9:3212–17
59. Ryan KA, Shapiro TA, Rauch CA, Englund PT. 1988. Replication of kinetoplast DNA in trypanosomes. *Annu. Rev. Microbiol.* 42:339–58
60. Saxowsky TT, Choudhary G, Klingbeil MM, Englund PT. 2003. *Trypanosoma brucei* has two distinct mitochondrial DNA polymerase β enzymes. *J. Biol. Chem.* 278:49095–101
61. Schnaufer A, Domingo GJ, Stuart K. 2002. Natural and induced dyskinetoplastic trypanosomatids: how to live without mitochondrial DNA. *Int. J. Parasitol.* 32:1071–84
62. Scocca JR, Shapiro TA. 2008. A mitochondrial topoisomerase IA essential for late theta structure resolution in African trypanosomes. *Mol. Microbiol.* 67:820–29
63. Sela D, Shlomai J. 2009. Regulation of UMSBP activities through redox-sensitive protein domains. *Nucleic Acids Res.* 37:279–88
64. Sela D, Yaffe N, Shlomai J. 2008. Enzymatic mechanism controls redox-mediated protein-DNA interactions at the replication origin of kinetoplast DNA minicircles. *J. Biol. Chem.* 283:32034–44
65. Shapiro TA. 1993. Kinetoplast DNA maxicircles: networks within networks. *Proc. Natl. Acad. Sci. USA* 90:7809–13
66. Shapiro TA, Englund PT. 1995. The structure and replication of kinetoplast DNA. *Annu. Rev. Microbiol.* 49:117–43
67. Shlomai J. 2004. The structure and replication of kinetoplast DNA. *Curr. Mol. Med.* 4:623–47
68. Shlomai J. 2010. Redox control of protein-DNA interactions: from molecular mechanisms to significance in signal transduction, gene expression, and DNA replication. *Antioxid. Redox Signal.* 13:1429–76
69. **Simpson AM, Simpson L. 1976. Pulse-labeling of kinetoplast DNA: localization of two sites of synthesis within the networks and kinetics of labeling of closed minicircles.** *J. Protozool.* **23:583–87**
70. Simpson L, Simpson AM, Wesley RD. 1974. Replication of the kinetoplast DNA of *Leishmania tarentolae* and *Crithidia fasciculata*. *Biochim. Biophys. Acta* 349:161–72
71. Souto-Padron T, de Souza W, Heuser JE. 1984. Quick-freeze, deep-etch rotary replication of *Trypanosoma cruzi* and *Herpetomonas megaseliae*. *J. Cell Sci.* 69:167–78
72. Strauss PR, Wang JC. 1990. The *TOP2* gene of *Trypanosoma brucei*: a single-copy gene that shares extensive homology with other *TOP2* genes encoding eukaryotic DNA topoisomerase II. *Mol. Biochem. Parasitol.* 38:141–50
73. Sundin O, Varshavsky A. 1981. Arrest of segregation leads to accumulation of highly intertwined catenated dimers: dissection of the final stages of SV40 DNA replication. *Cell* 25:659–69
74. Tzfati Y, Abeliovich H, Kapeller I, Shlomai J. 1992. A single-stranded DNA-binding protein from *Crithidia fasciculata* recognizes the nucleotide sequence at the origin of replication of kinetoplast DNA minicircles. *Proc. Natl. Acad. Sci. USA* 89:6891–95
75. Wang J, Englund PT, Jensen RE. 2012. TbPIF8, a *Trypanosoma brucei* protein related to the yeast Pif1 helicase, is essential for cell viability and mitochondrial genome maintenance. *Mol. Microbiol.* 471–85
76. Wang Z, Morris JC, Drew ME, Englund PT. 2000. Inhibition of *Trypanosoma brucei* gene expression by RNA interference using an integratable vector with opposing T7 promoters. *J. Biol. Chem.* 275:40174–79
77. Woodward R, Gull K. 1990. Timing of nuclear and kinetoplast DNA replication and early morphological events in the cell cycle of *Trypanosoma brucei*. *J. Cell Sci.* 95(Pt. 1):49–57
78. Zhao Z, Lindsay ME, Roy Chowdhury A, Robinson DR, Englund PT. 2008. p166, a link between the trypanosome mitochondrial DNA and flagellum, mediates genome segregation. *EMBO J.* 27:143–54
79. Ray DS. 1989. Conserved sequence blocks in kinetoplast minicircles from diverse species of trypanosomes. *Mol. Cell. Biol.* 9:1365–67

69. Presented remarkable autoradiography data that took 15 years to be explained with the rotating kinetoplast mechanism.

Pseudomonas aeruginosa Twitching Motility: Type IV Pili in Action

Lori L. Burrows

Department of Biochemistry and Biomedical Sciences, Michael G. DeGroote Institute for Infectious Disease Research, McMaster University, Hamilton, ON L8N 3Z5, Canada; email: burrowl@mcmaster.ca

Keywords

pilus, fimbriae, assembly system, secretin, type IVa, type IVb

Abstract

Type IV pili (T4P) are one of the most common forms of bacterial and archaeal surface structures, involved in adherence, motility, competence for DNA uptake, and pathogenesis. *Pseudomonas aeruginosa* has emerged as one of the key model systems for the investigation of T4P structure and function. Although its reputation as a serious nosocomial and opportunistic pathogen is well deserved, its facile growth requirements and the ready availability of molecular tools have allowed for rapid advances in our understanding of how T4P are assembled; their contributions to motility, biofilm formation and virulence; and their complex regulation. This review covers recent findings concerning the three different types of T4P found in *P. aeruginosa* (type IVa, type IVb, and Tad) and provides details about the modes of translocation mediated by T4aP, the architecture and function of the T4aP assembly system, and the complex regulation of T4aP biogenesis and function.

Contents

INTRODUCTION	494
T4a, T4b, AND Tad SUBCLASSES OF TYPE IV PILI	494
CRAWLING, WALKING, SWARMING, AND SLINGSHOT MOTILITIES	496
Other Forms of T4aP-Mediated Motility	497
TWITCHING MOTILITY: WHAT'S IT FOR?	499
Arrivals and Departures	499
Biofilm Development	499
Twitching Motility and Pathogenicity	500
T4P as Antennae?	501
THE TYPE IV PILI MACHINE AND ITS ENGINE	502
Overview of the T4P Assembly Machinery	502
How Does Twitching Happen?	506
WHERE ARE THEY GOING? REGULATION OF TWITCHING MOTILITY	508
Regulation of Pilin and Minor Pilin Expression by Two-Component Systems	509
cAMP-Dependent Regulation of T4aP Biogenesis	509
Cyclic-di-GMP Signaling and Twitching Motility	510
LOOKING FORWARD	511

INTRODUCTION

Watching *Pseudomonas aeruginosa* spread rapidly across a surface by twitching motility is endlessly fascinating (**Supplemental Movie 1**; follow the **Supplemental Material link** from the Annual Reviews home page at http://www.annualreviews.org). This form of bacterial movement was hypothesized by Bradley (29) to result from the repeated extension, tethering, and retraction of long protein fibers called type IV pili (T4P). He showed that the shortening of T4P after attachment of pilus-specific bacteriophages resulted in trafficking of phage particles to the bacterial cell surface (28). A wide variety of bacteria and archaea have since been found to use T4P for functions ranging from DNA uptake to electron transport (11, 141). Because they are genetically amenable human pathogens, *Neisseria* and *Pseudomonas* species—particularly *P. aeruginosa*—have emerged as the favored model organisms for T4P structure and function investigations. Other well-studied but substantially more complex systems (reviewed in Reference 113) include that of the soil bacterium *Myxococcus xanthus*. In this review, the rapid progress that has occurred in the field of twitching motility over the past decade since the last major coverage of this topic (107, 151) is discussed, with a particular focus on *P. aeruginosa*. For additional information, readers are directed to excellent reviews covering various aspects of T4P assembly, structure, and function (3, 11, 12, 30, 41, 42, 67, 69, 79, 120, 135, 137, 138).

T4a, T4b, AND Tad SUBCLASSES OF TYPE IV PILI

Like other kinds of pili and fimbriae, T4P provide the ability to adhere to chemically diverse surfaces—from glass and stainless steel to mammalian cells—and promote bacterial cell aggregation involved in microcolony formation and virulence. However, T4P are unique in mediating flagellum-independent motility (126). There are two major subfamilies of T4P, type IVa and

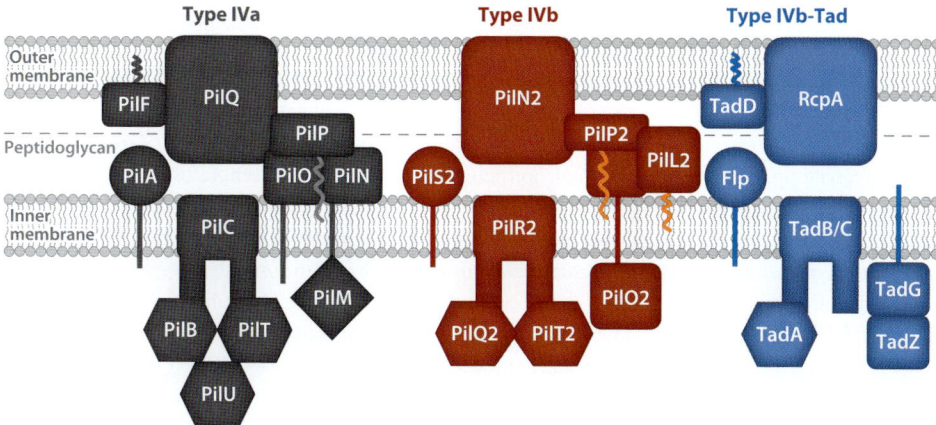

Figure 1

Type IV pili assembly systems in *Pseudomonas aeruginosa*. Putative organization in the cell envelope of potentially functionally equivalent components of the three assembly systems in *P. aeruginosa* is shown (12, 21, 31). PilA, PilS2, and Flp are the major pilin subunits; PilQ, PilN2, and RcpA are the secretins; PilF and TadD are putative pilotins; PilC, PilR2, TadB, and TadC are putative platform proteins; PilB, PilQ2, and TadA are putative pilin polymerases; and PilT, PilU, and PilT2 are putative pilin depolymerases. The PilD prepilin peptidase (not shown) is shared by the T4a and T4bP systems; the Tad system has a distinct FppA prepilin peptidase. Also not shown are the putative minor pilins FimU-PilVWXE for the T4a system, PilV2 for the T4b system, and TadF for the Tad system. Each system also has unique components of unknown function.

type IVb (**Figure 1**). Type IVa pili (T4aP) are a relatively homogeneous and broadly distributed subtype (135), whereas type IVb pili (T4bP) are a more heterogeneous group. The T4bP subfamily is common in enteric species, such as *Vibrio cholerae* (150) and enteropathogenic *Escherichia coli* (EPEC) (157), and on plasmids and other mobile genetic elements (86).

The T4b subfamily includes a monophyletic class called the tight adherence pili (Tad, or Flp pili), first described in *Aggregatibacter* (previously *Actinobacillus*) *actinomycetemcomitans* (77). Tad pili are found in a wide variety of environmental species such as *Caulobacter crescentus* (156). The Flp pilin subunit is much smaller (7–8 kDa) than the typical T4a or T4b pilin subunit (~15–20 kDa). Despite the differences in size, all three pilin subtypes have a highly conserved, hydrophobic α-helical N terminus (42) that contains a characteristic type III signal sequence (182). Unlike signal sequences recognized by signal peptidases I and II, both of which act at the exterior of the cytoplasmic membrane, the type III motif is cleaved by a dedicated prepilin peptidase (121) at the cytoplasmic face of the membrane. Although T4a and T4b pili were traditionally classified by subtle differences in their major pilins (41), it is now clear that each subclass has a distinct assembly system whose components are diagnostic (**Figure 1**) (and see below).

To date, *P. aeruginosa* is the only species in which T4aP, T4bP, and Tad systems in a single strain have been reported (21, 31, 49). The T4aP- and Tad-encoding genes are common to all *P. aeruginosa* genomes sequenced thus far (173), whereas T4bP are found in strains that have *P. aeruginosa* pathogenicity island 1 (PAPI-1) or related elements such as pKLC102 (31, 178). Although each of the three T4P subtypes has a separate assembly system (**Figure 1**), the *P. aeruginosa* T4bP system is dependent on the T4aP prepilin peptidase PilD for processing subunits prior to their assembly (31). The Tad system encodes its own prepilin peptidase, FppA, specific to the Flp subunit (49).

T4b and Tad pili are not typically associated with motility, and many systems lack an ortholog of the PilT ATPase involved in pilus retraction. However, at least some T4bP systems may have retraction capabilities associated with other phenotypes. In the case of *C. crescentus*, the disappearance of Tad pili that occurs with stalk biogenesis was suggested to occur via retraction, since neither pili nor their subunits were detected in the medium (96). EPEC uses T4bP to form bacterial aggregates on intestinal epithelia (the localized adherence, or LA, phenotype). Disaggregation of the microcolonies—presumably by pilus retraction—is important for virulence, as inactivation of the PilT ortholog BfpF reduced pathogenicity in human volunteers (24). A recent study also showed that BfpF function is important for disruption of host cell tight junctions and for promoting the close contact between bacterial and host cells needed for engagement of the type III secretion system. The T4bP system in *P. aeruginosa* has a PilT homolog (PilT2) (**Figure 1**) and is self-transmissible via conjugation (31); retraction of the T4b pilus may be required for productive mating pair formation.

CRAWLING, WALKING, SWARMING, AND SLINGSHOT MOTILITIES

Many (but not all) T4aP-expressing species, including *Neisseria* spp., *Dichelobacter nodosus*, and *P. aeruginosa*, exhibit pilus-mediated twitching motility (referred to as social motility in *M. xanthus*). Some T4aP-expressing species that lack macroscopic twitching motility when examined under the in vitro assay conditions used for *P. aeruginosa* may still have limited motility visible by light microscopy. Twitching occurs on moist surfaces of moderate viscosity, equivalent to that of 1% agar. Although individual cells are capable of movement, it is common to see them moving in rafts, groups of cells preferentially aligned along their long axes (149). Individual cells often snap into alignment with the group when they encounter a raft, contributing to the jerky appearance of twitching motility. It is surprising that even though cells typically move as a group, the exact requirements for an individual cell to participate have not been established. Does each cell need to have pili, or only a subset? Do all the cells need to be motile, or can some be nonmotile? These are simple questions, but important in understanding how twitching cells are coordinated.

The exact distance traveled by twitching cells depends on both extrinsic and intrinsic factors. Extrinsic factors include medium (nutrient) composition (75), viscosity, and hydrophobicity of the surface (149), and intrinsic factors include the amount of pili produced and their retraction rates, and the production of surface-tension-reducing surfactants (129). When a common 1% Luria-Bertani agar stab assay is used, *P. aeruginosa* strains can form a thin, radial twitching zone $\sim$2 cm in diameter around the point of inoculation after 20 h of incubation, a velocity of $\sim$1 mm h^{-1}. If the average *P. aeruginosa* cell is $\sim$2 μM in length, this translates to around 500 cell lengths per hour, although the actual distance traveled is likely to be considerably longer, as the bacteria do not travel in a straight line.

T4P are so thin (<8 nm in diameter) that they cannot be visualized readily by light microscopy, and thus it was initially difficult to conclusively link pilus retraction to motility. Evidence of pilus retraction by live cells was first provided by optical tweezers studies showing that *Neisseria* spp. connected by their pili to beads could reduce the distance between cells and beads while generating substantial forces in the $\sim$100 pN range (106, 111). Even greater forces ($\sim$150 pN) were found when similar studies were performed with *M. xanthus*, potentially because it has multiple retraction motors (37).

In 2001, Skerker & Berg (155) published a seminal study in which they used total internal reflectance microscopy to image live, nonflagellated *P. aeruginosa* labeled with a fluorescent, amino-specific dye. By using this method, it was possible to directly observe single, highly flexible pilus fibers extending from individual cells and exploring the adjacent surface before adhering to it via the pilus tip. Attached pili became taut and shortened, resulting in translocation of the cell body. In addition to supporting the hypothesis that twitching motility results from pilus retraction, the authors showed that retraction of single pilus fibers did not occur in a coordinated fashion; each fiber appeared to retract independently of the others. Also, newly extended pili appeared to be as brightly fluorescent as those already on the surface, implying that they were assembled—at least to some extent—from fluorescent subunits recycled from pili that were disassembled upon retraction after the initial labeling step.

Subsequent studies demonstrated that even though pili can be retracted individually, the cooperative retraction of several bundled fibers generates more substantial forces (in the nanonewton range, versus piconewton range, for a single pilus) (23) and allows cells to travel distances that exceed the length of individual pili. To correlate the number of pilus fibers per cell with the distance traveled by an individual cell, Holz et al. (73) used light microscopy to monitor the motility of individual live cells on electron microscopy grids, followed by rapid fixation, negative staining, and enumeration of the average number of pili per cell. By genetically manipulating the number of pili produced per cell, they showed that bacteria with the most pili traveled farther than bacteria with fewer pili, because of the sharing of forces among multiple fibers. The angles at which individual fibers bind—relative to the cell axis—and the comparative retraction forces on each ultimately affect the direction in which the cell body moves.

Other Forms of T4aP-Mediated Motility

Although twitching motility is the best-characterized type of movement associated with T4aP, a number of other T4aP-dependent modalities have been described for *P. aeruginosa*, including swarming, walking, and slingshot motilities. Swarming motility—a complex phenotypic adaptation (127, 180) that affects a number of traits—occurs on medium that is less viscous than standard twitching motility medium (0.4–0.7% agar versus 1% for twitching). Swarming motility is characterized by the formation of elaborate dendritic patterns by the swarming colony (117) (**Figure 2**). Under specific nutrient conditions, swarming motility requires T4aP (89, 128), but provision of specific carbon sources including glutamate, glucose, or succinate restores swarming—in some cases, hyperswarming—of pilin mutants (153). Swarming is controlled in part by the pilus-related chemotaxis system (see below), as point mutations in the chemosensory protein ChpA decrease or modulate swarming (98).

A novel type of T4aP-mediated walking motility was described recently (39, 57). The investigators developed customized particle-tracking algorithms to quantify the movement of large numbers of individual cells during videomicroscopy of *P. aeruginosa* in flow cells. Two distinct modes of T4aP-related motility were observed: crawling motility, the archetypal twitching mode in which cell bodies are oriented parallel to the surface, and walking motility, in which cell bodies are oriented at right angles to the surface plane and attached by pili extending from the surface-proximal pole. In crawling mode, the cells moved more slowly and were less likely to change direction (high directional persistence), whereas cells in walking mode moved more quickly and were more likely to ramble on the surface (low directional persistence) (**Figure 3**). The horizontal and vertical states were strongly favored over intermediate states, although cells frequently switched between the two orientations in a flow-independent manner. Use of a flow cell was key to observing walking motility, affording cells the opportunity to orient vertically.

The most recently identified mode of T4aP-mediated motility, slingshot motility, contributes significantly to the overall distance traveled by a cell and to its ability to change direction on a surface (80). By separately tracking the leading and trailing poles of cells crawling along a surface, Jin et al. (80) saw that most of the time during twitching (crawling) motility, the leading and

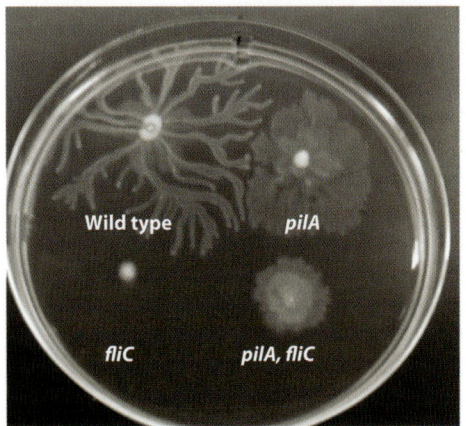

Figure 2

Role of Type IVa pili (T4aP) in swarming and sliding motility. On a medium of intermediate viscosity, wild-type *Pseudomonas aeruginosa* is capable of swarming motility, forming elaborate dendritic colonies (*top left*). Piliated mutants lacking flagella (*fliC*) are nonmotile under these conditions (*bottom left*), whereas flagellated mutants lacking type IV pili (T4P) (*pilA*) show aberrant swarming motility (*top right*). Mutants lacking both T4P and flagella (*pilA, fliC*) can move by sliding motility (*bottom right*), suggesting that expression of T4P hampers sliding in the *fliC* background. Figure reproduced from Reference 117 with permission.

trailing poles of a cell moved in the same direction, parallel to the long axis of the cell, at roughly similar speeds. However, about 5% of the time, and at regular intervals, they saw sudden, rapid lateral movements of the trailing pole that reoriented the cell body. The abrupt release of a single pilus—pulling at an angle relative to the long cell axis—was proposed to result in rotation of the cell body to a new position, dictated by the combined tensile forces generated by remaining adherent pili. Although this slingshot mode of motility occurred less frequently and for much shorter durations than crawling, it contributed equally to the overall trajectory of the cell owing to its 20-fold increase in relative velocity. This mode probably contributes to the classical jerky appearance of twitching motility.

Although T4aP clearly contribute to rapid movement across surfaces, the deployment of T4P inhibits other types of motility. This finding suggests that under specific environmental conditions there is competition between translocation modes that influences the types of surface interactions made by bacteria. For example, Murray & Kazmierczak (117) showed that *P. aeruginosa* lacking both flagella and T4aP could move by sliding motility on 0.5 to 1.0% agar plates (**Figure 2**). This

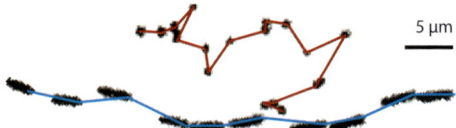

Figure 3

Type IV a pili (T4aP)-mediated crawling versus walking motility. Cells exhibiting crawling motility (*blue trace*), the mode traditionally associated with twitching, are oriented parallel to the surface and change direction infrequently. Cells exhibiting walking motility (*red trace*) are oriented perpendicular to the surface and change direction frequently. Figure reproduced from Reference 39 with permission.

mode was similar to swarming in terms of its complex regulation and dependence upon rhamnolipid surfactant production but was inhibited by T4aP expression. The authors suggested that flagella promote swarming by overcoming the adhesive interactions generated by T4aP. Similarly, Taylor & Buckling (162) showed that *P. aeruginosa* lacking T4aP traveled farther by swimming motility than isogenic mutants lacking the PilU retraction ATPase, presumably because the coexpression of nonretractile pili with flagella reduces the efficiency of flagellum function. Li et al. (100) showed recently that attachment of the polar Tad pili of swimmer cells of *C. crescentus* to surfaces resulted in the mechanical jamming of the flagellum at the same pole, a signaling event triggering the release of exopolysaccharide adhesins that subsequently promote irreversible surface interactions.

TWITCHING MOTILITY: WHAT'S IT FOR?

Although most bacterial pili and fimbriae are used for adherence, the ability to retract provides bonus functions: bringing the cell body into intimate contact with a surface, allowing migration along a surface away from the initial point of contact or toward an attractant, repositioning cells with respect to one another (differentiation within a biofilm), and helping cells efficiently escape from surfaces when necessary.

Arrivals and Departures

The primary function of twitching motility is likely exploration of surfaces, using the crawling, walking, swarming, and skidding modes described above. If the environment is unsuitable, however, a means of escape is required. In collaboration with flagella, T4aP are crucial to the dispersal of bacteria from surfaces in a coordinate series of events called a launch sequence by Conrad et al. (39). Cells tethered to the surface by their flagella and rotating rapidly moved from shallow ($\sim 30°$) to steeper ($\sim 70°$) angles in a T4aP-dependent manner, eventually breaking loose and swimming away. In contrast, flagellated cells lacking pili were unable to reorient themselves and detach. The inability of nonpiliated cells to detach efficiently from the surface following initial attachment resulted in profound differences in the architecture of biofilms arising from the colonizers (**Figure 4**).

Biofilm Development

T4P are important adhesins, promoting initial attachment to a variety of chemically diverse surfaces. They were one of the first factors identified in genetic screens looking for traits important for *P. aeruginosa* biofilm formation (124), participating in both surface attachment and microcolony formation. Chiang & Burrows (33) showed that although the adhesive capacity of T4aP was important in establishment of biofilms in flow cells, twitching motility played a role in development of biofilm architecture. Mutants lacking the retraction ATPase PilT formed thick, undifferentiated biofilms compared with those that were able to twitch. Those data are consistent with the study by Conrad et al. (39) in which nonpiliated cells formed aberrant biofilms, showing that fully functional T4aP are important for structuring a biofilm. It would be interesting to further study the contribution of twitching to biofilm architecture using strains in which PilT expression is inducible to determine whether its latent expression in preformed biofilms of abnormal architecture would lead to their remodeling.

Elegant studies by the Tolker-Nielsen group (15, 87, 88) shed further light on the specific role that T4aP play in biofilm differentiation. They grew biofilms of yellow fluorescent protein–labeled wild-type *P. aeruginosa* mixed with cyan-labeled nonpiliated cells. The resulting biofilms had the typical mushroom-like architecture of mature microcolonies separated by water channels, but

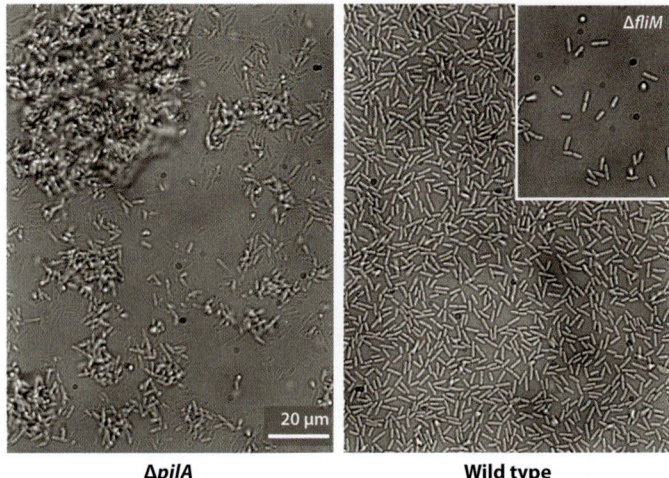

Figure 4

Effect of type IVa pili (T4aP)-mediated detachment on biofilm architecture. *Pseudomonas aeruginosa* uses its T4aP to orient the cells in such a way that they can detach effectively from surfaces using their flagella. Mutants lacking T4aP (*pilA* mutant, *left*) are proficient surface colonizers but deficient in detachment, forming large clumps of cells. In contrast, the detachment-proficient wild type (*right*) is more uniformly distributed on the surface. Bacteria lacking flagella (*fliM* mutant, *inset*) are poor surface colonizers. Figure reproduced from Reference 39 with permission.

there was a startling spatial separation between the two cell types. The nonpiliated mutants were concentrated in the stalk portion of the microcolonies, and the wild-type parent cells formed the majority of the cap portion (88), suggesting that the wild-type cells used twitching to climb a mound of their nonmotile brethren. Further studies using mutants in the Pil-Chp T4aP regulatory system (see below), as well as mutants lacking flagella or impaired in flagellar chemotaxis, showed that altering pilus function caused formation of atypical caps (in the case of *pilH* and *chpA* mutants) and that flagella are also necessary for formation of normal microcolony architecture (15). Although the authors concluded that twitching had only a minor role in cap development, they did not formally test its contribution using a *pilT* mutant.

A clear picture of twitching motility's participation in biofilm development is complicated by the fact that excess twitching can impair biofilm formation in the first place. This delicate balance is best illustrated by studies in which the nutrient content of the medium has been altered, affecting twitching motility. For example, Singh et al. (154) showed that chelation of iron by lactoferrin increased twitching motility, causing a commensurate decrease in biofilm formation. This phenotype is due partly to upregulation of rhamnolipid biosynthesis, a biosurfactant that increases twitching and alters biofilm architecture (61).

Twitching Motility and Pathogenicity

Because T4P help establish initial contact with a host, they are important virulence factors for many species including *P. aeruginosa* (65), an opportunistic pathogen of fungi, worms, plants, animals, and humans. T4P are deployed in the early or acute phase of infection but are frequently lost owing to downregulation or mutation in chronic infections such as cystic fibrosis (105). In the majority of reports, the specific contributions of T4P-mediated adhesion versus twitching motility have not been clarified, because it is common practice to compare the pathogenicity of wild-type strains to mutants lacking pili. Below, I focus on the particular role of twitching motility in virulence as demonstrated in studies using retraction-deficient mutants.

Comolli et al. (38) showed that the PilT and PilU retraction ATPases were important for *P. aeruginosa* cytotoxicity against epithelial cells and for virulence in vivo. Surprisingly, although such mutants are described as hyperpiliated, they found fewer bacteria associated with the epithelial cells compared with mutants lacking pili. This finding implies that additional adhesins on the cell surface need to be engaged to stabilize host-pathogen interactions, and that lack of pilus retraction may prevent those necessary contacts. The dependence of cytotoxicity on pilus retraction was based on the requirement for close apposition of the contact-dependent type III toxin secretion system with the host cell membrane, to allow efficient delivery of effectors (38). Strong evidence for the involvement of twitching motility in *P. aeruginosa* pathogenesis came from the Fleiszig group (181), who showed that mutants lacking PilT or PilU were impaired in corneal infection models. They subsequently demonstrated in vitro using multilayered corneal epithelium that twitching motility mutants were impaired in translocation across the epithelium and in escape from infected cells, phenotypes similar to mutants lacking pili altogether (2).

A recent in vitro evolution study looked at the effects of selecting *P. aeruginosa* strains for the traits of increased twitching or increased swimming motility (163). The two modes of motility were antagonistic, as strains with increased twitching had reduced swimming compared with the parent and vice versa. Strains selected for increased twitching motility (via undefined mechanisms) grew to higher densities in an acute waxworm infection model compared with strains selected for increased swimming. Because the mode of inoculation—injection—bypassed the initial steps of colonization in which T4P normally function, it is unclear whether increased twitching or decreased swimming (a metabolically expensive trait) provided a growth advantage in the host.

Beyond *P. aeruginosa*, a role for twitching motility, or at least for retractile T4P, in virulence is best supported by numerous studies using *Neisseria* retraction-deficient mutants. *N. gonorrhoeae* is an obligate human pathogen and a master manipulator of host signaling cascades. Formation of adherent microcolonies on human epithelia and subsequent T4P retraction, which generates forces in the 70 pN range (122), cause a number of changes in host cells, including recruitment of cytoskeletal and signaling proteins to a cortical plaque beneath the bacteria (54, 110, 168). *N. gonorrhoeae* downregulates proapoptotic and upregulates cytoprotective pathways in epithelial cells in a PilT-dependent manner (72, 74), and a recent study by Dietrich et al. (51) confirmed that the previously reported activation of NFκB by the pathogen was enhanced by T4P retraction. The phenotype was recapitulated in the presence of *pilT* mutants by imposing a shear force on the human cells, confirming early reports (110) that mechanical forces generated by pilus retraction modulate host signaling.

For the T4aP-expressing sheep pathogen, *Dichelobacter nodosus*, Han et al. (66) showed that mutants lacking either PilT or PilU were avirulent and that this deficiency was related specifically to loss of twitching motility. They noted an interesting dichotomy in the two mutants; although both were nonmotile and nonpathogenic, the *pilU* mutant continued to secrete proteases, a T4P-dependent trait in this species. Functional differences between the PilT and PilU ATPases have been noted in a number of species (22, 33–35, 132, 152, 172), suggesting that these proteins have discrete roles despite the apparent similarities between the proteins (see below) and some of their mutant phenotypes.

T4P as Antennae?

One intriguing but underexplored possibility is that in addition to providing motility, T4P can act as antennae to sense and respond to the bacterial environment. Because of their length (up to several microns) they can act at a distance from the cell body but are intimately connected to it via their assembly systems (see below). It is not hard to imagine that T4P could transduce to the bacterium a wealth of information about the physical and even chemical nature of the surfaces to which they become attached. The tactile response of a softer versus harder surface when pilus retraction occurs could provide clues that could inform bacterial behavior. In addition to the retraction-related *Neisseria* host-signaling manipulation studies described above, there is evidence that T4P of *M. xanthus* can detect bacterial molecules such as exopolysaccharides and respond by retracting (25). The T4P of *Geobacter sulfurreducens* can conduct electrons released from the respiratory chain to extracellular acceptors (104, 141), making it possible to envision the reverse scenario, in which bacteria use their T4P to receive information about the electrochemical properties of nearby surfaces.

THE TYPE IV PILI MACHINE AND ITS ENGINE

Overview of the T4P Assembly Machinery

From an evolutionary standpoint, the T4P machinery is ancient, present in gram-positive (including the oldest eubacterial class, *Clostridia*), gram-negative, and archaeal species (135, 164). It has been appreciated for some time that specific T4P system components share sequence similarity with those of the type II secretion system (T2SS) (134). However, recent structural studies of key components have revealed further unanticipated similarities, even between proteins with little to no sequence identity. Because of space limitations, a detailed discussion of structural findings is not possible here. For further information on the structural and functional similarities between T4P and T2S assembly systems, see References 12, 67, 69, 90, 134, and references therein.

In *P. aeruginosa*, most of the genes involved in pilus biogenesis were originally identified via mutant screens (48, 107, 108). Strains that exhibited aberrant motility were further examined for susceptibility to pilus-specific bacteriophages. Mutants deficient in pilus biogenesis or retraction typically exhibit resistance to phage killing. However, paradoxical phenotypes were occasionally observed, in which piliated but nonmotile mutants (e.g., *pilU*) (172) or even apparently nonpiliated mutants (e.g., *fimV*) (148) remained susceptible to pilus-specific phages, suggesting that there were intermediate phenotypes between the completely nonpiliated and the retraction-deficient hyperpiliated forms. These comprehensive surveys laid the foundation for detailed molecular, biochemical, and structural investigations of components involved in twitching motility in *P. aeruginosa*. The organization of the genes encoding the T4a assembly system and regulatory proteins is shown in **Figure 5**, and a list of proteins involved in biogenesis, twitching motility, and regulation of T4P function is provided in **Supplemental Table 1**. Although many of these components are conserved in other T4P-producing species (reviewed in Reference 135), others are unique to *P. aeruginosa*.

The T4a pilus and its assembly apparatus can be envisioned as four interdependent subcomplexes that together span the entire cell envelope, with most components located in the inner membrane (**Figure 6**). The outer membrane subcomplex, which is absent in gram-positive systems, consists of the secretin, a massive (>1 MDa), highly stable dodecamer of PilQ subunits, and its lipoprotein pilotin, PilF, composed entirely of protein-protein interaction motifs called tetratricopeptide repeats (91). PilF is responsible for correct localization and oligomerization of PilQ in the outer membrane, as removal of its lipidation site causes PilQ to multimerize aberrantly in the inner membrane (91). The secretin provides an outer membrane channel for the pilus fiber to exit the cell (175).

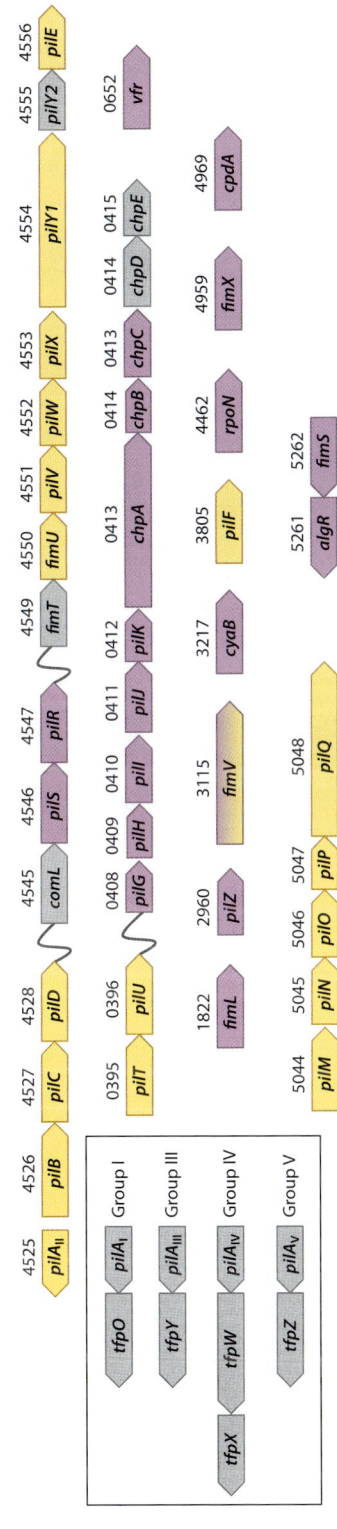

Figure 5

Genomic context of type IVa pili (T4aP) genes in reference strain PAO1. Gene names are shown on the arrows, and the corresponding four-digit PA number (PAxxxx) is shown above each gene (173). Genes encoding structural components are colored yellow, and genes encoding regulatory components are colored purple. Genes that are associated with known T4aP genes but lack a pilus-related phenotype are colored gray. PAO1 encodes a group II pilin; the four alternate pilins and their corresponding accessory genes (95) are boxed at the left.

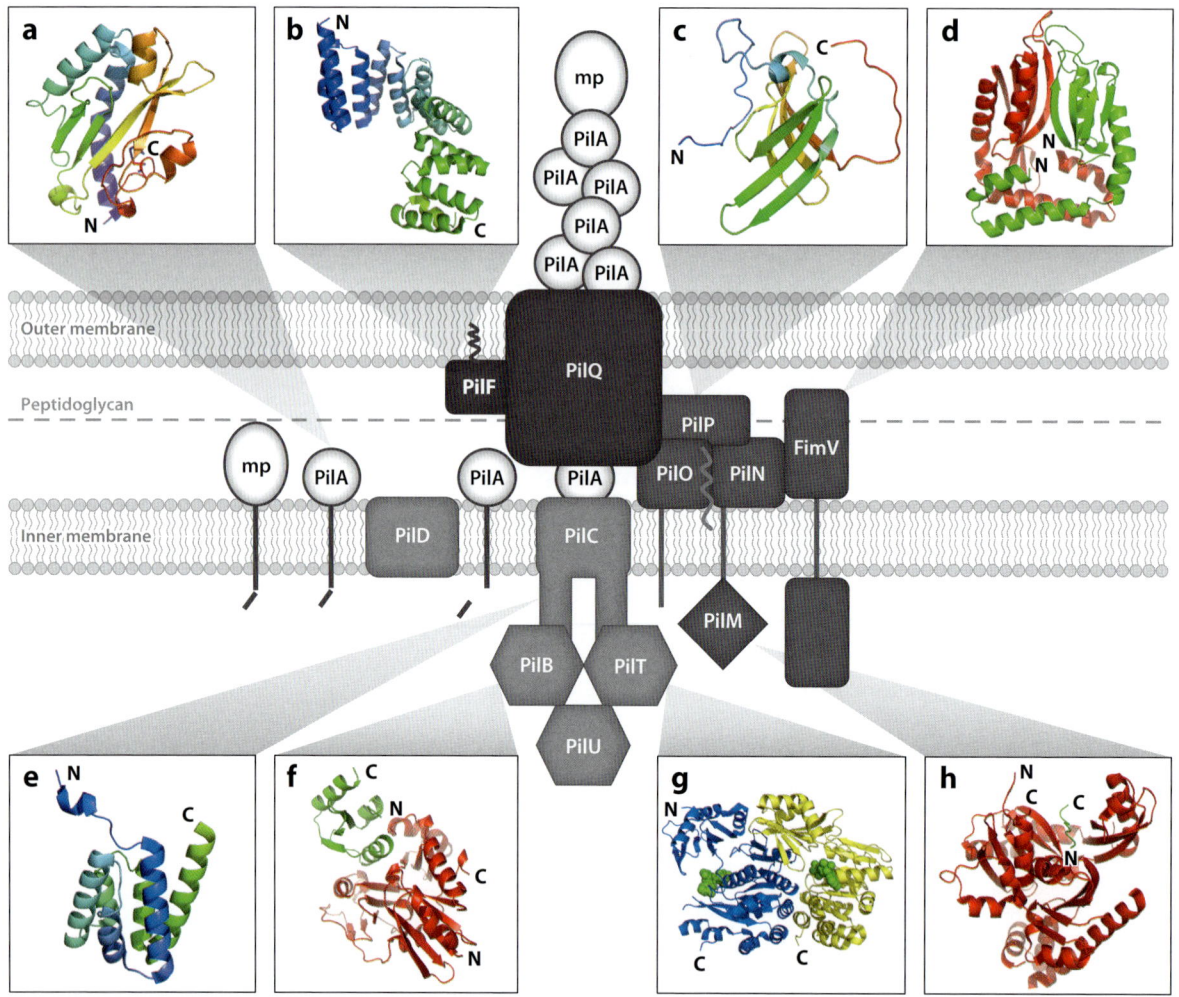

Figure 6

Structures of type IVa pili (T4aP) assembly subcomplex components. The T4aP assembly system comprises four interacting subcomplexes. The outer membrane subcomplex (PilF and PilQ) is shown in dark gray; the inner membrane motor subcomplex (PilB, PilC, PilD, PilT, and PilU) in light gray; the alignment subcomplex (PilM, PilN, PilO, PilP, and FimV) in medium gray; and the pilus subcomplex [the minor pilins (mp) FimU, PilV, PilW, PilX, and PilE and the major pilin, PilA] in white. Structural advances in the T4P and type II secretion systems reveal the architecture of specific components, shown in the surrounding panels. (*a*) The N-terminally truncated group V pilin from strain Pa80110594, PDB code 3JYZ (119). (*b*) Full-length PilF from strain PAO1, PDB code 2HO1 (91). (*c*) C-terminal fragment of PilP from strain PAO1, PDB code 2LC4 (161). (*d*) The periplasmic fragment of PilO (*red*, PDB code 2RJZ) modeled as a heterodimer with the predicted periplasmic fragment of PilN (*green*) (144). (*e*) The N-terminal cytoplasmic domain of PilC from *Thermus thermophilus*, PDB code 2WHN (83). (*f*) A cytoplasmic fragment of EspL (equivalent to PilM, *green*) and its interaction partner EspE (equivalent to PilB, *red*), PDB code 2BH1 (1). (*g*) Two monomers of PilT from strain PA103 (*yellow and blue*) with bound 5′-adenylyl (β,γ-methylene)diphosphonate (AMP-PCP) (*green*), PDB code 3JVV (116). (*h*) PilM (*red*) from *T. thermophilus* bound to an N-terminal peptide from PilN (*green*), PDB code 2YCH (82).

Both the motor and alignment subcomplexes are located in the inner membrane. In *P. aeruginosa*, the exact composition of the T4aP motor subcomplex has not been defined but, on the basis of available studies and evidence from related systems, potentially comprises PilB, PilC, PilT, and PilU (34, 170, 172) (**Figure 6**), as well as regulatory proteins (see below). PilB is a hexameric ATPase of the large VirB11 family (35) and is predicted to convert chemical energy from ATP hydrolysis to mechanical energy required for pilus assembly. PilT and PilU are PilB-like ATPases required for pilin depolymerization. PilC is a three-pass membrane protein with two large cytoplasmic domains, each predicted to contain a bundle of six α-helices (83). Aside from the pilins, PilB and PilC are arguably the most readily identifiable and highly conserved components in T4P, T2S, DNA-uptake, and archaeal motility systems, presumably composing the minimal functional unit of the motor (11, 134, 137). PilD, a membrane-bound aspartyl protease that processes the type III signal sequence at the pilins' N termini, and N-methylates the new terminus (160), might be transiently or permanently associated with the motor subcomplex on the basis of its genetic linkage to PilC in a number of systems and its essentiality for pilus biogenesis. Although the Tad pilus system lacks a retraction ATPase, it has two PilC homologs (21) (TadB and TadC) (**Figure 1**). It is possible that interaction of the sole TadA ATPase with one platform protein could promote pilin polymerization (see below), and interaction with the other could induce depolymerization.

The alignment subcomplex is, as the name implies, proposed to physically connect the inner membrane motor and outer membrane subcomplexes to ensure that the growing, highly flexible pilus is positioned correctly for egress as it begins to extend through the periplasm from its assembly site. In addition, evidence from the related T2SS suggests that components of the alignment subcomplex interact directly with subunits (63), which may increase their local concentration around the assembly site, and participate in gating of the secretin (12). The components of the *P. aeruginosa* T4aP alignment subcomplex include the broadly conserved PilMNOP proteins that are encoded with the PilQ secretin monomer (135), and potentially the peptidoglycan-binding protein FimV (166) (**Figure 6**). Interestingly, components of the alignment subcomplexes from the T4a, T4b, and Tad pilus systems (**Figure 1**) and the T2SS are the least conserved in terms of sequence and number, but emerging structural data have revealed some conservation in their architectures (12).

PilM is a cytoplasmic, actin-like protein bound to PilN via the latter's cytoplasmic N terminus (82). Together, PilM and PilN form the structural equivalent of GspL, an alignment complex protein in the T2SS (144). The PilN and PilO proteins have a similar domain organization, in that both are predicted to have short cytoplasmic N termini, a single transmembrane domain, and a periplasmic C terminus with a long coiled-coil region followed by a ferredoxin-like fold (144). The periplasmic domains of PilN and PilO form highly stable heterodimers in vitro and are dependent on one another for stability in vivo unless PilO is overexpressed, in which case they form stable but nonfunctional homodimers (13). Although interactions between PilMNO and PilC that would support the proposed alignment function have not been described, GspL and the T2S platform protein GspF interact with one another, as well as with GspM (equivalent to PilO) (144) and GspE (equivalent to PilB) (8, 139). PilP is an inner membrane lipoprotein with a long unstructured N terminus followed by a β-sandwich domain (62, 161). Recent biochemical studies showed that PilP forms a stable 1:1:1 heterotrimer with PilN and PilO, and that formation of this complex protects its proteolytically sensitive, unstructured N terminus from degradation (161). Formation of such a complex in vivo is also thought to explain the observation that the lipidation site of PilP can be mutated without loss of function. The isolated β-domain of PilP is unable to form a stable complex with PilNO, likely because it is the region of PilP that interacts with the secretin PilQ (14, 161).

The last member of the proposed alignment subcomplex is FimV, a large (97 kDa) protein containing a periplasmic domain with a peptidoglycan-binding LysM motif connected via a single transmembrane segment to a highly acidic cytoplasmic domain containing predicted tetratricopeptide repeat protein-protein interaction motifs (148). The peptidoglycan-binding function of FimV is required for correct formation of the PilQ secretin, as mutants deleted of the LysM motif have significantly fewer secretins and impaired motility (166). The protein may enable the massive secretin to pass through the covalently closed peptidoglycan layer and/or anchor the assembly system in the cell wall to brace it against the forces generated by pilus retraction (147). The function of the cytoplasmic domain of FimV is currently unknown, but it may participate in regulating pilus biogenesis (56) (see below).

The final subcomplex is, of course, the pilus itself. It is the most dynamic of the four, as it is repeatedly assembled and disassembled. It is composed of the major subunit PilA (130), as well as the minor (based on abundance) pilins FimU, PilVWX, and PilE (5, 6). Each *P. aeruginosa* strain expresses one of five PilA alleles, two of which are glycosylated, and one of two sets of minor pilins (32, 59, 94, 165). The major pilins can be exchanged between strains and function normally if accompanied by their cognate accessory genes that are linked to the pilin loci (9) (**Figure 5**); the exception is the group IV pilins, which also require their unique glycan biosynthetic machinery (68). The two sets of minor pilins are also interchangeable, with the exception of the PilX orthologs, which do not function in heterologous backgrounds (59).

Although the majority of the pilus fiber consists of PilA, the minor components are also present in surface-exposed pili (58, 174). Both major and minor pilins share the highly conserved N-terminal domain characteristic of T4P and T2S subunits, but their periplasmic C termini are divergent. Similar to PilA, minor pilins are processed by PilD, supporting the contention that they are assembled into the pilus (58, 174). The hydrophobic N termini of pilins act as transmembrane domains prior to assembly and as protein-protein interaction domains in the assembled fiber (43). Evidence from the T2SS suggests that the minor subunits form a complex that primes the subsequent assembly of major subunits below it, thereby forming the tip of a growing fiber (36, 52, 92). By analogy, the T4P minor pilins may form a similar initiation complex that is subsequently displayed at the distal ends of pili, though confirmation requires further studies. Immunogold microscopy showed infrequent labeling along pilus fibers, but it was not possible to distinguish whether the proteins were incorporated along the length of the pilus or present at the tips of bundled fibers of different lengths (58).

A large (~125 kDa) nonpilin protein called PilY1 is encoded within and coordinately regulated with the minor pilin operon (5, 26). PilY1 is important for pilus function, although its exact role is controversial (70, 93, 123). Its *Neisseria* homolog, PilC, is proposed to be a pilus tip adhesin, but to date PilY1 has not been convincingly associated with T4P or their tips. According to data from some studies (93, 123), PilY1 appears to have both direct and indirect regulatory effects on pilus function, including calcium-dependent control of pilus retraction. The question of whether PilA, minor pilins, or PilY1 is present at the tips of *P. aeruginosa* pili is a fundamental one, because the biochemical nature of the adhesin is likely to dictate the range and affinities of possible surface interactions. Although PilA was reported to be the T4P adhesin in *P. aeruginosa* (78), the contribution of other proteins must be examined to clarify their role.

How Does Twitching Happen?

In simplified terms, twitching is thought to result from removal of pilin subunits from the base of a previously assembled pilus fiber that is attached to a surface via its tip. Because both the pilus tip and the assembly system are fixed, shortening the pilus by disassembly pulls it increasingly taut. The forces thus generated eventually overcome the friction caused by interactions of the cell body with the surface, pulling the cell toward the adhered pilus tip. This scenario assumes that the contact between the cell body and the surface is the weakest link, so that movement happens before the pilus tip releases, the fiber breaks, or the pilus or assembly system tears free of the cell. The exact mechanisms by which the T4P system assembles and then disassembles a pilus fiber that is under tension are currently the subject of intense research, but biochemical, biophysical, structural, and molecular modeling studies have provided important clues that allow some speculation about the process.

The first stage is pilus assembly. According to current models, the minor pilins, oriented with their C termini in the periplasm, may form an initiation complex that, because of its size and shape, partly protrudes from and deforms the membrane, lowering the amount of energy necessary to extract the hydrophobic N termini (36). Molecular simulations suggest that both pilins and T2S pseudopilins adopt a slight tilt in the membrane (36, 99), which may further assist in their extraction. The low abundance of the minor pilins probably ensures that few initiation sites relative to the amount of major pilins are available for subsequent polymerization, so that each pilus is long enough to extend a useful distance from the cell. Changing the stoichiometry of minor subunits can cause pili of aberrant length to form (53, 58). Although priming by the minor pilins is important for efficient assembly, it is not essential; mutants lacking all the minor pilins still assemble surface pili that can be observed in a retraction-deficient background (59).

To further extract the initiation complex from the membrane by adding additional pilin subunits beneath it, mechanical energy generated by hydrolysis of ATP by PilB is thought to be necessary. Structural data suggest that monomers of both PilB and PilT ATPases have head and body domains connected by a flexible linker (143, 146). In the functional unit, a hexamer, the head of each subunit interacts with the body of the adjacent subunit (55) (**Figure 6**). Comparison of PilT crystal structures in unbound and nucleotide-bound states showed large domain movements of the head relative to the body around the flexible linker (116, 146) that are presumably translated to neighboring subunits and/or other interaction partners. How these motions in a cytoplasmic protein complex result in extraction of a pilin on the periplasmic side of the membrane remains unanswered.

There is evidence from both T2S and T4P systems that the polymerase interacts with the N-terminal cytoplasmic domain of the platform protein (8, 44, 139), although the stoichiometry of both the platform protein and the potential interaction remains murky. If the transmembrane regions of the platform protein simultaneously interact with the membrane-embedded hydrophobic N termini of pilins, a conformational change in the ATPase bound to the platform protein's cytoplasmic domain could shove it, its adjacent transmembrane domain(s), and an associated pilin subunit upward. The distance that a pilin needs to be pushed is ∼1 nm, estimated by fiber structural studies (43) and molecular simulations (99) to be far enough to allow it to establish stable interactions with the preceding subunit. This proposed mechanism is reminiscent of an upside-down version of the process hypothesized for transmission of signals across the membrane by chemotaxis receptor proteins. In that case, binding of chemical ligands to the periplasmic domain of chemoreceptors is thought to cause conformational changes that induce piston-like downward movements of transmembrane helices, generating corresponding movements in their cytoplasmic domains (125, 131).

As the assembly system pushes pilins out of the membrane, subunit-subunit interactions would stabilize them in the assembled fiber. A highly conserved, negatively charged Glu residue at +5 of the mature subunit forms a critical salt bridge with the methylated N terminus of the previous subunit, ensuring the correct registration of one pilin relative to the next (43). Only PilX and related minor (pseudo)pilins lack this key residue (5), which may reflect a role as one of the first proteins to move upward during initiation complex formation. A network of hydrophobic interactions between the pilins' N termini, as well as additional interactions between the loop regions of the C-terminal domains, provides the pilus with its impressive ability to withstand large forces generated upon retraction. The mechanism of pilus length control is unknown but is probably related to the number of available pilins per initiation site (58) and to the frequency of disassembly relative to assembly.

The next stage of twitching is tethering of the pilus tip (or possibly greater portions of the fiber) to a surface (97, 155). This step is poorly characterized, in terms of both the specific biochemical nature of the pilus tip (above) and how an ultimately reversible bond with the surface that is firm enough to withstand retraction forces is achieved. Early studies of PilA function suggested that a disulfide loop located at the C terminus of the protein was the adhesive portion (176), using main chain atoms to bind to receptors on epithelial cells (10). Peptides corresponding to that portion of the protein were also reported to block subsequent pilus-mediated adherence to abiotic surfaces (60). However, these studies did not consider the presence of other components that might modulate attachment. This is an area of T4P function that requires more attention to better explain the amazing binding capacity of these fibers.

The final stage of twitching is pilus retraction, which involves switching from fiber polymerization to depolymerization in response to signals that are not yet understood. The way in which the pilus is reeled into the cell at the blistering pace of $\sim 10^3$ subunits per second (155) while under tension remains a mystery. Although the PilT ATPase is required, exactly how it promotes pilin depolymerization is still unknown. Logically, one might imagine that it could use a strategy similar to that proposed for PilB, but in the opposite direction. Rather than pushing pilins from the membrane upward via interactions with PilC, it could pull them from the pilus downward. The two large cytoplasmic domains of the PilC platform protein have $\sim 34\%$ sequence identity in *P. aeruginosa* and potentially a similar α-helical bundle structure (83). If the N-terminal cytoplasmic domain interacts with PilB, the C-terminal domain might similarly interact with PilT. This proposed interaction is supported by a single piece of evidence showing that overexpression of this portion of the EPEC T4bP platform protein in *trans* mimics a retraction-deficient phenotype, potentially by titrating the retraction ATPase away from the system (44, 115). Alternatively, PilT could displace PilB in binding to the N-terminal cytoplasmic domain; however, there is currently no evidence for this interaction. Another possibility is that PilT interacts directly with the subunits (116), although how such an interaction could occur across the membrane to remove them from the pilus is not clear.

P. aeruginosa has two paralogous retraction ATPases, PilT and PilU (170, 172), that are not functionally interchangeable. Localization studies showed that whereas PilB and PilT are bipolar, PilU is unipolar (34), suggesting a possible role in regulating pilus function (see below), at one pole versus the other. *pilU* mutants are piliated but, unlike *pilT* mutants, remain susceptible to killing by pilus-specific bacteriophages (170, 172). This phenotype is consistent with a small amount of residual motility (66), implying that PilT continues to function, albeit inefficiently, in the *pilU* background. For now, the way in which PilU modulates PilT function is unknown.

WHERE ARE THEY GOING? REGULATION OF TWITCHING MOTILITY

Twitching allows bacteria to travel, but where are they going and how do they navigate? In *P. aeruginosa*, twitching is controlled by a number of regulatory systems that sense external signals—most of which are unknown—and transduce them to modulate pilus extension and retraction. Both physical (e.g., viscosity) and chemical (e.g., phospholipids, iron) signals that influence twitching have been identified (16, 22, 85, 154), as have several regulatory proteins and protein relays (4, 18, 19, 46, 75, 84, 159, 169). **Supplemental Figure 1** summarizes many of the regulatory inputs that control twitching, although a comprehensive understanding of the players and the extent and nature of their interconnectedness continues to evolve. The contributions of a subset of regulatory factors are discussed below.

Regulation of Pilin and Minor Pilin Expression by Two-Component Systems

Among the simplest mechanisms of regulation is expression (or not) of the major pilin subunit, PilA. In *P. aeruginosa*, as in many other T4aP-expressing bacteria, transcription of the *pilA* gene is controlled by the alternate sigma factor, RpoN (σ^{54}), and a two-component system, PilR-PilS (114, 159). PilS, the sensor kinase, has six membrane-spanning domains and is localized to both poles (27). Autophosphorylation of PilS and subsequent phosphorylation of its response regulator, PilR, upregulate *pilA* transcription. The exact signal sensed by PilS has not yet been determined but is potentially PilA itself. Bertrand et al. (22) showed that alteration of inner membrane PilA pools, either by inactivation of genes encoding PilA or by the ATPases PilB or PilT, affected *pilA* transcription. In a *pilB* mutant, where unassembled pilins accumulate in the inner membrane, *pilA* transcription was reduced. In *pilT* mutants, where pilin pools were depleted because of unopposed assembly in the absence of retraction, *pilA* transcription was increased. The ability of PilS to sense and respond to PilA levels in the inner membrane could occur directly, via protein-protein interactions between the highly conserved, membrane-embedded hydrophobic N terminus of PilA and the transmembrane segments of PilS. This hypothesis would account for the ability of PilS—which is identical among unrelated strains of *P. aeruginosa* (59)—to control the levels of pilins whose C-terminal sequences are divergent. Older studies of *M. xanthus* pilin expression (177) support this idea, as specific point mutations in the conserved N terminus of the pilin did not affect its stability but led to dysregulation of expression.

The minor pilin operon is under control of a separate two-component regulatory system, AlgR-FimS (20, 169). Although other genes are part of the AlgR regulon, its effects on pilus biogenesis are limited to control of minor pilin expression. Provision of the operon (but not individual open reading frames) in *trans* in an *algR* background is sufficient to restore pilus biogenesis (102). The alternate sigma factor AlgU (σ^{22}) controlling alginate synthesis also modulates twitching motility (169), although its effects appear to be indirect. AlgU controls the expression of the lectin LecB (17), required for PilJ expression (158), and PilJ controls pilus biogenesis (see below).

cAMP-Dependent Regulation of T4aP Biogenesis

Whereas regulation of pilin transcription is relatively straightforward, regulation of pilus assembly is more complex owing to the large number of components involved. Early in the investigation of *P. aeruginosa* T4aP biogenesis, a putative pilus-specific chemotaxis system (Pil-Chp) composed of components that share sequence similarity with those of the flagellar chemotaxis system of *E. coli* was identified (45–47, 171). The canonical flagellar system consists of membrane-bound methyl-accepting chemotaxis proteins (MCPs) that sense environmental stimuli and undergo a conformational change transmitted via an adaptor protein, CheW, to a histidine kinase, CheA. After autophosphorylation, CheA phosphorylates CheY, which binds to the flagellar switch complex to induce conformational changes that ultimately reverse the direction of flagellar motor rotation (133, 145). The extent of CheY phosphorylation is controlled by CheZ, and sensitivity of the MCPs to signals is controlled by their methylation and demethylation via CheR and CheB, respectively (142). Although the Pil-Chp system has identifiable homologs of many Che proteins, the extension-retraction mechanism of the pilus motor is clearly different from that of the rotary flagellum motor.

The Pil-Chp system has a single MCP, PilJ, localized to both poles of the cell (50). PilI is a CheW homolog and ChpA (PilL) is a more complicated version of CheA, with a combination of nine His, Ser, and Thr phosphotransfer domains and one C-terminal CheY-like domain (171). There are two CheY-like regulators, PilG and PilH, but no CheZ homolog. Fulcher and colleagues (56) proposed that in the absence of a phosphatase, PilH may act as a phosphate sink to control the extent of PilG phosphorylation. PilK and ChpB encode homologs of CheR and CheB, respectively. ChpC is predicted to be a second CheW homolog that may link other MCPs to ChpA (171). Similar chemotaxis systems can be identified in many other soil-, water-, and plant-associated bacteria that have T4aP, such as *Myxococcus*, *Xylella*, *Xanthomonas*, and *Ralstonia*, but not all T4aP-expressing bacteria encode such a system. Pathogens such as *Neisseria*, *Legionella*, and *Vibrio* lack Pil-Chp systems.

Recent analyses of *P. aeruginosa pil-chp* mutants confirmed that *pilG*, *pilI*, *pilJ*, and *chpA* are involved in pilus assembly, as mutants had little or no surface pili despite having normal levels of intracellular PilA (22, 56). Similarly, a *lecB* mutant, which has a substantial reduction in PilJ levels, does not assemble pili but has normal PilA levels (158). In contrast, *pilH*, *pilK*, *chpB*, and *chpC* mutants continued to twitch, although surface pilus levels were slightly altered in some strains; in particular, *pilH* and *chpB* mutants were reported to be hyperpiliated, suggesting their inactivation has effects on retraction. From these and other data, Bertrand et al. (22) proposed that PilG may control PilB activity and thus assembly, and PilH might affect PilT activity and therefore retraction. Although this is a simple and attractive hypothesis, further discoveries suggested that the regulatory picture is more complex (56, 76).

The expression of many *P. aeruginosa* virulence factors—including T4aP—is controlled in part by a cyclic AMP (cAMP)-binding protein called Vfr (virulence factor regulator) (167). cAMP is synthesized by two adenylate cyclases, CyaA and CyaB, with CyaB being the major contributor to cellular cAMP pools (56). While searching for factors that affected CyaB activity, Fulcher et al. (56) isolated mutants in the *pil-chp* genes, as well as others implicated in pilus biogenesis or function (*pilA*, *pilB*, *fimL*, *fimV*, *vfr*). Mutations that affected pilus biogenesis (*pilG*, *pilI*, *pilJ*, *chpA*, *fimL*, *fimV*) reduced cellular cAMP levels—similar to those of a *cyaAB* double mutant—even though CyaB protein levels were unaffected. Similarly, FimL appears to positively regulate CyaB activity, as mutants had normal amounts of CyaB but low cAMP levels (76). In contrast, *pilH*, *pilK*, and *chpB* mutants had high levels of cAMP and greater than wild-type levels of surface pili. From these data, PilG and PilH were proposed to modulate CyaB activity and thus the expression of Vfr-dependent proteins, including PilMNOPQ (81), among others. However, it was also apparent that the Pil-Chp system had cAMP-independent effects on pilus function, as supplying a *cyaAB-pilG* triple mutant with exogenous cAMP restored pilus biogenesis but not motility (56).

FimV appears to have dual roles in pilus biogenesis, consistent with its two-domain structure. Its periplasmic N terminus binds peptidoglycan and promotes PilQ secretin formation in the outer membrane (166), and its cytoplasmic C terminus may positively regulate CyaB function (56). The decreased levels of PilMNOP previously reported in *fimV* mutants (166) may be related to low cAMP levels and thus Vfr-dependent transcription (81). FimV affected function of the T2SS under specific growth conditions (112), also supporting a broader regulatory function.

Cyclic-di-GMP Signaling and Twitching Motility

Cyclic-di-GMP (cdG) is another key secondary messenger in bacteria that has received considerable attention because of its role in promoting biofilm formation (40, 71) and other phenotypes associated with chronic *P. aeruginosa* infections (103). Proteins involved in cdG metabolism have characteristic motifs GGDEF (found in diguanylate cyclases) and/or EAL (found in phosphodiesterases). In some proteins, degenerate forms of these motifs simply bind the molecule. In addition, a host of other cdG-binding motifs continue to be identified, including allosteric regulatory sites controlling enzymatic activity and RNA riboswitches controlling expression.

Two cdG-related proteins required for T4P biogenesis have been identified, PilZ and FimX. PilZ mutants express normal amounts of pilins but do not assemble pili and are resistant to pilus-specific phage (4). When identified, PilZ had a novel sequence that was subsequently identified as a archetypal cdG-binding adaptor domain, found in a large and diverse family of proteins (7). Ironically, although binding of cdG by many members of the PilZ domain family has since been demonstrated, PilZ itself (PA2960) does not bind the molecule (109). Structural studies (101) revealed that PilZ and its orthologs in other T4P systems have critical structural differences at the N terminus where members of the larger PilZ domain family typically bind cdG. Subsequent studies in *Xanthomonas* spp. showed that PilZ interacts with both PilB, the pilin polymerase, and the nonfunctional EAL domain of FimX, a cdG-binding protein (64). This work suggested that PilZ together with FimX might regulate PilB function in a cdG-dependent manner.

P. aeruginosa FimX is a unipolar signaling protein containing a CheY-like REC domain, a PAS sensor domain, and degenerate GGDEF (GDSIF) and EAL (EVL) domains (75, 84), implying that it senses and integrates environmental signals relevant to twitching to influence pilus function. Mutants lacking *fimX* have reduced twitching and respond only weakly to specific environmental cues (e.g., the presence of mucin) that markedly stimulate twitching of wild-type cells. The noncatalytic C-terminal EVL domain binds cdG with high affinity (K_d ~100 nM), causing a long-range (~7 nm) conformational change to be transduced to the N-terminal REC domain responsible for unipolar localization (118, 140). Deletion of the GDSIF or EVL motifs prevents localization and function (84), suggesting that integration of extracellular cues with intracellular levels of cdG is required to modulate twitching. *P. aeruginosa* has multiple diguanylate cyclases and phosphodiesterases that can modulate cdG levels. Many are linked to other signaling pathways, so it will be interesting to determine which ones connect to FimX, and how the circuits connecting cdG- and cAMP-based control of twitching are organized.

LOOKING FORWARD

The field of twitching motility has seen tremendous progress in many areas over the past decade. The discovery of the many ways beyond simple crawling motility in which T4P contribute to bacterial movement and biofilm development was made possible by development of elegant microscopy and biophysical techniques coupled with sophisticated particle-tracking algorithms. Other notable achievements include an increasingly detailed picture of the molecular structure of the T4P machinery and its potential mechanism. The rapid characterization of many of the previously identified genes that were implicated in T4P function but whose roles were enigmatic has been particularly interesting. Many of these mysterious *pil* and *fim* genes have turned out to be key regulators of pilus biogenesis or pilus function, and it will be exciting to add more pieces to this incomplete jigsaw puzzle.

Topics needing further attention include the similarities and differences between T4a, T4b, and Tad pili; their assembly systems; and their regulatory components. Most of our structural information comes from the T4a system, and with a single exception (179), there is almost nothing known about the structure of the Tad pilus and its machinery. A more complete understanding of how these systems evolved and how they function would be very useful. In particular, identification of key targets may aid in the development of type IV pilicides that could be used in antivirulence strategies. This approach has been used successfully for other types of bacterial fimbriae involved in pathogenesis (136). The stoichiometry and spatial organization of the components within the T4P complexes and the manner in which the systems are organized in the cell envelope, including the peptidoglycan layer, are poorly defined. Are they built in during cell division when new poles are formed, or retrofitted afterward? Do components of the assembly systems form stable complexes, or are specific subcomplexes (in addition to the pilus) dynamic? These areas and more await revelation in the next 10 years.

DISCLOSURE STATEMENT

The author is not aware of any affiliations, memberships, funding, or financial holdings that might be perceived as affecting the objectivity of this review.

ACKNOWLEDGMENTS

I am very grateful to the past and present members of my laboratory, and to my long-term collaborator Lynne Howell and her trainees, for their hard work and interesting ideas about all things pili. To those in the field, I'm sorry that space limitations made it impossible to cover all aspects with equal attention. Work in the Burrows laboratory on T4P is generously funded by the Canadian Institutes of Health Research.

LITERATURE CITED

1. Abendroth J, Murphy P, Sandkvist M, Bagdasarian M, Hol WG. 2005. The X-ray structure of the type II secretion system complex formed by the N-terminal domain of EpsE and the cytoplasmic domain of EpsL of *Vibrio cholerae*. *J. Mol. Biol.* 348:845–55
2. Alarcon I, Evans DJ, Fleiszig SM. 2009. The role of twitching motility in *Pseudomonas aeruginosa* exit from and translocation of corneal epithelial cells. *Invest. Ophthalmol. Vis. Sci.* 50:2237–44
3. Albers SV, Pohlschroder M. 2009. Diversity of archaeal type IV pilin-like structures. *Extremophiles* 13:403–10
4. Alm RA, Bodero AJ, Free PD, Mattick JS. 1996. Identification of a novel gene, pilZ, essential for type 4 fimbrial biogenesis in *Pseudomonas aeruginosa*. *J. Bacteriol.* 178:46–53
5. Alm RA, Hallinan JP, Watson AA, Mattick JS. 1996. Fimbrial biogenesis genes of *Pseudomonas aeruginosa*: *pilW* and *pilX* increase the similarity of type 4 fimbriae to the GSP protein-secretion systems and *pilY1* encodes a gonococcal PilC homologue. *Mol. Microbiol.* 22:161–73
6. Alm RA, Mattick JS. 1996. Identification of two genes with prepilin-like leader sequences involved in type 4 fimbrial biogenesis in *Pseudomonas aeruginosa*. *J. Bacteriol.* 178:3809–17
7. Amikam D, Galperin MY. 2006. PilZ domain is part of the bacterial c-di-GMP binding protein. *Bioinformatics* 22:3–6
8. Arts J, de Groot A, Ball G, Durand E, Khattabi ME, et al. 2007. Interaction domains in the *Pseudomonas aeruginosa* type II secretory apparatus component XcpS (GspF). *Microbiology* 153:1582–92
9. Asikyan ML, Kus JV, Burrows LL. 2008. Novel proteins that modulate type IV pilus retraction dynamics in *Pseudomonas aeruginosa*. *J. Bacteriol.* 190:7022–34
10. Audette GF, Irvin RT, Hazes B. 2004. Crystallographic analysis of the *Pseudomonas aeruginosa* strain K122-4 monomeric pilin reveals a conserved receptor-binding architecture. *Biochemistry* 43:11427–35
11. Averhoff B, Friedrich A. 2003. Type IV pili-related natural transformation systems: DNA transport in mesophilic and thermophilic bacteria. *Arch. Microbiol.* 180:385–93
12. Ayers M, Howell PL, Burrows LL. 2010. Architecture of the type II secretion and type IV pilus machineries. *Future Microbiol.* 5:1203–18

13. Ayers M, Sampaleanu LM, Tammam S, Koo J, Harvey H, et al. 2009. PilM/N/O/P proteins form an inner membrane complex that affects the stability of the *Pseudomonas aeruginosa* type IV pilus secretin. *J. Mol. Biol.* 394:128–42
14. Balasingham SV, Collins RF, Assalkhou R, Homberset H, Frye SA, et al. 2007. Interactions between the lipoprotein PilP and the secretin PilQ in *Neisseria meningitidis*. *J. Bacteriol.* 189:5716–27
15. Barken KB, Pamp SJ, Yang L, Gjermansen M, Bertrand JJ, et al. 2008. Roles of type IV pili, flagellum-mediated motility and extracellular DNA in the formation of mature multicellular structures in *Pseudomonas aeruginosa* biofilms. *Environ. Microbiol.* 10:2331–43
16. Barker AP, Vasil AI, Filloux A, Ball G, Wilderman PJ, Vasil ML. 2004. A novel extracellular phospholipase C of *Pseudomonas aeruginosa* is required for phospholipid chemotaxis. *Mol. Microbiol.* 53:1089–98
17. Bazire A, Shioya K, Soum-Soutera E, Bouffartigues E, Ryder C, et al. 2010. The sigma factor AlgU plays a key role in formation of robust biofilms by nonmucoid *Pseudomonas aeruginosa*. *J. Bacteriol.* 192:3001–10
18. Beatson SA, Whitchurch CB, Sargent JL, Levesque RC, Mattick JS. 2002. Differential regulation of twitching motility and elastase production by Vfr in *Pseudomonas aeruginosa*. *J. Bacteriol.* 184:3605–13
19. Beatson SA, Whitchurch CB, Semmler AB, Mattick JS. 2002. Quorum sensing is not required for twitching motility in *Pseudomonas aeruginosa*. *J. Bacteriol.* 184:3598–604
20. Belete B, Lu H, Wozniak DJ. 2008. *Pseudomonas aeruginosa* AlgR regulates type IV pilus biosynthesis by activating transcription of the *fimU-pilVWXY1Y2E* operon. *J. Bacteriol.* 190:2023–30
21. Bernard CS, Bordi C, Termine E, Filloux A, de Bentzmann S. 2009. Organization and PprB-dependent control of the *Pseudomonas aeruginosa tad* locus, involved in Flp pilus biology. *J. Bacteriol.* 191:1961–73
22. Bertrand JJ, West JT, Engel JN. 2010. Genetic analysis of the regulation of type IV pilus function by the Chp chemosensory system of *Pseudomonas aeruginosa*. *J. Bacteriol.* 192:994–1010
23. Biais N, Ladoux B, Higashi D, So M, Sheetz M. 2008. Cooperative retraction of bundled type IV pili enables nanonewton force generation. *PLoS Biol.* 6:e87
24. Bieber D, Ramer SW, Wu CY, Murray WJ, Tobe T, et al. 1998. Type IV pili, transient bacterial aggregates, and virulence of enteropathogenic *Escherichia coli*. *Science* 280:2114–18
25. Black WP, Xu Q, Yang Z. 2006. Type IV pili function upstream of the Dif chemotaxis pathway in *Myxococcus xanthus* EPS regulation. *Mol. Microbiol.* 61:447–56
26. Bohn YS, Brandes G, Rakhimova E, Horatzek S, Salunkhe P, et al. 2009. Multiple roles of *Pseudomonas aeruginosa* TBCF10839 PilY1 in motility, transport and infection. *Mol. Microbiol.* 71:730–47
27. Boyd JM. 2000. Localization of the histidine kinase PilS to the poles of *Pseudomonas aeruginosa* and identification of a localization domain. *Mol. Microbiol.* 36:153–62
28. Bradley DE. 1972. Evidence for the retraction of *Pseudomonas aeruginosa* RNA phage pili. *Biochem. Biophys. Res. Commun.* 47:142–49
29. Bradley DE. 1980. A function of *Pseudomonas aeruginosa* PAO polar pili: twitching motility. *Can. J. Microbiol.* 26:146–54
30. Burrows LL. 2005. Weapons of mass retraction. *Mol. Microbiol.* 57:878–88
31. Carter MQ, Chen J, Lory S. 2010. The *Pseudomonas aeruginosa* pathogenicity island PAPI-1 is transferred via a novel type IV pilus. *J. Bacteriol.* 192:3249–58
32. Castric P. 1995. *pilO*, a gene required for glycosylation of *Pseudomonas aeruginosa* 1244 pilin. *Microbiology* 141(Pt. 5):1247–54
33. Chiang P, Burrows LL. 2003. Biofilm formation by hyperpiliated mutants of *Pseudomonas aeruginosa*. *J. Bacteriol.* 185:2374–78
34. Chiang P, Habash M, Burrows LL. 2005. Disparate subcellular localization patterns of *Pseudomonas aeruginosa* type IV pilus ATPases involved in twitching motility. *J. Bacteriol.* 187:829–39
35. Chiang P, Sampaleanu LM, Ayers M, Pahuta M, Howell PL, Burrows LL. 2008. Functional role of conserved residues in the characteristic secretion NTPase motifs of the *Pseudomonas aeruginosa* type IV pilus motor proteins PilB, PilT and PilU. *Microbiology* 154:114–26
36. Cisneros DA, Bond PJ, Pugsley AP, Campos M, Francetic O. 2011. Minor pseudopilin self-assembly primes type II secretion pseudopilus elongation. *EMBO J.* 31:1041–53
37. Clausen M, Jakovljevic V, Sogaard-Andersen L, Maier B. 2009. High-force generation is a conserved property of type IV pilus systems. *J. Bacteriol.* 191:4633–38

38. Comolli JC, Hauser AR, Waite L, Whitchurch CB, Mattick JS, Engel JN. 1999. *Pseudomonas aeruginosa* gene products PilT and PilU are required for cytotoxicity in vitro and virulence in a mouse model of acute pneumonia. *Infect. Immun.* 67:3625–330
39. Conrad JC, Gibiansky ML, Jin F, Gordon VD, Motto DA, et al. 2011. Flagella and pili-mediated near-surface single-cell motility mechanisms in *P. aeruginosa*. *Biophys. J.* 100:1608–16
40. Cotter PA, Stibitz S. 2007. c-di-GMP-mediated regulation of virulence and biofilm formation. *Curr. Opin. Microbiol.* 10:17–23
41. Craig L, Li J. 2008. Type IV pili: paradoxes in form and function. *Curr. Opin. Struct. Biol.* 18:267–77
42. Craig L, Pique ME, Tainer JA. 2004. Type IV pilus structure and bacterial pathogenicity. *Nat. Rev. Microbiol.* 2:363–78
43. Craig L, Volkmann N, Arvai AS, Pique ME, Yeager M, et al. 2006. Type IV pilus structure by cryo-electron microscopy and crystallography: implications for pilus assembly and functions. *Mol. Cell* 23:651–62
44. Crowther LJ, Anantha RP, Donnenberg MS. 2004. The inner membrane subassembly of the enteropathogenic *Escherichia coli* bundle-forming pilus machine. *Mol. Microbiol.* 52:67–79
45. Darzins A. 1993. The *pilG* gene product, required for *Pseudomonas aeruginosa* pilus production and twitching motility, is homologous to the enteric, single-domain response regulator CheY. *J. Bacteriol.* 175:5934–44
46. Darzins A. 1994. Characterization of a *Pseudomonas aeruginosa* gene cluster involved in pilus biosynthesis and twitching motility: sequence similarity to the chemotaxis proteins of enterics and the gliding bacterium *Myxococcus xanthus*. *Mol. Microbiol.* 11:137–53
47. Darzins A. 1995. The *Pseudomonas aeruginosa pilK* gene encodes a chemotactic methyltransferase (CheR) homologue that is translationally regulated. *Mol. Microbiol.* 15:703–17
48. Darzins A, Russell MA. 1997. Molecular genetic analysis of type-4 pilus biogenesis and twitching motility using *Pseudomonas aeruginosa* as a model system—a review. *Gene* 192:109–15
49. de Bentzmann S, Aurouze M, Ball G, Filloux A. 2006. FppA, a novel *Pseudomonas aeruginosa* prepilin peptidase involved in assembly of type IVb pili. *J. Bacteriol.* 188:4851–60
50. DeLange PA, Collins TL, Pierce GE, Robinson JB. 2007. PilJ localizes to cell poles and is required for type IV pilus extension in *Pseudomonas aeruginosa*. *Curr. Microbiol.* 55:389–95
51. Dietrich M, Bartfeld S, Munke R, Lange C, Ogilvie LA, et al. 2011. Activation of NF-κB by *Neisseria gonorrhoeae* is associated with microcolony formation and type IV pilus retraction. *Cell. Microbiol.* 13:1168–82
52. Douzi B, Durand E, Bernard C, Alphonse S, Cambillau C, et al. 2009. The XcpV/GspI pseudopilin has a central role in the assembly of a quaternary complex within the T2SS pseudopilus. *J. Biol. Chem.* 284:34580–89
53. Durand E, Michel G, Voulhoux R, Kurner J, Bernadac A, Filloux A. 2005. XcpX controls biogenesis of the *Pseudomonas aeruginosa* XcpT-containing pseudopilus. *J. Biol. Chem.* 280:31378–89
54. Edwards JL, Shao JQ, Ault KA, Apicella MA. 2000. *Neisseria gonorrhoeae* elicits membrane ruffling and cytoskeletal rearrangements upon infection of primary human endocervical and ectocervical cells. *Infect. Immun.* 68:5354–63
55. Forest KT, Tainer JA. 1997. Type-4 pilus-structure: outside to inside and top to bottom—a minireview. *Gene* 192:165–69
56. Fulcher NB, Holliday PM, Klem E, Cann MJ, Wolfgang MC. 2010. The *Pseudomonas aeruginosa* Chp chemosensory system regulates intracellular cAMP levels by modulating adenylate cyclase activity. *Mol. Microbiol.* 76:889–904
57. Gibiansky ML, Conrad JC, Jin F, Gordon VD, Motto DA, et al. 2010. Bacteria use type IV pili to walk upright and detach from surfaces. *Science* 330:197
58. Giltner CL, Habash M, Burrows LL. 2010. *Pseudomonas aeruginosa* minor pilins are incorporated into the type IV pilus. *J. Mol. Biol.* 398:444–61
59. Giltner CL, Rana N, Lunardo MN, Hussain AQ, Burrows LL. 2011. Evolutionary and functional diversity of the *Pseudomonas* type IVa pilin island. *Environ. Microbiol.* 13:250–64

60. Giltner CL, van Schaik EJ, Audette GF, Kao D, Hodges RS, et al. 2006. The *Pseudomonas aeruginosa* type IV pilin receptor binding domain functions as an adhesin for both biotic and abiotic surfaces. *Mol. Microbiol.* 59:1083–96
61. Glick R, Gilmour C, Tremblay J, Satanower S, Avidan O, et al. 2010. Increase in rhamnolipid synthesis under iron-limiting conditions influences surface motility and biofilm formation in *Pseudomonas aeruginosa*. *J. Bacteriol.* 192:2973–80
62. Golovanov AP, Balasingham S, Tzitzilonis C, Goult BT, Lian LY, et al. 2006. The solution structure of a domain from the *Neisseria meningitidis* lipoprotein PilP reveals a new β-sandwich fold. *J. Mol. Biol.* 364:186–95
63. Gray MD, Bagdasarian M, Hol WG, Sandkvist M. 2011. In vivo cross-linking of EpsG to EpsL suggests a role for EpsL as an ATPase-pseudopilin coupling protein in the type II secretion system of *Vibrio cholerae*. *Mol. Microbiol.* 79:786–98
64. Guzzo CR, Salinas RK, Andrade MO, Farah CS. 2009. PILZ protein structure and interactions with PILB and the FIMX EAL domain: implications for control of type IV pilus biogenesis. *J. Mol. Biol.* 393:848–66
65. Hahn HP. 1997. The type-4 pilus is the major virulence-associated adhesin of *Pseudomonas aeruginosa*—a review. *Gene* 192:99–108
66. Han X, Kennan RM, Davies JK, Reddacliff LA, Dhungyel OP, et al. 2008. Twitching motility is essential for virulence in *Dichelobacter nodosus*. *J. Bacteriol.* 190:3323–35
67. Hansen JK, Forest KT. 2006. Type IV pilin structures: insights on shared architecture, fiber assembly, receptor binding and type II secretion. *J. Mol. Microbiol. Biotechnol.* 11:192–207
68. Harvey H, Kus JV, Tessier L, Kelly J, Burrows LL. 2011. *Pseudomonas aeruginosa* D-arabinofuranose biosynthetic pathway and its role in type IV pilus assembly. *J. Biol. Chem.* 286:28128–37
69. Hazes B, Frost L. 2008. Towards a systems biology approach to study type II/IV secretion systems. *Biochim. Biophys. Acta* 1778:1839–50
70. Heiniger RW, Winther-Larsen HC, Pickles RJ, Koomey M, Wolfgang MC. 2010. Infection of human mucosal tissue by *Pseudomonas aeruginosa* requires sequential and mutually dependent virulence factors and a novel pilus-associated adhesin. *Cell. Microbiol.* 12:1158–73
71. Hengge R. 2009. Principles of c-di-GMP signalling in bacteria. *Nat. Rev. Microbiol.* 7:263–73
72. Higashi DL, Lee SW, Snyder A, Weyand NJ, Bakke A, So M. 2007. Dynamics of *Neisseria gonorrhoeae* attachment: microcolony development, cortical plaque formation, and cytoprotection. *Infect. Immun.* 75:4743–53
73. Holz C, Opitz D, Greune L, Kurre R, Koomey M, et al. 2010. Multiple pilus motors cooperate for persistent bacterial movement in two dimensions. *Phys. Rev. Lett.* 104:178104
74. Howie HL, Shiflett SL, So M. 2008. Extracellular signal-regulated kinase activation by *Neisseria gonorrhoeae* downregulates epithelial cell proapoptotic proteins Bad and Bim. *Infect. Immun.* 76:2715–21
75. Huang B, Whitchurch CB, Mattick JS. 2003. FimX, a multidomain protein connecting environmental signals to twitching motility in *Pseudomonas aeruginosa*. *J. Bacteriol.* 185:7068–76
76. Inclan YF, Huseby MJ, Engel JN. 2011. FimL regulates cAMP synthesis in *Pseudomonas aeruginosa*. *PLoS One* 6:e15867
77. Inoue T, Tanimoto I, Ohta H, Kato K, Murayama Y, Fukui K. 1998. Molecular characterization of low-molecular-weight component protein, Flp, in *Actinobacillus actinomycetemcomitans* fimbriae. *Microbiol. Immunol.* 42:253–58
78. Irvin RT, Doig P, Lee KK, Sastry PA, Paranchych W, et al. 1989. Characterization of the *Pseudomonas aeruginosa* pilus adhesin: confirmation that the pilin structural protein subunit contains a human epithelial cell-binding domain. *Infect. Immun.* 57:3720–26
79. Jarrell KF, McBride MJ. 2008. The surprisingly diverse ways that prokaryotes move. *Nat. Rev. Microbiol.* 6:466–76
80. Jin F, Conrad JC, Gibiansky ML, Wong GC. 2011. Bacteria use type-IV pili to slingshot on surfaces. *Proc. Natl. Acad. Sci. USA* 108:12617–22
81. Kanack KJ, Runyen-Janecky LJ, Ferrell EP, Suh SJ, West SE. 2006. Characterization of DNA-binding specificity and analysis of binding sites of the *Pseudomonas aeruginosa* global regulator, Vfr, a homologue of the *Escherichia coli* cAMP receptor protein. *Microbiology* 152:3485–96

82. Karuppiah V, Derrick JP. 2011. Structure of the PilM-PilN inner membrane type IV pilus biogenesis complex from *Thermus thermophilus*. *J. Biol. Chem.* 286:24434–42
83. Karuppiah V, Hassan D, Saleem M, Derrick JP. 2010. Structure and oligomerization of the PilC type IV pilus biogenesis protein from *Thermus thermophilus*. *Proteins* 78:2049–57
84. Kazmierczak BI, Lebron MB, Murray TS. 2006. Analysis of FimX, a phosphodiesterase that governs twitching motility in *Pseudomonas aeruginosa*. *Mol. Microbiol.* 60:1026–43
85. Kearns DB, Robinson J, Shimkets LJ. 2001. *Pseudomonas aeruginosa* exhibits directed twitching motility up phosphatidylethanolamine gradients. *J. Bacteriol.* 183:763–67
86. Kim SR, Komano T. 1997. The plasmid R64 thin pilus identified as a type IV pilus. *J. Bacteriol.* 179:3594–603
87. Klausen M, Aaes-Jørgensen A, Molin S, Tolker-Nielsen T. 2003. Involvement of bacterial migration in the development of complex multicellular structures in *Pseudomonas aeruginosa* biofilms. *Mol. Microbiol.* 50:61–68
88. Klausen M, Heydorn A, Ragas P, Lambertsen L, Aaes-Jørgensen A, et al. 2003. Biofilm formation by *Pseudomonas aeruginosa* wild type, flagella and type IV pili mutants. *Mol. Microbiol.* 48:1511–24
89. Kohler T, Curty LK, Barja F, van Delden C, Pechere JC. 2000. Swarming of *Pseudomonas aeruginosa* is dependent on cell-to-cell signaling and requires flagella and pili. *J. Bacteriol.* 182:5990–96
90. Koo J, Burrows LL, Howell PL. 2011. Decoding the roles of pilotins and accessory proteins in secretin escort services. *FEMS Microbiol Lett.* 328:1–12
91. Koo J, Tammam S, Ku SY, Sampaleanu LM, Burrows LL, Howell PL. 2008. PilF is an outer membrane lipoprotein required for multimerization and localization of the *Pseudomonas aeruginosa* type IV pilus secretin. *J. Bacteriol.* 190:6961–69
92. Korotkov KV, Hol WG. 2008. Structure of the GspK-GspI-GspJ complex from the enterotoxigenic *Escherichia coli* type 2 secretion system. *Nat. Struct. Mol. Biol.* 15:462–68
93. Kuchma SL, Ballok AE, Merritt JH, Hammond JH, Lu W, et al. 2010. Cyclic-di-GMP-mediated repression of swarming motility by *Pseudomonas aeruginosa*: the *pilY1* gene and its impact on surface-associated behaviors. *J. Bacteriol.* 192:2950–64
94. Kus JV, Kelly J, Tessier L, Harvey H, Cvitkovitch DG, Burrows LL. 2008. Modification of *Pseudomonas aeruginosa* Pa5196 type IV pilins at multiple sites with D-Araf by a novel GT-C family arabinosyltransferase, TfpW. *J. Bacteriol.* 190:7464–78
95. Kus JV, Tullis E, Cvitkovitch DG, Burrows LL. 2004. Significant differences in type IV pilin allele distribution among *Pseudomonas aeruginosa* isolates from cystic fibrosis (CF) versus non-CF patients. *Microbiology* 150:1315–26
96. Lagenaur C, Agabian N. 1977. *Caulobacter crescentus* pili: structure and stage-specific expression. *J. Bacteriol.* 131:340–46
97. Lee KK, Sheth HB, Wong WY, Sherburne R, Paranchych W, et al. 1994. The binding of *Pseudomonas aeruginosa* pili to glycosphingolipids is a tip-associated event involving the C-terminal region of the structural pilin subunit. *Mol. Microbiol.* 11:705–13
98. Leech AJ, Mattick JS. 2006. Effect of site-specific mutations in different phosphotransfer domains of the chemosensory protein ChpA on *Pseudomonas aeruginosa* motility. *J. Bacteriol.* 188:8479–86
99. Lemkul JA, Bevan DR. 2011. Characterization of interactions between PilA from *Pseudomonas aeruginosa* strain K and a model membrane. *J. Phys. Chem. B* 115:8004–8
100. Li G, Brown PJ, Tang JX, Xu J, Quardokus EM, et al. 2012. Surface contact stimulates the just-in-time deployment of bacterial adhesins. *Mol. Microbiol.* 83:41–51
101. Li TN, Chin KH, Liu JH, Wang AH, Chou SH. 2009. XC1028 from *Xanthomonas campestris* adopts a PilZ domain-like structure without a c-di-GMP switch. *Proteins* 75:282–88
102. Lizewski SE, Schurr JR, Jackson DW, Frisk A, Carterson AJ, Schurr MJ. 2004. Identification of AlgR-regulated genes in *Pseudomonas aeruginosa* by use of microarray analysis. *J. Bacteriol.* 186:5672–84
103. Lory S, Merighi M, Hyodo M. 2009. Multiple activities of c-di-GMP in *Pseudomonas aeruginosa*. *Nucleic Acids Symp. Ser.* 2009:51–52
104. Lovley DR, Ueki T, Zhang T, Malvankar NS, Shrestha PM, et al. 2011. *Geobacter*: the microbe electric's physiology, ecology, and practical applications. *Adv. Microb. Physiol.* 59:1–100

105. Mahenthiralingam E, Campbell ME, Speert DP. 1994. Nonmotility and phagocytic resistance of *Pseudomonas aeruginosa* isolates from chronically colonized patients with cystic fibrosis. *Infect. Immun.* 62:596–605
106. Maier B, Potter L, So M, Seifert HS, Sheetz MP. 2002. Single pilus motor forces exceed 100 pN. *Proc. Natl. Acad. Sci. USA* 99:16012–17
107. Mattick JS. 2002. Type IV pili and twitching motility. *Annu. Rev. Microbiol.* 56:289–314
108. Mattick JS, Whitchurch CB, Alm RA. 1996. The molecular genetics of type-4 fimbriae in *Pseudomonas aeruginosa*—a review. *Gene* 179:147–55
109. Merighi M, Lee VT, Hyodo M, Hayakawa Y, Lory S. 2007. The second messenger bis-(3′-5′)-cyclic-GMP and its PilZ domain-containing receptor Alg44 are required for alginate biosynthesis in *Pseudomonas aeruginosa*. *Mol. Microbiol.* 65:876–95
110. Merz AJ, Enns CA, So M. 1999. Type IV pili of pathogenic *Neisseriae* elicit cortical plaque formation in epithelial cells. *Mol. Microbiol.* 32:1316–32
111. Merz AJ, So M, Sheetz MP. 2000. Pilus retraction powers bacterial twitching motility. *Nature* 407:98–102
112. Michel GP, Aguzzi A, Ball G, Soscia C, Bleves S, Voulhoux R. 2011. Role of fimV in type II secretion system-dependent protein secretion of *Pseudomonas aeruginosa* on solid medium. *Microbiology* 157:1945–54
113. Mignot T, Kirby JR. 2008. Genetic circuitry controlling motility behaviors of *Myxococcus xanthus*. *BioEssays* 30:733–43
114. Mikkelsen H, Sivaneson M, Filloux A. 2011. Key two-component regulatory systems that control biofilm formation in *Pseudomonas aeruginosa*. *Environ. Microbiol.* 13:1666–81
115. Milgotina EI, Lieberman JA, Donnenberg MS. 2011. The inner membrane subassembly of the enteropathogenic *Escherichia coli* bundle-forming pilus machine. *Mol. Microbiol.* 81:1125–27
116. Misic AM, Satyshur KA, Forest KT. 2010. *P. aeruginosa* PilT structures with and without nucleotide reveal a dynamic type IV pilus retraction motor. *J. Mol. Biol.* 400:1011–21
117. Murray TS, Kazmierczak BI. 2008. *Pseudomonas aeruginosa* exhibits sliding motility in the absence of type IV pili and flagella. *J. Bacteriol.* 190:2700–8
118. Navarro MV, De N, Bae N, Wang Q, Sondermann H. 2009. Structural analysis of the GGDEF-EAL domain-containing c-di-GMP receptor FimX. *Structure* 17:1104–16
119. Nguyen Y, Jackson SG, Aidoo F, Junop M, Burrows LL. 2010. Structural characterization of novel *Pseudomonas aeruginosa* type IV pilins. *J. Mol. Biol.* 395:491–503
120. Nudleman E, Kaiser D. 2004. Pulling together with type IV pili. *J. Mol. Microbiol. Biotechnol.* 7:52–62
121. Nunn DN, Lory S. 1991. Product of the *Pseudomonas aeruginosa* gene *pilD* is a prepilin leader peptidase. *Proc. Natl. Acad. Sci. USA* 88:3281–85
122. Opitz D, Clausen M, Maier B. 2009. Dynamics of gonococcal type IV pili during infection. *ChemPhysChem* 10:1614–18
123. Orans J, Johnson MD, Coggan KA, Sperlazza JR, Heiniger RW, et al. 2009. Crystal structure analysis reveals *Pseudomonas* PilY1 as an essential calcium-dependent regulator of bacterial surface motility. *Proc. Natl. Acad. Sci. USA* 107:1065–70
124. O'Toole GA, Kolter R. 1998. Flagellar and twitching motility are necessary for *Pseudomonas aeruginosa* biofilm development. *Mol. Microbiol.* 30:295–304
125. Ottemann KM, Xiao W, Shin YK, Koshland DE Jr. 1999. A piston model for transmembrane signaling of the aspartate receptor. *Science* 285:1751–54
126. Ottow JC. 1975. Ecology, physiology, and genetics of fimbriae and pili. *Annu. Rev. Microbiol.* 29:79–108
127. Overhage J, Bains M, Brazas MD, Hancock RE. 2008. Swarming of *Pseudomonas aeruginosa* is a complex adaptation leading to increased production of virulence factors and antibiotic resistance. *J. Bacteriol.* 190:2671–79
128. Overhage J, Lewenza S, Marr AK, Hancock RE. 2007. Identification of genes involved in swarming motility using a *Pseudomonas aeruginosa* PAO1 Mini-Tn5-*lux* mutant library. *J. Bacteriol.* 189:2164–69
129. Pamp SJ, Tolker-Nielsen T. 2007. Multiple roles of biosurfactants in structural biofilm development by *Pseudomonas aeruginosa*. *J. Bacteriol.* 189:2531–39
130. Paranchych W, Sastry PA, Frost LS, Carpenter M, Armstrong GD, Watts TH. 1979. Biochemical studies on pili isolated from *Pseudomonas aeruginosa* strain PAO. *Can. J. Microbiol.* 25:1175–81

131. Park H, Im W, Seok C. 2011. Transmembrane signaling of chemotaxis receptor Tar: insights from molecular dynamics simulation studies. *Biophys. J.* 100:2955–63
132. Park HS, Wolfgang M, Koomey M. 2002. Modification of type IV pilus-associated epithelial cell adherence and multicellular behavior by the PilU protein of *Neisseria gonorrhoeae*. *Infect. Immun.* 70:3891–903
133. Paul K, Brunstetter D, Titen S, Blair DF. 2011. A molecular mechanism of direction switching in the flagellar motor of *Escherichia coli*. *Proc. Natl. Acad. Sci. USA* 108:17171–76
134. Peabody CR, Chung YJ, Yen MR, Vidal-Ingigliardi D, Pugsley AP, Saier MH Jr. 2003. Type II protein secretion and its relationship to bacterial type IV pili and archaeal flagella. *Microbiology* 149:3051–72
135. Pelicic V. 2008. Type IV pili: e pluribus unum? *Mol. Microbiol.* 68:827–37
136. Pinkner JS, Remaut H, Buelens F, Miller E, Aberg V, et al. 2006. Rationally designed small compounds inhibit pilus biogenesis in uropathogenic bacteria. *Proc. Natl. Acad. Sci. USA* 103:17897–902
137. Pohlschroder M, Ghosh A, Tripepi M, Albers SV. 2011. Archaeal type IV pilus-like structures—evolutionarily conserved prokaryotic surface organelles. *Curr. Opin. Microbiol.* 14:357–63
138. Proft T, Baker EN. 2009. Pili in gram-negative and gram-positive bacteria: structure, assembly and their role in disease. *Cell. Mol. Life Sci.* 66:613–35
139. Py B, Loiseau L, Barras F. 2001. An inner membrane platform in the type II secretion machinery of gram-negative bacteria. *EMBO Rep.* 2:244–48
140. Qi Y, Chuah ML, Dong X, Xie K, Luo Z, et al. 2011. Binding of cyclic diguanylate in the non-catalytic EAL domain of FimX induces a long-range conformational change. *J. Biol. Chem.* 286:2910–17
141. Reguera G, McCarthy KD, Mehta T, Nicoll JS, Tuominen MT, Lovley DR. 2005. Extracellular electron transfer via microbial nanowires. *Nature* 435:1098–101
142. Roberts MA, Papachristodoulou A, Armitage JP. 2010. Adaptation and control circuits in bacterial chemotaxis. *Biochem. Soc. Trans.* 38:1265–69
143. Robien MA, Krumm BE, Sandkvist M, Hol WG. 2003. Crystal structure of the extracellular protein secretion NTPase EpsE of *Vibrio cholerae*. *J. Mol. Biol.* 333:657–74
144. Sampaleanu LM, Bonanno JB, Ayers M, Koo J, Tammam S, et al. 2009. Periplasmic domains of *Pseudomonas aeruginosa* PilN and PilO form a stable heterodimeric complex. *J. Mol. Biol.* 394:143–59
145. Sarkar MK, Paul K, Blair D. 2010. Chemotaxis signaling protein CheY binds to the rotor protein FliN to control the direction of flagellar rotation in *Escherichia coli*. *Proc. Natl. Acad. Sci. USA* 107:9370–75
146. Satyshur KA, Worzalla GA, Meyer LS, Heiniger EK, Aukema KG, et al. 2007. Crystal structures of the pilus retraction motor PilT suggest large domain movements and subunit cooperation drive motility. *Structure* 15:363–76
147. Scheurwater EM, Burrows LL. 2011. Maintaining network security: how macromolecular structures cross the peptidoglycan layer. *FEMS Microbiol. Lett.* 318:1–9
148. Semmler AB, Whitchurch CB, Leech AJ, Mattick JS. 2000. Identification of a novel gene, *fimV*, involved in twitching motility in *Pseudomonas aeruginosa*. *Microbiology* 146(Pt. 6):1321–32
149. Semmler AB, Whitchurch CB, Mattick JS. 1999. A re-examination of twitching motility in *Pseudomonas aeruginosa*. *Microbiology* 145(Pt. 10):2863–73
150. Shaw CE, Taylor RK. 1990. *Vibrio cholerae* O395 *tcpA* pilin gene sequence and comparison of predicted protein structural features to those of type 4 pilins. *Infect. Immun.* 58:3042–49
151. Shi W, Sun H. 2002. Type IV pilus-dependent motility and its possible role in bacterial pathogenesis. *Infect. Immun.* 70:1–4
152. Shi X, Bi J, Morse JG, Toscano NC, Cooksey DA. 2010. Differential expression of genes of *Xylella fastidiosa* in xylem fluid of citrus and grapevine. *FEMS Microbiol. Lett.* 304:82–88
153. Shrout JD, Chopp DL, Just CL, Hentzer M, Givskov M, Parsek MR. 2006. The impact of quorum sensing and swarming motility on *Pseudomonas aeruginosa* biofilm formation is nutritionally conditional. *Mol. Microbiol.* 62:1264–77
154. Singh PK, Parsek MR, Greenberg EP, Welsh MJ. 2002. A component of innate immunity prevents bacterial biofilm development. *Nature* 417:552–55
155. Skerker JM, Berg HC. 2001. Direct observation of extension and retraction of type IV pili. *Proc. Natl. Acad. Sci. USA* 98:6901–4
156. Skerker JM, Shapiro L. 2000. Identification and cell cycle control of a novel pilus system in *Caulobacter crescentus*. *EMBO J.* 19:3223–34

157. Sohel I, Puente JL, Murray WJ, Vuopio-Varkila J, Schoolnik GK. 1993. Cloning and characterization of the bundle-forming pilin gene of enteropathogenic *Escherichia coli* and its distribution in *Salmonella* serotypes. *Mol. Microbiol.* 7:563–75
158. Sonawane A, Jyot J, Ramphal R. 2006. *Pseudomonas aeruginosa* LecB is involved in pilus biogenesis and protease IV activity but not in adhesion to respiratory mucins. *Infect. Immun.* 74:7035–39
159. Strom MS, Lory S. 1993. Structure-function and biogenesis of the type IV pili. *Annu. Rev. Microbiol.* 47:565–96
160. Strom MS, Nunn DN, Lory S. 1993. A single bifunctional enzyme, PilD, catalyzes cleavage and N-methylation of proteins belonging to the type IV pilin family. *Proc. Natl. Acad. Sci. USA* 90:2404–8
161. Tammam S, Sampaleanu LM, Koo J, Sundaram P, Ayers M, et al. 2011. Characterization of the PilN, PilO and PilP type IVa pilus subcomplex. *Mol. Microbiol.* 82:1496–514
162. Taylor TB, Buckling A. 2010. Competition and dispersal in *Pseudomonas aeruginosa*. *Am. Nat.* 176:83–89
163. Taylor TB, Buckling A. 2011. Selection experiments reveal trade-offs between swimming and twitching motilities in *Pseudomonas aeruginosa*. *Evolution* 65:3060–69
164. Varga JJ, Nguyen V, O'Brien DK, Rodgers K, Walker RA, Melville SB. 2006. Type IV pili-dependent gliding motility in the gram-positive pathogen *Clostridium perfringens* and other clostridia. *Mol. Microbiol.* 62:680–94
165. Voisin S, Kus JV, Houliston S, St-Michael F, Watson D, et al. 2007. Glycosylation of *Pseudomonas aeruginosa* strain Pa5196 type IV pilins with mycobacterium-like α-1,5-linked D-Araf oligosaccharides. *J. Bacteriol.* 189:151–59
166. Wehbi H, Portillo E, Harvey H, Shimkoff AE, Scheurwater EM, et al. 2011. The peptidoglycan-binding protein FimV promotes assembly of the *Pseudomonas aeruginosa* type IV pilus secretin. *J. Bacteriol.* 193:540–50
167. West SE, Sample AK, Runyen-Janecky LJ. 1994. The *vfr* gene product, required for *Pseudomonas aeruginosa* exotoxin A and protease production, belongs to the cyclic AMP receptor protein family. *J. Bacteriol.* 176:7532–42
168. Weyand NJ, Lee SW, Higashi DL, Cawley D, Yoshihara P, So M. 2006. Monoclonal antibody detection of CD46 clustering beneath *Neisseria gonorrhoeae* microcolonies. *Infect. Immun.* 74:2428–35
169. Whitchurch CB, Alm RA, Mattick JS. 1996. The alginate regulator AlgR and an associated sensor FimS are required for twitching motility in *Pseudomonas aeruginosa*. *Proc. Natl. Acad. Sci. USA* 93:9839–43
170. Whitchurch CB, Hobbs M, Livingston SP, Krishnapillai V, Mattick JS. 1991. Characterisation of a *Pseudomonas aeruginosa* twitching motility gene and evidence for a specialised protein export system widespread in eubacteria. *Gene* 101:33–44
171. Whitchurch CB, Leech AJ, Young MD, Kennedy D, Sargent JL, et al. 2004. Characterization of a complex chemosensory signal transduction system which controls twitching motility in *Pseudomonas aeruginosa*. *Mol. Microbiol.* 52:873–93
172. Whitchurch CB, Mattick JS. 1994. Characterization of a gene, *pilU*, required for twitching motility but not phage sensitivity in *Pseudomonas aeruginosa*. *Mol. Microbiol.* 13:1079–191
173. Winsor GL, Van Rossum T, Lo R, Khaira B, Whiteside MD, et al. 2009. Pseudomonas Genome Database: facilitating user-friendly, comprehensive comparisons of microbial genomes. *Nucleic Acids Res.* 37(Database issue):D483–88
174. Winther-Larsen HC, Wolfgang M, Dunham S, van Putten JP, Dorward D, et al. 2005. A conserved set of pilin-like molecules controls type IV pilus dynamics and organelle-associated functions in *Neisseria gonorrhoeae*. *Mol. Microbiol.* 56:903–17
175. Wolfgang M, van Putten JP, Hayes SF, Dorward D, Koomey M. 2000. Components and dynamics of fiber formation define a ubiquitous biogenesis pathway for bacterial pili. *EMBO J.* 19:6408–18
176. Wong WY, Campbell AP, McInnes C, Sykes BD, Paranchych W, et al. 1995. Structure-function analysis of the adherence-binding domain on the pilin of *Pseudomonas aeruginosa* strains PAK and KB7. *Biochemistry* 34:12963–72
177. Wu SS, Kaiser D. 1997. Regulation of expression of the *pilA* gene in *Myxococcus xanthus*. *J. Bacteriol.* 179:7748–58
178. Wurdemann D, Tummler B. 2007. In silico comparison of pKLC102-like genomic islands of *Pseudomonas aeruginosa*. *FEMS Microbiol. Lett.* 275:244–49

179. Xu Q, Christen B, Chiu HJ, Jaroszewski L, Klock HE, et al. 2012. Structure of the pilus assembly protein TadZ from *Eubacterium rectale*: implications for polar localization. *Mol. Microbiol.* 83:712–27
180. Yeung AT, Torfs EC, Jamshidi F, Bains M, Wiegand I, et al. 2009. Swarming of *Pseudomonas aeruginosa* is controlled by a broad spectrum of transcriptional regulators, including MetR. *J. Bacteriol.* 191:5592–602
181. Zolfaghar I, Evans DJ, Fleiszig SM. 2003. Twitching motility contributes to the role of pili in corneal infection caused by *Pseudomonas aeruginosa*. *Infect. Immun.* 71:5389–93
182. Zolghadr B, Weber S, Szabó Z, Driessen AJ, Albers SV. 2007. Identification of a system required for the functional surface localization of sugar binding proteins with class III signal peptides in *Sulfolobus solfataricus*. *Mol. Microbiol.* 64:795–806

Postgenomic Approaches to Using Corynebacteria as Biocatalysts

Alain A. Vertès, Masayuki Inui, and Hideaki Yukawa

Research Institute of Innovative Technology for the Earth, Kizugawadai, Kizugawa, Kyoto 619-0292, Japan; email: avertes.sln2004@london.edu; mmg-lab@rite.or.jp

Keywords

reverse genetic engineering, genome engineering, primary and secondary metabolites, whole-cell biocatalysts

Abstract

Corynebacterium glutamicum exhibits numerous ideal intrinsic attributes as a factory of primary and secondary metabolites. The versatile capabilities of this organism have long been implemented at the industrial scale to produce an array of amino acids at high yields and conversion rates, thereby enabling the development of an entire industry. The postgenomic era provides a new technological platform not only to further optimize the intrinsic attributes of *C. glutamicum* whole cells as biocatalysts, but also to dramatically expand the product portfolio that can be manufactured by this organism, from amino acids to commodity chemicals. This review addresses the methods and strain optimization strategies enabled by genomic information and associated techniques. Their implementation has provided important additional incremental improvements to the economics of industry-scale manufacturing in which *C. glutamicum* and its episomal elements are used as a performing host-vector system.

Contents

INTRODUCTION	522
PHYSIOLOGY AND MOLECULAR BIOLOGY TECHNIQUES	523
Megabase-Scale Engineering of *C. glutamicum*	523
Comparative Genomics of the *C. glutamicum* Group of Bacteria	525
TRANSCRIPTOMIC, PROTEOMIC, METABOLOMIC, AND FLUXOMIC ANALYSES OF *CORYNEBACTERIUM GLUTAMICUM*	525
Transcriptomics	526
Proteomics	526
Metabolomics	528
Fluxomics	528
DEVELOPING THIRD-GENERATION INDUSTRIAL STRAINS	532
Inverse Metabolic Engineering	533
Systems Biology and Global Metabolic Design	534
Manipulating Microbial Function	534
Emerging *C. glutamicum* Systems Biology	535
Building Efficient Primary and Secondary Metabolite Whole-Cell Biocatalysts	537
C. glutamicum as a Manufacturing Workhorse	540
CONCLUSIONS	542

Commodity chemicals: chemicals without any attached composition of matter intellectual property rights that are manufactured at very large scales

Biorefinery: a manufacturing plant where an integrated array of operations is conducted to produce, from biomass and sustainable energy or materials, a highly interconnected line of products comprising fuels and chemicals

Rationale-based engineering: a targeted engineering approach in which, e.g., a specific mutation is introduced to alter in a predetermined manner the properties of an enzyme

INTRODUCTION

Since its discovery in 1957 as a natural glutamate producer (52, 108), *Corynebacterium glutamicum* has become an industrial microbial workhorse. *C. glutamicum* produces not only amino acids such as glutamate and lysine for the food and feed industries (each of which constitutes worldwide markets greater than $1.5 billion), but also nucleotides and vitamins (23, 24, 33, 35, 60, 62, 64, 95, 108, 129). Notably, *C. glutamicum* has been engineered to produce commodity chemicals, such as organic acids (e.g., lactate and succinate), poly-3-hydroxybutyrate, isobutanol, ethanol, and 1,2-propanediol, for the materials (plastics) and transportation industries (12, 39, 40, 42, 43, 73, 81, 97). Considering that the bio-based share of worldwide chemicals (commodity, specialty, fine, and polymer chemicals) is expected to grow from 1.8% in 2005, representing a market worth $21.2 billion, to 9.5–13% in 2010 ($130–180 billion), to 22–28% in 2025 ($480–610 billion) (2), the next iteration in the cycle of *C. glutamicum* technology is lignocellulosic biorefineries (27, 42, 46, 66, 69, 128). As a result, although investments in data acquisition and processing enable us to better understand and model the physiology of this organism, novel methods, including techniques to integrate or delete megabase-long DNA regions in a random or targeted manner, have been developed to manipulate the genetic material, from individual genes to chromosomes, of this bacterium (102, 103, 114, 135).

This review explores the physiology of *C. glutamicum* and the plasticity of its genome, as well as recent advances in the molecular biology techniques available to engineer it. This review also illustrates how the information made available by complete genome sequences enables the rationale-based engineering of the entire cellular metabolism, including how systems biology can improve the industrial attributes of corynebacteria as whole-cell biocatalysts.

The genus *Corynebacterium* is classified within the *Actinobacteria*; it is closely related to *Mycobacterium* and *Nocardia*, which collectively form the *Corynebacterineae* suborder of the

Eubacteria (64). *C. glutamicum* (as well as its close relative *C. efficiens*) is a gram-positive, non-spore-forming facultative anaerobic bacterium with moderate to high G+C content (64, 75, 76). Importantly, it is generally recognized as safe, for example, by the US Food and Drug Administration, a status that allows its use for the production of amino acids for the health, cosmetics, food, and feed industries. *C. glutamicum* is a fast-growing, nonmotile bacterium that was first isolated in Japan on the basis of its native ability to secrete glutamate under suitable conditions (52, 60, 95, 108). A biotin auxotroph, this saprophytic organism is typically found in soil or on the skin of fruits and vegetables; it grows under anaerobiosis in the presence of a terminal electron acceptor such as nitrate but reaches high cellular densities when cultivated under aerobic conditions (64, 75). *C. glutamicum* cannot sporulate but remains metabolically active when placed under growth-arrested conditions, e.g., incubation under oxygen-deprived conditions in the absence of a terminal electron acceptor (39, 40). Similarly, when subjected to stress such as biotin limitation, moderate heat shock, or penicillin G, ethambutol, or cerulenin treatment (95), *C. glutamicum* does not undergo autolysis either but rather secretes amino acids or ceases growth while keeping metabolically active, perhaps owing to a persistence and resuscitation mechanism as observed in the genus *Mycobacterium* (31, 47, 49, 107, 122). The cell envelope of *C. glutamicum* comprises a typical plasma membrane bilayer, a mycolic acid outer lipid layer that is anchored by an arabinogalactan-peptidoglycan polymer complex that forms a thick cell wall glycan core, and a crystalline surface S-layer composed primarily of high-molecular-weight glycans and arabinomannans as well as various proteins and lipids (6, 87).

Systems biology: the quantitative determination, mining, and integration of global genomics, transcriptomics, proteomics, metabolomics, and fluxomics data to develop in silico predictive models of physiology

Oxygen-deprived conditions: when aeration is stopped in liquid cultures without a terminal electron acceptor, the redox potential falls quickly and *C. glutamicum* ceases to grow but can be reacted at very large cell densities

Strain-directed evolution: an engineering protocol that makes use of evolution at the strain level

PHYSIOLOGY AND MOLECULAR BIOLOGY TECHNIQUES

Numerous genetic and physiological techniques have been developed to understand and modify corynebacterial biology. A complete molecular biology toolbox is available for manipulating *C. glutamicum*: efficient replicative and integrative transformation techniques, gene disruption and replacement, stable low- and high-copy-number vectors for gene cloning or expression, promoters of various strengths and regulatory controls, and transposon mutagenesis, as well as an array of reporter genes and positive selection markers such as antibiotic resistance genes (54, 72, 114). These tools complement conventional physiological analysis techniques and enable researchers to dissect metabolic pathways to understand their underlying mechanisms. Moreover, global methods—such as genome-scale engineering; strain-directed evolution or whole-genome shuffling; multivariate data analyses; and high-content analysis techniques, e.g., global transcriptomic, proteomic, metabolomic, or fluxomic analysis—have emerged in recent years to optimize the technological attributes of industrial fermenters (26, 28, 41, 44, 58, 88, 112, 114, 121, 123, 135). Complemented by advances in high-throughput analytic laboratory instruments and computing technologies, these techniques have in turn led to a new scientific discipline, systems biology, to model in silico the behavior of organisms and address biological questions at the systems level, rather than at the level of isolated parts of an organism. Surprisingly, systems biology was commercially applied initially to complex tasks in healthcare, for example, for the development of virtual patients to test the impact of new medical entities, rather than in industrial microbiology (34).

Megabase-Scale Engineering of *C. glutamicum*

Genome manipulation techniques have been developed to delete nonessential segments of the chromosome, chiefly as a means to optimize bioconversion yields or to identify essential and nonessential genes. This is possible given the plasticity of bacterial genomes that can accommodate megabase-sized rearrangements including deletions, insertions, duplications, and inversions

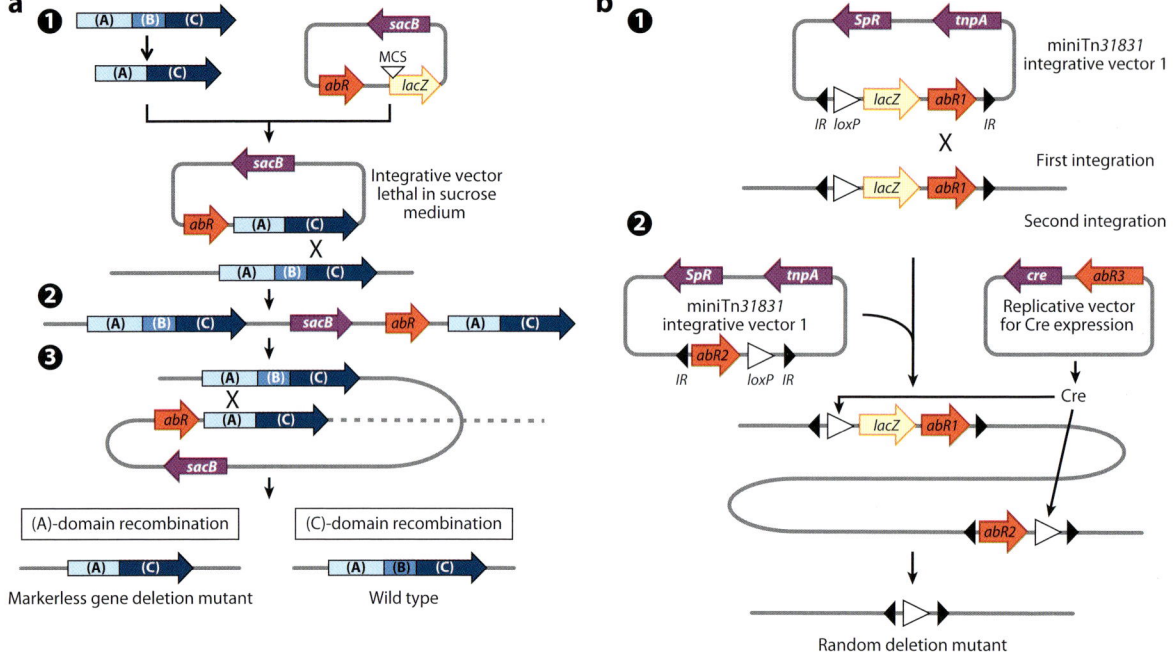

Figure 1

Schematic representation of markerless gene disruption and random genome deletion in *Corynebacterium glutamicum*. (*a*) The markerless gene disruption technique is an adaptation of the conventional gene disruption and replacement technique. ❶ A deletion mutant of the gene of interest is constructed in vitro and cloned into an integrative vector encoding the *sacB* gene, the expression of which is lethal to *C. glutamicum* grown on sucrose. ❷ Upon transformation under appropriate selective pressure, the plasmid integrates into the genome. ❸ Mutants resulting from single crossover events are grown on sucrose-containing medium that favors a second crossover event resulting in either a markerless deletion mutant or the reconstitution of the wild-type genotype (39). (*b*) The random genome deletion technique ❶ makes use of a transposon such as miniTn*31831* on an integrative plasmid to introduce two *loxP* sites into the chromosome; integrants are identified via a positive selection screen. ❷ The Cre recombinase protein that reacts with the *loxP* sites is provided in *trans* on a replicative plasmid that is introduced subsequently into the two transposition events; random deletion mutants are selected in a color test on X-gal-containing plates with deletion mutants having lost the *lacZ* gene (106). This technique is adaptable to targeted genome deletion when the *loxP* sites are integrated by gene disruption and replacement technique using known DNA fragments. Abbreviations: *abR*, antibiotic resistance; MCS, multiple cloning site; *tnpA*, transposase; SpR, spectinomycin resistance; IR, inverted repeat.

(114). The knowledge thus generated can be critical to the development of novel strains via synthetic biology (1). Random or directed deletions of several hundreds of kilobases can be achieved in *C. glutamicum* particularly by using a combination of transposon or gene disruption and replacement techniques and the Cre/*loxP* system. Basically, *loxP* recombination target sites are delivered to the chromosome via integrative vectors or transposons, and the *cre* gene coding for the corresponding *Escherichia coli* phage P1 Cre recombinase is provided by a replicative plasmid under the control of a strong inducible promoter such as p*lac* (**Figure 1**). Successive deletions are possible when using a Cre/mutant-*loxP* system combining a right-element mutant *lox* site and a left-element mutant *lox* site (102, 103, 106). This technique was demonstrated by deleting 331 open-reading frames (approximately 400 kb) from the genome of *C. glutamicum* R (106). Furthermore, the ability to randomly delete segments of the chromosome enables investigators to conduct strain-directed evolution experiments for reducing the metabolism to only the critical functions desired. Similarly, whole-genome shuffling recombines the chromosomes of different strains to accelerate the isolation of mutants with improved industrial attributes, especially when applied to a set of phenotypically selected strains (86, 137). This technique typically proceeds by way of protoplast fusion to combine genomes; it is thus reminiscent of natural sexual evolution processes (86). Furthermore, whole-genome shuffling can be recursive and iterative. Genome breeding, on the other hand, proceeds from the comparative genomic analysis of a high producer and its parent, the identification of beneficial mutations, and the assembly of selected mutations in a wild-type background (37). These technically simple methods have been applied for example to derive corynebacterial L-lysine producers with an improved glucose metabolic activity (48), or novel strains such as the *C. glutamicum* ATCC 13032 derivative strain AHP-3 that can produce L-lysine at 40°C (78). Moreover, random shuffling of DNA fragments was expanded to the genome level, which proved extremely useful in directed evolution approaches (21).

Comparative Genomics of the *C. glutamicum* Group of Bacteria

The genome sequences of several industrially important corynebacterial strains have been determined, including those of *C. glutamicum* R (135, 136), two variants of *C. glutamicum* ATCC 13032 (37, 45), and *C. efficiens* YS-314 (76). The availability of these sequences has facilitated the identification of insertion elements and transposons in *C. glutamicum*, as well as putative sigma factors, two-component systems and transcriptional regulators, secreted proteins, and enzymes involved in sugar metabolism (94, 136), among others. A relatively limited number of genes are present either in strain R or in strain ATCC 13032 but not in both (60 genes in strain R and 189 in strain ATCC 13032). Transposable elements aside, the most striking differences in the number of genes are found in amino acid transport and metabolism as well as secondary transport and metabolism (136). Dot blot hybridization used five such strain-specific islands to group *C. glutamicum* strain isolates into several distinct families, demonstrating that strains R, ATCC 13032, and ATCC 31831 exhibit significantly different hybridization patterns, which is consistent with the diversity observed among corynebacteria at the restriction and modification systems levels (RITE, unpublished data). Although *C. glutamicum* exhibits a broad metabolic diversity (37, 45, 136), determining the pan-genome of *C. glutamicum* would require the sequencing of a discrete number of additional corynebacterial genome sequences (136). Such projects have been carried out, for example, in *E. coli* (29 strains) to define core and variable (strain-specific) metabolism (119). The comparative analysis of the genomic sequences of *C. glutamicum* and *C. efficiens* was critical for understanding the determinants of thermostability of *C. efficiens*. The analysis showed that three substitutions (lysine to arginine, serine to alanine, serine to threonine) in the primary structure of key enzymes are particularly important for increased stability of *C. efficiens* (76).

TRANSCRIPTOMIC, PROTEOMIC, METABOLOMIC, AND FLUXOMIC ANALYSES OF *CORYNEBACTERIUM GLUTAMICUM*

The availability of complete information regarding the set of genes and corresponding deduced proteins expressed by *C. glutamicum* represents a discontinuity in corynebacterial research. It enabled researchers to determine for the first time global differences in genetic expression patterns when *C. glutamicum* cells are incubated under varying conditions. These data, combined with those arising from transcriptomics, metabolomics, and, in particular, fluxomics, can be integrated to enable the development of a systems biology platform for *C. glutamicum*. This data-driven approach has translated into practical biotechnological applications, as exemplified by redesigning the general metabolism to dramatically increase diaminopentane or lysine production (10, 51).

FROM GENOMIC SEQUENCE TO SYSTEMS BIOLOGY

The sequencing of complete genomes constitutes a milestone in biotechnology because research can now be conducted on fully defined living systems. The new biology does not, however, provide a simple path to biocatalysts improvement. In fact, there is a high level of compensatory mechanisms that exist within the primary molecular pathways and networks; this complexity results in metabolic fluxes exhibiting a tendency to be dependent on multiple enzymes. As a result of this redundancy, predicting in silico the behavior of living systems that exhibit five linked primary network dimensions (genome, transcriptome, proteome, metabolome, fluxome) remains a very complex undertaking. To be efficient and pave the way to optimized biomanufacturing processes, modeling and recombinant engineering need to take into account that, at steady state, cells regulate their overall metabolism to increase their fitness rather than improve product production, and do so by responding to changes dictated by their dynamic environment.

Transcriptomics

Global transcriptomics studies of *C. glutamicum* have been conducted under aerobic and oxygen-deprived conditions using whole-genome microarrays, and under different fermentation conditions. For example, an in-depth profile of lysine-producing *C. glutamicum* ATCC 13287 cells was carried out at different time points during aerobic lysine fermentation on minimum medium (growth: 5.6 h; phase shift: 6.2, 6.5, 6.8, and 6.9 h; lysine production: 7.0, 7.3, and 7.7 h) and revealed that the switch from growth to lysine production is characterized by a strong increase in glucose-uptake genes whereas those of the ATP synthase, PPP enzymes (glucose-6-phosphate dehydrogenase, 6-phosphogluconate dehydrogenase, transketolase, transaldolase) and tricarboxylic acid (TCA) cycle enzymes (citrate synthase, oxoglutarate dehydrogenase, membrane bound malate dehydrogenase) decrease (58). The parallel determination of transcriptomes, metabolomes, and fluxomes uncovered correlations that exist between the different dimensions that make the cellular blueprint (**Figure 2**) and provided evidence that the switch from biomass production to lysine production is accompanied by changes in the expression levels of genes of the central metabolism and changes in the corresponding metabolic fluxes (58). Global transcriptome analysis is also used to compare parent and mutant or recombinant strains with improved industrial attributes. For example, the underlying mechanisms of the superior lysine secretion performance of the industrial strain B-6 are attributable to the upregulation of *lysC* and other amino acid biosynthetic genes, thereby suggesting that a regulatory mechanism underpins it (32, 38). High-data-content analyses such as global transcriptome or proteome analyses provide clues regarding the molecular mechanisms responsible for yield improvements, which are typically not obvious in strains derived by random mutagenesis and selection (36, 38). Transcriptome analysis of *C. glutamicum* cells under oxygen-deprived conditions in the absence of a terminal electron acceptor such as nitrate revealed that genes coding for key enzymes of the glycolytic and organic acid production pathways (*gapA*, *pgk*, *tpi*, *ppc*, *ldhA*, *mdh*) are significantly upregulated, leading to increased rates of carbon source consumption (41). Furthermore, induction of the SOS response was demonstrated in *C. glutamicum* cells during nitrate respiration (74).

PPP: pentose phosphate pathway

TCA cycle: tricarboxylic acid cycle

Proteomics

Proteomics is used to measure the effective relative concentrations of enzymes of interest and thus can provide rationales for strain engineering or validate those attained by transcriptomic analyses (44). For example, a shotgun proteomics overview of the L-lysine-overproducing strain DM1730

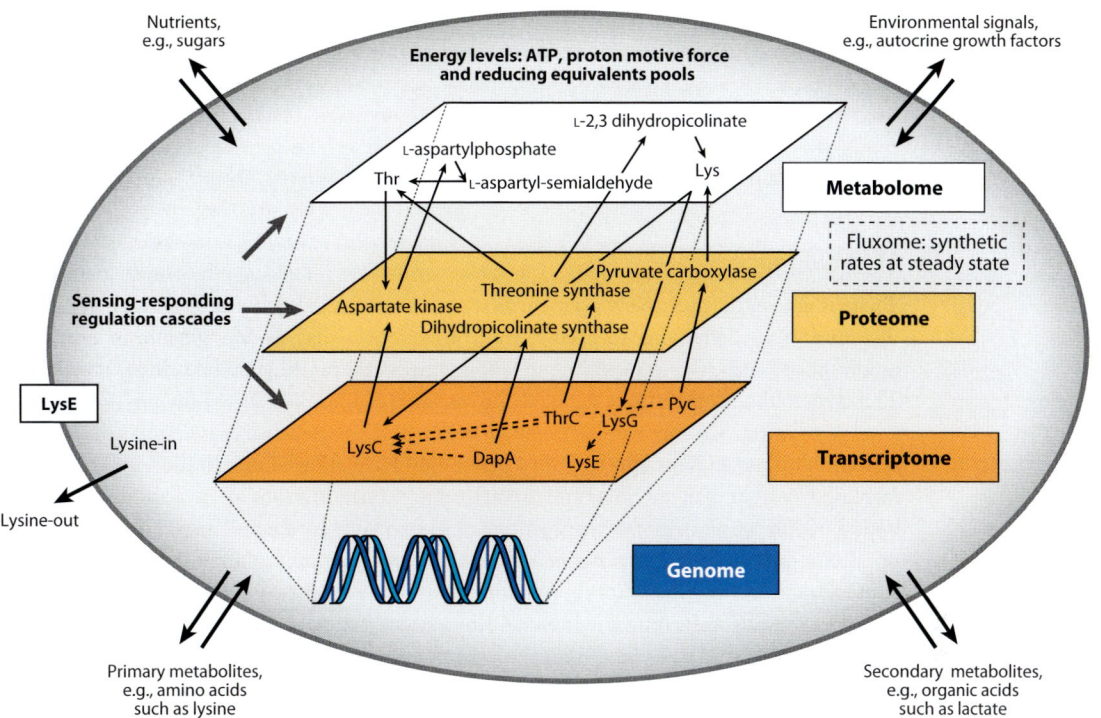

Figure 2

Links between genome and phenotype, interdependence of microbial biochemical networks, and the lysine biosynthetic pathway in *Corynebacterium glutamicum*. This simplified diagram illustrates the various biochemical networks, or dimensions, that regulate microbial metabolism: genome (DNA), transcriptome (mRNA), proteome (proteins), and metabolome (metabolites), with fluxome (fluxes calculated at steady state) constituting an additional level of complexity. These dimensions are not independent of one another as significant cross-talks occur between them, as shown by the anabolism of lysine and its transport to the extracellular space. The key enzyme of the lysine biosynthetic pathway, aspartate kinase (encoded by *lysC*), is under feedback inhibition of lysine and threonine (129). The complexity of the regulatory pathway makes simple recombinant engineering approaches typically ineffective, as exemplified by the limited improvement in yield attained with the overexpression of only one particular gene encoding an apparently rate-limiting enzyme of the pathway. This observation has been ascribed to metabolic flux imbalances that lead to physiological changes that are difficult to predict. On the other hand, expression of more than one gene may solve this hurdle. For example, the coordinated expression of the genes coding for pyruvate carboxylase and aspartate kinase results in a 250% increase in specific lysine productivity, whereas the overexpression of aspartate kinase alone leads to marginal increases in lysine productivity, and the overexpression of pyruvate carboxylase alone, the main anaplerotic enzyme for lysine biosynthesis, results in enhanced growth but reduced lysine productivity (56, 57). Lysine secretion by the exporter LysE under the control of the regulator LysG is another critical parameter for lysine manufacturing (120). Implementing a holistic approach to optimize the total intracellular molecular processes as a means to achieve maximum cell fitness in a particular production reaction is thus necessary to achieve significant improvement in yields (56, 126). The interactions between the cells and their environment also play a critical role via sensing and responding through regulatory cascades, as well as via primary and secondary metabolite excretion, autocrine growth factors, or cellular energetic levels including the pool of reducing equivalents (31, 105, 136). Virtual relationships between genes of the pathway are symbolized in the genome dimension by dashed arrows.

quantified 908 proteins (26). In this strain, the pyruvate carboxylase (PC) has a point mutation that enhances the supply of oxaloacetate as a lysine precursor. Interestingly, levels of this enzyme decrease when cells enter the lysine production phase. Similarly, levels of the glucose-6-phosphate dehydrogenase Zwf are reduced, which decreases the flux toward the PPP and thus increases that toward the TCA cycle, and as a result ultimately toward lysine synthesis (26). Furthermore, proteomics data can provide clues to process or fermentation medium improvements. For example,

PC: pyruvate carboxylase

PEP: phosphoenolpyruvate

compared with its parent strain ATCC 13032, strain DM1730 exhibits lower levels of iron-sulfur cluster biosynthesis and repair proteins as well as lower levels of iron storage proteins, whereas its catalase levels are increased 78-fold. These observations suggest that strain DM1730 suffers from iron deficiency and oxidative stress under the conditions tested (26). Remarkably, proteomes appear to be highly strain specific, indicating the need to base process improvement analyses on industrial strains rather than on the *C. glutamicum* type strain (63).

Metabolomics

There is in general a relatively poor correlation between transcriptome data and metabolic fluxes; therefore, it is challenging to predict the consequences that specific genetic or environmental modifications may have on specific metabolic fluxes and on the general cellular metabolism (4). The metabolome constitutes a detailed and sensitive characterization of the physiological state of a cell population. Advances, for example, in tracer studies by nuclear magnetic resonance, in chromatography, and in mass spectrometry techniques have allowed one to monitor the levels of an increased number of metabolites and their changing concentrations in a time series to compare the impact of differences in genetic background or in environmental conditions. Metabolome analyses of L-lysine-producing *C. glutamicum* ATCC 13287 cells, for example, demonstrated that the initiation of lysine production occurs approximately 30 min earlier than the extracellular accumulation of lysine, thereby indicating the induction of a lysine exporter. The analyses also showed that a dramatic change in the pools of intracellular amino acids parallels the switch from growth to lysine production, whereas these pools remain relatively constant during lysine production and until the end of the cultivation period (58). Similarly, metabolome studies combined with transcriptome analysis led to the discovery that nitrogen metabolism via the master regulator AmtR and L-lysine biosynthesis are connected via the regulation of the *dapD* gene coding for succinylase involved in *m*-diaminopimelate synthesis (17). In addition, nitrogen metabolism and carbon metabolism are connected via the regulation of the malic enzyme encoded by the *mez* gene (17).

Fluxomics

Flux determination provides quantitative information measured in vivo for individual pathways. Proceeding from the reactions of the central metabolism pathway, comprising both anabolic and catabolic reactions (glycolysis, TCA cycle, PPP, glyoxylate cycle) (**Figure 3**), to map the topology of the metabolic network, flux determination encompasses product formation and biomass formation (131). Moreover, flux coupling analysis identifies interdependencies that may exist between different reactions (19). Biomass precursors include glucose-6-phosphate, fructose-6-phosphate, ribose-5-phosphate, erythrose-4-phosphate, glyceraldehyde-3-phosphate, 3-phosphoglycerate, pyruvate, phosphoenolpyruvate (PEP), acetyl-CoA, 2-oxoglutarate, and oxaloacetate, with 16.4 mmol of NADPH used for the synthesis of 1 g of biomass; many of these biomass anabolic precursors are also precursors of products of interest and thus their metabolic imbalances lead to inferior yields (131). An appropriate supply of intracellular precursors, such as pyruvate and oxaloacetate for lysine production, is critical for attaining high industrial productivities. Optimizing their availability has thus become a central part of corynebacterial engineering. Intracellular fluxes are calculated by solving a set of linear equations derived from the established network topology inferred from the set of known metabolic reactions, and from extracellular metabolome analyses. The method encompasses substrate uptake and product secretion as well as biomass generation based on its known composition. Metabolite balancing was used, for example, to demonstrate that, in an aerobic batch fermentation process involving the threonine and methionine auxotroph *C. glutamicum* ATCC 21253, lysine production is limited neither by the availability of NADPH nor by that of pyruvate (109–111). This technique, which typically uses ^{13}C or ^{15}N and tracing by mass spectrometry or nuclear magnetic resonance, also defines the physiological steady state of cells in industrial fermentations; however, isotope chasing experiments are necessary to characterize in detail metabolic cycles, reversible reactions, or parallel pathways such as flux splitting between glycolysis and PPP or the anabolic reactions at the pyruvate node (131).

Flux measurements are particularly useful to assess via metabolic modeling in silico the impact of single gene knockouts, as these can be compensated by rerouting of the metabolism to alternative pathways. A canonical example is the compensation of a deficiency in pyruvate kinase, which converts PEP into pyruvate (**Figure 3**), by activation of a bypass involving PEP carboxylase, malate dehydrogenase, and malic enzyme that ultimately converts PEP into pyruvate while keeping other fluxes essentially unchanged (9). This approach can also lead to the discovery of novel intracellular reactions.

Flux studies have focused on elucidating the physiology of *C. glutamicum* grown on glucose or fructose, and importantly also on sucrose metabolism for lysine production, revealing that *C. glutamicum* has a large metabolic flexibility. This translates into a critical industrial production attribute as *C. glutamicum* cells adapt their metabolism to changing metabolic burdens when precursors are withdrawn for product production such as for amino acid synthesis (50, 68, 125, 132). Remarkably, metabolic flux analysis has defined NADPH balances under various culture conditions, including during lysine production (132). Here as predicted, increased lysine fluxes are linked with an increased PPP flux, because this pathway is the major pathway for NADPH supply (50, 132). Therefore, dual overexpression of the *zwf* gene coding for glucose-6-phosphate dehydrogenase and *fbp* (fructose 1,6-bisphosphatase) aimed to increase the PPP flux increased lysine production by a *lysC* feedback-resistant *C. glutamicum* ATCC 13032 mutant by 70% (7, 129). As pointed out above, the relatively large number of enzymes at the PEP/pyruvate node makes it a critical point of the metabolism given the anaplerotic reactions at this node to ensure the supply of the key TCA cycle intermediates oxaloacetate and malate (**Figure 3**). During the switch from growth to lysine production, *C. glutamicum* ATCC 13287 cells grown aerobically on glucose in minimum medium have an increased anaplerotic net flux around the pyruvate node, which corresponds to the high demand at that time of the fermentation for oxaloacetate to ensure biomass generation and lysine production (58). The flux at the pyruvate node for the replenishment of oxaloacetate is directly linked to that of lysine production, thus making the corresponding enzyme important engineering targets, including particularly PC (coded for by the gene *pyc*) and phosphoenolcarboxykinase (*pck*), which recycles two-thirds of anaplerotically synthesized oxaloacetate, a lysine precursor, into PEP (**Figure 3**) (22, 85, 92, 131). This hypothesis was further

Figure 3

The central metabolic pathways of *Corynebacterium glutamicum*. The central metabolic pathways of *C. glutamicum* have been represented under (*a*) aerobic and (*b*) anaerobic conditions. Under anaerobiosis, the synthesis of many enzymes of the tricarboxylic acid pathway is repressed and sugars are converted to pyruvate via glycolysis and the pentose phosphate pathway, and subsequently converted to L-lactate by lactate dehydrogenase. Simplified routes to amino acid biosynthesis are indicated in the aerobic panel (124). Some pathways have been engineered to achieve significant production of products of interest, such as L-serine from glucose under aerobic conditions as it results in higher yields as compared to the fermentative production from glycine (83) or isobutanol, 3-methyl-1-butanol and 2-methyl-1-butanol (97), or ethanol under oxygen-deprived conditions (39). Not all reactions are shown. Genes that are upregulated under oxygen-deprived conditions are indicated in red, genes that are downregulated in blue (41). Engineered pathways are indicated in orange. Cofactors are indicated for some, but not all, reactions. Abbreviations: PPP, pentose phosphate pathway (not represented in detail); TCA, tricarboxylic acid cycle; Ga3P, D-glyceraldehyde 3-phosphate.

a

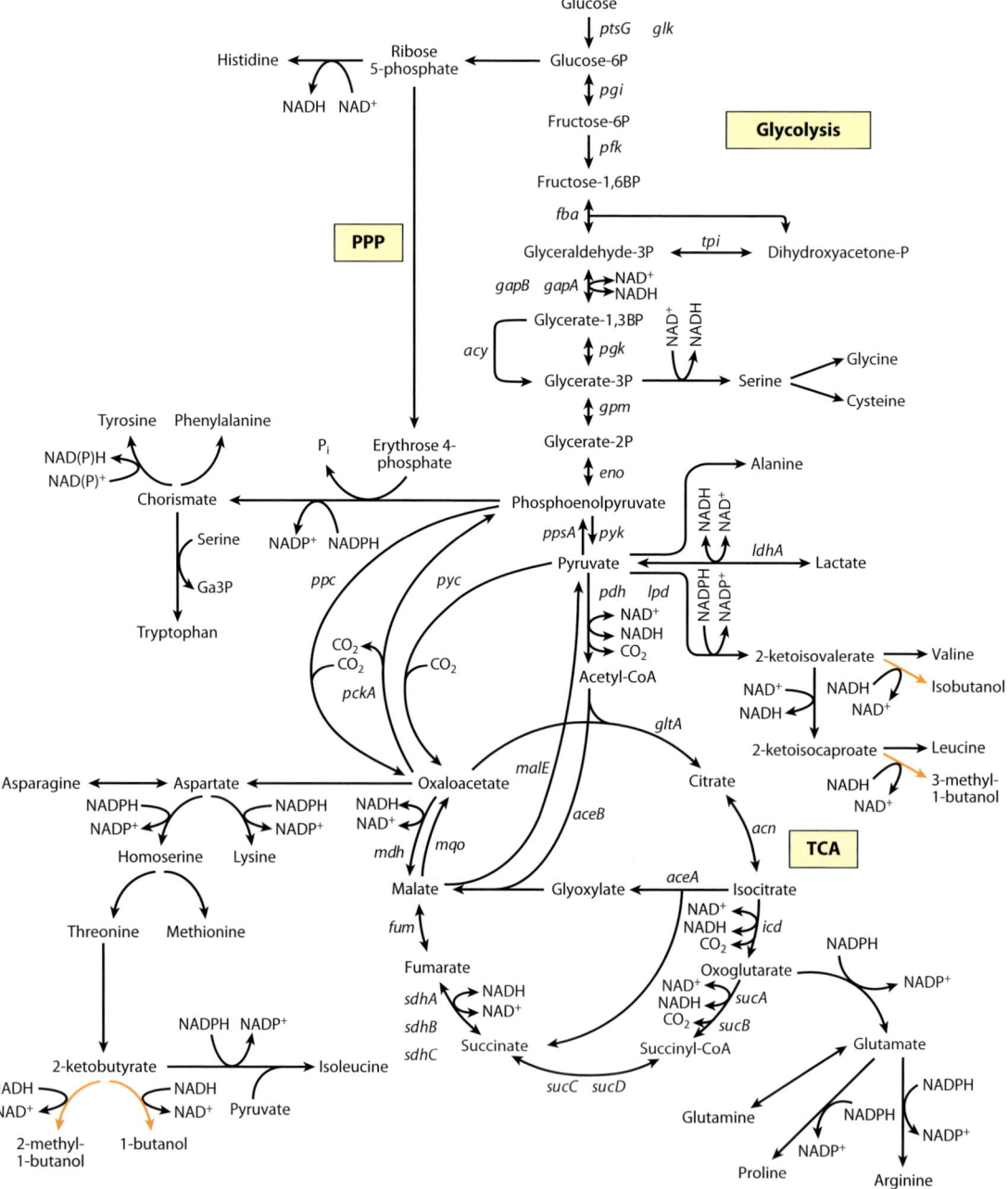

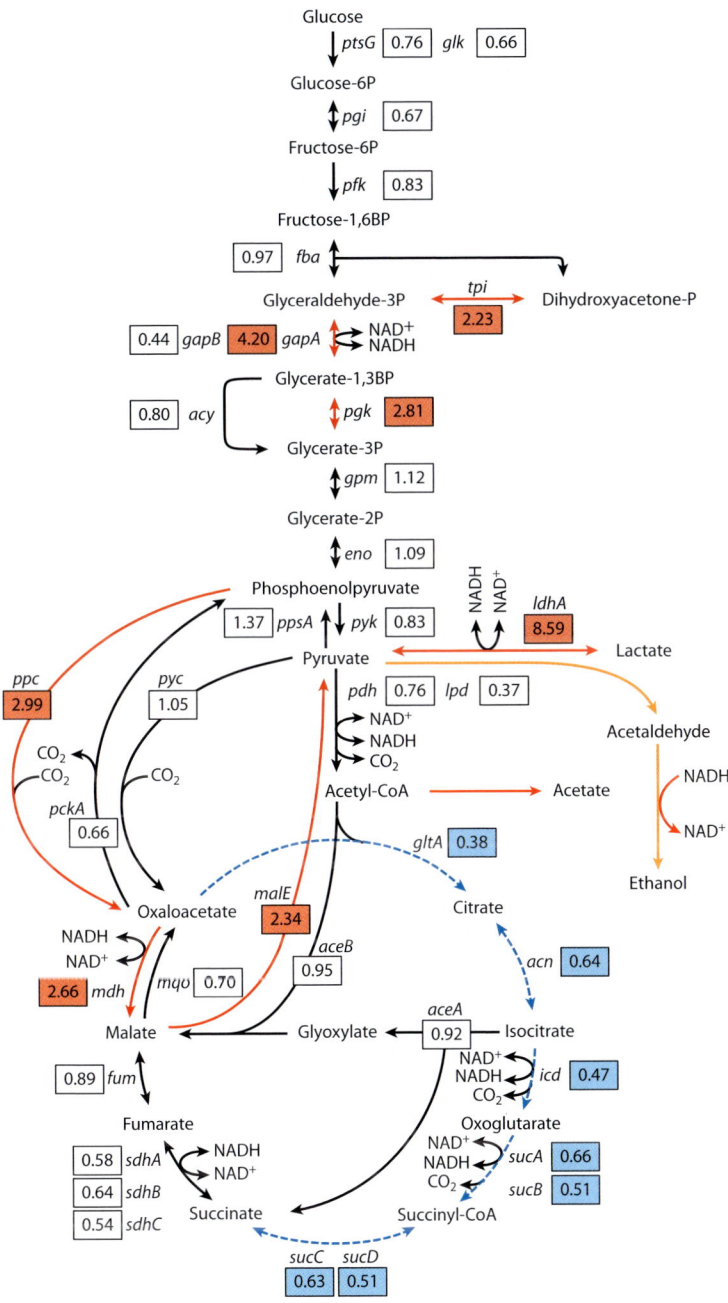

PEPC:
phosphoenolpyruvate carboxylase

confirmed by the observation with cells grown aerobically on glucose that inactivation of *pck* leads to 20% higher lysine accumulation in *C. glutamicum* MHB20-22B cells, or that attenuating pyruvate dehydrogenase complex activity leads to a 40% increase in substrate specific lysine yield by *C. glutamicum* DM1729 (15, 85). In *C. glutamicum*, the only two significant anaplerotic enzymes that can substitute for one another are PEP carboxylase (PEPC coded for by the *ppc* gene) and PC. Isotope chasing experiments demonstrated that in aerobically glucose-grown cells, the flux partition is 10/90 between PEPC and PC, but in contrast PEPC is the main anaplerotic enzyme when *C. glutamicum* cells are incubated in oxygen-deprived conditions (41, 84). Flux analysis with flux distribution maps throughout the growth curve thus constitutes a very powerful technique to identify metabolic bottlenecks and targets for metabolic engineering to improve product production in *C. glutamicum* (92, 104). Similarly, flux genealogy profiling of successive generations of industrial strains enables one to assess the impact of specific successive mutations in a given genetic background (130).

DEVELOPING THIRD-GENERATION INDUSTRIAL STRAINS

First-generation strains refer to the initial *C. glutamicum* wild-type isolates that were used industrially and to their metabolic analog-resistant derivatives. Second-generation corynebacterial producers refer to those strains derived by linear conventional methods for optimizing microbial bioconversion processes such as overexpressing the synthetic genes of the products of interest and relaxing their regulatory constraints, improving downstream processing yields by increasing compartmentalization (for example by improving secretion into the fermentation medium), and reducing product degradation reactions.

In this technological iteration step enabled by molecular biology techniques, attention has also been focused on achieving appropriate reducing equivalent balance and precursor supply. Although this approach has been largely successful, for example, for reaching cost-effective lysine production yields (129), its impact remains limited because it enables the practitioner to redesign only part of the metabolic machinery of industrial microorganisms.

By modeling cellular metabolism as an interdependent set of enzymes, regulatory factors, and metabolites to better grasp nonlinear systems effects, the derivation of efficient bioconverters by postgenomic tools proceeds from the two fundamental elements of rational design and evolutionary selection: deep physiological knowledge (generated by the implementation of "omics" techniques and integrated into systems biology, which provides a more holistic approach to fermenter engineering) and efficient genetic engineering technology platforms (114). As emphasized above, this third-generation strain development strategy has the advantage of breaking free from the constraints of manipulating only a few genes or a few parameters at a time. These particular constraints restricted advancement in the field up until the postgenomic era, since the complexity, stochasticity, and redundancy of regulatory mechanisms have often resulted in limited yield

OPTIMIZING CORYNEBACTERIA AS BIOCATALYSTS

The interconnectivity and complexity of biological networks restrict the impact of rational engineering based on the recognition of simple genotype-phenotype relationships. Instead, engineering a novel phenotype may require the concomitant modulation of several factors, some of which are potentially unknown or are from seemingly unrelated pathways. The engineering of novel or optimized pathways thus typically requires the balanced expression and activity of an array of enzymes and transcriptional or translational regulators.

improvement, as illustrated by the observation that overexpression of the eight genes of the glycolytic pathway in *Saccharomyces cerevisiae* does not result in any significant improvements in the production of ethanol despite 3.7- to 13.9-fold increases in the specific activities of these enzymes (93). Thus, manipulating fluxes and the intricate regulatory networks of *C. glutamicum*, rather than genes and protein levels, has become an important goal of postgenomic fermenter engineering (4, 8, 18). This can be exemplified by metabolic engineering of the TCA cycle to improve lysine production by 40% (8), or by overexpression and deregulation of the serine biosynthetic genes as promising engineering targets, which initially were insufficient to achieve serine production until the need to reduce hydroxymethyltransferase was discovered. This was optimally achieved by reducing folate supply (101). Strain-directed evolution for evolutionary engineering in conditions mimicking as closely as possible manufacturing conditions constitutes a positive-selection-based approach to optimize the industrial attributes of a host-vector system. Inverse metabolic engineering, comprising techniques such as whole-genome shuffling and genome breeding, and systems biology, comprising in silico biology and global metabolic design, represent complementary core technology platforms that define the new frontier enabled by the postgenomic era to resolve bottlenecks in productivity.

Inverse metabolic engineering: a strategy for manipulating biological function

Postgenomic era: a new era in biology research enabled by high-throughout, high-content, automated analytical techniques and characterized by large amounts of data that can be mined using bioinformatics tools

Inverse Metabolic Engineering

Inverse metabolic engineering is a global reverse engineering development approach (4, 5, 29). It aims to circumvent the shortcomings of rationale-based engineering in which the modification of a discrete number of target genes may not be sufficient to alter fluxes in order to achieve the desired compound production and yields (4). This methodology involves (*a*) defining a desired phenotype; (*b*) determining the genetic or growth conditions needed to achieve that phenotype, usually via identifying flux-limiting reactions; and (*c*) engineering that phenotype by implementing strain-directed evolution techniques or rational genetic modifications while placing the particular pathway of interest within the context of the global cellular metabolic network (5, 16, 29, 90, 121). Global analysis techniques such as comparative genomics or transcriptomics are automatable methods that can reveal the fine differences that exist between strains exhibiting varying industrial attributes.

For example, Ohnishi et al. (79) generated an optimized lysine producer: They identified beneficial mutations in the L-lysine-overproducing *C. glutamicum* strain B-6 from 16 genes encoding enzymes involved either in anaplerotic reactions, or in the L-lysine terminal biosynthetic pathway, or in lysine efflux. Strain B-6 is the result of multiple rounds of random mutagenesis and screening. However, although it overproduces lysine, it exhibits poor growth and sugar consumption, two characteristics likely ascribed to deleterious mutations introduced during the random mutagenesis and screening steps. Gene replacement therefore introduced selected beneficial mutations into the strain B-6 parent, the wild-type strain ATCC 13032 (79). This new approach constructs the optimized strain AHP-3 by introducing only favorable mutations. Introduction of the mutations present in either the *hom* or *lysC* gene of strain B-6 into strain ATCC 13032 resulted in L-lysine titers of 8 g liter^{-1} or 55 g liter^{-1}, respectively, whereas the double mutant exhibited synergistic effects with the accumulation of 75 g liter^{-1} of L-lysine (79). Introducing an optimized *pyc* gene further improved L-lysine titer to 80 g liter^{-1} and, compared with strain B-6, dramatically improved productivity of the new strain carrying only four well-defined mutations from 2.0 g liter^{-1} h^{-1} to 3.0 g liter^{-1} h^{-1} (79). Similar approaches will be utilized more frequently when genome-wide comparative analyses become routine (as the cost of whole-genome sequencing continues to decrease). On the other hand, this approach needs to be complemented by other global analysis techniques such as transcriptomics, metabolomics, and fluxomics (121). For example, although the engineered strain completes fermentation in approximately 30 h (whereas the classic producer requires 50 h), the final titers are lower than those achieved with the classically derived fermenter.

As illustrated above, global transcriptional profiling has revealed differences in the expression of the biosynthetic genes of several amino acids, including the *lysC*-encoded aspartokinase, thereby highlighting the existence of one or more global regulatory mechanisms and providing leads toward further rationale-based engineering of *C. glutamicum* (32, 38, 121). Global approaches help identify targets that synergistically generate the improved phenotype, a task that is virtually impossible when studying enzymes and their reactions in isolation. A similar comparative genomic approach was followed to develop an L-arginine and L-citrulline fermenter, in which the best mutation set selected via rationale-based design resulted in only one-third of the productivity achieved by the strain developed via classical mutagenesis. Again, global transcriptome analyses were instrumental in revealing that expression of the *arg* operon in the latter strain is dramatically higher, thus providing a clue that the observed low productivity was due to regulation of the key biosynthetic enzyme (36). An alternative to whole-genome sequencing is the analysis of plasmid libraries, which has proven useful for accelerating the identification of genes responsible for specific phenotypes (29). Moreover, whole-genome shuffling of the chromosomes of a library of strains that exhibit varying degrees of the desired phenotypic properties represents an evolutionary method that can be applied here as well. This is useful to incorporate mutations that result in incremental increases, and hence that are not obvious, but nevertheless these mutations collectively significantly contribute to the overall productivity of the system (86, 91). Other combinatorial techniques that enable inverse metabolic engineering include random knockout libraries (attained by transposon mutagenesis using mobile elements with low target specificities) and overexpression libraries (attained by transforming the cell with an episomal genomic library), as well as targeting the transcription and translation machineries and control to introduce global perturbations (90).

Genome-scale metabolic network: a mathematical reconstruction of the metabolic network of an organism, based primarily on its whole-genome sequence and systems constraints as validated by in vivo experimentation

Stoichiometric model: a model based on the stoichiometric properties of an enzyme

Kinetic model: a model based on the kinetic properties of enzymes and also on metabolite and enzyme concentrations

Systems Biology and Global Metabolic Design

Systems biology is perhaps best described as the quantitative analysis of biological systems through the use of predictive mathematical models of metabolisms derived from the analysis of global biological data (82) (**Figure 2**). It has become an inescapable tool for manipulating cellular function. This reconstruction of genome-scale metabolic networks to build or discover new properties of the cellular metabolic network, including its regulatory components, enables stoichiometric modeling (metabolites, directionality, enzymes, flux balance analyses, constraints, operating ranges) and kinetic modeling (metabolites, directionality, time-course of compound production and consumption). Flux balance analyses are used for predictive modeling when critical kinetic parameters such as forward and reverse reaction rates at physiologically relevant conditions are unknown (77).

Manipulating Microbial Function

The biochemical reaction space models thus defined ultimately facilitate the formulation of metabolic engineering strategies, the implementation of which makes use of genomic plasticity and typically requires a combination of rationale-based engineering approaches such as engineering of novel pathways and gene overexpression or silencing (in vitro and in silico, knowledge-based), and techniques such as genome shuffling or strain-directed evolution (in vivo, evolution-based) (82, 86, 91, 114). The resulting strain is subsequently tested (maximum specific rates, substrate consumption rates, products titers and yields, by-product formation) under conditions that closely mimic industry-scale conditions, and is characterized in detail by transcriptomic, proteomic, metabolomic, and fluxomic techniques. This cycle is typically repeated until the desired industrial attributes of the new process are reached (82).

Although it has long been envisaged that conducting experiments in silico leads to faster and cost-effective research and development, systems biology has been made possible only recently through major advances in high-throughput experimental techniques and bioinformatics. Nonetheless, systems biology represents a paradigm change and a new step in corynebacterial technology that stem not only from building models by assembling complementary knowledge as described above, but also from crunching high-throughput quantity data via multivariate analysis techniques to reduce data complexity and help in their visualization, further refining and validating the models thus generated. Practically, these models constitute invaluable tools to predict, at steady state or even at the dynamic phase, not only metabolic bottlenecks such as insufficient activity or dosage of a key enzyme (58), but also the physiological response of an organism of interest when subjected to a disruption. Such disruptions could be a change in environmental conditions, for example, a new manufacturing process or a change in a major manufacturing parameter, or a genetic change, for example, the deletion or overexpression of one or more genes. Likewise, network rigidities, that is, robustness or resistance of the system to change owing to redundant mechanisms, can be discovered when the applied changes only result in maintenance of the wild-type steady state. Particularly, such models enable researchers to map critical physiological differences that may exist between a wild-type strain and its industrial derivatives.

> **Network rigidities:** the phenomena whereby changes in metabolic fluxes are minimized by way of redundancy

Furthermore, metabolome data may also result in the in silico biology discovery of metabolic networks through reverse engineering (20). Fundamentally, as systems biology integrates functional information of a highly interlinked network of reactions and regulatory mechanisms in which multiple precursors are involved in multiple pathways, it permits researchers to define complex nonlinear relationships. In contrast, overexpression of one or more genes is not necessarily correlated directly with a resulting metabolite concentration due to cellular complexities, because one or more of the corresponding enzymes may not be rate-limiting or may be part of a tight regulatory pathway. Moreover, deeper meta-analysis enabled by enhanced data integration across heretofore unconnected data repositories, such as gene sequences, transcriptional and RNA profiles, as well as proteomics and metabolomics data spanning multiple industrial phylogenies and a wide phylogenetic distance, could be implemented to further strengthen the models.

Emerging *C. glutamicum* Systems Biology

To model complex nonlinear behaviors, systems biology proceeds from the definition of the networks of metabolites occurring in a cell, and from the definition of the networks of enzymatic reactions and their regulations leading to these metabolites. Stoichiometric metabolic network maps with flux analysis are useful not only to guide strain and process development, but also to calculate the maximum theoretical production potential of a particular metabolite, an important value for determining the long-term economic value of a new process and predicting the upper limit in economies of learning that can be gained through manufacturing experience (59, 100).

Several genome-scale models of the metabolic network of *C. glutamicum* ATCC 13032 have been constructed in silico; one such metabolic network model comprises 502 metabolic reactions and 423 metabolites (96), and another comprises 446 reactions and 411 metabolites (55). Flux balance analyses based on these simulations demonstrated a satisfactory correlation between simulated and biological data such as the PPP flow in the first of these two models. Fine-tuning is still needed because several discrepancies were observed, as exemplified by the prediction that 10–20% of carbon should be secreted under microaerobic or anaerobic conditions, whereas it was measured that only 5% of carbon is secreted as succinate (55, 96). Likewise, biochemical regulations need to be taken into account more precisely to more closely mimic biological events. In spite of these imperfections that should be resolved by incremental advances over time, the systems biology methodology has already been successfully tested and put to use for optimizing the corynebacterial metabolism. For example, Shinfuku et al. (96) predicted that, under oxygen-deprived conditions, the disruption of the succinate pathway, the disruption of some reactions of the PPP, or the disruption of H^+-ATPase activity leads to increased lactate overproduction. Likewise, disruption of the lactate production pathway was predicted to be effective in improving succinate production rates; this corroborates findings from biological experiments (40). Similarly, Kjeldsen & Nielsen (55) used in silico genome-scale reconstruction to demonstrate a correlation between a high PPP reaction rate to support NADPH generation and high lysine production; here again this is in agreement with in vivo observations (7, 129).

Furthermore, simulations to maximize lysine production while constraining biomass production at different levels generated several useful metabolic engineering development hypotheses. For example, the model predicts a maximum lysine yield of 75% when no biomass is formed [the maximum theoretical value for *C. glutamicum* is 75–82% (100, 129)], a state that can be approached under ATP-limiting conditions. Moreover, the model indicates that decreasing fluxes through the TCA cycle, which always results in high carbon loss through CO_2 formation, could lead to improved lysine production. This hypothesis was verified by downregulating by 70% the activity of isocitrate dehydrogenase, the TCA enzyme that is expressed the highest, by replacing the strong ATG start codon with the rare GTG codon in the *icd* gene; this simple modification increased lysine production by more than 40% (8). This latter example illustrates well the power of systems biology for industrial production optimization, with the TCA cycle engineering approach complementing conventional approaches already successfully deployed, such as biosynthetic pathway, pyruvate node, and reducing equivalent balance engineering (8, 129).

Kinetic models of reaction networks also have started to generate tangible results, as shown by the identification of enzyme targets for the optimization of a valine-producing *C. glutamicum* strain (65). To this end, a model of flux control through the valine-leucine biosynthetic pathway from pyruvate was constructed for a derivative of *C. glutamicum* ATCC 13032. A total of 47 data points per metabolite during a 25-s time period were measured from a cell culture at transient state as well as from a culture at steady state. A set of differential equations was subsequently assembled to describe the material balance of each metabolite, and the resulting model was used to assess the effect of enzyme overexpression or inactivation.

An interesting hypothesis suggested that a 20% increase in pyruvate availability could lead to a 150% increase of the valine formation rate, thereby highlighting several potential metabolic engineering targets including pyruvate dehydrogenase (*aceE*), pyruvate carboxylase (*pyc*), pyruvate:quinone oxidoreductase (*pqo*, formerly *poxB*), and pyruvate carboxykinase (*pyk*) (65). Deletion of the *aceE* gene alone in vivo significantly increased valine production from undetectable levels to 30–35 mM, whereas the additional deletion of *pqo*, *pgi* (coding for phosphoglucose isomerase), and *pyc*, and the overexpression of the *ilvBNCE* cluster, resulted in the production of 410 mM (47) g liter^{-1} of L-valine (i.e., 0.49–0.56 g valine per g glucose). These findings demonstrating that optimizing pyruvate supply, increasing NADPH intracellular pools, and increasing the gene dosage of the valine biosynthetic enzymes translate to the generation of a superior valine-producing strain (13, 14, 65). The model also suggested that the translocase responsible for valine export exerts strong control on the valine metabolic flux. As a result, the L-valine translocase also represents an interesting engineering target that could be addressed by process engineering methods, should genetic engineering or proteomic engineering ones fail, to lower the extracellular valine concentrations and thus reduce gradient effects (65).

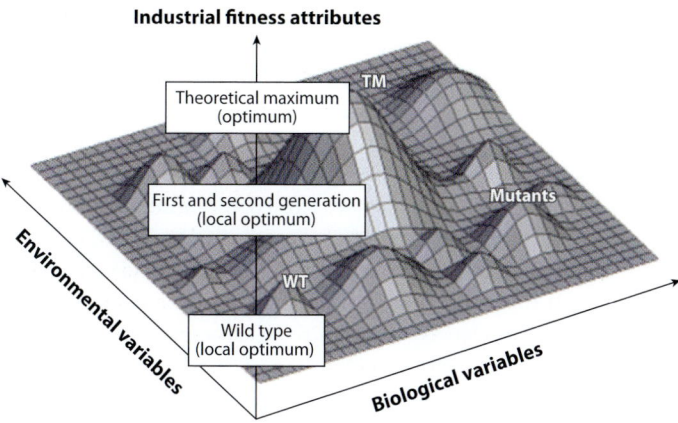

Figure 4

Metabolic fitness landscape. The industrial phenotype of a fermenter, or its industrial metabolic fitness, is a variable dependent on genetic and environmental variables. Under any given set of environmental conditions, there are in principle an optimal number of genetic changes that can redirect fluxes such that yields and titers of the product of interest are maximized. As in natural evolution, the key challenge in industrial strain engineering is establishing a path from a local optimum to paths that are closer to the global optimum, i.e., the maximum theoretical yield, such that process economics are maximized (assuming a correspondence between biological performance and economic performance, which may not necessarily be the case when comparing different environmental conditions with significantly different industrial operating variables). First- and second-generation fermenters have been developed that comprise multiple mutations, some of which may be detrimental. Postgenomic tools allow the creation of third-generation strains that are based on a blueprint that is more global, as though derived from a systems approach such that nonobvious synergistic changes can be implemented, and dependent on fewer genetic changes such that industrial robustness attributes are better preserved.

Building Efficient Primary and Secondary Metabolite Whole-Cell Biocatalysts

Building efficient whole-cell biocatalysts in essence refers to a process that defines the fitness landscape and conquers local fitness peaks (86, 91, 133) (**Figure 4**). Through use of the notion of phenotypic space (or metabolic network space), the elements of which comprise models of wild-type and mutant strains, to define fitness landscapes, the challenge is to design and evolve an industrial producer toward the global optimum of productivity, rather than toward a local optimum (86, 91, 133). As illustrated above, the overexpression of proteins or metabolites typically results in a metabolic burden to the cell. Flux optimization of *C. glutamicum* tailored to a specific production is therefore best achieved by changes at the main network nodes, resulting in a new balanced metabolism under process conditions (57, 100).

In summary, the engineering of third-generation microbial producers that exhibit superior industrial attributes via metabolic engineering has been enabled by four supportive technology domains: (*a*) data generation (whole-genome sequencing, global transcriptomics, metabolomics, fluxomics), (*b*) data analysis (bioinformatics, genome-scale modeling, and kinetic modeling), (*c*) biological reprogramming (recombinant DNA technology, protein engineering, and synthetic biology as defined below), and (*d*) biochemical engineering (manufacturing process optimization including product removal and downstream processing, and chemostats to model metabolism at steady state).

These techniques are useful in preserving the native industrial robustness of the wild-type strain while introducing a selected set of mutations to implement design-based microbial systems.

The best demonstration is perhaps provided by the increase in lysine production from 80 g liter^{-1} after 27 h at 3.0 g liter^{-1} h^{-1} (as achieved by genome breeding via the introduction of mutations in the *hom*, *lysC*, and *pyc* genes into wild-type *C. glutamicum* ATCC 13032) (79) to the industrial levels of 120 g liter^{-1} at 4.0 g liter^{-1} h^{-1} at a yield of 0.55 g lysine per g glucose (i.e., 68% molar yield) (rational systems engineering of 12 genes or their genetic signals) (10). The latter metabolic blueprint was determined by predicting the flux changes necessary to attain an optimal flux distribution scenario during lysine production (10). To this end, elementary flux model analysis was conducted with a genome-scale model of *C. glutamicum* for an optimal fermenter at zero growth. Under the tested environmental conditions, the optimal design included increasing flux through the lysine biosynthetic pathway, anaplerotic carboxylation, and PPP, and decreasing flux through counteracting decarboxylation reactions, the TCA cycle, and general anabolism (10). Importantly, the various changes implemented aim to collectively and synergistically optimize these key pathways. This is achieved by amplifying, mutating, or attenuating several target genes as follows (10):

- releasing the *lysC* gene product (aspartate kinase) from feedback inhibition,
- doubling the *ddh* gene dosage coding for diaminopimelate dehydrogenase,
- deleting the *pck* gene coding for PEP carboxykinase,
- overexpressing the *dapB* gene coding for dihydropicolinate reductase,
- doubling the *lysA* gene dosage (diaminopimelate decarboxylase),
- overexpressing *lysC* by promoter exchange,
- reducing the activity of homoserine dehydrogenase by mutation of the *hom* gene,
- increasing the kinetic properties of pyruvate carboxylase by rationale-based engineering of the *pyc* gene,
- replacing the native *pyc* promoter by a stronger promoter,
- decreasing the transcription of the isocitrate dehydrogenase encoding gene *icd* by replacing the start codon with a rare initiation codon,
- overexpressing the fructose 1,6-biphosphatase–encoding gene *fbp* by replacing its native promoter by a stronger one, and
- overexpressing the transketolase operon coding for glucose-6-phosphate dehydrogenase (genes *zwf-opcA*) as well as transketolase (*tkt*) and transaldolase (*tal*) by promoter exchange.

Remarkably, these results not only confirm but also extend the observations attained by conventional genetic engineering and physiological techniques that the simultaneous amplification of the activities of the *lysC* and *pyc* gene products dramatically improves lysine production without impacting growth, while in isolation these two mutations result in marginal overall process productivity improvement, if any (57). This nonobvious observation, in which more than one flux-controlling enzyme or its dosage is engineered, illustrates the need to address the complexity of cellular metabolism using advanced engineering tools where the cellular system itself is optimized via synergistic changes, rather than singular reactions, in order to increase both carbon conversion yields and titers while maintaining appropriate carbon source uptake and catabolism rates in conditions of reduced by-product formation. On the other hand, further productivity improvements seem possible because the theoretical maximal molar yield of lysine of the maximum fitness *C. glutamicum* network has been calculated to be up to 82% (129). Similar coordinated engineering of several genes within the context of the global metabolism, rather than of any single gene in isolation, was necessary to enable L-serine production up to 86 mM at a maximal specific productivity of 1.2 mmol^{-1} h^{-1} g dry weight (e.g., reduced expression of *glyA*, overexpression of a *serA* allele coding for an L-serine-insensitive 3-phosphoglycerate dehydrogenase, overexpression of *serB* and *serC*, and deletion of *sdaA*) (83).

Akin to that observed for the directed evolution of enzymes (25), and given the context dependence of biological responses, the conquest of metabolic fitness peaks to optimize the economic performance of a biotechnology manufacturing operation is ideally conducted by setting environmental conditions of the test design to mimic as closely as possible industry-scale conditions. Such a reverse engineering method first establishes physiologically relevant but economically optimal environmental conditions, and second, sets the corresponding optimal biological blueprint. Process engineering variables, directed by economic performance as measured by various financial indicators such as return on investment, thus define the appropriate environment in tune with the genetic background of the evolved strain in a manner in which biological performance is optimized.

Economical performance is not necessarily linear with biological performance, thus making mission critical the integration of economic variables into the design of a new strain or process. This is particularly true regarding the manufacture of commodity chemicals, including amino acids for feed use, in large quantities. This is perhaps best exemplified by the unexpected observation reported by Ohnishi et al. (78) that lysine production by the genome-breeding-derived *C. glutamicum* strain AHP-3 is temperature dependent, with higher production yields attained at 40°C than at 30°C even though lower biomass is generated at the elevated temperature. Nevertheless, additional optimization is still required to take advantage of this operational improvement, which reduces requirements of cooling utilities, because the final titer of L-lysine of 85 g liter^{-1} achieved by this strain does not reach the levels near 100 g liter^{-1} typically attained by the classical mutant strain B-6 used in conventional industrial production. Again, the use of transcriptomic analyses alone as a postgenomic tool enabled the identification of several differentially expressed genes as genetic engineering targets that could collectively support temperature resistance. These genes are involved in glycolysis (e.g., *gapA*), gluconeogenesis (e.g., *pck* and *malE*), the TCA cycle (e.g., *gltA*), or the glyoxylate shunt (78).

The key parameters that maximize the return on financial capital employed in bioconversions include three primary value axes. (*a*) Process robustness provides the fundamental industrialization basis (large operational temperature range, tolerance to growth inhibitors, contamination tolerance, tolerance to sugar concentration variations, tolerance to sugar mixture composition variations, salt resistance, high shear tolerance). (*b*) Economics provides investment incentives to build or retrofit manufacturing plants (high yields, high titers, end-product tolerance, limited by-products, by-products that can be valorized, broad range of substrates, low rate of carbohydrate consumption through vegetative functions, simple and efficient downstream processing, efficient product secretion, limited product degradation). (*c*) Industrialization enables cost-effective and reproducible manufacturing operations (scale-up feasibility, no genetic drift such that continuous production is possible, no diauxic phenomena, high volumetric productivity, optimal performance under a wide range of oxygen tensions, low pH tolerance, non-spore former, generally recognized as safe, adaptability of industrial equipment including reactors to conduct more than one process, simplicity in manufacturing design, and limited environmental footprint) (117).

These parameters are biologically answered by removing bottlenecks in the product biosynthetic pathway; optimizing central metabolism and precursor supply, including achieving reducing equivalent balance; optimizing uptake and secretion systems; reducing by-product formation; increasing tolerance to product; reducing product degradation; biocatalyst recycling; and enabling the use of cost-effective carbohydrates or their mixtures. In addition to conventional genetic engineering techniques, synthetic biology, which refers to engineering and synthesizing novel functions, represents an approach to either (*a*) achieve differential genetic expression, (*b*) generate novel biosynthetic routes to produce novel compounds by assembling modular biological components including particularly regulatory components, (*c*) manipulate genetic expression at the single

Chemical building blocks: chemical molecules that can be subsequently combined or converted into a number of more complex molecules or materials

Industrial process or strain robustness: the ability of an industrial process or strain to maintain acceptable industrial performance despite variability in manufacturing process operations

synthetic route or at the systems level, or (*d*) create sequences that do not exist in nature, for example, to generate novel enzymatic activities or to change signal transduction by altering ligand properties of membrane-bound extracellular biological sensors (1). Synthetic biology promises not only to develop bacterial cells with minimized genomes to reduce the quantity of energy dissipated in pathways other than those for product production, but also to generate biological modules to perform specific functions of industrial interest that are closer to their global fitness maximum than similar systems derived from naturally occurring proteins or modules would be. Synthetic functions can be established within well-defined metabolic boundaries in order to be replaced or adapted more easily (1, 67, 71). Furthermore, genome-scale and kinetic models enable combinatorial engineering by allowing the practitioner to test in silico, rapidly and at low cost, different permutations of these individual elements in order to generate the most appropriate and cost-effective metabolic network that can ultimately be validated in vivo (58, 130). Synthetic biology could be deployed, for example, to improve the production of D-amino acids in *C. glutamicum*, rather than the corresponding natural L-enantiomers (99), or to reprogram translational efficiency by replacing natural initiation codons or ribosome binding sites to achieve superior production (89).

C. glutamicum as a Manufacturing Workhorse

As discussed above, corynebacterial manufacturing processes constitute a mature technology that has been applied successfully to make amino acids available in large tonnage and at low cost for a range of feed and food supplement applications. Harnessing of the intrinsic properties of *C. glutamicum* to produce commodity chemicals other than amino acids is, however, more recent (40, 42, 117) and has only started to gain momentum (12, 97). The goal is to make use of renewable and cost-effective raw materials that do not significantly affect the food or feed supply chain, such as lignocellulosic biomass, by using conventional techniques and industrial practices of industrial biotechnology (also called white biotechnology) (53). As detailed above, it is remarkable that biotechnology-manufactured chemicals could represent more than 20% of all chemicals produced by 2020 and up to 28% by 2025, a significant segment considering that the worldwide chemistry market is forecasted to exceed $4 trillion by the next decade (27, 42, 46, 66, 69, 128).

The concept of deriving chemicals and chemical building blocks from biological raw materials was initially formulated as "chemurgy" (30). The first industrial example was implemented in the United States in 1883 for lactic acid production (11); another example is the production of acetone-butanol by *Clostridium acetobutylicum* during World War I (3). However, these processes were essentially replaced with the advent of the petrochemical industry during 1930–1950 (11, 98), with the exception of the production of higher-value-added compounds consisting of, in addition to amino acids, biologics such as enzymes and therapeutic proteins for which only biotechnological processes exist (24). However, the growing threats of global warming and peak oil phenomena have triggered a renewed interest in this fundamental technology platform enabling sustainable chemistry, which, in order to penetrate the market, has to overcome the economies of scale, scope, and learning that the petrochemical industry has gained over the years (113, 115).

A key factor enabling white biotechnology to significantly displace petrochemical-based production is the availability of cost-effective and robust industrial processes that optimally could be integrated into a "biorefinery" such that commodity chemicals including ethanol and isobutanol, as well as higher-value compounds including ethyl-acetate or isobutanol, are produced. Economies of scope could be leveraged further by designing manufacturing plants that make products for various industries, such as amino acids for feed use in addition to fuels, organic acids, or bioplastics, using the same fundamental corynebacterial technology and flexible manufacturing campaign schedules as a means to reduce capital expenses and thus the cost of the produced goods.

The higher oxygen content of biomass-based compounds makes possible the production of totally novel materials and chemicals with novel properties (66, 113, 127). This large difference in redox potential nevertheless constitutes a technical difficulty to avoid redox imbalances, as illustrated by the biosynthesis in *E. coli* of 1,4-butanediol, a highly reduced compound, achieved via a heterologous NADH-intensive pathway (134). A critical success factor for the biorefinery industry is therefore the availability of versatile, cost-effective, and robust industrial processes to manufacture an array of chemicals from an array of raw materials (e.g., sugars, starch, lignocellulosic hydrolysates, and sugar mixtures) (113, 117). Another critical success factor is the availability of host-vector systems that demonstrate superior operational robustness, industrialization, and economics (113, 117).

Beyond pharmaceuticals, among the highest-value-added products that can be manufactured using microbial cell factories are polymers, because these products can benefit from composition of matter patent protection, a key to ensure the generation of a sustainable high-profit-margin business, with such prospects triggering upstream investments in research and development (118). Manufacturing can be performed either by bioconversion alone or by a combination of biological and chemical processes, for example, to produce the necessary building blocks and to assemble them by condensation or addition. Polymer building blocks and polymers of economic interest include dicarboxylic acids to produce polyesters (polylactic acid, PLA; or polyethylene terephthalate, PET); diamines such as putrescine (1,4-diaminobutane) and cadaverine (1,5-diaminopentane) to produce nylon-4,6; or dienes such as 2-methyl-1,3-butadiene to produce polyisoprene (51, 61, 70).

The current state-of-the-art corynebacterial biotechnology already enables the production of a portfolio of compounds comprising various amino acids, organic acids such as succinate and lactate, alcohols such as ethanol and isobutanol, and several chemical building blocks (12, 39, 40, 43, 51, 60, 62, 70, 73, 80, 97). Integration of amino acid production and chemical production in a biorefinery has thus become possible.

Remarkably, pathways for the synthesis of aromatic amino acids could perhaps be redirected to synthesize aromatic chemicals in widespread use, for example, compounds with applications to the plastics industry such as terephthalate. This advancement is a fundamental one: The biorefinery concept aims to recreate for agricultural feedstock cost structures that are similar to those attained in the petrochemical industry, where commodity chemicals are produced as well as higher-value-added chemicals via processes that transform most of the carbon of the primary feedstock into useful product (113).

The corynebacterial biorefinery provides numerous ways to optimize the costs of products sold. This change is bound to have a far-reaching impact, because it proceeds from the integration of part of the agricultural value chain with the petrochemical value chain, thus colliding the transportation, polymer, and chemical value chains with other established value chains, particularly those of the food and feed industry (113). A fundamental advantage of utilizing *C. glutamicum* as a biotechnology workhorse in biorefineries is that, unlike *S. cerevisiae*, live *C. glutamicum* catalysts used under growth cessation conditions at high cellular concentrations can be prepared in advance in bulk quantities in dedicated centralized factories, and stored and transported at points of use. This translates to lower costs to transport bulky raw materials and to smaller reactor sizes, thus reducing capital expenses (116, 117). Postgenomic tools applied to *C. glutamicum* have allowed researchers to redesign the cellular metabolism to achieve efficiencies in industrial metabolic functions rather than physiological ones. As a result, they further optimize productivity performance, initiating a new iteration of the corynebacterial technology development cycle, and in turn further expanding technology frontiers to deliver, for the greatest benefits of the consumers, the promises of green chemistry and of a sustainable economic growth engine while limiting humankind's footprint on Nature.

CONCLUSIONS

Classical methods of strain engineering involve the use of mutagenesis to remove bottlenecks in product biosynthetic pathway, central metabolism, and precursor supply, reducing equivalent balance, or uptake and secretion systems. These methods have resulted in significant productivity and cost improvements. Nevertheless, the limitations of mutagenesis and recombinant DNA techniques have recently become apparent, which stem from the observation that microbial metabolic networks typically exhibit rigidities at their key nodes. These rigidities ensure robustness of cellular metabolism. Although it is a useful adaptive trait of self-organization whereby a system at steady state regulates itself in response to environmental changes, these resistances to change maintain microorganisms in their natural fitness peak as determined by natural evolution, therefore constituting a barrier to achieving optimal industrial applications, which typically correspond to a different fitness space determined by genetic and environmental dimensions. The mechanisms at play here include a highly redundant and interconnected metabolic network. There is in general no single rate-limiting step. Flux control tends to be distributed among multiple enzymes for most pathways. Classical methods are useful for delineating the behavior of single enzymes in vitro, but they are much less useful for understanding cellular behavior at the systems level. Many pathways are redundant, for example, via bypasses that become prominent when the natural predominant pathways are disrupted. Many biological mechanisms are interdependent, as exemplified by sugar uptake or product secretion, sensing/responding systems, glycolysis, and the TCA cycle, resulting in the embedment of biosynthetic pathways of various products such as amino acids into the central metabolism of biosynthetic pathways. The dimension of information exchange between a particular pathway and the rest of the metabolic network is thus a critical one to consider for attaining significant performance improvement. As a result, a more holistic approach is needed in order to develop superior productivity attributes by the coordinated manipulation of multiple genes for achieving systems-level synergies, i.e., by implementing an expert system that goes beyond conventional empirical approaches and beyond linear rationales.

Complete genome sequences are the foundation on which such a systems approach can be built, because complete genomic information makes possible global analyses to study gene and protein expression. Complemented by flux and metabolite analyses, the integration of the data derived from these high-content global experiments permits the optimal use of systems postgenomic tools including rationale-based design engineering and strain-directed evolution. Systems biology based on predictive genome-scale and kinetic models of industrial strains as well as proteome reference maps enable the assessment of novel genetic blueprints in silico prior to their translation into a biological system. Importantly, the approach is iterative, improving on the metabolic reconstruction using flux analysis to determine the metabolic engineering steps that need to be performed followed by testing in vivo. Notwithstanding the promises and initial successes of the new platform, there are still important limitations given remaining uncertainties and shortcomings in predictive power, which are due primarily to incomplete information and to the stochastic nature of biological processes.

Furthermore, although biological reconstruction is performed bottom-up, the new blueprints are designed in terms of economic efficiency, rather than in terms of biological efficiency. The key parameters affecting pathway redesign thus compose engineering and biological constraints under the hierarchy of process economics, as follows: process conditions, titer yield and productivity comprising carbon channeling and enzymatic efficiency, sustainable cofactor and reducing equivalent supply, industrial robustness, high-cell-density reactors, low operating variables, low environmental footprint or sustainability impact, and minimized capital expenses.

The biological parameters that can be altered to redesign a metabolic network for optimum industrial productivities include transcriptional control, posttranscriptional control, allosteric control, kinetic improvements, cofactor and reducing equivalent recycling, and transport reactions. These parameters can be altered by overproduction or disruption of relevant positive regulatory genes; overproduction or disruption of relevant negative regulatory genes; identification of rate-limiting steps; amplification of genes encoding rate-limiting steps; inactivation of competing secondary metabolite pathways; modulation of transcription and translation; appropriate precursor supply and balance by metabolic engineering; appropriate cofactor balance by metabolic engineering or enzyme replacement; deletion of product inhibition, e.g., by protein engineering; and construction of synthetic regulatory circuits to confer desired functions.

Postgenomic approaches encompass not only the activities of the enzymes directly related to the biosynthetic pathway of the product of interest, but also the pathways of the central metabolism and pleiotropic mechanisms of control of gene expression, as well as sensing and responding cascades that microorganisms use to best adapt to the environmental conditions they encounter. As such, robustness and homeostasis attributes are optimized in addition to biosynthetic attributes. The new approach aims to conquer fitness peaks under environmental conditions dictated by the specific industrial process being considered. The genetic blueprint of the microbial overproducer is amended, both within and outside the specific synthetic pathway of the product being manufactured. The systems approach thus enables the harnessing of nonobvious synergies that could not be attained before. These advances expand the technological frontier. Third-generation strains and associated processes thus developed pave the way for the large-scale production of commodity products, including not only amino acids or nucleotides for food or feed application, but also fuels and chemical building blocks for the development of a sustainable transportation and polymer industry, thereby further expanding the "economic power of the microbe," following the vision that Arnold Demain has set for industrial microbiology (24).

SUMMARY POINTS

1. Postgenomic tools, such as transcriptomics, proteomics, metabolomics, fluxomics, systems biology, and synthetic biology, allow researchers to analyze and engineer microbial strains at the systems level. The knowledge thus generated uncovers physiological insights and strain engineering strategies that cannot be attained when analyses are conducted solely at the gene or pathway level.

2. *C. glutamicum* has intrinsic robust industrial attributes and has become a biotechnology workhorse not only for amino acid production, but also for the production of commodity chemicals such as fuels or chemical building blocks. This is a critical development for the implementation of corynebacterial biorefineries.

3. The development of oxygen-deprived processes without growth due to the absence of a terminal electron acceptor enables the cost-effective production of reduced chemicals by *C. glutamicum*. Here cells are reacted at very high cell densities and with a limited share of carbon substrate being directed toward vegetative functions as opposed to product production.

4. *C. glutamicum* can concomitantly metabolize a large spectrum of carbon sources for which it is not subject to carbon catabolite repression, albeit this co-utilization is heavily regulated by a complex array of transcriptional regulators.

5. With the availability of genome-scale models, stoichiometric models, and kinetic models, in silico systems-level methods have been developed to engineer efficient corynebacterial host-vector systems or to further decipher the corynebacterial physiology. These models have already uncovered nonobvious engineering targets to conquer local fitness peaks and optimize industrial robustness, homeostasis, and biosynthetic power.

6. *C. glutamicum*, like many other industrial organisms, exhibits numerous metabolic rigidities at its key network nodes that result from a highly redundant and interconnected metabolic network.

7. A combination of rationale-based design and strain-directed evolution techniques applied to inverse metabolic engineering has already resulted in the construction of optimized industrial strains in which network rigidities are addressed with a minimum number of mutations, thus maximizing the overall fitness of industrial strains. These bioinformatics models are likely to become highly predictive once a more complete set of reactions is modeled.

8. Extensive manipulations of the microbial function of corynebacteria are made possible by a complete set of molecular biology tools. These comprise not only efficient electrotransformation techniques, high frequency and essentially random transposon mutagenesis or gene disruption and replacement tools, as well as constitutive or inducible promoters of various strengths, but also stable high- and medium-copy-number plasmids as well as megabase-scale engineering protocols.

DISCLOSURE STATEMENT

The authors are not aware of any affiliations, memberships, funding, or financial holdings that might be perceived as affecting the objectivity of this review.

LITERATURE CITED

1. Andrianantoandro E, Basu S, Karig DK, Weiss R. 2006. Synthetic biology: new engineering rules for an emerging discipline. *Mol. Syst. Biol.* 2:2006.0028
2. Anonymous. 2009. *The Bioeconomy to 2030*. Organisation for Economic Co-operation and Development. Paris, France
3. Awang GM, Jones GA, Ingledew WM. 1988. The acetone-butanol-ethanol fermentation. *Crit. Rev. Microbiol.* 15(Suppl. 1):S33–67
4. Bailey JE. 1999. Lessons from metabolic engineering for functional genomics and drug discovery. *Nat. Biotechnol.* 17:616–18
5. Bailey JE, Sburlati A, Hatzimanikatis V, Lee K, Renner WA, Tsai PS. 2002. Inverse metabolic engineering: a strategy for directed genetic engineering of useful phenotypes. *Biotechnol. Bioeng.* 79:568–79
6. Bayan N, Houssin C, Chami M, Leblon G. 2003. Mycomembrane and S-layer: two important structures of *Corynebacterium glutamicum* cell envelope with promising biotechnology applications. *J. Biotechnol.* 104:55–67
7. Becker J, Klopprogge C, Herold A, Zelder O, Bolten CJ, Wittmann C. 2007. Metabolic flux engineering of L-lysine production in *Corynebacterium glutamicum*—over expression and modification of G6P dehydrogenase. *J. Biotechnol.* 132:99–109
8. Becker J, Klopprogge C, Schröder H, Wittmann C. 2009. Metabolic engineering of the tricarboxylic acid cycle for improved lysine production by *Corynebacterium glutamicum*. *Appl. Environ. Microbiol.* 75:7866–69

9. Becker J, Klopprogge C, Wittmann C. 2008. Metabolic responses to pyruvate kinase deletion in lysine producing *Corynebacterium glutamicum*. *Microb. Cell Fact.* 7:8
10. Becker J, Zelder O, Häfner S, Schröder H, Wittmann C. 2011. From zero to hero—design-based systems metabolic engineering of *Corynebacterium glutamicum* for L-lysine production. *Metab. Eng.* 13:159–68

    > 10. Constitutes a seminal example of the application of postgenomic tools to the design and engineering of efficient industrial corynebacterial strains.

11. Benninga H. 1990. *A History of Lactic Acid Making*. Dordrecht, The Nether.: Kluwer
12. Blombach B, Riester T, Wieschalka S, Ziert C, Youn JW, et al. 2011. *Corynebacterium glutamicum* tailored for efficient isobutanol production. *Appl. Environ. Microbiol.* 77:3300–10
13. Blombach B, Schreiner ME, Bartek T, Oldiges M, Eikmanns BJ. 2008. *Corynebacterium glutamicum* tailored for high-yield L-valine production. *Appl. Microbiol. Biotechnol.* 79:471–79
14. Blombach B, Schreiner ME, Holatko J, Bartek T, Oldiges M, Eikmanns BJ. 2007. L-valine production with pyruvate dehydrogenase complex-deficient *Corynebacterium glutamicum*. *Appl. Environ. Microbiol.* 73:2079–84
15. Blombach B, Schreiner ME, Moch M, Oldiges M, Eikmanns BJ. 2007. Effect of pyruvate dehydrogenase complex deficiency on L-lysine production with *Corynebacterium glutamicum*. *Appl. Gen. Mol. Biotechnol.* 76:615–23
16. Bro C, Nielsen J. 2004. Impact of 'ome' analyses on inverse metabolic engineering. *Metab. Eng.* 6:204–11
17. Buchinger S, Strösser J, Rehm N, Hänßler E, Hans S, et al. 2009. A combination of metabolome and transcriptome analyses reveals new targets of the *Corynebacterium glutamicum* nitrogen regulator AmtR. *J. Biotechnol.* 140:68–74
18. Bulter T, Bernstein JR, Liao JC. 2003. A perspective of metabolic engineering strategies: moving up the systems hierarchy. *Biotechnol. Bioeng.* 84:815–21
19. Burgard AP, Nikolaev EV, Schilling CH, Maranas CD. 2004. Flux coupling analysis of genome-scale metabolic network reconstructions. *Genome Res.* 14:301–12
20. Cakir T, Hendriks MM, Westerhuis JA, Smilde AK. 2009. Metabolic network discovery through reverse engineering of metabolome data. *Metabolomics* 5:318–29
21. Crameri A, Raillard SA, Bermudez E, Stemmer WP. 1998. DNA shuffling of a family of genes from diverse species accelerates directed evolution. *Nature* 391:288–91
22. de Hollander JA. 1994. Potential metabolic limitations in lysine production by *Corynebacterium glutamicum* as revealed by metabolic network analysis. *Appl. Microbiol. Biotechnol.* 42:508–15
23. Demain AL. 2000. Small bugs, big business: the economic power of the microbe. *Biotechnol. Adv.* 18:499–514
24. Demain AL. 2007. The business of biotechnology. *Ind. Biotechnol.* 3:269–83
25. Dougherty MJ, Arnold FH. 2009. Directed evolution: new parts and optimized function. *Curr. Opin. Biotechnol.* 20:486–91
26. Fränzel B, Poetsch A, Trötschel C, Persicke M, Kalinowski J, Wolters DA. 2010. Quantitative proteomic overview on the *Corynebacterium glutamicum* L-lysine producing strain DM1730. *J. Proteomics* 73:2336–53
27. Gavrilescu M, Chisti Y. 2005. Biotechnology—a sustainable alternative for chemical industry. *Biotechnol. Adv.* 23:471–99
28. Gehlenborg N, O'Donoghue SI, Baliga NS, Goesmann A, Hibbs MA, et al. 2010. Visualization of omics data for systems biology. *Nat. Methods* 7:S56–68
29. Gill RT. 2003. Enabling inverse metabolic engineering through genomics. *Curr. Opin. Biotechnol.* 14:484–90
30. Hale WJ. 1934. *The Farm Chemurgic*. Boston, MA: Stratford
31. Hartmann M, Barsch A, Niehaus K, Pühler A, Tauch A, Kalinowski J. 2004. The glycosylated cell surface protein Rpf2, containing a resuscitation-promoting factor motif, is involved in intercellular communication of *Corynebacterium glutamicum*. *Arch. Microbiol.* 182:299–312
32. Hayashi M, Ohnishi J, Mitsuhashi S, Yonetani Y, Hashimoto S, Ikeda M. 2006. Transcriptome analysis reveals global expression changes in an industrial L-lysine producer of *Corynebacterium glutamicum*. *Biosci. Biotechnol. Biochem.* 70:546–50

33. Hermann T. 2003. Industrial production of amino acids by coryneform bacteria. *J. Biotechnol.* 104:155–72
34. Holford NH, Kimko HC, Monteleone JP, Peck CC. 2000. Simulation of clinical trials. *Annu. Rev. Pharmacol. Toxicol.* 40:209–34
35. Hüser AT, Chassagnole C, Lindley ND, Merkamm M, Guyonvarch A, et al. 2005. Rational design of a *Corynebacterium glutamicum* pantothenate production strain and its characterization by metabolic flux analysis and genome-wide transcriptional profiling. *Appl. Environ. Microbiol.* 71:3255–68
36. Ikeda M, Mitsuhashi S, Tanaka K, Hayashi M. 2009. Reengineering of a *Corynebacterium glutamicum* L-arginine and L-citrulline producer. *Appl. Environ. Microbiol.* 75:1635–41
37. Ikeda M, Nakagawa S. 2003. The *Corynebacterium glutamicum* genome: features and impacts on biotechnological processes. *Appl. Microbiol. Biotechnol.* 62:99–109

> 38. Uses the comparative genomic analysis of an industrial strain with its natural ancestor to identify a variety of mutations in genes associated with L-lysine production.

38. **Ikeda M, Ohnishi J, Hayashi M, Mitsuhashi S. 2006. A genome-based approach to create a minimally mutated *Corynebacterium glutamicum* strain for efficient L-lysine production. *J. Ind. Microbiol. Biotechnol.* 33:610–15**
39. Inui M, Kawaguchi H, Murakami S, Vertès AA, Yukawa H. 2004. Metabolic engineering of *Corynebacterium glutamicum* for fuel ethanol production under oxygen-deprivation conditions. *J. Mol. Microbiol. Biotechnol.* 8:243–54

> 40. Constitutes a paradigm shift in biotechnology whereby *C. glutamicum* is applied to the production of commodity chemicals.

40. **Inui M, Murakami S, Okino S, Kawaguchi H, Vertès AA, Yukawa H. 2004. Metabolic analysis of *Corynebacterium glutamicum* during lactate and succinate productions under oxygen deprivation conditions. *J. Mol. Microbiol. Biotechnol.* 7:182–96**
41. Inui M, Suda M, Okino S, Nonaka H, Puskas LG, et al. 2007. Transcriptional profiling of *Corynebacterium glutamicum* metabolism during organic acid production under oxygen deprivation conditions. *Microbiology* 153:2491–504
42. Inui M, Vertès AA, Yukawa H. 2010. Advanced fermentation technologies. In *Biomass to Biofuels: Strategies for Global Industries*, ed. AA Vertès, N Qureshi, HP Blaschek, H Yukawa, pp. 311–30. Chichester, UK: Wiley
43. Jo SJ, Maeda M, Ooi T, Taguchi S. 2006. Production system for biodegradable polyester polyhydroxybutyrate by *Corynebacterium glutamicum*. *J. Biosci. Bioeng.* 102:233–36
44. Josic D, Kovač S. 2008. Application of proteomics in biotechnology—microbial proteomics. *Biotechnol. J.* 3:496–509
45. Kalinowski J, Bathe B, Bartels D, Bischoff N, Bott M, et al. 2003. The complete *Corynebacterium glutamicum* ATCC 13032 genome sequence and its impact on the production of L-aspartate-derived amino acids and vitamins. *J. Biotechnol.* 104:5–25
46. Kamm B, Kamm M. 2007. Biorefineries—multi product processes. *Adv. Biochem. Eng. Biotechnol.* 105:175–204
47. Kana BD, Mizrahi V. 2010. Resuscitation-promoting factors as lytic enzymes for bacterial growth and signaling. *FEMS Immunol. Med. Microbiol.* 58:39–50
48. Karasawa M, Tosaka O, Ikeda S, Yoshii H. 1986. Application of protoplast fusion to the development of L-threonine and L-lysine producers. *Agric. Biol. Chem.* 50:339–46
49. Keep NH, Ward JM, Cohen-Gonsaud M, Henderson B. 2006. Wake up! Peptidoglycan lysis and bacterial non-growth states. *Trends Microbiol.* 14:271–76
50. Kiefer P, Heinzle E, Zelder O, Wittmann C. 2004. Comparative metabolic flux analysis of lysine-producing *Corynebacterium glutamicum* cultured on glucose or fructose. *Appl. Environ. Microbiol.* 70:229–39
51. Kind S, Jeong WK, Schröder H, Wittmann C. 2010. Systems-wide metabolic pathway engineering in *Corynebacterium glutamicum* for bio-based production of diaminopentane. *Metab. Eng.* 12:341–51
52. Kinoshita S, Udaka S, Shimono M. 1957. Studies on the amino acid fermentation. Part 1. Production of L-glutamic acid by various microorganisms. *J. Gen. Appl. Microbiol.* 3:193–205
53. Kircher M. 2006. White biotechnology: ready to partner and invest in. *Biotechnol. J.* 1:787–94
54. Kirchner O, Tauch A. 2003. Tools for genetic engineering in the amino acid-producing bacterium *Corynebacterium glutamicum*. *J. Biotechnol.* 104:287–99

> 55. Develops functional genome-scale metabolic models of *C. glutamicum* to demonstrate the biotechnological significance of annotated whole-genome sequences.

55. **Kjeldsen KR, Nielsen J. 2009. In silico genome-scale reconstruction and validation of the *Corynebacterium glutamicum* metabolic network. *Biotechnol. Bioeng.* 102:583–97**

56. Koffas M, Stephanopoulos G. 2005. Strain improvement by metabolic engineering: lysine production as a case study for systems biology. *Curr. Opin. Biotechnol.* 16:361–66

57. Koffas MA, Jung GY, Stephanopoulos G. 2003. Engineering metabolism and product formation in *Corynebacterium glutamicum* by coordinated gene overexpression. *Metab. Eng.* 5:32–41

58. Krömer JO, Sorgenfrei O, Klopprogge K, Heinzle E, Wittmann C. 2004. **In-depth profiling of lysine-producing *Corynebacterium glutamicum* by combined analysis of the transcriptome, metabolome, and fluxome.** *J. Bacteriol.* 186:1769–84.

58. Uses an integrated postgenomic approach to identify correlations that exist between gene expression and the in vivo activity of numerous enzymes.

59. Krömer JO, Wittmann C, Schröder H, Heinzle E. 2006. Metabolic pathway analysis for rational design of L-methionine production by *Escherichia coli* and *Corynebacterium glutamicum*. *Metab. Eng.* 8:353–69

60. Kumagai H. 2000. Microbial production of amino acids in Japan. *Adv. Biochem. Eng. Biotechnol.* 69:71–85

61. Lee JW, Kim HU, Choi S, Yi J, Lee SY. 2011. Microbial production of building block chemicals and polymers. *Curr. Opin. Biotechnol.* 22:1–10

62. Leuchtenberger W, Huthmacher K, Drauz K. 2005. Biotechnological production of amino acids and derivatives: current status and prospects. *Appl. Microbiol. Biotechnol.* 69:1–8

63. Li L, Wada M, Yokota A. 2007. Cytoplasmic proteome reference map for a glutamic acid-producing *Corynebacterium glutamicum* ATCC 14067. *Proteomics* 7:4317–22

64. Liebl W. 2006. *Corynebacterium* - nonmedical. In *The Prokaryotes*, ed. M Dworkin, S Falkow, E Rosenberg, KH Schleifer, E Stackebrandt, pp. 796–818. New York: Springer. 3rd ed.

65. Magnus JB, Oldiges M, Takors R. 2009. **The identification of enzyme targets for the optimization of a valine producing *Corynebacterium glutamicum* strain using a kinetic model.** *Biotechnol. Prog.* 25:754–62

65. Demonstrates the value of combining data-driven and model-based methods to generate a quantitative evaluation of flux control.

66. Marquardt W, Harwardt A, Hechinger M, Kraemer K, Viell J, Voll A. 2010. The biorenewables opportunity: toward next generation process and product systems. *AIChE J.* 56:2228–35

67. Martin CH, Nielsen DR, Solomon KV, Prather KL. 2009. Synthetic metabolism: engineering biology at the protein and pathway scales. *Chem. Biol.* 16:277–86

68. Marx A, Striegel K, de Graaf AA, Sahm H, Eggeling L. 1997. Response of the central metabolism of *Corynebacterium glutamicum* to different flux burdens. *Biotechnol. Bioeng.* 56:168–80

69. Meyer HP, Werbitzky O. 2011. How green can the industry become with biotechnology? In *Biocatalysis for Green Chemistry and Chemical Process Development*, ed. AT Junhua, R Kazlauskas, pp. 23–44. Hoboken, NJ: Wiley

70. Mimitsuka T, Sawai H, Hatsu M, Yamada K. 2007. Metabolic engineering of *Corynebacterium glutamicum* for cadaverine fermentation. *Biosci. Biotechnol. Biochem.* 71:2130–35

71. Moya A, Gil R, Latorre A, Peretó J, Pilar Garcillán-Barcia M, de la Cruz F. 2009. Toward minimal bacterial cells: evolution versus design. *FEMS Microbiol. Rev.* 33:225–35

72. Nešvera J, Pátek M. 2011. Tools for genetic manipulations in *Corynebacterium glutamicum* and their applications. *Appl. Microbiol. Biotechnol.* 90:1641–54

73. Niimi S, Suzuki N, Inui M, Yukawa H. 2011. Metabolic engineering of 1,2-propanediol pathways in *Corynebacterium glutamicum*. *Appl. Microbiol. Biotechnol.* 90:1721–29

74. Nishimura T, Teramoto H, Inui M, Yukawa H. 2011. Gene expression profiling of *Corynebacterium glutamicum* during anaerobic nitrate respiration: induction of the SOS response for cell survival. *J. Bacteriol.* 193:1327–33

75. Nishimura T, Vertès AA, Shinoda Y, Inui M, Yukawa H. 2007. Anaerobic growth of *Corynebacterium glutamicum* using nitrate as a terminal electron acceptor. *Appl. Microbiol. Biotechnol.* 75:889–97

76. Nishio Y, Nakamura Y, Kawarabayasi Y, Usuda Y, Kimura E, et al. 2003. Comparative complete genome sequence analysis of the amino acid replacements responsible for the thermostability of *Corynebacterium efficiens*. *Genome. Res.* 13:1572–79

77. Oberhardt MA, Palsson BO, Papin JA. 2009. Applications of genome-scale metabolic reconstructions. *Mol. Syst. Biol.* 5:320

78. Ohnishi J, Hayashi M, Mitsuhashi S, Ikeda M. 2003. Efficient 40°C fermentation of L-lysine by a new *Corynebacterium glutamicum* mutant developed by genome breeding. *Appl. Microbiol. Biotechnol.* 62:69–75

79. Ohnishi J, Mitsuhashi S, Hayashi M, Ando S, Yokoi H, et al. 2002. A novel methodology employing *Corynebacterium glutamicum* genome information to generate a new L-lysine-producing mutant. *Appl. Microbiol. Biotechnol.* 58:217–23
80. Okino S, Inui M, Yukawa H. 2005. Production of organic acids by *Corynebacterium glutamicum* under oxygen deprivation. *Appl. Microbiol. Biotechnol.* 68:475–80
81. Okino S, Noburyu R, Suda M, Jojima T, Inui M, Yukawa H. 2008. An efficient succinic acid production process in a metabolically engineered *Corynebacterium glutamicum* strain. *Appl. Microbiol. Biotechnol.* 81:459–64
82. Otero JM, Nielsen J. 2010. Industrial systems biology. *Biotechnol. Bioeng.* 105:439–60
83. Peters-Wendisch P, Stolz M, Etterich H, Kennerknecht N, Sahm H, Eggeling L. 2005. Metabolic engineering of *Corynebacterium glutamicum* for L-serine production. *Appl. Environ. Microbiol.* 71:7139–44
84. Petersen S, de Graaf AA, Eggeling L, Möllney M, Wiechert W, Sahm H. 2000. In vivo quantification of parallel and bidirectional fluxes in the anaplerosis of *Corynebacterium glutamicum*. *J. Biol. Chem.* 275:35932–41
85. Petersen S, Mack C, de Graaf AA, Riedel C, Eikmanns BJ, Sahm H. 2001. Metabolic consequences of altered phosphoenolpyruvate carboxykinase activity in *Corynebacterium glutamicum* reveal anaplerotic regulation mechanisms in vivo. *Metab. Eng.* 3:344–61
86. Petri R, Schmidt-Dannert C. 2004. Dealing with complexity: evolutionary engineering and genome shuffling. *Curr. Opin. Biotechnol.* 15:298–304
87. Puech V, Chami M, Lemassu A, Laneelle MA, Schiffler B, et al. 2001. Structure of the cell envelope of corynebacteria: importance of the non-covalently bound lipids in the formation of the cell wall permeability barrier and fracture plane. *Microbiology* 147:1365–82
88. Sahm H, Eggeling L, de Graaf AA. 2000. Pathway analysis and metabolic engineering in *Corynebacterium glutamicum*. *Biol. Chem.* 381:899–910
89. Salis HM, Mirsky EA, Voigt CA. 2009. Automated design of synthetic ribosome binding sites to control protein expression. *Nat. Biotechnol.* 27:946–50
90. Santos CN, Stephanopoulos G. 2008. Combinatorial engineering of microbes for optimizing cellular phenotype. *Curr. Opin. Chem. Biol.* 12:168–76
91. Sauer U. 2001. Evolutionary engineering of industrially important microbial phenotypes. *Adv. Biochem. Eng. Biotechnol.* 73:129–69
92. Sauer U, Eikmanns BJ. 2005. The PEP-pyruvate-oxaloacetate node as the switch point for carbon flux distribution in bacteria. *FEMS Microbiol. Rev.* 29:765–94
93. Schaaff I, Heinisch J, Zimmermann FK. 1989. Overproduction of glycolytic enzymes in yeast. *Yeast* 5:285–90
94. Schröder J, Tauch A. 2010. Transcriptional regulation of gene expression in *Corynebacterium glutamicum*: the role of global, master and local regulators in the modular and hierarchical gene regulatory network. *FEMS Microbiol. Rev.* 34:685–737
95. Shimizu H, Hirasawa T. 2007. Production of glutamate and glutamate-related amino acids: molecular mechanism analysis and metabolic engineering. In *Microbiology Monographs: Amino Acid Biosynthesis*, ed. VF Wendisch, pp. 1–38. Berlin: Springer-Verlag
96. Shinfuku Y, Sorpitiporn N, Sono M, Furusawa C, Hirasawa T, Shimizu H. 2009. Development and experimental verification of a genome-scale metabolic model for *Corynebacterium glutamicum*. *Microb. Cell Factories* 8:43
97. Smith KM, Cho KM, Liao JC. 2010. Engineering *Corynebacterium glutamicum* for isobutanol production. *Appl. Microbiol. Biotechnol.* 87:1045–55
98. Spitz PH. 1988. *Petrochemicals: The Rise of An Industry*. New York: Wiley
99. Stäbler N, Oikawa T, Bott M, Eggeling L. 2011. *Corynebacterium glutamicum* as a host for synthesis and export of D-amino acids. *J. Bacteriol.* 193:1702–9
100. Stephanopoulos G, Vallino JJ. 1991. Network rigidity and metabolic engineering in metabolite overproduction. *Science* 252:1675–81

101. Stolz M, Peters-Wendisch P, Etterich H, Gerharz T, Faurie R, et al. 2007. Reduced folate supply as a key to enhanced L-serine production by *Corynebacterium glutamicum*. *Appl. Environ. Microbiol.* 73:750–55
102. Suzuki N, Inui M, Yukawa H. 2008. Random genome deletion methods applicable to prokaryotes. *Appl. Microbiol. Biotechnol.* 79:519–26
103. Suzuki N, Okayama S, Nonaka H, Tsuge Y, Inui M, Yukawa H. 2005. Large-scale engineering of the *Corynebacterium glutamicum* genome. *Appl. Environ. Microbiol.* 71:3369–72
104. Takaç S, Çahk G, Mavituna F, Dervakos G. 1998. Metabolic flux distribution for the optimized production of L-glutamate. *Enzyme Microbiol. Technol.* 23:286–300
105. Takeno S, Murata R, Kobayashi R, Mitsuhashi S, Ikeda M. 2010. Engineering of *Corynebacterium glutamicum* with an NADPH-generating glycolytic pathway for L-lysine production. *Appl. Environ. Microbiol.* 76:7154–60
106. Tsuge Y, Suzuki N, Inui M, Yukawa H. 2007. Random segment deletion based on IS*31831* and Cre/*loxP* excision system in *Corynebacterium glutamicum*. *Appl. Microbiol. Biotechnol.* 74:1333–41
107. Tufariello JM, Jacobs WR Jr, Chan J. 2004. Individual *Mycobacterium tuberculosis* resuscitation-promoting factor homologues are dispensable for growth in vitro and in vivo. *Infect. Immun.* 72:515–26
108. Udaka S. 2008. The discovery of *Corynebacterium glutamicum* and birth of amino acid fermentation industry in Japan. In *Corynebacteria: Genomics and Molecular Biology*, ed. A Burkovski, pp. 1–6. Norfolk, UK: Caister Academic
109. Vallino JJ, Stephanopoulos G. 1993. Metabolic flux distributions in *Corynebacterium glutamicum* during growth and lysine overproduction. *Biotechnol. Bioeng.* 41:633–46
110. Vallino JJ, Stephanopoulos G. 1994. Carbon flux distributions at the glucose 6-phosphate branch point in *Corynebacterium glutamicum* during lysine overproduction. *Biotechnol. Prog.* 10:327–34
111. Vallino JJ, Stephanopoulos G. 1994. Carbon flux distributions at the pyruvate branch point in *Corynebacterium glutamicum* during lysine overproduction. *Biotechnol. Prog.* 10:320–26
112. van der Werf MJ, Jellema RH, Hankemeier T. 2005. Microbial metabolomics: replacing trial-and-error by the unbiased selection and ranking of targets. *J. Ind. Microbiol. Biotechnol.* 32:234–52
113. **Vertès AA. 2010. Axes of development in chemical process engineering for converting biomass into energy. In *Biomass to Biofuels: Strategies for Global Industries*, ed. AA Vertès, N Qureshi, HP Blaschek, H Yukawa, pp. 491–521. Chichester, UK: Wiley.**
114. Vertès AA, Inui M, Yukawa H. 2005. Manipulating corynebacteria, from individual genes to chromosomes. *Appl. Environ. Microbiol.* 71:7633–42
115. Vertès AA, Inui M, Yukawa H. 2006. Implementing biofuels on a global scale. *Nat. Biotechnol.* 24:761–64
116. Vertès AA, Inui M, Yukawa H. 2007. Alternative technologies for biotechnological fuel ethanol manufacturing. *J. Chem. Technol. Biotechnol.* 82:693–97
117. Vertès AA, Inui M, Yukawa H. 2008. Technological options for biological fuel ethanol. *J. Mol. Microbiol. Biotechnol.* 15:16–30
118. Vertès AA, Yochanan SB. 2010. Financing strategies for industrial-scale biofuel production and technology development start-ups. In *Biomass to Biofuels: Strategies for Global Industries*, ed. AA Vertès, N Qureshi, HP Blaschek, H Yukawa, pp. 523–45. Chichester, UK: Wiley
119. Vieira G, Sabarly V, Bourguignon PY, Durot M, Le Fèvre F, et al. 2011. Core and panmetabolism in *Escherichia coli*. *J. Bacteriol.* 193:1461–72
120. Vrljic M, Sahm H, Eggeling L. 1996. A new type of transporter with a new type of cellular function: L-lysine export from *Corynebacterium glutamicum*. *Mol. Microbiol.* 22:815–26
121. Warner JR, Patnaik R, Gill RT. 2009. Genomics enabled approaches in strain engineering. *Curr. Opin. Microbiol.* 12:223–30
122. Wayne LG, Sohaskey CD. 2001. Nonreplicating persistence of *Mycobacterium tuberculosis*. *Annu. Rev. Microbiol.* 55:139–63
123. Wendisch VF. 2003. Genome-wide expression analysis in *Corynebacterium glutamicum* using DNA microarrays. *J. Biotechnol.* 104:273–85
124. Wendisch VF. 2007. *Amino Acid Biosynthesis: Pathways, Regulation, and Metabolic Engineering*. Berlin: Springer. 413 pp.

113. Reviews the fundamental needs of the emerging sustainable chemical industry, including current hurdles and technological bottlenecks.

125. Wendisch VF, de Graaf AA, Sahm H, Eikmanns BJ. 2000. Quantitative determination of metabolic fluxes during coutilization of two carbon sources: comparative analyses with *Corynebacterium glutamicum* during growth on acetate and/or glucose. *J. Bacteriol.* 182:3088–96
126. Westerhoff HV, Kholodenko BN. 2004. Introduction. In *Metabolic Engineering in the Post Genomic Era*, ed. BN Kholodenko, HV Westerhoff, pp. 1–36. Wymondham, UK: Horizon Bioscience
127. **Weusthuis RA, Lamot I, van der Oost J, Sanders JPM. 2011. Microbial production of bulk chemicals: development of anaerobic processes. *Trends Biotechnol.* 29:153–57**
128. Wilke D. 1995. What should and what can biotechnology contribute to chemical bulk production? *FEMS Microbiol. Lett.* 16:89–100
129. **Wittmann C, Becker J. 2007. The L-lysine story: from metabolic pathways to industrial production. In *Microbiology Monographs: Amino Acid Biosynthesis*, ed. VF Wendisch, pp. 39–70. Heidelberg, Ger.: Springer**
130. Wittmann C, Heinzle E. 2002. Genealogy profiling through strain improvement by using metabolic network analysis: metabolic flux genealogy of several generations of lysine-producing corynebacteria. *Appl. Environ. Microbiol.* 68:5843–59
131. Wittmann C, Heinzle E. 2008. Metabolic network analysis and design in *Corynebacterium glutamicum*. In *Corynebacteria: Genomics and Molecular Biology*, ed. A Burkovski. pp. 79–112. Wymondham, UK: Caister Academic
132. Wittmann C, Kiefer P, Zelder O. 2004. Metabolic fluxes in *Corynebacterium glutamicum* during lysine production with sucrose as carbon source. *Appl. Environ. Microbiol.* 70:7277–87
133. Wright S. 1932. The roles of mutation, inbreeding, crossbreeding, and selection in evolution. *Proc. Sixth Int. Congr. Genet.* 1:356–66
134. Yim H, Haselbeck R, Niu W, Pujol-Baxley C, Burgard A, et al. 2011. Metabolic engineering of *Escherichia coli* for direct production of 1,4-butanediol. *Nat. Chem. Biol.* 7:445–52
135. Yukawa H, Inui M, Vertès AA. 2006. Genomes and genome-level engineering of amino acid-producing bacteria. In *Microbiology Monographs: Amino Acid Biosynthesis*, ed. VF Wendisch, pp. 349–401. Berlin: Springer
136. Yukawa H, Omumasaba CA, Nonaka H, Kós P, Okai N, et al. 2007. Comparative analysis of the *Corynebacterium glutamicum* group and complete genome sequence of strain R. *Microbiology* 153:1042–58
137. Zhang YX, Perry K, Vinci VA, Powell K, Stemmer WP, del Cardayré SB. 2002. Genome shuffling leads to rapid phenotypic improvement in bacteria. *Nature* 415:644–46

---

127. Describes the fundamental economic and biological factors to consider when developing bulk chemical microbial production technologies.

129. Gives an account of the history of L-lysine fermentation.

Cumulative Index

Contributing Authors, Volumes 62–66

A

Achtman M, 62:53–70
Ahlquist P, 64:241–56
Ajioka JW, 62:329–51
Aly AS, 63:195–221
An P, 66:213–35
Anca I, 63:363–83
Andersson JO, 63:178–93
Angert ER, 66:197–212
Atluri VL, 65:523–41

B

Babitzke P, 63:27–44
Baker CS, 63:27–44
Barbieri JT, 62:271–88
Barnard TJ, 64:43–60
Bartlett JG, 65:501–21
Battesti A, 65:189–213
Bedard DL, 62:253–70
Belser JA, 63:79–98
Best SM, 62:171–92
Bever JD, 66:265–83
Boetius A, 63:311–34
Bonfante P, 63:363–83
Borst P, 62:235–51
Brett PJ, 64:495–517
Britton RA, 63:155–76
Broughton WJ, 63:431–50
Brunet YR, 66:453–72
Buchanan SK, 64:43–60
Burrows LL, 66:493–520

Busby SJW, 66:125–52
Busch A, 64:539–59
Bush K, 65:455–78

C

Cannon GC, 64:391–408
Capra EJ, 66:325–47
Carroll KC, 65:501–21
Carter CA, 62:425–43
Casadevall A, 62:19–33
Cascales E, 66:453–72
Cashel M, 62:35–51
Chaconas G, 64:185–202
Chang S-S, 66:305–23
Chao MC, 64:293–311
Charon NW, 66:349–70
Chaurushiya MS, 64:61–81
Chistoserdova L, 63:477–99
Ciche T, 63:557–74
Clarke D, 63:557–74
Clausen T, 65:149–68
Cockburn A, 66:349–70
Crosson S, 65:261–86
Cullen BR, 64:123–41
Cushion MT, 64:431–52

D

de Jong MF, 65:523–41
de la Torre JR, 66:83–101
de Lorenzo V, 64:257–75

del Peso-Santos T, 65:37–55
Del Poeta M, 63:119–31
den Boon JA, 64:241–56
Deng Q, 62:271–88
den Hartigh AB, 65:455–78
DeShazer D, 64:495–517
Deveau H, 64:475–93
Doering TL, 63:223–47
Dorer MS, 65:329–48
Dunigan DD, 64:83–99

E

Ehrmann M, 65:149–68
Eisenstark A, 64:277–92
Emerson D, 64:561–83
Englund PT, 66:473–91
Erlich LS, 62:425–43

F

Falkow S, 62:1–18
Ferry JG, 64:453–73
Filter JJ, 62:211–33
Fischer R, 64:585–610
Fisher JF, 65:455–78
Fisher MC, 63:291–310
Fleming EJ, 64:561–83
Forney LJ, 66:371–89
Fournier P-E, 65:169–88
Fuchs G, 65:631–58

G

Galagan J, 63:385–409
Galyov EE, 64:495–517
Gao R, 63:133–54
Garneau JE, 64:475–93
Garner T, 63:291–310
Gerdes K, 66:103–23
Gilbert OM, 65:349–67
Goffeau A, 66:39–63
Gorbushina AA, 63:431–50
Göringer HU, 66:65–82
Gottesman S, 65:189–213
Green R, 62:353–73
Grimwade JE, 65:19–35
Guazzaroni M-E, 64:539–59
Guillier M, 64:43–60
Gunsalus RP, 66:429–52

H

Hackett JD, 65:369–87
Hammer ND, 65:129–47
Handelsman J, 62:375–401
Harris E, 62:71–92
Hatfull GF, 64:331–56
Hazelbauer GL, 66:285–303
Hedtke M, 64:585–610
Heinhorst S, 64:391–408
Heinzen RA, 65:111–28
Henderson IR, 62:153–69
Hendrixson DR, 65:389–410
Henry JT, 65:261–86
Hill KL, 63:335–62
Holmes EC, 62:307–28
Hood HM, 63:385–409
Horn M, 62:113–31
Huber R, 65:149–68
Hunstad DA, 64:203–21

I

Ibba M, 63:61–78
Inoue T, 65:287–305
Inui M, 66:521–50
Iwasaki A, 66:177–96

J

Jedd G, 66:237–63
Jensen RE, 66:473–91
Jogler C, 63:501–21
Johnson PJ, 64:409–29
Justice SS, 64:203–21

K

Kabututu ZP, 63:335–62
Kaiser M, 65:149–68
Kalyuzhnaya MG, 63:477–99
Kamilova F, 63:541–56
Kappe SH, 63:195–221
Kastner C, 64:585–610
Katz LA, 66:411–27
Keiler KC, 62:133–51
Kentner D, 64:373–90
Kerfeld CA, 64:391–408
Kirby JR, 63:45–59
Kleerebezem M, 63:269–90
Knittel K, 63:311–34
Kobryn K, 64:185–202
Kolstø A, 63:451–76
Kolter R, 63:99–118
Krell T, 64:539–59
Kuehn MJ, 64:163–84
Kuipers OP, 62:193–210
Kulp A, 64:163–84
Kung C, 64:313–29
Kyle JL, 62:71–92

L

Lacal J, 64:539–59
Lane LC, 64:83–99
Laub MT, 66:325–47
Law CJ, 62:289–305
Lee DJ, 66:125–52
Lengyel P, 66:27–38
Leonard AC, 65:19–35
Lertsethtakarn P, 65:389–410
Levasseur A, 65:57–69
Lewis K, 64:357–72
Ley RE, 65:411–29
Li C, 66:349–70
Lidstrom ME, 63:477–99
Lilley CE, 64:61–81
Ling J, 63:61–78
Little AEF, 62:375–401
Liu J, 66:349–70
Liu Y, 66:305–23
Ljungdahl LG, 63:1–25
Lopez-Rubio JJ, 62:445–70
Lovley DR, 66:391–409
Lowy DR, 64:23–41
Ludden PW, 62:93–111
Lugtenberg BJ, 63:541–56
Luján HD, 65:611–30

M

Ma B, 66:371–89
Maisonneuve E, 66:103–23
Majdalani N, 65:189–213
Maloney PC, 62:289–305
Martinac B, 64:313–29
Massana R, 65:91–110
Mather MW, 63:249–67
McBeth JM, 64:561–83
McConville MJ, 65:543–61
McDonald ME, 62:353–73
McInerney MJ, 66:429–52
Melehani JH, 63:335–62
Merdanovic M, 65:149–68
Miller KA, 66:349–70
Miller MR, 66:349–70
Minchin SD, 66:125–52
Miyata M, 64:519–37
Moineau S, 64:475–93
Moore P, 65:287–305
Morton ER, 66:265–83
Motaleb MA, 66:349–70
Mougous JD, 66:453–72
Müller S, 64:585–610
Murray CL, 65:307–27

N

Naderer T, 65:543–61
Naffakh N, 62:403–24
Neafsey DE, 63:385–409
Nikaido H, 65:1–18
Noinaj N, 64:43–60

O

Oberholzer M, 63:335–62
Økstad O, 63:451–76
Okuda S, 65:239–59
Omsland A, 65:111–28
Ornston LN, 64:1–22
Österberg S, 65:37–55
Ottemann KM, 65:389–410
Otto M, 64:143–62

P

Parkinson JS, 64:101–22
Paul AV, 65:583–609
Peterson SB, 62:375–401
Piel J, 65:431–53
Pieuchot L, 66:237–63
Pipas JM, 66:213–35
Platt TG, 66:265–83
Plattner F, 62:471–87
Potrykus K, 62:35–51
Prasad R, 66:39–63
Prucca CG, 65:611–30
Punt PJ, 65:57–69

Q

Queller DC, 65:349–67

R

Raffa KF, 62:375–401
Ralston KS, 63:335–62
Rameix-Welti M-A, 62:403–24
Ramos JL, 64:539–59
Raoult D, 65:169–88
Ravel J, 66:371–89
Record E, 65:57–69
Reynolds N, 63:61–78
Rhome R, 63:119–31
Rice CM, 65:307–27
Rivero FD, 65:611–30
Riviere L, 62:445–70
Roberts JW, 62:211–33
Robinson CJ, 62:375–401
Rodriguez-Romero J, 64:585–610
Romeo T, 63:27–44
Rubin EJ, 64:293–311
Rubio LM, 62:93–111

S

Sabatini R, 62:235–51
Sachs MS, 63:385–409
Sáenz Robles MT, 66:213–35
Salama NR, 65:329–59
Samuels DC, 65:479–99
Satchell KJF 65:71–90
Scherf A, 62:445–70
Schiller JT, 64:23–41
Schüler D, 63:501–21
Sessler TH, 65:329–48
Shankar S, 62:211–33
Shiflett AM, 64:409–29
Shingler V, 65:37–55
Sibley LD, 62:329–51
Sieber JR, 66:429–52
Silva-Jiménez H, 64:539–59
Silva-Rocha R, 64:257–75
Silverman JM, 66:453–72
Skaar EP, 65:129–47
Skalsky RL, 64:123–41
Smith G, 66:153–76
Smits WK, 62:193–210
Sockett R, 63:523–39
Soldati-Favre D, 62:471–87
Sourjik V, 64:373–90
Stahl DA, 66:83–101
Stock AM, 63:133–54
Straight PD, 63:99–118
Strassmann JE, 65:349–67
Stringer JR, 64:431–52
Sukharev S, 64:313–29

T

Tokuda H, 65:239–59
Tomoiu A, 62:403–24
Tourasse NJ, 63:451–76
Tsai B, 65:287–305
Tsolis RM, 65:455–78
Tumpey TM, 63:79–98

U

Ullmann A, 66:1–26

V

Vaidya AB, 63:249–67
van der Werf S, 62:403–24
van der Woude M, 62:153–69
Van Etten JL, 64:83–99
Vaughan AM, 63:195–221
Vaughan EE, 63:269–90
Veening J-W, 62:193–210
Velicer G, 63:599–623
Vertès AA, 66:521–50
Visser H, 65:57–69
Vos M, 63:599–623

W

Wagner M, 63:411–29
Walker SF, 63:291–310
Walter J, 65:411–29
Waterfield NR, 63:557–74
Wang D-N, 62:289–305
Weitzman MD, 64:61–81
Wery J, 65:57–69
Wimmer E, 65:583–609
Wisecaver JH, 65:369–87
Wolgemuth CW, 66:349–70
Wood JM, 65:215–38

X

Xavier MN, 65:523–41

Y

Yother J, 65:563–81
Young KD, 64:223–40
Youngman EM, 62:353–73
Yukawa H, 66:521–50

Z

Zhang Z, 66:305–23
Zuber P, 63:575–97

QR 1 .A5 v.66

Annual review of microbiology

	DATE DUE	

Concordia College Library
Bronxville, NY 10708